INSECT MICROBIOLOGY

BY EDWARD A. STEINHAUS

❖

Insect Microbiology

AN ACCOUNT OF THE MICROBES ASSOCIATED WITH INSECTS AND TICKS

WITH SPECIAL REFERENCE TO THE BIOLOGIC RELATIONSHIPS INVOLVED

By Edward A. Steinhaus

ASSISTANT PROFESSOR OF INSECT PATHOLOGY AND ASSISTANT INSECT PATHOLOGIST IN THE AGRICULTURAL EXPERIMENT STATION AT THE UNIVERSITY OF CALIFORNIA, BERKELEY. FORMERLY ASSOCIATE BACTERIOLOGIST AT THE ROCKY MOUNTAIN LABORATORY, U. S. PUBLIC HEALTH SERVICE

✦ ✦ ✦

ITHACA, NEW YORK
Comstock Publishing Company, Inc.
1947

Copyright 1946 by

Comstock Publishing Company, Inc.

FIRST PRINTED, AUGUST, 1946
SECOND PRINTING, SEPTEMBER, 1947

ALL RIGHTS RESERVED. THIS BOOK, OR PARTS THEREOF, MUST NOT BE REPRODUCED IN ANY FORM WITHOUT PERMISSION IN WRITING FROM THE PUBLISHER, EXCEPT BY A REVIEWER WHO WISHES TO QUOTE BRIEF PASSAGES IN A REVIEW.

❖

PRINTED IN THE UNITED STATES
OF AMERICA BY THE VAIL-BALLOU
PRESS, INC., BINGHAMTON, N. Y.

LIBRARY
LOS ANGELES COUNTY MUSEUM
EXPOSITION PARK

TO MY FATHER AND MOTHER

PREFACE

SINCE today's world is one of extremely rapid progress on all scientific fronts, there is a tendency to allow our "lines of communication" to become overextended, permitting our practical accomplishments to lose the sturdy support of our reserves of basic knowledge and information. Especially is this true with the biologic sciences in which the "practical" or applied accomplishments so frequently appear to overshadow the results of academic or so-called pure research. Until World War II necessitated the sending of our troops into all regions of the world, including the tropics, few Americans realized the really great part played by insects in the transmission of disease agents affecting man. America's part in producing ever-increasing quantities of food during this period has focused our attention as never before on the role of insects in the spread of diseases of animals and plants. Yet in the handling of most of the new problems concerned in this activity we were fortified with knowledge and progress gained during the prewar years in the peaceful pursuit of the answers to similar but less pressing problems. With the renewed realization of the importance of insects and the microorganisms associated with them, it is not only safe but wise to continue the study of the basic biologic relationships concerned in order to provide continuously a firm foundation upon which future accomplishments may be built securely. As concerns the field of insect microbiology, we seem to have got ahead of ourselves in many respects. There has been very little effort to group our forces, so to speak, and for the most part the available knowledge and information is not properly known, organized, or appreciated. Research workers in particular have long been handicapped by the widely scattered, unorganized, and often inaccessible literature of the field. If we hope to meet the opportunity and the challenge which the future of this field is offering, this situation must be corrected. To help accomplish this is the writer's modest purpose behind the presentation of this publication.

In general, this book is an attempt to treat the various associations and relationships existing between all types of microbes and insects (including ticks and mites) from a biologic standpoint, and, in a sense, to serve as a compendium of the data concerned. Many of these data have not been sufficiently studied to present adequate bases for discussions relating to the biologic relationships involved. While this is not a book on insect pathology or on medical entomology, an attempt has been made to include the biologic relationships existing between pathogenic agents and their arthropod hosts and vectors as well as all those between nonpathogenic agents and insects and ticks in general.

Although all types of microbes (bacteria, protozoa, yeasts, fungi, rickettsiae, viruses, and spirochetes) are concerned in the subject matter of the book, yet in no instance have we attempted to give the reader a treatment of any particular field of microbiology as such. In other words, this book does not include the fundamentals of bacteriology itself, or protozoology itself, etc. An effort has been made, however, to treat each of the groups of microbes in a fashion acceptable to the present authorities in each particular field, especially with reference to its association with insects. Brief treatments of the taxonomy of the various microbial groups are included both for the sake of order and to enable the reader to read in the group concerned even though he may not have an intimate knowledge of that particular branch of microbiology.

The writing of this book has been made difficult by the inaccessibleness at this time of many of the foreign references and by the fact that the nature of the subject matter is such that it calls for treatment in a variety of ways. For example, the bacteria could not very well be treated in the same manner as the protozoa, nor the intracellular bacterium- and yeastlike organisms in a manner similar to that used for the true rickettsiae. For this reason we have felt justified in considering each group of microorganisms in the way most suitable to that group, varying the method of treatment in different chapters of the book. In places, lists of microbes are given without much associated individual discussion. This has been done for two reasons: first, considering the dearth of information a discussion of each microbe is impossible; second, thinking such a listing desirable, the author tried to give at least the names of the microbes concerned. For example, hundreds of entomophilic protozoa have been discovered and named, but practically nothing is known of the nature of the biologic relationships between them and their hosts. A similar situation prevails with most of the entomogenous fungi. In the belief that the reader would like to have the names of these microorganisms, the writer has presented them at the risk of appearing to present parts of the book as merely annotated lists. A similar situation prevails concerning much of the general subject matter presented on these pages. In many cases an appreciable number of facts have been gathered, but very few generalizations or conclusions have been or can be made. The writer is fully aware that this is an unfortunate deficiency. It is hoped that with an increase in our knowledge of this field appropriate deductions, conclusions, and generalizations can be made.

The authorities for the specific names of the insects have been omitted in the text of the book but are given in the index. In the case of some of the microorganisms it occasionally has been expedient to include the names of their describers in the text.

Although the book's title specifies insects as the type of arthropods concerned, the microbiology of *ticks and mites* is also included. Hence, as it applies here, the term insect, when used in a general sense, includes the Acarina.

To the critical reviewer, who so often feels duty bound to find errors in a book, the author gives assurance that a detailed work of this kind is never entirely free from mistakes. The author has diligently tried to avoid them, however, and will welcome having them called to his attention. He has attempted to use the most recent names of insects and ticks but may not have done so in all cases. For the bacteria the systematics of the fifth edition of *Bergey's Manual of Determinative Bacteriology* have been followed, and similar standard works have been used for the other types of microorganisms. Cross references to synonyms have been kept to a minimum in the text. For the most part these are shown in the index. An effort has been made to list in the index the names of all the microbes associated with each insect.

The author has attempted to consult original papers and works as much as possible. Where the reference to the work would seem to be of special value to the reader, it has been cited. Although large numbers of references have been given, no attempt has been made to make the book completely bibliographic of the field of insect microbiology. The writer is greatly indebted to the many works read but not mentioned by name. The references cited in the text are listed together at the end of the book.

It is the earnest desire of the author that this volume shall be of use, not only as a reference book to workers in biologic research, but also as a textbook for the classroom. A specialized course in insect microbiology would serve as a basic study to those majoring in such fields as medical entomology, physiology of insects, insect pathology, plant pathology, bacteriology, or parasitology. Certainly there is abundant material available to support such a course.

The author is indebted to a great many people who have assisted him in the completion of this book. For thoughtful advice and inspiration I wish to express my appreciation to Professor Alvah Peterson of the Ohio State University and to my associates at the Rocky Mountain Laboratory and at the University of California. For kindly reading portions of the manuscript and for offering constructive criticisms on certain subjects, I am indebted to Prof. Lee Bonar of the University of California, Maj. Gordon E. Davis of the Army Medical

School, Prof. Harry Fitzpatrick of Cornell University, Prof. N. Paul Hudson of the Ohio State University, Dr. William L. Jellison of the Rocky Mountain Laboratory, Prof. Harold Kirby, Prof. K. F. Meyer, Prof. E. M. Mrak, and Dr. H. J. Phaff of the University of California, and Dr. R. R. Parker, Director of the Rocky Mountain Laboratory. I particularly wish to express my thanks to Mr. Nick J. Kramis of the Rocky Mountain Laboratory for preparing reproductions of most of the illustrations, and to the numerous persons who permitted me to use photographs and other illustrative material. I also wish to acknowledge the valuable assistance given me by Miss B. M. Nelson in preparing the index, and by Mrs. W. T. Ash and Miss Helen Spaulding in typing parts of the manuscript. To my wife, Mabry Clark Steinhaus, I am especially indebted for many hours of loyal assistance in reading and correcting manuscript and proofs.

Edward A. Steinhaus

Berkeley, California
September, 1945

CONTENTS

	Preface	vii
I	Introduction	1
II	Extracellular Bacteria and Insects	9
III	Specific Bacteria Associated with Insects	38
IV	Intracellular Bacteriumlike and Rickettsialike Symbiotes	188
V	Rickettsiae	256
VI	Yeasts and Insects	348
VII	Fungi and Insects	376
VIII	Viruses and Insects	413
IX	Spirochetes Associated with Insects and Ticks	451
X	Protozoa and Insects (Except Termites)	462
XI	Protozoa in Termites	526
XII	Immunity in Insects	555
XIII	Methods and Procedures	573
	References	603
	Author Index	693
	Subject Index	706

INSECT MICROBIOLOGY

❖

CHAPTER I

INTRODUCTION

TO ANYONE who gives the matter some thought it is readily apparent that the field embracing the microbiology of insects is one that needs to be introduced to the microbiologist and entomologist alike. In a sense, it is somewhat superfluous to begin this book with a chapter titled "Introduction" when in reality the purpose of the entire book is to introduce the field to those it may concern. An entire book is necessary for such an introduction since, in addition to entomology, at least five sciences are concerned: bacteriology, protozoology, mycology, pathology, and immunology. These sciences encompass the following types of microbes which may be associated with insects and ticks: bacteria, protozoa, yeasts, molds (fungi), rickettsiae, viruses, and spirochetes.

The present day of specialization has a tendency to demand that one's attention be limited to a single subject. The unfortunate part of this demand is that it neglects important borderline fields which are badly in need of investigation and study. Insect microbiology is just such a field. It is a rare entomologist who has had extensive training in bacteriology, protozoology, or mycology. Similarly, the microbiologist has seldom seen fit to acquire an entomologic background. Hence, problems involving the co-operation of these two groups of sciences are sorely neglected. There is probably no field of biologic endeavor more in need of interscience co-operation than that of the microbiology of insects and ticks. The entomologist and the microbiologist greatly need to be introduced to each other.

The need for the co-operative merging of abilities in entomology and microbiology and for adequate training of the individual in the fundamentals of both sciences has been well expressed by Leach (1940):

> The discovery of the nature of virus diseases of plants and the role of insects in the transmission of viruses has greatly stimulated the interest of both entomologist and pathologist in the general subject of insect transmission of plant diseases. The

appearance in Europe and America of the destructive and spectacular Dutch elm disease, which is so dependent upon insects for its spread and development, also served to focus the attention of both groups of workers on the problem. The necessity for cooperation between entomologist and plant pathologist in the solution of these borderline problems is now generally recognized by both groups, and the need has been expressed in the literature on numerous occasions. For various reasons, however, the present situation leaves much to be desired. It is one thing to talk of cooperation and another to cooperate. In fact, so much has been said and written in recent years about cooperation in research that the word, to some extent, has fallen into disrepute. Cooperation often works much better on paper than in practice. In order to avoid the difficulties of cooperation, we often plead for coordination of effort. This sounds much better but is often more difficult, for it requires a great deal of earnest cooperation before we can have successful coordination. But despite all these difficulties there is, and has been, a fair amount of real and successful cooperation in scientific research.

The failure or lack of cooperation may be caused by many different factors. The human or personal element is perhaps the most common of them all and the most difficult to overcome. Other causes may be administrative, political or, as indicated above, largely a matter of tradition. However, it is the author's belief that the greatest success in the solution of borderline problems cannot be achieved by the expedient of cooperation alone. Cooperative work is sometimes attempted on the principle of strict division of labor in which all "entomological" work is done by the entomologist and all "phytopathological" work by the phytopathologist. This type of cooperation nearly always is doomed to failure. In the study of the relation of insects to plant diseases, such strict division of labor is not practical. For the greatest success, the invisible, though very real, wall separating the two fields of research must be broken down. This may be rather difficult, but it can be done.

A first step in this direction would be a liberalization of the narrow professional viewpoint, which in effect often hangs out a sign reading, "This is the phytopathological field; entomologists encroach at their own risk" or "This is the entomological field; all phytopathologists keep off." Such a viewpoint may simplify some of the problems of organization, but it is not conducive to the solution of these neglected, but mutually important, problems. The necessity for well-defined fields of research with corresponding responsibility and authority is recognized. Such responsibility and authority are necessary, not only for efficient administration, but also for the existence of the guild spirit so important in scientific research. Nevertheless, when attempts are made to draw too sharp a line between related fields of activity, many problems of vital importance and significance usually are neglected.

A second step would be a modification of our educational procedure so that research workers would be given the viewpoint and training in techniques necessary for the solution of the problem in hand. The worker should have a thorough knowledge of the essentials of both entomological and microbiological techniques. Instead of placing the emphasis upon training entomologists or plant pathologists, some of the workers should be given the training and viewpoint necessary for the solution of this particular kind of problem, namely, the role of insects in the spread and development of plant diseases. It is not proposed that we train mental giants who can master both fields of knowledge, but rather workers who have a sufficient grasp of the essentials of both sciences for the solution of this particular kind of problem. When these qualifications are combined in one man, many of the difficulties of cooperation will be avoided. No claim is made for the novelty of the idea, for already it has met with considerable success in the solution of other borderline problems.[1]

Historical. Medical historians tell us that one of the first to link the spread of disease with the activities of insects was Mercurialis in 1577. He believed that the cause of bubonic plague was carried from the ill or the dead to the well by flies. Naturally Mercurialis had no conception of microbes in his day, but his idea that the virus of infection could be conveyed by insects was essentially correct, although in the case of plague the mode of transmission is usually not by way of flies. During the following three hundred years a few others put forth similar views, but most of these were purely conjecture. In some cases, however, certain observers came extremely close to the truth. Thus in 1848 Josiah Nott expressed his belief that mosquitoes were responsible for the occurrence of both yellow fever and malaria. Similarly, Beauperthuy in 1854 published a theory on the transmission of yellow fever and other diseases by mosquitoes.

It was not until several decades after the discovery of microbes that men's suspicions were fully aroused as to the possible connection between these microscopic forms and insects. Noteworthy in this connection are the experiments by Raimbert, who in 1869 showed by the inoculation of guinea pigs that flies could be contaminated with anthrax bacilli and very probably could disseminate them.

The greatest impetus given the study of microbes in relation to their arthropod hosts occurred during the years from 1890 to 1900. During this period several epoch-making discoveries were made. The first of these was Waite's (1891) discovery that bees and wasps were vectors of fire blight, a bacterial disease of pears and other orchard fruits. Then Smith and Kilbourne's (1893) discovery that the cattle tick, *Boöphilus annulatus,* is the invertebrate host of *Babesia bigemina,* the cause of Texas cattle fever, was of great fundamental importance. Of great significance also was their observation that the protozoan was transmitted to the next generation through the egg. In rapid succession were reported the discoveries that trypanosomes were carried by tsetse flies (Bruce, 1895), that mosquitoes carry the malaria parasite (Ross, 1897), that the plague bacillus may be transmitted to rats by infected fleas (Simond, 1898), and that the virus of yellow fever is transmitted by the mosquito *Aëdes aegypti* (Finlay, 1881; Reed, 1900).

---

[1] Quoted from J. G. Leach, *Insect Transmission of Plant Diseases,* by permission of McGraw-Hill Book Co., New York.

Similar discoveries followed throughout the decades after 1900 until the transmission of all types of microorganisms (bacteria, fungi, protozoa, spirochetes, rickettsiae, and viruses) by arthropods was clearly established.

Thus the discoveries concerning the arthropod-transmitted diseases of animals, man, and plants dramatically called attention to the association of microbes with insects and ticks. During the course of such investigations other entomophilic microbes, unrelated to disease, were also observed. In the hurry to work out the relationships of the microbes of medical importance, observations on the nonpathogenic microorganisms were largely neglected. Fortunately a few biologists, braving the wrath of those who scorn academic or "pure" science, laboriously sought the secrets of some of these supposedly less glamorous associations. As will be pointed out in later paragraphs, some of these little-known relationships may ultimately make greater contributions to our knowledge of life than have some of the more dramatic discoveries.

Fig. 1. Theobald Smith. One of the first investigators to show experimentally that an arthropod may transmit a disease agent. (Courtesy Dr. Carl Ten Broeck.)

The historical data relative to each of the groups of organisms constituting the subject matter of this book will be considered in the chapters dealing with these groups. This history consists of a continuing parade of many interesting personalities, institutions, discoveries, and mistakes.

Biologic Relationships. The reader may wonder where lies the essence of our story—what is the plot? Essentially it has to do with the phenomena concerned wherever and whenever microbe and insect happen to meet. We shall throughout be concerned chiefly with the *biologic relationships* existing between microbes and insects. We shall want to know the effects each of these forms of life has upon the other. We shall inquire into the adaptations and physiologic processes involved in these associations and their effect upon the respective ecologies and biologies.

To support such a plot, it will be necessary frequently to list or catalogue the microbes concerned. This, however, is done out of necessity, since without introducing the characters it would be difficult to carry out the theme, which, as we have just said, is concerned with the biologic relationships involved.

In chapter IV we shall have occasion to define in detail our acceptance of the terms "symbiosis," "mutualism," "parasitism," "commensalism," and the other terms used to denote manners in which organisms associate with one another. Suffice it to say here that we feel obligated to use the word "symbiosis" as its originator, De Bary, originally used it: as a general term referring simply to the living together of dissimilar organisms and not excluding parasitism or commensalism. Thus, in a sense, we may also say that our theme has to do with the factors involved in the symbiotic associations prevailing between insects and microorganisms, i.e., between macrosymbiote and microsymbiote. Regardless of the association between the insect and microbe—whether the latter is parasitic, mutualistic, or commensal; pathogenic or nonpathogenic to vertebrates or to invertebrates; necessary or beneficial to the life of the insect, or only an adventitious associate—it shall make no difference in our treatment of it. All relationships existing between insects and microbes will receive our equal consideration.

To some extent the type of association or relationship involved determines the location of the microbes with respect to the insect. The association may be one in which the insect may voluntarily bring about or encourage the microbes to grow (e.g., those insects having fungus gardens). Usually, however, there is no such freedom of choice for the insect. The microbes may be found in the interior of the insect or externally on its chitinous covering. They may live endogenously and penetrate to the outside of the insect, or they may originate or grow exogenously and penetrate to within the insect's body. Internally, the microbes may abide extracellularly in the alimentary tract or in the hemocoele, or they may live intracellularly in the epithelial lining of the alimentary tract, Malpighian tubes, salivary glands, or in the cells of other tissues of the insect's body. They may live extracellularly in definite tubes, pouches, or ceca, or they may inhabit only certain specialized cells in certain specialized organs.

The microbes themselves may be highly specialized organisms capable of living only in association with specific insects or groups of insects, or they may be common, saprophytic, adventitious microbes existing only in a fortuitous association with insects. They may be harmless, nonpathogenic forms, or they may be disease parasites of animals or of plants, the insect acting as a necessary or accidental carrying agent or vector. The microbes may parasitize the insect, causing it to become diseased, or they may actually benefit the insect by serving as food or as a source of enzymes and vitamins.

Thus it is apparent that almost every type of biologic relationship which is known to exist may be found among the many associations prevailing between microbes and insects. As concerns our knowledge of the extent of these associations—only the surface has been scratched!

Applications. The knowledge and information gained through the study of the microbes associated with insects and arachnids may be applied or put to use in numerous ways. These applications may fall into any one of three domains: agriculture, medicine, and general biology.

In *agriculture,* for instance, there exists one of the outstanding examples of the practical effects of co-operation between entomologist and microbiologist, namely the realization that insects (and ticks) transmit diseases of animals and plants. The insect-transmitted diseases, of the latter especially, are many and important, and the biologic relationships are as yet little understood. Another application of insect microbiology has to do with the biologic control of destructive insects by means of controlled bacterial and fungous diseases. The real possibilities of this application, in the past, have been considered mostly with a certain amount of dilettantism. Along with this goes the demand for still more knowledge of the microbial diseases which plague our useful insects such as the bee and the silkworm.

The applications of insect microbiology in *medicine* are likewise numerous and extremely important. An understanding of the biologic relationships between pathogenic microbes and their vectors forms the basis for the important field of medical entomology. Only through a thorough understanding of the general microbiology of insects and ticks can we acquire a clear picture of the manner in which these arthropods transmit disease-producing organisms. Our new interest in the arthropod-borne diseases of the tropics is making a heavy demand upon insect microbiology for fundamental information and data relating to the association of microbes and their tropical vectors.

In its applications to *general biology* the field of insect microbiology has an extremely promising future. It is now recognized that the academic and "pure" biology of today pays great dividends years and decades hence. It is not to be doubted, therefore, that present studies on the basic biologic principles underlying the many associations between microbes and insects will bring forth great scientific profit in the years to come. Who knows what biologic discoveries lie behind a knowledge of the mysterious ability of certain insects to harbor in their tissue cells large numbers of living microorganisms? Are certain of the intracellular symbiotes of today the pathogenic rickettsiae of tomorrow—or vice versa? The detailed facts relative to the phenomena of symbiosis (mutualism, commensalism, and parasitism) must certainly prove to be of great biologic significance when they become better known. The physiologist will in all probability be furnished with new knowledge which can be applied to the biology of man, other animals, and plants.

Insect microbiology enjoys the privilege of being able to contribute greatly to the realms of both the pure and the applied sciences. Realizing this, investigators in the field should content themselves in striving continuously ahead, accumulating and interpreting their newly discovered facts as best they can with the tools at hand. If the applications of their findings are not immediately apparent, they should recall the often-told story in which Gladstone, when shown the electromagnetic motor, is said to have asked, "What good is it?" Faraday was quick to reply, "What good is a baby?"

Nematode Parasites of Insects. The various groups of microorganisms with which we shall be concerned in this book have already been mentioned. In addition, there is a group of animals whose association with insects has recently been receiving considerable attention—we refer to the phylum Nemathelminthes (roundworms). The writer is of the opinion that for the time being a detailed discussion of the nematode parasites of insects is outside the scope of the present book. Nevertheless, it would be a mistake at least not to mention this group of metazoans and to express the hope that when more information is available a chapter on the biologic relationships involved may be included in a book of this kind. Among the facts which may be mentioned here in passing are those pertaining to the distribution of the nematodes in insects and the relation of their life cycles to those of their insect hosts.

The number of species of nematode worms associated with insects is known to be very great. Those which have received the most study, and with which insects are primary hosts, are included in the families Oxyuridae, Rhabditidae, Mermithidae, Anguillulinidae, and Gordiidae. The latter belongs to the Nematomorpha; the rest to the Nematoda. All known Gordiaceae (hairworms) are true parasites of insects.

The roundworms may enter their insect host through the body wall, mouth, and possibly the anus, although in most instances the portal of entry is unknown. Occasionally, as with the mermithids, the eggs of the worms are ingested by the insects while feeding on certain plants.

Nematodes have three stages (eggs, larvae, and adults) in their life cycle, and the eggs are always microscopic, although in some species the mature individuals may reach a length of several feet. Many species spend part of their life cycle in the insect host and part in another host (e.g., a vertebrate animal). Frequently the larvae may spend a short time in the free-living state, usually in aquatic surroundings. Most nematodes parasitic in insects overwinter in the host insect.

Recently, there has been considerable interest in the use of nematodes in the biologic control of certain insects, as for example, the use of *Neoaplectana glaseri* in the control of the Japanese beetle *Popillia japonica* (Glaser, 1932; Glaser and Farrell, 1935). This has been facilitated by the development of artificial culture methods by which the parasites can be reared in large numbers (McCoy and Glaser, 1936). For a discussion of some of the other nematode diseases of insects the reader is referred to Sweetman (1936).

Certain nematodes associated with insects are of medical importance. Chief of these is the filarial worm, *Wuchereria bancrofti,* causing filariasis of man, and transmitted by many species of mosquitoes, notably *Culex quinquefasciatus* and *Aëdes scutellaris*.

It is of interest to note that the presence of a tapeworm (a plerocercoid) has been recorded in the fatty tissue of a sandfly in India (Subramaniam and Naidu, 1944). The worm is similar but not identical to *Sparaganum proliferum*. Only a few other flatworms have been found associated with insects. Of these, one of the most noteworthy is the dog tapeworm, *Dipylidium caninum,* for which certain species of fleas serve as intermediate hosts.

CHAPTER II

EXTRACELLULAR BACTERIA AND INSECTS [1]

THERE are approximately 250 identified species of bacteria which have been found associated, in one way or another, with insects and ticks. This does not include the intracellular forms such as the rickettsiae and other intracellular "symbiotes." A large number of additional bacteria associated with arthropods have never been completely described nor correctly classified. In general, however, the bacteria found in insects are not characteristically different from most bacteria. Limited studies have shown that the bacteria isolated from insects consist roughly of 40 to 50 per cent gram-negative small rods, 15 to 25 per cent gram-positive sporeforming rods, 15 to 25 per cent gram-positive cocci, 10 to 12 per cent gram-positive small rods, and lower percentages of spirilla, "coccobacilli," and other forms. Bacteria isolated from ticks appear to be predominantly cocci (50 per cent or more in some species) with smaller numbers of other types. Percentages such as the above will probably have to be modified as more complete studies on the flora of insects are made.

External Bacterial Flora of Insects. The bacteria found on the external surfaces of insects are, for the most part, adventitious. This is especially the case with such insects as the housefly and cockroach, which frequent areas of filth and decomposing organic matter. Such insects acquire a great and ever-changing variety of bacteria, many of which are important to public health.

Because of peculiarities in body structure such as the housefly's bristles and sticky pads (fig. 2), many insects carry externally enormous numbers of microorganisms. Even the fly, however, usually has many more bacteria internally than externally. For example, Yao, Yuan, and Huie (1929) found an average of 3,683,000 bacteria (externally) a fly in the slum district of Peiping,

---

[1] Much of the information in this chapter is based on a section in *Catalogue of Bacteria Associated Extracellularly with Insects and Ticks* published for the author in 1942 by the Burgess Publishing Company, Minneapolis.

China, and 1,941,000 a fly in the cleanest district. The insides of the flies harbored from eight to ten times as many bacteria as the outsides. Similarly, Torrey (1912) found the bacteria in the intestine of the housefly to be 8.6 times as numerous as those occurring on the external surface of the insect.

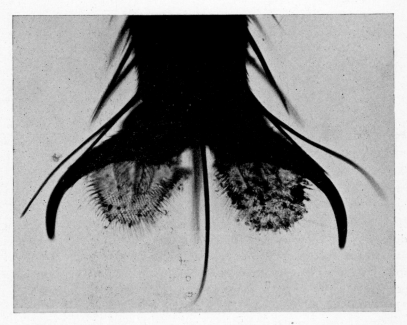

Fig. 2. A greatly magnified "foot" of a housefly showing pads (pulvilli) with glandular hairs and adherent detritus. (From Philip, 1937.)

Other insects, whose external body structures are not so complex, carry relatively few bacteria of any kind on their body surfaces. It is interesting to note that bees, whose body structures are so well adapted for carrying pollen, have relatively few bacteria on their external surfaces. White (1906) found only three species (*Bacillus A, Bacterium cyaneus,* and *Micrococcus C*) on adult bees from normal apiaries.

Bacterial Flora of Alimentary Tract Proper. There is extreme variation in the size and structure of the alimentary tracts of different species of insects. In some species the tract is merely a tube extending from one end of the body to the other. In other species it is a highly complex structure with various pouches, ceca, and diverticula. In most insects the tract is longer than the body and possesses three chief divisions: the fore-intestine, the mid-intestine, and the hind-intestine. The fore- and hind-intestines are invaginations of the body wall, and their chitinous lining is continuous with the body wall cuticula. The mid-intestine develops from an entodermal tube, the mesenteron.

Although very little comparative work has been done on the bacterial flora of main divisions of the digestive tracts of insects, Steinhaus (1941) reported that the milkweed bug, *Oncopeltus fasciatus*, appears to have a different flora in its pylorum and rectum from that in the four stomachs which precede them. The predominant bacterium found in the pylorum and rectum was named *Proteus recticolens*, but in the four stomachs the flora consisted mainly of *Proteus insecticolens*. *Eberthella insecticola* and occasionally *Streptococcus faecalis* were found throughout the entire tract.

The bacterial flora of the digestive tract may vary quantitatively as well as qualitatively. Whereas the tracts of some insects are packed with organisms, others have been found to be sterile. In the honeybee Hertig (1923) found the greater number of bacteria in the hind-intestine, particularly the rectum, and very few both in numbers and variety were found in the ventriculus, except at times of food accumulation. Occasionally when small sections of the gut wall and contents were inoculated into media, no growth at all resulted. Hertig explains the low bacterial content of the ventriculus as perhaps due to the fact that solid particles pass rapidly to the hind-intestine, and also to the fact that the contents of the ventriculus are at times rather acid. This might inhibit the multiplication of the bacteria. In the larvae of the olive fly, *Dacus oleae*, large numbers of bacteria are found in the mid-intestine, few in the hind-intestine, and none in the fore-intestine. In *Tephritis conura*, besides occurring in the intestinal lumen, bacteria are regularly found in the esophagus as far as the region of the mouth opening and even in the proboscis.

Stammer (1929), in a study of 37 species of trypetids, showed the presence of bacteria in all instances, but their manner of distribution varied with the stage and genus of the host. In the larvae and young adults the bacteria occurred diffusely or in clumps in the intestinal contents. In old adults they were always present in enormous numbers in the lumen of the intestine. In the case of *Agriotes mancus*, Melampy and MacLeod (1938) found the greatest number of bacteria in the hind-intestine. A similar condition in the petroleum fly, *Psilopa petrolei*, has been reported by Thorpe (1930).

The digestive tract of some arthropods, such as certain members of the bloodsucking group, is sterile. In other cases this sterility is restricted to certain parts of the tract. An example of this regional sterility is found in blowfly maggots used in the treatment of slow-healing lesions such as those in osteomyelitis. In the larvae of *Lucilia sericata*, for instance, the bacteria taken in with the food are destroyed in passing through the long, tubular stomach of the maggot. Bacteria may be found abundantly in the forestomach and occasionally in the intermediate area and the hind-stomach; none survive as far as the intestine (Robinson and Norwood, 1934).

Duncan (1926) found the gut contents of *Cimex lectularius, Argas persicus,* and *Ornithodoros moubata* to be consistently sterile. He quotes Breinl as finding the gut contents of lice to be invariably free of bacteria. Chapman (1924) examined the digestive tract of the confused flour beetle, *Tribolium confusum,* and found no living organisms. Nuttall (1899) found that the anthrax bacillus died in the stomach of the bedbug in 48 to 96 hours at 13° to 17° C. and in 24 to 28 hours at 37° C., and the feces of the bugs contained living bacilli during the first 24 hours after feeding.

During the years following 1905 considerable interest was aroused in regard to the bacterial flora of the housefly (*Musca domestica*) as well as that of several species of cockroaches (Longfellow, 1913; Barber, 1914). One of the first of these investigations was that of Cao who was interested in the ability of flies, cockroaches, and other insects to carry disease agents. This Italian worker published a series of three very interesting and important papers which are frequently overlooked by modern writers.[1] Cao (1906a,b) discovered many of the basic relationships known to exist between nonbiting flies and their microbial flora. He observed such things as the fact that fly larvae are able to take up bacteria with their food; that these bacteria, both those nonpathogenic and those pathogenic for man, may pass unharmed through the insect's alimentary tract and out with the feces; that some bacterial species are capable of persisting and even multiplying in the intestinal canal; and that bacteria taken up by larvae may survive through metamorphosis and be distributed by the adult flies. In 1907 Jackson found as many as 100,000 human fecal bacteria in a single fly, and recognized the fact that these bacteria might easily survive passage through the intestinal canal of the insect. Graham-Smith (1909) examined 148 flies caught in various parts of London and Cambridge, and 35 (24 per cent) possessed externally or internally, or both, bacteria belonging to the colon group. Later (1913) he reported that *Serratia marcescens* could be cultivated in large numbers from the contents of the crop and intestine of the housefly up to 4 or 5 days after inoculation and that it survived in the intestine up to 18 days. Graham-Smith also stated that although it seemed to have been proved that the spores of *Bacillus anthracis* may survive after being ingested by fly larvae, most observers agreed that such nonspore-forming pathogenic organisms as *Eberthella typhosa, Salmonella enteritidis,* and *Shigella dysenteriae,* derived from cultures and added to the food of the larvae, are not present in the flies which emerge, except under very special and

---

[1] The author is indebted to Dr. W. Dwight Pierce for original translations of the works of Cao (1898, 1906a,b).

TABLE I

KINDS OF BACTERIA ASSOCIATED WITH VARIOUS SPECIES OF TICKS

Species of Tick	Number of ticks examined according to stage						Kinds of bacteria present and numbers of strains of each isolated			
	Adults		Nymphs		Larvae		Cocci	Spore-formers	Gram-positive nonspore-forming rods	Gram-negative rods
	Fed	Unfed	Fed	Unfed	Fed	Unfed				
Haemaphysalis leporis-palustris	6	4	18	73	8	100	14	4		2
Amblyomma americanum	5,950		506	7			19	24*	1	7
Amblyomma maculatum	18	19	61		25		22	1		2
Amblyomma cajennense	3		1				1			
Ixodes dentatus	2						2			
Ixodes texanus	15						1			
Rhipicephalus sanguineus	14		33				6	2		
Dermacentor andersoni	486	2,016					47†	7	5	18
Dermacentor albipictus	15	125	10	20			2	2		7‡
Argas persicus	123			60			1		3	3
Ornithodoros rudis		17					4		1	
Ornithodoros coprophilus		5							1	
Otobius megnini	3						3			
Totals	6,635	2,186	629	160	33	100	122	40	11	39

* Mostly *Bacillus cereus*.
† And one strain of bacteriophage.
‡ And two strains of bacteriophage.

highly artificial conditions. However, as stated elsewhere, Bacot (1911) reported that when the food of newly hatched larvae of *Musca domestica* was inoculated with a culture of *Pseudomonas aeruginosa,* viable bacteria remained in the gut during metamorphosis.

Nicoll (1911) and Cox, Lewis, and Glynn (1912) also studied the numbers and varieties of bacteria associated with the housefly, finding large numbers of the coliform type. (See also Hewitt, 1914.)

The internal bacterial flora of many species of ticks consists principally of gram-positive cocci with smaller numbers of gram-positive and gram-negative rods. In a cursory examination of 13 species of ticks (8 genera) the writer found that the gram-positive cocci predominate in all species except perhaps *Amblyomma americanum* and *Argas persicus* (see table I). In a detailed study of *Dermacentor andersoni* it was found that of 2,016 unfed adult ticks, only 1.6 per cent harbored bacteria, but of 486 recently fed ticks, bacteria were found in a minimum percentage of 9.1. One possible explanation of the greater number of bacteria in recently fed ticks is the experimental finding that feeding ticks can ingest bacteria from the surface of skin to which they are attached. The evidence gained in this investigation suggested that the bacterial flora of *D. andersoni* is a fortuitous one consisting chiefly of adventitious organisms apparently acquired from its hosts (Steinhaus, 1942b).

In the years prior to 1900 several bacterial infections or toxic conditions were ascribed to the bites of ticks. For a detailed review of these early observations the reader is referred to a paper by Nuttall (1899, pp. 42-49).

Cecal Bacteria. In certain insects of the order Hemiptera peculiar saclike appendages are found opening into the posterior end of the midgut. These structures (fig. 3), called "ceca" or "bacterial crypts," are of various shapes and sizes and always harbor enormous numbers of bacteria morphologically characteristic for the particular species of insect sheltering them.

This bacteriologic-entomologic relationship was first studied in 1888 by Forbes (1892) during his investigations on contagious diseases of insects. He noted that certain appendages to the alimentary canal in members of the families Scutelleridae and Pentatomidae, and in some Lygaeidae and Coreidae, contained large numbers of bacteria. In the Coreidae and Lygaeidae the cecal structures are present in one genus and absent in another. While the higher Hemiptera (Pentatomidae, Scutelleridae, etc.) invariably possess them, they are always absent in the lower Hemiptera. According to Forbes, the gastric pouches of grasshoppers, cockroaches, and carabid beetles do not commonly contain bacteria. Earlier (1857) Leydig had apparently observed microorganisms in these ceca, but he was not aware of their true nature.

It was not until 1914 that Glasgow undertook a detailed study of this relationship and brought out many interesting facts concerning it. Among other things he observed that the bacteria from the different hosts vary a great deal in their morphology, though they are constant for each individual species of insect; that they range from minute coccuslike bacilli to huge spirochete-

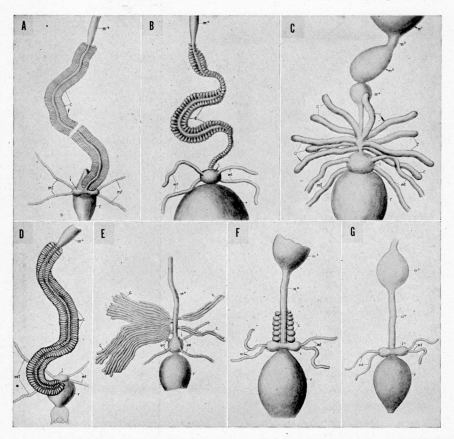

Fig. 3. Parts of the alimentary tracts of various species of Hemiptera showing the location, relative size, and shape of the gastric ceca. A, *Anasa tristis*; B, *Thyreocoris unicolor*; C, *Blissus leucopterus*; D, *Peribalus limbolarius*; E, *Myodocha serripes*; F, *Dysdercus suturellus*, female; G, *Dysdercus suturellus*, male (showing complete absence of ceca in this sex). (From Glasgow, 1914.)

like forms (see fig. 4); and that these bacteria are apparently passed from generation to generation through the egg, since they appear early in the alimentary tract of the developing embryo. This is one of the earliest recorded instances of the transovarial transmission of bacteria in insects. To study the constancy of the presence of bacteria in the ceca, Glasgow examined specimens of the harlequin cabbage bug, *Murgantia histrionica*, secured from widely separated points in the United States. In comparing the flora of these specimens he found that the peculiar, large, spirochetelike forms were constantly present in the ceca regardless of whether the specimen examined "was from California or Maryland." Few of the cecal bacteria could be grown on ordinary culture media, although those from *Anasa tristis* were cultured in nutrient broth.

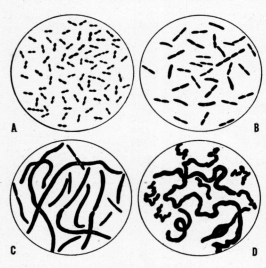

Fig. 4. A few of the morphologic types of cecal bacteria. *A*, bacteria from ceca of *Blissus leucopterus*; *B*, of *Euschistus servus*; *C*, of *Peribalus limbolarius*; *D*, of *Murgantia histrionica*. (From Glasgow, 1914.)

The midgut of these ceca-possessing insects is usually free of the invading bacteria and protozoa commonly present in many related insects. It appears that the normal ceca-inhabiting bacteria inhibit the development of, or exclude, these foreign organisms altogether. Glasgow believes that this is the chief function performed by the cecal bacteria in the life processes of the host. He assumes that the ceca merely provide a safe place for the multiplication of normal bacteria.

Kuskop (1924) lists 23 insects of the family Pentatomidae, 7 of the family Coreidae, and 4 of the family Lygaeidae which, she says, undoubtedly carry bacteria in their cecal appendages. She found the ceca to be as well filled with bacteria after the long period of winter rest as they were during the active summer season. Kuskop believes the bacteria play a symbiotic role, being essential to the insect's digestion.

In 1939 Rosenkranz examined over 30 species of Hemiptera for the presence of cecal bacteria. Of the four groups he investigated (Scutellerinae, Pentatominae, Acanthosominae, and Asopinae) only Asopinae was entirely without the bacteria. In the Scutellerinae he found the symbiotes to be more or less coccoid in shape, while in the Pentatominae they occurred as definite rods, although in both cases they apparently underwent several morphologic changes during their host's life cycle. Some species, such as *Stagonomus pusillus*, harbor two apparently different species of cecal bacteria. According to Rosenkranz, transfer of the symbiotes to the next generation is accomplished during the time of egg laying. The eggs are coated with the bacteria, which are later ingested by the larvae directly after hatching during a resting period in which the larvae stay on the egg surface.

In the Acanthosominae, Rosenkranz found the bacteria to live in saclike crypts, which were constricted from the midgut and which show a sexual size dimorphism. In members of this tribe, from the ninth abdominal segment into the abdomen on each side projects an epidermal fold which consists of numerous chitinous tubes filled with bacteria. This "Beschmierorgan," as Rosenkranz calls it, during the egg-laying process coats the eggs with the bacteria. Just how the bacteria get into this structure from the crypts is not known.

An interesting contribution to an understanding of the biologic relationships between the cecal bacteria and their insect hosts was made by Schneider (1940). This worker observed that in the plataspid *Coptosoma scutellatum,* part of the gut just behind the stomach forms a structure with lateral crypts filled with bacteria. When this insect lays its eggs,

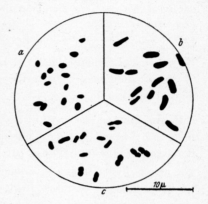

Fig. 5. A comparison of the transmitting forms of the symbiotes of (*a*) *Eurygaster maura* (Scutellerinae), (*b*) *Palomena prasina* (Pentatominae), and (*c*) *Acanthosoma haemorrhoidale* (Acanthosominae). (From Rosenkranz, 1939.)

the bacteria are deposited in packets among the eggs and the newly hatched larvae devour these first. The ingested bacteria then migrate to an appendage at the rear end of the yolk-filled larval gut, which eventually becomes the adult crypt gut. Bacteria-free larvae can be obtained if the bacterial packets are removed from among the eggs. These larvae, however, do not survive more than six to nine days! Whether or not this indicates that the organisms play a nutritional role has not been determined.

Besides the Hemiptera, certain of the chrysomelids in the order Coleoptera have intestinal ceca and vaginal pouches which contain bacteriumlike organisms. In some of these cases (e.g., *Bromius obscurus*) both intracellular and extracellular organisms are present. Stammer (1936) has described some of these relationships, which will be treated more fully in chapter IV.

Bacteria in the Blood. Several investigators have reported the presence of bacteria in the blood of apparently normal insects as well as in diseased individuals. In 1931 Lilly observed that the blood of the housefly, *Musca domes-*

tica, may normally contain one or more species of bacteria. He thought that the extent of this flora probably fluctuates with the age, nutrition, and environment of the fly. According to Paillot (1933), bacteria may occur freely in the blood of aphids. Cameron (1934) states:

> Bacteria are not infrequently found in the blood of normal caterpillars. Taking all precautions to prevent contamination I have cultured micro-organisms from the blood of *Graphiphora triangulum, Gonepteryx rhamni, Smerinthus ocellatus, Endromis versicolora, Graphiphora agathina, Ennomos autumnaria, Agrotis ashworthi, Euchloris vernaria,* and *Aproophyla nigra.*

Tauber and Griffiths (1942) isolated *Staphylococcus albus* from the blood of the roach *Blatta orientalis,* though in this case the insect may have been naturally diseased.

Factors Influencing Type of Bacterial Flora. It is obvious that the environment of a particular insect is important in determining the type of microbial flora associated with it. This is particularly true as concerns the external flora. On insects living in the soil one would expect to find soil bacteria; those living on animals would be likely to carry bacteria associated with the animals' skin or fur; and so on. Certain insects, such as houseflies and cockroaches, may carry one type of flora in an area of filth and quite a different type in "clean" areas.

Those insects which live as scavengers feeding on almost anything edible which comes their way are likely to have in and on their bodies a variety of bacteria. Those arthropods whose feeding habits demand foliage would be expected to have quite a different flora from those which feed only on blood or plant juices. Feeding habits also affect the numbers of bacteria found in a particular insect with the bloodsucking species usually having the fewest.

Both quantitatively and qualitatively the flora may vary according to the particular stage of the insect. This, of course, may be due simply to the fact that the feeding habits of each stage vary. It is easy to understand how the flora of a butterfly may differ from that of the caterpillar from which it came. To be sure, in the case of some insects the bacteria survive through metamorphosis into the adult stage. More frequently, however, the adult acquires a flora of its own.

The flora of some insects may change with the seasons of the year. This change may not necessarily be due to the effect of the season itself but rather to the change of habits of the insect during the year.

Torrey (1912) observed that flies examined up to the latter part of June were free from fecal bacteria of human origin and carried a homogeneous flora of coccal forms. During July and August, periods when the flies examined possessed several millions of bacteria alternated with periods in which the number of bacteria was reduced to hundreds. Bacteria of the colon type were first encountered in abundance during the early part of July. Another example of seasonal incidence has been observed in the case of the bacteria producing soft rot in potatoes. In this case the bacteria pass the winter in the digestive tract of the puparia of *Hylemya cilicrura* (Leach, 1933).

Fate of Bacteria During Metamorphosis. The fate of the bacteria harbored by the larva during the process of metamorphosis to the adult stage has been studied thoroughly in very few insects. More of such knowledge would be particularly valuable from the standpoint of public health. For example, housefly larvae may become a reservoir for bacteria pathogenic to man. Should these bacteria survive metamorphosis and be disseminated by the adult, the chances of spreading disease are great. That such does happen was first indicated by Cao (1906a,b), who observed that bacteria taken up by the larvae of nonbiting flies may survive through metamorphosis into the adult stage and for some time thereafter. The adult may then deposit the bacteria in its feces on food or other materials used by man. In 1909 (a,b) Faichnie suggested the possibility of the spread by adult flies of disease agents acquired in the larval stage. Hence in deciding what is the flora of an adult insect, one has to consider adventitious bacteria that have been acquired, not only by the imago itself, but by the larva as well.

Bacot (1911) found the pupae and imagines of *Musca domestica* bred from larvae infected with *Pseudomonas aeruginosa* under conditions which excluded the chance of reinfection in the pupal or imaginal period remained infected with the bacterium. Other authorities, however, are agreed that such nonsporeforming organisms as *Eberthella typhosa, Salmonella enteritidis,* and *Shigella dysenteriae* added to the food of fly larvae usually do not survive metamorphosis (Graham-Smith, 1913). Later Bacot (1914a) studied the bacteria of the alimentary canal of the flea during its metamorphosis and found that the alimentary canal of the flea larva may become "infected" with the following bacteria if they are mixed with its food: *Pseudomonas aeruginosa, Salmonella enteritidis, Staphylococcus aureus,* and *Staphylococcus albus.* These organisms may persist in the larval gut until the resting period of the larva in the cocoon, but there appears to be no satisfactory evidence that they can survive the pupal stage. In 1921 Pierce recorded at least 11 bacteria which have been shown to persist from larvae through metamorphosis to the adults. Very little recent work has been done to add to this number.

Brinley and Steinhaus (1942) observed that dipterous insects, such as those of the genera *Chironomus* and *Psychoda,* which live as larvae in sewage, are capable of acquiring bacteria during the larval stage and of disseminating them later during the adult stage. Although most of the bacteria in these insects are probably sporadic forms, Léger (1902b) observed three forms in *Chironomus plumosus* which he considered to be truly parasitic. These were a *Streptothrix,* a sporeforming *Bacillus,* and a spirochete similar to *"Sp. anserina."*

An interesting example of the survival of bacteria in an insect during metamorphosis has been shown by the work of Leach (1931, 1933) in the case of the bacterium *Erwinia carotovora,* causing potato blackleg, and the seed-corn maggot, *Hylemya cilicrura.* The maggots pick up the bacteria from their contaminated egg shells, from the soil, and probably from the surface of contaminated potato-seed pieces. After two or three weeks' development in the seed pieces the maggots leave this abode, enter the soil, and pupate. In this manner the bacteria survive the winter in the digestive tract of the puparia. Besides *Erwinia carotovora,* other bacteria regularly pass uninjured through the intestinal tract of both larvae and flies, overwinter in the puparia, and emerge with the adult fly. The most common of these are *Pseudomonas fluorescens* and *P. non-liquefaciens.* Leach (1933) found that the bacteria survive in the puparium in three different locations: in the cast-out linings of the hindgut, in the cast-out linings of the foregut, and in the lumen of the midgut of the pupa. The bacteria found surviving in the fore- and hindgut may be of several varieties since they are the ones that happen to be in these organs at the time of pupation. In the midgut, on the other hand, the bacteria appear to be of one species, resembling, but not identical to, *Pseudomonas fluorescens.* These bacteria become reduced to relatively small numbers during metamorphic histolysis, but just before the fly emerges from the puparium they appear to increase rapidly. According to Leach, there appears to be a selective action on the bacteria surviving in the midgut that is not operative on those surviving in the castoff linings of the fore- and hindguts.

Relationships similar to those just described exist between *Erwinia carotovora* and the cabbage maggot, *Hylemya brassicae. Xanthomonas (Phytomonas) savastanoi,* the cause of olive knot, is also known to survive in the puparium of the olive fly, *Dacus oleae* (Petri, 1910).

Transmission of Bacteria from Generation to Generation. Along with the discussion of the fate of bacteria during metamorphosis should be mentioned the phenomenon of transmission of bacteria from one generation to the next. Many instances of this process are known to occur with intracellular microorganisms.

An outstanding example of the perpetuation of extracellular bacteria through successive generations has been described by Petri (1909, 1910) in the case of *Xanthomonas (Phytomonas) savastanoi,* the cause of olive knot, and the nonpathogenic bacterium, *Ascobacterium luteum,* in the olive fly, *Dacus oleae.* These bacteria occur in the intestinal tract during all stages of the insect's development.

If one were to make a longitudinal section of the ovipositor of the olive fly, he would observe that the vagina and the anal tract unite at their posterior end forming a common opening. Peculiar saclike evaginations, filled with bacteria, occur in the wall of the anal tract near the point of union and open into the lumen (see fig. 6). A longitudinal slit in the membrane which

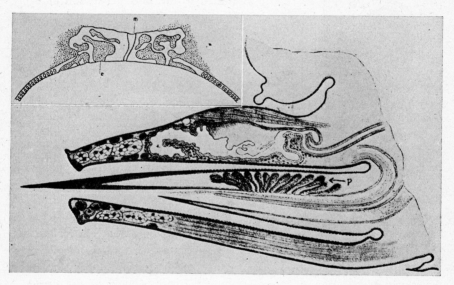

Fig. 6. A longitudinal section through the ovipositor of *Dacus oleae* showing longitudinal slit connecting the anal tract and the oviduct. On passing out, the eggs are pressed against the pockets (P) and are smeared with bacteria. (Inset) A section through the micropyle of an egg through which the bacteria enter the egg and "infect" the embryo. *m,* micropyle; *c,* air tubes filled with bacteria. (From Petri, 1910.)

separates the anal tract from the oviduct lies immediately opposite the opening of the evaginations. As the eggs pass along the vagina, the surface of each egg is pressed through this slit, against the openings. The bacteria contained within the evaginations are smeared over the surface of each egg, thence finding their way through the micropyle (fig. 6, inset) into the egg. The bacteria are then incorporated into the embryologic development of the insect.

The larvae which hatch from the eggs possess four spherical ceca near the fore part of the midgut. These ceca (fig. 7) contain the bacteria which may also be found throughout the lumen of the alimentary tract. During the pupal stage a bulblike diverticulum branches off the esophagus just in front of the brain. (A similar structure has been observed by Dean, 1933, 1935, in the apple maggot, *Rhagoletis pomonella*, but its possible relation to bacteria has not been determined.) The bacteria accumulate in this structure (fig. 8), from which they later (in the adult fly) spread throughout the alimentary tract, including the anal saclike evaginations. From this location the bacteria are transmitted to the next generation of eggs, and thus they are perpetuated. According to Stammer (1929a) and Allen and Riker (1932), similar bacterium-insect relationships exist in other species of Trypetidae. In *Tephritis heiseri* a similar aperture exists between the vagina and hindgut, but the latter does not possess the claviform protrusions of *Dacus oleae*. Instead, this area of the hindgut is differentiated into long-drawn-out channels which are narrowed in the direction of the opening. These are filled with bacteria which are applied to the eggs during oviposition. The freshly laid egg is covered with a layer of mucus in which the bacteria multiply until they enter the egg through the micropyle.

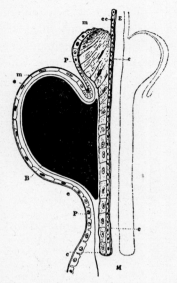

Fig. 7. A longitudinal section of the ceca of the anterior portion of the midintestine of the larva of *Dacus oleae*. *B*, bacteria colony; *c*, cuticula of proventriculus; *E*, esophagus; *ee*, esophageal epithelium; *e*, epithelium of the ceca; *M*, mid-intestine; *m*, musculature; *P*, peritrophic membrane. (From Petri, 1910.)

Among the ticks instances of generation-to-generation transmission of bacteria are not so well known. One important case in this regard, however, is the generation-to-generation and stage-to-stage transmission of *Pasteurella tularensis* in the Rocky Mountain wood tick, *Dermacentor andersoni*. This discovery was made by Parker and Spencer (1924). Philip and Jellison (1934) observed that a similar transovarial transmission takes place with this bacterium in *Dermacentor variabilis* but less frequently perhaps. *Salmonella enteritidis* may possibly be passed on occasionally through the egg of *D. andersoni* to the active stages of the next generation (Parker and Steinhaus, 1943c).

Bacteria and Insect Eggs. Atkin and Bacot (1917) and Bacot (1917) found that a great stimulus to the hatching of mosquito eggs (*Aëdes aegypti,* = *Stegomyia fasciata*) was the introduction into their environment of living yeasts or bacteria. The stimulus produced by killed cultures of bacteria and sterile watery extracts of brewers' yeast was more feeble, many of the eggs failing to hatch. Sterile filtrates of bacteria were less effective than killed cultures. The methods of experimentation were simple. Different species of living bacteria were introduced into tubes of sterile media, such as peptone water, in which the eggs had been lying dormant for 11 to 15, and in some cases 39, days. Upon inoculation with the bacteria all eggs hatched within 18 hours. Atkin and Bacot explain this phenomenon by supposing that the stimulus is of the nature of a "scent" which penetrates to the larvae lying dormant within the egg shells, causing them to make vigorous movements which result in the uncapping of the eggs. These results are similar to those obtained later by Roubaud and Colas-Belcour (1927). These workers believed that soluble ferments produced by the bacteria and yeasts were responsible for the hatching. They based this belief on the fact that certain enzymes such as pepsin, papainase, trypsin, and tyrosinase appeared to cause the eggs to hatch.

Fig. 8. Longitudinal section through the head of an adult of *Dacus oleae* showing the bulbous diverticulum off the esophagus (*E*). The diverticulum retains the bacteria during metamorphosis. Later the entire intestinal tract is recontaminated from this organ. (From Petri, 1910.)

On the other hand, Barber (1928) found "there was no indication that bacteria promoted hatching in either *C.* [*Culex*] *quinquefasciatus* or *A.* [*Aëdes*] *aegypti*. Eggs hatched out in water or in clear sterile media as promptly as in contaminated cultures. In a few cases bacteria seemed to encourage the hatching of eggs of *A.* [*Aëdes*] *sollicitans,* but they were surely not a necessary stimulus." Similarly, Hinman (1930) obtained results which were in direct opposition to those of Atkin and Bacot. Hinman "repeatedly found that eggs of this mosquito [*Aëdes aegypti*] (and also other species) hatched apparently as rapidly in sterile as in contaminated media."

Some of the differences of opinion on this phenomenon were reconciled by Rozeboom (1934), who found that a great deal depends on the age and condition of the egg. Of 240 old, dry eggs (*Aëdes aegypti*) only 4 hatched in sterile media, whereas 204 hatched within two days following inoculation of the media with bacteria. Of fresh, moist eggs, 35 per cent hatched in distilled water, and 82 per cent in water contaminated with bacteria.

Gjullin, Yates, and Stage (1939) found that tap-water infusions of dry cottonwood leaves, willow leaves, and grass gave consistently larger hatches of *Aëdes vexans* and *Aëdes aldrichi* (*A. lateralis*) eggs than either tap or river water alone. They concluded that the amino acids and proteins present in vegetation may be the stimulants which cause the eggs to hatch when flooded in nature. Later, however, Gjullin, Hegarty, and Bollen (1941) concluded that an important stimulus to hatching was a reduction of the dissolved oxygen in the medium containing the eggs. At least this was the case with the eggs of certain *Aëdes* mosquitoes. This condition may be brought about by several factors including the reducing action of microorganisms.

More recently Thomas (1943) reported that when *Aëdes aegypti* eggs were placed in sterile solutions they rarely hatched but that the introduction of bacteria and molds did stimulate the hatching of dormant eggs. He found the following heterotrophic bacteria and fungi capable of bringing about this stimulation: *Escherichia coli, Sarcina lutea, Pseudomonas aeruginosa, Serratia marcescens, Saccharomyces cerevisiae, Aspergillus niger, Penicillium* sp., *Rhizopus nigricans, Fusarium moniliforme,* and *Glomerella grossypii.* Thomas also observed that when the eggs were crowded together a relatively large percentage hatched, though not as many as when microorganisms were introduced. He thought perhaps this might explain why Barber (1928), MacGregor (1929), Hinman (1930), Rozeboom (1935), and Trager (1935a, 1937) did not observe any relationship between the presence of living microbes and the hatching of mosquito eggs.

In this connection an interesting observation of Hinman's (1932) should be mentioned. This worker found viable bacteria within the eggs of *Aëdes aegypti* and other mosquitoes. By both cultural and microscopic examinations he found cocci, bacilli, and yeast within the mosquito ova. The most common type of bacteria found in sections was the coccus, with bacilli rarely being encountered. As Hinman points out, probably only a relatively small percentage of eggs actually contains microorganisms.

According to Masera (1936d) the Russian workers Stieben, Kasarov, and Ziklauri isolated the following organisms from the eggs of normal silkworms, *Bombyx mori:* "*Bact. fermentationis, Bac. simplex, Bac. mesentericus, Bac. mesentericus ruber, Bac. novus, Bac. subtilis, M. candidus, M. perflavus, M. cinnebareus, Bact. turcosum.*"

Variation of Entomophytic Bacteria. It is well known that bacterial cells may change in shape, size, and structure. Some of these changes are due to changes in environment and are not inherited. In other cases the changes are more stable and are the result of artificial selection, and in still other instances distinct mutationlike changes occur.

Into which of these categories fall the various instances of variation among entomophytic bacteria it is difficult to say. Besides the occurrence of bacterial variation within the insect host itself, this phenomenon has also been observed in artificial cultures isolated from the host and in other insects artificially inoculated with the bacterium concerned.

As in the early history of bacteriology, variation of bacterial species in insects has caused considerable trouble and controversy among investigators. Typical of this is the case of *Bacillus alvei,* the cause of European foul brood of bees. Cheshire and Cheyne (1885) were the first to isolate this organism as the etiologic agent of the disease. Maassen (1907) believed that either *Bacillus alvei* or *Streptococcus apis* was the cause. White (1912, 1920b,c) was unable to produce typical European foul brood with *Bacillus alvei, Streptococcus apis,* or *Bacterium (Achromobacter) eurydice* and concluded that a new species, *Bacillus pluton,* was the real cause. Burnside (1934) attempted to bring some order out of this confusion by suggesting that *Bacillus pluton, Streptococcus apis,* and *Achromobacter eurydice* are variants or stages in the life history of *Bacillus alvei.* He found that "*Bacillus alvei* is capable of morphological, cultural, and biological transformation and is also capable of stabilization, at least temporarily, as a sporogenic rod, an asporogenic rod resembling *Bacterium eurydice,* or as a coccoid form resembling *Bacillus pluton."*

Paillot (1933) has found that the majority of coccobacilli isolated from insects change their form more or less according to the insect into which they are inoculated. The bacteria may undergo such minor changes as a slight elongation of the cell, or the changes may be of a more striking nature. Such variations have been observed with *Bacterium pieris liquefaciens alpha, Bacterium melolonthae liquefaciens gamma,* and *Bacterium lymantricola adiposus.* For instance, *Bacterium pieris liquefaciens alpha* is in the form of coccobacilli in the blood of the larvae of the cabbage butterfly, *Pieris brassicae.* In the blood of the larvae of *Vanessa urticae,* there is no appreciable difference. In the larvae of *Vanessa polychloros* and *Euproctis chrysorrhoea* the cells are considerably longer than in the first two species. In the blood of *Porthetria dispar,* however, the elongation is so great that the bacteria lose all aspects of coccobacilli and are transformed into sinuous filaments which may attain the length of 40 or 50 microns (see fig. 9). When inoculated back into the general cavity of the *Pieris brassicae* larvae, the cells return to their normal form.

Bacterium melolonthae liquefaciens gamma, which usually appears in the form of a coccobacillus, becomes elongated and thicker when inoculated into the larvae of the gypsy moth, *Porthetria dispar.* As the infection advances, a certain number of the bacteria show one or two median or polar swellings. According to Paillot, these swellings later become detached from the bacterial elongations and float freely in the blood, though they are not actively motile like the bacilli from which they originated. These forms resemble small true cells, possessing a central portion which may be taken for a true nucleus since it takes nuclear stains. When the insect dies these forms gradually disappear.

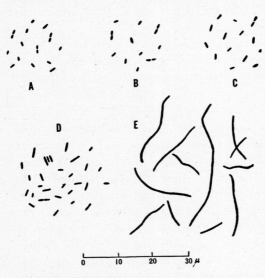

Fig. 9. Variations of *Bacterium pieris liquefaciens alpha*. A, in the blood of *Pieris brassicae*; B, in the blood of *Vanessa urticae*; C, in the blood of *V. polychloros*; D, in the blood of *Euproctis chrysorrhoea* (*Nygmia phaeorrhoea*), and E, in the blood of *Porthetria dispar*. (From Paillot, 1933.)

Similar "growth forms" have been observed in certain strains of *Bacterium lymantricola adiposus* inoculated into *Porthetria dispar*. In this case they originate from an enlargement of the central portion of the bacterium. These enlargements continue to elongate and often give rise to secondary elongations, the whole thing resembling a kind of mycelium. In old strains the bacteria and "growth forms" may be seen to possess a large central vacuole which disappears when the elongations reabsorb themselves into new rounded forms, which float freely in the blood. Similar forms of this bacterium have been described as developing in the blood of the larva of *Euxoa segetum*. Such forms of variation have also been observed in the case of *Bacterium pieris liquefaciens alpha*.

Paillot theorizes that these "growth forms" probably represent a degenerate stage of the bacteria. The increased ability of the organisms to give rise to these forms corresponded to a diminution in their original virulence for the insects. This hypothesis is similar to that which, according to some workers, characterizes the symbiotes of aphids. In the absence of detailed work in this field one wonders what types of variation might occur by the passage through insects of some of the better-known bacteria not usually found in association with insects. One indication of such interesting possibilities is that afforded by the work of Lal, Ghosal, and Mukherji (1939), who found that certain morphologic, metabolic, and chemical changes occurred in strains of *Vibrio comma* passed through houseflies (*Musca domestica*).

Involution forms occurring on artificial media frequently arise in cultures of bacteria isolated from insects. *Bacillus liparis* is normally a slightly elon-

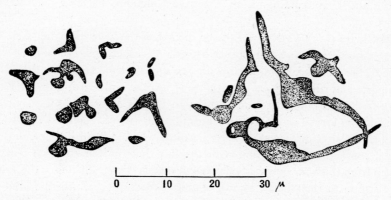

Fig. 10. Growth forms of *Bacterium lymantricola adiposus*. (From Paillot, 1933.)

gated, straight, or slightly bent rod, but when cultivated on carbohydrate media such as levulose agar, huge forms, swollen in clubs at one or both of their extremities, may be observed. These forms resemble very closely certain involution forms of the diphtheria bacillus.

Bacterium neurotomae, isolated from *Neurotoma nemoralis,* appears in the blood of various insects in the form of elongated but rarely filamentous rods. In young cultures on agar, however, some of the cells swell greatly and become more or less rounded. In the central part of these rounded cells is consolidated the chromatophilic substance, and the whole aspect is one of true nucleated cells. According to Paillot, they are without vitality and rapidly degenerate.

From the lyreman cicada, *Tibicen linnei,* Steinhaus (1941a) isolated a gram-positive bacterium (*Bacterium mutabile*) which normally has the form of a short rod. In fluid media such as tryptophane broth, bizarre, slightly branched forms appear.

THE ROLE OF BACTERIA
IN THE NUTRITION OF INSECTS

Inasmuch as most insects harbor large numbers of bacteria within their digestive tracts, it is apparent that these microorganisms may exert a profound effect upon insect physiology and nutrition. Despite this, very little investigation has been made of the actual function of bacteria and other microorganisms in such processes.

Bacteria as Food. Besides the possibility of being related to the food habits of an insect and aiding in its digestive functions, bacteria may serve as food.

Mitchell (1907) was one of the first to express the belief that the "wriggler" of *Aëdes aegypti* is pre-eminently a bacteria feeder, because the larvae develop rapidly in water contaminated with sewage. In later years her belief was supported by the work of Bacot (1916), Atkin and Bacot (1917), Barber (1928), Rozeboom (1935), and others. In Bacot's report the suggestion that the bacteria served as food for the mosquito larvae was based on the clearing action the latter displayed in water, originally turbid from its enormous bacterial content, in conjunction with the fact that the gut contents of larvae taken from this water showed relatively few bacteria. He attributed the scarcity of bacteria to their being rapidly digested. Barber (1927, 1928) found that algae alone, bacteria alone, or infusoria alone may serve as a sufficient source of food for *Anopheles* larvae but that a combination of bacteria with infusoria or with algae appeared to afford the best conditions for the growth of *Culex quinquefasciatus* and of *Aëdes aegypti*.

From the intestinal examination of over 600 mosquito larvae, Hinman (1930) concluded that the larvae ingest any material small enough to be taken in through the mouth. A considerable amount of this material appears to pass unchanged through the alimentary canal. Whereas larvae failed to develop in sterile, synthetic media or in autoclaved water taken from normal breeding places, the addition of certain types of bacteria to such water made it a suitable medium for complete larval development. Hinman (1933) later demonstrated the existence of a factor in bacteria which stimulated the growth of mosquito larvae, but he was unable to extract this vitaminlike substance from the bacteria with any regularity. Filtrates from these cultures failed to stimulate development. In 1935 Rozeboom studied the problem and concluded that bacteria, to a certain extent, can be utilized as food by mosquito larvae, though all kinds of bacteria are not equally suitable for such larvae. "Environmental bacteria," associated with the natural breeding places of mosquitoes, proved to support the best development of the larvae when bacteria were the only source of food. *Sarcina lutea* was of little value, and *Escherichia coli*, *Bacillus subtilis*, *Bacillus mycoides*, *Aerobacter aerogenes*, and *Pseudomonas fluorescens* were of equal value. In media inoculated with *Pseudomonas aeruginosa* (*P. pyocyanea*) the toxic products of this organism rapidly killed the larvae. Rozeboom's attempts to grow mosquito larvae in the absence of bacteria were unsuccessful. Trager (1935a,b) obtained normal development of the larvae of *Aëdes aegypti* in the absence of living microorganisms. He used a medium consisting of a standard, autoclaved, protein-free, liver extract with autoclaved yeast. He demonstrated that the larvae require two accessory food substances. One is present in yeast and aqueous yeast extracts, egg white, and wheat. It is heat- and alkali-stable and is not adsorbed by fuller's earth. The other is present in large amounts in purified liver extracts rich in the anti-anemia principle. It is heat-stable but cannot withstand the action of alkalis. In a slightly acid solution it is almost completely adsorbed by fuller's earth. Interestingly enough, it has been found (Trager, Miller, and Rhoads, 1938) that a substance, possibly flavine or a flavine compound, occurs in extracts prepared from urine of normal persons or patients with aplastic anemia or leukemia which enhances the growth of larvae of *Aëdes aegypti*.

A relationship similar to that of the mosquito larvae and the bacteria in contaminated water is suggested by Von Wolzogen Kühr (1932) between the larvae of *Chironomus plumosus*, which frequents sandfilters in the summer, and *Pseudomonas fermentans*. This was attributed to the presence in the filters of the bacterium, upon which the larvae supposedly fed. A similar situation was described by Dyson and Lloyd (1933) in sewage beds.

One of the first to advance the idea that bacteria are indispensable to the growth of certain insects was Bogdanow (1906), who found that the larvae of *Calliphora vomitoria* fail to develop in the absence of microorganisms. Later (1908) he stated that the larvae require a definite and fairly simple bacterial flora. Sterile larvae on sterile food never developed normally, although some of them reached the pupal stage. Weinland (1907), however, showed that the larvae of *Calliphora* are able to digest meat without the assistance of bacteria. Bogdanow also found that larvae of the housefly, *Musca domestica*, can be bred on starch paste or gelatin, but only in the presence of molds and bacteria. Wollman (1921) reported that microbe-free cultures of flies can be maintained indefinitely, as can also similar cultures of the moth *Galleria mellonella*. The work of Glaser (1924b) showed that the growing larvae of flies were dependent on certain accessory growth factors which may be obtained from bacteria and yeasts, but that microorganisms and their activities are not absolutely essential to the normal growth, development, and longevity of flies. Later (1938a) he developed a method whereby houseflies could be raised in sterile culture, free from microorganisms. Baumberger (1919) reported that the larvae of the fly *Desmometopa nigra* are probably dependent on microorganisms and that the larvae of the housefly probably feed on them. Trypetidae larvae can develop only when microorganisms are present, according to Stammer (1929a).

Bacteria and the Physiology of Insect Digestion. Considerable evidence has been advanced that bacteria may play a greater role in the nutrition of insects than merely serving as food. Bacteria are capable of producing proteolytic, lipolytic, saccharolytic, amylitic, and other enzymes, which no doubt exert considerable influence on the digestive processes of the insect harboring them.

The best-known examples of such a relationship are those concerned with the intestinal flagellates which take an active part in the digestion of cellulose in the gut of the termite and in the wood-feeding roach *Cryptocercus*. However, we shall limit ourselves here to a brief discussion of the bacteria which maintain similar relationships.

Petri (1905) was one of the earliest to assign to the bacteria a definite digestive role. The bacteria constantly present in the ceca of the olive fruit fly (*Dacus oleae*) were found to produce lipase. It was suggested that the activity of the bacteria in the digestion of fats must be very important for the larva, which feeds on the olive, a fruit rich in fats. In a later paper (1910) he asserted that partial digestion of the oil might be possible without the aid of bacteria, since many larvae living on seeds rich in oil do not possess intestinal bacteria. Bogdanow (1906) believed that the formation of ammonia during the larval development of *Calliphora vomitoria* is not a characteristic of protein digestion by the larvae but probably a result of bacterial activity. Weinland (1907), on the other hand, insisted that the ammonia is the result of larval metabolism. Wollman (1911, 1921) indicated that Weinland was mistaken as no ammonia is produced by sterile larvae, and its production, therefore, must be due to microorganisms. Weinland (1908) observed further that bacteria take no part in the process of fat formation in the larvae. Guyénot (1906, 1907) found that muscid larvae (mostly those of *Lucilia*) are unable to produce any digestive ferments which liquefy meat. He believed that this liquefaction is accomplished by bacteria. On the other hand, Wollman (1921) claimed that aseptically bred larvae liquefy gelatin, which indicates that they produce some proteolytic ferments.

In the intestine of the larvae of the wax moth, *Galleria mellonella*, Dickman (1933) found bacteria which may in part be responsible for the utilization of the beeswax on which this insect feeds. It is possible that from the wax these intestinal bacteria produce intermediate substances which in turn can be digested by the larvae.

One may expect to find almost any saccharolytic enzyme in the digestive tracts of insects if one considers the variety which may arise from the bacteria they harbor. The fermentative ability of bacteria isolated from the alimentary tracts of insects and ticks varies greatly. Some can ferment almost no carbohydrates, others 25 to 30. It is evident that in the insect the bacteria would not be called upon to produce most of these enzymes unless the appropriate substrates were included in the arthropod's food. In the case of the cattle grub, *Hypoderma lineata,* Simmons (1939) found the following enzymes to be present: lactase, maltase, invertase, glycogenase, lipase, trypsin, and erepsin. He believed the lactase, maltase, and invertase to be products of bacteria in the intestine of the larva. Brown (1928) believed that most of the enzymes found in the honeybee are produced by microorganisms. In most of the studies on the saccharolytic enzymes of insects too little attention has been given to the large amounts of these enzymes which bacteria are capable of producing.

Portier (1911) claims that leaf-mining larvae of *Nepticula malella* and *Gracilaria syringella* live under sterile conditions and do not harbor any microorganisms in their bodies. On the other hand, the normal leaf-feeding larva of the silkworm, *Bombyx mori,* has its digestive tube populated with bacteria, some of which destroy the wall of the leaf cell, while other bacteria thrive on the contents which are used directly as food. Glaser (1925a), however, reared large numbers of silkworms and rarely found many bacteria in the digestive system of normal worms. When bacteria became numerically high, the worms ailed and died. Hering (1926) criticized some of the views of Portier, stating that up to that time no true "symbionts" were known in leaf miners. Werner (1926a) found the digestive tract of the larva of *Potosia cuprea* to have a very rich microflora able to cause the fermentation of cellulose. A specific bacterium was isolated and named *Bacillus cellulosam fermentans*. Schütte (1921) found that cellulose is digested by the larva of *Hydromyza livens,* but apparently without the aid of bacteria.

It should be remembered, when one is considering the role of cellulose-fermenting bacteria in the nutrition of insects, that in most phytophagous insects the food passes through the gut so rapidly that no great amount of fermentation is likely to take place. The breakdown of cellulose by bacteria is usually too slow a process to be initiated and completed in the few hours during which food remains in the gut. On the other hand, cellulose-splitting bacteria are often associated with the food ingested by insects and for this reason cannot be completely ignored. Furthermore, certain insects, such as the lamellicorn larva, possess "fermentation chambers," which are probably used for such purposes.

In 1919 Roubaud asserted that adult tsetse flies were exclusively hemophagous. The blood ingested by the flies was digested only in the middle section of the gut where the epithelial cells included symbiotic organisms. According to Roubaud, these organisms play an important part in the digestion of the blood. Wigglesworth (1929) states, however, that there is no evidence that these organisms play any part in the digestion of the blood.

Fermentation Chambers. As has already been mentioned, the gut of certain insects, notably lamellicorn larvae, possess special sacs or chambers containing bacteria which are probably responsible for breaking down the cellulose ingested by the insect (fig. 11). Cuticular areas bearing branched spines occur on the walls of the chamber. The thin cuticle between these areas is pierced by fine canals. It appears that most of the digestion and absorption takes place within this chamber since the tiny particles of cellulose and wood are retained here for long periods of time and are acted upon by the cellulose-fermenting bacteria therein.

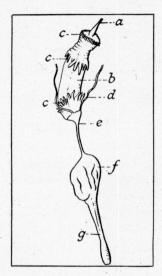

Fig. 11. The alimentary tract of an *Oryctes nasicornis* larva showing the bacterial fermentation chamber in the hindgut. *a*, esophagus; *b*, midgut; *c*, rings of midgut ceca; *d*, Malpighian tubes; *e*, hindgut; *f*, fermentation chamber; *g*, rectum. (From Wigglesworth's *The Principles of Insect Physiology*, 1939, E. P. Dutton and Co., New York. After Mingazzini.)

According to Werner (1926), larvae of *Potosia* (*Cetonia*), which feed on the decaying pine needles found in ant heaps, thrive only at those temperatures optimum for the cellulose-fermenting bacteria. Similar fermentation chambers are also possessed by certain lipulids. There are some insects, such as the larvae of *Dorcus* and *Osmoderma*, which possess "fermentation chambers" filled with bacteria which apparently do not break down the cellulose they ingest.

Bacteria as a Source of Vitamins and Growth-Accessory Substances. Portier (1919) was one of the first to suggest that the source of vitamins for the individual insect is the intracellular organisms it possesses. Wollman (1926) probably overlooked this possibility when he claimed that cockroaches (*Blattella germanica*) may dispense with vitamins and generalized that perhaps vitamins are not essential to insects. Though others (Frost, Herms, and Hoskins, 1936; Bowers and McCay, 1940) have shown that mosquitoes, cockroaches, and other insects can apparently do without certain vitamins, it has been definitely demonstrated that by and large insects need the essential growth substances as do higher animals. Some writers (Imms,

1937) have speculated that the chief functions of bacteria in insects are to supply growth-promoting substances and to liquefy the food.

Hobson (1933) found that the larvae of the blowfly, *Lucilia sericata,* were unable to develop aseptically on sterile blood owing to the lack of growth factors of the vitamin B type. The presence of bacteria improved growth, and yeast autolyzate allowed the larvae to grow at a normal rate. Later (1935) he reported that the natural flora must supply the necessary vitamins and that larvae grow readily on blood inoculated with pure cultures of various bacilli isolated from the gut and from blown meat. *Escherichia coli* proved equally effective in these experiments. The observations of Wigglesworth (1936) on *Rhodnius prolixus* support the view that symbiotic organisms in exclusively bloodsucking insects provide an endogenous source of vitamins.

At this point we may conclude that insect larvae can be reared on sterile media if they are supplied with all the necessary food factors. As stated by Wigglesworth (1939):

If these are deficient, infection with microorganisms (in the case of *Drosophila,* particularly the introduction of yeasts) improves the rate of growth. Sterile *Lucilia* larvae will grow on beef muscle; they fail to grow in guinea pig muscle; but if this is infected with *Bacillus coli* or if a yeast extract is added to it, normal growth takes place. . . . In these cases there is little doubt that the microorganisms are synthesizing the necessary vitamins of the 'B' group.

In connection with a discussion of growth-accessory substances might be mentioned the interesting discovery by Tatum (1939) that certain bacteria synthesize a "hormone" which can change the eye color from white to brown in *Drosophila* flies being reared on tryptophane. Neither the bacteria nor tryptophane separately have any influence on the production of eye pigment. In the presence of both the bacteria and tryptophane, however, eye pigmentation is greatly increased. According to Tatum, this shows that tryptophane is able to modify eye color only through the intermediation of microorganisms.

Bactericidal Principle Associated with Ticks and Insects. Through the work of Hindle and Duncan (1925) and Duncan (1926) it is known that certain arthropods possess a bactericidal principle in their alimentary tracts. These workers found that while *Bacillus anthracis, Bacillus subtilis,* and *Streptococcus faecalis* were able to survive in the alimentary canal of the fowl tick, *Argas persicus,* others, such as *Staphylococcus aureus,* died quickly after ingestion by the tick, and when tested *in vitro* the stomach contents were found to be definitely bactericidal to *Staphylococcus aureus, Bacillus anthracis,* and *Bacillus mycoides.* The results with *Pasteurella pestis* and *Bacillus subtilis* were inconclusive. *Eberthella typhosa*, *Serratia marcescens*, *Brucella abortus*, and *Streptococcus faecalis* were not affected. The inhibitory principle, the potency of which varied with the individual tick, was not inactivated by exposure to a temperature of 58° C. for thirty minutes.

Duncan (1926) investigated further the nature of the bactericidal action and its occurrence or nonoccurrence in the following arthropods: *Stomoxys calcitrans*, *Musca domestica*, *Anopheles bifurcatus*, *Aëdes cinereus*, *Cimex lectularius*, *Rhodnius prolixus*, *Argas persicus*, and *Ornithodoros moubata* He demonstrated a bactericidal principle in the gut contents of all of these, and with the exception of the last two (ticks), in the feces as well. Staphylococci and the sporeforming aerobes were the bacteria most affected by this bactericidal principle. These included *Staphylococcus aureus*, *Staphylococcus albus*, *Bacillus anthracis*, *Bacillus subtilis*, *Bacillus mesentericus*, and *Bacillus vulgatus*. Also inhibited by the gut contents of *Argas persicus* were *Neisseria catarrhalis* and *Streptococcus hemolyticus*. Apparently there is only one active principle in any given species of arthropod, but different groups of bacteria possess varying degrees of susceptibility to it. The widest range of action, in Duncan's tests, was exhibited by the gut contents of *Argas persicus* and *Stomoxys calcitrans* and the narrowest by those of the bugs. The sporeforming bacilli were the most susceptible to the gut contents of *S. calcitrans* and the staphylococci were more affected by the material from *A. persicus*.

As to the properties of the active principle, Duncan states:

. . . bactericidal action is greater and more rapid at 37° C. than at room temperature. This action is not accompanied by any visible bacteriolysis. The bactericidal principle retains its activity unimpaired for at least six months when kept in the dry state. It is very thermostable, resisting temperatures as high as 120° C. It is not destroyed by tryptic digestion. It is precipitated from solution with proteins by alcohol or acetone, but is not itself affected by these reagents. It is not soluble in the common fat-solvents, ether, chloroform, alcohol, or acetone. By allowing it to act upon repeated small doses of bacteria, it rapidly becomes exhausted and it can be inactivated, possibly through adsorption, by large doses of killed bacteria; even those species which are not destroyed by it. It may also be adsorbed in small amount by bibulous paper. It exhibits none of the properties of bacteriophage, and it differs from lysozyme.

Regarding the source of the active principle, there is no doubt that it is formed in the stomach, but whether as a secretion of the gastric cells or as a result of the processes of digestion is not clear. (Nuttall (1908) showed that the destruction of *Spirochaeta duttoni* in the gut of the bedbug was definitely related to digestion.)

Surgical Maggots. According to Livingston and Prince (1932), as early as the sixteenth century Paré observed that suppurating wounds in which blowflies had deposited their eggs healed with unusual rapidity. Larrey, the famous surgeon of Napoleon, observed that during the Syrian campaign the presence of larvae in the soldiers' wounds enhanced the healing processes. Other early physicians noticed the relationship between maggots and the healing of wounds. The real impetus to the study of this relationship came with the observations of Baer (1929, 1931) who, during the first World War, noticed that men wounded in battle and left unattended on the battlefield for as long as seven days before being taken to the dressing stations frequently had their wounds infested with maggots. These men had no fevers and did not develop infections nearly so often as did those who had received early treatment. On cleaning their wounds, he observed that, instead of pus and debris, they were filled with healthy, pink, granulation tissue. Baer concluded:

Maggots have been found to be a tremendously useful adjunct to thorough surgical treatment of chronic osteomyelitis. . . .

Maggots, by their digestive action, clear away the minute fragments of bone and tissue sloughs caused by operative trauma in a way not accomplished by any other means. This is a tremendously valuable asset in the healing of a wound.

Maggots cause wounds to become alkaline and in this way diminish growth of pathogenic bacteria.

Maggots seem to have other more subtle biochemical effects within the wound itself and perhaps cause also a constitutional reaction inimical to bacterial growth.

Though Baer did not live to investigate fully the "more subtle biochemical effects," subsequent investigation has shed more light on this phenomenon.

The following species of fly larvae were used in the early treatment of osteomyelitis: *Lucilia sericata, Lucilia caesar, Phormia regina,* and *Wohlfahrtia nuba. Lucilia sericata* was used most commonly. After working with this species, Stewart (1934) concluded:

L. sericata larvae are beneficial in osteomyelitis wounds because they ingest, by means of macerating mouth-hooks and excreted tryptase, acid forming and bacterial-growth-supporting necrotic tissue; because, most, if not all, of the bacteria ingested with the necrotic tissue and pus are killed by the acid in the middle portion of the mid-intestine; because they alkalize the wounds by means of excreted ammonia and calcium carbonate, and thereby reduce swelling, consequently increasing drainage and decreasing bone destruction and protect tissue cells from autolysis; because the exuded calcium carbonate stimulates phagocytosis; because the bacterial exotoxin is probably rendered inert by the acid condition of the middle region of the mid-gut; and because they promote the growth of healthy granulation tissue apparently by either raising the pH of the wound or by the activity of the exuded calcium ions, or both.

In 1935(c) Simmons obtained, from the excretions of the maggots, a thermostable bactericidal substance which would kill such bacteria as *Staphylococcus aureus,* hemolytic streptococci, and *Clostridium welchii.* In the same year Robinson (1935) isolated allantoin from maggot excretions.

Allantoin occurs naturally in animal and plant tissues as a metabolic product from the breakdown of cell nuclei. The amount of allantoin in the excretions of maggots is too small and the process of extraction too involved to make maggots a practical source for this substance (Robinson, 1937). Other methods of preparing the chemical were devised, and it was soon generally available. Upon hydrolysis, one of the side chains of allantoin is split off and goes to form urea. Although, according to Robinson (1937), it has not been shown that the effect of allantoin is due to this hydrolysis, it was soon found that urea (also present in the excretions of maggots) likewise has definite healing properties. Thus through the careful observation of maggots in human wounds the healing properties of allantoin and urea were rediscovered,[1] and subsequent reports (Robinson, 1940) indicate the gaining of ever greater rewards from this line of research.

Bacteria Associated with Termite Flagellates. Although themselves not directly associated with insects, there are certain bacteria and bacteriumlike

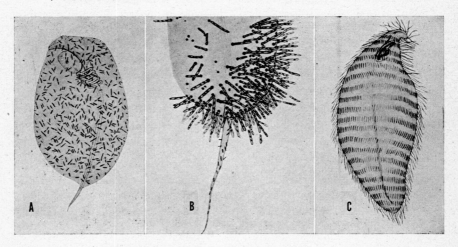

Fig. 12. Bacteriumlike organisms associated with protozoa of termites. *A*. Short rods adherent to the surface *Macrotrichomonas pulchra* from the termite *Glyptotermes taveuniensis*. *B*. Rods resting in cytoplasmic cuplike structures in the cytoplasm of *M. virgosa* from *Procryptotermes* sp. *C*. *Fusiformis*-like rods arranged in transverse bands on *Caduceia bugnioni* from *Neotermes greeni;* spirochetes shown at edge. (From Kirby, 1942a.)

forms which occur in and on certain protozoa found exclusively in the guts of termites (see chapter XI). These interesting forms have been seen in certain of the termite protozoa for many years, but only recently have they

---

[1] The original but forgotten discovery of the healing properties of allantoin was made by Macalister (1912).

gained the attention of investigators in this country, chief of whom is Kirby (1942a, 1944).

As shown in figures 12 and 13, these bacteria may be in the form of bacillary rods or of more or less spherical cocci. They have not, with certainty, been cultivated on artificial media; hence, practically nothing is known concerning their identity or physiologic characteristics.

Some of the coccoid organisms have been given names and have been placed in the family Micrococcaceae. The genus *Caryococcus* was established by Dangeard in 1902 for an organism (*Caryococcus hypertrophicus*) parasitizing the nucleus of *Euglena deses*. In this same genus Kirby (1944) has placed several of the nuclear parasites of termite flagellates. Examples are *Caryococcus nucleophagus* and *Caryococcus cretus* in *Trichonympha corbula, Caryococcus invadens* in *T. peplophora,* and *Caryococcus dilatator* in at least six species of *Trichonympha*. The termites, in most of these cases, belong to the genera *Kalotermes, Neotermes,* and *Procryptotermes*. The name *Micrococcus batrochorum* was given to a parasite of *Tritrichomonas batrachorum* by Yakimoff (1930). The taxonomic status of other similar parasites (*Caryoletira, Nucleophaga*) is still in doubt.

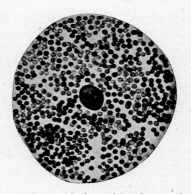

Fig. 13. Nucleus of *Trichonympha peplophora* greatly enlarged and filled with *Caryococcus dilatator*. (From Kirby, 1944.)

CHAPTER III

SPECIFIC BACTERIA
ASSOCIATED WITH INSECTS [1]

THERE has long been a need for an inventory and a systematic cataloguing of the bacteria associated extracellularly with insects and ticks. Much of the confusion which today prevents a thorough understanding of the biologic relationships existing between bacteria and these arthropods might be avoided if investigators were furnished with a comprehensive list of the entomophytic bacteria and a statement concerning their arthropod relationships. To supply these will be the chief objective of this chapter.

Nomenclature. A survey of the literature on bacteriologic-entomologic relationships indicates that there is an utter lack of conformity between the names given to certain bacteria by bacteriologists and those used by entomologists for the same bacteria. Entomologic literature is filled with names given to bacteria, references to which are not found anywhere in bacteriologic literature. Furthermore, there are numerous examples of the same bacterium being cited with different names by different authors. To add to the confusion, many investigators have not attempted to identify or to classify properly the bacteria they have isolated or found associated with insects. This has resulted in many taxonomic ghosts and ambiguities.

It would not be appropriate to discuss here all systematic inconsistencies which appear in the literature on this subject. On the other hand, it is perhaps advisable to point out a few of the more serious causes of confusion in the hope that by focusing attention upon them now they may be avoided in the future.

One of the principal inconsistencies has been in the choice of the generic names. Although taxonomic procedures were not very definite during the

---

[1] Much of the information in this chapter is based on material published by the author in *Catalogue of Bacteria Associated Extracellularly with Insects and Ticks,* Burgess Publishing Company, Minneapolis, 1942.

early days of bacteriology, nevertheless, the errors made were out of all proportion to what they excusably might have been. Today the rules of bacterial nomenclature have been fairly well established and are worthy of the attention of all those about to indulge in the naming of microorganisms. The indiscriminate use of the generic names *Bacillus* and *Bacterium* should definitely be discouraged. Present-day bacterial nomenclature reserves the genus *Bacillus* chiefly for the gram-positive, aerobic, sporeforming bacteria. In reviewing entomologic literature in particular, one repeatedly sees gram-negative, nonsporeforming short rods, and occasionally even cocci, referred to by the generic name *"Bacillus."* Similarly the generic name *Bacterium* has been used in referring to sporeforming bacilli.

The use of abbreviations, especially by early writers, also has been annoying. In many cases the only reference to a generic name throughout an entire paper has been a single letter *"B.,"* not indicating whether *Bacterium* or *Bacillus* is meant. Similarly, the abbreviation *"Bac."* is found all too frequently. Equally confusing is the abbreviation *"S.,"* which might be for *Staphylococcus, Streptococcus, Sarcina, Serratia, Salmonella,* or *Spirillum*.

Although the use of trinomials is being discouraged by modern taxonomists, early workers went so far as to use quadrinomials. Such names have little taxonomic value except from a historical viewpoint.

It is especially unfortunate that the literature contains the names of so many microorganisms for which no adequate description has been published. In most instances this is due to the neglect, on the part of the discoverers, to make adequate morphologic, cultural, and physiologic studies. It is not enough merely to observe that a microorganism is a short rod and then to give it a new specific name in the genus *Bacterium*. If an organism is worthy of a name it certainly should be worthy of an accurate, fairly complete description. Of course, some of the blame for this no doubt lies in the confusion still remaining in present-day bacterial taxonomy. As stated by De Bach and McOmie (1939): "The investigator sometimes does not know which of the characters studied is the more important. Often a valuable test is not performed because its importance is not realized, while a number of quite insignificant observations are laboriously made." On the other hand, if the investigator adheres to one or another of the more generally accepted methods of bacterial identification and classification, such as that set forth in *Bergey's Manual of Determinative Bacteriology,* he is not likely to go far wrong.

In using this catalogue the reader will find a large number of strange bacterial names which are not listed in the latest (5th) edition of *Bergey's Manual*. It is unfortunate that so many undescribed species of bacteria exist. Regardless of the limitations of the available descriptions of the bacteria themselves, the importance of these bacteria in the biologic relationships between them and insects should not be disregarded and overlooked. It is hoped that many of these bacteria will be reisolated and restudied, so as to secure for them a place in the accepted list of bacterial species. Perhaps the bringing together in one publication of these obscure names along with the names of the well-known species isolated from insects and ticks will lead to a uniformity, an enlargement, and an acceleration of study and publication in this field, and will make for a more thorough coverage of the literature of this subject.

PLAN AND METHOD OF USING CATALOGUE

The orders, families, tribes, genera, and species of the bacteria here catalogued are listed in alphabetical order in their respective classifications. More than 300 specific names of the bacteria which have in some way or other been associated with insects or ticks are listed. The system of bacterial classification followed is that used in *Bergey's Manual of Determinative Bacteriology* (5th ed., 1939) with a few exceptions as noted.

Under the appropriate order, family, tribe, and genus, the scientific name for each species of bacterium is given as a heading. When known, the specific name is followed by the authority. This is followed by the names of the arthropods with which the bacterium in question is associated. Next is a short abstract or synopsis of the nature of its arthropod relationship. This catalogue should not be considered bibliographic. The references cited are those which will give the reader the most readily available, complete, or pertinent citations on the bacterium in question. The index should be used to find any particular bacterium or insect.

The author has attempted to use the more recent names of the insects and ticks listed, but may not have done so in all cases. The authorities for the names of insects and ticks are given in the index. For the bacteria we have listed the name of the bacterium as designated in the original reference and its present name according to *Bergey's Manual*. Cross references to bacterial synonyms have been kept to a minimum in the text. For the most part these have been shown in the index.

ORDER OF CATALOGUE CONTENTS

	Page
Class: Schizomycetes	42
Order: Actinomycetales	42
Family: Mycobacteriaceae	42

SPECIFIC BACTERIA ASSOCIATED WITH INSECTS

	Page
Genus: Corynebacterium	42
Genus: Mycobacterium	46
Order: Eubacteriales	50
Family: Bacillaceae	50
Genus: Bacillus	50
Genus: Clostridium	94
Family: Bacteriaceae	97
Genus: Achromobacter	97
Genus: Bacterium	99
Genus: Bacteroides	114
Genus: Flavobacterium	114
Genus: Fusobacterium	115
Family: Enterobacteriaceae	116
Tribe: Erwineae	116
Genus: Erwinia	116
Tribe: Eschericheae	120
Genus: Aerobacter	120
Genus: Escherichia	121
Genus: Klebsiella	123
Tribe: Proteae	124
Genus: Proteus	124
Tribe: Salmonelleae	126
Genus: Eberthella	126
Genus: Salmonella	129
Genus: Shigella	131
Tribe: Serrateae	132
Genus: Serratia	132
Family: Lactobacteriaceae	136
Tribe: Streptococceae	136
Genus: Diplococcus	136
Genus: Streptococcus	138
Family: Micrococcaceae	144
Genus: Gaffkya	144
Genus: Micrococcus	144
Genus: Sarcina	151
Genus: Staphylococcus	152
Family: Neisseriaceae	157
Genus: Neisseria	157
Family: Parvobacteriaceae	158

	Page
Tribe: Brucelleae	158
Genus: Brucella	158
Tribe: Hemophileae	159
Genus: Hemophilus	159
Tribe: Pasteurelleae	159
Genus: Malleomyces	159
Genus: Pasteurella	160
Family: Pseudomonadaceae	169
Tribe: Pseudomonadeae	169
Genus: Pseudomonas	169
Genus: Xanthomonas	173
Tribe: Spirilleae	177
Genus: Vibrio	177
Family: Rhizobiaceae	180
Genus: Alcaligenes	180
Genus: Chromobacterium	180
Genera Incertae Sedis Vel Dubia	181
Genus: Ascobacterium	181
Genus: Coccobacillus	181
Genus: Diplobacillus	185
Genus: Enterococcus	186
Genus: Gyrococcus	186
Genus: Leptotrix	187

Class: SCHIZOMYCETES

Order: ACTINOMYCETALES

Family: MYCOBACTERIACEAE

Genus: Corynebacterium

Corynebacterium blattellae Glaser

Insect concerned: The roach, *Blattella germanica*.

   Glaser (1930b) claims to have cultivated on artificial media the intracellular organisms found in the fat body of the roach. To the diphtheroidal forms he isolated he gave the name *Corynebacterium blattellae*. Since this organism is presumably an intracellular one, it will be discussed in detail in chapter IV.

Corynebacterium diphtheriae (Flügge) Lehmann and Neumann

Insects concerned: The wax moth, *Galleria mellonella;* the housefly, *Musca domestica;* and the roach, *Blatta orientalis.*

Metalnikov (1920a) made a number of experiments to determine the susceptibility of the larvae of the wax moth, *Galleria mellonella,* to infection with *Corynebacterium diphtheriae* and found them to be completely immune. Cameron (1934) found that massive doses rapidly killed all larvae injected, whereas medium doses killed some larvae out of each batch. A very dilute dose did not kill. According to Huff (1940), Chorine found diphtheria toxin to be toxic for these caterpillars. He also was able to produce an immunity in the larvae by use of an "anatoxin."

The housefly has been suggested as a possible vector of diphtheria bacilli. Graham-Smith (1910) made a series of experiments which seemed to indicate that *Corynebacterium diphtheriae* does not remain alive for more than a few hours on the legs and wings of houseflies, but may live for 24 hours or longer in the intestinal tract. He recovered the organism from the feces 51 hours after the flies had fed on the bacilli. He states (1913), "There is no evidence that under natural conditions flies are concerned in the spread of this disease [diphtheria] . . . but, under suitable conditions, it is possible that the disease may occasionally be conveyed by them."

Longfellow (1913) cultivated the Westbrook type of diphtheria bacillus from the feces of roaches.

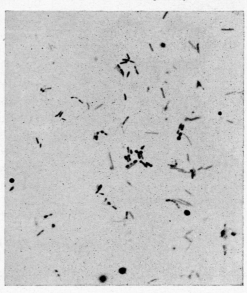

Fig. 14. *Corynebacterium diphtheriae.*

Corynebacterium lipoptenae Steinhaus

Insect concerned: The louse fly, *Lipoptena depressa*.

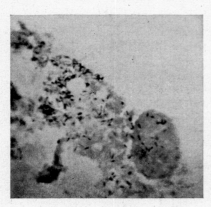

Fig. 15. *Corynebacterium lipoptenae* in the tissues of the louse fly *Lipoptena depressa*.

Corynebacterium lipoptenae was isolated by Steinhaus (1943) from crushed specimens of the louse fly. Every insect examined out of a lot of 10 harbored this organism, which occurred chiefly in the intestinal tract. It stained red with Macchiavello stain as do rickettsiae. The organism was cultured in six-day fertile eggs, and occurred most abundantly in the egg fluids and moderately in the yolk membrane. From the egg the organism was successfully cultured on leptospira medium, and later on glucose and sucrose agar plates. He suggested that the association of this bacterium with *Lipoptena depressa* may to some extent be similar to that between *Rickettsia melophagi* and the sheep ked, *Melophagus ovinus*.

Corynebacterium ovis Bergey et al.

Tick concerned: The winter tick, *Dermacentor albipictus*.

Humphreys and Gibbons (1942) isolated *Corynebacterium ovis*, the cause of caseous lymphadenitis in sheep, from four Rocky Mountain mule deer. They found that engorged *Dermacentor albipictus* females removed from the deer were infected with the organism, and larvae reared from these females also harbored *Corynebacterium ovis*, indicating "hereditary" transmission.

Corynebacterium paurometabolum Steinhaus

Insect concerned: The bedbug, *Cimex lectularius*.

While attempting to cultivate an intracellular symbiote from the ovaries and mycetome of the bedbug, Steinhaus (1941a) isolated a diphtheroid which he named *Corynebacterium paurometabolum*. The cultivated organism appeared very similar to the slender rod-shaped bacterium observed in the tissues of the insect and seemed to be constantly associated with it. In a later study numerous strains of the organism were isolated.

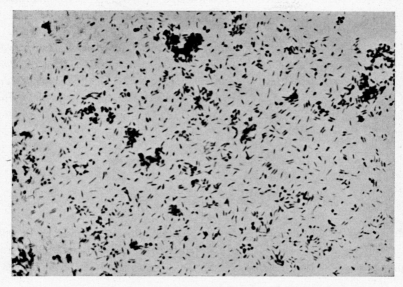

Fig. 16. *Corynebacterium paurometabolum* stained by Gram's method. See also Fig. 89. (Photo by N. J. Kramis.)

During the same investigation an unidentified diphtheroid was isolated from the alimentary tract of the larvae of the bagworm, *Thyridopteryx ephemeraeformis*. The two diphtheroids were similar in many respects, differing only in a few minor characteristics.

Corynebacterium periplaneta var. americana Glaser

Insect concerned: The American roach, *Periplaneta americana*.

From the fat body of the American roach, *Periplaneta americana*, Glaser (1930a) claims to have isolated and cultivated a diphtheroid bacterium which he named *Corynebacterium periplanetae* var. *americana*. Since this bacterium is presumably an intracellular one, it has been discussed elsewhere.

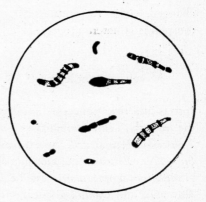

Fig. 17. *Corynebacterium periplaneta* var. *americana*. (From Glaser, 1930a.)

Corynebacterium pseudodiphthericum Lehmann and Neumann

Insects concerned: The roach, *Blatta orientalis;* *Tentyria* sp.; *Blaps mucronata;* *Pimelia bifurcata;* *Pimelia sardea;* and the wax moth, *Galleria mellonella.*

Cao (1898) found that *Bacillus pseudo-difterico* (*Corynebacterium pseudodiphthericum*) did not survive passage through the intestinal tract of *Blatta orientalis, Tentyria* sp., *Blaps mucronata, Pimelia bifurcata,* and *Pimelia sardea.* Cameron (1934) found that the bacillus of Hoffmann (*Corynebacterium pseudodiphthericum*) had no effect on the wax moth when injected.

Genus: MYCOBACTERIUM

Mycobacterium aquae

In referring to a thesis on immunity reactions in invertebrates by Porchet, Paillot (1933, p. 312) refers to *Mycobacterium aquae,* an acid-fast bacillus from tap water, as being pathogenic to various invertebrates, presumably insects.

Mycobacterium lacticola Lehmann and Neumann

Insect concerned: The wax moth, *Galleria mellonella.*

Mycobacterium smegmatis (*Mycobacterium lacticola*) upon injection was quickly phagocytosed by the phagocytes of the wax moth, but survived unchanged, having no injurious effect on the moth and persisting through metamorphosis (Cameron, 1934).

Mycobacterium leprae (G. A. Hansen) Lehmann and Neumann

Insects and ticks concerned: *Chlorops* (*Musca*) *leprae; Chlorops vomitoria;* the housefly, *Musca domestica; Sarcophaga pallinervis; Sarcophaga barbata; Volucella obesa; Lucilia* sp.; *Simulium pertinax; Phlebotomus intermedius; Stomoxys calcitrans;* the mosquitoes, *Aëdes aegypti, Anopheles albitarsis,* and *Anopheles tarsimaculatus;* the bedbug, *Cimex lectularius;* the mites, *Sarcoptes scabiei* and *Demodex* sp.; the roaches, *Periplaneta americana* and *Blabera* (?); the louse, *Pediculus humanus capitis;* the pubic louse, *Phthirius pubis;*

the fleas, *Pulex irritans* and *Pulex canis;* and the ticks, *Amblyomma cajennense* and *Boöphilus microplus.*

Mycobacterium leprae is the causative organism of leprosy. As early as 1872 Hansen observed small rod-shaped bacilli lying within the "lepra cells." The leprosy bacillus has never with certainty been cultivated on artificial media. Furthermore, very little is known concerning the method of its transmission. It is conceivable, however, that in some instances the bacilli may be transferred from one person to another by insects.

Rosenau (1927) writes concerning the role of insects in the transmission of leprosy:

> The evidence is reviewed by Nuttall [1899], who says: "It appears that Linnaeus and Rolander considered that *Chlorops (Musca) leprae* was able to cause leprosy by its bite." [1] Blanchard and Corrodor tell of flies in connection with leprosy. Flies frequently gather in great numbers on the leprous ulcers and then visit and bite other persons. An observation by Boek of the presence of *Sarcoptes scabiei* in a case of cutaneous leprosy led Joly to conclude that these parasites might at times serve as carriers of the infection. . . . Carrasquillo of Bogotá found the bacillus of Hansen in the intestinal contents of flies. The British Leprosy Commission investigated the possible role played by insects with entirely negative results. Wherry . . . found the fly *Chlorops vomitoria* took up enormous numbers of lepra bacilli from the carcass of a leper rat and deposited them with their feces, but the bacilli apparently do not multiply in flies, as the latter are clear of bacilli in less than 48 hours. Larvae of *Chlorops vomitoria* hatched out in the carcass of a leper rat become heavily infested with lepra bacilli. If such larvae are removed and fed on uninfected meat they soon rid themselves of most of the lepra bacilli. A fly, *Musca domestica,* caught on the face of a human leper was found to be infected with lepra-like bacilli. . . . Lepra-like bacilli have been found in bedbugs and these insects have long been associated with the spread of the disease.

In 1909 Ehlers (see Pierce, 1921) found leprosy bacilli in bedbugs of the West Indies. Sanders found the bacillus in 20 out of 75 bedbugs fed on leprous patients. The organisms were in the proboscis up to the fifth day, in the digestive tract to the sixteenth day, and were also found in the feces. In 1911 Long allowed two bedbugs (*Cimex lectularius*) to bite lepers near leprous nodules. On examining the alimentary tract of the bugs he found them to harbor acid-fast bacilli.

Currie (1910), in experiments with mosquitoes, found little reason to believe that they were transmitters of the infection. However, Vedder (Riley

---

[1] Dr. W. D. Pierce has informed the writer that Linnaeus quoted Rolander in naming *Musca leprae,* stating that it was found associated with elephantiasis of American Negroes. At one time elephantiasis was considered a form of leprosy, which fact probably accounts for the Linnean name.

and Johannsen, 1938) found acid-fast bacilli in 41 per cent of the mosquitoes (*Aëdes aegypti*) which he fed on lesions of leprosy. The leprosy bacillus was found by St. John, Simmons, and Reynolds (1930) to survive in the gut of *Aëdes aegypti* for at least 24 hours, but they could not be demonstrated after an interval of 7 or more days.

Currie found that *Musca domestica, Sarcophaga pallinervis, Sarcophaga barbata, Volencella obesa,* and *Lucilia* sp. may contain leprosy bacilli in their intestinal tracts and feces for several days after feeding on leprous fluids. Honeij and Parker (1914) concluded from their experiments that *Stomoxys calcitrans* potentially plays an important role as a carrier of the acid-fast bacilli of leprosy. They also found acid-fast bacilli in *Musca domestica*.

McCoy and Clegg (1912) found acid-fast bacilli in two head lice out of many examined from patients having leprosy. Macfie (Riley and Johannsen, 1932) found that *M. leprae* passed through the cockroach intestine unharmed. Tejera (1926) claimed to have found leprosy bacilli in cockroaches (*Blabera?*) collected in a leper colony.

Dr. H. Souza-Araujo and others have shown that leprosy may perhaps be transmitted through the bite of the ticks *Amblyomma cajennense* and *Boöphilus microplus.* (See *J. Am. Med. Assoc.*, 1942, **120**, 1053.) In 1943 this worker reported that mosquitoes (*Anopheles albitarsis* and *A. tarsimaculatus*), phlebotomus flies (*Phlebotomus intermedius*), and a species of *Simulium* (probably *S. pertinax*) could acquire leprosy bacilli by biting leprous patients. *Phthirius pubis* was also found capable of carrying the bacteria.

Muñoz Rivas (1943) has shown experimentally that the fleas, *Pulex irritans* and *Pulex canis,* can acquire leprosy organisms both from patients and from earth in the native homes of patients, and concludes that fleas transmit leprosy.

Many insects other than those discussed above have been thought to be associated with *Mycobacterium leprae,* but most of the evidence is unconvincing.

Mycobacterium tuberculosis (Schroeter) Lehmann and Neumann

Insects concerned: The wax moth, *Galleria mellonella;* the housefly, *Musca domestica; Achraca grissella;* the roaches, *Periplaneta americana* and *Blatta orientalis; Tentyria* sp.; *Blaps mucronata; Pimelia bifurcata; Pimelia sardea;* and possibly the bedbug, *Cimex lectularius.*

Metalnikov (1914) found that the larvae of *Achraca grissella,* when kept at room temperature, were susceptible to infection with piscine strains of *Mycobacterium tuberculosis.*

Spielman and Haushalter (1887) appear to have been the first to express belief that houseflies that have fed on tubercular sputum may serve as carriers. They found *Mycobacterium tuberculosis* in the intestinal contents and feces of flies fed on tubercular sputum. Others (Hofmann, 1888; Cellí, 1888;

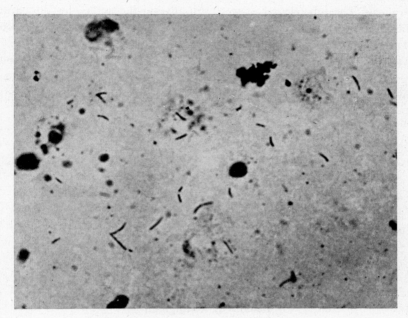

Fig. 18. *Mycobacterium tuberculosis*. (Photo by N. J. Kramis.)

André, 1908; Graham-Smith, 1913) have made similar observations. Riley and Johannsen (1938) express the opinion that laboratory and epidemiologic evidence indicates that houseflies play a definite role in the dissemination of tuberculosis.

Mycobacterium tuberculosis was found by Cao (1898) to pass unchanged through the intestinal tracts of the roach, *Blatta orientalis,* several Coleoptera, *Tentyria* sp., *Blaps mucronata, Pimelia bifurcata,* and *Pimelia sardea*. Küster (1902) fed tubercle bacilli to the oriental roach and later recovered the organism from the feces. Macfie (Riley and Johannsen, 1932) fed *Periplaneta americana* on tubercular sputum, and normal tubercle bacilli were isolated from the feces on the second to fifth day. To prove their virulence, guinea pigs were injected, which subsequently became infected. Tejera (1926) also reported that the tubercle bacillus passes through the cockroach intestine unharmed.

Metalnikov (1920a,b) found that the wax moth, *Galleria mellonella,* was not susceptible to infection with *M. tuberculosis*. Cameron (1934) injected large numbers of human, bovine, and avian strains into the body cavity of the wax-moth larvae with no ill effects. The organisms were not killed, but were segregated chiefly in the pericardial cells and persisted there through metamorphosis. The tubercle bacilli were still alive in the adults, for the bacilli could be grown on culture media, and guinea pigs were infected upon inoculation. The adults appeared perfectly normal.

Dewèvre (1892b) has reported the isolation of *M. tuberculosis* from bedbugs (*Cimex lectularius*) collected in a room occupied by a tubercular patient. However, there are reasons for doubting these observations (Nuttall, 1899).

Order: EUBACTERIALES
Family: BACILLACEAE
Genus: BACILLUS

Bacillus A Ledingham

Insect concerned: The housefly, *Musca domestica*.

A non-lactose-fermenting bacillus isolated from the feces of children has been found by Tebbutt (1913) to be normal to the housefly (*Musca domestica*). It was present on the ova and in the larvae and adults. When the bacillus was fed to the larvae, it survived through the metamorphosis to the adult fly.

Bacillus A White

Insect concerned: The adult honeybee, *Apis mellifera*.

White (1906) isolated *Bacillus A* from the body of a healthy bee and from the combs. He indicates that the organism may be the same as *B. mesentericus*.

Bacillus aerifaciens Steinhaus

Insect concerned: The cabbage butterfly, *Pieris rapae*.

Steinhaus (1941a) isolated this bacillus from triturated specimens of the white cabbage butterfly. It probably belongs to the aerobacillus group of the genus *Bacillus* since the original cultures produced large amounts of gas in glucose, sucrose, and maltose.

Bacillus agilis Mattes

Insects concerned: The flour moth, *Ephestia kühniella*, and the honeybee, *Apis mellifera*.

Mattes (1927) described *Bacillus agilis* as a very motile, short bacillus with pointed ends, resembling *Bacillus lanceolatus* and causing a mild form of foul brood in bees. He found it to be pathogenic for *Ephestia kühniella* under certain conditions.

Bacillus agilis larvae Toumanoff

Insect concerned: The honeybee, *Apis mellifera*.

While studying the microbial flora of the larvae of diseased bees, Toumanoff (1927) isolated and described this sporeforming bacillus. *Bergey's Manual* lists it as a synonym for *Bacillus agilis* Hauduroy *et al*. Apparently it has no relation to *Bacillus agilis* Mattes.

Bacillus agrotidis typhoides Pospelov

Insect concerned: *Euxoa (Agrotis) segetum*.

In Russia the larvae of *Euxoa segetum* were found to be killed by a bacterial disease caused by a mixed flora of bacteria which included *Bacillus agrotidis typhoides* (Pospelov, 1927).

Bacillus alacer

Insect concerned: The nun moth, *Lymantria monacha*.

Eckstein (1894) found *Bacillus alacer* associated with the eggs of the nun moth.

Bacillus alvei Cheshire and Cheyne

Insects concerned: The honeybee, *Apis mellifera*, and *Polia oleracea*.

Cheshire and Cheyne (1885) first described *Bacillus alvei* and thought it was the exciting cause of European foul brood of bees. The etiology of this disease has been the subject of considerable controversy and is still undecided. Maassen (1907) believed that it was caused by either *Streptococcus apis* or *Bacillus alvei*. White (1912, 1920b, 1920c) was unable to produce the disease experimentally with *Bacillus alvei* and concluded that a new species, *Bacillus pluton*, which he was unable to cultivate on artificial media, was the cause. He assumed that *Bacillus alvei* was a secondary invader. Sturtevant (1925) thinks that even in a secondary role it has a marked influence upon the course of the disease.

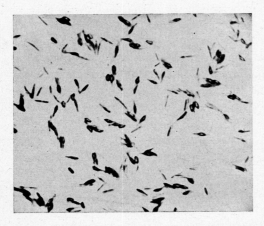

Fig. 19. Sporeforming rods of *Bacillus alvei*. (Courtesy Dr. C. E. Burnside.)

Lochhead (1928c) showed that *Bacillus alvei* possesses a coccoid stage (when grown on sugar-containing media for several weeks), which looks like *Bacillus pluton*.

Burnside (1934) found that "*Bacillus alvei* is capable of morphological, cultural, and biological transformation and is also capable of stabilization, at least temporarily, as a sporogenic rod, an asporogenic rod resembling *Bacterium [Achromobacter] eurydice,* or as a coccoid form resembling *Bacillus pluton.*" He suggests that *Bacillus pluton, Streptococcus apis,* and *Achromobacter eurydice* are variants or stages in the life history of *Bacillus alvei.*

Serbinow (1912a) attributed the cause of "blackbrood in bees" (American foul brood) to be due to *Bacillus alvei*.

The pupae of *Polia oleracea* were found to be killed by *Bacillus alvei* by Zorin and Zorina (1928).

Bacillus alveolaris

Insect concerned: The honeybee, *Apis mellifera*.

Ksenjoposky (1916) states that bees suffer from a disease caused by *Bacillus alveolaris*.

Bacillus anthracis Cohn emend. Koch

Insects and ticks concerned: The biting stablefly, *Stomoxys calcitrans;* the horseflies, *Tabanus striatus, Tabanus atratus, Tabanus rubidus, Tabanus bovinus,* and *Tabanus* sp., near *nigrovittatus;* the horn fly, *Haematobia irritans;* *Haematopota pluvialis;* the blowflies, *Calliphora erythrocephala,* and *Calliphora vomitoria;* the housefly, *Musca domestica;* *Lucilia caesar;* *Sarcophaga carnaria;* *Chrysops caecutiens;* the sheep ked, *Melophagus ovinus;* the mosquitoes, *Psorophora sayi* (= *Janthinosoma sayi*) and *Aëdes vexans* (= *A. sylvestris*); the bedbug, *Cimex lectularius;* the roach, *Blatta orientalis;* the ticks, *Argas persicus, Ixodes ricinus, Otobius megnini* and *Boöphilus decoloratus;* the hide beetle, *Dermestes vulpinus; Attagenus pellio; Anthrenus museorum; Ptinus* sp.; *Tentyria* sp.; *Blaps mucronata; Pimelia bifurcata;* and *Pimelia sardea*.

The beginning of modern bacteriology was marked by Robert Koch's demonstration, in 1876, of the causal relationship of *Bacillus anthracis* to anthrax. Earlier, in 1869, Raimbert had shown experimentally that anthrax could be disseminated by flies. (See also Davaine, 1868, 1870a,b) Bollinger (1874) is cited by Nuttall as having captured flies on a cow dead from anthrax and having seen the bacilli in preparations made from the stomachs and intestines of the insects. Two rabbits inoculated with the insects died of anthrax. For a detailed early history of the relation of insects to the spread of anthrax, the reader is referred to a paper by Nuttall (1899).

Cao (1906b) isolated virulent *B. carbonchio* (*B. anthracis*) from the glutinous secretion surrounding the eggs of *Calliphora vomitoria*. He placed eggs, sterilized externally, of *Musca domestica, Calliphora vomitoria, Lucilia caesar,* and *Sarcophaga carnaria* on the flesh of animals dead from anthrax, and from day to day he dissected the larvae feeding on this flesh and always was able to isolate the anthrax bacillus. He also showed that these larvae retained the organisms from pupation to maturity and for at least nine days after maturity. Graham-Smith (1912) found that a large proportion of houseflies which develop from larvae fed on infected material are themselves infected. He also found that when *Calliphora erythrocephala* and *Lucilia caesar* emerged from larvae fed on anthrax meat, they were infected for 15 days or more after reaching maturity.

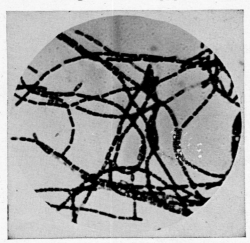

Fig. 20. *Bacillus anthracis*. (From Hagan, 1943.)

In 1912 Schuberg and Kuhn found that *Stomoxys calcitrans* fed on the cadaver of an animal dead from anthrax would transmit the infection. They also found viable anthrax bacilli in the guts and feces of the flies for considerable periods after an infective feeding. Mitzmain (1914), working with *Tabanus striatus* and *Stomoxys calcitrans*, showed that anthrax could be mechanically transmitted to guinea pigs by the bites of both species. In 1918 Morris found that the horn fly, *Haematobia irritans*, the horsefly, *Tabanus* sp., and the mosquitoes, *Psorophora* (*Janthinosoma*) *sayi* and *Aëdes vexans*, are capable of transmitting anthrax after biting an infected animal. However, Duncan (1926) says the anthrax bacillus is very susceptible to a bactericidal principle in the gut contents of *Stomoxys calcitrans*.

According to Bequaert (1942), in 1939 Zumpt quoted an apparently unpublished report of experiments by Bongert to the effect that transmission of *B. anthracis* from infected to healthy sheep was accomplished through the bite of the sheep ked, *Melophagus ovinus*. Furthermore, *B. anthracis* was found in the gut of keds taken from sheep dead of anthrax. It does not seem likely, however, that this insect has ever been important in the spread of the disease.

Nieschulz (1935) has reported experimental transmission of anthrax by the bedbug, *Cimex lectularius*. (However, see Nuttall 1898.)

Proust in 1894 found virulent anthrax bacilli in the excrements of the hide beetle, *Dermestes vulpinus,* taken from goatskins, as well as in the eggs and larvae. Similarly, in 1894 Heim found larvae of *Attagenus pellio, Anthrenus museorum,* and *Ptinus* to harbor virulent anthrax spores on their surfaces and in their excreta.

Küster (1902) fed *Bacillus anthracis* to *Blatta orientalis* and later recovered it from the insect's feces. Cao (1898) found that *Bacillus anthracis* even multiplied in the intestines of *Blatta orientalis* and of several Coleoptera (*Tentyria* sp., *Blaps mucronata; Pimelia bifurcata,* and *Pimelia sardea*).

As for ticks, Cao (1898) carried on a series of experiments to find out what part *Ixodes ricinus* played in the spread of anthrax and other diseases. He states that *Ixodes* had been accused of inoculating anthrax, but there was no direct proof. He concluded that ticks can contain for a certain time the bacteria circulating in the blood stream of the animals to which they are attached. Martinaglia (1932) found anthrax bacilli to be still viable 24 hours after ingestation by the blue tick, *Boöphilus decoloratus,* but the bacilli eventually disappeared. Hindle and Duncan (1925) found that *Bacillus anthracis* not only persists in *Argas persicus* indefinitely but is also passed in the feces at least up to the hundredth day after an infective feeding. An instance of the actual transmission of anthrax to man through the bite of *Argas persicus* has been recorded by Delpy and Kaweh (1937).

For descriptions of other experiments on the role of insects in the transmission of anthrax see Graham-Smith (1913), Nieschulz (1929), and Pierce (1921, p. 475).

Bacillus apisepticus Burnside

Insect concerned: The honeybee, *Apis mellifera.*

Bacillus apisepticus is the cause of a septicemia in adult honeybees. The disease is only slightly infectious and is spread by soil and water that has come in contact with diseased bees. Food does not seem to be a method by which the disease is spread. (See Burnside and Sturtevant, 1936.)

Bacillus aureus

Insects concerned: *Vanessa polychloros; Vanessa urticae; Stilpnotia salicis;* and *Liparis auriflua.*

While studying the infectivity of certain bacteria for various larvae, Eckstein (1894) found that he was able to infect *Vanessa polychloros* with *Bacillus aureus* and that he was unable to infect *Vanessa urticae, Stilpnotia salicis*, and *Liparis auriflua*.

Two organisms are mentioned by the name *Bacillus aureus* in *Bergey's Manual* (5th ed., pp. 629 and 661). Both were described before 1894, so it is difficult to know with which of these, if either, Eckstein worked.

Bacillus B Hofmann

Insect concerned: The nun moth, *Lymantria monacha*.

This bacillus was isolated in 1891 by Hofmann and thought to be the cause of a polyhedral wilt disease (*Wipfelkrankheit*) of the nun moth. Later experiments have proved the disease is due to a virus. Eckstein (1894) considered *Bacillus B* and *Bacterium monachae* of Von Tubeuf (1892a,b) to be the same.

Bacillus B White

Insect concerned: The honeybee, *Apis mellifera*.

White (1906) found that there occurred very constantly in the pollen and intestines of adult honeybees a species of bacteria he referred to as "*Bacillus B*."

Bacillus barbitistes Statelov

Insect concerned: *Isophya (Barbitistes) amplipennis*.

This bacillus was isolated from the tettigonid listed above. An outbreak of an infectious disease caused by this organism occurred for the first time in Bulgaria in the spring of 1930. Statelov (1932) has described its cultural characteristics.

Bacillus bombycis *auctt.*

Insects concerned: The silkworm, *Bombyx mori*, and *Bothynoderes punctiventris*.

The name *Bacillus bombycis* is usually used in referring to a sporeforming bacillus originally observed by Pasteur and now believed to be an important the cause of an epidemic in the laboratory among mature larvae of the southern armyworm. The organism was isolated from the blood of diseased larvae and upon injection into healthy armyworms caused a septicemia with 100 per cent mortality. Cockroaches injected with this strain of *Bacillus cereus* died in 96 hours. *B. cereus* has also been found in diseased *Plodia*.

Hatcher (1939) isolated *Bacillus albolactis,* which is considered a variant of *Bacillus cereus,* from the oral cavity and feces of the American cockroach, *Periplaneta americana*. *Bacillus cereus* was isolated on numerous occasions from triturated ticks, *Amblyomma americanum,* collected in Texas (Steinhaus, 1944).

Bacillus circulans "Group"

Insect concerned: The cecropia moth larva, *Samia cecropia*.

Steinhaus (1941a) isolated a sporeforming bacillus which probably belongs to the *Bacillus circulans* group from the caterpillar of the cecropia moth.

Bacillus cleoni Picard

Insects concerned: *Temnorrhinus (Cleonus) mendicus; Conorrhynchus mendicus*.

While conducting his investigations on "coccobacilli" as insect parasites, Picard (1913) found a bacterium, which he provisionally named *Bacillus cleoni,* in the diseased larvae of *Temnorrhinus mendicus*. He was uncertain whether or not it was different from *Bacillus (Coccobacillus) cajae* (isolated from *Arctia caja* by Picard and Blanc, 1913a). Later Picard (1914) stated that *Bacillus cleoni* resembled *Escherichia coli* but that it differed in its power to liquefy gelatin.

Bacillus coeruleus

Insect concerned: The nun moth, *Lymantria monacha*.

Eckstein (1894) cultivated this organism from larvae of the nun moth. The fifth edition of *Bergey's Manual* (pp. 94 and 630) mentions two organisms by this name as members of groups of organisms which have not received sufficient comparative study to justify definite classification. Inasmuch as these bacteria were described before 1894, Eckstein could have considered his organism to be either one of these.

Bacillus colisimile

Insects concerned: The housefly, *Musca domestica; Calliphora vomitoria, Lucilia caesar;* and *Sarcophaga carnaria.*

Cao (1906b) fed *Bacillus colisimile* to larvae of *Musca domestica, Calliphora vomitoria, Lucilia caesar,* and *Sarcophaga carnaria* and later recovered it from the feces of these insects.

Bacillus cubonianus Macchiati

Insect concerned: The silkworm, *Bombyx mori.*

Cuboni and Garbini (1890) thought this bacillus was the cause of flacherie in silkworms. However, Paillot (1928b) points out that it is possible to infect silkworms with the bacillus artificially and that when this is done the symptoms characteristic of flacherie are not present.

Bacillus cuenoti Mercier

Insect concerned: The oriental roach, *Blatta orientalis.*

Mercier (1906, 1907) studied in considerable detail the "bacteroids" found in the adipose tissues of the oriental cockroach. He isolated a sporeforming rod which he thought was an intracellular microorganism and named it *Bacillus cuenoti.* More recent work, however, has shown that this bacillus is probably a contaminant and not identical with the intracellular microorganism found in the fat body of the cockroach. (See Hertig, 1921; Glaser, 1930a; and Gier as quoted by Steinhaus, 1940a.)

Bacillus decolor

Insect concerned: *Vanessa urticae.*

While studying the bacteria associated with the nun moth, *Lymantria monacha,* Eckstein (1894) found this bacillus in the larva of *Vanessa urticae.*

Bacillus dobelli Duboscq and Grassé

Insect concerned: The termite, *Glyptotermes iridipennis*.

According to Dougherty (1942), Duboscq and Grassé (1927) isolated three bacteria from *Calotermes* (*Glyptotermes*) *iridipennis*: *Fusiformis termitidis*, *Fusiformis hilli*, and *Bacillus* (*Flexilis*) *dobelli*. Dougherty states that for this last bacterium Duboscq and Grassé "proposed the group name (subgenus?) *Flexilis* to include certain bacilli characterized by a considerable length (up to 250 microns)." *Bacillus dobelli* is probably the same as *Bacillus flexilis* Dobell.

Bacillus E White

Insect concerned: The honeybee, *Apis mellifera*.

White (1906) isolated *Bacillus E* from the honeybee while studying the latter's intestinal flora.

Bacillus ellenbachi
(*Bacillus ellenbachensis*)

Insect concerned: The silkworm, *Bombyx mori*.

Sawamura (1906) lists this bacillus as one which produced "flacherie by multiplying in the body of the silkworm." *Bacillus ellenbachi* (*B. ellenbachensis*) is probably a synonym for *Bacillus cereus*.

Bacillus entomotoxicon Duggar

Insects concerned: The squash bug, *Anasa tristis;* the chinch bug, *Blissus leucopterus;* the box-elder bug, *Leptocoris trivittatus;* larvae of the white-lined morning sphinx, *Celerio lineata;* the May beetle, *Phyllophaga fusca;* and a tomato hornworm, *Protoparce*.

Bacillus entomotoxicon was found by Duggar (1896) to be the cause of a disease of the squash bug, *Anasa tristis*, which was first observed among the insects in laboratory breeding cages. Both laboratory and field experiments showed the disease to be easily transmissible to healthy squash bugs by contact with pure cultures of the organism from the fluids of infected insects, nymphs being more readily infected than adults. Infusions made from the growth of this organism on agar contained an active principle which "kills many insects after a very short period of immersion." While young chinch bugs, *Blissus leucopterus,* were also susceptible to the infection, adult chinch bugs were strongly resistant, as were the grubs and larvae of the box-elder bug, the white-lined morning sphinx moth, a tomato hornworm, and *Phyllophaga fusca.*

Bacillus equidistans Noguchi

Tick concerned: The wood tick, *Dermacentor andersoni.*

Noguchi (1926a) isolated this bacillus and two others from the spotted fever tick or wood tick, *Dermacentor andersoni.* He described its cultural characteristics in detail.

Bacillus ferrugenus

Insect concerned: The silkworm, *Bombyx mori.*

Sawamura (1906) lists *Bacillus ferrugenus* as an organism experimentally pathogenic to the silkworm. Whether this is the same as *Bacillus ferrugineus,* a cellulose destroyer, now called *Cellulomonas ferruginea,* is not known.

Bacillus flavus

Insect concerned: *Vanessa polychloros.*

While studying the bacteria associated with various insects, Eckstein (1894) found *Bacillus flavus* [Fuhrmann?] in dead larvae of *Vanessa polychloros.*

Bacillus flexilis Dobell

Insect concerned: The crane fly, *Tipula* sp.

In his examinations of crane fly larvae for protozoa, Mackinnon (1912) observed large numbers of bacteria. He states: "Chief among them is a large sinuous form resembling *Bacillus flexilis* Dobell." This may be the same as *Bacillus dobelli*.

Bacillus foetidus

Insect concerned: *Vanessa urticae*.

Eckstein (1894) found *Bacillus foetidus* along with *Bacillus lineatus* and *Bacillus similis* in dead larvae of *Vanessa urticae*.

B. foetidus, according to Lehmann, Neumann, and Breed (1931, p. 659) is the same as *Bacillus verrucosus*.

Bacillus fuchsinus Boekhaut and De Vries

Insect concerned: The silkworm, *Bombyx mori*.

Sawamura (1906) lists this organism as one experimentally pathogenic to the silkworm. It is probably synonymous with *Serratia fuchsina*.

Bacillus gasoformans nonliquefaciens

Insect concerned: The housefly, *Musca domestica*.

Nicoll (1911) isolated *Bacillus gasoformans nonliquefaciens* from the surface and intestinal tract of houseflies.

Bacillus gaytoni Cheshire

Insect concerned: The honeybee, *Apis mellifera*.

White (1906) refers to *Bacillus gaytoni* by stating: "It is believed by some bee keepers that *Bacillus gaytoni* of Cheshire is the cause of bee paralysis, but this is not claimed by Cheshire, and the belief is not grounded on bacteriological findings."

Bacillus gibsoni (Chorine)

Insect concerned: The corn borer, *Pyrausta nubilalis*.

According to Paillot (1933), Chorine isolated this organism from the corn borer in 1929 and found it to be pathogenic. This is apparently the same organism that Chorine isolated in 1929 and called *Coccobacillus gibsoni* (which see). Paillot says that the organism is a sporeformer, but in his original articles (1929a,b), Chorine states that the organism does not form spores.

Bacillus gigas Van der Goot

Insect concerned: *Adoretus compressus*.

According to Van der Goot (1915), *Adoretus compressus* is attacked both in the larval and adult stages by *Bacillus gigas*. He found that it destroyed large numbers of the larvae in the insectary.

Bacillus gortynae Paillot

Insects concerned: *Gortyna ochracea* and the gypsy moth, *Porthetria dispar*.

Paillot (1913) isolated *Bacillus gortynae* from diseased caterpillars of *Gortyna ochracea*. The organism had caused an epidemic among the insects, death being due to a septicemia. *Bacillus gortynae* is one of the many so-called "coccobacilli" that have been isolated from insects.

Bacillus grünthal

Insect concerned: The housefly, *Musca domestica*.

Nicoll (1911) isolated *Bacillus grünthal* (*Bacillus gruenthali* Morgan?) from the surface and intestinal tract of houseflies.

Bacillus gryllotalpae Metalnikov and Meng

Insect concerned: *Gryllotalpa gryllotalpa* (*G. vulgaris*).

Bacillus gryllotalpae was one of two bacteria which were found by Metalnikov and Meng (1935) to be the cause of the death of *Gryllotalpa gryllotalpa* in the laboratory. *Bacterium gryllotalpae* was the other organism.

Bacillus hoplosternus Paillot

Insects concerned: The brown-tail moth, *Nygmia phaeorrhoea* (*Euproctis chrysorrhoea*); the cockchafer, *Melolontha melolontha; Malacosoma neustria; Arctia caja; Vanessa urticae;* and *Porthetria dispar*.

Paillot (1919d) found *Bacillus hoplosternus*, which he had isolated from diseased cockchafers, to be very pathogenic for *Nygmia phaeorrhoea*. The insects died within 24 hours after being inoculated. He found the same thing to be true with *Malacosoma neustria, Arctia caja,* and *Vanessa urticae*. In the cases of the last two insects, Paillot found the blood at death contained few bacteria and concluded that the bacillus is chiefly pathogenic because of a toxin it secretes. *Porthetria dispar* showed a decided immunity to the bacillus. Paillot (1933, pp. 154-156) has discussed in some detail the cytology of *Bacillus hoplosternus*.

Bacillus immobilis Steinhaus

Insect concerned: The catalpa sphinx, *Ceratomia catalpae*.

From the rectum of larvae of *Ceratomia catalpae,* Steinhaus (1941a) isolated this nonmotile, sporeforming bacillus which he called *Bacillus immobilis*.

Bacillus intrapallens Forbes

Insects concerned: The cabbage worm, *Pieris rapae;* the silkworm, *Bombyx mori;* the yellow-necked caterpillar, *Datana ministra; Datana angusii;* and the zebra caterpillar, *Ceramica picta* (= *Mamestra picta*).

Bacillus intrapallens, a short, broad bacillus, was isolated by Forbes (1880) from a number of insects. Since it was similar to *Bacillus subtilis,* he was uncertain whether or not to consider it a new species. He thought that *Bacillus intrapallens* displaced the micrococci in cultures obtained from diseased *Pieris rapae* larvae and silkworms. He states that this suggested "an alternation of certain forms." He also isolated this bacillus from diseased larvae of *Datana ministra, D. angusii,* and *Mamestra picta*.

Bacillus lanceolatus Maassen

Insect concerned: The honeybee, *Apis mellifera*.

Maassen (1913) isolated *Bacillus lanceolatus* from bee larvae affected with European foul brood, and stated that it resembled the larger forms of *Bacillus pluton*. The organism is lancet-shaped (Lehmann, Neumann, and Breed, p. 234), hence the name. Sturtevant (1923, 1924) isolated an organism similar to *Bacillus lanceolatus* from larvae affected with European foul brood that differed from *Streptococcus apis* and *Bacillus pluton* in being gram-negative and that grew best on media containing 10 per cent dextrose. It has been suggested by Mattes (1927) that *Bacillus lanceolatus* was similar to *Bacillus agilis,* an organism that is supposed to cause a mild form of foul brood in bees.

Since the etiology of European foul brood is still a subject of controversy, the part, if any, that *Bacillus lanceolatus* has in the disease is not known.

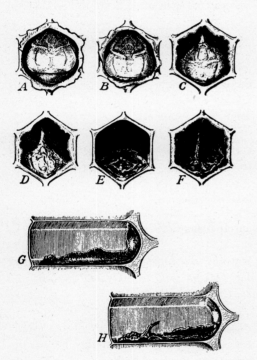

Fig. 21. Larva and pupae of *Apis mellifera* showing symptoms of American foul brood. *A,* healthy pupa; *B–F,* stages in decay and drying of pupae; *G,* scale of dead larva, lateral view; *H,* scale of dead pupa, lateral view. (From Burnside and Sturtevant, 1936; U.S.D.A. Circular 392.)

Bacillus larvae White

Insect concerned: The honeybee, *Apis mellifera*.

Bacillus larvae is the cause of American foul brood in bees. The infectious nature of foul brood had been recognized for a long time, but it was not until White (1904) succeeded in cultivating a sporeforming bacterium on artificial media that the etiology was clearly established. White temporarily called the sporeforming *Bacillus* "X," later (1905, 1906) naming it *Bacillus larvae*.

Burri (1904) and Maassen

(1908), working on American foul brood independently of each other and of White, established the fact that a sporeforming organism difficult to cultivate was the cause of the disease. Maassen called the organism he cultivated *Bacillus brandenburgiensis,* and Cowan (1911) has referred to Burri's organism as *Bacillus burrii*. Both names are now considered synonyms for *Bacillus larvae,* although one will find *Bacillus brandenburgiensis* used a great deal in earlier literature. Serbinow (1913) thought *Bacillus brandenburgiensis* was the cause of European foul brood, a disease distinct from American foul brood.

White (1920a), Sturtevant (1924), and Lochhead (1928b, 1937, 1942) have made studies of the cultural characteristics of *Bacillus larvae,* and Dr. E. C. Holst has found the bacillus to yield an antibiotic.

Bacillus lasiocampa Brown

Insect concerned: The eastern tent-caterpillar moth, *Malacosoma americana*.

While studying the bacteria found in certain insects, Brown (1927) found *Bacillus lasiocampa* throughout the entire female genital system (ovaries and egg tubes) of the tent-caterpillar moth, and he readily cultivated it from the dissected organs. The females infected with this sporeforming bacillus seemed unable to deposit their eggs; although the organism was not found in the eggs, another organism, *Alcaligenes stevensae* (which see) was.

Bacillus lentimorbus Dutky

Insect concerned: The Japanese beetle, *Popillia japonica*.

Bacillus lentimorbus is the cause of type B milky disease of Japanese-beetle larvae, type A being caused by *Bacillus popilliae*. Dutky (1940) describes both organisms and the diseases they cause. Type A and type B milky diseases are very similar, and in the early spring and summer it is difficult to distinguish macroscopically between the two since the larvae in both cases are milky-white in appearance. However, the overwintering diseased larvae affected with type B are chocolate-brown to black because of the formation of blood clots. Both diseases are easily produced in healthy larvae by the injection of heated or unheated suspensions of the blood of diseased larvae, both organisms being sporeformers. *Bacillus popilliae* has centrally located spores, a re-

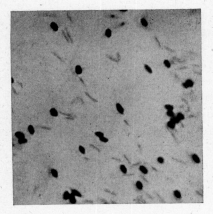

Fig. 22. *Bacillus lentimorbus*, showing both rods and spores. (From Dutky, 1940; courtesy Dr. C. H. Hadley, U.S.D.A.)

fractile body about half the size of the spore, and staining characteristics similar to *Bacillus lentimorbus,* which has no refractile body.

The milky diseases have become increasingly important as natural controls for the Japanese beetle (White, 1940, 1941; White and Dutky, 1942), *Bacillus popilliae* being more common than *Bacillus lentimorbus.* The Bureau of Entomology and Plant Quarantine of the United States Department of Agriculture has developed a method for economically producing large quantities of the milky-disease spores and for distributing them in field treatment.

Bacillus leptinotarse White

Insect concerned: The Colorado potato beetle, *Leptinotarsa decemlineata.*

White (1928) isolated this organism from diseased larvae of *Leptinotarsa decemlineata.* The disease is characterized by a septicemia, and the bacillus may easily be found in the larval blood. The infected larvae are found clinging to the potato plant; the dead ones are usually on the ground. In a later report White (1935) stated that this organism was similar to *Bacillus sphingidis* and *Bacillus noctuarum.*

Bacillus lineatus

Insects concerned: The nun moth, *Lymantria monacha; Vanessa urticae; Porthesia auriflua;* and *Stilpnotia salicis.*

This organism was among those found in larvae of the nun moth by Eckstein (1894). He also isolated this organism from dead *Vanessa urticae, Porthesia auriflua,* and *Stilpnotia salicis.*

Bacillus liparis Paillot

Insect concerned: The gypsy moth, *Porthetria dispar*.

Paillot (1917) isolated *Bacillus liparis* from larvae of *Porthetria dispar*. He found it to be very pleomorphic and to resemble the diphtheria bacillus in morphology. The bacillus appeared to be of little pathogenic importance.

Bacillus lutzae Brown

Insects concerned: The greenbottle fly, *Lucilia sericata*, and the housefly, *Musca domestica*.

Brown (1927) isolated this bacillus from dying and dead greenbottle flies (*Lucilia sericata*) and found it to be pathogenic for the housefly also. Dying individuals yielded pure cultures, and those that had just died, colonies of *Bacillus lutzae* and two cocci. (See *Micrococcus rushmori* and *Neisseria luciliarum*.) This organism has subsequently been tentatively assigned to the genus *Flavobacterium*.

Bacillus lymantriae Picard and Blanc

Insect concerned: The gypsy moth, *Porthetria dispar*, and *Anoxia australis*.

Picard and Blanc (1913b) discovered that a fatal septicemia occurring in the larva of the gypsy moth was caused by an organism they called *Bacillus lymantriae* (*Coccobacillus lymantriae* by some authors). The bacillus was also pathogenic for *Anoxia australis*.

Bacillus lymantriae beta Paillot

Insect concerned: The gypsy moth, *Porthetria dispar*.

Paillot (1919a) isolated *Bacillus lymantriae beta* and *Bacillus lymantricola adiposus* from larvae of *Porthetria* (*Lymantria*) *dispar* sick with a septicemia. He states that one is not to confuse *Bacillus lymantriae beta* with the *Bacillus lymantriae* (see above) of Picard and Blanc, which he says, should be called

Bacillus lymantriae alpha. Paillot pointed out that the cultural characteristics of the two organisms were different.

Bacillus lymantricola adiposus Paillot

Insects concerned: The gypsy moth, *Porthetria dispar; Vanessa urticae;* the brown-tail moth, *Nygmia phaeorrhoea* (*Euproctis chrysorrhoea*); and the silkworm, *Bombyx mori.*

Paillot (1919a) isolated *Bacillus lymantricola adiposus* from the diseased larvae of *Porthetria dispar,* which presented the external symptoms of *grasserie* and flacherie. He was able to reproduce the same symptoms experimentally in the caterpillars of *Vanessa urticae* and *Nygmia phaeorrhoea.* He named this bacterium, which he says is a coccobacillus, *Bacillus lymantricola adiposus,* because of its specific "désorganisation" action on the adipose tissue of *Porthetria dispar.* Paillot (1933, p. 138) observed that this species of bacterium underwent, after its isolation, profound morphologic changes. During the year the bacterium was isolated, he observed only typical coccobacilli in *P. dispar* which had been artificially infected, with only a tendency for the organism to elongate and to swell in the middle portion. Later this tendency was much developed, and few of the original coccobacilli remained.

Bacillus megaterium

Insect concerned: The silkworm, *Bombyx mori.*

Sawamura (1906) lists *Bacillus megaterium* as one of the organisms found to produce flacherie by multiplying in the body of the silkworm.

Regarding the name *Bacillus megaterium,* Breed (1929) states: "In the publication of De Bary's original description of the common spore forming organism, *Bacillus megatherium,* a typographical error occurs which has frequently caused trouble. The incorrect spelling is *Bacillus megaterium.*"

Bacillus megaterium [megatherium?] bombycis Sawamura

Insect concerned: The silkworm, *Bombyx mori.*

Sawamura (1905, 1906) cites this organism as one artificially pathogenic to the silkworm. He assumes that it is a variant of *Bacillus megatherium* De Bary.

Bacillus megatherium De Bary

Insect concerned: *Lecanium corni.*

Benedek and Specht (1933) found the secondary "symbiont" in diseased Lecaniidae to be *Bacillus megatherium,* the main "symbiont" being a fungus, *Torula lecanii cornii.* Both organisms were found free in the hemolymph of the host.

Bacillus megatherium Ravenel

Insect concerned: The cockroach, *Blatta orientalis.*

According to Pierce (1921), "*Bacillus megatherium* Ravenel, a chromogenic organism found in soil, was isolated by Cao from the feces of a *Blatta orientalis* in a single series of experiments."

Bacillus melolonthae Chatton

Insects concerned: The cockchafer, *Melolontha melolontha;* the silkworm, *Bombyx mori;* the gypsy moth, *Porthetria dispar;* and *Vanessa urticae.*

In experimenting with *Coccobacillus acridiorum* on cockchafers, Chatton (1913a) noted a septicemia independent of *Coccobacillus acridiorum* which he found to be due to *Bacillus melolonthae.* The organism resembles *Coccobacillus acridiorum* but is different in that it imparts a fluorescence to the medium in five or six days. This organism behaved the same way in both the silkworm and the cockchafer, being virulent when injected and innocuous when taken into the alimentary tract.

Paillot (1916) studied the effect of *Bacillus melolontha* on *Porthetria dispar* and *Vanessa urticae.*

Bacillus melolonthae liquefaciens Paillot

Insects concerned: The cockchafer, *Melolontha melolontha;* the gypsy moth, *Porthetria dispar;* the brown-tail moth, *Nygmia phaeorrhoea* (*Euproctis chrysorrhoea*); and *Euxoa segetum.*

Paillot (1918, 1922) isolated three strains (*alpha, beta,* and *gamma*) of *Bacillus melolonthae liquefaciens* from diseased cockchafers. At various times he isolated eight other bacteria from the cockchafer, the organism causing the infection varying with the locality. The insects die from a septicemia, characterized by noncoagulable blood, which also becomes more or less turbid. The three strains of *Bacillus melolonthae liquefaciens* are all gram-negative. There is considerable variation (Paillot, 1933, p. 137) in the morphology of the organisms, depending upon the insect in which they are encountered. *Bacillus melolonthae liquefaciens gamma* has the characteristic coccobacilli form in the blood of the cockchafer and the larvae of *Euproctis chrysorrhoea;* in *Euxoa segetum* it is a long, slender rod; and in *Porthetria dispar* there is a general elongation of the bacilli with a perceptible increase in thickness and the ends drawn out to points. The *alpha* and *beta* forms also vary. There was considerable variation in the virulence of the strains injected into the insects depending upon a number of factors such as age of the insect, temperature, route of inoculation, and changes in the degree of virulence of the cultures themselves.

Bacillus melolonthae non liquefaciens Paillot

Insects concerned: The cockchafer, *Melolontha melolontha;* the gypsy moth, *Porthetria dispar;* the brown-tail moth, *Nygmia phaeorrhoea* (*Euproctis chrysorrhoea*); and *Euxoa segetum*.

Paillot (1918) isolated *Bacillus melolonthae non liquefaciens alpha,* along with the *beta, gamma, delta,* and *epsilon* strains, from diseased cockchafers. He (1919a) found that the gypsy moth and the brown-tail moth were both immune to infection with this organism. Paillot differentiated between *Bacillus melolonthae liquefaciens* and strains and *Bacillus melolonthae non liquefaciens* and strains by their ability to liquefy gelatin.

Bacillus mesentericus Trevisan

Insects and ticks concerned: The honeybee, *Apis mellifera;* the wax moth, *Galleria mellonella;* the bedbug, *Cimex lectularius;* the stablefly, *Stomoxys calcitrans; Rhodnius prolixus;* and the ticks, *Argas persicus* and *Ornithodoros moubata*.

In attempting to determine the cause of black brood in bees, Serbinow

(1912a) isolated *Bacillus mesentericus*, which is widely distributed in soil and dust, and thought it was the cause of the disease.

Duncan (1926) found "*B. mesentericus*" and "*B. vulgatus*" [1] to be susceptible to the bactericidal principle in the gut contents of *Argas persicus, Ornithodoros moubata, Stomoxys calcitrans, Cimex lectularius,* and *Rhodnius prolixus.*

Cameron (1934) isolated several strains of *Bacillus mesentericus* from the alimentary tract of a normal wax moth and found them to be extremely virulent upon reinjection, killing in 24 hours.

Bacillus milii Howard

Insect concerned: The honeybee, *Apis mellifera.*

Howard (1900) reported a new bee disease and called it "New York bee disease" or "black brood." He gave as its cause an organism which he called *Bacillus milii.*

White (1906) was of the opinion that "New York bee disease" was not a new disease. He stated: "In our investigations of this diseased condition [New York bee disease] which have covered five years, we have not found an organism corresponding to *Bacillus milii* in any of the specimens that we have received; but we have found *Bacillus alvei.* . . ."

Bacillus minimus

Insects concerned: The nun moth, *Lymantria monacha,* and *Stilpnotia salicis.*

Eckstein (1894) found this bacillus, which he isolated from the larva of the nun moth, to be pathogenic for the larvae of *Stilpnotia salicis.*

Bacillus monachae (Von Tubeuf) Eckstein
(See also *Bacterium monachae* and *Bacillus B*)

Insects concerned: The nun moth, *Lymantria monacha; Vanessa urticae; Porthesia auriflua; Trachia piniperda; Pieris brassicae;* the satin moth, *Stilpnotia salicis;* and *Hyponomeuta evonymella.*

---

[1] According to *Bergey's Manual* (5th ed., pp. 647–649), *Bacillus vulgatus* is considered synonymous with *Bacillus mesentericus.*

Eckstein (1894) believed this sporeforming organism to be identical to *Bacterium monachae* of Von Tubeuf (1892a,b) and *Bacillus B* of Hofmann (1891). Eckstein isolated *Bacillus monachae* from sick nun-moth larvae. He found it to be pathogenic to the larvae of *Vanessa urticae* and *Porthesia auriflua*, and to be occasionally pathogenic for the larvae of *Trachea piniperda*, *Pieris brassicae*, and *Stilpnotia* (*Liparis*) *salicis*. It was not pathogenic for *Hyponomeuta evonymella*.

Bacillus mycoides Flügge

Insects and ticks concerned: The silkworm, *Bombyx mori;* the honeybee, *Apis mellifera; Orgyia pudibunda;* the stablefly, *Stomoxys calcitrans; Cimex lectularius; Rhodnius prolixus;* the wax moth, *Galleria mellonella;* and the ticks, *Argas persicus* and *Ornithodoros moubata*.

This sporeforming bacillus was originally described by Flügge and is widely distributed in the soil. Eckstein (1894) isolated it from soil containing dead larvae of *Orgyia pudibunda* which had been stored for a period of two years.

White (1906) isolated a *Bacterium mycoides* from the intestine of a healthy honeybee. Since the organism he isolated was a sporeformer, it was probably *Bacillus mycoides*.

Cameron (1934) found *Bacillus mycoides* to be extremely virulent for the wax-moth larvae, killing in 12 to 24 hours. The insects seem to have no resistance whatever.

Sawamura (1906) listed *Bacillus mycoides* as artificially pathogenic to the silkworm, *Bombyx mori*.

After feeding *B. mycoides* to *Argas persicus*, Hindle and Duncan (1925) found that the bacillus neither survived long in the tick nor appeared in its feces. Unlike *Bacillus subtilis*, *B. mycoides* was either not affected or only slightly affected by the bactericidal gut contents of *Argas persicus, Ornithodoros moubata, Stomoxys calcitrans, Cimex lectularius,* and *Rhodnius prolixus* (Duncan, 1926).

Bacillus neurotomae Paillot

Insect concerned: *Neurotoma nemoralis*.

This bacillus was isolated from diseased *Neurotoma nemoralis* insects by

Paillot (1924a). It was found not to be of any practical importance in checking their numbers in nature.

Bacillus noctuarum White

Insects concerned: Cutworms of the genera *Feltia, Agrotis,* and *Prodenia;* larvae of the silkworm, *Bombyx mori;* the tobacco hornworm, *Protoparce sexta;* the tomato hornworm, *Protoparce quinquemaculata;* the larvae of the catalpa moth, *Ceratomia catalpae;* grasshoppers; and the red-backed cutworm, *Euxoa ochrogaster.*

White (1923b) isolated *Bacillus noctuarum* from diseased cutworms, and suggested "cutworm septicemia" as the name to be used for the disease. The blood of the larvae of the sick or dead of this disease contained pure or almost pure cultures of the organism. The septicemia is not produced easily by feeding but is readily so by puncture inoculations, the mortality being nearly 100 per cent. No important changes in the virulence of *B. noctuarum* were observed when it was kept on artificial media for four years.

The larvae of silkworms, hornworms, catalpa moths, and grasshoppers were found experimentally to be susceptible by the puncture method, the mortality being about 100 per cent.

King and Atkinson (1928) found a similarity between cutworm septicemia caused by *Bacillus noctuarum* and a septicemia of the red-backed cutworm, *Euxoa ochrogaster.*

Economically, cutworm septicemia is of little importance. White states that comparatively few cutworms die of this disease in the field during the active growing season of the crops on which these larvae feed.

The morphology, cultural characteristics, and pathogenesis of *Bacillus noctuarum, Bacillus sphingidis,* and *Bacillus acridiorum* are rather similar, but serologically the three species are quite different.

Bacillus noctuarum has also been called *Escherichia noctuarii* (*Bergey's Manual,* 3d ed., 1930, p. 327) and *Proteus noctuarum* (*Bergey's Manual,* 4th ed., 1934, p. 363).

Bacillus oblongus

Insect concerned: *Hyponomeuta evonymella.*

Eckstein (1894) found *Bacillus oblongus* present in the larvae of *Hyponomeuta evonymella.*

Bacillus ocniriae

Insect concerned: *Loxostege sticticalis*.

During their experiments on microbiologic methods of combating caterpillars of *Loxostege sticticalis*, Drobotjko, Martchouk, Eisenman, and Sirotskaya (1938) found *Bacillus ocniriae* to be slightly pathogenic when fed to the insects. They obtained similar results with an organism designated as "*B. ephestia.*" (See also Martchouk, 1934.) Both of these sporeforming bacteria were obtained from Metalnikov.

Bacillus ontarioni Chorine
(See also *Bacterium ontarioni*)

Insect concerned: The corn borer, *Pyrausta nubilalis*.

Chorine (1929a,b) isolated an organism from diseased corn borers which he called *Bacterium ontarioni*, but Paillot (1933) called it *Bacillus ontarioni*. Since it is a sporeforming bacillus, the correct generic name should be *Bacillus*.

Bacillus orpheus White

Insects concerned: The honeybee, *Apis mellifera*, and the silkworm, *Bombyx mori*.

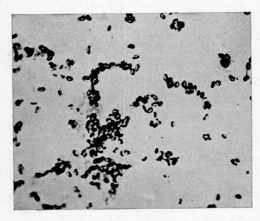

Fig. 23. Spores of *Bacillus orpheus*. (Courtesy Dr. C. E. Burnside.)

White (1912) found a bacterium, which he called *Bacillus orpheus*, occasionally associated with European foul brood. He (1920b) considered it to be one of the secondary invaders encountered in the disease. In one instance he found it very widely distributed in an apiary that had suffered heavy losses due to European foul brood. In this case the larvae when dry were "stone-like." Usually the organism is not so prevalent. Bee larvae are not susceptible to *Bacillus orpheus* by feeding, but silkworms are both by feeding and by puncture. McCray (1917) has made a thorough study of the characteristics of the bacillus.

Bacillus para-alvei Burnside

Insect concerned: The honeybee, *Apis mellifera*.

Burnside (1932) and Burnside and Foster (1935) first found *Bacillus para-alvei* in diseased colonies of bees in southeastern United States. It is the cause of the disease parafoulbrood which affects the larval and sometimes the pupal stages but never the adults. The organism lives in the digestive tract (Burnside and Sturtevant, 1936). *Bacillus para-alvei* resembles the bacteria of European foul brood.

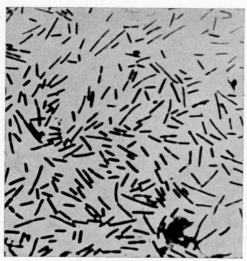

Fig. 24. Vegetative rods of *Bacillus para-alvei*. (Courtesy Dr. C. E. Burnside.)

Bacillus pectinophorae White and Noble

Insect concerned: The pink bollworm, *Pectinophora gossypiella*.

In the summer of 1932 White and Noble (1936) encountered a septicemia among laboratory-reared pink-bollworm larvae which they found to be due to a bacillus which they called *Bacillus pectinophorae*. Pink-bollworm septicemia was observed only under laboratory conditions. Fifty larvae were inoculated with a pure culture of the bacillus, and the mortality was 100 per cent in 24 hours. To prevent death from pink-bollworm septicemia, the larvae should not be crowded in containers, for they produce excessive moisture, which is unfavorable, and also they may bite each other, allowing easy entrance of the bacillus. *Bacillus,* as a generic name, is not correct inasmuch as the bacterium is a small, nonsporeforming rod.

Bacillus pediculi Arkwright and Bacot

Insect concerned: The louse, *Pediculus humanus.*

While Arkwright, Bacot, and Duncan (1919) were working on the association of rickettsiae with trench fever, they noted a bacillary infection of the excreta and gut of the human louse. Later, Arkwright and Bacot (1921) had occasion to study the causative organism in more detail. They found it to be a gram-negative coccobacillus which they called *Bacillus pediculi,* concluding that it was a parasite of the copulatory apparatus. It occurred in the folds of the *vesica penis* of the male, and in the vaginal orifice and passage leading to the ovaries of the female. There was no evidence that the gut was parasitized.

Bacillus pestiformis apis Malden

Insect concerned: The honeybee, *Apis mellifera.*

Bullamore (1922) states that Malden was the first to associate a definite causal organism with Isle of Wight disease of honeybees. From the chyle stomach of bees Malden observed an organism similar to *Bacillus pestis* and suggested there might be some relationship between the organism and the disease. He proposed the name *Bacillus pestiformis apis.* The view that the organism causes the disease has since been abandoned.

Near Bacillus pestis (?)

Insect concerned: *Lachnosterna smithi.*

Bourne (1921) found that many larvae of *Lachnosterna smithi* were killed by an organism which resembled *"Bacillus pestis."* Just which species this may be is difficult to ascertain. It probably was not the true *Pasteurella pestis*, the cause of plague.

Bacillus pieris agilis Paillot

Insect concerned: The white cabbage butterfly, *Pieris brassicae*.

Paillot (1919b) isolated *Bacillus pieris agilis* from infected caterpillars of the white cabbage butterfly. At the same time he isolated eight other bacteria from the same source. He considered these organisms as secondary invaders in causing the death of the larvae, the parasite *Apanteles glomeratus* being the predisposing factor to bacterial infection. The other bacteria isolated were *Bacillus pieris liquefaciens alpha* and *beta*, *Bacillus pieris non liquefaciens alpha* and *beta*, *Bacillus pieris fluorescens*, *Diplococcus pieris*, *Diplobacillus pieris*, and *Bacillus proteidis*.

Bacillus pieris fluorescens Paillot

Insect concerned: The white cabbage butterfly, *Pieris brassicae*.

Bacillus pieris fluorescens was isolated by Paillot (1919b) from diseased white-cabbage-butterfly larvae.

Bacillus pieris liquefaciens Paillot

Insect concerned: The white cabbage butterfly, *Pieris brassicae*.

From diseased white-cabbage-butterfly larvae Paillot (1919b) isolated *Bacillus pieris liquefaciens alpha* and *beta*. (See *Bacillus pieris agilis*.)

Bacillus pieris non liquefaciens Paillot

Insect concerned: The white cabbage butterfly, *Pieris brassicae*.

Paillot (1919b) isolated *Bacillus pieris non liquefaciens alpha* and *beta* from infected white-cabbage-butterfly larvae. (See *Bacillus pieris agilis*.)

Bacillus piocianemus

Insect concerned: The fly, probably *Musca domestica*.

According to Cao (1898), Charrin found *Bacillus piocianemus* (*Bacillus pyocyaneus?*) still living in the body of a fly six to eight days after feeding.

Bacillus pirenei
(See *Bacterium pyrenei*)

Insects concerned: *Pieris* spp. and the European corn borer, *Pyrausta nubilalis*.

In laboratory and field experiments carried out near Leningrad and in Moldavia, Pospelov (1936) found that cultures of *Bacillus pirenei* proved to be very virulent to various lepidopterous larvae. When applied in sprays against *Pieris* spp. on cabbage during sunny weather, the bacilli gave 10 to 100 per cent mortality and remained effective for 20 days. They also killed 25 per cent of the larvae of *Pyrausta nubilalis* on maize.

It is fairly probable that this organism is the *Bacterium pyrenei* of Metalnikov, Ermolaev, and Skobaltzyn (1930) which they isolated from diseased corn borers.

Bacillus pluton White

Insect concerned: The honeybee, *Apis mellifera*.

Bacillus pluton was thought by White (1920b) to be the cause of European foul brood, a disease of bees. He first (1912) referred to the organism as Bacillus "Y" and claimed that it caused European foul brood. Lochhead (1928a) says,

The claim of White, however, may be said to be based on purely indirect evidence, on the basis of microscopical examination and inoculation tests with impure cultures, as it is said to be incapable of cultivation on laboratory media. The selection of *B. pluton* as etiological factor was derived by a process of elimination whereby the other organisms, associated with European foulbrood, including *B. alvei*, *Str. apis*, and others, were excluded by a series of inoculation tests.

Wharton (1928) claimed to have successfully cultivated *Bacillus pluton*. However, according to Lochhead (1928c), Wharton did not culture *Bacillus pluton*, but instead an organism closely related, if not identical, to *Streptococ-*

cus apis. He pointed out the similarity between certain stages of *Streptococcus apis* in pure culture and those of *Bacillus pluton* in diseased material, and raised doubts as to whether *Bacillus pluton* can be said to exist at all, since it has never been known to be obtained in pure culture. Lochhead also showed that *Bacillus alvei* possesses a coccoid stage which looks like *Bacillus pluton*. It has been suggested by Burnside (1934) that *Bacillus pluton* was one of the variants or stages in the life history of *Bacillus alvei*.

Borchert (1935a,b) does not think *Bacillus pluton* is the cause of the disease, but rather that *Bacillus orpheus*, *Streptococcus apis*, and *Bacillus alvei* all possess pathogenic properties, and the symptoms of the disease depend on the organism. Some workers believe *Bacillus pluton* and *Bacterium (Achromobacter) eurydice* to be the same.

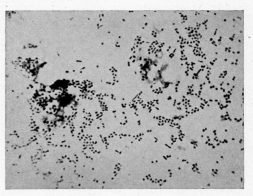

Fig. 25. *"Bacillus pluton."* Smear from the stomach of a bee larva in the advanced stage of European foul brood. (Courtesy Dr. C. E. Burnside.)

Tarr (1935) considers *Bacillus pluton* to be the cause of European foul brood and to be a strict parasite.

Bergey's Manual seems to accept Burnside's theory that *Bacillus pluton*, *Streptococcus apis*, and *Achromobacter eurydice* are stages in the life history of *Bacillus alvei*, although *Achromobacter eurydice* is given a separate description. It is apparent that the question of the etiology of European foul brood is still not satisfactorily settled.

Bacillus pluton was called *Diplococcus pluton* in the second edition of *Bergey's Manual*.

Bacillus poncei Glaser

Insects concerned: The red-legged grasshopper, *Melanoplus femur-rubrum;* *Encoptolophus sordidus;* and the cricket, *Gryllus assimilis* (= *G. pennsylvanicus*).

In 1918 (a) Glaser made a study of the organisms distributed under the name of *Coccobacillus acridiorum* D'Herelle. In carrying out these studies, he obtained from Dr. Ponce of Honduras an organism which was not a "coccobacillus" at all but a bacillus heretofore not described. He studied its cultural characteristics and named it *Bacillus poncei*. The organism was pathogenic to *Melanoplus femur-rubrum, Encoptolophus sordidus,* and *Gryllus pennsylvanicus*. In most cases attempts to recover the bacillus from the blood, the alimentary tract, or the feces failed.

Glaser (1918b) immunized grasshoppers with *Bacillus poncei* and found that the blood showed a high degree of immunity toward the bacteria. He found agglutinins in the immune grasshopper blood.

Bacillus popilliae Dutky
(See also *Bacillus lentimorbus*)

Insects concerned: The Japanese beetle, *Popillia japonica;* the oriental beetle, *Anomala orientalis;* the Asiatic garden beetle, *Autoserica castanea; Cyclocephala borealis; Phyllophaga anxia; Phyllophaga bipartita; Phyllophaga ephilida; Phyllophaga fusca; Phyllophaga rugosa; Strigoderma arboricola; Cotinis nitida;* and *Macrodactylus subspinosus*.

Dutky (1940) described two sporeforming bacteria, which he named *Bacillus popilliae* and *Bacillus lentimorbus* (which see), as the causative agents, respectively, of types A and B milky disease of the larvae of the Japanese beetle, *Bacillus popilliae* being the more important of the two. The milky diseases are playing an important role in the control of Japanese beetle larvae in eastern United States. Larval reductions of over 90 per cent have been observed in some areas upon the introduction of the disease organisms. It has been possible to hasten the reduction of the number of larvae by applying the disease agent in areas where it already occurred, and it has also been possible to prevent larval population increases in areas where there was no disease

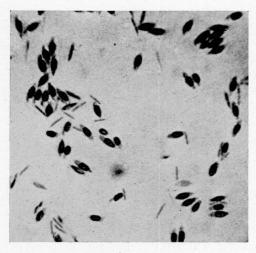

Fig. 26. *Bacillus popilliae* in blood of Japanese beetle larva showing rod and spore forms. (From Dutky, 1940; courtesy Dr. C. H. Hadley, U.S.D.A.)

before its introduction. Once the soil becomes highly infected with *Bacillus popilliae,* no substantial population of Japanese beetles can live. The organism survives in the soil despite adverse climatic conditions.

Since no artificial medium has yet been found which will allow *Bacillus popilliae* to produce spores, living Japanese beetle larvae are used as a culture medium. After the larvae are inoculated they are kept at 87° F. for 10 to 12 days. By that time the contents of each larva average from one to three billion spores. The larvae are then introduced into field plots or are ground up into suspensions for dispersion.

Birds and insects have been proved to be important agents of dispersion of this bacillus. Ants also are probably important in its spread as they have been seen dragging the dead diseased grubs.

According to Beard (1944), the virulence of *Bacillus popilliae* in causing milky disease of Japanese beetle larvae is far less than that of *Bacillus larvae* in causing foul brood of honeybee larvae. Only about 25 *Bacillus larvae* spores are required to produce foul brood in 50 per cent of the bee larvae, whereas 11,000 *Bacillus popilliae* spores are required to cause milky disease in 50 per cent of the beetle larvae when the spores are injected parenterally.

Dutky (1941) found the following scarabaeid species to be susceptible to milky disease: *Anomala orientalis, Autoserica castanea, Cyclocephala (Ochrosidia) borealis, Phyllophaga anxia, P. bipartita, P. ephilida, P. fusca, P. rugosa, Strigoderma arboricola,* and *Strigodermella pygmaea*. On the other hand, *Cotinis nitida* and *Macrodactylus subspinosus* were not susceptible.

Bacillus proteidis Paillot

Insect concerned: The white cabbage butterfly, *Pieris brassicae.*

Paillot (1922) isolated *Bacillus proteidis* from larvae of diseased cabbage butterflies. From the same source, he isolated eight other bacteria. (See *Bacillus pieris agilis.*)

Bacillus proteisimile Cao

Insects concerned: The roach, *Blatta orientalis; Tentyria* sp.; *Blaps mucronata; Pimelia bifurcata;* and *Pimelia sardea.*

Bacillus proteisimile was isolated by Cao (1898, 1906a) from the feces of the oriental roach and from *Blaps mucronata, Pimelia bifurcata, Pimelia sardea,* and *Tentyria* sp. He describes the organism rather completely. He found that passage through a series of roaches given a variation in diet increased the pathogenicity of the organism for guinea pigs.

Bacillus pseudoxerosis Noguchi

Tick concerned: The wood tick, *Dermacentor andersoni*.

Noguchi (1926a) isolated *Bacillus pseudoxerosis* from the wood tick. It was one of three microorganisms he found in this arthropod. These organisms morphologically resembled the rickettsia causing spotted fever. However, they were found to be nonpathogenic for laboratory animals, and immunologically they were not related to the spotted fever rickettsia.

Bacillus pseudoxerosis is a nonmotile, slender, pleomorphic bacillus. It is apparently nonsporeforming, hence the generic name *Bacillus* is not acceptable according to present rules of nomenclature.

Bacillus punctatus Zimmermann

Insect concerned: *Locusta migratoria*.

In experimental infection of *Locusta migratoria* using a mixture of *Bacillus fluorescens liquefaciens*, *Bacillus punctatus*, and *Coccobacillus acridiorum*, Shulguina and Kalinicker (1927) found the mixture gave a low mortality.

Bacillus pyrameis Paillot

Insect concerned: *Vanessa cardui* (= *Pyrameis cardui*).

Paillot (1913) isolated two different "coccobacilli" from the tissues and blood of caterpillars of *Vanessa cardui*. He named these *Bacillus pyrameis I* and *Bacillus pyrameis II*.

Bacillus radiciformis

Insects concerned: The roach, *Blatta orientalis;* the housefly, *Musca domestica; Calliphora vomitoria; Lucilia caesar; Sarcophaga carnaria; Tentyria* sp.; *Blaps mucronata; Pimelia bifurcata;* and *Pimelia sardea*.

Cao (1898, 1906a) fed *Bacillus radiciformis* to the oriental cockroach, the common housefly, several other species of flies, and several species of Coleoptera, among them being a *Tentyria* sp., *Blaps mucronata*, and *Pimelia sardea*. In the case of each insect, the organism survived passage through the intestinal tract and was isolated from the feces.

Bacillus rickettsiformis Noguchi

Tick concerned: The wood tick, *Dermacentor andersoni*.

Bacillus rickettsiformis was isolated by Noguchi (1926a) from the wood tick. This bacillus and two other microorganisms which Noguchi isolated from the same source morphologically resembled the rickettsia which causes spotted fever. *Bacillus rickettsiformis* was more frequently isolated than the other two. They were found to be nonpathogenic for laboratory animals and immunologically not related to the spotted fever rickettsia.

Bacillus rickettsiformis is moderately motile, and lanceolate, fusiform, or rod-shaped. In old cultures there is considerable pleomorphism. Since the organism is a gram-negative, apparently nonsporeforming rod, the generic name *Bacillus* is not correct according to accepted nomenclature.

Bacillus rotans Roberts

Insect concerned: Termites in Texas. (Identity not stated in original paper.)

During an investigation of the intestinal flora of termites of central Texas, Roberts (1935) isolated an organism which seemed to be uniformly present in the termite intestine. He named the organism *Bacillus rotans* because of the rotary motility of young colonies.

Bacillus rubefaciens Zimmermann

Insect concerned: The silkworm, *Bombyx mori*.

This bacterium was listed by Sawamura (1906) as pathogenic to silkworm larvae when experimentally infected.

Bacillus salutarius Metchnikoff

Insect concerned: *Anisoplia austriaca.*

Paillot (1933, p. 123) referred to this bacillus as having been found by Metchnikoff in 1879 while studying the green muscardine of *Anisoplia austriaca*. Metchnikoff found this organism in many of the dying larvae. This is one of the earliest observations of the fact that bacteria may cause diseases of insects.

Bacillus septicaemiae lophyri Shiperovich

Insect concerned: The sawfly, *Diprion sertifer.*

In 1925 Shiperovich observed a bacterial disease among the larvae of sawflies that he attributed to an organism which he named *Bacillus septicaemiae lophyri*. Schwerdtfeger (1936) says that it probably caused mass mortality among the larvae of sawflies in nature.

Bacillus septicus insectorum Krassilstschik

Insect concerned: The cockchafer, *Melolontha melolontha.*

In 1893 Krassilstschik isolated this organism from cockchafers and found it to be the cause of a septicemia among the larvae. The organism seemed to gain entrance into the larvae by way of the skin; for Krassilstschik (1916) stated that the larvae are not cannibals in the sense that they devour each other, but they often attack and wound one another, and these wounds, however slight, often prove fatal, as they provide means of entry of noxious bacteria.

Northrup (1914a) found a gas-producing bacillus associated with *Micrococcus nigrofaciens* which she thought might be *B. septicus insectorum*.

Bacillus similcarbonchio Cao

Insects concerned: The roach, *Blatta orientalis;* the housefly, *Musca domestica; Lucilia caesar; Sarcophaga carnaria; Calliphora vomitoria;* and *Melolontha vulgaris.*

Bacillus similcarbonchio was isolated by Cao (1898) from the feces of a series of oriental roaches. He found it to be pathogenic for guinea pigs, conies, and doves. The virulence of the organism was increased by continued passage through roaches which were given a varied diet. He (1906b) also isolated *Bacillus similcarbonchio* from the intestinal contents of larvae of *Melolontha vulgaris,* and from the larvae of several flesh-eating flies. The organisms pass alive and virulent through the intestinal tract and are deposited on the external envelope of the eggs of the flies.

Bacillus similis Eckstein

Insects concerned: The nun moth, *Lymantria monacha;* the white cabbage butterfly, *Pieris brassicae; Porthesia auriflua;* and *Vanessa urticae.*

This sporeforming bacillus was found by Eckstein (1894) to be pathogenic for the larvae of the above-listed insects which had been experimentally infected. Eckstein asserts that this organism is similar to the well-known *Bacillus megatherium.*

Bacillus similtubercolari Cao

Insect concerned: *Melolontha vulgaris.*

Cao (1906b) isolated a *"Bacillus similtubercolari"* from the intestinal contents of the larvae of *Melolontha vulgaris,* a burrowing beetle.

Bacillus sotto Ischivata

Insects concerned: The silkworm, *Bombyx mori,* and the corn borer, *Pyrausta nubilalis.*

Beginning with Joly's observation (about 1858) of *"Vibrio Aglaiae,"* several sporeforming organisms have been isolated from diseased silkworms. *Bacillus sotto* is one of these. Knowledge of their relation to each other and to flacherie is still very confused.

Paillot (1928b) refers to *"Bacillus sotto* of Ischivata" whereas Metalnikov and Chorine (1928b) refer to "Ischivata's *Bacterium sotto."* The two terms probably refer to the same organisms, which is a sporeformer and rightly should be called *Bacillus*.

According to Aoki and Chigasaki (1915a,b), Ischivata stated that it was the cause of a severe epidemic of flacherie in 1902 among silkworms of Japan, and they too found it to be pathogenic to the silkworm. Paillot (1928), however, states that *Bacillus sotto* has been "erroneously considered the cause of flacherie" by the Japanese authors.

Paillot (1928) carried out experiments using *Bacillus sotto* against larvae of *Pyrausta nubilalis*. He found the organism was not very effective against the insect, as only three out of the ten corn borers infected showed signs of disease at the end of three days. Metalnikov and Chorine (1928b) were not able to infect the corn borer by mouth with *Bacillus sotto*.

Fig. 27. Hornworm two days after inoculation with *Bacillus sphingidis*, hanging by hook of proleg from leaf of tobacco plant on which it had been feeding. (From White, 1923a; courtesy Bureau of Entomology, U.S.D.A.)

Bacillus spermatozoides

Insect concerned: *Hyponomeuta evonymella*.

While studying the bacteria associated with the nun moth, *Lymantria monacha*, Eckstein (1894) cultivated this bacillus from dead *Hyponomeuta evonymella*.

Bacillus sphingidis White
(See also *Bacillus noctuarum* White)

Insects concerned: The tomato hornworm, *Protoparce quinquemaculata;* the tobacco hornworm, *Protoparce sexta;* the catalpa-moth caterpillar, *Ceratomia catalpae;* the silkworm, *Bombyx mori;* cutworms; and grasshoppers.

White (1923a) observed a septicemia among laboratory-reared hornworms that was so severe that it ruined his experiments. He isolated an organism which he called *Bacillus sphingidis*. He found that all hornworms in all instars were susceptible to infection by the puncture method, although infection was more likely to occur during the fifth instar. The mortality was practically 100 per cent. However, only a small percentage became infected by the feeding method. White thought that it was likely that few hornworms die of hornworm septicemia in nature.

The larvae of silkworms and catalpa moths as well as cutworms and grasshoppers were found to be susceptible to infection by the puncture method.

Bergey's Manual (5th ed., p. 605) indicates that *Bacillus sphingidis* is very similar if not identical to *Bacillus noctuarum*. White, however, says that though their cultural, morphologic, and pathogenic characters are very similar, serologically they are quite different. *Bacillus sphingidis* has also been called *Escherichia sphingidis* (*Bergey's Manual,* 3d ed., p. 327) and *Proteus sphingidis* (*ibid.,* 4th ed., p. 366). Inasmuch as the organism is a nonsporeformer, it does not belong in the genus *Bacillus*.

Bacillus subgastricus White

Insect concerned: The honeybee, *Apis mellifera*.

White (1906) isolated *Bacillus subgastricus* from the intestine of a healthy honeybee. He gives a complete description of it. This organism may be a variant of *Bacillus gastricus* Ford (see *Bergey's Manual,* 5th ed., p. 603), though the two differ in oxygen requirements, indole production, and nitrate reduction.

Bacillus subtilis Cohn *emend*. Prazmowski

Insects and ticks concerned: The wood-digesting roach, *Cryptocercus punctulatus;* the cockroach, *Blatta orientalis;* the wax moth, *Galleria mellonella;* the cecropia moth, *Samia cecropia;* the catalpa moth, *Ceratomia catalpae; Sinea diadema;* the tarnished plant bug, *Lygus pratensis; Conocephalus fasciatus* var. *fasciatus;* a species each in the families Chrysomelidae and Curculionidae; the Alpine rock crawler, *Grylloblatta campodeiformis campodeiformis;* the stablefly, *Stomoxys calcitrans; Cimex lectularius; Rhodnius prolixus;* the mealworm, *Tenebrio molitor;* and the ticks, *Argas persicus* and *Ornithodoros moubata*.

Bacillus subtilis, the common "hay bacillus," is widely distributed in nature in the air, soil, and decomposing organic materials. For this reason it is perhaps not surprising that it should be found associated with insects. It seems very probable that some of the early named sporeforming organisms described as being associated with insects may have been *Bacillus subtilis* or closely related species.

In 1906 Cao isolated *Bacillus subtilis* from the feces of *Blatta orientalis*. He also fed cultures to larvae of *Musca domestica, Lucilia caesar, Sarcophaga carnaria,* and *Calliphora vomitoria* and isolated the organism from the feces of the larvae.

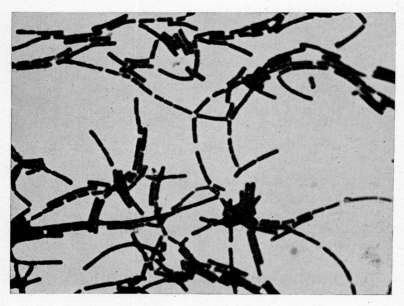

Fig. 28. *Bacillus subtilis.* (Photo by N. J. Kramis.)

Metalnikov (1920a) found *Galleria mellonella* to be very susceptible to infection with *Bacillus subtilis,* the insects dying from extremely small doses. Cameron (1934) isolated several strains of *B. subtilis* from normal larval blood and the alimentary tract of the wax moth, and found the strains to be extremely virulent, the larvae dying in 12 to 24 hours. Other strains of *B. subtilis* had no effect on the larvae even in enormous doses. Larvae that survived had spores of the bacillus in their body cavities for a long time. Masera (1936) has reported the occurrence of a septicemia among larvae of *Tenebrio molitor* caused by *B. subtilis.* Inoculation of healthy larvae at unfavorable temperatures caused a septicemia similar to flacherie of silkworms. Only slight pathogenicity was observed when the bacilli were ingested. Hatcher (1939) found *B. subtilis* in the colon of the roach, *Cryptocercus punctulatus.* Steinhaus (1941a) found bacilli of this group in the alimentary canals of normal larvae of *Samia cecropia, Ceratomia catalpae,* in normal adults of *Lygus pratensis, Conocephalus fasciatus,* in normal nymphs of *Sinea diadema,* and in a species each of Chrysomelidae and Curculionidae. Burroughs (1941) found several strains of *Bacillus subtilis* in the alimentary tract of a grylloblattid.

Hindle and Duncan (1925) noted that *Bacillus subtilis* survives "in the stomach of *Argas persicus* for a time, and is passed in the feces, thus behaving similarly to *B. anthracis*." Duncan (1926) found the gut contents of *Argas persicus*, as well as of *Ornithodoros moubata*, *Stomoxys calcitrans*, *Cimex lectularius*, and *Rhodnius prolixus*, to be bactericidal to *B. subtilis*.

Bacillus tenax

Insect concerned: The nun moth, *Lymantria monacha*.

Eckstein (1894) isolated this sporeforming bacillus from the larva of the nun moth.

Bacillus thoracis Howard

Insect concerned: The honeybee, *Apis mellifera*.

Howard (1900) found an organism which he named *Bacillus thoracis* in the thorax, air passages, and spiracles of the honeybee. This bacillus was usually in association with *Bacillus milii*, which he believed was the cause of the "New York bee disease." In some cases *Bacillus thoracis* itself appeared to cause the death of the bees.

Bacillus thuringiensis Berliner

Insects concerned: The flour moth, *Ephestia kühniella;* the corn borer, *Pyrausta nubilalis;* the cabbage butterfly, *Pieris brassicae; Echocerus cornutus;* the gypsy moth, *Porthetria dispar; Vanessa urticae; Aporia crataegi;* the grasshoppers, *Chorthippus pulvinatus, Chorthippus dorsatus, Chorthippus biguttulus,* and *Calliptamus italicus;* the mosquitoes, *Anopheles maculipennis* and *Culex pipiens;* and the beetle, *Acanthoscelides obtectus*.

In 1915 Berliner isolated *Bacillus thuringiensis* from diseased larvae of the flour moth, *Ephestia kühniella*. Experiments showed that the infection occurred through ingestion and developed in the intestinal tract. Shepherd (1924) states that the organism has been used for the control of *Echocerus cornutus*.

Husz (1927) first infected corn borers with *Bacillus thuringiensis* and found it to be one of the most pathogenic bacteria known to this insect, experimental infection through the mouth giving 100 per cent fatality in one and one-half days. In later experiments he (1929, 1930) has shown that this bacterium may be applied successfully in combating the corn borer, and that the dusting method gives as good results as spraying. Metalnikov and Chorine

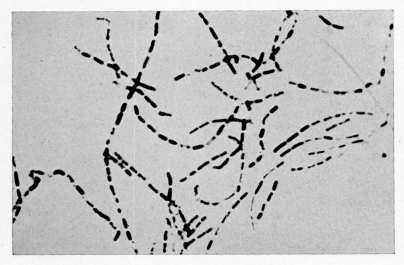

Fig. 29. *Bacillus thuringiensis.* (Photo by N. J. Kramis.)

(1929c) stated that the corn-borer larvae found on corn plants which had been sprayed with cultures of *Bacterium* [*Bacillus*] *thuringiensis* were all poorly developed and less than half the size of the healthy larvae, and that the corn plants yielded more and larger ears than the check plots.

Metalnikov and Chorine (1929d) found *Bacterium thuringiensis* to be very pathogenic for *Porthetria dispar,* and larvae of *Aporia crataegi* and *Vanessa urticae.* Four genera of grasshoppers, two of mosquitoes, and one species of beetle were found to be resistant to infection by mouth. Metalnikov and Chorine thought from their experiments that *Bacterium thuringiensis* was specific in its pathogenicity for the larvae of Lepidoptera.

Metalnikov and Chorine and others have referred to *Bacillus thuringiensis* in most cases as *Bacterium thuringiensis;* but it is presumed that they are identical, for the authors give Berliner credit for having discovered *Bacterium thuringiensis.* *Bacillus* is correct since the organism is a sporeformer. Metalnikov and Chorine (1929a,e) designated two bacteria isolated from diseased *Ephestia kühniella* as *Bacterium ephestiae* No. 1 and No. 2. At that time *Bacillus thuringiensis* had not been sufficiently studied to permit comparison, so Ellinger and Chorine (1930a,b) undertook a comparative study of the three bacteria and found that their strains of *Bacterium ephestiae* were the same as *Bacillus thuringiensis* isolated by Berliner from diseased *Ephestia*.

Bacillus tifosimili

Insects concerned: The roach, *Blatta orientalis;* the housefly, *Musca domestica; Calliphora vomitoria; Lucilia caesar;* and *Melolontha vulgaris.*

Cao (1898) isolated several strains of *Bacillus "Tifosimili"* a typhoidlike organism from roaches, flies, and larvae of *Melolontha vulgaris*. The organisms live for a long time in the intestines of the roach, but finally become exhausted after two or three months. The organism also lives for quite a time in larvae of *Melolontha vulgaris,* even multiplying. Cao (1906a) also refers to a *Bacillus typhosimili,* giving its cultural characteristics; it is probably the same as *Bacillus tifosimili.*

Bacillus tingens

Insects concerned: *Orgyia pudibunda* and the nun moth, *Lymantria monacha.*

Eckstein (1894) found *Bacillus tingens* in the dead larvae of *Orgyia pudibunda*. It was experimentally infective for the nun moth.

Bacillus tracheïtis sive [or] graphitosis Krassilstschik

Insect concerned: The cockchafer, *Melolontha melolontha.*

Krassilstschik (1893), working on the diseases of the larvae of the cockchafer, isolated a sporeforming motile bacillus which he called *Bacillus tracheïtis sive graphitosis*. The disease, "graphitose," was characterized by a lead color of the body at the time of the insect's death.

Bacillus tropicus Heaslip

Tick concerned: The tick, *Haemaphysalis humerosa.*

Heaslip (1940, 1941) reported that a bacillus of the anthrax group, *Bacillus tropicus*, is the cause of certain cases of pyrexia that have been called "coastal fever." The fever may also occur as a double infection with typhus. The disease has a distinct clinical entity, and the organism is always present in the blood of the patient. Mice which were injected with the contents of an engorged tick taken from an infected bandicoot (*Isoodon torosus*) were affected the same way as those which had been injected with blood of human patients. *Haemaphysalis humerosa* transmitted the bacillus from infected to healthy mice by biting. Mosquitoes and mites have also been suspected of being vectors.

Bacillus vesiculosis

Insect concerned: The housefly, *Musca domestica*.

Nicoll (1911) isolated *Bacillus vesiculosis* (*Bacterium vesiculosum* Henrici?) from the surface of flies.

Bacillus viridans

Insect concerned: The silkworm, *Bombyx mori*.

This organism was listed by Sawamura (1906) as pathogenic to silkworm larvae when the latter were artificially infected.

Genus: CLOSTRIDIUM

Clostridium botulinum Type C Bengtson

Insects concerned: The greenbottle fly, *Lucilia caesar*, and the water beetle, *Enochrus hamiltoni*.

Bengtson (1922) isolated an anaerobic sporeforming organism from the larvae of *Lucilia caesar* which caused limber-neck in chickens. She did not give a name to the organism, but in *Bergey's Manual* (3d ed., 1930) the organism is referred to as *Clostridium luciliae* Bengtson. In the fifth edition (1939, p. 755) the organism is called *Clostridium botulinum* Type C Bengtson.

Gunderson (1935) reported finding *Clostridium botulinum* Type C in both larvae and cocoons of the water beetle, *Enochrus hamiltoni*, and attributed an epizootic among sandpipers to the presence of this organism in the larvae, the sandpiper feeding upon them.

Clostridium chauvoei (Arloing et al.) Holland

Insects concerned: The roach, *Blatta orientalis; Blaps mucronata; Tentyria* sp.; *Pimelia bifurcata;* and *Pimelia sardea.*

Cao (1898) found that the bacillus of symptomatic anthrax or blackleg (*Clostridium chauvoei*) passed alive and virulent through the intestinal tract of *Blatta orientalis, Blaps mucronata, Pimelia bifurcata, Pimelia sardea,* and *Tentyria* sp.

Clostridium novyi (Migula) Bergey et al.

Insect concerned: The wax moth, *Galleria mellonella.*

Metalnikov (1927) and Cameron (1934) found that *Clostridium oedematiens* (*Clostridium novyi*) varied in its pathogenicity for the wax moth, the result depending upon the dosage.

Clostridium perfringens (Veillen and Zuber) Hauduroy et al.

Insect concerned: The wax moth, *Galleria mellonella.*

Cameron (1934) found that *Clostridium welchii* (*Clostridium perfringens*) varied in its pathogenicity for the wax moth, the result depending on the dosage.

Clostridium sporogenes (Metchnikoff) Bergey et al.

Insects concerned: The roach, *Blatta orientalis; Blaps mucronata; Pimelia bifurcata; Pimelia sardea; Tentyria* sp.; a burrowing beetle, *Melolontha vulgaris;* and the wax moth, *Galleria mellonella.*

The bacillus of a malignant edema, *Clostridium sporogenes* was found by Cao (1898) to pass in the spore form through the intestines of the oriental roach and several Coleoptera (*Blaps, Pimelia,* and *Tentyria* sp.) and was alive and virulent in the feces.

Cao (1906b) also isolated *Clostridium sporogenes* from the intestinal contents of larvae of *Melolontha vulgaris*.

Metalnikov (1920a) and Cameron (1934) found that *Clostridium sporogenes* had no lethal effect on the wax moth when injected.

Clostridium tetani (Nicolaier) Holland

Insects and ticks concerned: A burrowing beetle, *Melolontha vulgaris; Tentyria* sp.; *Blaps mucronata; Pimelia bifurcata; Pimelia sardea;* the roach, *Blatta orientalis;* the wax moth, *Galleria mellonella;* the flea, *Tunga penetrans* (=*Dermatophilus penetrans*); and the tick, *Argas persicus*.

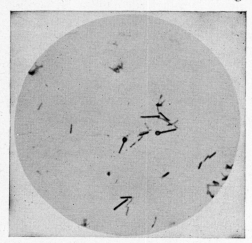

Fig. 30. *Clostridium tetani*. (From Hagan, 1943.)

Clostridium tetani was isolated by Cao (1906b) from the intestinal contents of the larvae of *Melolontha vulgaris*, and was found by him to pass alive and virulent through the intestines of *Blatta orientalis* and several Coleoptera (*Tentyria* sp., *Blaps mucronata, Pimelia sardea,* and *Pimelia bifurcata*). According to Cao (1898), *Argas persicus* has been accused of disseminating the bacillus of tetanus.

Pierce (1921) states that the attack of the flea, *Tunga penetrans*, occasionally leads to infection with *Clostridium tetani*.

Metalnikov (1920a) made a number of experiments to determine the immunity of the larvae of the wax moth to various classes of microorganisms. He found the larvae to be completely immune to infection with the tetanus bacillus. Cameron (1934) confirmed his results.

Clostridium werneri Bergey et al.

Insect concerned: *Potosia cuprea*.

Werner (1926a) isolated a cellulose-fermenting bacillus from the digestive tract of the larva of *Potosia cuprea*, which feeds chiefly on spruce and pine needles. He called the organism *Bacillus cellulosam fermentans,* and it was so designated in the fourth edition of *Bergey's Manual*. In the fifth edition, 1939 (see p. 785) the name was changed to *Clostridium werneri*.

Werner was unable to decide whether the larva used the products of fermentation of cellulose directly for metabolic purposes, or whether the fermentation only destroyed the cell walls and made the contents of the cell available as food for the larva.

Family: BACTERIACEAE

Genus: ACHROMOBACTER

Achromobacter delicatulum (Jordan) Bergey *et al.*
(See also *Achromobacter hyalinum*)

Insect concerned: The Colorado potato beetle, *Leptinotarsa decemlineata*.

Steinhaus (1941a) isolated this bacterium from the normal alimentary tract of the Colorado potato beetle.

Achromobater eurydice
(White) Bergey *et al.*
(See also *Bacillus alvei* and *Bacillus pluton*)

Insect concerned: The honey-bee, *Apis mellifera*.

White (1912) named this organism *Bacterium eurydice*. Bergey *et al.* (see *Bergey's Manual,* 2d ed., 1935, p. 170; also 5th ed., 1939, p. 517), renamed it *Achromobacter eurydice*.

White was unable to produce European foul brood in

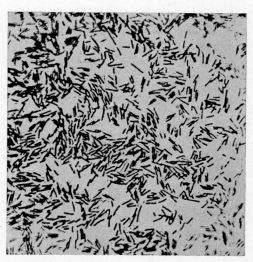

Fig. 31. Asporogenic form of *Bacillus alvei*, morphologically closely resembling White's *Achromobacter eurydice*. (From Burnside, 1934.)

bees with *Bacterium eurydice,* although it is apparently a secondary invader in the disease. Burnside (1934) suggested that *Bacillus pluton, Streptococcus apis,* and *Achromobacter (Bacterium) eurydice* are variants, or stages, in the life history of *Bacillus alvei.*

Achromobacter hyalinum Bergey *et al.*

Insect concerned: The American cockroach, *Periplaneta americana.*

Hatcher (1939) states that *"Acromobacter hyalinum* (Jordan)" was one of the species of bacteria she isolated from the fecal matter of *Periplaneta.* The generic name of this organism was probably misspelled and no doubt was meant to be *Achromobacter.* According to the fifth edition of *Bergey's Manual* (1939, p. 505), *Bacillus hyalinus* Jordan (*Achromobacter hyalinum* Bergey *et al.*) is possibly a synonym for *Achromobacter delicatulum* (Jordan) Bergey *et al.* Probably it is this organism which Hatcher found in the roach, *Periplaneta americana.*

Achromobacter larvae (Stutzer and Wsorow) Bergey *et al.*

Insect concerned: *Euxoa segetum.*

This organism, isolated by Stutzer and Wsorow (1927) from the intestinal tract of normal caterpillars of *Euxoa segetum,* was named by its isolators *Enterobacillus larvae,* and was so designated in the third edition of *Bergey's Manual,* 1930, page 227. However, in the fifth edition of *Bergey's Manual,* 1939, page 517, it is described under the name *Achromobacter larvae.* Besides this species several other species of bacteria were found in the intestinal tracts of the normal caterpillars.

Achromobacter superficiale (Jordan) Bergey *et al.*

Insect concerned: *Urographis fasciata.*

In normal larvae of *Urographis fasciata,* Steinhaus (1941a) found a bacterium which corresponded fairly well to the incomplete description given in *Bergey's Manual* (5th ed., 1939, p. 511) for *Achromobacter superficiale.* However, the growth on agar of the organism isolated was more abundant than that of *Achromobacter superficiale,* which is described as being "limited."

Genus: BACTERIUM

Bacterium sp. De Bach and McOmie

Insect concerned: The termite, *Zoötermopsis angusticollis.*

De Bach and McOmie (1939) found their laboratory stock of the termite, *Zoötermopsis angusticollis,* to be afflicted with two bacterial diseases. One of these diseases was caused by *Serratia marcescens* (which see), and the other was caused by a bacterium which they designated as *Bacterium* sp. The disease caused by the latter organism was less common than that caused by *Serratia marcescens.*

De Bach and McOmie found *Bacterium* sp. to be similar to *Bacterium neapolitanum* according to the classification system of Kluyver and Van Niel (1936).

Bacterium acidiformans Sternberg

Insect concerned: The honeybee, *Apis mellifera.*

White (1906) isolated *Bacterium acidiformans* from the scrapings of propalis and wax from honey frames and hives of healthy colonies of honeybees.

Bacterium agrigenum (Trevisan) Migula

Insect concerned: A fly, probably *Musca domestica.*

Marpmann (1897) fed flies on cultures of *Bacillus septicus agrigenus* (*Bacterium agrigenum*), and after 12 hours the contents of the flies were inoculated into mice, the majority dying. However, he concluded that in some cases the organisms were digested or attenuated in the flies' guts.

Bacterium canadensis Chorine

Insects concerned: The corn borer, *Pyrausta nubilalis;* the wax moth, *Galleria mellonella;* and the flour moth, *Ephestia kühniella.*

Chorine (1929a,b) isolated *Bacterium canadensis* from diseased Canadian corn-borer larvae, and found it to be very virulent to the larvae experimentally. Ninety per cent or more of the borers died when they were fed pure cultures. The larvae of *Galleria mellonella* and *Ephestia kühniella* died when injected with very minute doses.

According to Chorine, in its general characteristics *Bacterium canadensis* resembles *Bacillus megatherium* and certain bacteria, such as *Bacillus thuringiensis, Bacillus sotto,* and *Bacillus hoplosternus,* isolated from insects.

Paillot refers to this organism as *Bacillus canadensis*. Since the organism is a sporeformer, the generic name *Bacillus* is preferable to that of *Bacterium*.

Bacterium cazaubon Metalnikov *et al.*

Insects concerned: The corn borer, *Pyrausta nubilalis;* the gypsy moth, *Porthetria dispar; Vanessa urticae;* the flour moth, *Ephestia kühniella; Stilpnotia salicis; Loxostege sticticalis;* and *Aporia crataegi.*

Metalnikov, Ermolaev, and Skobaltzyn (1930) isolated *Bacterium cazaubon* No. I and *Bacterium cazaubon* No. II from diseased larvae of the corn borer that came from the Pyrenees in the environs of Cazaubon. They found that the organism was one of the most virulent bacteria known for the corn borer, killing the larvae in 10 to 15 hours after infection through the mouth. The infected larvae usually became black in color. Metalnikov (1930) also found it to be pathogenic for the gypsy moth, the flour moth, *Vanessa urticae, Stilpnotia salicis,* and *Aporia crataegi.*

Bacterium cazaubon No. I and *Bacterium cazaubon* No. II differ only in insignificant characters, being very similar both in morphologic and physiologic characteristics. Presumably in later publications the authors made no distinction between Nos. I and II and just referred to *Bacterium cazaubon*. The preferable generic name would be *Bacillus,* since the organism is a sporeformer.

Bacterium cellulosum

Insect concerned: *Cetonia floricola.*

According to Jepson (1937), Werner thinks that *Bacterium cellulosum* plays an essential role in the nutrition of the inquiline larva of *Cetonia floricola.*

Bacterium christiei Chorine

Insect concerned: The corn borer, *Pyrausta nubilalis*.

Chorine (1929a) isolated this organism from diseased Canadian larvae of *Pyrausta nubilalis*. The organism, in its biologic characteristics, resembles *Bacterium ontarioni*, though not in its morphologic characters and in the development of the colonies. It is also more virulent than *Bacterium ontarioni*, but it was not so virulent as *Bacterium canadensis*, *Bacterium cazaubon*, and others isolated from diseased borers. The organism is a sporeformer, so it should have been placed in the genus *Bacillus*.

Bacterium cleonusum Beltiukova and Romanevich

Insect concerned: "The beet weevil" (species not given).

While investigating the bacterial diseases affecting the beet weevil in the Ukraine, Beltiukova and Romanevich (1940) isolated and named this bacterium which they described as one of a group of bacteria causing a "rot" in these larvae. They also observed infections in this insect caused by *"B. prodigiosus"* and by *"B. gallerie* No. 3 (Metalnikov)."

Bacterium coli apium Serbinow

Insect concerned: The honeybee, *Apis mellifera*.

Serbinow (1915) states that *Bacterium coli apium* and *Proteus alveicola* were the cause of an infectious diarrhea among honeybees during the spring of the year. The organisms were found to be infectious to mice by peritoneal inoculation.

Bacterium conjunctivitidis (Kruse) Migula

Insect concerned: The housefly, *Musca domestica*.

According to Patton (1931), Wollman, working in Tunis, placed houseflies in tubes containing cultures of *Bacillus aegyptius* (*Bacterium conjunctivitidis*) and the Morax-Axenfeld bacillus (see *Hemophilus duplex*) of subacute conjunctivitis, and observed that whereas they became infective immediately afterwards, they were not infective after an interval of three and a half hours.

Bacterium cyaneus

Insect concerned: The honeybee, *Apis mellifera*.

White (1906) isolated *Bacterium cyaneus* from pollen and from the bodies of healthy honeybees.

Bacterium D White

Insect concerned: The honeybee, *Apis mellifera*.

White (1906) frequently found *Bacterium D* present in the intestines of normal honeybees. The body temperature of the honeybee and that of warm-blooded animals is about the same, and he found that the bacterial flora of the honeybee intestine was very similar to that of man and animals.

Bacterium delendae-muscae Roubaud and Descazeaux

Insects concerned: The stablefly, *Stomoxys calcitrans;* the housefly, *Musca domestica;* blowflies; and flesh flies.

In 1923 Roubaud and Descazeaux discovered a bacterial disease of fly larvae caused by *Bacterium delendae-muscae*. They state that it is the first microorganism described in literature that causes a specific bacterial infection of house and other flies. The disease is primarily a larval infection, the larvae becoming infected by ingesting the bacteria with their food and dying after 2 to 30 days. Flesh flies die at the commencement of the pupae stage, houseflies at the end of that stage, and *Stomoxys* after emergence.

"Bacterium elbvibrionen"

Insect concerned: The yellow mealworm, *Tenebrio molitor*.

"*Bacterium elbvibrionen*" is a misnomer, the correct name being *Vibrio albensis* Lehmann and Neumann. It was found pathogenic for the mealworm by Pfeiffer and Stammer (1930).

Bacterium ephestiae No. 1 and No. 2 Metalnikov and Chorine
(See also *Bacillus thuringiensis*)

Insects concerned: The corn borer, *Pyrausta nubilalis;* the flour moth, *Ephestia kühniella;* and *Pectinophora gossypiella*.

Bacterium ephestiae No. 1 and No. 2 were isolated by Metalnikov and Chorine (Chorine, 1929a) from *Ephestia kühniella* and used to combat corn borers. These organisms are identical with *Bacillus thuringiensis* as shown by Ellinger and Chorine (1930a).

Bacterium galleriae Metalnikov and Chorine

Insects concerned: The wax moth, *Galleria mellonella;* the flour moth, *Ephestia kühniella;* and the corn borer, *Pyrausta nubilalis*.

Metalnikov (1922) isolated *Bacterium galleriae* from wax moths. Later Kitajima and Metalnikov (1923) isolated it from the same insects during an epizoötic.

This organism is described as being an elongated, rod-shaped, sporeforming organism, so the preferable generic name would be *Bacillus*.

Bacterium galleriae No. 2 and No. 3 Metalnikov and Chorine

Insects concerned: the corn borer, *Pyrausta nubilalis,* and the wax moth, *Galleria mellonella*.

Metalnikov and Chorine (1928b) used *Bacterium galleriae* No. 2 at first rather effectively in combating the corn borer, but later results (Chorine 1930a,b) were poor when compared with those of some other bacteria. The insects turn black a few hours after death. Old cultures are as virulent as fresh cultures.

Bacterium galleriae No. 3 was not very effective in killing corn-borer larvae, as it was noninfective by mouth and only 20 to 50 per cent effective upon injection.

Bacterium gelechiae No. 1, No. 2, and No. 5 Metalnikov and Metalnikov

Insects concerned: The corn borer, *Pyrausta nubilalis; Gelechia gossypiella;* and *Prodenia litura.*

Metalnikov and Metalnikov (1932) isolated from dead or dying larvae of *Platyedra (Gelechia) gossypiella* two forms of *Bacterium gelechiae* which they called Nos. 1 and 2. Both organisms produced death within 24 to 48 hours. Later (1933) they also isolated strain No. 5 from dead larvae.

Bacterium gryllotalpae Metalnikov and Meng

Insect concerned: *Gryllotalpa gryllotalpa.*

Bacterium gryllotalpae was one of two organisms isolated by Metalnikov and Meng (1935) from the diseased larvae of *Gryllotalpa gryllotalpa* during an epidemic that occurred in the laboratory. *Bacillus gryllotalpae* was the other organism.

Bacterium hebetisiccus Steinhaus

Insect concerned: The walkingstick, *Diapheromera femorata.*

Steinhaus (1941a) isolated this bacterium from the walkingstick while studying its normal bacterial flora.

Bacterium hemophosphoreum Pfeiffer and Stammer

Insects concerned: *Mamestra oleracea;* the yellow mealworm, *Tenebrio molitor;* the cabbage butterfly, *Pieris rapae;* and *Agrotis* sp.

Pfeiffer and Stammer (1930) isolated this phosphorescent organism from the larvae of *Mamestra oleracea.* According to them, the organism is not identical with any of the known light-producing bacterial strains pathogenic for invertebrates. Laboratory infection was possible only when the bacterial masses from the hemolymph were inoculated. Attempts to infect by mouth were not successful. The organism was found to be pathogenic for other insects also.

Bacterium imperiale Steinhaus

Insect concerned: The imperial moth, *Eacles imperialis*.

Steinhaus (1941a) isolated this bacterium from the alimentary tract of a normal *Eacles imperialis*. It is a gram-positive short rod (fig. 32).

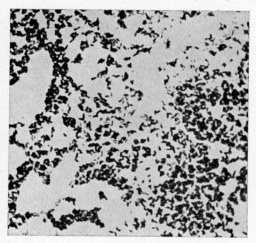

Fig. 32. *Bacterium imperiale*.

Fig. 33. *Tibicen linnei*. (From Comstock, 1936.)

Bacterium incertum Steinhaus

Insect concerned: The lyreman cicada, *Tibicen linnei* (fig. 33).

While studying the natural flora of the lyreman cicada, Steinhaus (1941a) isolated from the ovaries of this insect a bacterium which in some respects closely resembled bacteria of the genus *Listeria* (*Listerella*). The pathogenic characteristics were not the same as those of *Listeria monocytogenes*, however, so the bacterium was tentatively placed in the genus *Bacterium* and given the specific epithet *incertum*. (See fig. 34.)

Bacterium insectiphilium Steinhaus

Insect concerned: The bagworm, *Thyridopteryx ephemeraeformis*.

Bacterium insectiphilium was isolated by Steinhaus (1941a) from the body wall of the bagworm. It is a gram-positive, nonsporeforming rod which hydrolyzes starch. (See fig. 35.)

Bacterium intrinsectum Steinhaus

Insect concerned: An unidentified leaf beetle (Chrysomelidae).

From the alimentary tract of an unidentified normal leaf beetle, Steinhaus (1941a) isolated this bacterium, a gram-negative small rod.

Bacterium italicum No. 2 Metalnikov *et al.*

Insect concerned: The corn borer, *Pyrausta nubilalis.*

Bacterium italicum No. 2 was isolated from diseased larvae of the corn borer from Italy by Metalnikov, Ermolaev, and Skobaltzyn (1930). It is a very virulent organism for corn-borer larvae, killing in 20 to 24 hours.

Since it is a large, sporeforming rod, a more suitable generic name would be that of *Bacillus.*

Bacterium knipowitchii

Insect concerned: The mealworm, *Tenebrio molitor.*

Bacterium knipowitchii, a phosphorescent bacterium, was found to be pathogenic for the mealworm by Pfeiffer and Stammer (1930).

Bacterium lymantricola adiposus Paillot

Insects concerned: *Euxoa segetum;* the gypsy moth, *Porthetria dispar;* and the silkworm, *Bombyx mori.*

Paillot isolated *Bacterium lymantricola adiposus* in 1917 from *Porthetria dispar.* He (1933) found that the bacillus underwent considerable morphological change by 1923. The organism which he thought was a typical coccobacillus had elongated considerably, and very few of the original coccobacillus forms remained. The organism presented various involution forms in the blood of *Porthetria dispar, Bombyx mori,* and *Euxoa segetum.*

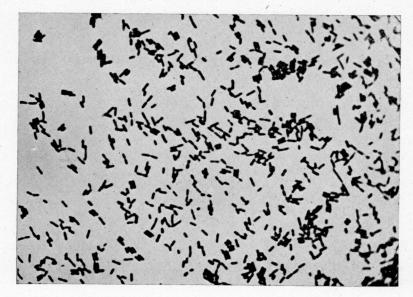

Fig. 34. *Bacterium incertum*. (Photo by N. J. Kramis.)

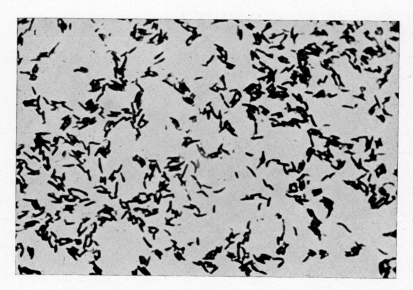

Fig. 35. *Bacterium insectiphilium*. (Photo by N. J. Kramis.)

Bacterium melolonthae liquefaciens Paillot
(See also *Bacillus melolonthae liquefaciens*)

Insects concerned: *Melolontha vulgaris; Vanessa urticae;* and the cockchafer, *Melolontha melolontha*.

Paillot (1916) isolated *Bacterium melolonthae liquefaciens* from the cockchafer, and found it pathogenic for other insects as well.

Bacterium melolonthae liquefaciens gamma Paillot

Insects concerned: The gypsy moth, *Porthetria dispar; Nygmia phaeorrhoea* (=*Euproctis chrysorrhoea*); *Euxoa segetum;* and the cockchafer, *Melolontha melolontha*.

Paillot (1933) speaks of the involution forms of this organism in the blood of *Euproctis chrysorrhoea, Porthetria dispar, Euxoa segetum,* and *Melolontha melolontha*.

Bacterium melolonthae liquefaciens gamma is probably a variant of *Bacterium melolonthae liquefaciens*.

Bacterium minutiferula Steinhaus

Insect concerned: The mud-dauber wasp, *Sceliphron caementarium*.

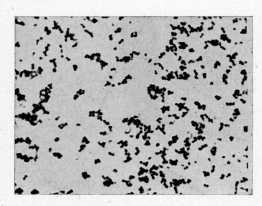

Fig. 36. *Bacterium minutiferula.* (Photo by N. J. Kramis.)

Steinhaus (1941a) isolated this bacterium from a triturated specimen of the mud-dauber wasp, *Sceliphron caementarium*.

Bacterium monachae Von Tubeuf
(See also *Bacillus B* and *Bacillus monachae*)

Insect concerned: The nun moth, *Lymantria monacha*.

Von Tubeuf (1892a,b) isolated this organism from the intestinal tract of the nun moth, considering it to be the causative agent of the "wilt" disease

SPECIFIC BACTERIA ASSOCIATED WITH INSECTS

(*Wipfelkrankheit*). Later Von Tubeuf (1911) reversed his opinion on the specific cause of the disease and concluded that it was due to a variety of intestinal bacteria becoming dominant. Since then Wahl (1909 to 1912) has shown that the disease is due to a virus.

Eckstein (1894) considered *Bacterium monachae, Bacillus B* of Hofmann (1891), and *Bacillus monachae* to be identical organisms.

Bacterium mutabile Steinhaus

Insect concerned: The lyreman cicada, *Tibicen linnei*.

This pleomorphic bacterium was isolated by Steinhaus (1941a) from the alimentary tract of the cicada while studying the natural flora of this insect.

Bacterium mycoides

Insect concerned: The honeybee, *Apis mellifera*.

White (1906) isolated *Bacterium mycoides* from the intestine of a healthy honeybee. His strain, differing from *Bacterium mycoides* Migula in that spores are formed and pigment is not, is probably identical with *Bacillus mycoides*.

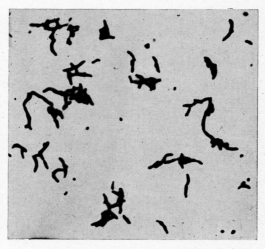

Fig. 37. *Bacterium mutabile*. (Photo by N. J. Kramis.)

Bacterium neurotomae Paillot

Insect concerned: *Neurotoma nemoralis*.

This organism was isolated in 1922 by Paillot (1933) from the larvae of *Neurotoma nemoralis*. The organism is very pleomorphic, often presenting elongated forms but rarely filamented forms. Paillot (1933) states that on gelatin a certain number of the organisms swell abnormally and transform themselves into masses more or less rounded, of which the "chromatophile" is generally condensed in the central part. The masses present what appear to be veritable nucleated cells, but they are deprived of their vitality and rapidly degenerate.

This organism is probably the same as *Bacillus neurotomae* isolated by Paillot (1924a).

Bacterium noctuarum White
(See also *Bacillus noctuarum* and *Bacillus sphingidis*)

Insect concerned: The honeybee, *Apis mellifera*.

Metalnikov and Chorine (1928b) refer to *Bacterium noctuarum* White, saying that it resembles *Coccobacillus ellingeri*. They probably mean *Bacillus noctuarum* White.

Bacterium ochraceum (Zimmermann)

Insect concerned: *Euxoa segetum*.

Stutzer and Wsorow (1927) isolated *Bacterium ochraceum* from the intestine of healthy larvae of *Euxoa segetum*.

Bacterium ochraceum is probably the same as *Flavobacterium ochraceum* (Zimmermann) Bergey *et al.* found in water. Zimmermann called it *Bacillus ochraceus*.

Bacterium ontarioni Chorine

Insects concerned: The corn borer, *Pyrausta nubilalis;* the flour moth, *Ephestia kühniella;* and *Galleria mellonella*.

This organism was isolated from diseased Canadian corn borers by Chorine (1929a,b), and experimentally it has been shown to be pathogenic to larvae of *Ephestia kühniella* and *Galleria mellonella* as well as to the corn-borer larvae, about 50 per cent dying when the organisms were added to the feed. This organism is a sporeformer, so the correct name should be *Bacillus ontarioni*.

Bacterium paracoli Gilbert and Lion

Insect concerned: *Euxoa segetem*.

Stutzer and Wsorow (1927) isolated *Bacterium paracoli* (*Bacillus bruneus* Maschek) from the intestines of healthy larvae of *Euxoa segetum*.

Bacterium pieris liquefaciens alpha Paillot

Insects concerned: *Pieris brassicae; Vanessa urticae; Vanessa polyschloros; Nygmia phaeorrhoea* (= *Euproctis chrysorrhoea*); and *Porthetria dispar*.

Paillot (1933) found *Bacterium pieris liquefaciens alpha* to be very pleomorphic, its shape depending upon the insect it parasitized. In the blood of the larvae of *Pieris brassicae,* from which it was isolated, the bacterial elements were in the form of coccobacilli. In the larvae of *Vanessa polyschloros,* the longer elements were considerably longer than in the above-mentioned larvae. In the larvae of *Porthetria dispar* the organism lost all aspects of a coccobacillus and was transformed into veritable filaments, some as long as 40 microns. These morphological changes did not affect the vital and biochemical properties of the organism.

Bacterium pityocampae Dufrenoy

Insect concerned: The processionary caterpillars of pines, *Cnethocampa pityocampa*.

Dufrenoy (1919) isolated three pathogenic organisms from *Cnethocampa pityocampa*: *Bacterium pityocampae* and *Streptococcus pityocampae alpha* and *beta*. *Bacterium pityocampae* according to Paillot (1933) is a gram-negative, encapsulated organism.

Bacterium prodeniae Metalnikov and Metalnikov

Insect concerned: *Prodenia litura*.

Metalnikov and Metalnikov (1932) isolated this organism from larvae of

Prodenia litura while studying diseases of this insect and those of *Gelechia gossypiella*.

Bacterium pyraustae Nos. 1–7 Metalnikov and Chorine

Insect concerned: The corn borer, *Pyrausta nubilalis*.

Metalnikov and Chorine (1928b) isolated these seven bacteria from diseased corn borers which had been collected from mugwort plants around Paris. The authors considered them to be of little importance because they were not able to infect corn borers by mouth, so the organisms were merely numbered. Nos. 1 and 2 were found to be very virulent for corn-borer larvae when injected, killing 75 to 100 per cent, while No. 6 only killed from 1 to 20 per cent.

Bacterium pyrenei No. 1, No. 2, and No. 3 Metalnikov *et al.*

Insects concerned: The corn borer, *Pyrausta nubilalis;* the white cabbage butterfly, *Pieris brassicae;* the gypsy moth, *Porthetria dispar; Vanessa urticae; Stilpnotia salicis;* the flour moth, *Ephestia kühniella;* and *Aporia crataegi*.

Bacterium pyrenei No. 1 was isolated by Metalnikov, Ermolaev, and Skobaltzyn (1930) from dead, black larvae of corn borers that had been received from the Pyrenees. Two other organisms, *Bacterium pyrenei* No. 2 and *Bacterium pyrenei* No. 3, were isolated from the same larvae. The authors, pending complete analysis, gave the bacteria provisional names. All three strains were very pathogenic for the corn-borer larvae and for other Lepidoptera larvae, which often died in 10 to 15 hours when infected by mouth with a virulent strain. The present author was unable to find subsequent references to these organisms other than Metalnikov (1930) and Sweetman (1936), who refer only to *Bacterium pyrenei*. Metalnikov found that *Bacterium pyrenei* was pathogenic for the larvae of the gypsy moth, the flour moth, the white cabbage butterfly, *Vanessa urticae, Stilpnotia salicis,* and *Aporia crataegi*.

The three strains are all sporeforming, so the generic name should be *Bacillus*.

Bacterium qualis Steinhaus

Insect concerned: The tarnished plant bug, *Lygus pratensis*.

Steinhaus (1941a) isolated this bacterium from the alimentary tract of a normal tarnished plant bug.

Bacterium rubrum

Insect concerned: *Platyedra (Gelechia) gossypiella.*

Metalnikov and Metalnikov (1932) isolated this organism from the larvae of *Platyedra (Gelechia) gossypiella.*

Bacterium tegumenticola Steinhaus

Insect concerned: The bedbug, *Cimex lectularius.*

The integuments of several specimens of normal *Cimex lectularius* were found by Steinhaus (1941a) to harbor this bacterium.

Bacterium termo Gillette

Insects concerned: The grasshoppers, *Melanoplus bivittatus* and *Melanoplus femur-rubrum.*

Gillette (1896) isolated this organism from grasshoppers in Colorado. According to Smith (1933), the hoppers generally died on the ground, their bodies turning black and their vital organs disintegrating. According to Jordan and Burrows (1941), *Bacterium termo* is synonymous with *Proteus vulgaris.*

Bacterium tumefaciens Smith and Townsend

Insect concerned: The nun moth, *Galleria mellonella.*

This organism was found in the larvae of *Galleria mellonella* by Metalnikov, Kostritsky, and Toumanoff (1924). Probably it is identical with *Xanthomonas tumefaciens.*

Bacterium viscosum non liquefaciens

Insect concerned: *Euxoa segetum.*

Stutzer and Wsorow (1927) isolated *Bacterium viscosum non liquefaciens* from a normal pupa of *Euxoa segetum.*

Genus: BACTEROIDES

Actinomyces necrophorus (Flügge) Lehmann and Neumann

Mite and ticks concerned: The mite, *Demodex folliculorum*, and ticks of the genus *Amblyomma*.

According to Pierce (1921), the blackhead mite, *Demodex folliculorum*, causes an irritation giving rise to papules which become infected with *Bacillus necrophorus* (*Actinomyces necrophorus*). Similar infections are sometimes caused by bites of ticks of the genus *Amblyomma*.

The classification of this organism is uncertain. *Bergey's Manual* appends it to the genus *Bacteroides;* others place it in the genus *Fusiformis*.

Genus: FLAVOBACTERIUM

Flavobacterium acidificum Steinhaus

Insects concerned: The grasshopper, *Conocephalus fasciatus* var. *fasciatus;* the Colorado potato beetle, *Leptinotarsa decemlineata;* the cabbage butterfly, *Pieris rapae;* and an unidentified lady-beetle larva.

Steinhaus (1941a) isolated *Flavobacterium acidificum* from a grasshopper, a potato beetle, a cabbage butterfly, and an unidentified lady-beetle larva while studying their normal bacterial floras.

Flavobacterium chlorum Steinhaus

Insect concerned: The nine-spotted lady beetle, *Coccinella novemnotata*.

This bacterium was isolated from the alimentary tract of a normal nine-spotted lady beetle by Steinhaus (1941a).

Flavobacterium devorans (Zimmermann) Bergey et al.

Insect concerned: *Coccinella novemnotata*.

Steinhaus (1941a) found a species of bacteria very similar to *Flavobacterium devorans* in the alimentary tract of normal *Coccinella novemnotata* collected in nature.

Flavobacterium maris Harrison

Insect concerned: The catalpa sphinx, *Ceratomia catalpae*.

From the midgut of normal larva of *Ceratomia catalpae,* Steinhaus (1941a) isolated a gram-positive bacterium which in many of its characteristics was similar to *Flavobacterium maris*. The two organisms may not be identical (they differed in a few of their physiologic reactions), but until this group is better defined they may be considered the same.

Flavobacterium rheni (Chester) Bergey et al.

Insect concerned: The walkingstick, *Diapheromera femorata*.

Steinhaus (1941a) isolated from walkingstick eggs about to be laid an organism similar to *Flavobacterium rheni*.

Genus: FUSOBACTERIUM
(*Fusiformis* Prévot)

Fusiformis hilli Duboscq and Grassé

Insect concerned: The termite, *Glyptotermes iridipennis*.

Duboscq and Grassé (1927) isolated an organism, which they named *Fusiformis hilli*, from the termite *Glyptotermes iridipennis*.

Fusiformis termitidis Hoelling

Insects concerned: Termites, including *Glyptotermes iridipennis*.

According to Dougherty (1942), Hoelling (1910) described a fusiform bacterium, which he called *Fusiformis termitidis,* from smears of the intestinal tracts of termites. Duboscq and Grassé (1927) found this microorganism in *Calotermes* (*Glyptotermes*) *iridipennis*.

Family: ENTEROBACTERIACEAE
Tribe: ERWINEAE
Genus: ERWINIA

Erwinia amylovora (Burrill) Winslow et al.

Insects concerned: The rapid plant bug, *Adelphocoris rapidus; Campylomma verbasci;* the tarnished plant bug, *Lygus pratensis; Orthotylus flavosparsus; Plagiognathus politus;* the apple leafhopper, *Empoasca maligna; Aphis avenae;* the apple aphid, *Aphis pomi;* the convergent lady beetle, *Hippodamia convergens;* the western spotted cucumber beetle, *Diabrotica soror; Orsodacne atra;* the black carpet beetle, *Attagenus piceus; Anthrenus* sp.; *Glischrochilus fasciatus;* the Oregon wireworm, *Melanotus oregonensis;* the shot-hole borer, *Scolytus rugulosus;* the codling moth, *Carpocapsa pomonella; Bibio albipennis;* the onion maggot, *Hylemya antiqua; Hylemya lipsia; Pegomyia calyptrata; Cynomya cadaverina;* the silky ant, *Formica fusca subsericea; Formica pallide-fulva schaufussi* var. *incerta;* the cornfield ant, *Lasius niger americanus; Prenolepsis imparis; Polistes* sp.; *Vespula* sp.; the housefly, *Musca domestica; Muscina assimilis; Muscina stabulans;* the pomace fly, *Drosophila funebris; Drosophila ampelopila* (*Drosophila melanogaster*); the greenbottle fly, *Lucilia sericata;* and bees.

Burrill (1881) was the first to prove that bacteria can be the cause of a plant disease, and it was he (1882) who isolated the causative organism of fire blight, *Micrococcus amylovorus*. Forbes in 1884 observed fire-blight lesions associated with the feeding of the tarnished plant bug and thought that this insect was acting as a vector of the disease. Waite (1891) was one of the first who offered experimental proof that insects were important vectors of plant diseases when he showed that bees and wasps were transmitters of fire blight.

According to Leach (1940), fire blight is principally a disease of pears and apples, although other orchard fruits as well as many ornamental plants are often attacked. It affects primarily the blossoms and young twigs of apple and pear trees, although more than 90 species of plants, mostly in the family Rosaceae, are hosts to the disease. The literature on this disease is contradictory as well as voluminous. Concerning this state of affairs, Leach writes: "It is somewhat ironical that the first association between insect and plant disease to be established should, after 50 years, remain in such an uncertain and unsatisfactory state." Many theories of insect transmission have been advanced and many insects have been incriminated, in some cases without adequate proof.

Two of the recent workers on the association of this disease with insects, Ark and Thomas (1936), have shown that *Erwinia amylovora* may live for several days in the intestinal tract of *Drosophila melanogaster, Lucilia sericata,* and *Musca domestica.* The eggs of *Musca domestica,* which had been laid by contaminated females, were found to harbor externally the pathogen. Furthermore, bacteria fed to the larvae of *Drosophila melanogaster* and *Musca domestica* persisted through the puparia and could be found associated with the emerging adult.

Erwinia cacticida (Johnston and Hitchcock) Hauduroy *et al.*

Insects concerned: *Melitara prodenialis; Olyca junctoliniella; Mimorista flavidissimalis;* a longicorn beetle, *Moneilema* sp.; several members of the Drosophilidae; *Cactoblastis cactorum;* and *Cactoblastis bucyrus.*

Erwinia cacticida, which causes a disease of the prickly pear (*Opuntia* spp.), was described and named *Bacillus cacticidus* by Johnston and Hitchcock (1923).

According to Leach, it has been shown experimentally that the larvae of *Melitara prodenialis, Olyca junctoliniella,* and *Mimorista flavidissimalis* are effective vectors in the spread of the disease, and it was predicted that *Cactoblastis cactorum* and *C. bucyrus* would also prove to be effective. The larvae of longicorn beetles and several members of the Drosophilidae were found experimentally to be able to spread the infection. The disease of itself does not spread unaided from one segment to another; insects not only spread the disease from plant to plant, but from one segment to another of the same plant.

Erwinia cacticida is regarded as a synonym of *Erwinia aroideae* (*Bergey's Manual*, 5th ed., p. 412).

Erwinia carnegieana Standring

Insect concerned: *Cactobrosis fernaldiales*.

Erwinia carnegieana, the cause of necrosis of the giant cactus in Arizona, was cultured from the external surfaces and from the intestinal tracts of larvae of a nocturnal moth, *Cactobrosis fernaldiales*. The bacteria were also isolated from the body of the adult moth and from its eggs. This insect is apparently an important vector of the disease organism (J. G. Brown and associates, *55th Ann. Rept., Agr. Expt. Sta., Univ. Arizona*, 1944).

Erwinia carotovora (Jones) Holland

Insects concerned: The seed-corn maggot, *Hylemya cilicrura*; *Hylemya trichodactyla*; the cabbage maggot, *Hylemya brassicae*; *Elachiptera costata*; *Scaptomyza graminum*; *Phorbia fusciceps*; the tarnished plant bug, *Lygus pratensis*; and the carrot rust fly, *Psila rosae*.

Erwinia carotovora causes a soft rot in carrot, celery, eggplant, cabbage, cucumber, iris, muskmelon, hyacinth, turnip, tomato, potato, onion, radish, parsnip, pepper, and other plants. The three principal types of disease caused by this bacterium are potato blackleg, soft rot of crucifers, and heart rot of celery. Jones first isolated the causative organism in 1901 and named it *Bacillus carotovorus*.

The seed-corn maggot is often found in diseased potatoes, and Leach (1926) has shown that it is an important agent in the spread of blackleg. The burrowing habit of the larva serves as a good means of inoculation. Leach (1931, 1933) studied the symbiotic relationship between bacteria and the corn maggot and found that *Erwinia carotovora* frequently but not always occurred in the intestinal tract of both the larvae and adult flies. The organism apparently propagates within the adults and larvae and may survive the winter within the insect puparia, and upon emerging in the spring, the adults are infected. Thus begins another cycle of infection. The organism has also been isolated from the eggs of *H. cilicrura*.

The cabbage maggot has been incriminated in the spread of soft rot, and *Scaptomyza graminum, Elachiptera costata,* and *Lygus pratensis* have been found to be vectors of the heart rot of celery (Leach, 1927).

Bonde (1939a) isolated *Erwinia carotovora* from a carrot infested with the larvae of the carrot rust fly.

Harding and Morse (1909) considered *Erwinia aroideae* and *Erwinia carotovora* to be the same, but Link and Taliaferro (1928) found them to be different serologically.

Erwinia lathyri (Manns and Taubenhaus) Magrou

Insect concerned: The bean aphid, *Aphis rumicis.*

Needham (1937) isolated from diseased *Aphis rumicis* an organism culturally resembling *Bacillus lathyri.* The same bacillus was not found in uninfected aphids.

Bacillus lathyri was first isolated from diseased sweet peas. It is now called *Erwinia lathyri.*

Erwinia salicis (Day) comb. nov.

Insects concerned: The gall midge, *Rhabdophaga saliciperda,* and the sawfly, *Euura atra.*

A survey of the insects associated with willow trees was made to determine, if possible, the insects that were responsible for the transmission of *Bacterium salicis* (*Erwinia salicis*), the cause of a watermark disease of willows in England. The larvae of the gall midge and the sawfly were commonly found in the wood and were strongly suspected of transmitting the bacterium.

Erwinia tracheiphila (Erw. Smith) Holland

Insects concerned: The twelve-spotted cucumber beetle, *Diabrotica duodecimpunctata,* and the striped cucumber beetle, *Diabrotica vittata;* the squash bug, *Anasa tristis;* the squash beetle, *Epilachna borealis;* the melon aphis, *Aphis gossypii;* the honeybee, *Apis mellifera;* and the potato flea beetle, *Epitrix cucumeris.*

Erwinia tracheiphila, the cause of a wilt of cucumbers, cantaloupes, muskmelons, pumpkins, and squashes, was isolated by Smith in 1893, and later he suggested that insects, especially the striped cucumber beetle, might be responsible for the spread of the disease. In 1915 Rand advanced experimental evidence incriminating the twelve-spotted cucumber beetle. These two beetles are the only known vectors of the disease by natural means. Rand and Cash (1920) have shown that the organism may overwinter in the digestive system of *Diabrotica vittata*. In fact, no other method of survival of *Erwinia tracheiphila* in nature is known. Rand and Enlows (1916) have shown that the squash bug, the squash beetle, the melon aphis, the honeybee, and the potato flea beetle are not vectors.

Tribe: ESCHERICHEAE
Genus: AEROBACTER

Aerobacter aerogenes (Kruse) Beijerinck

Insects concerned: The housefly, *Musca domestica; Euxoa segetum;* the cabbage butterfly, *Pieris rapae;* Pentatomidae (probably *Loxa variegata*); the potato beetle, *Leptinotarsa decemlineata;* the meadow grasshopper, *Conocephalus fasciatus* var. *fasciatus;* the cricket, *Nemobius fasciatus* var. *fasciatus;* and *Urographis fasciata*.

Cox, Lewis, and Glynn (1912) isolated *B. lactis aerogenes* (*Aerobacter aerogenes*) from Liverpool flies, 46 per cent of the colon organisms isolated belonging to this group. Torrey (1912) isolated *B. aerogenes* (*Aerobacter aerogenes*) from the intestinal contents of *Musca domestica* caught in a poor tenement section of New York City. Nicoll (1911) isolated an organism which he called *Bacillus oxytocus perniciosus* (*A. aerogenes*) from intestines of flies.

Stutzer and Wsorow (1927) found *Bacillus lactis aerogenes* (*Aerobacter aerogenes*) to be a part of the normal flora of the intestines of healthy caterpillars of *Euxoa segetum*. Steinhaus (1941a) isolated it from the intestinal tracts of *Pieris rapae,* a Pentatomidae, *Leptinotarsa decemlineata, Conocephalus fasciatus* var. *fasciatus, Nemobius fasciatus* var. *fasciatus,* and *Urographis fasciata*.

Aerobacter cloacae (Jordan) Bergey *et al.*

Insects concerned: The German roach, *Blattella germanica;* the honeybee, *Apis mellifera;* the cabbage butterfly, *Pieris rapae;* the blister beetles, *Epicauta pennsylvanica* and *Epicauta cinerea marginata;* and the housefly, *Musca domestica.*

White (1906) isolated *Bacillus cloacae* (*Aerobacter cloacae*) from the intestine of the healthy honeybee and Nicoll (1911) from the alimentary tract of the housefly.

During a survey of the natural bacterial flora of 30 species of insects, Steinhaus (1941a) found coliform bacteria of the *Aerobacter cloacae* type in the alimentary tracts of the German roach, cabbage butterfly, and two species of blister beetles.

Genus: ESCHERICHIA

Escherichia coli (Migula) Castellani and Chalmers
(*Bacterium coli* and *Bacillus coli communis*)

Insects concerned: The silkworm, *Bombyx mori;* the honeybee, *Apis mellifera;* the wax moth, *Galleria mellonella;* the oriental roach, *Blatta orientalis;* the German roach, *Blattella germanica;* the cecropia-moth larva, *Samia cecropia; Paria canella gilvipes;* the lady beetle, *Coccinella novemnotata;* the cricket, *Nemobius fasciatus* var. *fasciatus; Calliphora vomitoria; Lucilia caesar; Sarcophaga carnaria;* the housefly, *Musca domestica;* and the fly, *Chrysomyia megacephala.*

Cao (1898, 1906b) found *Bacillus coli* (*Escherichia coli*) constantly present in the feces of all flies he examined. However, he found the contents of eggs at the time of oviposition

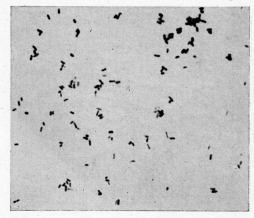

Fig. 38. *Escherichia coli.*

to be sterile, although the external envelope contained *E. coli.* He succeeded in demonstrating the transmission of bacteria from the surroundings of adult flies to their eggs, through metamorphosis to the mature flies, which deposited the bacteria with their feces. Cao's studies dealt with *Musca domestica, Calli-*

phora vomitoria, Sarcophaga carnaria, and *Lucilia caesar. Bacillus coli communis* (*E. coli*) has also been isolated from the body and intestinal contents of *Musca domestica* by Nicoll (1911), Cox, Lewis, and Glynn (1912), Torrey (1912), and Scott (1917). Nicoll (1911) also isolated *Bacillus schafferi,* which is now called *E. coli,* from flies. Ostrolenk *et al.* (1939, 1942) have shown that as many as 100,000 *Escherichia coli* may be deposited by flies on pecans exposed for four hours, and that long periods of storage decrease the number of bacteria. From 80 to 100 per cent of the *Chrysomyia megacephala* caught in Peiping by Chow (1940) were found to have *Bacterium coli* (*E. coli*) on their surfaces.

Bacillus coli communior (*E. coli* var. *communior*) was isolated by Torrey (1912) and Scott (1917); *B. acidi lactici* (*E. coli* var. *acidilactici*) by Torrey (1912) and Cox, Lewis, and Glynn (1912); *B. neapolitanus* (*E. coli* var. *neapolitana*) by Cox, Lewis, and Glynn (1912); and *Bacillus coli mutabilus* (*Bacterium coli mutabile,* a variant of *E. coli*) by Nicoll (1911), all from flies.

Cao (1898, 1906a,b) also isolated *Bacillus coli* from the feces of cockroaches (*Blatta orientalis*). Even after long fasting, the roach retained the organism in the intestine. It may also carry the organism on its legs, according to Longfellow (1913). Steinhaus (1941a) also isolated *E. coli* from the intestinal tract of the German roach.

Sawamura (1906) lists *Bacillus coli* as one of the organisms capable of experimentally producing flacherie in silkworms. White (1906) isolated this colon bacillus from honeybees, and it has also been isolated from cecropia-moth larvae, *Paria canella* var. *gilvipes,* a lady beetle, and *Nemobius fasciatus* var. *fasciatus* by Steinhaus (1941a).

E. coli has been used by several investigators in studying the immunity of various insects to it, especially the wax moth. Both Metalnikov (1927) and Cameron (1934) found that the wax-moth larvae were very susceptible to infection with *E. coli*. The larvae died in 12 to 24 hours.

Escherichia freundii (Braak) Bergey *et al.*

Insects concerned: The cecropia-moth caterpillar, *Samia cecropia,* and *Paria canella* var. *gilvipes.*

Coliform bacteria similar to *Escherichia freundii* were isolated from the alimentary tracts of the above-listed normal insects by Steinhaus (1941a).

Escherichia paradoxa (Toumanoff) Haudroy *et al.*

Insects concerned: The honeybee, *Apis mellifera,* and *Galleria mellonella.*

While studying the microbial flora of diseased bee larvae, Toumanoff (1927) isolated a bacterium which he named *Colibacillus paradoxus* (*Escherichia paradoxa*). He believed that in many respects it was similar to *E. coli.* It should be noted, however, that the organism he described has only feeble capacities for fermenting carbohydrates and it did not ferment lactose at all. Experimentally, the bacterium was pathogenic for *Galleria mellonella* as well as for bees.

Genus: KLEBSIELLA

Klebsiella capsulata (Sternberg) Bergey *et al.*

Insect concerned: The armyworm, *Barathra configurata.*

Munro (1936) studied an outbreak of septicemia among Bertha armyworms in North Dakota in collaboration with Professor C. I. Nelson, who isolated an organism which he identified as *Klebsiella capsulata.*

Klebsiella ozaenae (Abel) Bergey *et al.*

Insect and tick concerned: *Rhodnius prolixus* and *Ornithodoros furcosus.*

Olarte (1942) reports the isolation of two strains of *Klebsiella ozaenae* from two guinea pigs that were inoculated 35 and 49 days before with *Ornithodoros furcosus* and *Rhodnius prolixus.* However, Olarte thought the long period of incubation indicated that the bacteria may not have had the arthropods as their true source.

Klebsiella paralytica Cahn, Wallace, and Thomas

Tick concerned: The winter tick, *Dermacentor albipictus.*

In 1932 Thomas and Cahn described a disease among moose, *Alces ameri-*

cana var. *americana*, which occurred in northeastern Minnesota and the adjacent region of Ontario, Canada. The moose in this area were heavily infested by the tick *Dermacentor albipictus*, the final stage of the tick appearing in the spring coincident with the appearance of the disease. Ticks taken from moose dying of the disease transmitted it to guinea pigs and rabbits in the laboratory. A bacterium, which Cahn, Wallace, and Thomas (1932) named *Klebsiella paralytica*, was isolated from these ticks taken from diseased moose. When this bacterium was injected into animals, symptoms were produced similar to those in the tick-infested laboratory animals and in the diseased moose. Summarizing their experiments, Wallace, Cahn, and Thomas (1933) state: "While we have not proved that *Klebsiella paralytica* is the cause of moose disease, we have presented a series of observations which strongly indicate that it may be the cause."

Klebsiella pneumoniae (Schroeter) Trevisan

Insects concerned: The roach, *Blatta orientalis; Blaps mucronata; Pimelia bifurcata; Pimelia sardea;* and *Tentyria* sp.

Cao (1898) found that Friedländer's bacillus (*Klebsiella pneumoniae*) passes alive and virulent through the feces of the oriental roach and several Coleoptera (*Blaps, Pimelia,* and *Tentyria* sp.).

Tribe: PROTEAE

Genus: PROTEUS

Proteus alveicola Serbinow

Insect concerned: The honeybee, *Apis mellifera*.

Serbinow (1915) attributed the cause of infectious diarrhea of honeybees to *Proteus alveicola* and *Bacterium coli apium*. Both organisms were infectious for mice by peritoneal inoculation.

Proteus bombycis Bergey et al.

Insect concerned: The silkworm, *Bombyx mori*.

Glaser (1924a) isolated and described but did not name an organism from diseased silkworms. The organism was isolated from the feces, blood, and various tissues of the diseased insects. The normal worms were infected by ingestion of food that was contaminated by the feces and body fluids of diseased worms. The organism was not very resistant to its environment and did not survive very long on the outside of eggs. Some evidence was presented to show that the organism did not pass from one generation to another through the egg.

Bergey *et al.* at first named the organism *Aerobacter bombycis* (*Bergey's Manual,* 3d and 4th eds.) and later called it *Proteus bombycis* (*ibid.,* 5th ed.). Lehmann, Neumann, and Breed named it *Bacterium bombycivorum*.

Proteus insecticolens Steinhaus

Insect concerned: The milkweed bug, *Oncopeltus fasciatus.*

Steinhaus (1941a) isolated *Proteus insecticolens* consistently from the four stomachs of the milkweed bug. It resembled *Proteus ammoniae* in many of its fermentation reactions, but culturally it was quite different.

Proteus morganii (Winslow *et al.*) Rauss

Insect concerned: The housefly, *Musca domestica.*

Morgan and Ledingham (1909), Nicoll (1911), and Cox, Lewis, and Glynn (1912) have isolated various strains of Morgan's No. 1 Bacillus (*Proteus morganii*) from flies caught in infected and uninfected areas. Graham-Smith (1912) has shown that the organism when fed to the larvae of *Musca domestica* survives metamorphosis to the mature fly.

Proteus photuris Brown

Insect concerned: The firefly, *Photuris pennsylvanica.*

Brown (1927) isolated this organism from the luminous organ of fireflies. He tried to prove by a number of experiments that these bacteria were responsible for the light, but without success. He states that the organism appeared to be a normal symbiotic inhabitant of that organ.

Proteus recticolens Steinhaus

Insect concerned: The milkweed bug, *Oncopeltus fasciatus*.

The pylorus and rectum of the milkweed bug were found by Steinhaus (1941a) to contain large numbers of *Proteus recticolens* as part of the normal flora. In a few cases the organism was isolated from seeds in the cages, but these seeds were probably contaminated by the fecal droppings.

Proteus vulgaris Hauser

Insects concerned: The roach, probably *Blatta orientalis;* the wax moth, *Galleria mellonella;* the housefly, *Musca domestica; Calliphora vomitoria; Lucilia caesar; Sarcophaga carnaria;* and the corn borer, *Pyrausta nubilalis*.

Longfellow (1913) found that cockroaches carried *Proteus vulgaris* on their legs.

Cao (1906b) fed larvae of *Musca domestica, Calliphora vomitoria, Lucilia caesar,* and *Sarcophaga carnaria* on *Bacillus proteus vulgaris* (*Proteus vulgaris*) and isolated it from the feces. Scott (1917) also isolated this organism from *Musca domestica* caught in Washington, D.C.

Metalnikov (1920a), in carrying out a series of experiments to determine the immunity of the wax moth to certain organisms, found that the insect larva offered no resistance at all to even small doses of *Bacillus proteus* (*Proteus vulgaris*). However, Metalnikov and Gaschen (1921) found they could produce immunity by injection of a vaccine, which carried over into the adult stage. Cameron (1934) also found that *Proteus vulgaris* was extremely virulent for wax-moth larvae.

Tribe: SALMONELLEAE

Genus: EBERTHELLA

Eberthella belfastiensis (Weldin and Levine) Bergey *et al.*

Insect concerned: The housefly, *Musca domestica*.

Bacterium coli anaerogenes (*Eberthella belfastiensis*) was isolated by Scott (1917) from *Musca domestica* caught in Washington, D.C.

Eberthella insecticola Steinhaus

Insects concerned: The meadow grasshopper, *Conocephalus fasciatus* var. *fasciatus;* the milkweed bug, *Oncopeltus fasciatus;* and the stinkbug, probably *Loxa variegata.*

Steinhaus (1941) isolated *Eberthella insecticola* from the alimentary tracts of the meadow grasshopper, the milkweed bug, and the stinkbug. Relatively large inocula of some of the strains were pathogenic to mice.

Eberthella pyogenes (Migula) Bergey et al.

Insect concerned: The cricket, *Nemobius fasciatus* var. *fasciatus.*

An organism similar to *Eberthella pyogenes* was isolated by Steinhaus (1941a) from the alimentary tracts of crickets that had been collected in nature.

Eberthella typhosa (Zopf) Weldin

Insects concerned: The housefly, *Musca domestica;* the fly, *Chrysomyia megacephala;* the wax moth, *Galleria mellonella;* the roaches, *Blatta orientalis* and *Periplaneta americana;* the lice, *Pediculus humanus capitis* and *Pediculus humanus corporis;* the mosquito, *Aëdes aegypti;* and ants.

A great deal has been written concerning the transmission of *Eberthella typhosa* by insects. The insect most frequently accused is the common housefly, *Musca domestica.* The typhoid bacilli may contaminate the body and appendages of the fly, or they may occur in the contents of the intestinal tract or in the feces of the insect, thereby contaminating food and drink.

One of the earliest definite reports incriminating the housefly in this regard was made by the Army Commission appointed to investigate the cause of epidemics of enteric fever in the army camps of the southern United States during the Spanish-American War (Reed, Vaughan, and Shakespeare, 1904). This Commission attributed the wide spread of typhoid fever to transmission by flies. Cellí (1888) showed that *Eberthella typhosa* could pass through the intestines of flies into the feces. Hamilton (1903) isolated typhoid bacilli from flies caught in houses in which patients were ill with typhoid fever, and proved that an outbreak of typhoid fever was due to flies acting as carriers. Ficker (1903) allowed flies to feed on pure cultures of *Eberthella typhosa* and was able to culture the bacilli from the crushed flies 5 to 23 days afterwards. Faichnie (1909a,b) concluded from his experiments that typhoid bacilli are not as readily transmitted via the legs of flies as by the excrement of flies.

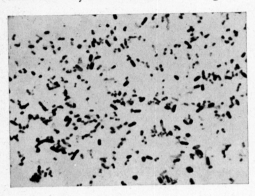

Fig. 39. *Eberthella typhosa.* (Photo by N. J. Kramis.)

Veeder (1898), Klein (1908), Graham-Smith (1909), Bertarelle (1910), Ledingham (1911), Howard (1911), Cochrane (1912), Bahr and Comb (1914), and Manson-Bahr (1919) have also incriminated the fly as a carrier of *Eberthella typhosa.*

Experimentally, Chow (1940) found that *Eberthella typhosa* as well as *Shigella dysenteriae* survives for five or six days in or outside the body of artificially infected *Chrysomyia megacephala.* However, he did not find *E. typhosa* associated with flies of this species which he caught in Peiping.

Tebbutt (1913) found that *Eberthella typhosa* failed to persist in the metamorphosis of *Musca domestica* from larva to imago in every case examined, though Ledingham (1911) recovered a few organisms from one pupa under similar experimental conditions.

Morischila and Tsuchimochi (Riley and Johannsen, 1932) found that *E. typhosa* passed through the cockroach intestine apparently unharmed. In 1922 Macfie (Riley and Johannsen, 1932) had negative results.

In 1907 Abe recovered the typhoid bacillus from head and body lice fed on typhoid fever patients in 75 per cent of the insects examined.

Darling (Wheeler, 1914) performed a series of experiments to determine whether ants carry *E. typhosa* on the surface of their bodies or in their intestinal tract, with negative results. He concluded that the formation of formic acid killed and inhibited the bacteria. Wheeler (1914) thinks that this was an erroneous conclusion, but that it is very likely that ants, because of their habits, do not often spread pathogenic bacteria.

The possible germicidal action of the gastrointestinal secretions of the yellow fever mosquito (*Aëdes aegypti*) on *E. typhosa* and *Serratia marcescens* was found to be negative by St. John, Simmons, and Reynolds (1930).

Metalnikov (1927) found that the wax moth, *Galleria mellonella,* varied in its susceptibility to infection with *Eberthella typhosa,* while Cameron (1934) found that the organism was always nonpathogenic for the wax-moth larvae.

Genus: SALMONELLA

Salmonella choleraesuis Weldin

Insects concerned: The honeybee, *Apis mellifera;* the housefly, *Musca domestica;* and the flea, *Pulex irritans.*

In 1906 White isolated *Bacillus cholerae suis* from the intestine of the honeybee. Probably this is the same organism that is now known by the name *Salmonella choleraesuis.* Scott (1917) isolated *B. suipestifer* (*Salmonella choleraesuis*) from flies caught in Washington. Messerlin and Couzi (1942) transmitted *S. suipestifer* (*S. choleraesuis*) from guinea pig to guinea pig by *Pulex irritans.*

Salmonella enteritidis (Gaertner) Gastellani and Chalmers

Insects and tick concerned: The housefly, *Musca domestica;* the bluebottle fly, *Calliphora erythrocephala;* the wax moth, *Galleria mellonella;* a louse; and the tick, *Dermacentor andersoni.*

Ficker (1903), Hamilton (1903), Graham-Smith (1909), Ledingham (1911), and Bahr and Comb (1914) isolated *Salmonella enteritidis* from the intestinal tract of houseflies; and Cox, Lewis, and Glynn (1912) found the organism on the bodies of flies caught in Liverpool. The organism that Cox *et al.* isolated, however, gave negative serological tests with *S. enteritidis.* Ostrolenk and Welch (1942) have shown that *Musca domestica* artificially contaminated with *Salmonella enteritidis* carried it for the duration of the life of the fly, and was capable of depositing large numbers of the organisms on food in a very short time. Peppler (1944) found that flies breeding in a sewage pond were contaminated with *Salmonella enteritidis,* and were the vectors of the infection, during an outbreak of gastroenteritis in an army camp.

Graham-Smith (1912) fed larvae of *Calliphora erythrocephala* and *Musca domestica* with food infected with *Salmonella enteritidis* but failed to recover it from the matured adults.

Huang, Chang, and Lieu (1937), during their studies on 17 cases of systemic infection of *Salmonella enteritidis* not associated with food poisoning, found that finely ground lice from their patients gave a growth of *Salmonella enteritidis* when cultured.

Parker and Steinhaus (1943c) found that when *Salmonella enteritidis* was ingested by the wood tick, *Dermacentor andersoni,* it survived in the tick and was occasionally passed through the egg to the next generation. The tick feces were found to be infective. The fact that the tick was infective for as long as 35 days suggested that possibly the transmission was not mechanical. It was also found that *Salmonella enteritidis* was often fatal to the ticks.

Zernoff used heated Danysz bacilli (*S. enteritidis*) to produce immunity in wax moths (see Huff, 1940).

Salmonella paratyphi (Kayser) Bergey *et al.*

Insects concerned: The housefly, *Musca domestica; Euxoa segetum;* and the bedbug, *Cimex lectularius.*

Torrey (1912) isolated *B. paratyphus* Type A (*Salmonella paratyphi*) from the intestinal tract of houseflies caught in a poor tenement section of New York City. Stutzer and Wsorow (1927) isolated *Bacterium paratyphi* (*Salmonella paratyphi*) from the intestines of larvae of *Euxoa segetem*. Caspari (1939) found that bedbugs (presumably *Cimex lectularius*) fed on mice infected with paratyphoid bacilli retained the organisms in their stomachs for two to three weeks, but transmission to healthy animals by biting could not be conclusively demonstrated.

Salmonella schottmülleri (Winslow *et al.*) Bergey *et al.*

Insects concerned: The housefly, *Musca domestica,* and the wax moth, *Galleria mellonella.*

Nicoll (1911) isolated "*B. para-typhosus*" (*Salmonella schottmülleri*) from the body and intestinal tract of *Musca domestica* caught in London, with the evidence that it had been carried by the flies for at least 11 days.

Metalnikov (1927) and Cameron (1934) found that *Salmonella schottmülleri* was not pathogenic for the wax moth. However, Zernoff and Ajolo (1939) found that lethal doses of *Bacillus paratyphi* B (*S. schottmülleri*) did not kill the wax moths if administered simultaneously with para-aminophenylsulfonamide.

Salmonella schottmülleri var. **alvei** Hauduroy and Ehringer

Insect concerned: The honeybee, *Apis mellifera*.

In the vicinity of Copenhagen an acute enteritis of bees was found to be due to *Bacillus paratyphi alvei* (*Salmonella schottmülleri* var. *alvei*) by Bahr (1919). The bees usually died in 25 hours to a few days.

Genus: SHIGELLA

Shigella spp.

Insect concerned: The housefly, *Musca domestica*.

Ficker (1903), Hamilton (1903), Graham-Smith (1909), Ledingham (1911), and Nicoll (1911) isolated several *Shigella* strains from the intestinal tract of flies.

Shigella dysenteriae (Shiga) Castellani and Chalmers

Insects concerned: The fly, *Chrysomyia megacephala;* a species of fly of Macedonia; and the wax moth, *Galleria mellonella*.

Shigella dysenteriae was isolated by Dudgeon (1919) from flies in Macedonia, and he established a decided correlation between the incidence of flies and dysentery.

Chow (1940) found that 8 per cent of the flies (*Chrysomyia megacephala*) caught by him in Peiping harbored *Shigella dysenteriae*. Experimentally both *Shigella dysenteriae* and *Eberthella typhosa* survived in or outside the fly's body for five to six days.

Metalnikov (1927) found that *Shigella dysenteriae* varied in its pathogenicity for the wax-moth larva, whereas Cameron (1934) found that it was not at all pathogenic.

Shigella paradysenteriae (Collins) Weldin

Insects concerned: Flies, probably *Musca domestica,* and the fire ant, *Solenopsis geminata*.

Graham-Smith (1909) noted the presence of an organism which corresponded "in cultural reactions with *B. dysenteriae (Flexner)*" (*Shigella paradysenteriae*) on flies on two different occasions. Kuhns and Anderson (1944) reported a fly-borne bacillary dysentery epidemic during which they isolated *Shigella paradysenteriae* from nine lots of flies collected in military kitchens and latrines.

Griffitts (1942) found that ants could carry *Shigella flexner V* (*Shigella paradysenteriae*) on their feet from one place to another for at least 24 hours after feeding on or traversing infected material.

Tribe: SERRATEAE

Genus: SERRATIA

Serratia kielensis (Lehmann and Neumann) Bergey *et al.*

Insects concerned: The housefly, *Musca domestica; Lucilia caesar; Calliphora vomitoria;* and *Sarcophaga carnaria.*

Cao (1906b) placed fly eggs on pieces of flesh polluted with the "red bacillus of Kiel" (either *Serratia kielensis* or *Bacterium kiliense*) and later recovered the organism from the feces of the larvae. Larvae fed on flesh which was artificially contaminated with the red-pigment-forming bacillus were allowed to mature, and all the adult flies yielded cultures of the red bacillus up to the ninth day after emergence. The flies died after nine days.

According to Lehmann, Neumann, and Breed, *Bacterium kiliense* and *Serratia marcescens* differ only in that *Serratia marcescens* has a stronger acid reaction, which affects the brilliance of the pigment produced. There seem to be two organisms that have been called the Kiel bacillus (see *Bergey's Manual,* 5th ed., p. 426).

Serratia marcescens Bizio

Insects concerned: The silkworm, *Bombyx mori; Pseudococcus citri;* the wax moth, *Galleria mellonella;* the corn borer, *Pyrausta nubilalis;* the gypsy moth, *Porthetria dispar;* the potato-tuber moth, *Gnorimoschema operculella; Schistocerca gregaria; Tenebrio molitor;* the roaches, probably *Blatta orientalis* and *Rhyparobia maderae;* the termite, *Zoötermopsis angusticollis;* the mosquito, *Aëdes aegypti;* the milkweed bug, *Oncopeltus fasciatus;* the housefly, *Musca domestica;* the stablefly, *Stomoxys calcitrans; Lucilia caesar; Calliphora vomitoria; Sarcophaga carnaria;* a burrowing beetle, *Melolontha vulgaris;* the white fringed beetle, *Pantomorus peregrinus; Chironomus* sp.; *Loxostege sticticalis;* and the Rocky Mountain wood tick, *Dermacentor andersoni.*

Cao (1906b) fed adult flies of *Musca domestica, Calliphora vomitoria, Lucilia caesar,* and *Sarcophaga carnaria* on cultures of *Bacillus prodigiosus* (Ser-

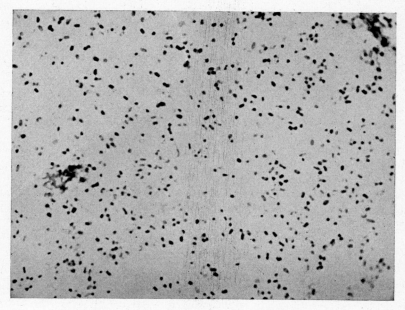

Fig. 40. *Serratia marcescens.* (Photo by N. J. Kramis.)

ratia marcescens) and recovered it from their feces and eggs 24 hours later. The organism also survived from the larval stage through metamorphosis to the adult fly and could be recovered from the intestinal tract nine days after emergence. Graham-Smith (1911) confirmed Cao's findings regarding the persistence of this organism throughout the metamorphosis of *Musca domestica.*

Serratia marcescens was also found by Cao to live and multiply 30 to 35 days in the intestines of *Melolontha vulgaris,* a burrowing beetle.

The history of the associations between other insects and *Serratia marcescens* is very interesting. In stating it briefly we quote from De Bach and McOmie (1939):

Masera (1936a) in a comprehensive treatment of the subject states that as early as 1817 Rozier noticed a red coloration forming in the dead bodies of silkworms. This was again noticed by Pollini and Vasco in 1819, Re and Ascolese in 1837, etc. However, the credit for the actual isolation in 1886 of *Bacillus prodigiosus* from a silkworm larva is due to Perroncito. Bandelli about the same time isolated *Bacillus prodigiosus* from the exterior of silkworms (*Bombyx mori*) and later stated that the red pigment did not appear until after the death of the larvae.

Metalnikov (1930) isolated a red-pigment former like *Bacillus prodigiosus* from the larvae of the gypsy moth, *Lymantria dispar* (L.) which was very virulent. [Earlier (1920a) this worker found the wax moth, *Galleria mellonella*, to offer no resistance to *Serratia marcescens*, dying from very small doses.]

Zernoff (1931) has reported that this bacillus is very virulent to the larvae of the wax moth, *Galleria mellonella* (L.), by inoculation but not by ingestion, whereas both methods result in infection in the larvae of the European corn borer, *Pyrausta nubilalis* (Hbn.).

Masera has recently (1934a and b, 1936a,b and c) published an extensive series of papers dealing with experimental studies of the pathogenicity of *Bacillus prodigiosus*. He found it to be fatal to *Pyrausta nubilalis* by inoculation and by ingestion; while to *Bombyx mori* fatal when inoculated but not necessarily fatal when ingested, depending particularly on the age of the larvae. In the case of *Galleria mellonella* he found it to be fatal only by inoculation, never by ingestion, and finally he found it to be nonpathogenic to the larvae of *Tenebrio molitor* L. [According to Masera, the immunity of the insects may be due to their food, which normally contains microorganisms.]

Lepesme (1937a,b) has reported the occurrence of an epizoötic in laboratory-bred *Schistocerca gregari* Forsk. caused by *Bacillus prodigiosus* and has shown that inoculation produces death in one or two days with the usual red coloration occurring (in this case mainly in the abdomen) while ingestion produces death only occasionally. [This worker (1938) also found it to be a secondary invader to the infestation of *Schistocerca gregaria* by the fungus *Aspergillus flavus*.]

De Bach and McOmie (1939) found their laboratory stock of the termite, *Zoötermopsis angusticollis*, to be suffering from two bacterial diseases. One of these diseases was due to *Bacterium* sp. (which see); the other to *Serratia marcescens*. The latter organism causes the head and appendages of the dead termites to turn red.

Longfellow (1913) was able to cultivate *Bacillus prodigiosus* from the feces of the common house roach, and Barber (1912) presumably isolated the organism from the roach *Rhyparobia maderae*. Duncan (1926) found "*B. prodigiosus*" several times in the gut contents of *Stomoxys calcitrans*. Pospelov (1936) isolated *Bacterium prodigiosum* from *Pseudococcus citri* and found it to be virulent for several species of *Pseudococcus*.

In 1927 Dr. Breed (1940) received 32 cultures of *Serratia marcescens* from Professor E. Hiratsuka, Japan, which had been isolated from silkworm larvae and cocoons. Metalnikov and Chorine (1928a) found that an organism which Tateiwa had isolated from silkworms with flacherie and had called *Bacterium prodigiosus* (probably *prodigiosum*) produced 20 to 30 per cent mortality in corn borers infected by mouth.

The writer (Steinhaus, 1939) attempted to establish *Serratia marcescens* as part of the bacterial flora in the gut of the milkweed bug, *Oncopeltus fasciatus,* by experimental feedings but was unsuccessful. Later (1942b) he used this bacterium to demonstrate that the tick, *Dermacentor andersoni,* could acquire bacteria from the skins of experimental animals and pass these bacteria from larvae to adults.

Brinley and Steinhaus (1942) using *Chironomus* larvae attempted to trace a known organism from sewage to the emerged adult. A 24-hour broth culture of *Serratia marcescens* was added to the sewage in which were living the *Chironomus* larvae. Several adults yielded pure cultures of *Serratia marcescens.* From their results, these workers concluded that larvae living in sewage are, at least occasionally, capable of acquiring microorganisms during the larval stage from the sewage and later of disseminating them during the adult stage.

The possible germicidal action of the gastrointestinal secretions of the yellow fever mosquito, *Aëdes aegypti,* on *Serratia marcescens* was found to be negative by St. John, Simmons, and Reynolds (1930). It was further found that this bacterium, as well as others, survived for at least 24 hours in the gut of the mosquito but that it could not be demonstrated after an interval of seven or more days.

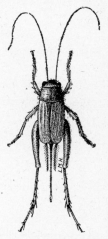

Fig. 41. *Nemobius fasciatus.* (From Comstock, 1936; after Lugger.)

Serratia plymouthensis (Migula) Bergey *et al.*

Insect concerned: The cricket, *Nemobius fasciatus* var. *fasciatus.*

The growth of *Serratia plymouthensis* is characterized by a bright red pigment similar to *Serratia marcescens,* but is a visible gas-producing type.

This bacterium was found by Steinhaus (1941a) to occur in the alimentary canal of nearly every specimen of the species of cricket which he examined. The crickets were collected in nature in a meadow near Columbus, Ohio.

Family: LACTOBACTERIACEAE
Tribe: STREPTOCOCCEAE
Genus: DIPLOCOCCUS

Diplococcus bombycis Paillot

Insects concerned: The silkworm, *Bombyx mori; Nygmia phaeorrhoea;* the gypsy moth, *Porthetria dispar; Eriogaster lanestris;* and *Vanessa urticae*.

Paillot (1922) isolated from diseased silkworms a gram-positive elongated coccus which he called *Diplococcus bombycis*. He states (1933) that in this organism there are constantly present transverse, double bands of "chromatophiles." The larvae of both *Porthetria dispar* and *Nygmia phaeorrhoea* were found to be very resistant to infection with the organism, whereas the larvae of *E. lanestris* and *Vanessa urticae* were easily infected.

Diplococcus liparis Paillot

Insect concerned: The gypsy moth, *Porthetria dispar*.

Diplococcus liparis was isolated by Paillot (1917) from larvae of the gypsy moth. The organism appeared to be of little pathogenic importance.

Diplococcus lymantriae Paillot

Insect concerned: The gypsy moth, *Porthetria dispar*.

Because of the confusion in the literature, it has been rather difficult to determine whether or not *Diplococcus lymantriae, Bacillus lymantriae,* and *Coccobacillus lymantriae* are the same organism. It appears likely, however, that *Diplococcus lymantriae* is distinct, since Paillot (1917) states that the microbes parasitic in the larvae of *Porthetria dispar* described in his paper were coccobacilli provisionally identified as *Bacillus lymantriae, Diplococcus lymantriae,* sp. n., which is only slightly pathogenic to the caterpillars, and *Bacillus liparis* sp. n.

Diplococcus melolonthae Paillot

Insects concerned: The cockchafer, *Melolontha melolontha; Vanessa urticae; Rhizotrogus solstitialis; Eriogaster lanestris;* the silkworm, *Bombyx mori;* and the gypsy moth, *Porthetria dispar.*

Paillot (1917) found that a coccobacillus and *Diplococcus melolonthae* killed cockchafers and other insects in 24 hours, whereas the diplococcus alone was only slightly pathogenic. The larvae of *Porthetria dispar* and the silkworm offered marked resistance to infection with the latter organism. Paillot (1933, p. 139) mentions *Diplococcus melolonthae* as an example of bacterial variation.

Diplococcus pemphigocontagiosus
(*Diplococcus pemphigi contagiosi* Manson)

Insect concerned: Louse.

Wardle (1929) refers to the transmission of *Diplococcus pemphigocontagiosus* by lice. Pierce (1921) lists "*Diplococcus pemphigi contagiosi* Manson" as being carried by lice.

Diplococcus pieris Paillot

Insects concerned: The cabbage butterfly, *Pieris brassicae,* and *Vanessa urticae.*

Paillot (1919b) isolated *Diplococcus pieris* from larvae of diseased white cabbage butterflies. From the same source he isolated eight other bacteria. (See *Bacillus pieris agilis.*) He considered these bacteria to be secondary invaders, the primary invader, *Apanteles glomeratus,* a parasitic wasp, being the predisposing factor in the bacterial infection of the white cabbage butterfly.

Diplococcus pneumoniae Weichselbaum

Insects concerned: The roach, *Blatta orientalis; Blaps mucronata; Pimelia bifurcata; Pimelia sardea; Tentyria* sp.; lice; the wax moth, *Galleria mellonella;* and the flea, *Pulex irritans.*

"The diplo bacillus of Frankel" (*Diplococcus pneumoniae*) was found by Cao (1898) to be destroyed in the intestines of the oriental roach and of several Coleoptera (*Tentyria, Blaps,* and *Pimelia*).

According to Pierce (1921), in 1915 Widmann fed lice on mice in which he had established a pneumococcic septicemia. Widmann was unable to infect other mice by the louse bites, but he did find the louse feces to be infective during the first 24 hours.

Metalnikov (1927) and Cameron (1934) found the pneumococcus to be noninfective for the wax moth.

Pinto (1930) lists the organism "Pneumococco sp." as being cultivated from the digestive tract of *Pulex irritans* by Da Silva (1916). Presumably the pneumococcus (*Diplococcus pneumoniae*) is meant, but since the original article was not available to us, this could not be ascertained.

Genus: STREPTOCOCCUS

Streptococcus sp.

Insect concerned: *Stomoxys calcitrans*.

Schuberg and Böing (1914) found an unidentified streptococcus in *Stomoxys calcitrans*.

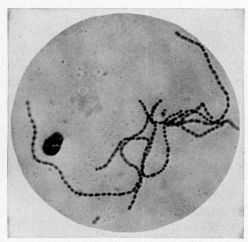

Fig. 42. *Streptococcus agalactiae.* (From Hagan, 1943.)

Streptococcus agalactiae Lehmann and Neumann

Insects concerned: The housefly, *Musca domestica,* and hippelates flies.

Saunders (1904a,b) concludes that the housefly and hippelates flies serve as vectors of bovine mastitis. Since this worker failed to state definitely the name of the microorganism concerned in his experiments, it is assumed he refers chiefly to *Streptococcus agalactiae*.

Ewing (1942) thinks that it is possible for the housefly to carry *Streptococcus agalactiae*, but doubts its importance as an agent in the spread of the infection.

Streptococcus apis Maassen

(See also *Bacillus alvei, Bacillus pluton, Bacterium eurydice,* and *Streptococcus liquefaciens*)

Insect concerned: The honeybee, *Apis mellifera*.

Maassen (1908) believed that the etiology of European foul brood is not uniform but that the disease is caused chiefly by *Streptococcus apis* and *Bacillus alvei*. Maassen (1908, 1913) stated that the disease could not be experimentally produced by pure cultures of these organisms. When bees were fed on triturated sick or dead larvae mixed with honey, they became sick with a mild form of foul brood from which they ultimately recovered. Numerous attempts by other workers to produce European foul brood by inoculation with pure cultures of *Streptococcus apis* have been largely unsuccessful.

Concerning this organism, Burnside (1934) states:

There seems to be insufficient reason for assuming that the lancet-shaped bacterial cell, *B. pluton,* found in late stages of infection in European foulbrood, is of different genus and species from the similar form *Streptococcus apis,* which is readily obtained in culture from sick larvae.

The identity of *Streptococcus apis* and *Bacillus pluton* is suggested by morphological similarity.

Burnside goes on to suggest that *Bacillus pluton* and *Streptococcus apis* are variants, or stages, in the life history of *Bacillus alvei*.

Bergey's Manual (5th ed., 1939, p. 339) lists *Streptococcus apis* as a synonym for *Streptococcus liquefaciens*. This is done on the basis of the work of Hucker (1932) who found these two gelatin-liquefying streptococci to be culturally similar. Agglutination studies also indicated that these two organisms were the same.

Streptococcus bombycis (Pasteur) Flügge
(See also *Streptococcus pastorianus*)

Insect concerned: The silkworm, *Bombyx mori*.

Silkworms are susceptible to infection by several pathogenic bacteria. One of the worst of these plagues of sericulture is gattine, with which *Streptococcus bombycis* is associated. The primary cause is probably a virus (see chapter VIII). *Streptococcus bombycis* is a gram-positive coccus of the enterococcus type, belonging to the serologic Group D. It forms chains from 5 to 12 microns long, and is a facultative anaerobe.

According to Wardle (1929), Paillot states that the term "flacherie" represents a group of three distinct maladies, one of which appears to be associated with *Streptococcus bombycis*. These are:

(1) A disorder associated with an abnormal abundance, in the intestine of the silkworm, of a sporulating bacillus morphologically identical with that described by Pasteur.

(2) An acute form of flacherie or "flacherie typique" associated apparently with a filterable virus.

(3) A chronic type of flacherie, the "gattine" of French workers, and "macilenze" of Italian workers, probably associated with *Streptococcus bombycis*.

In the last-mentioned disease there is generally a swelling of the anterior body wall, which becomes more or less translucent, and *Streptococcus bombycis* is isolated from the intestinal contents.

According to Paillot (1928b), *Streptococcus pastorianus* is synonymous with *Streptococcus bombycis*.

Streptococcus disparis Glaser

Insects concerned: The Japanese race of the gypsy moth, *Porthetria dispar;* the silkworm, *Bombyx mori;* and the armyworm, *Cirphis unipuncta*.

Glaser (1918c) described an infectious disease of caterpillars of the Japanese race of the gypsy moth which spread to cultures of the American race. It is clinically, pathologically, and etiologically distinct from wilt (a filterable virus disease). A streptococcus, which he named *Streptococcus disparis,* was found to be the cause of the disease.

The streptococcus is ingested with contaminated food. During the latter stages of the disease and after death it invades practically all the tissues. The symptoms are diarrhea and loss of appetite and of muscular co-ordination. The skin of the dead insect does not rupture as in the case of wilt, though the larva hangs in a flaccid condition by its prolegs and has the appearance of a caterpillar dead from wilt.

Streptococcus disparis is a gram-positive, nonmotile, encapsulated organism with a diameter of less than one micron.

Successful field experiments were conducted with *Streptococcus disparis* in sections of the gypsy-moth-infected territory in the United States. In two places severe epizoötics were produced. The organism is not pathogenic to silkworms (*Bombyx mori*) nor to the armyworm (*Cirphis unipuncta*). It is not pathogenic to human beings, guinea pigs, or rabbits.

Streptococcus equinus Andrewes and Horder

Insect concerned: The housefly, *Musca domestica*.

Torrey (1912) states that "to a certain extent the favorite breeding place of flies, viz., horse manure, is revealed by the presence of *Str. equinus* on the surface of the flies." He was able to isolate this streptococcus from flies from June through August in moderate to large numbers (42,000 to 3,850,000).

Streptococcus faecalis Andrewes and Horder

Insects concerned: The German roach, *Blattella germanica;* the webworm, *Hyphantria cunea;* the milkweed bug, *Oncopeltus fasciatus;* the bagworm, *Thyridopteryx ephemeraeformis;* the lyreman cicada, *Tibicen linnei;* the housefly, *Musca domestica; Triatoma infestans;* and the wax moth, *Galleria mellonella*.

Streptococcus faecalis has been isolated by Cox, Lewis, and Glynn (1912), Torrey (1912), and Scott (1917) from the intestinal tract of *Musca domestica*. According to Torrey, the presence of this organism in the intestinal contents of these insects indicates that their selected food is human dejecta, an important item in sanitation.

This streptococcus was isolated by Steinhaus (1941a) from the alimentary tracts of the German roach, a webworm, a cicada, a bagworm, and the milkweed bug.

Cameron (1934) found *Streptococcus faecalis* to be pathogenic for the wax moth, *Galleria mellonella*.

Brecher and Wigglesworth (1944) on several occasions isolated *Streptococcus faecalis* from *Triatoma infestans*.

Streptococcus galleriae Metalnikov

Insect concerned: The wax moth, *Galleria mellonella*.

Metalnikov (1927) isolated *Streptococcus galleriae* from the wax moth. Cameron (1934) isolated it from the larval blood and intestinal tract of the wax moth. He found that it was not pathogenic.

Streptococcus lactis (Lister) Löhnis

Insect concerned: The housefly, *Musca domestica*.

Torrey (1912) isolated what he called *B. lactis acidi* from the surface of city-caught flies. According to *Bergey's Manual* (5th ed.), *B. lactis acidi* could be either *Streptococcus lactis* or *Lactobacillus lactis* or neither, depending on whether the *B.* stands for *Bacterium* or *Bacillus*. In this case *Streptococcus lactis* is probably the organism meant.

Streptococcus pastorianus Krassilstschik
(See also *Streptococcus bombycis*)

Insect concerned: The silkworm, *Bombyx mori*.

This organism was isolated in 1896 by Krassilstschik. Paillot (1928b) thinks that *Streptococcus pastorianus* is the same as *Streptococcus bombycis*.

Streptococcus pityocampae Dufrenoy

Insect concerned: The processionary-moth caterpillar, *Cnethocampa pityocampa*.

Dufrenoy (1919) found that three organisms attacked the processionary-moth caterpillar: *Bacterium pityocampae, Streptococcus pityocampae alpha,* and *Streptococcus pityocampae beta*. When the caterpillars are inoculated with *Streptococcus pityocampae alpha,* they die in two to four days. The muscles become infiltrated with the coccus, and the fibers degenerate and lose their striation.

Paillot (1933) states that he is surprised that the *beta* strain of this organism was given the same generic name as the *alpha* strain since the *beta* strain is gram-negative. Furthermore, Paillot indicates that the *alpha* type is ill-placed in the *Streptococcus* genus because it is motile.

Streptococcus pyogenes Rosenbach

Insects and ticks concerned: The housefly, *Musca domestica;* the wax moth, *Galleria mellonella;* and the ticks, *Argas persicus* and *Ornithodoros moubata*.

Scott (1917) isolated *Streptococcus pyogenes* from *Musca domestica* caught in Washington, D.C., and indicates that the common housefly may be a factor in the dissemination of the suppurative processes. Torrey (1912) in a similar survey of a congested part of New York City isolated 15 cultures of streptococci, but none were *Streptococcus pyogenes*. From the exteriors of 9 out of 27 flies caught in hospital wards with cases of streptococcus infection, Shooter and Waterworth (1944) isolated cultures of hemolytic streptococci. Flies caught in other rooms did not yield these organisms.

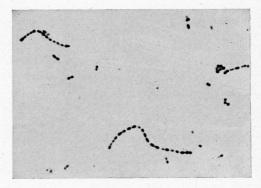

Fig. 43. *Streptococcus pyogenes.*

Duncan (1926) found that *Streptococcus haemolyticus* (*Streptococcus pyogenes*) was inhibited by the bactericidal principle found in the gut contents of *Argas persicus,* but a similar principle was not demonstrated in *Ornithodoros moubata*.

Metalnikov (1927) found *Streptococcus pyogenes* nonpathogenic for the wax moth, though Cameron (1934) found that large doses were pathogenic.

Streptococcus salivarius Andrewes and Horder

Insect concerned: The housefly, *Musca domestica*.

Torrey (1912) isolated *Streptococcus salivarius,* an organism found in the human mouth, throat, and nasopharynx, from the intestinal contents of houseflies, caught in New York City; and Cox, Lewis, and Glynn (1912) from the surface of flies caught in Liverpool.

Family: MICROCOCCACEAE

Genus: GAFFKYA

Gaffkya tetragena (Gaffky) Trevisan

Insect concerned: The housefly, *Musca domestica*.

Micrococcus tetragenus (*Gaffkya tetragena*) was isolated from *Musca domestica* by Scott (1917).

Genus: MICROCOCCUS

Micrococcus C White

Insect concerned: The honeybee, *Apis mellifera*.

White (1906) isolated *Micrococcus C* from the body of a healthy honeybee.

Micrococcus chersonesia Corbet

Insect concerned: An unidentified long-horned grasshopper of the family Tettigoniidae.

During studies on the bacterial flora of normal insects, Steinhaus (1941a) isolated a gram-positive coccus from a long-horned grasshopper. The coccus at times appeared almost as a very short rod. This characteristic is shared by an organism which *Bergey's Manual* (5th ed., 1939, p. 258) describes under the name *Micrococcus chersonesia,* isolated from the latex of a rubber tree. The growth of this organism is described as being "dull," whereas that of the coccus isolated from the grasshopper is glistening. Otherwise the physiologic, morphologic, and cultural characteristics of the two organisms agree fairly well.

Micrococcus conglomeratus Migula

Insect concerned: The bedbug, *Cimex lectularius*.

An organism similar to *Micrococcus conglomeratus* was isolated from the bedbug by Steinhaus (1941a).

Micrococcus curtissi Chorine

Insects concerned: The corn borer, *Pyrausta nubilalis;* the flour moth, *Ephestia kühniella;* and the wax moth, *Galleria mellonella*.

In July, 1928, Chorine (1929a) observed a very high mortality among the young corn-borer larvae caused by an organism which he called *Micrococcus curtissi*. The organism also proved to be very virulent toward full-grown borers when injected but less so when fed. By injection it was virulent to the larvae of *Ephestia kühniella,* though the larvae of *Galleria mellonella* proved to be more resistant.

Micrococcus ephestiae Mattes

Insect concerned: The flour moth, *Ephestia kühniella*.

According to Mattes (1927), *Micrococcus ephestiae* is apparently a nonpathogenic inhabitant of the intestinal tract of flour-moth larvae. It is an encapsulated organism.

Micrococcus epidermidis (Kligler) Hucker

Insect concerned: *Grylloblatta campodeiformis campodeiformis*.

Burroughs (1941) isolated three strains of this coccus from the alimentary tract of the above-mentioned grylloblattid.

Micrococcus flaccidifex danai Brown
(See also *Gyrococcus flaccidifex*)

Insect concerned: The monarch butterfly, *Danaus plexippus* (= *Danaus archippus*).

Brown (1927) considered this organism to be the cause of "wilt" disease in monarch-butterfly larvae. Smears showed the body fluid teaming with "motile cocci." Brown states:

One cannot be certain from the meager cultural details given by Glaser and Chapman [1912] of [*Gryococcus*] *flaccidifex* whether the present organism is specifically distinct. On account of its pathological effect it seems to be closely related to *flaccidifex* and I am naming it as a new subspecies. I failed to find any undue gyrating of the cocci in the hanging drop upon which they erect their genus *Gyrococcus*.

Micrococcus flavus Lehmann and Neumann

Insect concerned: The housefly, *Musca domestica*.

Micrococcus flavus, an air and milk organism, was isolated by Torrey (1912) from the intestinal contents and the body surfaces of flies.

Micrococcus freudenreichii Guillebeau

Insects concerned: The bagworm, *Thyridopteryx ephemeraeformis*, and *Grylloblatta campodeiformis campodeiformis*.

Steinhaus (1941a) found *Micrococcus freudenreichii* to be part of the bacterial flora of the alimentary canal of the larva of the bagworm moth. Burroughs (1941) isolated a strain of this micrococcus from the alimentary tract of a grylloblattid.

Micrococcus galleriae No. 3

Insect concerned: The wax moth, *Galleria mellonella*.

Ishimori and Metalnikov (1924) immunized larvae of the wax moth with heated cultures of *Micrococcus galleriae* No. 3 and produced an immunity against cholera in the insect.

Micrococcus insectorum Burrill

Insect concerned: The chinch bug, *Blissus leucopterus*.

Forbes (1882) found a micrococcus in the chinch bug which occurred primarily in the cecal organs. He concluded it was normal to these organs, being exceedingly abundant in all those examined. Burrill (1883) subsequently made a technical study of this organism and gave it the name *Micrococcus insectorum*.

Smith (1933, p. 54) refers to *"Bacillus insectorum* Burrell" as a cause of a bacterial disease of the chinch bug. Later (p. 59) he refers to *Micrococcus insectorum* Burrill as being similar to organisms which cause the silkworm diseases. He probably means the same organism.

Fig. 44. The chinch bug, *Blissus leucopterus*. (From Comstock, 1936.)

Micrococcus lardarius Krassilstschik

Insect concerned: The silkworm, *Bombyx mori*.

Krassilstschik (1896) found this organism in the intestines and body cavities of silkworms and thought it to be the cause of *grasserie*. He showed that the organism was distinctly different from *Streptococcus bombycis*.

Micrococcus luteus liquefaciens Adametz

Insect concerned: The honeybee, *Apis mellifera*.

While studying the microbial flora of diseased honeybee larvae, Toumanoff (1927) isolated a bacterium which he designated as *Micrococcus luteus liquefaciens* var. *larvae*. In many of its characteristics this organism was similar to *Micrococcus luteus* (Schroeter) Cohn of which *M. luteus liquefaciens* is possibly a variety.

Micrococcus major

Insects concerned: The nun moth, *Lymantria monacha,* and *Hyponomeuta* sp.

Eckstein (1894) in working with bacteria associated with nun-moth larvae isolated *Micrococcus major*. He found it pathogenic also for a *Hyponomeuta* species.

Micrococcus neurotomae Paillot

Insects concerned: *Neurotoma nemoralis; Euxoa segetum;* and *Agrotis pronubana*.

Paillot (1924a) isolated this bacterium from diseased *Neurotoma nemoralis* larvae, but he did not find it of any great help in checking the insect in nature. He (1933) mentions the pathogenicity of the organism to *Euxoa segetum* and *Agrotis pronubana*.

If the organism is gram-negative, as Paillot suggests, it should be placed in the genus *Neisseria*.

Micrococcus nigrofaciens Northrup

Insects concerned: *Phyllophaga* (*Lachnosterna*) spp.; *Allorrhina* spp.; the American cockroach, *Periplaneta americana;* the green June beetle, *Cotinis nitida* (= *Allorrhina nitida*); *Malacosoma americana;* the cockchafer, *Melolontha melolontha;* the rhinoceros beetle, *Strategus utanus;* and the May beetle, *Phyllophaga vandinei*.

Fig. 45. A white grub infected with *Micrococcus nigrofaciens*. One leg may be seen with only a blackened stump remaining; another in an earlier stage with only its end blackened. One of the spiracles is another site of infection. (From Sweetman, 1936; after Davis.)

In 1914 Northrup (1914a,b) described a bacterial disease of June-beetle larvae, *Lachnosterna* spp., caused by *Micrococcus nigrofaciens*. She found that this micrococcus was almost always accompanied by a putrefactive organism which she regarded as probably being *Bacillus septicus insectorum* Krassilstschik. Northrup concluded from her observations that the organisms exist in the soil and that the diseased larvae become infected through surface injury. Experiments on the infection of soils showed that excessive moisture in the soil greatly favored the spread of the micrococcus. The common cockroach, *Periplaneta americana*, was also found to be attacked by the micrococcus, but the infection apparently limited itself to the legs.

Smyth (1917, 1920) reported a high mortality among the grubs of May beetles and of rhinoceros beetles in experimental boxes caused by *Micrococcus nigrofaciens*. Du Porte (1915) reported it pathogenic for *Malacosoma americana*, and Davis and Luginbill (1921) for the green June beetle.

Micrococcus nitrificans Bergey et al.

Insect concerned: The lyreman cicada, *Tibicen linnei*.

Steinhaus (1941a) isolated an organism similar to *Micrococcus nitrificans* from a normal lyreman cicada.

Micrococcus nonfermentans Steinhaus

Insects concerned: The lyreman cicada, *Tibicen linnei;* and an unidentified damsel fly (Coenagrionidae).

Micrococcus nonfermentans was isolated by Steinhaus (1941a) from the alimentary tracts of the above-mentioned insects.

Micrococcus ochraceus Rosenthal

Insects concerned: The imperial moth, *Eacles imperialis,* and the nine-spotted lady beetle, *Coccinella novemnotata.*

Micrococcus ochraceus was isolated from the alimentary tract of the nine-spotted lady beetle and the hindgut of the imperial moth by Steinhaus (1941a).

Micrococcus ovatus

Insect concerned: The silkworm, *Bombyx mori.*

According to Northrup (1914a), pébrine of silkworms, now known to be caused by a protozoan, was at one time supposed to be due to *Micrococcus ovatus.*

Micrococcus parvulus Bergey *et al.*

Insect concerned: The American roach, *Periplaneta americana.*

Hatcher (1939) isolated *Micrococcus parvulus* from the feces of the American roach. Veillon and Zuber originally isolated this bacterium from human appendices, buccal cavities, and lungs, and named it *Staphylococcus parvulus.* In the third edition of *Bergey's Manual* (1930, p. 92) this bacterium was called *Micrococcus parvulus.* In the fifth edition (p. 285) the name has been changed to *Veillonella parvula.*

Micrococcus pieridis Burrill

Insect concerned: The cabbage butterfly, *Pieris rapae*.

Chittenden (1926) states, "In some seasons the larvae are destroyed in large numbers by a contagious bacterial disease, caused by *Micrococcus pieridis*." As a rule, however, the larvae were not very susceptible.

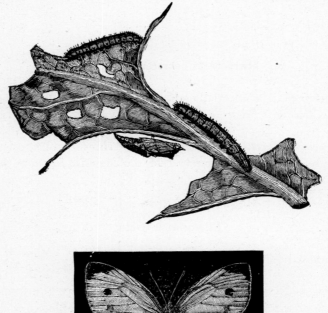

Fig. 46. *Pieris rapae*. (From Comstock, 1936.)

Micrococcus rushmori Brown

Insect concerned: *Lucilia sericata*.

Brown (1927) found *Micrococcus rushmori* to be a secondary invader in a disease of flies primarily caused by *Bacillus lutzae*.

Micrococcus saccatus Migula

Insect concerned: *Euxoa segetum*.

Pospelov (1927) isolated this coccus, along with *Bacillus agrotidis typhoides* and *Bacillus subtilis*, from dead larvae of *Euxoa segetum*. The larvae had died with symptoms of a septicemia.

Micrococcus subflavus Bumm

Insect concerned: *Grylloblatta campodeiformis campodeiformis*.

Micrococcus subflavus was isolated by Burroughs (1941) from the alimentary tract of a grylloblattid.

Micrococcus vulgaris

Insects concerned: The nun moth, *Lymantria monacha*; *Vanessa urticae*; *Pieris brassicae*; and *Stilpnotia salicis* (= *Liparis salicis*).

Eckstein (1894) isolated this organism from sick nun moths and from experimentally dead larvae of *Vanessa urticae*, *Pieris brassicae*, and *Stilpnotia salicis*.

Genus: SARCINA

Sarcina alba Zimmermann

Insects concerned: The roach, *Blatta orientalis*; *Pimelia sardea*, *Pimelia bifurcata*; *Blaps mucronata*; and *Tentyria* sp.

Cao (1898, 1906a) isolated both pathogenic and nonpathogenic strains of *Sarcina alba* from the feces of the oriental roach. In one instance Cao (1898) refers to the pathogenic strain as *Sarcina alba* "*patogena*." Experimentally Cao found that *Sarcina alba* survived in the intestines and passed out in the feces of the oriental roach and the following Coleoptera: *Pimelia sardea*, *Pimelia bifurcata*, *Blaps mucronata*, and a *Tentyria* sp.

Sarcina aurantiaca Flügge

Insects concerned: The honeybee, *Apis mellifera;* the oriental roach, *Blatta orientalis;* and the flies, *Musca domestica, Calliphora vomitoria, Sarcophaga carnaria,* and *Lucilia caesar.*

Cao (1906a) isolated *Sarcina aurantiaca* from the feces of the oriental cockroach, and upon refeeding found, by varying the type of food fed with the organism, that it became slightly pathogenic. He (1906b) also found that the organism may remain in the bodies of *Musca domestica, Calliphora vomitoria, Sarcophaga carnaria,* and *Lucilia caesar* throughout the metamorphosis of the insects. The organism fed to flies was recovered from the feces.

Serbinow (1912b) found this organism together with *Sarcina lutea* and *Bacillus mesentericus* in the dead larvae of honeybees.

Sarcina flava De Bary

Insects concerned: *Euxoa segetum,* and the bedbug, *Cimex lectularius.*

Stutzer and Wsorow (1927) isolated *Sarcina flava* from normal pupae of *Euxoa segetum.* Steinhaus (1941a) on one occasion found this organism as a fortuitous associate of the bedbug.

Sarcina lutea Schroeter

Insects concerned: The honeybee, *Apis mellifera;* the roach, *Blatta orientalis;* and the silkworm, *Bombyx mori.*

Serbinow (1912a) in trying to determine the cause of black brood among bees isolated *Sarcina lutea* from the dead larvae. This organism has also been isolated from oriental cockroaches and from the eggs of silkworms. (See also *Sarcina aurantiaca.*)

Genus: STAPHYLOCOCCUS

Staphylococcus spp.

Insects and tick concerned: The housefly, *Musca domestica;* the lice, *Pediculus humanus capitis* and *Pediculus humanus corporis;* and the tick, *Ixodes dentatus.*

Torrey (1912) isolated several strains of staphylococci, which he called *Albococcus* and *Aurococcus,* from the outer surfaces of houseflies caught in a tenement section of New York. Cox, Lewis, and Glynn (1912) also isolated six strains from flies caught in Liverpool.

According to Pierce (1921),

Staphylococcus aureus and *albus,* the cause of impetigo contagiosa, an acute contagious pustular inflammation of the skin, can be carried by head lice, as was proven by Dewèvre (1892 [a]) by removing lice from impetigo cases and placing them on the heads of healthy children, who some days later developed this disease. This claim has been supported by various authors. Widmann (1915) attempted to transmit *Staphylococcus septicaemia* by louse bite and failed, although he recovered living cocci from the louse feces after 60 hours, but not later. In view of recent findings with other louse-borne diseases, we may expect that infection could have been obtained by slightly abrading the surface on which the lice had defecated.

Bell and Chalgren (1943) report that the tick *Ixodes dentatus* may be responsible in furnishing portals of entry for abscess-producing staphylococci in cottontail rabbits.

Staphylococcus acridicida Kuffernath

Insects concerned: The cabbage butterfly, *Pieris rapae,* and the locust, *Locusta viridissima.*

In 1913 Kuffernath (1921) received locusts from Greece which had been part of a disastrous invasion. He found the locusts infected with an organism which he isolated and named *Staphylococcus acridicida,* finding it to be closely allied to *Staphylococcus pyogenes* (*Staphylococcus aureus*).

In the title of his paper he refers to the organism as *Micrococcus* (*Staphylococcus*) *acridicida.*

Staphylococcus albus Rosenbach

Insects and ticks concerned: The roaches, *Blatta orientalis* and *Blattella germanica;* a burrowing beetle, *Melolontha vulgaris;* the wax moth, *Galleria mellonella;* the stablefly, *Stomoxys calcitrans; Tabanus* sp.; *Sarcophaga carnaria; Lucilia caesar; Calliphora vomitoria;* the housefly, *Musca domestica;* the bedbug, *Cimex lectularius,* the mosquitoes, *Aëdes cinereus* and *Anopheles claviger; Rhodnius prolixus;* and the ticks, *Argas persicus* and *Ornithodoros moubata.*

Joly (1898) isolated *Staphylococcus albus* from a tabanid taken from a heifer; Cao (1906b) from the mucilaginous envelope covering the eggs of *Musca domestica, Sarcophaga carnaria, Lucilia caesar,* and *Calliphora vomitoria*; and Torrey (1912) and Scott (1917) from the surface of city-caught flies. Torrey refers to the organism as *Albococcus pyogenes*. Others have used

Fig. 47. *Blattella germanica. a,* first instar; *b,* second instar; *c,* third instar; *d,* fourth instar; *e,* adult; *f,* adult female with egg case; *g,* egg case, enlarged; *h,* adult with wings spread. (From Comstock, 1936; after Howard and Marlatt.)

the name *Staphylococcus pyogenes albus*. Cao also isolated *Staphylococcus albus* from the intestinal tract of *Melolontha vulgaris*, a burrowing beetle.

Herms (1939) found the roach, *Blattella germanica*, to be a carrier of *Staphylococcus albus*, and he has shown that food on which the insect feeds, or with which it comes in contact, becomes contaminated. Tauber (1940) found in the hemolymph of the oriental roach two organisms which were pathogenic for the insect. One was an unidentified rod, and the other was *Staphylococcus albus*. Just how the bacteria made their entrance into the hemolymph of normal roaches was not clear. He suggested, however, that after the insect molts, the exoskeleton is very soft and is easily injured. Then the uninfected roaches come in contact with the infected ones, and the bacteria penetrate the delicate, newly exposed exoskeleton, or pass through breaks in the surfaces and so into the hemolymph.

Metalnikov (1927) found *Staphylococcus albus* to be nonpathogenic for the wax-moth larvae, whereas Cameron (1934) found some larvae died within 24 hours after injection.

Duncan (1926) found the gut contents of *Argas persicus, Ornithodoros moubata,* and *Stomoxys calcitrans*, but not of *Cimex lectularius, Rhodnius prolixus, Musca domestica, Aëdes cinereus,* and *Anopheles claviger*, to be bactericidal to *Staphylococcus albus*.

Staphylococcus aureus Rosenbach

Insects and ticks concerned: *Melolontha vulgaris;* the silkworm, *Bombyx mori;* the wax moth, *Galleria mellonella;* the roaches, *Blattella germanica* and *Blatta orientalis; Sarcophagula* sp.; the stablefly, *Stomoxys calcitrans; Tabanus* sp.; the housefly, *Musca domestica;* the bedbug, *Cimex lectularius; Rhodnius prolixus;* the mosquitoes, *Aëdes aegypti, Aëdes cinereus,* and *Anopheles claviger;* and the ticks, *Argas persicus, Argas reflexus, Ornithodoros moubata,* and *Ixodes ricinus.*

In 1898 Joly obtained *Staphylococcus aureus* and *Staphylococcus albus* from a tabanid taken from a heifer. Sawamura (1906) lists *Micrococcus pyogenes aureus (Staphylococcus aureus)* as being one of the organisms experimentally pathogenic to the silkworm. Longfellow (1913) found that roaches, probably *Blatta orientalis,* carried *Staphylococcus aureus* on their legs, and Herms (1939) says that the bacterium has been isolated from the antennae and feet, as well as from the stomach contents, of the roach, *Blattella germanica.* Cox, Lewis, and Glynn (1912) isolated several strains of *Staphylococcus* from houseflies, and Scott (1917) isolated strains of *Staphylococcus aureus* from the feet of flies. Cellí (1888) showed that *Staphylococcus aureus* retained its virulence when it passed through the intestines of flies. Nicholls (1912) found that larvae of the *Sarcophagula* raised in material infected with *Staphylococcus aureus* were free from these bacteria in the pupae if they were removed from their infected surroundings.

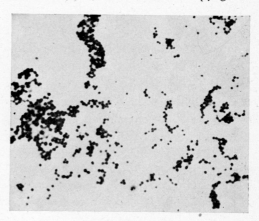

Fig. 48. *Staphylococcus aureus.*

Cao (1906b) isolated a strain of *Staphylococcus (pyogenes) aureus* from the intestinal tract of *Melolontha vulgaris.*

Galli-Valerio (1907) states that *Staphylococcus pyogenes aureus* has been spread by *Argas reflexus,* and *Ixodes ricinus* has also been similarly indicted. Hindle and Duncan (1925) found that *Staphylococcus aureus* dies out quickly after ingestion by the tick, *Argas persicus.* Duncan (1926) supported these results by the observation that this coccus was greatly inhibited by the gut contents of the tick. The same bactericidal effect against this organism was found by Duncan to be characteristic of the gut contents of *Ornithodoros moubata* and *Stomoxys calcitrans,* but not of *Cimex lectularius, Rhodnius prolixus, Musca domestica, Aëdes cinereus,* and *Anopheles claviger.*

St. John, Simmons, and Reynolds (1930) found *Staphylococcus aureus* to survive for at least 24 hours in the gut of the mosquito, *Aëdes aegypti,* though it could not be demonstrated after an interval of seven or more days.

Metalnikov found that the pathogenicity of *Staphylococcus aureus* for waxmoth larvae varied, but Cameron (1934) found it nonpathogenic. Zernoff and Ajolo (1939) found that the wax moth, *Galleria mellonella,* survived several lethal doses of *Staphylococcus aureus* if an injection of para-aminophenylsulfonamide was given simultaneously with the bacteria.

Staphylococcus citreus (Migula) Bergey et al.

Insects concerned: The cockroach, *Blatta orientalis;* the flies, *Musca domestica, Calliphora vomitoria, Lucilia caesar, Sarcophaga carnaria;* and the burrowing beetle, *Melolontha vulgaris.*

Longfellow (1913) found that cockroaches carried *Staphylococcus citreus* on their legs. Scott (1917) isolated it (*Staphylococcus pyogenes citreus*) from bodies of houseflies caught in Washington, D.C. Cao (1906b) fed larvae of *Musca domestica, Calliphora vomitoria, Lucilia caesar,* and *Sarcophaga carnaria* on food contaminated with *Staphylococcus citreus* and recovered it from feces of the adults nine days after the adult flies emerged. Cao also isolated *Staphylococcus citreus* from the intestinal tract of a burrowing beetle, *Melolontha vulgaris.*

Staphylococcus insectorum Krassilstschik

Insect concerned: The silkworm, *Bombyx mori.*

Krassilstschik (according to Paillot, 1928b) isolated and named *Staphylococcus insectorum,* and thought it was found normally in the intestinal tract of the silkworm.

Staphylococcus muscae Glaser

Insect concerned: The housefly, *Musca domestica*.

Glaser (1924c) isolated and named *Staphylococcus muscae*. He found it to be the cause of a fatal infection in the housefly. The disease is rather sporadic and never assumes the form of an epizoötic, only about 50 per cent of adult flies contracting the infection when experimentally infected. Males were more susceptible than females.

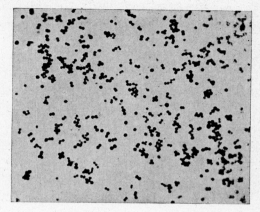

Fig. 49. *Staphylococcus muscae*. (Courtesy Dr. R. W. Glaser.)

Family: NEISSERIACEAE
Genus: NEISSERIA

Neisseria catarrhalis (Frosch and Kolle) Holland

Tick concerned: *Argas persicus*.

Duncan (1926) found that the gut contents of *Argas persicus* possessed a bactericidal principle active against *Micrococcus catarrhalis* (*Neisseria catarrhalis*) and other bacteria.

Neisseria gonorrhoeae Trevisan

Insects concerned: A fly, probably *Musca domestica,* and the wax moth, *Galleria mellonella*.

Welander (1896) found that *Diplococcus gonorrhoeae* (*Neisseria gonorrhoeae*) was carried on the feet of a fly for three hours after being soiled with human secretions.

Metalnikov (1927) found that *Neisseria gonorrhoeae* varied in its pathogenicity for the wax-moth larvae, while Cameron (1934) found it nonpathogenic.

Neisseria intracellularis (Lehmann and Neumann) Holland

Insect concerned: A fly, probably *Musca domestica*.

Diplococcus intracellularis (*Neisseria intracellularis*) is probably carried by flies, according to MacGregor (1917).

Neisseria luciliarum Brown   prob. a Micrococcus

Insect concerned: The greenbottle fly, *Lucilia sericata*.

Brown (1927) isolated this gram-negative, motile coccus from dead *Lucilia sericata* which had been killed by *Bacillus lutzae*. Taxonomically, according to Brown, this organism should be placed near *Neisseria perflava* Bergey *et al.*

Family: PARVOBACTERIACEAE
Tribe: BRUCELLEAE
Genus: BRUCELLA

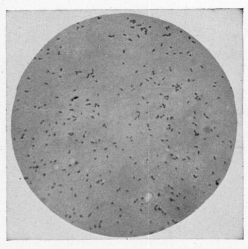

Fig. 50. *Brucella abortus*. (From Hagan, 1943.)

Brucella abortus (Schmidt and Weis) Meyer and Shaw

Insects concerned: The cockroach, *Periplaneta americana*, and the flies, *Musca domestica*, *Muscina stabulans*, *Stomoxys calcitrans*, *Calliphora* sp., and *Lucilia* sp.

Ruhland and Huddleson (1941) in attempting to account for the appearance of brucellosis in noninfected cattle kept under ideal conditions fed cockroaches and flies for two hours on a virulent strain of *Brucella abortus*, after which the insects were examined bacteriologically. In the 110 cockroaches tested the bacterium did not remain alive in their intestinal tracts for more than 24 hours. "Data obtained by culturing the droplets was heavier and more free from contamination 48 hours after exposure than at earlier periods. Although no flies were cultured later than 96 hours after exposure, it is possible that they carry the organism for a considerable period of time."

According to Patton (1931), Wollman found that flies kept in contact with *Brucella abortus* for 48 hours, and then placed in a tube containing a culture medium, remained infective for 24 hours but not later.

Tribe: HEMOPHILEAE
Genus: HEMOPHILUS

Hemophilus duplex (Lehmann and Neumann) *comb. nov.*

Insect concerned: The housefly, *Musca domestica*.

According to Patton (1931), Wollman placed houseflies in tubes containing cultures of *Bacillus aegyptius* and the Morax-Axenfeld bacillus (*Hemophilus duplex*) of subacute conjunctivitis and observed that they became infective immediately afterwards, though they were not infective after three and one-half hours.

Tribe: PASTEURELLEAE
Genus: MALLEOMYCES

Malleomyces mallei (Flügge) Pribram

Insects concerned: *Stomoxys calcitrans; Blatta orientalis; Tentyria* sp.; *Blaps mucronata; Pimelia bifurcata;* and *Pimelia sardea*.

According to Osborn (1896), *Stomoxys calcitrans* has been blamed for transmitting *M. mallei,* the cause of glanders, to horses. As early as 1898 Cao observed that this bacillus ("bacillus della morva") passed unchanged through the intestines of *Blatta orientalis, Tentyria* sp., *Blaps mucronata, Pimelia bifurcata,* and *Pimelia sardea*. Using these same insects Cao found that "bacillus della barbone bufalino" also passed through the intestinal tracts unharmed.

Malleomyces pseudomallei (Whitmore)

Insects concerned: The rat flea, *Xenopsylla cheopis,* and the mosquito, *Aëdes aegypti.*

Blanc and Baltazard (1941) found that fleas that had fed on guinea pigs infected with Whitmore's bacillus (*Malleomyces pseudomallei*) infected healthy guinea pigs. One test of fleas remained infective for 50 days. The organism was also recovered from the feces of the fleas. Later Blanc and Baltazard (1942b) reported that *Aëdes aegypti* also readily transmitted the infection from one guinea pig to another. Twenty-one guinea pigs were infected by rat fleas and two by *Aëdes aegypti.*

Genus: Pasteurella

Pasteurella avicida (Gamaleia) Trevisan
(*Pasteurella aviseptica* [Kitt] Shütze)

Insect concerned: The cockroach, *Blatta orientalis.*

Küster (1902) and also Cao (1906a) fed *Bacterium cholerae gallinarum* (*Pasteurella avicida*) to *Blatta orientalis* and recovered it from the insects' feces. Cao found that when an attenuated culture of the organism was fed to starved roaches, it passed through the intestines without an increase in virulence for guinea pigs. When fed in conjunction with bread and putrid beef liver infusion, it partially regained its lost virulence.

Pasteurella bollingeri Trevisan

Insects concerned: The flies, *Tabanus rubidus, Tabanus striatus, Chrysops dispar, Stomoxys calcitrans, Lyperosia exigua,* and *Musca inferior,* and the mosquitoes, *Aëdes aegypti, Anopheles fuliginosus,* and *Armigeres obturbans.*

In 1929 Nieschulz and Kranefeld reported on a series of transmission experiments using the above-named insects to transmit *Pasteurella bollingeri* (also known as *Pasteurella bubaliseptica*), the cause of hemorrhagic septicemia of buffaloes, to rabbits. Transmission was most successful with the tabanids. *Tabanus rubidus* remained infective for four days and *Chrysops dispar* for three days; *Stomoxys calcitrans, Lyperosia exigua,* and *Musca inferior* were infective for 24 hours after biting an infected rabbit. Of the mosquitoes, *Aëdes aegypti* was infective up to two days, *Anopheles fulignosus* up to 6 hours, and *Armigeres obturbans* up to 24 hours after an infective feeding.

Pasteurella cuniculicida (Flügge) Trevisan

Insect concerned: The housefly, *Musca domestica.*

Scott (1917) isolated *Bacillus cuniculicida* from a common housefly caught in the animal room of the Army Medical School. He thought that outbreaks of rabbit septicemia among his laboratory rabbits and guinea pigs might be due to houseflies at times acting as carriers of the infection from pens of infected animals to pens of healthy animals. The organism isolated from the housefly was very pathogenic for rabbits—killing in 36 to 48 hours.

Scott gives, as a synonym for *Bacillus cuniculicida, Bacterium cholerae gallinarum* Flügge, which makes it difficult to determine to what organism he refers. According to *Bergey's Manual* (5th ed., pp. 290, 292), *Bacillus cuniculicida* is a synonym for *Pasteurella cuniculicida,* and *Bacterium cholerae gallinarum* is a synonym for *Pasteurella avicida.* Scott's description of his *Bacillus cuniculicida* does not fit the description of *Pasteurella avicida,* and a complete description of *Pasteurella cuniculicida* is not given. Both organisms produce septicemia in rabbits, so his *Bacillus cuniculicida* could be either one of these two *Pasteurella* species.

Pasteurella pestis (Lehmann and Neumann) Bergey et al.

Insects and ticks concerned: Plague bacilli have been cited as associated with an extensive number of arthropods. The species listed below are perhaps the most important and representative ones from the standpoint of plague transmission.

Fleas: *Xenopsylla cheopis; Xenopsylla astia; Xenopsylla brasiliensis; Diamanus montanus* (= *Ceratophyllus acutus*); *Nosopsyllus fasciatus* (= *Ceratophyllus fasciatus*); *Monopsyllus anisus* (= *Ceratophyllus anisus*); *Oropsylla silantiewi; Citellophilus tesquorum* (= *Ceratophyllus tesquorum*); *Dinopsyllus lypusus; Leptopsylla segnis* (= *L. musculi*); *Ctenophthalmus agyrtes;*

Pulex irritans; Ctenocephalides canis; Ctenocephalides felis; Ceratophyllus gallinae; Synopsylla fonquernii; and others. Some fleas, such as *Malaraeus telchinum,* while not efficient vectors, will, nevertheless, transmit the bacillus experimentally under the right conditions (Burroughs, 1944).

The reader is referred to the work of Eskey and Haas (1940) for a list of other species of fleas which are potential vectors of plague, especially among rodents. (See also Brown, 1944.)

Lice: *Haematopinus columbianus; Linognathoides citelli;* and *Pediculus humanus.* (Plague bacilli have been found associated with the latter species, but there is no evidence that plague can be transmitted by it.)

Many of the following have been implicated in the spread of plague only incidentally and largely on the basis of experimental findings:

Ticks: *Ixodes autumnalis; Rhipicephalus schulzei; Argas persicus; Hyalomma volgense;* and *Dermacentor silvarum.*

Flies: *Musca domestica* and *Stomoxys calcitrans.*

Beetles: *Necrophorus dauricus* and others.

Mosquitoes: *Culex pipiens* and *Aëdes aegypti.*

Ants (*Monomorium vastator*), roaches, bedbugs, and mites have also been suspected of carrying plague bacilli.

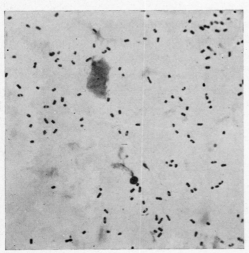

Fig. 51. *Pasteurella pestis* in a smear of the liver of an infected guinea pig. (Courtesy Dr. K. F. Meyer.)

The plague bacillus was independently discovered in 1894 by Kitasato and by Yersin. Three years later Ogata (1897) and soon thereafter Simond (1898) pointed out, on epidemiologic and experimental grounds, the role of the flea in the transmission of plague bacilli. However, it was the English Plague Research Commission that clarified and established the essential part that fleas play in the spread of plague, especially among rodents. (See also Meyer, 1938.)

Although the number of fleas known to transmit the plague bacillus is large and ever-increasing, the principal vector is the rat flea, *Xenopsylla cheopis*. It is present on wild rats in many parts of the world and is the predominant species in India, Java, Egypt, and most parts of China. Besides its widespread occurrence, there are other factors which make this species an efficient vector. As stated by Eskey (1938):

When compared with all other species studied, *Xenopsylla cheopis* are considered the most efficient transmitting agents because they are more readily infected when fed on septicemic blood, and they transmitted the disease to many more guinea pigs. They also tend to become blocked earlier and to remain infectious for a longer time than other fleas. Blocked cheopis are very persistent in their efforts to obtain blood. . . .

On the other hand, Douglas and Wheeler (1943) have shown that under laboratory conditions the California ground-squirrel flea, *Diamanus montanus,* is approximately twice as efficient a vector as *X. cheopis*. (Wheeler and Douglas, 1945, have developed a method by which the vector efficiency of a given flea species may be determined.)

According to Lien-teh, Chun, Pollitzer, and Wu (1936), *Xenopsylla astia* is not as important as *X. cheopis* in the transmission of plague though it may be the responsible vector in certain circumscribed and isolated outbreaks. *X. brasiliensis* is the predominant rat flea of Uganda, Kenya, and Nigeria, and has been known to transmit plague from rat to rat and from rat to man. *Nosopsyllus fasciatus,* the so-called "European rat flea," attacks man less readily than *X. cheopis* and does not predominate on rats in plague areas. *Leptopsylla segnis* likewise does not readily bite man and is of minor importance in the spread of the disease.

The biologic relationships between *Pasteurella pestis* and its flea host are interesting and important from the standpoint of the bacterium's transmission. Simond (1898) thought that infection resulted from the contamination of the bite wound with the infective feces of the flea. A few years later Verjbitski (1908) showed that the bacteria could be transmitted directly through the infected insect's bite. Bacot and Martin (1914) were able to prove definitely that *P. pestis* multiplied in the flea's alimentary tract, especially in the proventriculus and esophagus. Thus it was learned that the gut lumen became blocked by the multiplying organisms, as a result of which the flea would regurgitate the bacteria when it attempted to feed. Some workers (Blanc and Baltazard, 1942a) discredit the importance of this blocking mechanism in fleas believing the chief factor in the flea's ability to transmit the infection to be the rapid multiplication of the bacilli in the fresh blood taken up in frequent feedings. Working with the fleas *Xenopsylla cheopis* and *Diamanus montanus,* Douglas and Wheeler (1943) observed that the location of the bacterial plug is to some extent a characteristic of the species, and is probably controlled by the anatomical structures and the physiologic activities of the insect's alimentary tract. Therefore, the ability of a species of flea to become infected is no proof that it will eventually become infective. These workers also showed that when the hungry infected flea attempts to suck blood the esophagus becomes distended with fresh blood, which becomes contaminated with *P. pestis* and which is driven back into the wound by the recoil of the esophageal and pharyngeal walls.

Other facts of biologic interest observed by Douglas and Wheeler were: (1) Under optimum conditions either *X. cheopis* or *D. montanus* is capable of ingesting an average of 3×10^5 plague bacilli. (2) Approximately 2 per cent of infected *X. cheopis* and 60 per cent of *D. montanus* appear able to free their alimentary tracts of *P. pestis* after a single infectious blood meal. This may occur within 24 hours after infection in *D. montanus*, but in *X. cheopis* it requires more than 48 hours to take place. (3) Occlusion of the gut lumen by the multiplying bacteria requires an average of 16 days in *X. cheopis* and 10 days in *D. montanus*. (4) Both fleas excrete *P. pestis* very irregularly. Of daily fecal samples, 25 per cent of those of *X. cheopis* and 56 per cent of those of *D. montanus* contained viable *P. pestis*. The bacteria are present in very small numbers in the feces of infected fleas.

In some parts of the world plague is harbored in animals other than the rat. In such cases the ectoparasites of these animals are of importance in maintaining the disease. Chief of these are *Diamanus montanus*, found on the ground squirrels of California; *Oropsylla silantiewi*, found on the tarbagan of Mongolia; *Citellophilus tesquorum*, carried by the suslik or marmot of southeastern Russia; and *Dinopsyllus lypusus*, found on the gerbil of Africa. (See also Meyer, 1942.)

As indicated in the list of arthropods above, other insects besides fleas, including flies, ants, beetles, mosquitoes, bedbugs, and roaches, have been suspected of carrying plague bacilli. As to bedbugs, *Cimex lectularius*, Walker (see Pierce, 1921) found that 22 per cent of the bugs in the huts of natives in India harbor the plague bacillus. Cao (1898) stated that Nuttall demonstrated that bedbugs placed on the body of a dead plague rat became infected by the bacillus more or less rapidly according to the temperature of the corpse. Regarding ants, Cao declared that Hankin (1897) isolated the plague bacillus from the feces of ants, *Monomorium vastator*, which had fed on the bodies of rats dead of plague and suggested that in Bombay the ant might have been a means of spreading the infection. (See also Nuttall, 1899.)

Data concerning the possible role of ticks in the transmission of plague is very meager. Skorodumoff in 1928 (quoted by Lien-teh *et al.*, 1936) was among the first to obtain positive experimental results. He infected a wild mouse and a guinea pig from crushed suspension of ixodid ticks. In 1929 Tichomirova and Nikanoroff (Lien-teh *et al.*, 1936) found three ticks (*Ixodes autumnalis*) upon the carcass of an experimentally infected tarbagan. The tissues of these ticks yielded positive cultures of the plague bacillus and were infective to guinea pigs. Faddeeva (1932) fed *Argas persicus* on infected guinea pigs when the bacteremia was most marked. Both the inoculation and the culture experiments gave positive results. Borzenkov and Donskov (1933) reported the finding of plague bacilli in *Ixodes autumnalis* and *Rhipicephalus schulzei* taken from animals infected with plague. These workers also found that *Hyalomma volgense*, from larval to adult stages, can be infected with plague by feeding on infected animals. These authors also state that "by the direct bites the plague infected adults may cause an infection and death in healthy animals." Sassuchin and Tichomirova (1936) have reported studies on the survival of plague bacilli in the larvae and nymphs of *Dermacentor silvarum*. Over 60 per cent of both nymphs and larvae died during the experiments, suggesting that *Pasteurella pestis* was pathogenic to them. From time to time the United States Public Health Service has reported the isolation of *Pasteurella pestis* from ticks and fleas pooled together. However, the *Public Health Reports* for October 2, 1942 (57, 1514), report the isolation of this bacterium from a pool of four unidentified ticks taken from one cottontail, *Sylvilagus* sp., from the Fort Ord Military Reservation, Area A.

Pasteurella tularensis (McCoy and Chapin) Bergey et al.[1]

Insects and ticks concerned: The rabbit louse, *Haemodipsus ventricosus;* the mouse louse, *Polyplax serratus;* the louse, *Neohaematopinus laeviusculus;* the bedbug, *Cimex lectularius;* the fleas, *Spilopsyllus cuniculi, Cediopsylla simplex, Ctenophthalmus pollex, Ctenophthalmus assimilis, Ctenophthalmus orientalis, Nosopsyllus fasciatus* (= *Ceratophyllus fasciatus*), *Ceratophyllus*

---

[1] Considerable disagreement persists as to the correct classification and naming of this bacterium. Serologically it is allied to the members of the genus *Brucella*, and hence is called *Brucella tularensis* by English writers. However, the importance of this serologic similarity as a taxonomic criterion means little if the findings of Mallman (1930) concerning the common antigenicity of *Brucella, Pasteurella,* and *Pfeifferella* are correct. Although originally named *Bacterium tularense* by McCoy and Chapin, more recent classifications have placed the organism in the genus *Pasteurella* because of its similarities with other members of the genus, especially *Pasteurella pestis*. These organisms agree to a large extent in pathogenicity, pathologic manifestations, morphology, fermentation reactions, selective affinity for rodent and human hosts, and in insect transmission. The tularemia organism differs from other members of the genus in the difficulty with which it is grown on ordinary bacteriologic media. Inasmuch as we have arbitrarily followed the systematics of *Bergey's Manual*, we have used the name *Pasteurella tularensis* for the sake of consistency. Recent morphologic evidence (Hesselbrock and Foshay, 1945) indicates that the tularemia organism may in fact belong to none of the above genera.

rectangulatus, and *Diamanus montanus* (= *Ceratophyllus acutus*); the flies, *Chrysops discalis, Musca domestica, Tabanus autumnalis, Tabanus agrestis, Tabanus bromius, Tabanus erberi, Tabanus flavoguttatus, Tabanus solstitialis, Tabanus peculiaris, Tabanus karybenthinus, Tabanus turkestanus, Tabanus septentrionalis, Tabanus rupestris, Chrysops noctifer, Chrysops relicta, Chrysozona turkestanica, Simulium decorum katmai, Stomoxys calcitrans;* certain mites of the family Gamasidae; the ticks, *Dermacentor andersoni, Dermacentor albipictus, Dermacentor occidentalis, Dermacentor variabilis, Dermacentor parumapertus* (= *Dermacentor parumapertus marginatus*), *Dermacentor silvarum, Haemaphysalis leporis-palustris, Haemaphysalis cinnabarina,* and *Ixodes ricinus californicus* (now *I. pacificus*); *Ornithodoros turicata, Ornithodoros parkeri, Ornithodoros lahorensis, Rhipicephalus sanguineus,* and *Amblyomma americanum;* and the mosquitoes, *Aëdes nearcticus, Aëdes vexans, Aëdes dorsalis, Aëdes stimulans, Aëdes canadensis, Aëdes aegypti, Aëdes caspius, Aëdes cinereus, Theobaldia incidens, Culex tarsalis, Culex apicalis,* and *Anopheles hyrcanus.*

In 1911 McCoy described a plaguelike disease of ground squirrels (*Citellus beecheyi*) of California. The following year McCoy and Chapin (1912) described *"Bacterium tularense"* as the causative agent. The disease was later named "tularemia" by Francis. During his original investigations McCoy (1911) found it possible to reproduce the infection in guinea pigs by the subcutaneous inoculation of crushed squirrel fleas (*Ceratophyllus acutus*) which had been removed from recently dead rodents. Ten years later Francis and Mayne (1921) reported experimental transmission of tularemia by the bloodsucking deer fly, *Chrysops discalis*. This was followed by similar reports by Francis and Lake (1921, 1922a,b) involving the rabbit louse, *Haemodipsus ventricosus,* the bedbug, *Cimex lectularius,* and the mouse louse, *Polyplax serratus.*

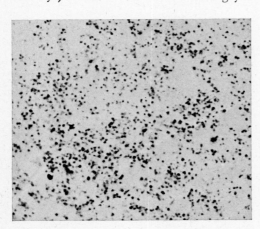

Fig. 52. *Pasteurella tularensis.* (Photo by N. J. Kramis.)

During the years 1922 and 1923 Parker, Spencer, and Francis (1924) made observations which indicated the spontaneous occurrence of the tularemia organism in the tick, *Dermacentor andersoni*. They also demonstrated stage-to-stage transmission of the bacterium from larva to adult tick. Later Parker and Spencer (1926) proved the "hereditary" transmission of *Pasteurella tularensis* (*Bacterium tularense*) in *Dermacentor andersoni*. This appears to have been the first recorded instance of the transovarial transmission of a known pathogenic bacterium by an arthropod.

Parker and his associates (Davis, Philip, and Jellison) have made numerous other studies regarding the association of arthropods and the tularemia bacterium. As to their findings, we quote from Parker (1933):

(a) The demonstration of the survival of *Bact. tularense* from the larvae through to the adults of both *H. leporis-palustris* and *D. variabilis*, its transmission by the successive stages involved, and generation-to-generation transmission from infected females to their progeny.

(b) Larval to adult survival in, and transmission by, the successive stages of the brown dog tick, *R. sanguineus*, and the lone-star tick, *A. americanum*, within a single generation.

(c) Survival in, and later transmission by, adult rabbit dermacentor, *D. parumapertus marginatus*, and the Pacific Coast tick, *D. occidentalis*, that had previously ingested virus.

(d) Mechanical transmission to guinea pigs by *Tabanus septentrionalis*, by another species of horsefly tentatively identified as *T. rupestris*, and by *Chrysops noctifer*, and the survival of viable *Bact. tularense* in *C. noctifer* for a period of over a month. Survival only was also shown in an undetermined species of Ceratopogonidae. [Philip, unpublished experiments.]

(e) Failure to transmit infection by species of fleas occurring in the native Montana fauna.

(f) Mechanical transmission by immediate interrupted feeding by the black fly, *Simulium decorum katmai*. [Philip, unpublished experiments.]

(g) Transmission by the sucking louse, *Neohaematopinus laeviusculus*, of the Columbian ground squirrel.

(h) The repeated recovery of *Bact. tularense* from specimens of *H. leporis-palustris* collected from rabbits and grouse in Morrison County, Minnesota, in 1931 and 1932. (These tests were made in conjunction with Dr. R. G. Green of the University of Minnesota Medical School.) [Parker also reports the isolation of the tularemia organism from *Haemaphysalis cinnabarina* taken from sage hens in eastern Montana.]

According to other reports, *Pasteurella tularensis* has been found spontaneously in *Dermacentor occidentalis* (Parker, Brooks, and Marsh, 1929), *Ixodes ricinus californicus* (Davis and Kohls, 1937), and *Dermacentor variabilis* (Green, 1931). Kamil and Bilal (1938a) have reported the transmission by *Ornithodoros lahorensis*. Zasukhin (1936a) mentions that in 1934 Golov showed that *Dermacentor silvarum* can be infected with *Pasteurella tularensis* and that it does not lose the infection from one stage to the next. Davis (1940a) has found that the tularemia bacterium may survive in the tissues of *Ornithodoros turicata* and *Ornithodoros parkeri,* but is not transmitted by the ticks during feeding.

In a study of the infection of ticks (*Dermacentor variabilis*) by *P. tularensis,* Bell (1945) concluded that "hereditary" transmission may be the exception rather than the rule in nature. He also observed that the fecundity of infected ticks was not diminished nor was viability adversely affected. Of particular significance was Bell's demonstration that

infected ticks feeding on immune or normal hosts lose their infection, presumably as a result of the stimulating effect of the blood meal upon a normal bactericidal function of the tick's gut but before losing their infection as a result of feeding, infected ticks may inoculate the host, whereupon the host, if it is not immune, will develop septicemia and infect all ticks feeding upon it. Infected ticks feeding on immune hosts, on the other hand, permanently lost their infections and the bacteria were not transmitted to normal ticks feeding concurrently.

Volferz, Kolpakov, and Flegontoff (1934) have reported the survival of *Pasteurella tularensis* in mites of the family Gamasidae and in the fleas, *Ctenophthalmus pollex* and possibly *Ctenophthalmus orientalis.* Green and Evans (1938) isolated the bacterium of tularemia from fleas (*Spilopsyllus cuniculi*) removed from snowshoe hares and from cottontail rabbits. Waller (1940) recovered *Pasteurella tularensis* from *Cediopsylla simplex* collected from a sick cottontail rabbit. Davis (1935) isolated the organism from lice (*Neohaematopinus laeviusculus*) taken from white-tailed prairie dogs. Philip, Davis, and Parker (1932) using *Aëdes nearcticus, A. vexans, A. dorsalis, A. stimulans, A. canadensis, Theobaldia incidens,* and *Culex tarsalis* showed experimentally that the mosquito could be a significant factor in the epidemiology of tularemia by infecting persons mechanically (1) by biting, (2) by being crushed on the skin, and (3) by depositing excrement on the skin. Bozhenko (1936) demonstrated that *Pasteurella tularensis* may survive in *Culex apicalis* and in the feces of this insect. This worker (1935) has also found that transmission of tularemia by the bites of infected bedbugs was successful 15 hours after an infective feeding and that organisms remain virulent in the bugs for as long as 136 days after feeding. Working with bedbugs, Francis and Lake (1922a) and Francis (1927) reported the infection of mice by interrupted feedings of the bugs. Kamil and Bilal (1938a) were able to transmit tularemia to guinea pigs through the bites of bedbugs, by rubbing triturated bugs into the shaven hide, and by feeding infected bugs to the guinea pigs. On the other hand, Davis (1943d) concluded that bedbugs fail to transmit *P. tularensis* to experimental animals when a "clean" method of feeding was used, eliminating fecal contamination of the host. Davis was unable to confirm the transovarial transmission of the organism as reported by Bilal (1941). Infected bedbugs do not live as long as uninfected ones.

In recent years Olsufiev and Golov (1936) and Olsufiev (1939a,b,c, 1940a,b) have gathered considerable data on the transmission of tularemia implicating horseflies, the stablefly, the rainfly, mosquitoes, and other insects. Steinhaus (unpublished experiments) isolated *Pasteurella tularensis* from flies collected in an animal room which contained guinea pigs infected with tularemia.

Family: PSEUDOMONADACEAE

Tribe: PSEUDOMONADEAE

Genus: Pseudomonas

Pseudomonas aeruginosa (Schroeter) Migula

Insects concerned: The grasshopper, *Schistocerca gregaria;* the wax moth, *Galleria mellonella;* the silkworm, *Bombyx mori;* the housefly, *Musca domestica; Stomoxys calcitrans; Calliphora vomitoria; Lucilia caesar; Sarcophaga carnaria;* and a burrowing beetle, *Melolontha vulgaris.*

Bacot (1911) and Ledingham (1911) by carefully controlled experiments have shown that *Musca domestica* larvae fed on food infected with *Bacillus pyocyaneus* (*Pseudomonas aeruginosa*) retain the organism in their intestinal tracts through metamorphosis.

Cao (1898) isolated *Bacillus pyocyaneus* from the intestinal tract of the larva of *Melolontha vulgaris*. Later (1906b) he fed the larvae of *Musca domestica, Calliphora vomitoria, Sarcophaga carnaria,* and *Lucilia caesar* on flesh containing *Bacillus pyocyaneus* (*Pseudomonas aeruginosa*) and found it to be the predominant organism in the feces of the larvae. The organ-

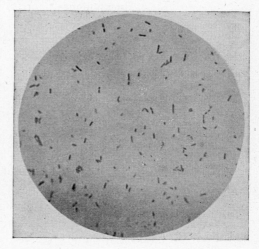

Fig. 53. *Pseudomonas aeruginosa.* (From Hagan, 1943.)

ism taken up in the larval stage persisted through the pupal stage and could be obtained from the feces of the mature flies. The organism was also isolated from the outside of the eggs of the flies.

Metalnikov (1920a, 1927), in a series of experiments to determine the immunity of the wax moth to certain classes of microorganisms, found that the wax moth was very susceptible even to small doses of *Bacillus pyocyaneus*. Cameron (1934) also found that the organism was very pathogenic for the wax moth.

Sawamura (1906) includes *Pseudomonas aeruginosa* among the organisms that experimentally have been shown to be pathogenic for the silkworm. However, Couvreur and Chahovitch (1921) found that the blood and digestive juices of the larvae and pupae of *Bombyx mori* destroyed the organism.

Duncan (1926) isolated it from the gut contents of one lot of *Stomoxys calcitrans*.

During a laboratory epidemic among *Schistocerca gregaria*, Lepesme (1937a) isolated *Pseudomonas aeruginosa* from the body fluid. Experimentally his strain caused death in one to two days. Later (1938) he found that *Pseudomonas aeruginosa* was a secondary invader to the infestation of the fungus, *Aspergillus flavus*, in this grasshopper.

Pseudomonas fluorescens Migula

Insects concerned: The honeybee, *Apis mellifera;* the roach, *Blatta orientalis;* the seed-corn maggot, *Hylemya cilicrura;* the housefly, *Musca domestica; Calliphora vomitoria; Sarcophaga carnaria; Lucilia caesar; Melolontha vulgaris; Blaps mucronata; Euxoa segetum; Pimelia bifurcata;* and *Pimelia sardea.*

Cao (1898, 1906a,b) isolated *Bacillus fluorescens liquefaciens* (*Pseudomonas fluorescens*) from the feces of a series of oriental roaches. The strains isolated varied in virulence for guinea pigs, most of them being avirulent. The strains reisolated from the guinea pigs were fed to another series of roaches which had been without food for about 45 days. The feces of these roaches yielded strains of *Pseudomonas fluorescens* which killed guinea pigs in five to seven days. Cao also found that the diet of the roaches caused a variation in the virulence of the strains of the bacteria recovered from the feces.

Cultures of *Pseudomonas fluorescens* were also isolated by Cao from the feces of several Coleoptera, including *Blaps mucronata, Pimelia bifurcata, Pimelia sardea,* and *Melolontha vulgaris,* a burrowing beetle. He found a constant relationship between the various bacteria existing in the earth and the intestinal contents of the larvae of *Melolontha vulgaris*.

Cao (1906b) fed the larvae of *Musca domestica, Calliphora vomitoria, Sarcophaga carnaria,* and *Lucilia caesar* on flesh containing *Bacillus fluorescens liquefaciens* (*P. fluorescens*) and found it to be the predominant organism in the feces of the larvae. The organism taken up in the larval stage persisted through the pupal stage and could be obtained from the feces of the mature flies. The organism was also isolated from the outside of the eggs of the flies.

According to Leach (1931), the bacterial flora of normal larvae and adult *Hylemya cilicrura* frequently includes bacteria closely resembling *Pseudomonas fluorescens*.

White (1906) isolated *Pseudomonas fluorescens liquefaciens* (presumably *Pseudomonas fluorescens*) from the intestine of the normal honeybee.

Pospelov (1927) isolated *Bacillus fluorescens liquefaciens* (*Pseudomonas fluorescens*) and *Bacillus fluorescens non liquefaciens* (*Pseudomonas non-liquefaciens*) from the larvae of cutworms, *Euxoa segetum,* which had died with symptoms of a septicemia.

Pseudomonas jaegeri Migula

Insects concerned: The housefly, *Musca domestica; Lucilia caesar; Calliphora vomitoria;* and *Sarcophaga carnaria.*

Bacillus proteus fluorescens (*Pseudomonas jaegeri*) was isolated by Cao (1906b) from the intestines of several different fly larvae. His strain possessed a notable pathogenicity for guinea pigs.

Pseudomonas non-liquefaciens Bergey et al.

Insects concerned: The roach, *Blatta orientalis;* the seed-corn maggot, *Hylemya cilicrura;* the housefly, *Musca domestica; Calliphora vomitoria; Lucilia caesar; Sarcophaga carnaria; Blaps mucronata; Euxoa segetum; Pimelia bifurcata;* and *Pimelia sardea.*

Cao (1898, 1906a) isolated several strains of *Bacillus fluorescens non-liquefaciens* (*Pseudomonas non-liquefaciens*) from the feces of *Blatta orientalis.* He studied the variation of the virulence of the organism after passing it through the cockroach a second time. In some cases it became virulent for guinea pigs, killing in 48 to 54 hours, while in others the organism remained avirulent. He also varied the diet of the roach to see what effect was had on the virulence.

Pseudomonas non-liquefaciens was fed by Cao (1898, 1906a,b) to larvae of *Musca domestica*, *Calliphora vomitoria*, *Lucilia caesar*, *Sarcophaga carnaria*, *Blaps mucronata*, *Pimelia bifurcata*, and *Pimelia sardea*. In all cases the organism was isolated from the feces.

Leach (1931) found that the bacterial flora of normal larvae and adult *Hylemya cilicrura* frequently includes bacteria closely resembling *Pseudomonas non-liquefaciens*.

Pseudomonas ovalis Chester

Insect concerned: The Colorado potato beetle, *Leptinotarsa decemlineata*.

Steinhaus (1941a) found this bacterium in the alimentary canal of the larvae of the Colorado potato beetle. This organism is frequently found in the soil and hence may easily become associated with this insect.

Pseudomonas septica Bergey *et al.*

Insects concerned: *Euxoa segetum;* the firefly, *Photinus pyralis;* and the Colorado potato beetle, *Leptinotarsa decemlineata*.

Stutzer and Wsorow (1927) isolated from *Euxoa segetum* an organism which they named *Bacillus fluorescens septicus*. They thought it was one of two agents that caused a "spring disease" among the caterpillars in 1925. Experimentally they were able to produce the disease by infecting the insects through the damaged integument. It was thought that the infection was brought about in a similar manner when the caterpillars were in the earth.

Steinhaus (1941a) found an organism similar to *Pseudomonas septica* in normal fireflies and potato beetles. It was the only organism he cultured from both the alimentary tract and ground-up specimens of the firefly. He isolated it from the alimentary tract of the potato beetle.

Genus: XANTHOMONAS (= PHYTOMONAS) [1]

Xanthomonas angulata (Fromme and Murray)
(*Bacterium angulatum*)

Insects concerned: Flea beetles, garden leafhoppers, and probably the tobacco hornworm, *Protoparce sexta*.

Although the true means of the dissemination of angular leaf spot of tobacco has not been entirely ascertained, the causative bacterium (*Xanthomonas angulata*) has been isolated from "flea beetles" and "garden leaf hoppers" by Valleau and Johnson (1936). Some evidence was obtained by Johnson (1934) that caterpillars of *Protoparce sexta* might spread it mechanically on a leaf.

Xanthomonas campestris (Pammel)

Insect concerned: The cabbage looper, *Autographa brassicae* (= *Plusia brassicae*).

Pammel (1895) first isolated *Bacillus campestris* (*Xanthomonas campestris*) from diseased rutabagas. Two years later Smith (1897) showed experimentally that this organism, which causes black rot of crucifers, could be transmitted by slugs, *Agriolimax agrestis*, and the cabbage looper, *Plusia brassicae*. According to Leach (1940), no actual proof of the insect transmission of this disease in the field has, as yet, been forthcoming.

Xanthomonas coronofaciens (Elliott)

Insect concerned: Aphids.

Bacterium coronofaciens (*Xanthomonas coronofaciens*) is the cause of a halo spot on oats. Aphids have been suspected of carrying the disease from

---

[1] The generic name *Phytomonas*, although in the past widely used for this group of bacteria, was earlier used for a genus of protozoa. For this reason many workers now prefer the name *Xanthomonas* as proposed by Dowson (1939).

infected plants to healthy ones. It is probable, however, that these insects play only a minor part in the spread of the disease.

Xanthomonas maculicola (McCulloch)

Insect concerned: The red-bordered stinkbug, *Euryophthalmus convivus*.

Xanthomonas maculicola, the cause of spot disease of cauliflower, was isolated by Goldsworthy (1926) from the extremities of the red-bordered stinkbug. He concluded that the insect was a common agent of dissemination, but he gave no experimental proof.

Xanthomonas medicaginis var. phaseolicola (Burkholder)

Insect concerned: The thrips, *Hercinothrips femoralis* (= *Heliothrips femoralis*).

This bacterium was isolated from leaves, pods, and stems of beans afflicted with halo blight. Buchanan (1932) showed that this bacterial disease of beans was transmitted by the thrips *Heliothrips femoralis*. The bacterial lesions on the plants are always associated with the feeding wounds of the insect. According to Leach (1940), the transmission appears to be incidental and entirely mechanical, and under field conditions the insect is probably of little importance as a vector of the disease.

Xanthomonas melophthora (Allen and Riker)

Insect concerned: The apple maggot, *Rhagoletis pomonella*.

Xanthomonas melophthora causes decay of apples and is found associated with the apple maggot, *Rhagoletis pomonella* (Allen and Riker, 1932). Besides being associated with the larvae, this bacterium has been found both in and on male and female flies. The organism also has been found on, but not in, the eggs, and no bacteria have been isolated from the inside of the puparia, although they are abundant on the surface.

Leach (1940) suggested that perhaps the development of the larvae may depend on the bacterial growth in the apples, for the maggots prefer the decayed tissue.

Xanthomonas pseudotsugae (Hansen and Smith)

Insect concerned: *Chermes cooleyi*.

Xanthomonas pseudotsugae was first isolated from galls on Douglas fir (*Pseudotsuga taxifolia*) in California by Hansen and Smith (1937) and called *Bacterium pseudotsugae*. The infection depends on deep wounds, which fact suggests transmission by an insect vector. Strong circumstantial evidence incriminates *Chermes cooleyi*, a sucking insect.

Xanthomonas saliciperda (Lindeijer)

Insect concerned: The willow borer, *Cryptorhynchus lapathi*.

Lindeijer (1932) described a bacterial disease of willows (*Salix* spp.) caused by *Pseudomonas saliciperda* (*Xanthomonas saliciperda*).

According to Leach (1940), the disease causes a wilt of the branches followed by early defoliation and death of the affected limbs. Natural infections most frequently originate at the site of wounds made by the willow borer, which, after having been contaminated with the bacteria, infects the tree. The disease has been experimentally produced by allowing infected insects to feed on willow twigs.

Xanthomonas savastanoi (Smith)

Insect concerned: The olive fly, *Dacus oleae*.

Xanthomonas savastanoi gives rise to a disease of olive trees on which "knots" or galls result from the infection. The disease is prevalent in Italy and other southern European countries, and Smith isolated cultures from olive galls collected in California where the disease has been known since 1898.

In Italy there appears to be a close association between the olive fly, *Dacus oleae*, and the spread of the disease. Petri (1909, 1910) studied the relationships between this insect and bacteria found in the intestinal tract of the insect. *Ascobacterium luteum* (which see) is one of the nonpathogenic bacteria which Petri found occurring as a "symbiote" in the olive fly. From the four blind appendages of the middle stomach of the larvae of *Dacus oleae*, he (1910) isolated *Bacterium savastanoi* (*Xanthomonas savastanoi*). According to Buchner (1930), the latter organism is the real symbiote associated with the olive fly. The bacteria are transmitted through the egg and persist in the puparium. Petri suggested that the physiological role of the bacteria in the digestive tract is probably connected with the feeding habit of the larva, which bores in olives, a food which is rich in fats. The larva has to ingest very large quantities of oil in order to extract enough nitrogenous substances for its development. The bacteria in the digestive tract may be useful in breaking down the fats and releasing the nitrogen.

Xanthomonas sepedonica (Spieckermann)

Insects concerned: The potato beetle, *Leptinotarsa decemlineata;* the differential grasshopper, *Melanoplus differentialis;* and the black blister beetle, *Epicauta pennsylvanica.*

Xanthomonas sepedonica is the cause of ring rot of potatoes, and insects have been suspected as carriers of the infection. List and Kreutzer (1942) made a series of studies and obtained evidence that the potato beetle, grasshoppers, and the black blister beetle can transmit the infection, though the results were not taken as definite proof that insects are a factor in the field spread of ring rot. It has been recently proposed that the name of this organism be changed to *Corynebacterium sepedonicum.*

Xanthomonas solanacearum (Smith)

Insect concerned: The potato beetle, *Leptinotarsa decemlineata.*

In warm, moist climates this bacterium attacks potatoes, tobacco, tomatoes, peppers, and other related plants. According to Leach (1940), the bacteria are found first in the vascular bundles, but eventually they enter the parenchyma cells of the cortex and pith. After spreading through the vascular bundles of the stolons, they reach the tubers where they decay the storage tissues.

E. F. Smith (1896) incriminated the Colorado potato beetle as a disseminator of the disease on the basis of greenhouse experiments. He named the bacterium *Bacillus solanacearum,* though it is now placed in the genus *Xanthomonas.*

Xanthomonas stewarti (Smith)

Insects concerned: The flea beetles, *Chaetocnema denticulata, Chaetocnema confinis,* and *Chaetocnema pulicaria;* the corn rootworms, *Diabrotica duodecimpunctata* and *Diabrotica longicornis; Phyllophaga* sp.; *Disonycha glabrata; Stilbula apicalis; Euscelis bicolor; Thamnotettix nigrifrons;* and *Macrosiphum solanifolii* (= *Illinoia solanifolii*).

Xanthomonas stewarti (often called *Aplanobacter stewarti*) is a bacterium which gives rise to a wilt in corn (*Zea mays*). Sweet corn seems to be the most susceptible, although, according to Leach (1940), teosinti (*Euchlaena mexicana*) and *Tripsacum dactyloides* are known to be susceptible. The organism is essentially a vascular pathogen, although other tissues are frequently affected.

Rand and Cash (1933) during 1920–1923 were the first to show that bacterial wilt of corn could be spread by insects. They demonstrated that two species of flea beetles, *Chaetocnema pulicaria* and *C. denticulata,* were responsible for the secondary spread of the wilt in midsummer, and later showed that early seasonal occurrence of the disease was due to insects such as the corn rootworms working at the base or roots of the corn.

Poos and Elliott (1936) collected 7,338 insects representing 40 species in 33 genera in and around corn fields and tested them for the presence of *Xanthomonas stewarti*. The organism was isolated from the following eight species: *Chaetocnema pulicaria, C. denticulata, C. confinis, Disonycha glabrata, Stilbula apicalis, Euscelis bicolor, Thamnotettix nigrifrons,* and *Macrosiphum solanifolii*. Excepting the *Chaetocnema* spp., *Xanthomonas stewarti* was obtained only once from the insects. Elliott and Poos (1940) later found that *Chaetocnema pulicaria* was the only insect of any importance in harboring the bacterium over winter and in spreading it during the growing season, though *C. denticulata* ranked next in the number of adults that yielded cultures of the organism.

Tribe: SPIRILLEAE

Genus: VIBRIO

Vibrio colerigenes

Insects concerned: The roach, *Blatta orientalis; Blaps mucronata; Pimelia bifurcata; Pimelia sardea;* and *Tentyria* sp.

Cao (1898) found that *Vibrio colerigenes* passed unharmed through the intestines of *Blatta orientalis, Pimelia bifurcata, Pimelia sardea, Blaps mucronata,* and *Tentyria* sp.

Vibrio comma (Schroeter) Bergey *et al.*

Insects concerned: The housefly, *Musca domestica;* the bluebottle fly, *Calliphora vomitoria;* the drone fly, *Eristalis tenax;* the cockroach, *Periplaneta americana;* the wax moth, *Galleria mellonella;* and the fly, *Aphiochaeta ferruginea.*

The specific relationship of *Vibrio comma* to cholera was discovered by Robert Koch in 1883, when he isolated the organism from the intestinal contents of cholera patients. Nicholas (1873) was the first to connect the prevalence of cholera with flies. Maddox (1885) found the cholera organisms in the feces of *Eristalis tenax* and *Calliphora vomitoria* after having fed the flies on cultures of the vibrios. Cattani (1886) caught flies in cholera wards and after several hours obtained cultures of *Vibrio cholera* (see also Tizzoni and Cattani, 1886). During the following 25 years several investigators (Simmonds, 1892; Macrae, 1895; Hamilton, 1903; Faichnie, 1909a; Ledingham, 1911; and Nicoll, 1911) conducted similar experiments, and in nearly every case they found the bacteria to be taken up by flies (including *Musca domestica*). Graham-Smith (1913) found that flies fed on old laboratory cultures passed infected feces for 30 hours. The bacteria soon died on the legs and wings, however, and after 48 hours cultures made from the intestines of the flies were negative. For a discussion of other early experiments on flies as vectors of cholera see Nuttall (1899), Howard (1911), and Graham-Smith (1913). According to Herms (1939), Gill and Lal in 1931 found evidence that possibly one phase of the life cycle of the cholera vibrio is passed in the body of the housefly. According to these workers, the bacteria disappear from the body of the fly after approximately 24 hours but reappear about the fifth day,

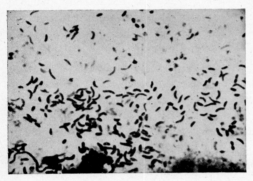

Fig. 54. *Vibrio comma.* (Photo by N. J. Kramis.)

when the fly is capable of contaminating food with its feces. This work, however, has not been definitely confirmed by other workers.

Roberg (1915) has pointed out the danger of the spread of cholera vibrios by minute flies (e.g., *Aphiochaeta ferruginea*) which breed in human feces. He also observed that cholera vibrios under certain circumstances may be transmitted from larvae, through pupae, into the emerging imagines.

Barber (1914) and Toda (1923) have shown that cockroaches may be a factor in the spread of cholera.

Metalnikov (1920a, 1927) found that *Vibrio comma* was sometimes pathogenic for the wax moth. Cameron (1934), however, found that it was non-pathogenic in all his experiments with this insect.

Vibrio leonardi Metalnikov and Chorine

Insects concerned: The wax moth, *Galleria mellonella,* and the corn borer, *Pyrausta nubilalis.*

Metalnikov and Chorine (1928a) isolated *Vibrio leonardi* from diseased corn-borer larvae, and found it to be very pathogenic to both corn-borer and wax-moth larvae. They found the insects very susceptible to infection by mouth, dying within 24 hours.

Vibrio metschnikovii Gamaleia

Insect concerned: The cockroach, *Blatta orientalis.*

Cao (1906a) experimented with this organism, feeding it to roaches to determine variability in its virulence for guinea pigs. He found that it passed unchanged through the intestines of the roach.

Vibrio pieris Paillot

Insect concerned: *Pieris brassicae.*

Paillot (1933) refers to this organism as having been frequently encountered in the caterpillars of *Pieris brassicae* which had been parasitized by larvae of *Apanteles glomeratus.*

Family: RHIZOBIACEAE

Genus: ALCALIGENES

Alcaligenes ammoniagenes Castellani and Chalmers

Insects concerned: The imperial moth, *Eacles imperialis;* the blister beetle, *Epicauta pennsylvanica;* and *Urographis fasciata.*

Bacteria very similar to *Alcaligenes ammoniagenes* were found in the imperial moth, the blister beetle, and *Urographis fasciata* by Steinhaus (1941).

Alcaligenes faecalis Castellani and Chalmers

Insect concerned: A fly, probably the housefly, *Musca domestica.*

Torrey (1912) isolated *Bacillus faecalis alkaligenes* (*Alcaligenes faecalis*) from the intestinal tract of city-caught flies.

Alcaligenes stevensae Brown

Insect concerned: *Malacosoma americana.*

Brown (1927) isolated this organism from crushed egg masses of the moth, *Malacosoma americana.* According to Brown, the organism is probably allied to *Alcaligenes bronchisepticus.*

Genus: CHROMOBACTERIUM

Chromobacterium violaceum (Schroeter) Bergonzini

Insect concerned: The roach, *Blatta orientalis.*

Longfellow (1913) was able to cultivate *Bacillus violaceus* (*Chromobacterium violaceum*) from the feces of the oriental roach.

Genera Incertae Sedis Vel Dubia

Genus: Ascobacterium

Ascobacterium luteum Babes

Insect concerned: The olive fly, *Dacus oleae*.

Petri (1910) found *Ascobacterium luteum* in the intestinal tract of the larva of the olive fly. He suggested that the presence of the bacteria might be favorable for the digestion of the olive oil since *Dacus oleae* larvae bore into olives and necessarily ingest very large quantities of oil. This may be especially true of *Ascobacterium luteum*, which is found in large numbers of larvae living on ripe and therefore oily fruit.

Apparently, the following organisms have no relation to the organism which Petri isolated. *Bacteridium luteum* was described about 1872 and is commonly found in air and water; it is now known as *Micrococcus luteus*. Adametz described in 1885 *Bacterium luteum*, which he isolated from the stomach contents of sheep and from water.

Genus: Coccobacillus

Coccobacillus acridiorum D'Herelle

Insects concerned: *Schistocerca pallens; Schistocerca peregrina; Schistocerca paranensis; Melanoplus femur-rubrum; Melanoplus bivittatus; Melanoplus atlantis; Tmetis muricatus; Oedalens nigrofasciatus; Arcyptera flavicosta; Locusta migratoroides; Atta sexdens; Dociostaurus maroccanus; Pezomachus botrana; Stauronotus maroccanus; Solenopsis gemminata; Bombyx mori; Zonocerus elegans; Caloptenus* sp.; and *Tropidacris dux*.

Coccobacillus acridiorum was first isolated by D'Herelle (1911) in Yucatan, Mexico, from a species of locust, *Schistocerca pallens* (fig. 55). While in Mexico, D'Herelle noticed a heavy mortality occurring in these destructive and very abundant insects, which had arrived from the borders of Guatemala. From 1909 to 1911 the epizoötic occurred so extensively that by 1912 it had reduced the number of locusts to such an extent that no invasion into Mexico occurred. D'Herelle succeeded in isolating the bacterium respon-

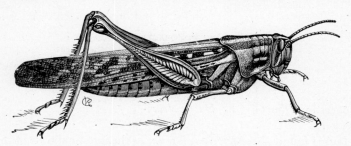

Fig. 55. *Schistocerca pallens*. (From Comstock, 1936; after Riley.)

sible for the epizoötic from the intestinal contents of dead locusts, the same organism not occurring in healthy specimens. D'Herelle found further that locusts artificially inoculated with this organism died with the characteristic symptoms. However, Mereshkovsky (1925), Pospelov (1926), and others have expressed the belief that *Coccobacillus acridiorum* is a normal "symbiont" of the blood of locusts, which can, under certain conditions of temperature and humidity, become a virulent parasite.

Coccobacillus acridiorum is a pleomorphic bacillus with coccoid forms (0.4 to 0.6 microns) and bacillary forms (0.9 to 1.5 microns) appearing in the same culture. In young cultures and in the intestinal contents of the locust the coccoid forms are the most abundant. Many cultures have been considered to be *Coccobacillus acridiorum* when in reality they were not. This fact has been brought out by Glaser (1918a) in a systematic study of the organisms placed under the name of *Coccobacillus acridiorum* D'Herelle.

A careful study of the published bacteriologic descriptions of *Coccobacillus acridiorum* suggests that in reality the bacterium is one of a group which modern systematists designate as coliform bacteria. Some strains which have been considered to be *C. acridiorum* may be paracolon intermediates.

Since D'Herelle's early successes some investigators have been able to confirm his results while others have not. For one thing, there seems to be a difference in susceptibility of different locusts to the disease. The bacteria appear to be more effective against locusts belonging to the genus *Schistocerca* than against other genera of the Locustidae. Species of the genus *Caloptenus* in Argentina and *Stauronotus maroccanus* in Algeria have been found to be experimentally susceptible. D'Herelle apparently was successful in combating plagues of *Schistocerca paranensis* in Argentina. In 1915 D'Herelle controlled an outbreak of *Schistocerca peregrina* in Tunisia by a combination of mechanical methods and the use of his organism. Experimentally, the organism was found to be pathogenic for *Schistocerca paranensis* and *Tropidacris dux* in Trinidad (Rorer, 1915). Glaser (1918a) found that a certain strain of *Coccobacillus acridiorum* ("Souche Cham") was pathogenic to *Melanoplus atlantis,* and to *Melanoplus bivittatus* and *Melanoplus femur-rubrum* to a lesser degree. Another strain ("Souche Sidi") was less pathogenic to *Melanoplus atlantis* and *Melanoplus bivittatus* than the "Souche Cham" strain.

Among the insects which were slightly or not at all susceptible to wholesale destruction by *Coccobacillus acridiorum* were: *Locusta migratoroides* (Mackie, 1913), *Zonocerus elegans* (Lounsbury, 1913), and *Oedalens nigrofasciatus* (Barber and Jones, 1915).

Among other insects susceptible to this bacterium are several species of ants and crickets. Fowls, guinea pigs, rabbits, cows, sheep, and man are refractory to infection, but the sewer rat dies in three to four days after a subcutaneous inoculation with *Coccobacillus acridiorum*.

Coccobacillus cajae Picard and Blanc

Insects concerned: *Arctia caja; Poecilus koyi; Epacromia strepens;* the oriental roach, *Blatta orientalis; Eurydema ornata; Temnorrhinus mendicus* (= *Cleonus mendicus*); *Chrysomela sanguinolenta; Anoxia australis; Melolontha vulgaris; Opartum sabulosum; Cetonia aurata; Porthesia chrysorrhoea;* the silkworm, *Bombyx mori; Acridium aegyptium; Hydrophilus* sp.; *Dytiscus* sp.; *Cybister* sp.; *Notonecta* sp.; *Nepa* sp.; and *Ranatra* sp.

The caterpillars of *Arctia caja,* which were extremely abundant in the vineyards of southern France, were almost completely wiped out by two diseases. One disease was caused by the fungus, *Empusa aulicae,* the other by a bacterium, *Coccobacillus cajae*. The organism, which is apparently allied to *Coccobacillus acridiorum* D'Herelle, was obtained by Picard and Blanc (1913a) from the blood of diseased caterpillars, and the disease was experimentally reproduced.

Coccobacillus cajae is motile, gram-negative, and shows bipolar staining. Unlike *Coccobacillus acridiorum,* it is a parasite of the blood of the caterpillar and not of the digestive tube. When caterpillars are given by ingestion a few drops of a culture by means of a pipette introduced into the pharynx, they die after a few hours from a septicemia.

Besides the larvae of *Arctia caja,* Picard and Blanc (1913b) found *Cocco-*

bacillus cajae to be pathogenic to the following insects: Coleoptera—*Poecilus koyi, Opatrum sabulosum, Cetonia aurata, Melolontha vulgaris, Anoxia australis, Chrysomela sanguinolenta,* and *Temnorrhinus mendicus;* Hemiptera—*Eurydema ornata;* Orthoptera—*Blatta orientalis, Epacromia strepens,* and *Acridium aegyptium;* Lepidoptera—*Porthesia chrysorrhoea* and *Bombyx mori.* According to Picard and Blanc, this list could probably be extended indefinitely since it is likely that the majority of insects are killed by *Coccobacillus cajae.* It is interesting to note, however, that the aquatic beetles are among those which are immune: Coleoptera—*Hydrophilus, Dytiscus,* and *Cybister;* Hemiptera—*Notonecta, Nepa,* and *Ranatra.* While the white rat is immune, the tree frog, *Hyla arborea,* dies of septicemia in about two days, its blood containing numerous *Coccobacilli.*

Various workers have referred to this bacterium as *"Bacillus cajae"* (Mareschal, 1914; Picard, 1914; Paillot, 1933).

Coccobacillus ellingeri Metalnikov and Chorine

Insects concerned: The European corn borer, *Pyrausta nubilalis;* the wax moth, *Galleria mellonella;* and *Ceratitis capitata.*

This organism was first isolated in November, 1927, by Metalnikov and Chorine (1928a,b), who found it repeatedly in diseased corn borers. Larvae of *Pyrausta nubilalis* and *Galleria mellonella* inoculated with a "tiny dose" of *Coccobacillus ellingeri* died in the course of 2 to 12 hours. Corn borer larvae were very susceptible to infection through the intestinal tract, the bacteria passing through the wall of the intestine into the blood, where they were found in great numbers. The bacterium has no pathogenic effect on guinea pigs or rabbits.

According to its discoverers, *"Coccobacillus ellingeri* somewhat resembles *Bacterium sphingidis* White, *Bacterium noctuarum* White, and *Bacterium melolonthae liquefaciens alpha* Paillot. It differs, however, from these three species by being non-motile. There are also other minor differences."

Citrus fruits infested by *Ceratitis capitata* in Sicily were found by Ciferri (1934) to be "infected" by a bacterium, apparently *Escherichia (Coccobacillus) ellingeri.* It appeared to be a "symbiont" but could become pathogenic to the larva under certain conditions.

In the third edition of *Bergey's Manual* (1930, p. 330) this organism is referred to as *Escherichia ellingeri* Bergey *et al.* In the fifth edition (1939, p. 606), however, it was designated by its original name, *Coccobacillus ellingeri.*

Coccobacillus gibsoni Chorine
(See also *Bacillus gibsoni*)

Insect concerned: The corn borer, *Pyrausta nubilalis*.

Chorine (1929a,b) isolated this bacterium from sick corn-borer larvae which he had received from Canada. He described this organism as being a very pleomorphic, motile, nonsporeforming, gram-negative bacterium. The organism was extremely virulent for the corn-borer larvae. The borers could be infected by injection and *per os*.

Coccobacillus insectorum var. malacosomae Hollande and Vernier

Insects concerned: *Malacosoma castrensis; Malacosoma neustris;* and *Vanessa urticae*.

Hollande and Vernier (1920) found the blood of 50 per cent of the caterpillars of *Malacosoma castrensis* examined to be infected with a new organism which they named *Coccobacillus insectorum* var. *malacosomae*.

Genus: DIPLOBACILLUS

Diplobacillus melolonthae Paillot

Insects concerned: The cockchafer, *Melolontha melolontha;* the gypsy moth, *Porthetria dispar;* the silkworm, *Bombyx mori; Nygmia phaeorrhoea;* and *Vanessa urticae*.

Diplobacillus melolonthae was isolated by Paillot (1917) from the larvae of the cockchafers. He (1922) describes it as being a small, nonmotile, gram-positive rod occurring singly, in pairs, and in short chains. In the blood of the cockchafer it becomes abnormally elongated. While the bacterium is pathogenic for the larvae of *Melolontha melolontha* and *Vanessa urticae*, those of *Porthetria dispar, Nygmia phaeorrhoea,* and *Bombyx mori* offer it a considerable amount of resistance.

Diplobacillus pieris Paillot

Insect concerned: The white cabbage butterfly, *Pieris brassicae*.

Paillot (1919b) isolated *Diplobacillus pieris* from the larvae of diseased white cabbage butterflies. From the same source he isolated eight other bacteria. (See *Bacillus pieris agilis*.) He considered these bacteria to be secondary invaders, the parasite, *Apanteles glomeratus*, being the predisposing factor in the bacterial infection of the white cabbage butterfly.

Genus: ENTEROCOCCUS

Enterococcus citreus Stutzer and Wsorow

Insect concerned: *Euxoa segetum*.

Stutzer and Wsorow (1927) isolated this organism from normal pupae of *Euxoa segetum*.

Genus: GYROCOCCUS

Gyrococcus flaccidifex Glaser and Chapman

Insect concerned: The gypsy moth, *Porthetria dispar*.

Glaser and Chapman (1912) isolated this organism from caterpillars of the gypsy moth during their studies on the cause and nature of the wilt disease occurring in this insect. These investigators originally believed that this organism was the cause of the wilt disease. It was later discovered, however, that nearly all the insects used for experimentation had become "accidentally infected" with this bacterium. In 1913 Glaser and Chapman (see Glaser, 1928) corrected their mistake regarding the cause of wilt disease and showed it to be caused by a filterable virus.

Gyrococcus flaccidifex was described as a small (0.5 to 0.85 microns), gram-negative, encapsulated organism, resembling the pneumococcus more than any other form except that it was motile, progressing in a "gyrating" manner. The generic name was derived from the latter characteristic.

Genus: LEPTOTRIX
(*Leptothrix*)

Leptotrix buccalis
(*Leptothrix buccalis?*)

Insect concerned: The mosquito, *Anopheles maculipennis*.

According to Keilin (1921c), who quoted Howard, Dyar, and Knab (1912), Perroncito in 1899 discovered a bacterial parasite of *Anopheles maculipennis* resembling *"Leptotrix buccalis."* "The parasite infests the larva, passes into the pupa, and destroys the imago soon after it emerges" (Keilin). From the information given it is difficult to ascertain whether this organism is the same as the one now known as *Leptotrichia buccalis* or not.

CHAPTER IV

INTRACELLULAR BACTERIUMLIKE AND RICKETTSIALIKE SYMBIOTES

I. GENERAL ASPECTS

STRANGE as it may seem, the tissue cells of many normal insects and ticks harbor living microorganisms which for the most part exert no harmful effects upon these cells. In fact, some of them may be distinctly beneficial to their hosts, carrying out their part of a mutually helpful relationship. To such microorganisms many authors have limited the use of the words "symbionts" or "symbiotes." Actually the words may have a much broader meaning.

Before going further a definition of terms is necessary. We refer particularly to "symbiosis," "symbiont," and "symbiote."

Webster's Dictionary defines "symbiosis" as the "living together in intimate association or even close union of two dissimilar organisms"; and points out that the term is ordinarily used where the association is advantageous. Following the recommendation of the Committee on Terminology of the American Society of Parasitologists (*J. Parasitol.*, 1937, **23**, 326–329), we have chosen to use the word as De Bary (1879) originally used it, that is, as a general term referring simply to the *living together of dissimilar organisms,* not excluding parasitism or commensalism. The term *mutualism* will be used to refer to the relationship between organisms which live in a mutually advantageous association.

As to the use of "symbiont" and "symbiote," the Committee on Terminology states:

Symbiont is the form coined by de Bary, Webster's Dictionary gives *symbiont* as the preferred form (derived from a participle of the corresponding Greek verb), whereas symbiote is listed as a synonym or variant. Meyer (1925) and others have maintained that symbiote is the correct form. The Committee has consulted a member of the Department of the Classics, Harvard University, who has informed us that *symbiote* is derived from the Greek *sumbiotes,* meaning "one who lives with," "companion," "partner," whereas symbiont has no Greek original. He has stated further that the philologist would prefer symbiote which has a definite Greek original and is correctly formed in English. It may be pointed out that symbiote is the common form in French literature, whereas most German, British and American writers use de Bary's symbiont, which possesses whatever advantage priority is able to cast into the etymological balance. The matter is apparently one of taste and usage rather than correctness.

The Committee adds:

The terms symbiont or symbiote are applied to the members of a symbiotic association and may properly be used for either member, though it has become the custom to refer to the smaller as the *symbiont* or *symbiote* and to the larger as the *host*. [The term *microsymbiote* is sometimes used to avoid confusion and to designate the microorganism.]

It may be pointed out that in the literature of the intracellular symbionts of insects, which comprise a large number of unnamed and unclassified microorganisms, there is a tendency to consider *symbiont* as possessing systematic connotations. For example, one author has objected to the designation of the rickettsia-like organisms of the bedbug as symbionts, maintaining that they cannot be rickettsiae and symbionts at the same time. It is obvious that symbiont connotes only a mode of life.

After careful consideration of the Committee's statements, the writer is inclined to accept the preference of the philologist and use the term "symbiote" except, of course, in quoting from the works of those who use the term "symbiont." In other words, *for our purpose, in referring to the microorganisms associated with insects and ticks,* we shall use "symbiote" as meaning a microorganism which lives in association with an insect or a tick. It may include bacteria, rickettsiae, protozoa, or other forms of microbial life. The term usually refers to a microorganism living in an intimate and rather constant or continuous association with its host. Used in the broad sense, however, it may include even temporary associations.[1]

Early History. One of the first to recognize the possible presence of intracellular microorganisms in insects was Leydig (1850, 1854), who found small lanceolate bodies in the lymph of *Lecanium hesperidum*. Though he observed the multiplication of these yeastlike forms by budding, he apparently did not realize their significance. Similarly, Putnam (1880) observed the symbiote (a

---

[1] Moshkovsky (1945) uses the term "cytotrope" or "cytotropic agent" to designate those microorganisms and viruses which live in an intimate and obligatory relationship with the cells of their hosts. In some respects this term is probably preferable to the word "symbiote," as it applies to most of the microorganisms considered in this chapter, since it avoids the implication of a definite mutually beneficial relationship.

fungus) in *Pulvinaria innumerabilis,* though he was not entirely convinced of its organismal nature. In 1886 Blochmann noticed that bacterialike bodies were present in both the eggs and the follicular membranes of ants and wasps. He suspected that these forms were bacteria. Later (1887, 1888) he expressed greater confidence in his belief as to the nature of these bodies, which he also found in the fat tissue and eggs of *Blatta orientalis* and *Blattella germanica.* Blochmann's contention that they were bacteria was based upon their reactions to various reagents and stains, their multiplication by fission, and their method of infection through the eggs. His attempts to cultivate the "organisms" in beef-peptone-gelatin agar were unsuccessful. Shortly after Blochmann reported his findings, Wheeler (1889) observed intracellular bacterialike bodies in tissues of *Blattella germanica,* and Cholodkowsky (1891) and Heymons (1895) confirmed these observations.

Among the first insects studied with respect to their intracellular organisms were aphids. In 1850 Leydig observed certain organs in aphids which have subsequently been called "symbiotic organs," "pseudovitelli," "green bodies," "bacteriotomes," and "mycetoms" or "mycetomes." Following Leydig's report, the nature, origin, and development of these structures were described by many workers including Huxley (1858), Balbiani (1866 to 1871), and Tannreuther (1907). These "organs" are made up of cells usually known as "mycetocytes" or "bacteriocytes," and it is in these that the intracellular forms are usually located. The terms "mycetome" and "mycetocyte" or "bacteriocyte" are now also applied to similar structures in other insects.

One of the first to be interested in the intracellular organisms in aphids was Krassilstschik (1889, 1890). The work of this investigator and others has been briefly reviewed by Uichanco (1924). According to the latter, Krassilstschik's investigations were concerned with 20 species of aphids. He observed bacilli in the insects' tissues and suspected that they might have some relation to the "symbiotic organs." He noted that the "organisms" were located only between the layer of fat cells and the pseudovitellus. Apparently he did not recognize the cytoplasmic inclusions of the mycetocytes as microorganisms. Henneguy (1904) stated that the green, granular mass in the eggs of aphids consisted of microorganisms. However, the organismal nature of these bodies was not definitely recognized until reports appeared by Pierantoni (1909, 1910a,b, 1911) and Šulc (1910a). Principally on the basis of morphologic characteristics, they believed the organisms to be related to yeasts.

About the same time that these observations were being made on aphids, investigations were in progress concerning the symbiotes in other insects. Prominent among these were the investigations of Buchner (1912, 1921a,b,c) concerning symbiosis in aleyrodids, coccids, aphids, psyllids, cicadellids, blattids, Cicada, Hymenoptera, Hemiptera, and Coleoptera. Later (1930) Buchner enlarged upon this with an excellent treatise covering the entire field up to that time. Other reviews of the subject have been presented by Glaser (1930c), Paillot (1933), and Steinhaus (1940a). Comprehensive work on the intracellular microorganisms (excluding the true rickettsiae) in ticks began with that of Cowdry (1923, 1925a) and Mudrow (1932), though early observations on ticks of the genus *Rhipicephalus* were made by Robert Koch.

Despite the numerous but scattered investigations, this subject is still largely a virgin field. The intimate relationships existing between the intracellular microorganisms and their hosts offer a challenge to the investigator. The early investigations were marked by numerous unsuccessful attempts to cultivate the microorganisms. Though some progress has been made in this direction, it perhaps is still the outstanding "thing to be done."

The Insect Host. Intracellular microorganisms have so far been found to occur in approximately a dozen orders of insects and in the ticks and mites. Most of the knowledge concerning these organisms has been gained through a study of their anatomical relationships with their hosts, and less through a study of the microorganisms themselves. In fact, a large amount of work has been done concerning the relationship of the microorganisms to the embryonic development of the insect hosts. For this reason it is imperative that one understand the essentials of insect anatomy and embryology as they relate to the intracellular symbiotes. The following discussion is more or less cursory in nature, and accepted textbooks should be consulted for a more comprehensive treatment of the subject.

General Morphology of the Sex Organs. The reproductive system of insects consists of paired ovaries in the female and paired testes in the male. In most insects reproduction is bisexual, that is, complete development of the female egg cell will take place only after fertilization with the male spermatozoa. Exceptions to this occur in some insects such as aphids, in which the egg cell may develop without fertilization by the male. This type of asexual reproduction is known as "parthenogenesis." With respect to insects having this type of reproduction, the contrasting term "amphigonous" is sometimes used to refer to those eggs which have been fertilized by the male.

The female reproductive system (fig. 56) in general consists of a pair of ovaries, the accompanying genital ducts, and accessory structures. Each ovary is made up of a more or less parallel series of cylindrical egg tubes or ovarioles varying in number from one to more than two thousand in certain insects. The ovarioles converge anteriorly, forming a threadlike suspensory ligament or terminal filament which frequently serves as a means of attachment to the body wall or dorsal diaphragm. Posteriorly the ovarioles frequently converge through the pedicel and upon the calyx, entering into the lateral oviduct, which unites with the lateral oviduct from the other ovary forming the main oviduct, which in turn joins the vagina.

Each ovariole consists of a chain of developing egg cells and is usually covered with a structureless membrane called the *tunica propria*. There may be an epithelial sheath outside of this, though it is frequently absent in adult insects. Inside the *tunica propria*, the egg tube, which usually refers to the middle section of the ovariole, is lined by a layer of follicular epithelium. In addition to the terminal filament the two main divisions of the ovariole are the germarium and the vitellarium. The germarium is the region just posterior to the terminal filament which contains the primordial germ cells or oögonia, which became differentiated into oöcytes and nutritive nurse cells or trophocytes. The vitellarium follows the germarium and is the region in which the developing egg cells grow and reach their mature size. As the oöcytes grow, they distend the vitellarium into a series of progressively larger egg chambers or follicles. The follicular epithelium invaginates between the developing eggs so that each egg becomes fairly well enclosed. Although the egg tube thus tends to be stripped of its epithelium, a new one is always being formed. The chief function of the follicular membrane appears to be that of forming the egg shell.

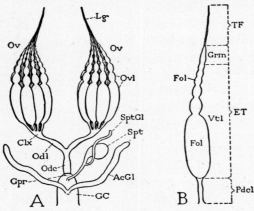

Fig. 56. *A*. Diagram of the essential parts of the reproductive organs of a female insect. *B*. Diagram of an ovariole. *AcGl*, accessory gland; *Clx*, calyx; *ET*, egg tube; *Fol*, follicle, or egg chamber; *GC*, genital chamber (vagina); *Gpr*, gonopore; *Grm*, germarium; *Lg*, ovarial ligament; *Odc*, oviductus communis; *Odl*, oviductus lateralis; *Ov*, ovary; *Ovl*, ovariole; *Pdcl*, ovariole pedicel; *Spt*, spermatheca; *SptGl*, spermathecal gland; *TF*, terminal filament; *Vtl*, vitellarium. (From Snodgrass' *Principles of Insect Morphology*, 1935, McGraw-Hill Book Co., New York.)

Figure 57 shows diagrammatically the three principal types of ovarioles. In the panoistic type there are no special cells, the nutritive function being assumed solely by the follicular epithelium. In the polytrophic type the oöcytes alternate with a group of nurse cells, so that each oöcyte has a number of nurse cells with it. By the time an oöcyte becomes mature, its nurse cells are exhausted and have degenerated. In the egg tube of the acrotrophic type the nurse cells remain in the apex of each ovariole and are connected to the oöcytes in the early stages of their development by long, plasmatic strands or nutritive cords.

In addition to the structures already discussed, the ovaries usually have associated with them certain accessory structures. The posterior oviduct exits through the female gonopore into the genital chamber or bursa copulatrix, at the anterior end of which are attached the ducts of the spermatheca and accessory glands. In some insects the genital chamber is like an internal pouch or tubular passage, in which case it may be called the vagina, the external opening of which is the vulva. The bursa copulatrix serves as a pouch, or diverticulum of the vagina, which receives the sperm from the male. From this region the sperm is accumulated and stored in the spermatheca or *receptaculum seminis*. The female accessory or colleterial glands are usually a pair of glands which have some function concerning the laying of the eggs such as secreting an adhesive substance or material forming a covering or case (oötheca) for the eggs.

The principal parts of the male reproductive system are a pair of testes, a pair of lateral ducts or *vasa deferentia*, which may include a section differentiated into a sperm reservoir (*vesicula seminalis*) or a convoluted coil (epididymis), an exit tube or *ductus ejaculatorius* and its opening (male gonopore), which is in the penis (or *phallus*), and the male accessory glands (see fig. 58). Each testis is made up of a group of sperm tubes or testicular follicles. Each tube is divided into a series of zones, similar to the female ovarioles, and in these the spermatozoa develop.

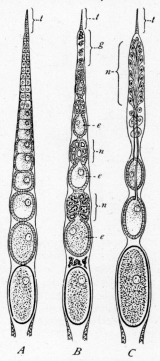

Fig. 57. The principal types of ovarian tubes. *A*. Panoistic type. *B*. Polytrophic type. *C*. Acrotrophic type. *e*, egg; *n*, nurse cells. (From Comstock, 1936; after Berlese.)

Development of the Egg. Fertilization of the egg usually takes place just before it is laid (oviposition), which may occur long after copulation by one or more spermatozoa passing into the egg through the micropyle, which is an opening or group of openings in the anterior end of the shell or chorion (fig. 59). When this happens the nucleus of the egg is stimulated to maturation, which takes place by the migration of the nucleus to the periphery of the egg, there undergoing two maturation divisions and at the same time throwing off two, or occasionally three, polar bodies. One of these divisions reduces the number of chromosomes to half (haploid). As the egg nucleus migrates back toward the center of the egg, it unites with the spermatozoon nucleus (pronucleus), producing the so-called fusion nucleus, which now contains the diploid number of chromosomes again. In most insects this zygote nucleus then undergoes mitotic division, and the several daughter nuclei migrate through the yolk toward the periphery of the egg (meroblastic cleavage). An increasing number of cleavage cells is thus produced. The cells arrange themselves in a single layer at the periphery of the egg just beneath the vitelline membrane, which was the delicate cell wall of the ovum. This stage of the embryo is equivalent to the blastula stage of a holoblastic egg. Within this blastoderm or primary epithelium one may find the yolk mass and a few primary yolk cells (vitellus) together with some other cells proliferated from the blastoderm that are called vitellophages. Some embryologists claim these cells to be endoderm cells which partially digest the yolk before the embryonic stomach is formed.

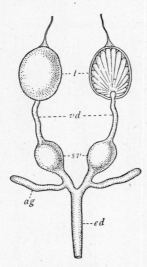

Fig. 58. Reproductive organs of a male insect, one testis shown in section. *ag*, accessory glands; *ed*, ejaculatory duct; *sv*, seminal vesicles; *t*, testis; *vd*, vasa deferentia. (From Comstock, 1936.)

During the time the cleavage cells are migrating to the egg periphery, certain of them on their way to the posterior pole pass through a protoplasmic substance known as germ-track determinant or oösome. As the cells pass through this oösome, it breaks up into fragments which accompany the cells, forming a characteristic clump at the posterior pole of the egg. This group of cells is known as the "germ cells." Eventually in the development of the insect, the germ cells move into the interior of the embryo and finally come to rest in that part of the mesoderm that gives rise to the ovaries and testes.

The first differentiation of the blastoderm is the thickening of the cells on the ventral side, and the surface appearance of this region is that of an opaque oval or elongated area. This is known as the "germ band," "primitive streak," or "germinal disc" (fig. 60). This germ band soon develops a head lobe at one end, while the other, or caudal, end moves into the yolk, taking with it the adjoining primary epithelium. Subsequently there develops the formation

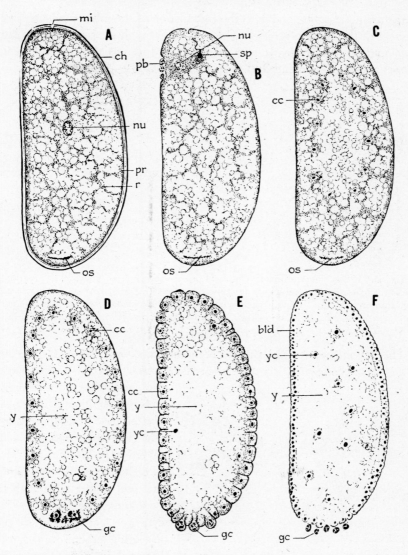

Fig. 59. Various stages in the continuous development of the egg of an insect. *A*. Unfertilized egg. *B*. Maturation. *C*. Cleavage. *D*. Cleavage stage. *E*. Blasteme stage. *F*. Blastoderm formation. *bld*, blastoderm; *cc*, cleavage cells; *ch*, chorion; *gc*, germ cell; *mi*, micropyle; *nu*, nucleus of egg; *os*, oösome; *pb*, polar bodies; *pr*, periplasm; *r*, protoplasmic reticulum; *sp*, sperm; *y*, yolk; *yc*, yolk cell. The chorion is shown only in *A*. (From Johannsen and Butt's *Embryology of Insects and Myriapods*, 1941, McGraw-Hill Book Co., New York.)

of the amniotic cavity, and with the gastrulation of the germ band the development of the embryo proper begins. For a thorough discussion of the later developments the reader is referred to such works as those of Snodgrass (1935) and Johannsen and Butt (1941).

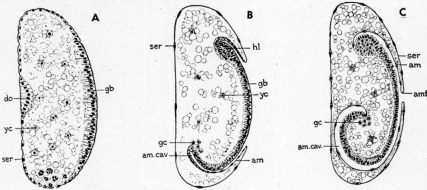

Fig. 60. Further stages in the development of the embryo of an insect. *A.* Formation of the germ band. *B.* Formation of embryonic envelopes. *C.* Saggital section of germ band. *am*, amnion; *am. cav*, amniotic cavity; *amf*, amniotic folds; *do*, dorsal organ; *gb*, germ band; *gc*, germ cells; *hl*, head lobe; *ser*, serosa; *yc*, yolk cells. (From Johannsen and Butt's *Embryology of Insects and Myriapods*, 1941, McGraw-Hill Book Co., New York.)

The Mycetome. As has already been indicated, *mycetome* is the name given to that "organ" or structure, the cells of which, in many insects, harbor the intracellular microorganisms or symbiotes. The name was proposed by Šulc (1910a), who also suggested the name mycetocytes for the individual cells making up the mycetome that are usually filled with the symbiotes. The term "mycetocyte" (or "bacteriocyte" for cells harboring organisms definitely bacterial in nature) may also refer to single cells containing symbiotes, even though they are not grouped together in a mycetome. The mycetomes of various insects are shown in several figures in the present chapter and in chapter VI; chief of these are figures 61, 62, 63, 64, 65, 80, 82, 93, 140, 141, and 142.

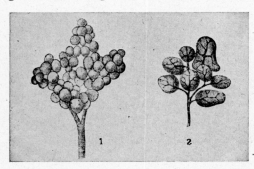

Fig. 61. Mycetome from the larva of *Cicada orni*. 1. A freed mycetome in its entirety. 2. An end group of mycetocytes showing the branching of the tracheae. (From Šulc, 1910b.)

Since these terms were apparently first used in reference to aphids, it might be well to consider here the mycetome and its development in this insect, even though some investigators believe the symbiotes found in the aphid mycetocytes to be yeasts. Leydig (1850) was the first to describe the symbiotic "organ" or mycetome in aphids, though he did not give it a name. A short time later Huxley (1858) observed this structure, which in its formation appeared as a central mass in the parthenogenetic egg, but he did not describe it in the amphigonous form. He called the structure "pseudovitellus" because "it completely simulates the vitellus of an impregnated ovum." Metchnikoff (1866) was one of the next to work on the histology and ontogeny of the "organ." He called it a secondary yolk, suggesting that its function was nutritive. Balbiani (1869 to 1871) traced the origin to the "organ," designating it "masse polaire," and discounted the secondary yolk theory of Metchnikoff. Apparently Balbiani was the first to report the presence of this structure in the amphigonous generation.

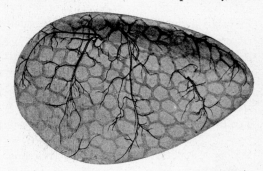

Fig. 62. The unpaired mycetome of *Pseudococcus citri* as seen from the side. (From Buchner's *Tier und Pflanze in Symbiose,* 1930, Gebrüder Borntraeger, Berlin.)

Witlaczil (1882, 1884) traced the origin of the aphid mycetome from the follicular epithelium. He also observed the presence of the *tunica propria* which envelopes the mycetome. Subsequent work to clarify the details of this study was conducted by numerous investigators, particularly Will (1889), Henneguy (1904), Flögel (1905), Stevens (1905), Tannreuther (1907), Šulc (1910a,b), and Buchner (1912).

Perhaps one of the most thorough treatments of the development of the mycetome in aphids is that by Uichanco (1924). He worked

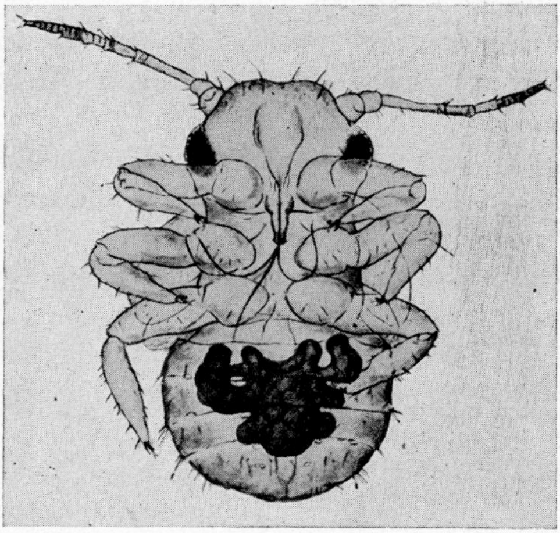

Fig. 63. A nymph of *Psylla buxi* showing the location of the mycetome in the abdomen. (From Buchner's *Tier und Pflanze in Symbiose,* 1930, Gebrüder Borntraeger, Berlin.)

principally with one species, *Macrosiphum tanaceti,* though he also studied about twenty others. According to Uichanco, the mycetome of the newly born nymph is a longitudinally bipartite organ, extending from about the second to about the sixth or seventh abdominal segment. Each lobe occupies the cavity on either side of the alimentary tract. As the aphid develops (to about the fourth instar) the mycetome is reduced to isolated groups of two, three, or more mycetocytes located anteriorly and posteriorly with respect to the ovarioles, in the interstices formed by the adjacent portions of the latter. A membrane enveloping most of the mycetome persists throughout the life of the aphid. Tracheoles about 7 microns in diameter supply the mycetome. The mycetocytes of the newly born nymphs measure in size from about 42 to 45 microns. After the birth of the aphid the mycetocytes apparently do not undergo further cell division but increase in size only. The number of mycetocytes present in each individual aphid ranges from about 60 to 70. When the adult stage is reached, the symbiotes have increased in number, and the mycetocytes measure from 72 to 108 microns in size, the average being about 92 microns. The symbiotes never invade the nuclei of the mycetocytes. After reaching the adult stage the mycetocytes degenerate, one by one, until toward the end of the aphid's life very few of these cells are left. The mycetome is present in amphigonous as well as parthenogenetic forms, occurring in both males and females, and is present in every individual.

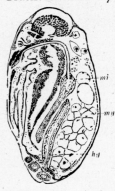

Fig. 64. A sagittal section of a full-grown embryo of *Macrosiphum tanaceti* showing the position of the mycetome (*my*) in relation to the midintestine (*mi*) and the hindgut (*hg*). (From Uichanco, 1924.)

During the early stages of the aphid's development the follicular epithelium becomes "infected" with the symbiotes, apparently harboring them in a dormant condition. As the egg forms in the adult, the symbiotes in the follicles come in contact with the yolk. This contact apparently stimulates the dormant symbiotes, bringing about their rapid multiplication and causing considerable swelling in the portion of the follicular epithelium adjoining the posterior pole. The symbiotes soon break through the thin epithelial cells and invade the posterior portion of the egg through the posterior opening in the periplasm. Thus the symbiotes find their way from the adult mother to the beginning of the next generation. The vitellophages or "mycetoblasts" show a marked attraction to the symbiotes working toward the posterior half of the egg cavity. The mycetoblasts form a subglobular mass together with the cytoplasmic network, within which they are in syncytium, and the enclosed symbiotes. Thus begins the formation of the mycetome. Cellular differentiation of the structure takes place after an active mitotic division of the mycetoblasts. The mycetome first exists as a single mass of mycetocytes, then divides into two lateral halves, and finally becomes the size and shape we have described as occurring in the adult aphid.

We have discussed the development of the mycetome in the aphid at some length because it illustrates the typical close relationship between this structure and the host insect. The development of the mycetome in other insects is very similar, though there are many modifications. In species of *Glossina, Camponotus,* and the order Anoplura, the mycetome is found in the wall of the intestinal tract. In some species of *Apion* (Curculionidae) and *Donacia* (Chrysomelidae), modified Malpighian tubes form the mycetome. Strings or masses of modified cells become mycetomes in coccids, aleyrodids, cicadids, and others. In some roaches mycetocytes are scattered throughout the fat body. Mycetomes or mycetocytes associated with the fat body or gonads and not with the gut occur in species of *Cimex, Lyctus, Calandra,* and *Rhizopertha* and in *Formica fusca, Stegobium paniceum,* and *Oryzaephilus surinamensis.* The hog louse, *Haematopinus suis,* has a well-developed mycetome in each oviduct, though it also has mycetocytes on the intestinal wall.

The mycetomes of the bedbug, *Cimex lectularius,* have been described in detail by Buchner (1923). They are paired structures lying on either side in about the third abdominal segment, near the gonads, and usually among lobes of the fat body. They occur in every individual of both sexes. Their opaque-to-transparent, oval or pear-shaped form is sharply outlined, though the size and shape are not constant. In the male each mycetome is connected with the base of the corresponding testis by a delicate strand or thread. The mycetome in each sex is supplied with a tracheal branch from the fourth abdominal spiracle.

In *Rhizopertha dominica* the mycetomes are in the form of two spheres, one on each side of the alimentary tract near the anterior end of the proctodaeum. In *Sinoxylon ceratoniae* and *Bostrychoplites zickeli* the mycetomes are also paired and occur in similar positions. Instead of being spherical, however, they are long, convoluted masses (Mansour, 1934a). The structure of the mycetome in these three species is similar. Each has a peripheral covering of a single layer of epithelial cells. In *Rhizopertha* the mycetocytes are grouped in small syncytial masses (fig. 65), whereas in the other species they are mostly separated from each other, though syncytial masses do occur. The mycetomes undergo an interesting period of growth during metamorphosis. According to Mansour, the nuclei first emit chromatin globules, which become dispersed all over the cytoplasm.

The whole mass is invaded by blood corpuscles (amebocytes), each mycetocyte receiving three to five of these invaders. The nuclei of the mycetocytes, which are by this time gigantic in size, emit a great deal of their chromatin, and in a later stage look like collapsed rubber balls. This stage is then followed by a direct, and in many cases unequal, division of the nuclei, and thus the number of nuclei and consequently of the cells of the mycetomes, is remarkably increased. Meanwhile the covering of the mycetome has been increasing in capacity by the active mitotic division of its cells, thus coping with the increase in bulk from inside.

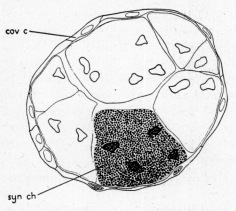

Fig. 65. Portion of a section through a mycetome of a first instar larva of *Rhizopertha dominica*. *cov c*, covering cell; *syn ch*, syncytial chamber of mycetome. (From Mansour, 1934a.)

By the time the advanced pupal stage is reached the nuclei are fully distended and the cytoplasm contains practically no amebocytes and chromatin extrusions. There is no essential difference between the mycetomes in males and in females.

It should be mentioned that apparently the mycetome may be present but contain no symbiotes. In certain ants and in the weevil, *Sitophilus granarius* (= *Calandra granaria*), this is apparently the situation, though more study is necessary in order to explain these apparent inconsistencies. It is also interesting that Koch (1931) succeeded in rearing generations of a beetle (*Oryzaephilus*) free from its intracellular symbiotes.

In a series of experiments Ries (1932a) transplanted the mycetomes or other symbiote-containing tissues of *Blatta orientalis, Psylla buxi, Pseudococcus citri, Pseudococcus adonidum, Oryzaephilus surinamensis, Stegobium paniceum* and *Eomenacanthus stramineus* (= *Menopon biseriatum*) into the meal worm (presumably *Tenebrio molitor*); of *Blattella germanica* and *Stegobium paniceum* (= *Sitodrepa panicea*) into *Blatta orientalis;* and of *Blatta orientalis* into the larva of *Ephestia kühniella*. The transplantations were successful as a whole. There were marked changes in the morphology of most of the symbiotes that Ries believed were not due to physical causes. In most cases there were no changes of structure of the implants caused by autolysis or the fermentative processes of the new hosts, though these did occur a few times. Several days after the operation a darkening or melanization of the implants and the symbiotes usually took place. This apparently set in directly after the death of the tissues. Ries assumed this to indicate a detoxication and rendering harmless of the implants and their metabolic products. Another reaction of the hosts was the increased production of lymphocytes, which tended to isolate the implants.

It should be emphasized that intracellular symbiotes do not always occur in mycetomes or even in specialized mycetocytes. In certain species of insects and ticks they may be regularly present in the epithelial cells lining the alimentary tract, the cells of Malpighian tubes, or they may be found in the cells of the salivary glands, ovaries, testes, or muscles. Occasionally the symbiotes are localized both in a mycetome and in other cells in other parts of the insect's body.

Functions of the Mycetome. There have been attempts to ascribe to the mycetome certain functions in the maintenance of the insect's life processes. Balbiani (1866) connected the mycetomes with the sex of the insect, postulating male and female mycetomes. He thought the function of these in some way would explain parthenogenesis. Metchnikoff (1866) thought that the function of the mycetome in aphids was probably nutritive, supplying food material for the developing embryo. This belief was discredited because instead of becoming smaller as a yolk would do during the embryo's development it became larger. Witlaczil (1882) believed that the mycetome took the place of the Malpighian tubes in their excretory function, but he later abandoned this idea. Portier (1918) thought the mycetome of aphids was "an organ of synthesis," transforming certain substances in the imbibed plant sap into substances utilizable by the insect. As will be pointed out later, the work of Aschner (1932) on the symbiotes of lice (*Pediculus*) indicates that the mycetome has no function of its own directly related to the metabolism of the insect, its main role being that of housing the symbiotes. It has been conjectured that, originally at least, the mycetomes were analogous to plant galls produced by the host as a response to irritation by a foreign inhabitant. When the beetle, *Oryzaephilus surinamensis,* is deprived of its symbiotes by being held at an increased temperature (36° C.), the deserted mycetomes continue to be formed the same as ever in subsequent generations (Koch, 1936b).

As with the symbiotes themselves, knowledge concerning the function of the mycetomes, if they have one other than that of harboring the symbiotes, will have to await further research.

The Nature of Intracellular Symbiotes (Microsymbiotes). The term "symbiote" or "microsymbiote" was defined at the beginning of this chapter. In general, as it concerns the field under discussion, it refers to any microorganism (bacterium, yeast, rickettsia, protozoan, etc.) which usually lives in a more or less constant association with its insect host. It is most frequently used in reference to the intracellular forms in these animals, but it may also apply to extracellular microorganisms. Some writers employ the term "endosymbiote" in referring to those which live intracellularly, and "exosymbiote" to those which live extracellularly.

The intracellular yeasts and fungi, the protozoa, and the true rickettsiae will be discussed in other chapters. To separate with certainty the true bacteria, the bacteriumlike organisms, and the rickettsialike organisms which live intracellularly in arthropods is at the present time impossible, or at least extremely difficult. Those microorganisms which have been placed in the genus *Rickettsia,* even though temporarily, will be discussed in the next chapter. For the present we shall concern ourselves with the intracellular bacteria and the bacteriumlike and rickettsialike organisms associated with insects and ticks.

Not all investigators have been willing to concede the organismal nature of these cellular inclusions. Some workers believe them to be mitochondria or waste products, as did Cuénot (1896) and Henneguy (1904). Several have recently claimed successful cultivation of the mitochondria. If these inclusions are readily cultivable on artificial media, they would immediately be removed from the category of metabolic or waste products.

Some biologists, notably Portier (1918), Meves (1918), and Wallin (1922, 1927), have been proponents of the belief that mitochondria are in reality symbiotic microorganisms. Wallin (1927) "has arrived at the unqualified conviction that these bodies [mitochondria] in the cell are bacterial in nature." Recently (1943) Wallin has reaffirmed his contentions as to the bacterial nature of mitochondria, which he believes are as much a part of the cell as the nucleus but which have their origin in originally free-living bacteria. At any rate, it is certainly difficult at times to establish by ordinary techniques whether certain obscure intracellular elements are to be considered as mitochondria or microorganisms. That recognized symbiotes are not mitochondria is supported by the claim that these bodies can usually be distinguished from microorganisms by the application of various methods of differentiation such as staining reactions and responses to lipoid solvents (Cowdry and Olitsky, 1922; Cowdry, 1923). However, others (Wallin, 1922; Miller, 1937) have maintained that such characteristics as these are not sufficient to distinguish bacteria from mitochondria. In recent years cytologists have been able to separate the mitochondria from tissues into pure suspensions (Bensley, 1942). It would be interesting to see if similar techniques could be applied to the so-called symbiotes in insects.

In the mycetocytes of some insects "mitochondria" have been seen lying beside the "symbiotes" as, for example, in the roaches (Milovidov, 1928).

Evidence that the symbiotes are actually foreign elements is advanced by Neukomm (1932), who ran complement fixation tests with the symbiotes of the roach *Blattella germanica*. This worker made suspensions of the symbiotes alone, and the roach fat body alone, and inoculated these antigens into rabbits to obtain the antisera. He found the complement fixation reaction to be positive only in homologous combinations; symbiote antigen in symbiote antiserum, and fat body antigen in fat body antiserum. Since there is generally no specificity of antibodies for various tissues of the same organism, the symbiotes may be considered as foreign to their hosts.

Perhaps the first real proof that microorganisms may live harmlessly within tissue cells was the demonstration by Beijerinck and others that the rodlike elements in certain plant tissues could be cultivated in artificial media and would infect young plants grown in sterile soil. It is interesting to note the similarities between the intracellular bacteria in the tissue cells of leguminous plants and those which occur in the mycetocytes of insects. The root nodules harboring the plant symbiotes (*Rhizobium*) to some extent may be likened to the mycetome of insects. The various involution forms seen in insects remind one of the life cycle of the *Rhizobium*. Unfortunately attempts to grow the intracellular microorganisms of insects on culture media have not yet met with the success that has accompanied similar attempts with plant symbiotes.

The development of the belief in the organismal nature of the intracellular inclusions in insects has been outlined in the historical section of this chapter.

Apparently considering them in the narrow sense, Gier (1936) sought to avoid such terms as "symbiont" (symbiote), "symbiosis," "parasite," and "parasitism," and the implications arising from their usage. For this reason and because of the wide disagreement as to the true physiologic nature of these intracellular organisms, he designated the symbiotes in roaches as "bacteroids." No doubt there is some justification for applying such a term in the sense "bacteriumlike." However, care must be taken in using a designation which has already a definite meaning in soil microbiology. The term "bacteroid" was first used by Brunchorst (1885) in referring to the root nodule organisms in the legume plants. As defined by Fred, Baldwin, and McCoy (1932), this term is used to designate "the enlarged, frequently club-shaped or branched, vacuolated or banded forms of the root nodule bacteria, both as they occur in the nodule and in culture media." Merely because comparable forms are found associated with insects is no valid reason why these cannot be called "bacteroids." Nevertheless, the same objection may be raised in this instance as that made in the case of the root nodule bacteria by Löhnis and Smith (1916). They objected to the use of the term "bacteroid" because it is now generally recognized that these forms are true bacteria and not plant products resembling bacteria. Similarly, when the intracellular forms in insects are more thoroughly studied, and perhaps some placed among the bacteria, they then will deserve more than the mere designation of "bacteroid."

Cultivation of Bacteriumlike and Some Rickettsialike Symbiotes. Perhaps the greatest need in the study of the intracellular organisms associated with insects and ticks is for adequate methods of cultivating these microsymbiotes. Much of the controversy that prevails as to the organismal nature of many of these bodies could be definitely settled if reliable culture methods were perfected. On the other hand, our failure to cultivate some of these forms should not be taken as an irrevocable indication that they are not microorganisms. The "fault" does not lie with the organisms but rather with the inadequacies of our methods. No one doubted that the leprosy bacillus was a living organism, yet it has defied ready cultivation. It must be admitted, however, that not until these bodies have been independently cultivated can they always be definitely differentiated from such elements as mitochondria.

Unfortunately, many of the early reports on the cultivation of intracellular organisms in arthropods are unreliable. In many instances contaminants of similar morphology were accepted as the species under consideration. This was no doubt the case with *Bacillus cuenoti*, isolated by Mercier (1906) from the fat body of the roach *Blatta orientalis*. He believed this sporeforming rod to be the symbiote found regularly in the roach. Hertig (1921) was unable to culture the organisms and decided that Mercier's bacillus was a contaminant. Later, however, Glaser (1930a) reported the cultivation of a microorganism from the American roach, *Periplaneta americana*. To the diphtheroidal forms isolated he gave the name *Corynebacterium periplanetae* var. *americana*. In the same year Glaser (1930b) cultivated *Corynebacterium blattellae* from the German roach, *Blattella germanica*. On the other hand, Gier (1936) indicated that only negative results were obtained in his attempts to cultivate the symbiotes. In a personal communication Gier asserts that he made hundreds of attempts to cultivate these forms on many kinds of media, but always without success. He believes that Glaser's diphtheroids are slow-growing contaminants of rather unusual character (see also Gier, 1937). He considers the results of Gropengiesser (1925) and of Bode (1936), as well as those of Mercier (1906, 1907), to be due to poor techniques. Gier was unable to get any signs of growth on chick chorio-allantoic membranes or in amniotic fluid either with the living chick or in a test tube. Glaser (personal communication) explains the failure of other workers to duplicate his results as due to faulty technique. He maintains that repeated transfers from a medium which has been inoculated with these microorganisms, but which is apparently sterile, are necessary before the growth of the organisms finally appears as tiny pin-point colonies. According to Glaser, a period of slow adaptation to artificial conditions apparently is necessary. He also believes that successful cultivation may depend, not only on the age of the roach, but on the season of the year as well. In 1942 (e) Steinhaus isolated an unidentified diphtheroid, possibly a contaminant, from fertile chicken eggs into which the fat bodies of *Blattella germanica* had been inoculated. Glaser examined this bacterium and noted that in many respects it was very similar to *Corynebacterium blattellae*. Positive, final proof of the cultivation of the symbiotes of roaches is apparently still lacking.

The controversy which has prevailed concerning the cultivation of the symbiotes of cockroaches is more or less typical of the situation in regard to many other intracellular bacteriumlike microorganisms in arthropods. By far the majority of these have not been cultivated either on artificial media or in tissue cultures. In general, the same situation exists with these unidentified microrganisms as with the known rickettsiae. Furthermore, there is a tendency to consider many of these forms as rickettsiae, especially when the organism will grow in chick embryos or in tissue cultures but not on artificial media. These aspects of the problem will be more completely discussed in the next chapter, though a certain amount of overlapping cannot be avoided.

Mahdihassan (1939, 1941) has offered an explanation concerning the unsuccessful attempts to cultivate intracellular symbiotes by postulating the presence of a substance which controls microbial growth in the mycetome tissues. When this substance comes in contact with the air it becomes oxidized and acts as a strong germicide. Hence, Mahdihassan believes that when these tissues are smeared over a medium, certain substances ("aldehydes of unsaturated fatty acids called Plasmal"), on being exposed to air, are oxidized to acids which kill the microorganisms. This theory, however, has not yet found general acceptance and demands more experimental proof.

Methods for the cultivation of certain extracellular microorganisms appear to be a good starting point toward the cultivation of intracellular symbiotes. Nöller (1917a) succeeded in cultivating *Rickettsia melophagi* on an enriched blood medium, Jungmann (1918) and Hertig and Wolbach (1924) confirming his results. Kligler and Aschner (1931) cultivated the extracellular forms in *Lipoptena caprina, Hippobosca equina, Hippobosca capensis,* and *Melophagus ovinus,* and are mentioned here because of the methods employed. Kligler and Aschner's procedure was briefly as follows: The insects were dipped in a 5 per cent tincture of iodine for five to ten seconds, then washed in 95 per cent alcohol for the same length of time, and then rinsed in sterile saline. The insect's abdomen was cut off and the intestinal contents pressed out into a drop of sterile saline. The abdomen was then transferred to another sterile slide where the intestines and sex organs were separated. The dissected parts were placed on Nöller blood-agar plates or in a fluid medium. Cultures were incubated at 26° C. and observed for a period of at least two weeks. The three media found most successful were (1) peptone-gelatin-blood medium, (2) Locke semisolid, and (3) Nöller blood-agar medium. The organisms these workers obtained in their cultures were minute gram-negative, coccoidal rods resembling the forms seen in the guts of the insects. Successful results were obtained especially when liquid media were used, the peptone-gelatin-blood medium being the best. Growth was extremely slow (7 to 14 days). Transfers had to be made every two weeks.

Steinhaus (1943) isolated a small diphtheroid (*Corynebacterium lipoptenae*) from a number of *Lipoptena depressa* by using the chick embryo technique for the original isolation. In an earlier study, however, this method seemed to be of no avail in attempting to cultivate a rickettsialike organism occurring in the tick *Argas persicus*.

Generation-to-Generation Transmission. One of the most interesting aspects of the relationship between insects and their symbiotes is the mode of transmission of these organisms from parent to offspring. This subject has already been discussed with respect to the extracellular bacteria. Similarly ingenious methods of transfer occur with the intracellular organisms. They are sometimes known as "cyclic endosymbioses."

In those insects in which the symbiotes are harbored in mycetocytes or in mycetomes within the body cavity, the mode of transmission is usually rather complicated. This fact has already been mentioned in the discussion of the mycetome. In most of these cases the eggs become internally "infected" before oviposition by way of the nurse cells. For example, in bedbugs (*Cimex*) and in certain weevils (*Apion*) the symbiotes invade the nurse cells and are carried by them to the oöcytes. A similar arrangement prevails in the scale insects. As has already been described, Uichanco (1924) traced the path of transmission of the symbiotes from the follicular epithelium of the parent to the definitely formed mycetome in the offspring. In some ants (e.g., *Camponotus*) the symbiotes in the follicle give rise to an invasion of the egg plasma. In certain Coleoptera (*Oryzaephilus* and *Lyctus*) and Anoplura the posterior pole of the egg is invaded shortly before the secretion of the chorion. In some Pupipara and in *Glossina*, which nourish the larva in the uterus, the symbiotes are transmitted through the milk glands.

The modes of symbiote transmission in roaches and some bostrychid beetles are particularly interesting and merit more detailed descriptions. Heymons (1895), Buchner (1912), and Gier (1936) have made detailed studies of the "life cycle" of the symbiotes of roaches. In general, the symbiotes behave similarly in all species of roaches, though there are differences in their number and in the structure of the ovary. All migrations of the symbiotes appear to be passive. The following description is taken largely from a paper by Gier (1936) who worked principally with *Periplaneta americana*.

In both the male and female roach the cavity around the abdominal viscera is rather well filled with a lobated fat body well supplied with tracheae. Each fat-body lobe consists largely of two types of cells—a central row of mycetocytes whose cytoplasm is filled with symbiotes, and an outer layer of cells supplied with fat globules and urate crystals. The latter cells completely surround the mycetocytes (fig. 66). Thicker portions of the fat body may have several

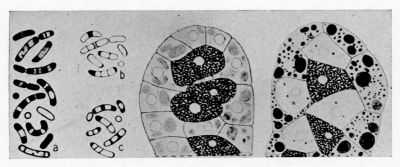

Fig. 66. Symbiotes in the fat body of cockroaches. (Left, *a*, *b*, and *c*.) Typical symbiotes from around oöcytes. (Center) Section of fat-body lobe from *Blatta orientalis*. (Right) Fat-body lobe of *Periplaneta americana* with urate cells around the mycetocytes. (From Gier, 1936.)

rows of mycetocytes. In *Periplaneta*, *Eurycotis*, and *Blatta*, many lobes are thin with only one row of mycetocytes. Thicker lobes occur in *Blattella*, *Parcoblatta*, and *Cryptocercus* with from 2 to 20 mycetocytes in cross section.

Mycetocytes, not associated with fat cells, are present in the connective tissue covering both ovaries and testes, but no symbiotes have ever been found within the testes themselves. In the ovaries, however, a layer of symbiotes is present around the larger oöcytes. The number of symbiotes between the oöcyte membrane and the follicle cells increases until there is a uniform layer two or three organisms thick. Before the egg is oviposited, the original oöcyte membrane breaks down and permits the symbiotes to enter the cytoplasm. This introduces the organisms into the embryo in which, accompanied by nuclei similar to the vitellophag nuclei, they move in masses to the center of the yolk. As the embryo develops, a few organisms from this central mass move posteriorly between the yolk granules through the incomplete margins of the gut epithelium into the body cavity. From here most of them are taken up by the cells of the lateral lobes of the fat bodies while a few are caught between the cells of the developing ovaries. The symbiotes within the ovariole appear to lie dormant for several weeks, after which they rapidly multiply and spread over the surfaces of the enlarging oöcytes. In this manner transmission from one generation to the next is accomplished.

It is interesting to note that Gier found the "bacteroids" doubling their numbers in the embryos and young nymphs in about ten days. The numbers diminished as the animals neared maturity, and apparently decreased in adults except in the ovaries.

Mansour (1934a), working with certain bostrychid beetles (*Rhizopertha dominica, Sinoxylon ceratoniae,* and *Bostrychoplites zickeli*), reported the transmission of symbiotes from one generation of the host to the next thus: The microorganisms from the mycetomes invade the testis lobes, multiply, and mix with the sperm. The symbiotes then pass with the sperm during copulation into the bursa copulatrix of the female. From this region they pass through the micropyle of the fully formed eggs during their passage to the outside. (See figs. 67 and 68).

Mansour did not determine the state at which the male genitalia are in-

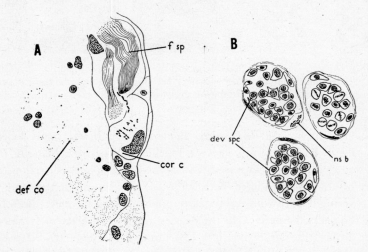

Fig. 67. Symbiotes of bostrychid beetles in testes. *A.* Portion of longitudinal section through testis lobe of *Rhizopertha dominica* showing symbiotes in cortical cells and their progression into the inside of the testis lobes. *B.* Portion of a section through a prepupa of *Sinoxylon ceratoniae,* showing three developing testis lobes, one of which has a nest of microorganismlike structures in its cortex. *cor c,* cortical cell; *def co,* deformed core; *dev spc,* developing spermatids; *f sp,* fully formed spermatozoa; *ns b,* nest of bacteriumlike microorganisms. (From Mansour, 1934a.)

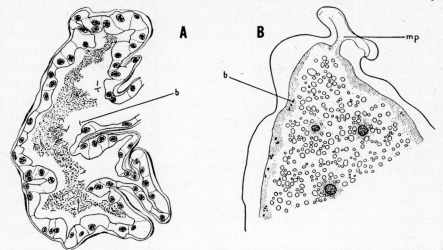

Fig. 68. Symbiotes of bostrychid beetles. *A*. Portion of a section through the genitalia of an egg-laying *Rhizopertha dominica*, showing the bursa copulatrix with symbiotes mixed with debris. *B*. The micropyle end of a segmenting egg of *Sinoxylon ceratoniae*, showing the symbiotes near the micropyle embedded in the peripheral protoplasmic layer. *b*, bacteriumlike symbiotes; *mp*, micropyle. (From Mansour, 1934a.)

vaded by the symbiotes. He felt fairly certain, however, that the "infection" first starts in the outer cover of the testis lobes. Although in some lobes it takes place during the pupal or prepupal stages, in others it may occur during the active period of the males. How the symbiotes leave the mycetomes and are carried to the testes region is not known with certainty, though they may leave the mycetomes through a rupture in the outer cover as they do in the females of *Oryzaephilus, Lyctus, Camponotus,* and *Formica fusca* (Koch, 1931).

Origin of Intracellular Symbiotes. Before discussing the role of microsymbiotes in relation to their hosts, we may well consider the possible origin of this group of microbes.

It is not difficult to imagine that in their initial association these microorganisms were actually pathogenic parasites, later assuming a more or less commensal relationship, until finally a definite mutualistic association was established between the host and the invaders. Certainly the association of these two forms of life must have been an extended one, especially in view of the congenital transmission of the organisms and the intimate connection between most insects and these forms. It should be pointed out, however, that in certain ants and in the weevil *Sitophilus granarius* the mycetome is present, but apparently contains no symbiotes.

Some writers (e.g., Meyer, 1925) have enlarged upon this idea that the intracellular forms have but reached a certain stage in an evolutionary process of adaptation. Originally, the organisms were probably parasites producing pathologic conditions and disease. Later, acquired immunity to the microorganisms became inherited, and gradually the invading microbes were brought under control. Transmission from generation to generation became established, and the microorganisms remained in constant contact with their host. They gradually lost their original harmful characteristics until finally they produced enzymes of use to the host and became essential to the insect's existence. Presumably such organisms as *Rickettsia prowazeki*, which is to some extent pathogenic for its host, the louse, are not so far along in their evolutionary process of adaptation as *Dermacentroxenus rickettsi*, which exerts no apparent deleterious influence on its tick host.

All such reasoning is, of course, highly speculative. There is very little evidence to indicate that this is the usual manner by which the intracellular symbiotes have become so intimately associated with insects.

As to the more immediate origin of the symbiotes, Paillot (1933) has maintained that the processes involved are essentially those concerned in natural antimicrobial immunity. For example, in the case of the aphid *Drepanosiphum platanoidea* the bacteria "normally" found in aphids are found free in the blood or attached to the mycetocytes in more or less large clumps. Cellular and humoral reactions act to limit the number of these microbes and to cause them to clump or agglutinate in contact with the mycetocytes and eventually to penetrate into the cytoplasms of these cells. At the same time the bacteria in the blood are phagocytosed. The phagocytes filled with the bacteria may be incorporated into the mycetocytes, which thus play the role of macrophages. According to Paillot, in all infected cells it is possible to detect normal bacilli as well as the growth forms which represent the transition forms between the normal bacilli and the symbiotes. In other words, the intracellular symbiotes (which are frequently large rounded or globular forms) originate from certain "normal" bacteria associated with the aphid.

Role of Intracellular Symbiotes. When we wish to consider the true role or actual function of the intracellular symbiotes we find ourselves in a sea of uncertain speculation with only a few small fragments of possible explanations. There is a trend, in much of the literature, toward considering these cellular inhabitants as assuming a true mutualistic relationship with their hosts. Others feel that they are commensals or some sort of tolerated parasites. They have also been regarded as true parasites to which the insect has developed an immunity.

It is often difficult to make a definite distinction between mutualism and parasitism, especially in borderline cases. It is generally recognized that there are degrees of mutualism, that is, mutualism may be obligate or facultative just as the broader relationship, symbiosis, may be obligate or facultative. In the case of the intracellular organisms in insects, very little work has been done to determine whether or not the association between these two forms is indispensable to both. It is probable that both the obligate and the facultative types of mutualism exist between insects and their symbiotes. In any case, it is sometimes difficult to imagine just what constitutes the "mutual advantage" to the two forms of life concerned. Although there are some experimental data on the benefits secured by the host through such a relationship, the evidence for the advantage to the symbiotes is considerably more uncertain and hypothetical. This is indicated in a statement by Meyer (1925): "Buchner's suggestion that the intracellular organisms are benefited by being protected within the host from the drastic atmospheric influences of heat, cold, desiccation, etc., is a trifle unreasonable." This much may be said, however; if the symbiote is unable to live outside the body or egg of the insect host, then a disruption of the association would very likely mean extinction of the symbiote species. To such an extent, at least, the host is benefiting the microbe. The providing of a place for a highly specialized organism to multiply and live must certainly be considered as a benefit furnished by the host.

The reasons why these microorganisms have been generally considered to be truly mutualistic in nature rather than parasitic have been set forth by Glaser (1920) as follows:

(a) Every individual of a species is infected.
(b) The infection produces changes in the host cells but these are harmless.
(c) The infection routes and methods of localization, while different in different hosts and symbionts, follow very definite courses within a species.
(d) The microorganisms are numerically controlled by the host, never increasing up to a point where they prove fatal.
(e) The microorganisms within the insects obtain nourishment and protection from drastic temperature and drought conditions.

To these reasons may be added the experimental evidence presented by Aschner (1932, 1934) and Aschner and Ries (1933), who obtained results which, in their opinion, warranted the conclusion that the symbiotes play an essential role in the life of the body louse, *Pediculus humanus*. It was found that if the louse was deprived of its symbiotes by operative removal of the mycetome, or through elimination of the symbiotes by centrifugalization of the egg, its powers of nutrition and reproduction were greatly impaired. Specimens from which the symbiotes were removed did not live so long as did normal insects. The female did not lay eggs, or at the most only two or three during the remainder of her life. These workers maintained that the harmful effects noted were not due to experimental injury but to the absence of the symbiotes.

Aschner conducted his experiments on female larvae of lice one or two days before the third molt. In these cases, the removal of the mycetome does not prevent "infection" of the adults by symbiotes which have reached the ovaries a short time before the molt. These females deprived of mycetomes but nevertheless provided with symbiotes behave the same as normal females in respect to their longevity and their reproductive ability. Partial removal of the mycetomes resulted in a decrease in the number of symbiotes in the eggs and embryos. Females which no longer possessed their complete mycetomes were much less fertile than normal females or females operated on after the passage of the symbiotes into the ovary. According to Aschner, these facts show that the physiologic effects noted do not depend on secretions of the mycetomes, but are due exclusively to the action of the symbiotes themselves. The harmful effects caused by the absence of symbiotes could be partially reduced by the rectal injection of yeast extract.

In the male lice the symbiotes more or less degenerate after the third molt, indicating that the role of the symbiotes in the males is not as important as in the females. The removal of the mycetomes is better withstood by the males, certain ones being able to live as long as three weeks after being separated from the symbiotes.

Aschner concludes that such results as these justify the belief that the relationship here between symbiote and host is one of true symbiosis (mutualism).

A view somewhat similar to that of Aschner's is taken by Brues and Dunn (1945) with respect to the gram-positive symbiotes occurring in the fat body of the roach *Blaberus cranifer*. These workers administered penicillin to the roaches and observed the symbiotes to reduce greatly in numbers or to be entirely destroyed in which case the insects died several days after the treatment. According to Brues and Dunn, this indicated that the symbiotes are mutualistic and not parasitic microorganisms.

Some workers have thought that in certain insects the constant occurrence of the symbiotes in, and their close association with, the intestinal tract indicated that the microorganisms might aid in digestion. This belief has been particularly strong in the case of such bloodsucking insects as the tsetse fly (*Glossina*). In this case, at least, it has been shown that there is no digestion and no digestion enzymes in that segment of the intestine where the symbiotes are located (Wigglesworth, 1929). Pierantoni (1910a) thought that since sap-sucking insects such as aphids absorb great quantities of sugar, the symbiotes break down this substance, making it available for the insect's utilization. Concerning exclusively bloodsucking insects, Wigglesworth (1936) suggests the possibility that the symbiotes provide an endogenous source of vitamins or accessory factors which enable their host to live permanently on a restricted or highly specialized diet deficient in some respects. Fraenkel and Blewett (1943) have conducted experiments on the larvae of the beetles *Lasioderma serricorne* and *Stegobium paniceum* and have presented evidence to support the suggestion that the intracellular organisms provide a source of vitamins to these nonbloodsucking insects.

The intimate association of the symbiotes of some insects with the reproductive organs has lent support to the view that the microorganisms influence the process of reproduction. This possibility has already been pointed out in our discussion of the experimental work of Aschner and Ries.

Of great interest are the experiments of Koch (1936b), who was able to deprive the beetle *Oryzaephilus surinamensis* of its symbiotes by increasing the surrounding temperature to 36° C. In this insect the bacteriumlike symbiotes are located in four definitely formed mycetomes. Under the influence of increased temperatures the symbiotes develop involution forms, which undergo a gradual process of degeneration and disappear. The highest temperature at which the beetle will live is 38° C. The lethal threshold for the symbiotes is between 32° and 33° C., and at 30° C. no apparent damage is done the symbiotes. The symbiote-free beetles in turn produce symbiote-free offspring. In these descendants the deserted mycetomes continue to be formed the same as ever and remain intact through at least 25 generations. In no case did the symbiote-free individual appear to suffer as the result of the lack of symbiotes, even when kept on a diet of only starch. There were no changes in either its feeding habits or in its powers of reproduction. Koch was quick to point out, however, that the fact that an organ is not necessary for life does not exclude its usefulness, and that in this case we have not been in position to determine the field of usefulness concerned.

The whole question as to the role of intracellular symbiotes in the life of their insect hosts is obviously far from a solution. Here, indeed, is a field awaiting further investigation!

II. SELECTED EXAMPLES OF BACTERIUMLIKE AND RICKETTSIALIKE MICROSYMBIOTES ARRANGED ACCORDING TO ARTHROPOD HOSTS

Most of the studies on the intracellular microorganisms in insects and ticks have been made by European investigators. It is hoped that more workers in the Americas will soon be aware of the interesting possibilities offered by this field and will similarly study the arthropods of this hemisphere. The entire field is still largely unexplored. To be sure, many interesting facts and observations have been recorded, but we still do not know what most of these facts mean, and we certainly are not fully aware of the applications this knowledge might have.

At this stage of our progress it is impossible to make satisfactory generalizations or definite conclusions. For example, at one time it was supposed that intracellular symbiotes were limited largely to the bloodsucking arthropods. It is now apparent, however, that insects with much different feeding habits (e.g., Coleoptera) harbor these microorganisms to an extent almost as great as the hemophagous insects. The time is not yet here when the true significance of the intracellular symbiotes can be written down with any degree of certainty.

Studies of the relationships between the intracellular symbiotes and insects from a phylogenetic standpoint have been attempted in a limited manner only. More data on the symbiotes of certain groups of insects are needed before any generalization as to possible correlations with evolutionary development can safely be made. Some groups (e.g., Lepidoptera) appear to have very few examples of intracellular symbiosis, but it should be remembered that these insects have had much less study from this standpoint than have those of most other orders. If the Lepidoptera do in fact harbor few intracellular symbiotes, the significance of this is far from understood. Both the more primitive and the more highly evolved groups of insects include examples of the type of association under consideration. Perhaps when the results of more observations have been compiled, we may be able to see more clearly some phylogenetic relationship between arthropods and their intracellular microorganisms.

The discussion to follow is intended principally for the purpose of orientation and for an indication of some of the possibilities this field holds for further research. The examples discussed are by no means the only ones known, though most of the known types of associations concerned are included. The microsymbiotes are discussed according to their arthropod hosts and the insect orders are arranged in the manner of Comstock's (1936) classification.

	Page
Class: Hexapoda	215
Order: Thysanura	215
Order: Orthoptera	215
Order: Neuroptera	218
Order: Corrodentia	218

BACTERIUMLIKE AND RICKETTSIALIKE SYMBIOTES

	Page
Order: Mallophaga	219
Order: Anoplura	220
Order: Homoptera	226
Order: Hemiptera	233
Order: Coleoptera	236
Order: Diptera	243
Order: Siphonaptera	247
Order: Hymenoptera	247
Class: Arachnida	249
Order: Acarina	249
(a) Superfamily: Ixodoidea (Ticks)	249
(b) Mites	254

Class: HEXAPODA

THYSANURA

Lepismatidae. Cowdry (1923) found gram-negative rickettsialike organisms to occur within the intestinal epithelial cells and fat cells and between the fibers of nerve ganglia in a male silver fish, *Lepisma saccharina*. The microorganisms were straight rods with rounded ends and measured 0.2 by 0.2 to 0.7 microns in size.

ORTHOPTERA

Blattidae (Cockroaches). All species of blattids harbor symbiotes in certain of their tissue cells. Some aspects of the biologic relationships between the cockroaches and their intracellular microorganisms have already been used as examples in the discussion on the preceding pages (see pp. 204 and 207.

After Blochmann's (1887) early reports on the bacteriumlike bodies in certain cells of the fat bodies of *Blattella germanica* and *Blatta orientalis,* Wheeler (1889), Cholodokowsky (1891), and Heymons (1892) confirmed their presence. It was not immediately accepted that these were

Fig. 69. *Lepisma saccharina.* (From Comstock, 1936; after Lubbock.)

living organisms, however, since such investigators as Cuénot (1896), Schneider (1902), Prenant, Bouin, and Maillard (1904), and Henneguy (1904) believed them to be accumulations of metabolic products or other materials. In 1906 Mercier claimed to have cultivated the inclusions, obtaining a sporeforming bacillus (*Bacillus cuenoti*) which later workers (Javelly, 1914; Hertig, 1921; Wollman, 1926; Glaser, 1930a,b; and Gier, 1936) were unable to isolate. Some of them were convinced that Mercier was working with a contaminant. Mercier's work is supported by Gropengiesser (1925), who isolated a bacillus similar to *Bacillus cuenoti*. As has already been mentioned, recent work has tended to discount Mercier's organism as being that occurring in the cockroach bacteriocytes. Glaser (1930a,b) isolated diphtheroids from *Periplaneta americana* and *Blattella germanica* and named them *Corynebacterium periplanetae* var. *americana* and *Corynebacterium blattellae*. However, Gier (1936, 1937) has been unable to confirm any of these isolations.

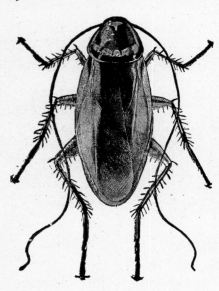

Fig. 70. *Periplaneta americana*. (From Comstock, 1936; after Howard and Marlatt.)

Nevertheless, most authorities in this field feel that the symbiotes of roaches are perhaps living organisms and should be considered as such. Furthermore, they have been observed in large numbers of blattids including *Periplaneta americana, Periplaneta australasie, Blatta orientalis, Blatta lapponica, Blatta aethiopica, Blattella germanica, Ectobia livida, Heterogomia aegiptica, Epilampra grisea, Nauphoeta cinerea, Derocalymma stigmosa, Platyzosteria armata, Homalo demascruralis, Loboptera decipiens, Parcoblatta virginica, Parcoblatta pennsylvanica,* and *Blabera* sp.

The mycetocytes or bacteriocytes containing the symbiotes are arranged in various ways in the different species. The simplest occurs in *Blatta aethiopica* where the mycetocytes occur at random throughout the fat tissue. Some species (*Eurycotis*) have a single row of mycetocytes lying longitudinally in the fat body. Some (*Nauphoeta cinerea*) have two rows. Others have three, four, and up to twenty rows. Whatever the arrangement, there is always at least one layer of fat cells on all sides of the mycetocytes. As to their location in the lobes of the fat body, it seems that there are rarely, if ever, any mycetocytes in the peripheral fat lobes. According to Wolf (1924a,b), in the fat lobes nearest the male and female gonads the mycetocytes frequently may be observed on the immediate periphery. In this connection it should be remembered that mycetocytes are also present in the connective tissue covering both the testes and ovaries and surrounding the oöcytes.

The microsymbiotes themselves vary with respect to their hosts. In adult roaches, particularly the female, the organisms are considerably longer than those in the immature stages. Fraenkel (1921) found the sizes of the symbiotes in the different species to vary thus: In *Blatta aethiopica* they measured 1.6 microns in length, in *Homalo* 2.2 microns, in *Blattella germanica* 2.6 microns, in *Platyzosteria* 4.6 microns, in *Periplaneta* 2.7 to 5.3 microns, in *Derocalymma* 5.9 microns, and in *Heterogomia* 5.3 to 9.0 microns. Other workers have arrived at different figures, but these apparent contradictions may be explained by the fact that in some cases at least, the symbiotes become much smaller when the insect host is starved. Furthermore, according to Mercier (1906, 1907), the shape and structure of the symbiotes may also change upon starvation of the host. Normally the organisms are oval to rod-shaped, slightly curved, occurring singly and occasionally in short chains or threads, the endoplasm usually containing vacuoles. On the other hand, in starving insects the ends of the rods stain more deeply, the centers stain lightly, and the ends may enlarge. Mercier claims they may even go over into spore formation, but this has not been confirmed. Long-continued starvation may also bring about a decrease in the number of symbiotes. Gier (personal communication) found that this decrease could also be brought about by sublethal doses of X-rays or ultraviolet light, and by the injection of crystal violet, hexylresorcinol, and metaphen.

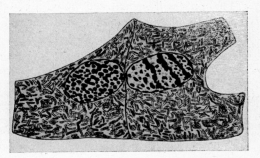

Fig. 71. Two bacteriocytes from the fat body of *Blatta orientalis*. The cells are filled with bacteriumlike symbiotes. (From Mercier, 1907.)

The embryologic considerations of this relationship have already been discussed to some extent (see p. 207). Suffice it to say that as the result of the investigations of Blochmann (1887), Mercier (1906, 1907), Buchner (1912), Fraenkel (1921), and Gier (1936) the "life cycles" of the symbiotes of roaches are fairly well known. In brief, the oöcytes in the ovaries are surrounded by one or several layers of symbiotes. As the egg develops the oöcyte membrane breaks, allowing the symbiotes to enter the cytoplasm. This introduces them into the embryo, and another generation-to-generation transmission has been accomplished. Apparently the organisms in the eggs are always smaller than those in the adult. The behavior of the symbiotes is similar in all roaches though variations do occur, especially in the formation of the primary mycetome in the egg. The reader is referred to Buchner (1930) for a more complete discussion of these variations.

Though it will be discussed more fully in the section on intracellular yeasts, the fact should be mentioned here that some workers (Mercier, 1906, 1907; Gropengiesser, 1925) have observed yeastlike parasites in roaches. These forms (probably *Torula*) crowd the regular intracellular organisms and apparently may replace them in the mycetocytes altogether.

NEUROPTERA

Chrysopidae. In 1923 Cowdry observed gram-negative, rickettsialike organisms within the pigmented fat cells of *Chrysopa oculata*. The microorganisms were rod-shaped and measured 0.2 by 0.5 to 1.2 microns in size.

CORRODENTIA

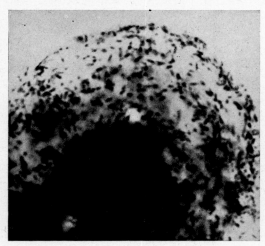

Fig. 72. Smear preparation showing rickettsialike microorganisms in the Malpighian tubes of the book louse *Dorypteryx pallida*. (From Hertig and Wolbach, 1924.)

Psocidae. A small, rickettsialike organism was observed by Sikora (1918, 1920) on the stomach epithelium of the dust louse, *Psocus* sp. Every individual specimen examined harbored the organism, which was transmitted to the next generation through the egg.

Atropidae (Book Lice). Hertig and Wolbach (1924) found a tiny, rickettsialike organism in the cells of the Malpighian tubes and ovaries of *Dorypteryx pallida,* both nymphs and adults (fig. 72). These organisms are gram-negative, coccoid and rodlike forms measuring 0.25 to 0.3 by 0.5 to 0.6 microns in size.

MALLOPHAGA

Philopteridae. The pigeon louse, *Columbicola columbae* (= *Lipeurus baculus*), passes its symbiotes to the next generation via the egg. Ries (1931) found that the symbiotes of the adults migrate individually from the mycetocytes up to the ovariole and into a depression at the posterior end of the egg. Although the symbiotes are not as closely associated with the germ band as those in *Pediculus*, like the latter they are carried more deeply into the yolk with the invagination of the germ band. Yolk cells mingle with the mass of symbiotes, becoming the nucleus of the mycetocytes, which are surrounded by membranes. In the male embryos these mycetocytes penetrate the epithelium of the midgut and migrate into the fat bodies. In the female embryos they pass into the ampullae of the reproductive organs. In the adult the mycetocytes have two or three nuclei. The symbiotes themselves are very small and do not stain easily. Sikora (1922) found rickettsialike organisms in the intestinal lumina of *Columbicola columbae* and *Trinoton* sp. (Menoponidae).

In *Lipeurus lacteus* the mature female retains some mycetocytes under the hypodermis as well as in the ampullae of the reproductive organs, and in both locations the symbiotes tend to be spherical. Similarly, in *Lipeurus frater* the symbiotes are large, round, and sausage-shaped.

Very similar to *Lipeurus* in their symbiotic arrangements are the species of *Nirmus*. These include *Nirmus merulensis, N. fuscus,* and *N. subtilis*.

The symbiotes of *Docophorus leontodon* are very thin rods which stain with difficulty. In the immature females the mycetocytes gather about the strands connecting the ovarioles and oviduct, and penetrate them as entire mycetocytes. Similar symbiotes occur also in *Docophorus icterodes*.

Other philopterids which are known to harbor symbiotes include *Goniocotes compar, Goniocotes hologaster,* and *Goniodes damicornis*. In this last species the mycetocytes are not situated between the fat lobes but are enclosed in them in much the same way as they are in the roaches. According to Buchner (1930), in an unidentified species of *Goniodes*, unlike any of the other symbiote-carrying lice, no ampulla is formed, and in both sexes and at all stages the symbiotes are either in scattered mycetocytes or entirely free.

Trichodectidae. Buchner (1930) states that no symbiotes have been found in the family Trichodectidae of which Ries examined a large number of specimens, or in the whole suborder Liotheidae, as indicated by the examination of *Gyropus, Gliricola, Physostomum, Menopon, Nitzschia,* and others. It should be mentioned, however, that rickettsialike organisms have been observed in some of these groups. For example, Arkwright (1923) found such an organism in the intestine of the goat louse, *Trichodectes caprae,* and earlier Hindle (1921) found 8 per cent of his specimens of the horse louse, *Trichodectes pilosus, to* harbor in their alimentary tracts an organism he named *Rickettsia trichodectae.* Cowdry (1923) discovered that the intestinal lumen of the hen louse, *Menopon gallinae* (= *M. pallidum*) contained gram-positive and gram-negative diplococci, and Hertig and Wolbach (1924) observed great numbers of a characteristic small bacterium in their specimens. In one specimen they found masses of rickettsialike organisms near the ovaries. In two other specimens small rickettsialike organisms were found in the mid- and hindgut, and in the Malpighian tubes (fig. 73).

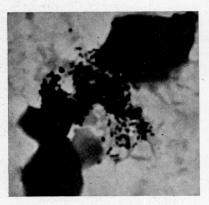

Fig. 73. Smear preparation of a Malpighian tube of the hen louse, *Menopon gallinae,* showing rickettsialike microorganisms. (From Hertig and Wolbach, 1924.)

Haematomyzidae. In the elephant louse, *Haematomyzus elephantis,* the adult male has paired mycetomes (fig. 74). The female carries its symbiotes in the oviducts where the mycetocytes are large and multinuclear as though fused with one another. Although Ries (1931), in studying the symbiotes of this insect, considered it an Anoplura, Ferris (1931) has shown that it is not a sucking louse and has placed it with the Mallophaga.

ANOPLURA

The rickettsial organisms associated with the true lice have been studied in considerable detail. However, since those organisms which have had taxonomic consideration and have already been placed in the family Rickettsiaceae are being considered in another chapter, our discussion here will exclude such organisms as *Rickettsia prowazeki, R. rocha-limae, R. pediculi,* and *R. linognathi.*

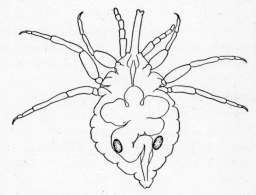

Fig. 74. *Haematomyzus elephantis.* Schematic section through an adult male. The paired mycetomes are located in the fat tissue without relation to the intestinal tract. (From Ries, 1931.)

BACTERIUMLIKE AND RICKETTSIALIKE SYMBIOTES

Pediculidae. Many early investigators observed and partially described the mycetomes of lice without the slightest idea as to their significance. Of course, one can hardly have expected them to be able to do so back in 1752 and earlier when Hooke and Schwammerdam (see Buchner, 1930) saw in the bodies of lice a yellowish structure which Hooke felt was some kind of liver and Schwammerdam thought was a gastric gland. Other early writers who mention this peculiar "organ" include Landois (1864), Graber (1872), Cholodkowski (1904), and Müller (1915). Sikora (1916) at first did not seriously consider the inclusions of this structure in the body louse to be microorganisms. Later (1919), however, she recognized their true significance as did Buchner (1919). Most of the details of the symbiote's life cycle in the host were worked out by Ries (1931).

In the larva of the human head louse, *Pediculus humanus capitis,* the symbiotes are located in a small "organ" known as the "stomach disk" (the *Magenscheibe* of the German workers). This structure is round or oval in shape and is attached to the outer ventral midgut where it lies exactly in the midventral line of the body, slightly nearer the anal region than the head. It consists of cells derived from the midgut and is covered with an irregular outer layer of mesodermal tissue. The enclosed spherical cavity is formed as a part of the stomach lumen during the louse's embryonic development. After the larva is hatched, however, this structure has no open connection with the gut lumen. The symbiotes are enclosed in the 12 to 16 chambers into which the stomach disk is radially divided. For further description of this structure we quote from Aschner (1934):

> The whole organ has an opaque, yellowish colour, produced by certain granules of high refraction index, enclosed in the inner cells. Because of its colour, which stands out sharply against the dark red background of the blood-filled stomach, it can easily be recognised in the living animal even with an ordinary hand lens.
>
> Prior to the formation of the stomach disc in the embryo the symbionts are enclosed in a group of cells which float freely in the yolk inside the lumen of the midgut to the secondary mycetome, the stomach disc, in the following manner. The primary mycetome migrates to the place on the ventral midgut wall where later the stomach disc is formed. Here the wall of the midgut is pushed out in a hernia-like fashion so that a pocket formation results which includes the primary mycetome with its symbionts. This pocket is closed later in such a way that the symbionts are retained in the newly formed cavity and the cells of the primary mycetome are pushed back into the yolk where they degenerate and are finally dissolved. The newly formed cavity containing the symbionts is then divided by radial septa into chambers and thus the previously described stomach disc results [see fig. 75]. In the male host the symbionts remain in the stomach disc from the larva stage until its death, while in the female host a few days before the imago stage all the symbionts emigrate from the stomach disc and infect a certain region in the oviducts. This region later develops into the so-called ovarian ampules.

In the adult, according to Buchner (1930), the structure consists of a fibrous cortical layer, a central syncytium, and a mass of tissue which occupies the space between the two and in which are imbedded the large vacuoles which shelter the symbiotes. In the female louse this structure, even during the development of the immature stages, is completely without symbiotes. Instead, the microorganisms are present in the ovarian ampules, and eventually the female mycetome becomes a degenerate cell mass. In the male lice, on the other hand, the mycetome frequently remains intact possessing symbiotes, though there are usually some internal signs of degeneration. Preparatory to transmission to the next generation, the symbiotes invade the ovaries and gain entrance to the egg through the egg stem. At the time of deposition a mass of symbiotes may be found at the posterior end of the egg.

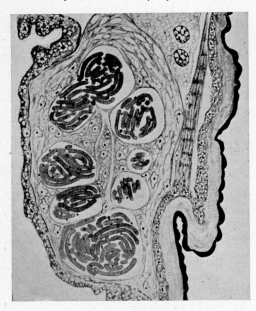

Fig. 75. A sagittal section of the stomach disk of *Pediculus humanus capitis* showing the symbiotes which lie in groups separated by septa. The midgut epithelium is on the left. (From Ries, 1931.)

According to Johannsen and Butt (1941), these become imbedded in a depression in the thickened germ band which invaginates into the yolk carrying the mass of symbiotes at its upper or caudal end. The symbiotes are enclosed in a syncitial envelope or mycetome which passes out of the germ band into the yolk at the anterior end of the egg. Later the mycetome is carried posteriorly near the region of the future midgut. Johannsen and Butt (see also Schölzel, 1937) describe its subsequent development thus:

As soon as the anlage of the midgut is formed, the mycetom is pushed into the still irregularly formed mid-gut epithelium, developing there a pocket, or diverticulum. The diverticulum at first encloses some yolk in addition to the mycetom. Its connection with the mid-gut becomes so constricted that only a narrow canal unites them, through which the syncitium composing the mycetom and the yolk is discharged, leaving the symbionts behind in the pocket. The canal then closes, the syncitium later degenerating in the yolk. The diverticulum, or pocket, containing the symbionts, enveloped by a layer of mesoderm cells, may now be designated as the "second mycetom" and forms the structure that in the larval louse has been called the "stomach disk."

Apparently the body louse, *Pediculus humanus corporis,* has a relationship to its symbiotes very similar to that of *Pediculus humanus capitis.* According to Buchner (1930), the pubic or crab louse, *Phthirius pubis,* is also very similar in this respect except that the mycetome or stomach disk consists of 20 to 24 chambers instead of 12 to 16 as in *Pediculus.* The pubic louse also was found by Guimarais (1922) to harbor extracellularly a rickettsialike organism in its intestine.

In the larvae and adults of both sexes of *Pedicinus rhesi* numerous mycetocytes may be found between the intestinal epithelium and the basal membrane (fig. 76). They form a limited zone around the gut like a girdle, causing a constriction. The symbiotes leave the cells and are carried by the blood stream back to the oviducts which they invade (Buchner 1930). Transmission to the next generation is through the egg.

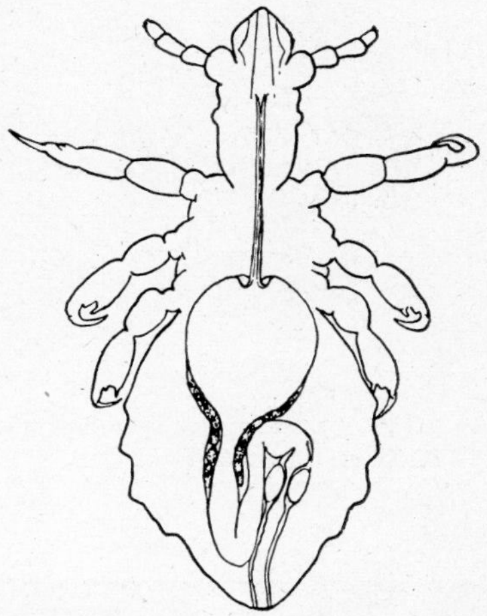

Fig. 76. *Pedicinus rhesi.* Schematic section through an adult male showing the mycetome located between the midgut epithelium and the basement membrane. (From Ries, 1931.)

Haematopinidae. In 1919, independently of each other, Sikora and Buchner observed an intracellular symbiote in the hog louse, *Haematopinus suis.* Hindle (1921) mentions having seen this organism in specimens he examined, and the studies of Florence (1924) have also added to our knowledge of this association.

According to Florence, the symbiotes are rather large, very pleomorphic gram-negative rods, approximately 1 micron wide and 2.5 microns or more long (fig. 77). The longer forms vary considerably in width. With Giemsa's or Wright's stain they stain homogeneously violet with no internal structure discernible. All attempts to cultivate them have been unsuccessful.

The mycetocytes containing these symbiotes are located on the wall of the mid-intestine (fig. 78). Florence believes they are modified epithelial cells which have been invaded by the symbiotes. On the other hand, Buchner (1919) believes them to be special cells, lying between the muscle layers and the epithelium, that push their way through the epithelial cells into the lumen of the mid-intestine. In any case, the cytoplasm of the host cells is pushed toward the periphery, and the balls of filamentous symbiotes occupy the center. Upon reaching the limits of its growth the cell ruptures, freeing the

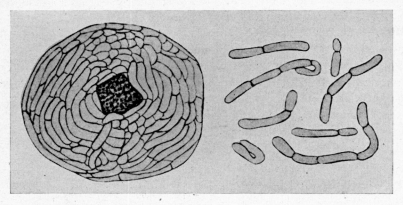

Fig. 77. A mycetocyte and the symbiotes of *Haematopinus suis*. (From Buchner, 1930.)

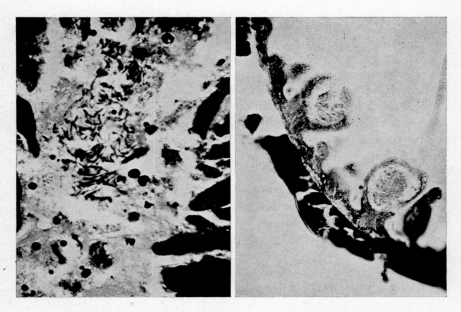

Fig. 78. Symbiote of the hog louse, *Haemotopinus suis*. (Right) Transverse section through the stomach of a mature louse three hours after feeding showing symbiotes in epithelial cells. (Left) Longitudinal section through the stomach of a mature louse showing symbiotes lying free in the lumen. (From Florence, 1924.)

symbiotes into the lumen where the filaments break up into rods and short chains. Transmission to the next generation takes place through the egg, there being a mycetome in each oviduct.

Florence supported the hypothesis that the symbiotes were in some way connected with the physiology of digestion.

The behavior of the symbiotes in the short-nosed cattle louse, *Haematopinus eurysternus,* and in the horse louse, *Haematopinus asini,* is very similar to that

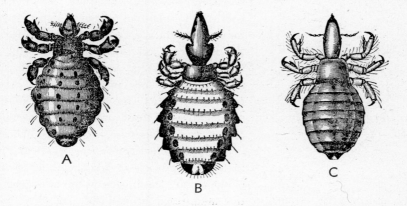

Fig. 79. *A*. The short-nosed ox louse, *Haematopinus eurysternus*. *B*. The horse louse, *Haematopinus asini*. *C*. The hog louse, *Haematopinus suis*. (From Comstock, 1936; after Law.)

in *Haematopinus suis*. According to Buchner (1930) describing the work of Ries, the intestines of male and female *Haematopinus eurysternus* are similarly furnished with mycetocytes which they retain throughout life. In the female a large number of the mycetocytes pass through the intestinal epithelium into the dorsal region and form three provisional azygous mycetomes in the fatty tissue. Ries and Buchner called these "depot mycetocytes" because they have the function of storing the symbiotes for future use. Just prior to the third molting the cell boundaries disappear, and as the last larval cuticula comes off, many symbiotes pass in masses through the loose hypodermis to the space between the nymphal and imaginal chitin, which is filled with molting fluid. In the meantime the sex apparatus has formed a wide opening at one of its ends while the other terminates in a peculiar nest of cells which receive the incoming symbiotes. These nests of cells, one on either side, then become loose and migrate to the place where the ampullae are to develop. Here the symbiotes leave their conducting or transporting cells and invade new cells at the rear of the ovarian ducts, which become the definitive mycetocytes of the ovarian ampulla.

It should, perhaps, be mentioned here that this interesting shifting of the symbiote from one place to another varies in different lice. For example, in *Linognathus* and *Polyplax* the location of the symbiotes is changed only once and then only in the female. In *Pediculus* they are shifted twice in the male and three times in the female. And, as has just been described, in the female *Haematopinus* they are shifted from yolk cells to intestinal cells or to depot mycetomes, to conducting cells, and to ampullae; in the male, from yolk cells to intestinal cells.

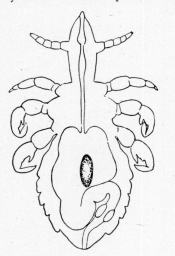

Fig. 80. *Linognathus vituli*. A schematic section through an adult male showing the mycetome ventral to the midgut. (From Ries, 1931.)

Buchner (1930) has observed the presence of azygous mycetomes in *Linognathus vituli* (= *L. tenuirostris*) (fig. 80) and in *Linognathus piliferus*. Histologically, the mycetome of the latter, a dog louse, is composed of well-defined, mononuclear mycetocytes. In the female the symbiotes invade the walls of the oviduct, and transmission to the next generation takes place through the eggs.

The rat louse, *Polyplax spinulosus*, has a similar azygous mycetome arrangement (fig. 82). As in the previous cases, the mycetome persists in the male, but in the female the symbiotes are found in the sex organs.

HOMOPTERA

Cicadellidae (Leafhoppers). Swezy and Severin (1930) investigated a rickettsialike organism they found both extracellularly and intracellularly in the sugar-beet leafhopper, *Eutettix tenellus*. In their intracellular position the symbiotes are found in the cytoplasm of the cells of the mid-intestine. Sometimes the cells are distended with microorganisms, and at other times only a few of the latter may be seen. The fat cells, cells in the posterior portion of the esophagus near the esophageal valve, and the blood may also show the presence of symbiotes.

The organisms are gram-negative diplobacilli, with somewhat rounded ends. from 0.24 to 3 microns in length.

Fig. 81. The dog louse, *Linognathus piliferus*. (From Comstock, 1936; after Law.)

Swezy and Severin conclude that *Eutettix tenellus* harbors two different organisms, the difference being that one is filterable while the other is not. The connection either of these may have with curly-top disease of sugar beets is not conclusively demonstrated.

Aphididae (The Aphids). As we shall have occasion to point out again, there has been considerable controversy among European investigators as to the true nature of some of the intracellular symbiotes in aphids or plant lice. Although early workers believed them to be yeasts or yeastlike organisms, the present trend is to consider them very pleomorphic intracellular bacteria. Paillot (1933) is one of the leaders of those who are convinced of their bacterial nature, and he has presented much evidence supporting his views. This worker believes that symbiosis should be considered as a particular case of

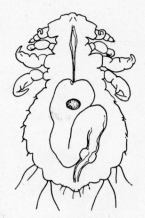

Fig. 82. *Polyplax spinulosus*. Schematic section through an adult male showing the mycetome ventral to the midgut. (From Ries, 1931.)

antimicrobial immunity and that the various large forms found in aphids are merely the result of humoral reactions against the regular bacteria associated with aphids. Paillot feels that their manner of multiplication (by fission) is also evidence of their bacterial nature.

Fig. 83. A group of aphids. (From Comstock, 1936.)

Careful study of these pleomorphic forms, according to Paillot, reveals that they contain no true nucleus. With iron-hematoxylin stain most of them appear homogeneous in structure though the center is frequently stained more deeply than the periphery. Giemsa stain brings out chromatin and basophilic granules. The mycetomes in which the aphids harbor these symbiotes have already been described in an earlier section of this chapter.

When the various forms occur together in the same insect specimen, they frequently appear as different species of microorganisms. An example of this is found in *Eriosoma lanigerum,* in which large rounded forms may appear along with filamentous or typical short bipolar bacilli (fig. 85). Some of the larger forms, which Paillot calls "growth forms," contain what appear to be nuclei but which Paillot calls "pseudonuclei," and which divide by simple fission when the cell undergoes division. All these forms presumably arise from the one simple bacterial form.

Tetraneura ulmi has two forms of the same symbiote in its mycetome. One is barred, is more or less elongated, and usually has one end pointed. The other is large, not vacuolated, and has irregular contours. Similarly, *Chaitophorus aceris* has two types of symbiotes. One is large and round, has a granular protoplasm, and occurs in groups or masses. The other occurs in isolated elements which are paired and coffee-bean-shaped. *Chaitophorus lyropictus* has one form which appears as a typical bipolar coccobacillus and another large, round form, 0.5 to 1.2 by 1 to 5 microns in size (fig. 86). The symbiotes of *Drepanosiphum platanoides* occur in two greatly contrasting morphologic forms (fig. 87). One is a typical bipolar bacillus frequently considerably elongated. The second form is a large, very pleomorphic, elongated cell varying considerably in width and considerably enlarged and swollen at points. These forms resemble fungi but originate from the smaller bacterial forms. Other species of aphids having similarly contrasting forms include *Macrosiphum tanaceti*, *Macrosiphum jaceae*, *Macrosiphum rosae*, *Aphis atriplicis*, *Aphis rumicis*, *Aphis pomi*, *Aphis forbesi*, *Pterocallis juglandicola*, and *Rhopalosiphum ribis*.

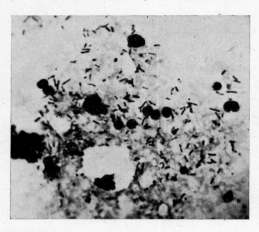

Fig. 84. Bacteriumlike symbiote in tissues of the cabbage aphid, *Brevicoryne brassicae*. (Photo by N. J. Kramis.)

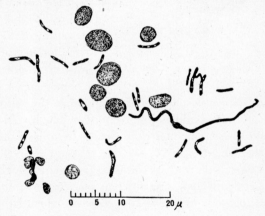

Fig. 85. Symbiotes of the aphid *Eriosoma lanigerum* showing both the typical short bacillary forms and the large round forms. (From Paillot, 1933.)

Klevenhusen (1927) discusses cases of mono-, di-, and tri-symbiosis and the manner in which the eggs and embryo become "infected" with one, two, or three symbiotes. What Klevenhusen describes as primary symbiotes are found in all aphids. Although characteristic for the particular species of aphid, these primary forms are invariably spherical in shape, never degenerate, and are located in specialized tissues. Phylogenetically these symbiotes are considered the oldest. The secondary symbiotes are often seen degenerating and are usually rod- or sausage-shaped. These forms are not as well localized as the primary ones and may occur free in the body cavity, blood cells, connective tissue cells, oenocytes, or syncytial cell masses.

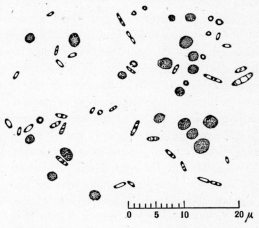

Fig. 86. Symbiotes of *Chaitophorus lyropictus* showing the typical bipolar coccobacillus and the large round forms. (From Paillot, 1933.)

Attempts to cultivate the symbiotes of aphids have generally been unsuccessful. Pierantoni (1910a) and Peklo (1912, 1916) have reported success in some cases, but their work remains to be confirmed. It is very possible, however, that if this work were repeated, using some of the newer culture methods and techniques, definite success would be forthcoming. This would also aid considerably in establishing the true nature of these microorganisms.

Adelgidae. In most of the Adelgidae the mycetome consists of two strands or elongated groups of cells alongside the gut. *Adelges laricis, Sacchiphantes viridis,* and *S. abietis* have a single type of elongated symbiote located in mononucleate mycetocytes. *Pineus pini, Pineus orientalis, Pineus strobus, Dreyfusia nordmannianae,* and *Dreyfusia picea* harbor two kinds of symbiotes. One is spherical, of constant size, and lies inside mononucleate cells; the other is oval, of varying size and shape, and is situated in multinucleate cells. In *Pineus pineoides* the mononucleate mycetocytes are absent, and the

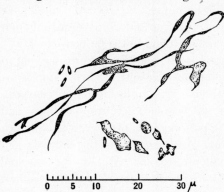

Fig. 87. Symbiotes of *Drepanosiphum platanoides* showing the short bacillary cells and the large pleomorphic forms. (From Paillot, 1933.)

rod-shaped symbiotes are located in more or less round groups in the cells of the fat body. The exact nature (bacteria or yeasts) of the symbiotes has not been determined.

According to Profft (1937), who has studied the symbiotes of this group of insects in some detail, transmission of the organisms to the next generation occurs through the egg. "Infection" takes place by the accumulation of symbiotes between the follicle cells around the eggs. They pass into each egg and into definite cells at the time of invagination of the embryo. In those insects which carry two kinds of symbiotes the organisms separate and the cells enclosing one kind remain mononucleate, those enclosing the other kind become multinucleate. In each insect species the mycetome and symbiotes are alike in all generations. In sexual females the entire symbiote population passes into the eggs. There is no essential difference in the behavior of the symbiotes between the holocyclic and the anholocyclic species.

Profft also examined three species of Phylloxeridae (*Dactylosphaera vitifoli, Acanthochermes quercus,* and *Phylloxera coccinea*) but found no symbiotes of any kind present.

Chermidae (Psyllidae), (The Jumping Plant Lice). In 1937 Profft studied the mycetomes and symbiotes of 22 species of jumping plant lice of the genera *Psylla, Arytaena, Strophingia, Aphalara, Phyllopsis, Trioza,* and *Trichochermes.* In most of these insects the mycetome consists of a single structure in the young and a paired structure in the adult, and during ontogeny varies in its shape and location. The mycetome is composed of a central syncytium and mononucleate mycetocytes, which more or less make up the border. In *Strophingia ericae* and in an unidentified *Trizoa,* unlike most of the other species, there are no symbiotes present in the syncytium, although in the latter the corresponding symbiotes are found in the fat body.

The symbiotes in this group of insects are elongated organisms varying in shape from short rods to filaments. Tinctorially and in their cytoplasmic structure the syncytial symbiotes differ somewhat from those in the mycetocytes. Both types of symbiotes enter the ovarian eggs simultaneously and are taken up by the follicle cells at the posterior pole of the egg from which place they pass into the egg itself. Upon the formation of the embryo's germinal streak, the two types of symbiotes separate. At first the syncytial symbiotes occupy mononucleate cells in the mycetome's periphery—these cells later lose their walls and become the central syncytium. The peripheral cells of the final mycetome originate in the center of the embryonic syncytial mycetome, which forms cell walls, and become the mononucleate mycetocytes.

Ortheziidae (Ensign Coccids). Buchner (1921c) found intracellular bacteria in peculiar fat cells scattered throughout the tissues of *Orthezia insignis*. The symbiotes are slender rod-shaped organisms which frequently fill the cytoplasm of the cell and lie parallel to each other, thus simulating bands or fascicles of constant width. The fascicles seem to be held together by some sort of a mucous secretion. The size of the symbiotes vary from short to very long rods, but those of each cell are usually uniform in size.

Transmission to the following generation takes place via the egg. In invading the egg the symbiotes first arrange themselves in a peculiar tactical formation between the follicular membrane and the anterior pole of the egg. At the proper time the egg plasma envelopes this mass of symbiotes and pulls it completely within itself.

Eriococcidae. The symbiotes of the citrus mealy bug, *Pseudococcus citri*, are considered by many to be yeasts or yeastlike organisms. Buchner (1930) finally came to the conclusion that the symbiotes of *Pseudococcus citri*, like those of *Pseudococcus adonidum*, were bacteria and not yeasts. The symbiotes of *Pseudococcus citri* may be spherical, oval, or sausage-shaped and usually have blunt, rounded ends. The protoplasm contains small granules but apparently no distinct nucleus.

Fig. 88. *Pseudococcus citri*. (From Comstock, 1936.)

The mycetome of *Pseudococcus citri* was first observed in 1893 by Berlese and was described in more detail later by Pierantoni (1910a, 1913). It lies below the intestine of the insect and is about a third of the length of the host. In the mycetocytes the symbiotes are contained within peculiar balls or spheres, which have a very interesting relationship to their contents. This and other aspects of the symbiotic arrangements of this insect will be considered in chapter VI.

As with all other species of *Pseudococcus* examined, the mycetome of *Pseudococcus brevipes* is unpaired and egg-shaped. Carter (1935) found it to lie in a median, somewhat posterior, position and close to and practically encircled by the midgut. Though the mycetome is usually light brown in color, in some specimens it has been found to be pale creamy white. The mycetome is enclosed in a mass of tracheae and grows by the mitotic division of the mycetocytes.

Carter considers the insect to be bisymbiotic since he observed two forms to occur in its mycetocytes. Probably Paillot would consider them to be two forms of the same microorganism. In any case, one of the forms is a rather large, bizarre, pleomorphic, yeastlike organism with homogeneous protoplasm, which sometimes shows extensions similar to buds. Some individuals are extremely vacuolated. As the insect approaches maturity, the forms are smaller and more spherical, probably being the stage in which they are passed to the embryo.

The other symbiote is a smaller, rod-shaped, bacteriumlike organism, which apparently occurs only under certain conditions of the host's nutrition. According to Carter (1936), the presence of the symbiote conditions the insect's oral secretions, and it is always found in mealybugs whose feeding produces green spots on pineapple leaves. By special techniques Carter found that this rod-shaped form disappears from the mycetome coincident with the loss of the green-spotting capacity. In its place appears an intermediate coccus-rod form which is a stable form in non-green-spotting insects.

The symbiotes of *Pseudococcus brevipes* are transmitted to the following generation via the egg in a manner similar to that of *Pseudococcus citri*. As the ovaries develop, the contents of certain mycetocytes aggregate into small, very dense, and deeply staining bodies. These migrate from the mycetome and enter the egg at its junction with the nurse cell. Apparently these dense, deeply staining masses take care of both types of symbiotes since both forms are found in the newly born larvae.

Other Homoptera. Most of the intracellular symbiotes associated with species of Homoptera come under the category of yeasts and yeastlike organisms. (See chapter VI.) On the other hand, as we have just seen, an occasional species is encountered in which the symbiotes have more of the appearance of bacteria than they do of yeasts. At the present time, however, it is not safe to make dogmatic statements concerning most of these. Frequently, as we have seen in the case of the aphids, what appears as two species of microorganisms occurring together in the same insect may actually be two widely divergent forms of the same species. Such, for example, may be the case with the symbiotes found in *Asterolecanium variolosum,* in which small, round to oval inclusions occur together with typical rod-shaped, motile bacteria.

The periodical cicada or seventeen-year locust, *Magicicada septendecim,* was found by Cowdry (1925a) to be heavily infested with a rickettsialike organism in eleven out of fifty specimens he examined. The distribution of the organisms was limited to the lumina of the alimentary tract and the Malpighian tubes. Cowdry seldom observed them in cells. It is possible that the organisms are transmitted transovarially.

HEMIPTERA

Most of the studies pertaining to symbiotes in Hemiptera have to do with the bacteria-filled ceca in members that feed on plant juices. For the most part, their symbiotes are located extracellularly. There are, however, some instances of intracellular symbiosis in this order.

Cimicidae (Bedbugs). One of the outstanding examples of the mycetome type of intracellular symbiosis is that of the bedbug, *Cimex lectularius*. The symbiote has been placed in the rickettsial group and named *Rickettsia lectularia*; its relationship will therefore be discussed in the chapter on rickettsiae. It should be mentioned, however, that some workers (Pfeiffer, 1931) believe that a bacterial symbiote also occurs in the bedbug. That this is at least an occasional possibility is indicated by the fact that Steinhaus (1941a) repeatedly isolated a diphtheroid organism (*Corynebacterium paurometabolum*) from the ovaries and mycetomes of this insect. He did not, however, claim it to be identical with the intracellular symbiotes.

Reduviidae (Assassin Bugs). Perhaps the first to report the presence of intracellular symbiotes in this group of insects was Wigglesworth (1936), who worked with *Rhodnius prolixus*, although in 1926 Duncan reported the presence in this insect of a gram-positive bacillus which may have been the symbiote. In 1937 Dias reported a similar organism in *Triatoma megista* and stated that symbiotes have been found in at least eight other species of blood-sucking reduviids (*Rhodnius prolixus, Triatoma rubrofasciata, Triatoma infestans, Triatoma sordida, Triatoma brasiliensis, Triatoma protracta, Eutriatoma flavida,* and *Psammolestes coreodes*).

Fig. 89. *Corynebacterium paurometabolum*. See also fig. 16. (Photo by N. J. Kramis.)

According to Wigglesworth, the intestinal lumen of newly hatched *Rhodnius prolixus* contains no microorganisms although the cells lining the anterior narrow segment of the midgut are packed with intracellular symbiotes. When the insect molts these microorganisms are set free into the lumen by means of vesicular swellings which extrude through the cell walls (fig. 90). Here they multiply in the undigested blood meal, and most of them are eventually digested. They have been observed to persist both extra- and intracellularly throughout the life of the insect.

At first Wigglesworth (1936) thought that the organism was transmitted to the next generation through the egg. However, in 1944 Brecher and Wig-

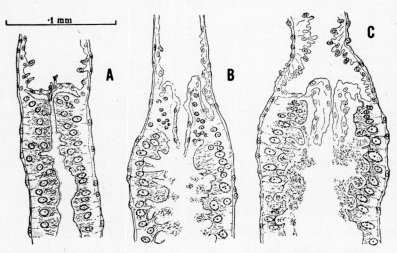

Fig. 90. Longitudinal sections of the junction between foregut and midgut of *Rhodnius prolixus* showing the intracellular bacteria and their extrusion into the gut lumen. *A*, an unfed, recently hatched, first-stage nymph; *B*, first-stage nymph, 7 days after feeding; *C*, second-stage nymph, 3 weeks after feeding (10 days after molting). (From Wigglesworth, 1936.)

glesworth reported that this is not the case, that the organism is taken up by the young nymph from the environment. It appears that the symbiotes may be acquired in several ways, chiefly from the dry excreta of other members of the species, but sometimes from the contaminated surface of the egg.

The symbiote itself was in 1936 reported by Wigglesworth as having the characters of a diphtheroid bacillus, and from the insect, but not the egg, he cultured such an organism on glucose agar. Intracellularly in the insect it stained poorly, though its diphtheroid character could be observed. Extracellularly it stained more deeply, resembling a typical diphtheroid. Involution forms occurred, becoming so beaded as to resemble "strings of cocci" which break up into clusters of coccoid bodies. The cultural forms were grampositive and non-acid-fast. Then in 1944 Brecher and Wigglesworth discounted some of the latter's earlier observations and reported the repeated isolation of *Actinomyces rhodnii*, an organism first reported and isolated by Erikson (1935) from the same insect (*Rhodnius prolixus*). Their medium was 5 per cent blood agar, on which at 30° C. the white to muddy-colored colonies were visible in 48 hours. On 10 to 20 per cent blood agar the colonies were dark red. They isolated what appeared to be the same organism from *Triatoma infestans* reared in their laboratory alongside *R. prolixus*.

Brecher and Wigglesworth's (1944) paper is of special significance since by rearing *Rhodnius prolixus* free from *Actinomyces rhodnii* they found that in the absence of the symbiote the insects grow and molt normally until the fourth or fifth instar only. Then, in spite of repeated blood meals, molting is delayed, or may fail completely, very few becoming adults. If reinfected with *Actinomyces rhodnii*, normal growth, molting, and egg production are resumed.

Definite proof is lacking as to the exact nature of this phenomenon and the benefit resulting from the presence of the symbiote. In 1936 Wigglesworth described certain limited experiments by Hobson, which, in the opinion of both men, supported the view that the organisms provide vitamin B to the insect's blood. Brecher and Wigglesworth (1944) similarly suggested that the cessation of the insect's development may be explained by some vitamin deficiency which may be corrected by the presence of *Actinomyces rhodnii*.

In *Triatoma rubrofasciata* rickettsialike bodies were found by Webb (1940) to occur both intracytoplasmically and intranuclearly in cells of the gut, salivary glands, ovaries, Malpighian tubes, and muscles. Very few organisms were present in the lumen of the intestine, although the lumen of the Malpighian tubes contained a "fair number." They were observed in every stage of the insect, including the egg.

The organisms themselves appeared in numerous forms: minute coccoid and paired coccoid bodies staining a brilliant red with Giemsa; minute bacillary bodies staining a light blue; slender, small, paired, lanceolate forms staining purple; filamentous organisms, occasionally branched, staining homogeneously blue, pinkish blue or purple; and rod-shaped organisms purplish red in color with darker-stained polar granules. The latter organisms Webb found to predominate in all of his smears. All forms were gram-negative. Attempts to cultivate the organism on ordinary culture media were unsuccessful.

Webb carried out some interesting animal experiments with this microorganism. Infections in small laboratory animals were sometimes fatal. White mice were more susceptible than were wild rats and guinea pigs. The latter animals were definitely susceptible when fed on a vitamin-deficient diet. Pathological lesions were produced, particularly in the brain.

Other Hemiptera. In addition to *Cimex lectularius* and the reduviids, a few other Hemiptera have been found to possess symbiote-containing mycetomes. Schneider (1940), for example, found such structures in the lygaeids, *Ischno-*

demus sabuleti (fig. 91), *Nysius senecionis, Nysius punctipennis, Nysius thymi, Nysius lineolatus, Ischnorrhynchus resedae,* and *Ischnorrhynchus ericae.* The genera *Ischnodemus* and *Nysius* have paired mycetomes, and *Ischnorrhynchus* has an unpaired mycetome. In the last two genera the structure is colored red. Transmission of the symbiotes to the next generation takes place through the egg.

COLEOPTERA

Intracellular symbiotes are known to occur in several families of Coleoptera, including Anobiidae, Cerambycidae, Chrysomelidae, Lyctidae, Curculionidae, Tenebrionidae, Cucujidae, and Bostrichidae. In some families, such as the Anobiidae, the symbiotes are in the nature of yeasts; in others, such as Bostrichidae, Curculionidae, Lyctidae, and Cucujidae, they are bacteriumlike.

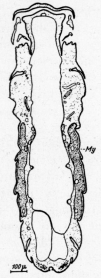

Fig. 91. *Ischnodemus sabuleti.* Schematic section of a young female adult showing the location of the mycetome (*My*). (From Schneider, 1940.)

Bostrichidae (Powder-Post Beetles). Three members of this family have been studied by Mansour (1934a,b) with respect to their intracellular symbiotes: the lesser grain borer, *Rhizopertha dominica,* and the wood borers, *Sinoxylon ceratoniae* and *Bostrychoplites zickeli.*

The embryologic development of the mycetomes in this group of insects has already been discussed on page 199 and will not be considered further here. The symbiotes themselves are pleomorphic bacteriumlike microorganisms. Their wide variation in size during the insect's life cycle is very interesting and has been determined by Mansour as follows:

Infection form in ♀ genitalia	5–9 × 1	microns
In segmenting egg	5 × 1.5	"
In developing embryo (fig. 92)	2.5 × 1.2	"
In larva	2.5 × 0.5	"
In pupa	2 × 0.5	"
In advanced adults in mycetomes	1.5 × 0.5	"
In testes (first invading forms)	1.5 × 0.5	"
In testes (growing form)	10–15 × 1	"

Lyctidae. Only a few species of this family have been studied with respect to their symbiotes. In 1926 Gambetta found that the larva of *Lyctus linearis* possessed mycetomes embedded between the fatty tissue and the gonads on both sides of the mid-intestine. These mycetomes are cylindrical, slightly curved structures. In the adult they are similar in appearance and lie near the first abdominal stigmata.

The symbiotes are oval- or rod-shaped organisms and are transmitted to the next generation through the egg. The symbiotes leave the mycetome and penetrate the follicle, where they lie between it and the surface of the developing egg.

In 1936 (a) Koch reported his observations on the mycetome and symbiotes of this same beetle (*Lyctus linearis*). According to this worker, larva, pupa, and adult each have two mycetomes which are isolated from the intestine and the Malpighian tubes and lie embedded in the fat body. Each mycetome consists of three main parts: a central layer of 7 to 12 large multinuclear mycetocytes, a group of 8 to 14 multinuclear, cortical mycetocytes located caplike on the central layer, and a thin-walled epithelial covering surrounding the whole organ.

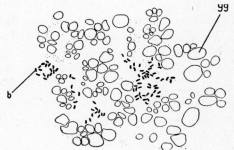

Fig. 92. Portion of a section through an advanced embryo of *Sinoxylon ceratoniae*, showing the symbiotes (*b*) between the yolk globules (*yg*). (From Mansour, 1934a.)

The mycetomes are inhabited by two different symbiotes. One symbiote, which occurs in rosettes, is located in the cortical layer. The other symbiote, round to oval in shape, is found in the central layer. Transmission of both symbiotes takes place through ovarian infection, and all eggs receive the organisms. The behavior of the symbiotes during embryogenesis is described by Koch in detail. Both sexes carry the symbiotes, which Koch feels belong to the bacteria. This worker also believes there is a possibility that the symbiotes are a source of vitamins to the beetle, or are somehow concerned with its nitrogen supply, since the main diet of the insect is the starch of the wood it ingests.

Curculionidae (Snout Beetles). The presence of bacteriumlike symbiotes is widespread in this family and Buchner (1930) suggests that all snout beetles may harbor them. In *Hylobius abietis* larvae the mycetomes adhere closely to the outside walls of the intestines at the juncture of the fore- and mid-intestines where they appear as a girdle of roundish humps. In a way these structures resemble the ceca of the more primitive types of these symbiotic relationships. As the old larvae near pupation, these humps become irregular and less numerous. The mid-intestine, which has appeared somewhat constricted just behind the mycetome girdle, now becomes distended. This distention is due to a layer of mycetocytes pushing itself backward between the newly formed mid-intestine epithelium and the muscularis. Following this comes a phase during which the mycetocytes in this enlarged section of the intestine penetrate deep in between the new intestinal epithelium, where they surround the groups of embryonic crypt cells. According to Buchner (1930), in the older, mating, and egg-laying beetles this zone of mycetocytes disappears and only a few isolated mycetocytes are left. He assumes that the mycetocytes have been gradually expelled into the intestinal lumen, where they disintegrate, freeing the symbiotes. How the bacteriumlike organisms are carried over to the next generation is not known with certainty. Buchner thinks that the symbiotes probably get into the female sex apparatus from the anus. At any rate, when the *Hylobius* larva emerges from its eggshell, it already possesses its symbiote-filled mycetomes. (See also Scheinert, 1933.)

Other snout beetles known to have symbiotic organs similar to, though somewhat divergent from, the larvae of *Hylobius abietis* include *Sternochetus lapathi, Molytes germanus, Miarus campanulae, Sibina pellucens, Brachycerus apterus, Brachyrhinus inflatus, Brachyrhinus gemmatus, Pissodes pini, Pissodes notatus, Odioporus glabricollus,* and *Boris granulipennis.*

Sitophilus granarius and *Sitophilus oryzae* have a characteristic relationship with their mycetomes that differs from that of *Hylobius abietis.* In *Sitophilus* the unpaired mycetome mass lies on the mid-intestine in the form of about twenty "ceca." After the disintegration of the mid-intestine has begun in old larvae, the mycetome penetrates between the new intestinal epithelium and the muscularis until eventually a layer of mycetocytes (or "accessory cell mass") surrounds the intestine. After this the formation of the ceca begins. According to Mansour (1934b), however, the accessory cell mass of *Sitophilus granarius,* which is much smaller than that in *Sitophilus oryzae,* contains no symbiotes.

The arrangement in *Rhynchophorus ferrugineus,* although similar to that of *Sitophilus,* differs in that whereas in *Hylobius* and *Sitophilus* the larval mycetomes do not persist until the adult stage, in *Rhynchophorus* they remain unchanged and in the same location. Similar in this respect are some of the Apioninae (species of *Aponida, Oxytoma, Erytrapion,* and *Protapion*) in which the symbiotic organs are found in the adults as well as in the larvae. In the Apioninae the mycetomes are oval, paired structures located on each side in the last larval segment. Each mycetome is made up of comparatively few, but very large, mycetocytes which contain symbiotes of extremely variable sizes, and "bubble-like" objects of unknown identification (Buchner, 1930).

Buchner has observed other interesting mycetocyte types in snout beetles. *Omphalapion laevigatum* has an extensive but sharply defined region in the fatty tissue which is uniformly filled with bacteriumlike symbiotes. In adult species of *Aspidapion* and *Perapion* there are no definite mycetomes, but the cells of the Malpighian tubes are thickly "infected." In *Smicronyx cuscutae*, *Ceutorhynchus punctiger*, and probably in *Rhyncolus lignarius* the mycetomes are in the posterior region of the larvae instead of up near the junction of the fore- and mid-intestines.

In 1933 Buchner listed those European species of curculionids which up to that time had been found to harbor symbiotes. According to the tribe or subfamily this list included the following species:

Otiorhynchinae

Brachyrhinus sulcatus
Brachyrhinus gemmatus
Brachyrhinus inflatus
Brachyrhinus corruptor
Brachyrhinus sulphurifer
Brachyrhinus cardiniger
Brachyrhinus niger
Brachyrhinus ventricola

Cleoninae

Coniocleonus glaucus var. *turbatus*
Bothynoderes punctiventris
Bothynoderes angullicollis
Bothynoderes meridionalis
Mecaspis alternans
Chromoderus fasciatus
Cyphocleonus tigrinus
Leucochromus imperialis
Leucochromus lehmanni
Neocleonus vittiger
Stephanocleonus superciliosus
Pseudocleonus cinereus
Cleonus piger
Cleonus verrucosus
Cleonus exanthineaticus
Cleonus maculatus
Cleonus strabus
Cleonus clathratus

Cleonus ophthalmicus
Adosomus roridus
Leucosomus pedestris
Pachycerus madidus
Eulixus myagri
Eulixus iridis
Eulixus subtilis
Dilixellus barbanae
Ortholixus sanguineus
Compsolixus ascanii
Lixochelus elongatus
Lixus paraplecticus
Lixus frater
Lixus fimbriolatus
Lixus cavipennis
Lixus coarctatus
Lixus truncatulus
Lixus praegrandus
Lixus plicaticollis
Lixus sturmi
Lixus rhomboidalis
Lixus nitidicollis
Lixus nebulofasciatus
Lixus massaicus
Lixus bisoleatus
Lixus bidentatus
Larinus latus
Larinus scolymus
Larinus cynarae
Larinus flavescens

Larinus jaceae
Larinus sturnus
Larinus turbinatus
Larinus planus
Rhinocyllus comicus
Rhinocyllus olivieri
Bangasternus orientalis

Curculionae

Phytonomus arator
Alophus pericarpus
Alophus triguttatus
Hylobius abietis
Liparus germanus

Brachycerinae

Brachycerus apterus
Brachycerus undatus

Calandrinae

Pissodes notatus
Magdalis armigera
Rhyncolus lignarius
Sternochetus lapathi
Stenocorus cardui
Sirocalus pulvinatus
Ceutorhynchus punctiger
Ceutorhynchus constrictus
Ceutorhynchus alliariae
Ceutorhynchus symphyti
Ceutorhynchus sulcicollis
Ceutorhynchus pleurostigma
Orobitis cyaneus
Cidnorrhinus quadrimaculatus
Rhinoncus bruchoides
Coryssomerus capucinus
Baris morio
Baris laticollis
Baris maculipennis
Baris chlorizans

Sitophilus oryzae
Sitophilus granarius
Rhynchophorus ferrugineus
Balaninus nucorum
Balaninus glandium
Dorytomus longimanus
Dorytomus melanophthalmus
Notaris acridulus
Smicronyx jungermanniae
Bagous binodulus
Elleschus bipunctatus
Tychius meliloti
Tychius flavicollis
Sibinia pellucens
Gymnetron antirrhini
Gymnetron villosulum
Gymnetron collinum
Miarus campanulae
Miarus sp.
Cionus hortulanus
Cionus scrophulariae
Cionus thaspi

Apioninae

Oxystoma craccae
Aspidapion aeneum
Omphalapion laevigatum
Protapion aestivum
Protapion assimile
Protapion varipes
Protapion nigritarse
Protapion flavipes
Erytrapion miniatum
Perapion violaceum
Perapion curtirostre
Eutrichapion loti
Eutrichapion virens
Exapion fuscirostre
Apion pisi
Catapion seniculus

Cucujidae (Flat Beetles). The presence of symbiotes in the cucujids was discovered by Pierantoni (1930), who worked with *Oryzaephilus surinamensis*.

The larva of this beetle has two large egg-shaped mycetomes lying dorsally over the intestine in the first and second segments and two lying ventrally in the third and fourth segments. The mycetocytes comprising the mycetome have a central polymorphous nucleus and a cytoplasm which is divided into cell-like regions containing rather large vermiform symbiotes (fig. 93). It is interesting that the center of these cell-like regions contains small, irregularly shaped nuclei.

As the insect goes into the pupal stage, it retains the four mycetomes, though the foremost two draw nearer together and lie over the intestine, and the two posterior mycetomes move farther apart. In the adult they become irregular, distorted, and considerably smaller. Transmission of the symbiotes occurs through the egg, which acquires the organisms from the follicles.

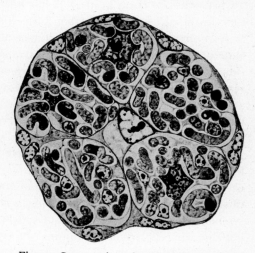

Fig. 93. Cross section of one of the four mycetomes of *Oryzaephilus surinamensis*. (From Koch, 1936b.)

Koch (1931, 1936b) has also studied the mycetomes and symbiotes of *O. surinamensis* and has been particularly concerned with the nature of their relationship to the host. By changing various conditions, such as diet and temperature, he has been able to free the beetle of its symbiotes. By holding the insects under a temperature of 36° C., Koch found the symbiotes to develop involution forms and then undergo a gradual process of degeneration. This injurious influence does not affect the symbiotes uniformly in all stages of the beetle's development. Only those stages that are thermosensitive show the short, compact, "infection" forms.

Along with the disintegration of the symbiotes, the cortical layer of the mycetome hypertrophies and finally makes up the main mass of the mycetome. The nuclei of the mycetocytes remain demonstrable in the sterile mycetome.

The descendants of these symbiote-free individuals are likewise without symbiotes, and this lack apparently causes no visible ill effects. Koch reared the beetle through 25 generations, and the sterile mycetome developed regularly in each.

A symbiotic arrangement similar to that of *O. surinamensis* is known to occur in *Oryzaephilus unidentatus*.

Chrysomelidae (Leaf Beetles). A few members of several known genera of chrysomelids are known to harbor intracellular symbiotes. In fact some, such as *Bromius obscurus,* have both intracellular and extracellular symbiotes. Stammer (1935, 1936) has studied the symbiotes of certain species of *Cassida, Bromius,* and *Donacia.* These insects have intestinal ceca and vaginal pouches very similar to the bacteria-containing ceca occurring in certain Hemiptera as described in chapter II. In particular, Stammer studied *Bromius obscurus,* which harbors intracellular symbiotes in 15 to 20 ceca located at the beginning of the midgut, and extracellular symbiotes in longer ceca attached to the posterior midgut. Passage of the symbiotes from one generation of beetles to the next is insured by two clublike vaginal pouches which provide the eggs with a superficially smeared layer of symbiotes. Upon hatching, the larva eats part of the eggshell thus acquiring the symbiotes. In the larva, both symbiotes live intracellularly in four small ceca at the beginning of the midgut and extracellularly in the lumina of the six Malpighian tubes. During the pupal stage some of the symbiotes are cast out into the intestinal lumen and colonize the newly formed imaginal organs.

Stammer found several species of *Cassida* in Germany to contain symbiotes in specialized ceca and vaginal pouches. *Cassida viridis* has four intestinal ceca and two groups of three vaginal pouches, *C. hemisphaerica* has four ceca and two groups of three vaginal pouches, *C. rubiginosa* and *C. vibex* have two intestinal ceca each and two groups of three vaginal pouches each, and *C. nobilis* has two ceca and two groups of two vaginal pouches. *C. nebulosa* and *C. flaveola* have no ceca, vaginal pouches, nor symbiotes. Unlike *Bromius obscurus,* the symbiote-containing species of *Cassida* harbor only one species of symbiote. The manner of transmission is similar, however, the larva acquiring the organisms from the eggshell. In the *Cassida* species the symbiotes are placed in a well-circumscribed heap at the upper pole of the eggs. In the intestinal ceca of both larval and adult *Cassida* insects the symbiotes live intracellularly.

In the females of all the *Donacia* species which Stammer (1935) studied, the bacteriumlike symbiotes were found in the cells of greatly enlarged U- or S-shaped loops of two of the six Malpighian tubes. At the time the eggs are about to be laid, the symbiotes pass out into the hindgut and from there are deposited in a mass at the upper pole of the egg in the foamy secretion which surrounds the eggs. When the larvae hatch, they ingest these symbiotes, which then locate themselves in the four ceca attached at the beginning of the larval midgut. Later the organisms migrate to the two Malpighian tubes of both sexes. When the insect pupates, the ceca and their symbiotes degenerate, but the organisms in the Malpighian tubes multiply abundantly in the females while in the males the symbiotes degenerate. In the adult beetles the symbiotes are short rods which become oval or almost spherical at the time of egg laying. In the ceca of the larvae they appear as long filaments which break up into short fragments.

Tenebrionidae (Darkling Beetles). Though Pierantoni (1929) found no symbiotes in *Tribolium confusum,* Koch (see Buchner, 1930) observed mycetocytes dispersed throughout the fatty tissue of the adult. Contained in these cells were numerous small rods or filaments. In the larvae, these probable symbiotes are not as definite in appearance.

DIPTERA

Ceratopogonidae (Biting Midges). Keilin (1921d) was the first to observe four large bodies in the thorax of larva, pupa, and adult *Dasyhelea obscura.* He noted that these bodies (mycetomes) were filled with bacteriumlike forms. Stammer (Buchner, 1930) studied *Dasyhelea versicolor* (fig. 94) and either *Dasyhelea flavifrons* or *Dasyhelea brevitibialis* and found similar symbiotes. *Dasyhelea longipalpis* was apparently devoid of symbiotes.

In 9 out of 27 specimens of *Culicoides sanguisuga* Hertig and Wolbach (1924) found tiny, rickettsialike cocci, diplococci, and very short rods varying in dimensions from 0.3 to 0.4 by 0.4 to 0.5 microns. They

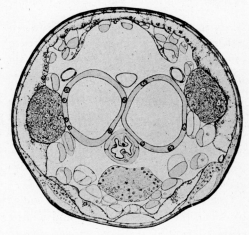

Fig. 94. Cross section through a larva of *Dasyhelea versicolor* showing the two mycetomes, one on either side, filled with rod-shaped organisms. (From Buchner's *Tier und Pflanze in Symbiose,* 1930, Gebrüder Borntraeger, Berlin.)

occurred free in smears of the abdomen and occasionally in the fat body. None were found in the intestine or Malpighian tubes.

Tabanidae (Horseflies). Hertig and Wolbach (1924) examined several species of *Tabanus* for the presence of rickettsialike microorganisms. They found them present in *Tabanus pumilus* and *Tabanus costalis,* but could not find them in *Tabanus lineola* or in *Tabanus abdominalis*. *Tabanus trispilus* showed no rickettsialike microorganisms but did show large numbers of diplococci, apparently within the intestinal epithelial cells.

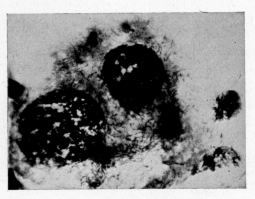

Fig. 95. Smear preparation showing rickettsialike microorganisms in pericardial cells of *Tabanus pumilus.* (From Hertig and Wolbach, 1924.)

In a single specimen of *Tabanus pumilus* Hertig and Wolbach found the Malpighian tubes and the pericardial cells to contain "large numbers of delicate rods and filaments" (fig. 95). The rods were curved, frequently having tapering ends and measuring 0.25 to 0.3 by 1.0 to 3.0 microns in size. The filaments were curved and beaded.

At least two intracellular organisms were found in *Tabanus costalis*. Rickettsialike organisms which morphologically resembled the rickettsia of Rocky Mountain spotted fever were found in the pericardial cells of one specimen. In the same and other specimens large coccobacilli were observed in cells of the rectal gland.

Pupipara (Hippoboscidae, Nycteribiidae, Streblidae). This group of insects is peculiar in that the larvae attain their full growth in the body of the female. Furthermore, though they are flies, in most cases the adults live parasitically, like lice, upon birds and mammals. For these reasons early interest was manifested in their symbiotes, particularly in the case of the sheep ked, *Melophagus ovinus*.

One of the symbiotes (*Rickettsia melophagi*) of the sheep ked has been placed in the rickettsial group and will be discussed in the chapter on rickettsiae. The other symbiote, universally present, is larger and definitely intracellular. The exact nature of these symbiotes is not known: some workers consider them to be yeasts, others think of them as forms of bacteria, and some have even regarded them as developmental stages of *Rickettsia melophagi*. At any rate, they occur in the cytoplasm of mycetocytes located in the epithelial lining of the posterior portion of the midgut. The mycetome here formed is always situated at the same sharp bend of the gut. The symbiotes are transmitted by the female to the young of the next generation through the secretion from the milk glands. A third symbiote has been found in certain specimens of *M. ovinus*. Zacharias (1928) observed this organism in the ordinary flat epithelial cells of the midgut and occasionally in the cells of the hindgut and Malpighian tubes. These organisms are also transferred through the milk glands.

Organisms similar to *Rickettsia melophagi* and found in other species of pupipara, particularly in the Hippoboscidae, have not been named and will be discussed at this time, though there is probably no good reason why they should be differentiated from rickettsiae.

In 1919 Roubaud reported the presence of symbiotes in *Lipoptena cervi*, *Hippobosca equina*, and *Melophagus ovinus*, though a year earlier Sikora (1918) had reported their presence in the last-named insect. Kligler and Aschner (1931) studied eight species of pupipara: *Melophagus ovinus, Lipoptena caprina, Hippobosca equina, Hippobosca capensis, Nycteribia kollari, Nycteribia biarticulata, Nycteribia blasii,* and *Pseudolynchia canariensis*. In some of these species the rickettsialike symbiotes have been observed only in an extracellular position. Kligler and Aschner found intracellular forms in *Nycteribia kollari, Nycteribia biarticulata, Nycteribia blasii,* and *Pseudolynchia canariensis*. In these cases most of the microorganisms occurred in the epithelial cells of the gut and in the cells of the Malpighian tubes. In addition, extracellular forms were present in the lumen of the gut massed on the epithelial lining. Buchner (1930) discusses the rickettsialike organisms, as well as the larger intracellular symbiotes, in this group of insects in considerable detail.

Kligler and Aschner were able to cultivate in artificial media the extracellular forms occurring in *Melophagus ovinus*, the sheep ked; *Lipoptena caprina*, a parasite of goats; *Hippobosca equina*, a parasite of horses; and *H. capensis*, a parasite of dogs. In general, these organisms were minute, gram-negative, coccoidal rods measuring from 0.3 to 0.5 microns in length.

Zacharias (1928) was the first to discover that the extracellular symbiotes in this group of insects do not invade the egg as they do in so many of the other instances we have discussed. Working with *Melophagus ovinus* and other pupipara, he found that the larvae, which develop within the female, become infected through the milk glands of the mother. According to Kligler and Aschner (1931), the intracellular forms they studied are transmitted through the eggs.

Though it was not an intracellular bacterium, Steinhaus (1943) isolated a diphtheroid (*Corynebacterium lipoptenae*) from the louse fly, *Lipoptena depressa*, collected from deer (fig. 15). The organism was present in every insect examined out of a lot of ten.

In *Ornithomyia avicularia* Zacharias (1928) found long, threadlike symbiotes located between the midgut epithelium and the muscularis. Similar organisms were found in the milk glands.

Referring to some early work of Aschner's, Buchner (1930) states that the symbiotic arrangement in the family Nycteribiidae differs considerably from the *Melophagus* type. In *Nycteribia blasii, Nycteribia biarticulata,* and *Nycteribia blainvillei* the mycetomes are not in close connection with the intestinal canal. *Nycteribia biarticulata* possesses two elongated masses of cells lying dorsally in the abdomen on either side of the rectal sac. The symbiotes are small spherical and threadlike organisms. In addition, as has already been mentioned, rickettsialike microorganisms are found in the intestines and milk glands.

Muscidae. Most of the investigations concerning intracellular microorganisms in this family have had to do with species of the genus *Glossina*. It is interesting that perhaps the first to observe these symbiotes was the eminent bacteriologist, Robert Koch, who spoke of them to Stuhlmann (1907). The latter discovered the "giant cell" zone which harbors the symbiotes. The first detailed investigation as to the nature of these organisms was made by Roubaud (1919) who believed that they play an important role in the digestion of blood. This worker also found that the giant cell zone appears in the pupa as a ring around the intestine. In the adult this ring has become more oblique and eventually forms definite longitudinal bands. In 1929 Wigglesworth, working with *Glossina submorsitans* and *Glossina tachinoides,* discounted the digestive role of the symbionts and studied further the anatomy, histology, and digestive enzymes of the tsetse flies. Wigglesworth stated that no material differences were detected between the two species just mentioned and *Glossina palpalis*.

Although Roubaud thought the symbiotes to be yeastlike organisms, the evidence seems to be in favor of their bacterial nature. Wigglesworth described them as "pleomorphic, bacteriumlike, gram-negative intracellular microorganisms, 3-10 microns in length."

Many aspects of the behavior of the symbiotes, such as the manner of transmission to the next generation, have not been adequately investigated. Their close association and adaptation to the host is indicated by Roubaud's finding that the cells in the gut wall which eventually harbor the symbiotes enlarge considerably over the other cells during the pupal stage. This happens before the invasion by the symbiotes takes place. It is possible that transmission to the next generation takes place through the secretion of the milk glands. Definite proof of this, however, is lacking. Although Roubaud maintained that the giant cells (mycetocytes) discharged their symbiotes into the intestinal lumen, Wigglesworth claimed that they did not do this regularly.

SIPHONAPTERA

Identified rickettsiae have been found associated with some species of fleas. These relationships will be covered in chapter V.

Pulicidae. Cowdry (1923) found in serial sections of a male dog flea, *Ctenocephalides canis,* small gram-negative rods measuring 0.2 to 0.3 by 0.4 to 1.5 microns in size. They were distributed intracellularly in the following tissues: hypoblast, salivary glands, muscle, fat cells, testis, intestinal epithelium, and the Malpighian tubes.

In sections of one *Pulex irritans* female out of 7 males and 18 females, Cowdry found filamentous organisms (0.3 to 0.6 by 0.6 to 4 microns) within the body cavity and in the smaller fat cells. He also observed tiny gram-negative rods in the intestinal lumen.

Fig. 96. The dog flea, *Ctenocephalides canis,* and its larva. (From Comstock, 1936.)

Hystrichopsyllidae. Faasch (1935) found the mole flea, *Hystrichopsylla talpae,* to harbor small rod-shaped symbiotes in the cells of the gut and in the intestinal lumen. Many of the cells are packed with them.

Dolichopsyllidae. The intestinal epithelial cells of the rat flea, *Nosopsyllus fasciatus,* were found by Faasch (1935) to harbor uniformly short rod-shaped microorganisms which also occurred in the intestinal lumen. Apparently the symbiotes multiply until the cells burst, discharging the organisms into the intestine. As Peus (1938) suggests, perhaps they are thus passed out with the feces and ingested by the flea larvae which feed greedily on the fecal material of the adults, thus passing to the next generation.

HYMENOPTERA

Formicidae (Ants). Knowledge concerning intracellular microorganisms in ants appears to be limited chiefly to their behavior during the embryological development of these insects. Blochmann (1884, 1887) initiated the interest taken in this group by Adlerz (1890), Strindberg (1913), Tanquary (1913), Hegner (1915), Buchner (1918), Hecht (1924), and Lilienstern (1932). Buchner (1930) did much to systematize and correlate the early available information.

Most of the study has been done on two genera of ants, *Camponotus*

(*C. ligniperdus, C. senex, C. maculatus, C. atramentarius* var. *liocnemsis, C. rectangularis, C. rubroniger*) and *Formica* (*F. fusca, F. sanguinea, F. rufa*). In *Formica sanguinea, Formica rufa,* and *Formica fusca* var. *glebaria,* apparently no symbiotes are present, though there are evidences of mycetocytes (Lilienstern, 1932).

The bacteriumlike symbiotes of *Camponotus* and *Formica* are very similar. Lilienstern (1932) compares those of *Camponotus ligniperdus* and *Formica fusca:* The organisms are gram-negative, non-motile, and usually slightly curved. With Giemsa they stain dark red to dark violet. Involution forms occur. Those of *C. ligniperdus* are long rods, 1 micron wide and from 5 to 18 microns long (fig. 97). Those of *F. fusca* are short thick rods 1 by 4 to 8 microns in size.

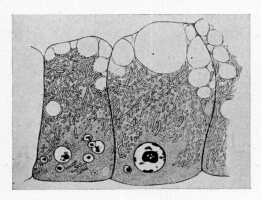

Fig. 97. Blastoderm cells of *Camponotus ligniperdus* showing the symbiotes during the insect's embryologic development. (From Buchner, 1930.)

All species of *Camponotus* so far examined have symbiotes. Furthermore, each caste, queen, male, or worker, of each species appears to harbor them. For the most part they are localized in the epithelium of the midintestine. As explained by Buchner (1930), this part of the intestinal tract at first glance appears to be composed of two different kinds of cells. Among the large epithelial cells are others which also lie on the basement membrane. The cytoplasm of these latter cells is usually completely filled with bundles of threadlike symbiotes which surround the nucleus. This arrangement presents a very peculiar appearance since, unlike the symbiotes in most mycetocytes, these long rods do not lie irregularly in the cytoplasm but instead run parallel to each other, forming rings about the nucleus. Sometimes one can also see these bundles crosswise.

How the symbiotes get from the intestinal epithelium to the developing eggs is not known with certainty. It is known that the symbiotes enter the follicular cells very early in the insect's development. From these cells the symbiotes migrate to the enlarging egg cell, the follicle almost completely freeing itself of the organisms. In turn, the developing egg cell is now packed with them as though it had no other purpose than to hold the symbiotes. Eventually, however, the host cell gains the upper hand, and the microorganisms in the primary mycetocytes begin to be segregated toward the rear of the egg. Later, numerous small, angular cells appear among the primary mycetocytes, which may be distinguished by their accessory nuclei. These new cells grow and somehow suddenly become filled with the symbiotes, while the primary mycetocytes with only a few remaining symbiotes begin to deteriorate. The secondary mycetocytes go on to find their proper places in the embryonic mid-intestinal cells.

The location of the mycetocytes in *Formica fusca* is somewhat different from that in the genus *Camponotus*. In *F. fusca* these specialized cells do not lie in the intestinal epithelium but behind it in a loose, long-drawn-out layer. In the larva the mycetocyte group lies in the bend between the midgut and the hindgut. The behavior of the symbiotes during the embryologic development of *Formica* is very similar to that in *Camponotus* except that it perhaps is not quite so complicated as in the latter genus. An interesting finding of Lilienstern's was that in queenless colonies of *F. fusca,* where the workers lay the eggs, the progeny are either very poorly "infected" or harbor no intracellular symbiotes at all, even though the workers are well supplied with them.

Although termites belong to the order Isoptera, it may be of comparative interest to mention here that a few intracellular bacteriumlike organisms have been seen in the tissues of termites but these instances have not been adequately studied.

Ichneumonidae (The Ichneumon Flies). In 1923 Cowdry reported the presence of gram-negative rickettsialike organisms within the intestinal epithelial cells and body cavity of *Casinaria infesta*.

Class: **ARACHNIDA**

Order: **ACARINA**

(a) Superfamily: IXODOIDEA (Ticks)

In some respects the relationships between arachnids and their symbiotes are very similar to those between insects and theirs. Some differences, however, are noteworthy. The outstanding difference appears to be the absence of mycetomes in the ticks, though some mites have these structures. Furthermore, most of the tick symbiotes occur in the Malpighian tubes and ovaries instead of in the alimentary tract, though this may not be true in all cases or for certain of the rickettsiae (*Dermacentroxenus rickettsi* and *Rickettsia dermacentrophila*) which will be discussed in another chapter.

The two families of ticks (Ixodidae and Argasidae) are similar with respect to their symbiotes. In both families the same organs are associated with the microorganisms. The two families appear to differ, however, in the manner of ovarial infection. This will be brought out in the discussion which follows.

Ixodidae (Scutate Ticks). Intracellular and rickettsialike organisms have been found in at least twenty species of ixodid ticks.

The symbiotes of the genus *Ixodes* were first comprehensively studied by Buchner (1922, 1926, 1930). This investigator worked with *Ixodes hexagonus*

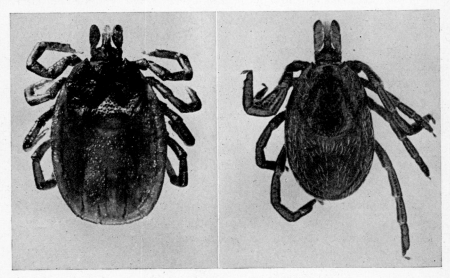

Fig. 98. Females of *Ixodes hexagonus* (left) and *Ixodes ricinus* (right). (Courtesy Dr. R. A. Cooley.)

from hedgehogs, and *Ixodes ricinus* from dogs. The symbiotes of the first-named species are threadlike forms, and within the host cell they arrange themselves vertically to the base of the cell. Like most other ticks, they invade only the cells of the Malpighian tubes and ovaries. According to Mudrow (1932), the Malpighian tubes become very much distended and nearly every cell is heavily colonized. Only those parts of the Malpighian tubes bordering directly on the rectal vesicle were free from symbiotes. Buchner found the symbiotes in his specimens to occur in "pillar-like bundles," though Mudrow did not notice very much of this. Both workers acknowledged that these threadlike forms actually consisted of rows or chains of rods and granules. When occurring freely, the symbiotes usually appeared as masses of tiny rods.

Transmission to the next generation occurs through the egg, the symbiotes passing from the heavily invaded cells of the ovary to the developing oöcytes.

At first the symbiotes are gathered at either pole of the egg. Later, however, the symbiotes increase greatly in numbers, and as the storing of the yolk commences, a ball of symbiotes begins to form between the nucleus and the periphery until the mass symbiotes appear as a secondary nucleus. Still later, in the older eggs, this mass appears to break down into its individual components again. The bundles and long threads of symbiotes found in the Malpighian tubes are not met with in the egg. According to Buchner (1930), Rondelli found the relationship of *Ixodes* to its symbiotes to be so different that it is difficult to reconcile her observations with those of other workers. Rondelli found the small, rod-shaped microorganisms in the lumen of the esophagus, at the beginning of the intestine, salivary glands, and other tissues. Occasionally they also occurred intracellularly in these regions. With regard to the Malpighian tubes, she found organisms in the lumen but none in the cells. Mudrow (1932) was unable to find any symbiotes in *Ixodes ricinus*.

In 1922 Godoy and Pinto reported the presence of microorganisms, which they called "Ixodisymbionte" in the ovaries of *Amblyomma agamum* (now *Amblyomma rotundatum*) and *Amblyomma cajennense*. A year later Cowdry (1923) reported the occurrence of intracellular microorganisms in *Amblyomma americanum, Amblyomma hebraeum*, and *Amblyomma maculatum*, though one female specimen of *Amblyomma tuberculatum* was negative. Cowdry (1925d) also found symbiotes in the ixodid ticks *Hyalomma aegyptium* (see also Jaschke, 1933), *Haemaphysalis leachi, Boöphilus annulatus, Boöphilus annulatus australis, Boöphilus decoloratus, Rhipicephalus sanguineus, Rhipicephalus appendiculatus, Rhipicephalus evertsi, Rhipicephalus simus, Dermacentor andersoni,* and *Dermacentor variabilis*. He did not find any to be present in *Haemaphysalis leporis-palustris, Margaropus winthemi, Rhipicephalus capensis,* and *Dermacentor albipictus*.

Godoy and Pinto (1922) describe the symbiotes in the ovaries of *Margaropus microplis* (*Boöphilus annulatus microplis*) as curved, rod-shaped organisms measuring 4.6 by 0.9 microns. When stained by Giemsa's method, they are seen to be violet-colored granules.

As has already been mentioned, Cowdry (1925d) found symbiotes in several species of *Rhipicephalus* (*R. sanguineus, R. evertsi, R. appendiculatus,* and *R. simus*). Mudrow (1932) worked with *Rhipicephalus sanguineus* and found the behavior of the symbiotes in this tick's embryonic stages to be the same as that of those in *Boöphilus annulatus*. Jaschke (1933) also reported observations on *Rhipicephalus sanguineus* as well as other ticks. According to Mudrow, the stage already mentioned in the case of *Ixodes,* of diffused distribution of the symbiotes in the egg yolk, is followed by another collecting of the symbiotes near the egg's posterior pole. This occurs shortly before oviposition. With the continued division of the first cells the symbiotes are crowded inward, and with the development of the germ band the symbiotes are assigned to the future location of the excretory organs. The Malpighian tubes and the sex cells originate side by side from the entoderm. From the very first these structures are provided with symbiotes.

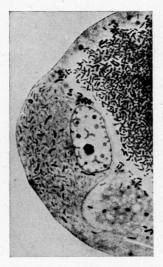

Fig. 99. Part of a cross section of a Malpighian tube of *Rhipicephalus sanguineus* showing symbiotes. (From Cowdry, 1923.)

Mudrow found the Malpighian tubes of *Rhipicephalus, Boöphilus,* and *Dermacentor* to harbor the symbiotes in their anterior ends exclusively. She supposes this to be due to the fact that newly forming cells originate at the base of the tubes, pushing the originally infected anterior portion ahead. In these ticks the symbiotes apparently propagate only in the cells originally invaded. This applies to the cells of the sex organs as well as to those of the Malpighian tubes. It is interesting that only the females show symbiotes in their sex cells. The male sex cells apparently are always free of the microorganisms.

The symbiotes of *Rhipicephalus sanguineus* [1] are nonmotile, gram-negative, straight or slightly curved short rods with an average width of 0.5 microns. Numerous tiny granulelike forms also occur. These short rods and granules are also present in chains or rows forming the filamentous bundles already described in the case of *Ixodes hexagonus*.

The symbiotes of *Boöphilus annulatus* occur in various groupings with respect to the stage of the tick. In the larvae they appear as small granules and in thicker, threadlike formations composed of granules. There seems to be a mucilaginous foundation in which the organisms lie. In the nymphal stage the filaments become longer and may lie in double rows. They become still longer in the adult ticks and sometimes lie in several parallel rows. On cross section the symbiotes may occur in ring-shaped colonies. Other pleomorphic arrangements may also be observed. In replete females some of the groupings, or colonies, may measure as large as 2.5 by 15 microns.

The symbiotes of *Dermacentor reticulatus* consist of granules which usually arrange themselves in round, compact groups. In general they do not stain as

---

[1] Probably the first to observe the symbiotes in this tick was Robert Koch, who spoke of them to Stuhlmann (1907), who in turn reported them.

deeply with Giemsa's solution as do the symbiotes of *Rhipicephalus sanguineus*. In old females (after oviposition) one may find certain "degeneration forms," which take on irregular shapes. Mudrow (1932) reports the presence of what may be another intracellular symbiote, which she found in the same specimens in which she studied the regular symbiotes. In this case, however, they appear as extremely fine rods in that portion of the Malpighian tubes just posterior to the middle. Their relation to the first-described symbiotes is not known.

The intestinal cells of engorging *Dermacentor andersoni* ticks contain certain globoid cytoplasmic inclusions, but it is doubtful if these are living organisms. A discussion of these and other inclusion bodies found in this tick is given by Gregson (1938). Similar bodies have been observed by Trager (1939) in *Dermacentor variabilis*. In larvae and nymphs they are found in the epithelial cells of the anterior tip of the anterior diverticula of the alimentary tract. In the adults they may be found in the epithelial cells of any part of the alimentary tract, and in the ovaries and eggs. Trager believes these bodies, and similar ones in the larvae of *D. andersoni*, are probably microorganisms.

Argasidae (Nonscutate Ticks). In 1923 Cowdry found symbiotes in *Ornithodoros turicata* but not in *Argas persicus* (= *A. miniatus*). In 1925 (d) he reported that intracellular organisms were present in *Argas persicus* and *Otobius megnini* but not in *Ornithodoros moubata* larvae. In 1924, however, Hertig and Wolbach had found rickettsialike organisms in the salivary glands of *Ornithodoros moubata,* and in 1932 Mudrow considered the symbiotes of this species in more detail.

Ornithodoros moubata, unlike the ixodid ticks, does not possess symbiotes in the cells at the anterior ends of the Malpighian tubes. Instead, they occupy the region just posterior to this, which takes in about one fifth of the entire length of each tube. One the other hand, the cells of Malpighian tubes of *Argas persicus* are filled with symbiotes their whole length, although a few cells at their posterior ends near the rectal vesicle may be free of them. In this tick Jaschke (1933) observed the symbiotes to occur intracellularly in peculiar groups or masses as large as 5 microns in diameter and containing as many as 40 individual organisms each.

Apparently the manner the symbiotes have of invading the egg differs in the Argasidae from that in the Ixodidae. That is, in the argasid tick the symbiotes may not invade the first sex cells but are probably transferred from the Malpighian tubes to the ovaries at a later stage. This is borne out by Mudrow's observation that the eggs closest to the points of contact between the Malpighian tubes and ovaries were always well provided with symbiotes.

The symbiotes of the Argasidae do not appear to be as pleomorphic as those of the Ixodidae. They usually are of the simple rod or coccus type, though they do group themselves in spherical groups or colonies, which seem to be held together in gelatinous masses. No rows or chains of granules nor any filamentous bundles are seen.

In no case have any of the symbiotes associated with ticks been cultivated on artificial media. Steinhaus (1940b) made several attempts to cultivate the rickettsialike organisms found in *Argas persicus* in the fluids and tissues of the developing chick, but without definite success.

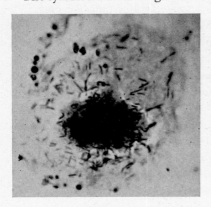

Fig. 100. Rickettsialike organism in the tissues of *Argas persicus*.

(b) Mites

Dermanyssidae. In 1922 Reichenow reported the presence of intracellular symbiotes in two species of mites, *Liponyssus saurarum* and *Ceratonyssus musculi*. Unlike the ticks, both of these mites have mycetomes closely associated with their alimentary tracts. Of particular interest is the fact that Reichenow found not one, but six, different symbiotes associated with *Liponyssus saurarum*. Generally one type was found in one host. When more than one occurred in a single host, each type was always in an individual mycetocyte. To illustrate, specimens collected in Spain harbored large, slightly curved rods similar to those occurring in roaches. Others had long, threadlike symbiotes. Those from Italy contained organisms which differed from any of the others, and so on.

Transmission to the next generation takes place via the egg. The symbiotes remain enclosed in the yolk mass through the larval stage of the mite. When the host reaches the nymphal stage, the intestinal epithelium is formed, the yolk is absorbed, and the mycetomes are formed.

Unlike *Liponyssus saurarum,* which has three mycetomes, *Ceratonyssus musculi* has only one, which lies further toward the head of the mite than those in the first-named species. It is possible that in this case the ovaries are also invaded (Buchner, 1930).

According to Hertig and Wolbach (1924), rickettsialike organisms have been found in *Dermanyssus avium* by Nöller, and in *Dermanyssus* sp. (from doves) by Reichenow.

Trombidiidae (Harvest Mites). Cowdry (1923) reported the occurrence of intracellular rickettsialike organisms in the hypoblastic cells of *Atomus* sp. and in the intestinal epithelial cells of *Lucoppia curviseta*.

CHAPTER V

RICKETTSIAE

ONE of the most interesting groups of microorganisms intimately associated, for the most part, with insects, ticks, and mites is comprised of organisms known as "rickettsiae" or "rickettsias" (singular: "rickettsia"). These microbes, like some of the entomophytic microorganisms we have discussed, are so well adapted to their arthropod hosts that they are generally considered to be peculiar to arthropods, although there are apparent exceptions to this relationship. The discovery of rickettsiae and most of their notoriety have arisen from the disease-producing properties of some of them. It is important to remember, however, that a study of only the pathogenic members of the group does not in itself reveal all the interesting biologic relationships involved between these organisms and their arthropod hosts.

The generic name *Rickettsia* was suggested by Da Rocha-Lima (1916) in memory of the American investigator, Howard Taylor Ricketts, who in 1910 died of typhus fever while studying its etiology in Mexico. The name of the type species, *prowazeki,* honors Von Prowazek, another scientist who gave his life in the investigation of typhus. *Rickettsia prowazeki* is, therefore, the genotype species (by monotypy) and forms a convenient nucleus about which any discussion of the group in general, and of the genus *Rickettsia* in particular, should be centered. In other words, *Rickettsia* is the type genus of the family Rickettsiaceae, and *R. prowazeki* the type species of the genus *Rickettsia*.

What Are Rickettsiae? Based upon a definition by Cowdry (1923, 1926), the characterization frequently given rickettsiae is similar to that given by Pinkerton (1942):

> Small, often pleomorphic, gram-negative, bacteriumlike organisms, living and multiplying in arthropod tissues, behaving as obligate intracellular parasites, and staining lightly with aniline dyes. With few exceptions, criteria adequate for classi-

fication on the basis of biological properties are available only for those members of the group that are pathogenic for mammals.

This definition is one of the best to date, but, as will be pointed out later, it would seem that such a characterization is more applicable to a single genus or two (*Rickettsia* and *Dermacentroxenus*) than to the entire group of microorganisms that might be included in the family Rickettsiaceae. Many writers consider the term "rickettsia" synonymous with the genus *Rickettsia*, whereas it is more logical that it be synonymous with the family Rickettsiaceae.

Sikora (1942) made a careful study of the morphology of various rickettsiae and described forms which she considered as corresponding to different stages in the growth and multiplication of the organisms. These stages may be listed as follows: (1) Rounded or oval noncellular granular forms are found. Each of these elongates and becomes (2) a short rod- or barrel-shaped cellular body with a granule at each end. These bodies may divide giving rise to two granules, but normally they elongate further with a subdivision of the granules, becoming (3) bacilliform bodies, each containing three or four granules. Each bacilliform body divides into (4) two plump bodies, each of which has two polar granules. The plump bodies elongate, become slender and then dumbbell-shaped. When the intermediate substance between the granules disappears, the cycle is complete and four daughter granules are produced from the parent granular form. Using these morphologic characters, Sikora has defined rickettsiae as microorganisms which even under favorable environmental conditions maintain a granular form, becoming bacilliform only during periods of multiplication.

Fig. 101. Howard T. Ricketts, the American investigator after whose name the rickettsial group of organisms is known. (U.S.P.H.S. Photo.)

Some writers are inclined to maintain the position that if an organism is not pathogenic to higher animals it cannot be a rickettsia. Considered from a biologic or taxonomic standpoint, the absurdity of such a conception can be realized when one thinks of the chaos that would result if it were applied to bacteria and protozoa in general. It should be remembered that pathogenicity is merely an incidental or acquired physiologic characteristic and should at the most be considered no more than a specific characterization. Systematically there is no reason why pathogenic and nonpathogenic species cannot be included in the same genus, let alone in the same family. Admittedly in the case of rickettsiae, it may for the time being be convenient to consider their pathogenic properties in grouping them; but care must be taken to prevent that which is now convenient to the medical mind from overriding or disrupting that which is biologically sound.

Other writers have advised the exclusion of rickettsialike organisms which do not occur intracellularly, which do not have a known arthropod host, or which can be cultivated on lifeless media. What has just been said regarding the property of pathogenicity applies to these arbitrary criteria. A true bacterium, for example, is no less a bacterium when it occurs extracellularly even though in animals it normally occurs intracellularly (e.g., *Neisseria intracellularis*). Some organisms described as rickettsiae have no known arthropod host, and there is evidence that others (*Rickettsia burneti*) which do have arthropod hosts may give rise to outbreaks of disease (e.g., Q fever pneumonia) in which apparently no arthropod has been involved. Yet there is very little reason to believe that these agents are anything else but rickettsiae. The mere fact that an organism is cultivated on cell-free media seems a very poor biologic reason for claiming that it is no longer a rickettsia. Would a spirochete be any less a spirochete simply because man learned how to grow it in a cell-free medium? Or, if we suddenly were able to cultivate filterable viruses on nonliving media, would we be forced to deny their designation as viruses? Our unsuccessful attempts to cultivate most rickettsiae on all cell-free media is very likely a transitory state of affairs and speaks of man's inabilities rather than the organism's reluctance to be cultivated.

The whole question as to which organisms should and which should not be included in the rickettsial group must await further study, and for the time being most assignments should be considered tentative. For the present the writer is inclined to accept provisionally as a rickettsia any adequately studied organism which is usually, or at times, intimately associated with an arthropod, which is usually small, often pleomorphic, and gram-negative, which in tissues stains lightly with most aniline dyes, and which is usually difficult to cultivate on cell-free media or which has not been so cultivated.[1] Such a conception as this will, no doubt, have to be altered in the future. For

---

[1] The latter would, of course, be a temporary characteristic, since it is possible that when techniques are sufficiently improved, rickettsiae may well be cultivated with ease on such media. However, the present situation is such that the difficulty with which rickettsiae grow on artificial media is a property useful in differentiating them from other microbes.

instance, there may some day be found a gram-positive organism that is definitely a rickettsia, and so it might be with the other characteristics.

Many writers have thought of rickettsiae as being in a position somewhere between viruses on the one hand and bacteria on the other. This was a convenient academic designation, but in recent years there has been a tendency to consider them as a distinct group of bacteria. Pinkerton (1936, 1942) considered them as such when he erected the bacterial family Rickettsiaceae. He pointed out that typical rickettsiae resemble bacteria in their visibility and morphology, and in most cases in their apparent nonfilterability. As Pinkerton (1942) says, a more satisfactory concept than the "missing-link" idea, stated at the beginning of this paragraph,

> would appear to be that of an unbroken series, with free-living bacteria at one end, the smallest viruses at the other end, and a large number of organisms and elementary bodies intermediate at many points between these two groups. On the whole, therefore, it seems most logical to regard the rickettsiae as bacteria which have become adapted to intracellular life in arthropod tissues.

Studying the morphologic structure of rickettsiae with the electron microscope, Plotz and his co-workers (1943) concluded that they were similar to bacteria and, like bacteria, appeared to have a limiting membrane which surrounds a protoplasmic substance with a number of dense granules embedded in it. For all practical purposes then we may consider the rickettsiae as a specialized group of bacteria of which the exact relationship to other bacterial families is, for the present, unknown.

For a comparison of rickettsiae with viruses and with bacteria, the reader is referred to a paper by Macchiavello (1938), who also favors the view that rickettsiae are bacteria. This writer discusses the three most generally held concepts as to the nature of rickettsiae: (1) that rickettsiae are nonspecific organisms which accompany a specific filterable virus; (2) that they have an evolutionary cycle with an ultramicroscopic or filterable phase; and (3) that rickettsiae are, in themselves, specific organisms, always identical to their visible microscopic forms, pleomorphic, without any invisible and filterable stage as a consequence of ultramicroscopic evolution. His conclusions include the acceptance of the last hypothesis, and he gives fairly adequate reasons for discarding the first two.

Evolution of Pathogenic Rickettsiae. Considerable speculation has arisen as to how the parasitic habits of the various pathogenic rickettsiae have evolved. Burnet (1942) has adequately outlined the present conception of the evolution of rickettsiae. It is generally thought that parasitic microorganisms have evolved from free-living forms. In the case of the rickettsiae, however, there appears to be represented a secondary evolution from a group of ancient, well-adapted symbiotes or mutualists. According to Burnet, the first stage was probably a colonization of the gut of primitive insects by saprophytic bacteria. Because of the ready availability of food in this location, some bacterial species came to live almost exclusively in the guts of insects. Some exploited their opportunities and soon began to derive nutriment from the living cells lining the gut, thus becoming true parasites causing damage to the host. After many centuries or geologic periods the host began to tolerate the bacteria as symbiotes or mutualists within the cells. However, the potential pathogenicity of these bacteria may still become active when brought in contact with other types of living cells. This contact may arise when the insect host begins living on the blood of a vertebrate which the bacterium is able to parasitize. Thus *Dermacentroxenus rickettsi* is a harmless symbiote of the wood tick, but in man and other vertebrates it causes the highly fatal Rocky Mountain spotted fever. Its contact with man is purely a biologic accident. Of course, when the rickettsiae invade a new host, arthropods other than the original one may become transmitters, and these secondary vectors may markedly change the epidemiologic features of the disease.

Baker (1943) uses several of the rickettsial diseases as examples to explain his conception of "the typical epidemic series." According to Baker, diseases which may be thought of as originating in arthropods seem to specialize in a series of steps, eventually reaching their peak in the typical epidemic. Thus, in typhus fever, the flea *Xenopsylla cheopis* (or possibly the mite *Liponyssus bacoti*) is the primary arthropod which originally was diseased by *Rickettsia prowazeki*. The second step in the series is the animal disease, represented in this case by the rat. The flea transmits the infection sporadically to man. This part of the series is at present represented by the endemic or murine typhus of the southern United States. Next, a secondary arthropod, which is always a close associate of man, enters the series. In this case, it is the louse, and Mexican intermediate typhus is the disease caused. In the next, or fifth, step of the series, we have the typical epidemic human disease, by which time the rat and the flea have dropped out of the picture, and the virulent, epidemic disease is maintained by a specialized reversible transmission between man and louse. The sixth or last class of diseases in the series is the human diease, in the case of typhus probably represented by Brill's disease in which the louse has also dropped out. After this, Baker presumes, the series dies out. In the case of Rocky Mountain spotted fever the disease has not advanced beyond the third class of the series, the endemic human disease. A secondary arthropod, similar to the louse in typhus fever, has not as yet definitely made its appearance unless the first evidence of this is the fact that in Brazilian spotted fever the louse becomes naturally infected; in this transitory infection, however, the louse is not able to transmit the disease. Baker similarly applies his theory to relapsing fevers and to yellow fever.

General Characteristics of Rickettsiae. Although the particular characteristics of each species of rickettsia will be discussed in detail later in this chapter, it may be well at this point to refer briefly to characteristics of rickettsiae in general.

Rickettsiae are usually smaller than most bacteria, frequently less than half a micron in diameter. They usually occur as short rods or cocci, though filamentous forms are frequently observed. Short rods or lanceolate coccoid forms often appear in pairs. Most of the rickettsiae so far studied are not motile, though forms seen in the bedbug have been observed by some investigators to be motile. Rickettsiae in tissues stain characteristically poorly with most aniline dyes, though in pure suspensions they stain well. Giemsa's, Castañeda's, and Macchiavello's staining methods are the most frequently used in rickettsial work. (See also Gracian, 1942, and Fielding, 1943.) All rickettsiae described so far are gram-negative.

Most attempts to cultivate rickettsiae on cell-free media have been without success, but *Rickettsia melophagi* and possibly *Rickettsia rocha-limae* have been grown on such media. *In vitro* tissue cultures and the tissues of embryonic chicks have proved to be successful media for the rearing of most rickettsiae. Other methods of cultivation will be described under *R. prowazeki*. The majority of these organisms are rather easily inactivated by heat, desiccation, and chemical agents. With the exception of *Rickettsia burneti*, rickettsiae are generally considered to be nonfilterable.

For a discussion of the various techniques used in the investigation of rickettsiae, the reader is referred to chapter XIII.

Rickettsiae and Other Symbiotes of Arthropods. As defined in chapter IV the term "symbiote," as it applies to insects and ticks, refers to any microorganism which lives in intimate and more or less constant association with such arthropods. Thus, rickettsiae are very much symbiotes, though many writers tend to disassociate the two.

The differences between rickettsiae and other symbiotes living in insects and ticks in some cases are very apparent; in other cases they are very inapparent. It should be kept in mind that from a biologic standpoint the relationships between rickettsiae and their arthropod hosts may be similar or identical to many of those existing between the other symbiotes and their arthropod hosts. The nonrickettsial symbiotes may be fungi, protozoa, bacteria, or other microbes. Most of these are well differentiated structurally and are larger than are rickettsiae. As pointed out in chapter IV, however, there is a large number of unnamed "rickettsialike" organisms (Hertig and Wolbach, 1924; Cowdry, 1925d, 1926; Buchner, 1930) which resemble nonpathogenic rickettsiae so closely that there is no good reason for excluding them from the rickettsial group. Such organisms await further cytologic and physiologic study and naming.

Some microorganisms, such as the so-called "bacteroids" of roaches, are definitely organisms which belong in different bacteriologic families than do the rickettsiae. Biologic differences are also indicated by the fact that unlike pathogenic rickettsiae, these organisms infest all members of the insect species concerned. However, there are nonpathogenic rickettsiae which also have this characteristic. Furthermore, rickettsiae do not seem to be as dependent upon special cellular structures (mycetomes) as some of the intracellular organisms which seem to be rickettsial in nature. The fact that many of these organisms are gram-positive and stain well with the ordinary aniline dyes also aids in differentiating them from rickettsiae.

Nomenclature and Classification of Rickettsiae. Although in general the rickettsiae appear to have more in common with one another than with any other group, there are some organisms, now in other taxonomic groups, that do have features strikingly similar to rickettsiae. The genera *Bartonella, Grahamella, Anaplasma,* and *Eperythrozoön* include such organisms whose points of similarity have been pointed out by several writers (Neitz, Alexander, and Du Toit, 1934; Strong, 1940; Pinkerton, 1942). The bacterium *Pasteurella tularensis* also possesses some properties in common with rickettsiae. The agents of psittacosis, trachoma, and climatic bubo or lymphogranuloma venereum, have been grouped with the rickettsiae by some writers. In fact, they have been given rickettsial names: *Rickettsia psittaci* (Lillie, 1930), *Rickettsia trachomae* (Busacca, 1935) (and *Rickettsia trachomitis,* Foley and Parrot, 1937), and *Miyagawanella lymphogranulomatosis* (Brumpt, 1938). Acceptance of these agents as true rickettsiae has not been general and the usual practice is to consider them as viruses or other types of infectious agents. However, the morphologic resemblance of some of the so-called inclusion bodies of certain of these diseases to known rickettsiae is striking, and for the present it is perhaps best to withhold opinion concerning the taxonomy of this group. When further knowledge is available it is possible that these agents may be included with the rickettsiae.

In 1936 Pinkerton suggested that the rickettsiae be considered as members of a bacterial family, the Rickettsiaceae. He (1942) also pointed out that the term "rickettsiae," spelled with a small letter, may be used loosely as a synonym for the Rickettsiaceae, just as we use the term "actinomycetes" for various species of Actinomycetaceae. Subsequent to Pinkerton's familial designation for the group, the suggestion was made (Buchanan and Buchanan, 1938, and Gieszczykiewicz, 1939) that the rickettsiae be included in the order Rickettsiales in the class Schizomycetes. These proposals were logical steps forward in the establishment of a systematic niche for the rickettsiae. It is entirely possible that eventually the order Rickettsiales may be included within the boundaries of one of the existing orders.

In the study of the rickettsial group of microorganisms it soon becomes apparent that they are too diverse to be included in a single genus. Out of necessity most of the rickettsiae described have been, at least tentatively, placed in the genus *Rickettsia*. Fortunately, there has been an effort in recent years to get away from this, and at least four generic names in addition to *Rickettsia* have been suggested: *Dermacentroxenus* Wolbach (1919), *Rickettsioides* Da Rocha-Lima (1930), *Miyagawanella* Brumpt (1938), and *Wolbachia* Hertig (1936). In addition, two subgeneric names have been erected: *Ehrlichia* Moshkovsky (1937) and *Coxiella* Philip (1943). It is very likely, however, that some of these will not stand. Brumpt (1938) suggested the generic name *Ixodisymbiotes* for the "symbiotes" specific for the ixodids. Philip (1943) has seen fit to reduce the generic name *Dermacentroxenus* to subgeneric standing because "there do not appear at present sufficient grounds for taking them out of the genus *Rickettsia*." This appears to be a matter of opinion, and the writer is inclined to favor a broader concept of the family Rickettsiaceae, and to accept Wolbach's (1919) and Pinkerton's (1936) view of the existence of definite generic differences between *R. prowazeki* and *D. rickettsi*, the type species of the genera *Rickettsia* and *Dermacentroxenus*, respectively. In fact, such a concept would also agree with the tendency to group the rickettsial diseases into four large groups, viz: the typhus group, the spotted fever group, the tsutsugamushi disease group, and the Q fever group. The members of each of these groups appear to deserve generic categories. Thus, the typhus fever group bears the generic name *Rickettsia*, the spotted fever group the name *Dermacentroxenus*, etc. *R. tsutsugamushi* appears to have characteristics sufficiently different from the type species *R. prowazeki* to warrant giving it a generic differentiation. Furthermore, the writer favors the lifting of the subgenus *Coxiella* to full generic status, since, as will be shown later, the Q fever group of rickettsiae have distinct characteristics of their own. In a like manner the other pathogenic as well as the nonpathogenic rickettsiae will have to be assigned to their proper genera. Certainly all these species are not sufficiently alike in their characteristics to be conveniently dumped into the one genus *Rickettsia*. Since such regrouping and naming of the genera will necessitate extensive and detailed systematic treatment, we shall not attempt to do more than suggest it here. With respect to the arrangement of the rickettsial species on the following pages we shall use only three taxonomically legitimate generic names: *Rickettsia, Dermacentroxenus,* and *Wolbachia.*

For an excellent treatment of the nomenclature of the pathogenic rickettsiae the reader is referred to a paper by Philip (1943) in which much of the confusion and uncertainty concerning the proper names of these agents is explained and clarified. Do Amaral and Monteiro (1932) have attempted to classify the various rickettsial diseases, and such a systematization is a valuable background for the taxonomic consideration of the causative agents.

One of the chief reasons for the absence of a system or key of classification of the rickettsiae is the lack of any good single basis for such a classification. Various criteria have been proposed, but each has been inadequate when the entire rickettsial group is considered. Classifications using the arthropod host, type of cells infected, serology, pathogenicity, morphology, or other characters as their bases have all proved inadequate, although the *diseases* caused by certain of the rickettsiae have been to some extent satisfactorily grouped on such characteristics, e.g., louse typhus, flea typhus, tick typhus, and mite typhus. With this taxonomic insufficiency in mind, Macchiavello (1938) concluded that at present there exists no certain criteria which permits the inclusion of all the rickettsiae in a single group, although sufficient data has been assembled to suppose that they cannot be separated from the bacteria. He presents a classification of groups, which, he says, has no other object than to bring out the differences and points of similarity of organisms which may or may not have a common origin. A summary of Macchiavello's grouping follows:

Rickettsialike Organisms
GROUP I SYMBIOTES

Microorganisms with perfect functional adaptation to the arthropod host and which live in specialized organs called "mycetomes." Hereditary transmission perfect. Morphology rickettsiaform.

(All the fungoid, bacteroid, or similar forms of yeasts or molds have been excluded from this group, leaving only those of rickettsial form. Also excluded have been the gram-positive bacteriumlike organisms which are of distinct and definite shape, and which stain intensely with ordinary aniline dyes.)

GROUP II RICKETTSIAE
(Family Rickettsiaceae?)

Bacteriumlike organisms, usually less than 0.5 microns in diameter, gram-negative, not precise in shape, staining characteristically with certain dyes, often intracellular in arthropods with preference for certain well-defined tissues, sometimes pathogenic.

Subgroup 1. Genus: *Wolbachia* Hertig

Microorganisms with characteristics of rickettsiae, small, bacteriumlike, and large forms which contain very small individuals. Slight damage to some host cells. Not localized exclusively in mycetomes. Hereditary transmission perfect. Not pathogenic.

Type species: *Wolbachia pipientis* Hertig (1936), the rickettsia of *Culex pipiens*. In this subgroup or genus might be included *R. lectularia*, which has some similar characteristics.

Subgroup 2. Unnamed

Rickettsialike microorganisms. Do no damage to host. Not present in mycetomes. Hereditary transmission through egg probably perfect. Possibility of being cultivable.

Example: *Rickettsia melophagi*.

Subgroup 3. Genus: *Dermacentroxenus* Wolbach

Rickettsialike microorganisms almost completely adapted to the host. Hereditary transmission in infected individuals but not necessarily characteristic for species as such.

1. Slightly pathogenic for host.
2. Nonpathogenic for host, but pathogenic for animals.
   Examples: *Dermacentroxenus rickettsi, R. conori,* and *R. brasiliensis*.

Subgroup 4. Genus: *Rickettsia* Da Rocha-Lima

Rickettsialike microorganisms incompletely or not at all adapted to the host. Hereditary transmission does not occur.

1. Nonpathogenic for arthropod host and for animals.
   Example: *Rickettsia rocha-limae*.
2. Pathogenic for animals but not for arthropod host.
   Examples: *Rickettsia ruminantium* and *R. wolhynica*.
3. Pathogenic for at least some of the arthropod hosts and for animals.
   Example: *Rickettsia prowazeki*.
4. Pathogenic for animals. Vector not known or studied.
   Example: *Rickettsia conjunctivae*.

TENTATIVE ORIENTATION OF RICKETTSIACEAE

Since Macchiavello's arrangement characterized only groups and did not differentiate species, we suggest the following tentative orientation. The object of this list is merely to bring out certain points of similarity and difference between the various species, and the writer disclaims any attempt at classification or systematic arrangement. This listing, or key, is purely arbitrary; others of equal merit could no doubt be prepared along different lines.

I. Pathogenic for vertebrates
  A. Exclusively intracellular
    1. Intracytoplasmic only
      a. Parasitize various tissues
        (1) High susceptibility of experimental animals (e.g., the guinea pig).
          1. Rickettsia prowazeki vars. *prowazeki* and *typhi* (pp. 268 and 280).
        (2) Relatively low susceptibility of certain experimental animals. (Strains may vary, however.)
          (a) Distributed throughout cytoplasm of infected cell. Typical morphologic characteristics.
            2. Rickettsia tsutsugamushi (p. 290).
          (b) In aggregate, intact masses.
            3. Rickettsia ruminantium (p. 302).
          (c) Arrangement in cells, unreported.
            4. Rickettsia suis (p. 304).
      b. Parasitize monocytes of circulating blood.
        (1) Parasitizes dogs.
          5. Rickettsia canis (p. 306).
        (2) Parasitizes sheep.
          6. Rickettsia ovina (p. 310).
        (3) Parasitizes cattle.
          7. Rickettsia bovis (p. 308).
        (4) Parasitizes birds.
          8. Rickettsia avium (p. 310).
        (5) Parasitizes fish.
          9. Rickettsia pisces (p. 311).
        (6) Parasitizes guinea pigs.
          10. Rickettsia (Ehrlichia) kurlovi (probably not a microorganism, p. 324).
      c. Parasitize conjunctival epithelium.
        (1) Parasitizes sheep, goats.
          11. Rickettsia conjunctivae (p. 304).
        (2) Parasitizes fowl.
          12. Rickettsia conjunctivae-galli (p. 305).
        (3) Parasitizes cattle.
          13. Rickettsia conjunctivae-bovis (p. 305).

(4) Parasitizes swine.
   14. *Rickettsia lestoquardi* (p. 306).
 2. Intranuclear and intracytoplasmic.
   a. No ulcer produced at site of tick bites.
      15. *Dermacentroxenus rickettsi* (p. 328).
   b. Ulcer usually produced at site of tick bite and accompanying adenopathy.
      16. *Dermacentroxenus conori* (p. 339).
B. Occasionally extracellular.
 1. Filterable.
      17. *Rickettsia (Coxiella) burneti* (p. 295).
C. Apparently exclusively extracellular.
      18. *Rickettsia wolhynica* (p. 287).
      19. *Rickettsia weigli* (p. 311).
II. Nonpathogenic for vertebrates.
 A. Hexapoda hosts.
  1. Small and large forms. Within the body of the large forms are contained several smaller individuals. Exceptionally pleomorphic.
    a. Enlarged forms contain individuals appearing as rings, curved or bent rods, or compact aggregations of rod and coccoid forms.
       20. *Wolbachia pipientis* (p. 325).
  2. Small forms; some thread forms with internal granules. Sometimes filamentous.
    a. Harbored in mycetome and other tissues.
       21. *Rickettsia lectularia* (p. 316).
  3. Small forms with no apparent internal structure. Rarely filamentous.
    a. Extracellular in insect host.
      (1) Cultivable on nonliving media.
         22. *Rickettsia melophagi* (p. 312).
      (2) Have not been cultivated on nonliving media.
         (a) Associated with the biting louse, *Trichodectes pilosus*.
            23. *Rickettsia trichodectae* (p. 321).
         (b) Associated with the goat-louse, *Linognathus stenopsis*.
            24. *Rickettsia linognathi* (p. 321).
         (c) Associated with the cat flea, *Ctenocephalides felis*.
            25. *Rickettsia ctenocephali* (p. 320).
    b. Intracellular (possibly sometimes extracellular) in insect host.
       26. *Rickettsia rocha-limae* (p. 319).
    c. Intracellular in insect host.
       27. *Rickettsia culicis* (p. 321).

B. Arachnida hosts.
1. Intracytoplasmic, never intranuclear.
28. *Rickettsia dermacentrophila* (p. 322).
III. Species of doubtful rickettsial nature.
1. *Rickettsia psitaci*.
2. *Rickettsia trachomitis*.
3. *Miyagawanella lymphogranulomatosis*.

CONSIDERATION OF VARIOUS RICKETTSIAL SPECIES

No attempt to understand rickettsiae as a whole can hope to be successful without a detailed understanding of each of the various species concerned. In the pages which follow, an effort will be made to present the information on each of the species in a brief and condensed form. In some cases, aspects unrelated to the relationship of the rickettsia to its arthropod host have been included for the sake of completeness. Since rickettsiae are characteristically associated with insects and ticks, a consideration of any of their properties should be of value to one interested in the microbiology of insects and ticks. In general, the data and information are presented from the standpoint of one interested in the biologic relationships which exist between the rickettsiae and their hosts, particularly their arthropod hosts. To complete this understanding, the properties and characteristics of the rickettsiae themselves are of prime importance. Questions as to synonymy will arise, and the writer is ready to admit that in some cases the proper synonymy is in doubt and will probably be changed in the future.

It should be kept in mind that knowledge concerning rickettsiae as a group is still extremely limited and that great developments are to be expected in the future, thus changing our present concept.

RICKETTSIA PROWAZEKI Da Rocha-Lima (1916)
var. *PROWAZEKI* Pinkerton (1936)

Synonym: *Rickettsia exanthematotyphi* Kodama (1932).

Varieties. As will be brought out in the pages which follow, two varieties of *Rickettsia prowazeki*[1] are recognized: *R. prowazeki* var. *prowazeki* and *R. prowazeki* var. *typhi* (sometimes called "var. *mooseri*"), the agents of classical typhus and murine typhus, respectively. The latter will be discussed in detail in the next section. The differences between these two varieties may

---
[1] In Da Rocha-Lima's original paper (1916) the specific name of this rickettsia was spelled *prowazekii*. Many authorities drop the last "i" and accept the name as *Rickettsia prowazeki*.

be noted serologically, pathologically in experimental animals and in humans, and in the modes of transmission. In general, they also occur in different parts of the world, though these geographic areas may overlap, and the two diseases may even occur together in the same area. The situation as it exists in Mexico is a good example of this. At one time it was thought that only murine typhus was present in that country. Through the work of Castañeda (1930, 1939), Mooser, Varela, and Pilz (1934), and others, it is now known that strains indistinguishable from typical European typhus, as well as several types of intermediate strains, occur in this part of the world. Transformation of certain Europeanlike strains into murinelike strains was accomplished by animal passage. The latter fact would seem to indicate the possible fundamental unity of the typhus rickettsiae.

Disease Caused. *Rickettsia prowazeki* var. *prowazeki* is the etiologic agent of classical European typhus, also known as "epidemic typhus fever," or as "human typhus." In the past, this disease has been one of the world's great scourges, appearing wherever large numbers of people congregated under unsanitary conditions such as those which frequently occur during wars and times of famine. For a thrilling history of this disease, the reader is referred to Hans Zinsser's book, *Rats, Lice, and History* (1934).

The case mortality in different epidemics varies, but it is usually about 12 to 20 per cent, though it may be much higher. In humans the onset of typhus may be sudden, beginning with chills, fever, headache, flushed face, dry tongue, and suffused eyes. The fever may reach 104° F. or higher, and usually lasts about fourteen days. Three to six days after onset a spotted rash appears on the lower chest and upper abdomen; later it is found on the extremities, though seldom on the soles, palms, or face. Sore throat, bronchitis, or an unproductive cough frequently develop in the later stages; delirium is common. Early observers confused typhoid ("typhuslike") fever with typhus. Ulceration of the bowels is common late in typhoid fever, but is not a symptom of typhus. Also, typhus is usually of shorter duration.

The nature of the causative agent of typhus was clouded in doubt and superstition until recent times. In 1910 Ricketts and Wilder described small bacillary bodies in the blood of individuals ill with typhus and in lice that had fed on typhus patients. These microorganisms were similar in appearance to those described earlier by Ricketts (1909) as occurring in the blood of patients with Rocky Mountain spotted fever. Similar observations were made by Hegler and Von Prowazek (1913), and by Sergent, Foley, and Vialette (1914). A thorough study of the organism was then made by Da Rocha-Lima (1916) who, in honor of Ricketts and Prowazek, both of whom died while investigating typhus fever, gave it the name *Rickettsia prowazekii*.

Da Rocha-Lima's observations were confirmed by Nöller (1916), Töpfer and Schüssler (1916), Otto and Dietrich (1917), and others.

Insects Concerned. Several early writers postulated that various vermin such as fleas or lice may play a role in the spread of typhus. It was Nicolle, Comte, and Conseil (1909), however, who first indisputably proved that lice were the vectors of European, or classical, typhus. The lice concerned are the so-called "body louse" *Pediculus humanus corporis,* and the "head louse" *Pediculus humanus capitis.*

Fig. 102. *Pediculus humanus corporis.* (From Comstock, Comstock, and Herrick, 1938.)

Experimentally, other insects have been found capable of harboring and transmitting *R. prowazeki* var. *prowazeki*. Atkin and Bacot (1922) found this true for the monkey louse *Pedicinus longiceps,* and Dyer (1934a) for the rat flea *Xenopsylla cheopis.* Blanc and Woodward (1945) found that the louse *Pedicinus albidus* may be infected on monkeys. Positive transmission from rat to rat has been obtained by *Polyplax spinulosus*. Although it does not transmit the disease, the bedbug, *Cimex lectularius,* is capable of harboring the rickettsiae alive in its coelomic cavity for as long as 26 days (Nauck and Zumpt, 1941).

Korshunova (1943), in Russia, has reported the occurrence of *Dermacentor nuttalli* naturally infected with what he calls "tick typhus" (apparently not spotted fever) because its symptoms in man are similar to epidemic typhus. This agent has also been isolated from ground squirrels (*Citellus*) and has been transmitted experimentally to rodents by *Dermacentor silvarum* and *Haemaphysalis concinna.*

Biologic Relationships. Weigl (1920) has reported that after periods of prolonged feeding the lice infected with the typhus rickettsia are distended and have a characteristic pinkish color. Wolbach, Todd, and Palfrey (1922) were not able to confirm these observations as to color, although lice which remained swollen after forty-eight hours' fasting were found to be almost invariably heavily infected with *Rickettsia prowazeki*. In heavily infected lice digestion is suspended or greatly retarded owing to the loss in function of practically all of the cells of the midgut, due to the packing of their cytoplasm with the rickettsia. The swollen appearance after fasting proved to be the only reliable gross evidence of infection; many lice which were moribund, swollen, or discolored at the time of their removal, proved not to be infected.

Infection of the louse probably begins in the anterior part of the midgut, perhaps in or near the diverticulae. Although Sikora (1920) reported finding rickettsiae in the salivary glands of lice, Wolbach, Todd, and Palfrey (1922) were never able to find the organisms in any organs other than the alimentary tract. Rickettsiae are also present in the feces of the louse. On the basis of the evidence (see Sparrow and Mareschal, 1942) that infected feces are the prime means of infection in man, Löffler and Mooser (1942) have listed the possible modes of transmission of *R. prowazeki* under natural conditions as follows: (1) The rickettsiae contained in the feces and crushed bodies of lice may enter through abrasions of the skin; (2) they may be introduced by the mouth parts of the louse which have been soiled with its infected feces; (3) rickettsiae contained in the feces or bodies of lice may be introduced into the eye by fingers which have been contaminated through crushing the insects; and (4) dried feces of infected lice result in dust particles which may be inhaled. It has also been suggested that infected louse feces may constitute the reservoir of infection during interepidemic periods.

Studying the typhus virus in the louse, Breinl (1924) observed certain differences between it and animal-tissue virus. Guinea pigs inoculated intraperitoneally with louse virus underwent a shorter period of latency, a more irregular fever, and had a higher death rate than animals inoculated with tissue virus. Other investigators have noted the virulence of louse strains and have found that the intestine of an infected louse may contain up to one billion infecting doses for rats.

So far as is known, *Rickettsia prowazeki* is not transmitted transovarially by the louse *Pediculus humanus*.

Dyer (1934a) was unable to find any evidence of change in the European typhus rickettsia after it was passed three times through rat fleas (*Xenopsylla cheopis*).

Morphology and Staining. *Rickettsia prowazeki* var. *prowazeki* is usually described as a small, pleomorphic, bacteriumlike microorganism, ovoid or elliptical in shape, usually occurring in pairs. It ranges from minute, just visible "coccobacilli" to longer filaments of sometimes extraordinary size. The ovoid or coccoid form in pairs or the diplobacillary form is usually regarded as the type form. The two elements composing the pairs stain deeply with Giemsa's stain, taking a reddish-purple coloration, with the material between the two a pale blue. Where the organism occurs singly there is often a zone of pale blue material surrounding or tapering from the deeply staining, reddish, ovoid body. The minute paired forms, usually consisting of ovoid elements, are occasionally composed of two pyriform elements, which may be unequal in size, joined at their narrow extremities, and lying in a straight line or forming an obtuse angle. With Giemsa's stain they color blue and contain in their distal extremities red- or purplish-stained, round, ovoid, or ellipsoidal bodies.

With Castañeda's stain the rickettsiae take on a light blue appearance. With

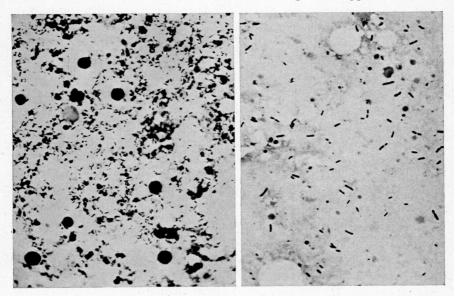

Fig. 103. Two views of *Rickettsia prowazeki* var. *prowazeki* growing in the yolk sac of fertile hens' eggs. (Photos by N. J. Kramis.)

Macchiavello's stain, they are colored bright red in contrast with the tissue and cellular elements, which stain blue.

The original description given by Da Rocha-Lima (1916) in naming the organism is as follows:

The basic form is elliptic, very short, almost spherical at the moment of origin, later stretching and dividing longitudinally into two short bodies. During and after division an external sheath which stains faintly is visible. Even after division the young bodies remain connected by means of a fine filament, thus giving rise to the characteristic form found in Giemsa preparations—the biscuit or dumb-bell form. When such forms are examined by means of dark-field illumination, two short elliptic bodies superimposed on one another can be detected. In cyanoquin preparations these elements appear as comparatively long rods. A similar appearance is given by Löffler's flagellar staining method, except that the bodies seem to be rather broader rods of varying length. With this method it is possible to stain the small olive-shaped bodies very intensely, while the external substance stands out as a pale sheath. In Giemsa preparations there may also be developmental forms which resemble polar stained bacilli. Evidently these are small bodies which have already divided, but whose sheaths have not been completely constructed. A phenomenon, which is not absolutely regular but which is common and to a certain extent characteristic, is the appearance of aggregations of these bodies scattered in large numbers throughout the preparations.

Since many of the bacteria which have most improbably been regarded as causal agents are Gram-positive, these little structures have been carefully tested for their reaction to the Gram stain. Even after incomplete decolorization of preparations which were in the first instance allowed to react for an especially long time with gentian violet, the subsequent contrast staining proved ineffective; the structures, however, can easily be seen in large numbers by means of dark-ground illumination.

As regards the dimensions of the bodies, I have not been able to make any more accurate measurements than could be effected with a 3000-fold magnification. The small bodies are about 0.3 by 0.4, and the biscuit-shaped double structures approximately 0.3 by 0.9. In practice comparison with known bacteria yields results which are at least as valuable as measurements. In concentrated carbol fuchsin, carbol gentian violet, and Giemsa preparations the bodies are undoubtedly smaller than specimens of *Micrococcus melitensis* and of *Bacillus prodigiosus* which are stained with diluted carbol fuchsin or thionin. The greater dimensions of the bacteria used for comparative purposes can be recognized when the smear preparations are examined on a dark ground.

According to Wolbach, Todd, and Palfrey (1922), dark-field preparations from heavily infected lice show *R. prowazeki* as paired and single ovoid bodies without motility but exhibiting Brownian movement. The double contour seen with most bacteria under dark field does not appear with typhus rickettsiae.

The following description of *R. prowazeki* in smear preparations from lice is taken largely from Wolbach, Todd, and Palfrey (1922).

In general, the minute coccoid and paired coccoid forms predominate over the short and long rods. The smallest single elements range from 0.25 by 0.4 to 0.3 by 0.45 microns. The paired forms range from 0.25 by 0.7 to 0.3 by 1.1 microns. Frequently small numbers of slightly larger, more deeply staining, and more uniformly shaped coccoid paired bodies, somewhat lanceolate in shape, may be seen.

Lice may also harbor delicate filamentous forms, usually curved, sometimes sharply flexed in one or several places. In size they range from 0.3 to 0.4 by 10 to 40 microns. The threads stain blue with Giemsa's stain and the outlines seem unbroken though varying slightly in width in the same filament. The blue-stained filaments contain red-stained bodies in pairs and chains corresponding in dimensions to the free-lying rickettsiae. Some of the filaments stain uniformly pale blue. It is interesting to note that the filamentous forms are frequently found in sections of lice, lying curled within nonswollen cells of the midgut at a time when very few cells are infected with the coccoid rickettsiae.

In the lightest and probably the earliest infections in lice, the rickettsiae are bacilliform. When this form is present in large numbers, they distend the cell and the rod elements seem to grow in chains arranged parallel in coils within the cell.

When large numbers of cells are infected, the minute coccoid form predominates. The cells containing these are usually swollen to many times their normal size, and the multiplication of the rickettsiae leads finally to the rupture of the cells.

All these forms have also been found to occur freely in the lumen of the gut.

As already mentioned, infection of the louse very probably begins near the diverticulae of the midgut. This initial infection is characterized by the fact that the filamentous and bacillary forms precede the appearance of the small coccoid forms. In fact, Wolbach, Todd, and Palfrey (1922) believe that

Rickettsia prowazeki undergoes a developmental cycle in the louse in which the first stages are the thread or filamentous and bacillary forms. The minute paired rickettsia bodies or coccoid forms probably represent the stage of most active multiplication. The deep stained lanceolate form is possibly the latest stage of development in the louse and a quiescent stage of slightly greater resistance to external agents.

These authors distinguished with certainty only two forms of *Rickettsia prowazeki* in human tissue: in endothelial cells, a relatively large, deeply staining, paired, lanceolate form and a smaller, lightly stained, paired, coccoid form in globular masses or diffusely distributed. Occasionally they also observed short, straight or curved, rodlike forms in association with other forms.

Ricketts and Wilder's original description of the organisms they suspected of being the cause of typhus fever includes the following:

In the stained (Giemsa) preparation of the blood of patients, taken on from the seventh to the twelfth day of the disease, we invariably have found a short bacillus which has roughly the morphology of those which belong to the "hemorrhagic septicaemia group." Usually it appears to stain solidly but on minute examination an unstained or faintly stained bar is seen to extend across the middle. Occasionally two organisms are seen end to end. Exact measurements have not been made, but when compared with the size of the erythrocyte, their length is estimated hardly more than two micromillimeters, and their diameter at about one-third this figure. Certain other bodies, the identity of which is not so clear, may represent degeneration or involution forms of the above. They consist of two stained granules, connected by an "intermediate substance" which is stained faintly blue or not at all. Frequently one of these granules or "rods" is larger than the other and stained a deep purple, whereas the small one takes a faint blue color.

In moist preparations of the blood of patients, bacillary bodies, with a structure like that mentioned above, have been encountered in all cases. The differentiation of the forms into two bodies, separated by a line or narrow zone of a substance of different refractive power may be observed. They possess no active motility, but vibrate more or less rapidly.

In guinea pigs scattered paired coccoid bodies can often be found in swollen endothelia in blood vessels of the skin of the scrotum, testes, and brain.

Some workers (Giroud and Panthier, 1942) have maintained that each morphologic or structural phase of *R. prowazeki* represents the result of a special type of reaction of the host to the invading rickettsiae. The actively multiplying and highly pathogenic rickettsiae are bacilliform in shape. Although these are very pathogenic, nevertheless, they offer little resistance to adverse influences. The granular forms are produced when the host is offering resistance to the organisms, and when the host has overpowered them, the rickettsiae appear as homogeneous bodies without structural identity. In this connection, Begg, Fulton, and Van den Ende (1944) have reported that if mouse-lung-adapted strains of typhus rickettsiae are passed into the lungs of rabbits, no bacillary forms are found at first, but only certain homogeneous inclusion bodies. By repeated passage the strain can be so adapted that well-defined intracytoplasmic colonies of rickettsiae, which they call "morulae," can be found in some cells. Eventually these morulae apparently break down into the bacillary rickettsiae. More work of this type needs to be done in order to determine whether or not most intracellular rickettsiae multiply in this way or go through a stage such as this when being adapted to a new host.

With the electron microscope *R. prowazeki* morphologically appears bacteriumlike, with a limiting membrane surrounding an inner protoplasmic substance. For the various forms seen with the electron microscope, the reader may consult such papers as those by Plotz, Smadel, Anderson, and Chambers (1943), Weiss (1943), and Eyer and Ruska (1944).

Cultivation. *Rickettsia prowazeki* has never with certainty been cultivated in the absence of living cells. In 1927 Kuczynski claimed to have cultivated microorganisms of morphology similar to *R. prowazeki* on media consisting chiefly of digested animal proteins and heated ascitic fluid. Weigl (1923), Fejgin (1927), and Anigstein and Amzel (1927) claim to have obtained growth *in vitro*. The last two workers later (1930) reported a series of experiments confirming Kuczynski. Recent investigations have not as yet supported these claims of the continuous *in vitro* cultivation of rickettsiae.

Wolbach, Pinkerton, and Schlesinger (1923) and Wolbach and Schlesinger (1923) were the first to cultivate rickettsiae outside the animal body. They used brain tissue cultures and found that primary cultures remained alive and virulent for 1 to 2 weeks and generation cultures for 2 to 4 weeks. The method of growing viruses on the chorio-allantoic membrane of the developing chick embryo was found by Da Cunha (1934) to be applicable to rickettsiae. Later Cox (1938a) found the yolk sac to support luxuriant growth of *R. prowazeki* as well as other rickettsiae.

Weigl (1924) devised the ingenious method of cultivating *R. prowazeki* by the intrarectal injection of body lice with infected material. Lice infected in this manner were also used by this worker for the production of a vaccine against typhus fever.

Other methods of cultivating rickettsiae have been devised by Zinsser and Castañeda (1932), Zilber and Dosser (1933), Zinsser and Macchiavello (1936), Aschner and Kligler (1936), Zinsser, Wei, and Fitzpatrick (1937), Hitz-Green (1938), Nauck and Weyer (1941b), and Singer (1941). The optimum conditions for the cultivation of rickettsiae will be considered in the discussion of their physiology.

Physiology. Zinsser and Macchiavello (1936) state:

In tissue cultures set up in the ordinary manner with guinea pig serum—tyrode solution and tunica vaginalis material from medium-sized guinea pigs, the oxygen consumption reaches its maximum and flattens out at the end of 40 to 46 hours. The potential distinctly and consistently rises, also reaching its maximum at about 40 hours. The pH usually shows a gradual change toward the acid side from an initial pH of about 7.8 to 7 to 7.2 at the end of 6 or 7 days. Thus, although the changes in the factors mentioned slow down and become more or less stabilized in such cultures before the end of the first 48 hours, the most active growth of viruses usually occurs after the 2nd day and that of Rickettsiae does not seem to take place until after the 5th to 7th days. It would seem, therefore, that the conditions favoring growth depended upon the preliminary establishment of some sort of equilibrium.

In other words, the most luxuriant growth of rickettsiae occurs after metabolism has slowed down or ceased.

In tissue cultures maintained at 37.5° C. the rickettsiae rapidly disappear from their host cells, and after about the tenth day the cultures will no longer infect guinea pigs. It should be pointed out that it is possible to maintain strains in fertile eggs incubated at 37.5° C. According to Pinkerton (1942), the optimum temperature in tissue cultures is 32° C., 27° C. being too low and 42° C. too high for continued growth. He points out that the organisms grow luxuriantly at 32° C. not because they prefer that temperature, but because that temperature is most effective in bringing about suitable growth conditions within their host cells. Alterations in oxygen tension and pH apparently have little effect on the intracellular growth of the rickettsiae unless they are so great as to kill the host cells. However, Macchiavello

(1939) considers the rickettsiae to be aerobic and to grow best at a pH of 7.4 to 7.5.

Aschner and Kligler (1936) found that infected guinea pig testicles kept 24 hours in a sterile petri dish at room temperature still gave positive cultures, and an emulsion of rickettsiae-infected tissue or culture in saline lost its infectivity for guinea pigs after 24 hours at 37° C. The duration of the life of the rickettsiae in their tissue cultures was from 15 to 20 days. After this time subcultures or infection of guinea pigs no longer gave uniform results, even though numerous rickettsiae could be seen in stained preparations.

Rickettsia prowazeki is susceptible to the action of many chemical and physical agents. In blood or tissue suspensions which have been dried, it is inactivated in a few hours at room or incubator temperatures. In dried louse feces it remains alive for at least as long as ten days and probably longer. (*R. prowazeki* var. *typhi* has been reported to retain its virulence in flea feces for as long as 651 days.) A temperature of 50° C. kills the organism in 15 to 30 minutes. Pinkerton (1942) reports that at minus 20° C. a brain or spleen from an infected guinea pig remains infectious for periods ranging up to eight months. Anderson (1944) found suspensions of *R. prowazeki* to survive well in skim milk and Elford and Van den Ende (1944) found the same to hold true for serum broth. The rickettsia remains virulent for months when preserved by the lyophile apparatus or when stored at low temperatures (-25° to -77° C.). Infected tissues suspended in glycerin retain their virulence for several months, probably through its dehydrating action on the host cells. In considering most of such data it should be remembered that the majority of these tests were made under conditions in which the rickettsiae were afforded some protection by the cytoplasm of the host cells or the medium in which they were contained. In certain instances, such as the inhibition of rickettsial growth in yolk sacs and in animals by penicillin and by para-aminobenzoic acid (paba), the physiologic effect may be more direct.

Pathogenicity. Experimental typhus can infect apes, monkeys, guinea pigs, rats, especially X-rayed rats, and gerbils. Inapparent infections have been described in the white mouse and rabbit. The young eastern cotton rat has been reported as being very susceptible. Infection in birds is slight and inconsistent. The horse, sheep, and other large animals are generally considered to be entirely resistant; in the donkey inapparent and even febrile infections have been described.

In guinea pigs the incubation period is generally 6 to 14 days, but may extend to 26 days; it is longer after subcutaneous than after intraperitoneal injection. The main characteristic of the disease is the febrile reaction. The temperature rises to between 39.6° and 41° C. for 3 to 14 days and then returns to normal. There is a mononuclear leucocytosis reaching its maximum as the fever declines. The animals recover and are subsequently immune to fresh inoculations. If they are killed and autopsied, one observes a slight enlargement and darkening of the spleen, and sometimes slight congestion of the testicles, which may be covered with a gelatinous exudate. Microscopically the main lesions are found in the blood and capillaries, especially those in the skin, skeletal muscles, and central nervous system. The primary lesion is in the endothelial cells lining the capillaries. Infected guinea pigs do not give positive Weil-Felix reactions.

In monkeys there is an incubation period of about a week followed by a gradual rise in temperature. The fever is maintained from 7 to 10 days and then falls rapidly. A period of hypothermia may occur just prior to a return to normal temperature. Accompanying the fever there are general constitutional symptoms such as anorexia, ruffled coat, and conjunctival congestion. On the third or fourth day a rash sometimes breaks out on the face. Death may occur. During the early part of the fever there is a leucopenia followed by a return to normal; the leucocytes continue to increase, passing above normal during convalescence and returning to normal about a month after inoculation.

Although the rabbit is not highly susceptible to *R. prowazeki,* it is known that rickettsiae may remain viable in the blood stream of this animal for at least three days after intravenous inoculation. This fact was made use of by Snyder and Wheeler (1945) in developing methods to experimentally infect lice easily and rapidly.

One of the most interesting characteristics of the rickettsia of European typhus is that, unlike that of murine typhus, it will not survive continuous passage through mice.

The course of disease in man may be found in textbooks on the clinical aspects of the disease.

Immunologic Aspects. Infection with typhus fever, as with most of the rickettsial diseases, usually leaves the individual with a long-lasting immunity. Nonpathogenic immunizing strains of *R. prowazeki* have been produced experimentally, but their practical use as a living vaccine has not been perfected. Dead vaccines against the disease have been prepared from infected lice, animal tissues (rat, goat, and others), tissue cultures, and from the yolk sac of the developing chick embryo. Passive immunization has been attained in guinea pigs through the use of immune horse serum.

The sera of patients suffering from typhus and typhuslike diseases frequently agglutinate certain strains of *Proteus vulgaris,* designated as *Proteus* OX-19, OX-2, or OX-K. This is a nonspecific reaction since none of the *Proteus* strains have anything to do with the etiology of typhus. Typhus immune serum usually agglutinates *Proteus* OX-19 in a high titer and *Proteus* OX-K in a low titer. This aids in differentiating between typhus and tsutsugamushi disease, which gives a high OX-K and a low OX-19 titer. In 1933 White obtained evidence of the existence in *Proteus* OX-19 of two distinct somatic receptors: (1) an alkali-labile receptor (Castañeda's P factor) which is mainly responsible for the agglutination of this organism by its own antiserum; (2) an alkali-stable receptor (Castañeda's X factor) which is responsible for the reaction of this organism with sera of typhus patients. Later, Castañeda (1934, 1935) succeeded in extracting soluble specific substances of a polysaccharide nature from *Proteus* OX-19 and *R. prowazeki.* From his work it appears that *Proteus* OX-19 and *R. prowazeki* possess a common alkali-stable carbohydrate antigenic factor which is responsible for the Weil-Felix reaction. Normal guinea pigs contain an antigen which gives rise to the Weil-Felix reaction and may partially explain the difficulty in obtaining the Weil-Felix reaction with the immune sera of these animals.

Other diagnostic tests have been devised using pure suspensions of rickettsiae in agglutination and complement fixation reactions. With the latter reaction it is possible to differentiate the serum of a case of European typhus from that of murine typhus. A diagnostic test has been developed in which the urine from typhus patients causes a precipitation of antityphus human and rabbit serum (Leon, 1942). Demonstrable agglutinins to *Proteus* OX-19 have also been detected in the urine of typhus patients (Shkorbatov, 1943). Other serologic tests have been perfected for use in making bedside diagnoses, and the "dry-blood test" has been reported by German workers as being a successful diagnostic test.

Castañeda and Silva (1941) have demonstrated certain immunologic relationships between spotted fever and typhus, and they believe not only that the biologic relationship between spotted fever and typhus rickettsiae is one of similar general properties, but that the antigenic similarities are greater than generally supposed. It would seem likely, however, that the degree of immunologic overlapping might depend on the particular strains used.

R. prowazeki var. *prowazeki* produces a toxin or toxic substance when grown in yolk sac cultures and in animal tissues. This property was first observed with *R. prowazeki* var. *typhi* and hence will be discussed more fully in the next section.

RICKETTSIA PROWAZEKI Da Rocha-Lima (1916) var. *TYPHI* (Wolbach and Todd, 1920) [1]

Synonyms: *Dermacentroxenus typhi* Wolbach and Todd (1920); *Rickettsia manchuriae* Kodama, Takahashi, and Kono (1931); *Rickettsia mooseri* Monteiro (1931); *Rickettsia prowazeki* var. *mooseri* Pinkerton (1936); *Rickettsia exanthematofebri* Kodama (1932); *Rickettsia muricola* Monteiro and Fonseca (1932); *Rickettsia murina* Megaw (1935); *Rickettsia fletcheri* Megaw (1935); *Rickettsia prowazeki* subsp. *typhi* (Wolbach and Todd) Philip (1943).

Disease Caused. The form of typhus under consideration is frequently called "murine typhus" because it has a natural reservoir in rats. In most countries where it occurs it is endemic, hence it is also designated by the name "endemic typhus." However, since this form of typhus may also occur in epidemics and since it can pass from man to man by the louse, the name "murine" is probably more satisfactory than the name "endemic."

Murine typhus is usually differentiated from the classic epidemic or European typhus by its reactions in rats and mice. The murine type produces a severe febrile reaction and is generally more virulent for rats than is the European type. Murine strains may be maintained indefinitely in mice while the European type usually cannot be carried beyond the third or fourth passage in these animals. Intraperitoneal inoculation of the rickettsia of murine typhus into guinea pigs produces typical scrotal swelling and testicular lesions, whereas these signs are seldom present in guinea pigs inoculated with the European virus. The first to report these scrotal lesions was Neill (1917) working with a strain of "Mexican typhus." Though strains vary slightly, the temperature rise in guinea pigs usually occurs earlier and may be higher with the murine than with the classical type.

Apparently several schools of thought exist as to the identity of some of the typhus diseases. Dyer and some of his co-workers (1932) have maintained, for instance, that Brill's disease is endemic typhus while Mooser (1929), Zinsser (1937), and others believe the disease described by Brill (1898) is a sporadic form of epidemic typhus. This latter belief is supported by certain

---

[1] Many writers use the name *Rickettsia mooseri* Monteiro (1931), or *Rickettsia prowazeki* var. *mooseri* (Pinkerton, 1936) to designate the causative agent of murine typhus. As pointed out by Philip (1943), these as well as several other names are antedated by *Dermacentroxenus typhi* Wolbach and Todd (1920). The latter workers proposed this name for the agent of "Mexican typhus" which, for various reasons, Dr. Wolbach (personal communication) believes was probably of the murine type. Since intranuclear organisms were not seen during Wolbach and Todd's pathologic studies in guinea pigs, the name *typhi* may be removed from the genus *Dermacentroxenus* and transferred to the genus *Rickettsia,* where, becoming a varietal name, it results in the designation *Rickettsia prowazeki* var. *typhi.*

serologic findings (see Plotz, 1943). The identity of some of the Mexican strains of typhus is likewise difficult. Mooser, Varela, and Pilz (1934) have divided the Mexican typhus into "endemic" and "epidemic" strains. Castañeda and Silva (1939), however, do not believe that such designations based on the clinical aspects of the guinea pig infection are correct. They state that typical murinelike strains have been found during epidemics, and that Europeanlike strains have been isolated during a nonepidemic period. Most authorities consider the typhus fever of Mexico (*tabardillo*), which is usually a louse-borne epidemic disease, to belong to the murine type, which may occasionally occur in an epidemic as well as an endemic form. In Mexico the typhus rickettsiae may be carried from man to man by lice as in the case of typhus in Europe, though it is usually transmitted from rat to rat by rat lice and rat fleas and from rat to man by rat fleas. It has also been thought that the disease in some cases is contracted through the inhalation of the rickettsiae from dried rat-flea feces (Burnet, 1942).

In man, murine typhus is similar clinically to European typhus. The murine type is usually somewhat milder and there is a low death rate. When the murine type occurs in epidemics, however, it may attain a greater severity than that usually associated with it.

Insects and Ticks Concerned. In 1926 Maxcy completed an epidemiologic study which excluded transmission of murine typhus by lice. Because this disease occurs largely during the season when rat fleas are most prevalent, Dyer, Rumreich, and Badger (1931a) examined fleas from rats trapped in Baltimore for the presence of the murine virus. From a mixture of two species of these fleas (*Xenopsylla cheopis* and *Nosopsyllus fasciatus*, both now known to be vectors) they recovered a strain of this rickettsia. In the same year (1931b) they also isolated a strain from a mixture of rat fleas, *Xenopsylla cheopis* and *Leptopsylla segnis* in Savannah. About this same time Mooser, Castañeda, and Zinsser (1931a) isolated the virus of the similar Mexican typhus from the brains of rats trapped in the Belem prison in Mexico City. Subsequent investigations have proved that rat fleas are vectors of *Rickettsia prowazeki* var. *typhi* from the rat to man. In 1944 Irons, Bohls, Thurman, and McGregor recovered the agent from suspensions of the cat flea, *Ctenocephalides felis,* collected in Texas. Other fleas capable of harboring the rickettsia, at least experimentally, include *Ctenocephalides canis*, *Xenopsylla astia,* and *Pulex irritans*. Brigham (1941) found the chicken flea *Echidnophaga gallinacea* taken from rats to be naturally infected.

In addition to the rat fleas, the rickettsia of murine typhus can be transmitted from rat to rat by the rat louse, *Polyplax spinulosus* (Mooser, Castañeda, and Zinsser, 1931b). Once *Rickettsia prowazeki* var. *typhi* has gained entrance into man from rats via the rat flea, the human body louse, *Pediculus humanus corporis,* is capable of transmitting the organism to man. This possibility was first indicated by Mooser and Dummer's (1930a) demonstration of experimental transmission of the rickettsia of murine typhus by the body louse.

Numerous other species of fleas and lice have been tested experimentally to determine their potentialities as vectors, but in most cases only negative results were obtained.

Castañeda and Zinsser (1930) found virulent *Rickettsia prowazeki* var. *typhi* capable of surviving in bedbugs (*Cimex lectularius*) for 10 days after intracoelomic injection with tunica material from infected guinea pigs. They also succeeded in infecting bedbugs by feeding them on infected benzolized rats. They were unable, however, to infect guinea pigs by feeding infected bedbugs on them or by rubbing the feces into the uninjured skin.

Mites and ticks have also been considered as potential vectors of the disease. Dove and Shelmire (1931) found the tropical rat mite, *Liponyssus bacoti,* capable of transmitting the rickettsiae from guinea pig to guinea pig and from guinea pig to rat. Kodama and Kono (1933) were able to infect the mite *Liponyssus nagayoi* on rats. The rickettsiae were first isolated from spontaneously infected mites (*Liponyssus bacoti*) by Pang (1941) during an outbreak of murine typhus in a Chinese orphanage. The true importance of this mite as a vector of *R. prowazeki* var. *typhi* remains to be determined.

With regard to ticks, Zinsser and Castañeda (1931) were able to infect *Dermacentor andersoni, Otocenter nitens,* and *Amblyomma* sp, as demonstrated by animal inoculation experiments. The ticks were infected by rectal injection after which the rickettsiae remained alive in the ticks for at least 12 days.

Biologic Relationships. The biologic relationships existing between *Rickettsia prowazeki* var. *typhi* and its arthropod vectors are not thoroughly understood.

According to Blanc and Baltazard (1940), a single infectious feeding is sufficient to infect a flea (*Xenopsylla cheopis*) for the remainder of its life, during which time it voids virulent rickettsiae in its feces. Ceder, Dyer, Rumreich, and Badger (1931) also believe this flea to remain infective for life, and, experimentally, they show that the rickettsiae may remain virulent in the flea for as long as 36 days after the last infecting feeding. These workers also found the feces to be infectious. Attempts to recover the virus from fleas hatched from eggs of infected fleas were negative. Unlike *R. prowazeki* var. *prowazeki* which is pathogenic to the louse, *R. prowazeki* var. *typhi* apparently does not harm the flea.

As yet rickettsiae have not been seen in the salivary glands of the flea, and all experiments to demonstrate their transmission by the bite of the flea alone have been unsuccessful. As in the case of *R. prowazeki* var. *prowazeki,* the rickettsiae of murine typhus remain highly infectious in the feces of fleas for considerable lengths of time, and it is likely that transmission occurs through this medium in a manner similar to that which occurs in European typhus.

In the intestine of the louse, *Pediculus humanus,* murine typhus rickettsiae multiply to such an extent that $1/1,000,000,000,000$ of an infected intestine caused infection in rats (Sparrow, 1940). When introduced into the louse *per anum* the rickettsiae kill the insect in a few days. On the other hand, when introduced into the louse by this same method, the rickettsiae of classical typhus require about three weeks to kill the insect regardless of the number of viable organisms injected.

During their passage through fleas or lice, the rickettsiae are not known to undergo any significant change. Lewthwaite and Savoor (1938), however, report an instance of the apparent mutation of a murine strain of typhus from the *Proteus* X-19 type to the X-K type in passing through and being transmitted by *X. cheopis.*

Experiments by Dyer, Workman, Ceder, Badger, and Rumreich (1932) have shown that *R. prowazeki* var. *typhi* undergoes an enormous multiplication in the rat flea.

Morphology and Staining. *Rickettsia prowazeki* var. *typhi* is similar in all morphologic and tinctorial characteristics to *Rickettsia prowazeki* var. *prowazeki.* Mooser (1928a,b) was the first to observe the rickettsia of murine typhus. Smears of scrapings of the *tunica vaginalis* stained with Giemsa's solution show swollen endothelial cells containing small diplobacilli. Mooser described them as consisting of two extremely small segments joined end to end in a straight line, rarely at an obtuse angle. Usually they are separated by a barely perceptible unstained intermediate space or connected by a pale-blue-staining intermediate part. The two parts comprising the pair stain reddish or purplish.

Mooser and Dummer (1930b) fed lice, *Pediculus humanus corporis,* on monkeys infected with Mexican (murine) typhus and then examined them for the presence of rickettsiae. The microorganisms stained pale blue or purple with Giemsa's solution, showed polar staining, and were bacillary, diplobacillary, and coccoid in shape. As infection in the lice progressed, the number of bacillary forms decreased, while the tiny red or purple staining diplobacillary and coccoid forms increased in number. According to Mooser and Dummer, this phenomenon is exactly what occurs in the tunica of rats

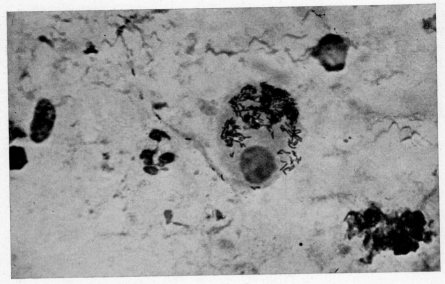

Fig. 104. *Rickettsia prowazeki* var. *typhi* in a smear of the *tunica vaginalis* of an infected guinea pig. (Courtesy Col. H. Plotz, Army Medical School. Photo by Army Medical Museum.)

and guinea pigs. It appeared to them that the minute forms, which are liberated from heavily infected endothelial cells, take on the larger bacillary form after they invade a fresh cell of the louse or infected mammal. Very few large forms are seen in heavily infected cells, the cytoplasm of which is completely filled with small forms. Similarly, small colonies of closely packed organisms in large endothelial cells usually consist of the minute forms, and loosely packed colonies generally consist of the larger, bipolar staining forms. These larger forms probably represent the stage of active multiplication, and the small coccoid forms may be more of a resting stage.

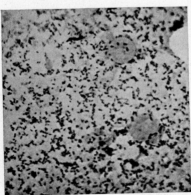

Fig. 105. *Rickettsia prowazeki* var. *typhi*.

Pathogenicity. According to Mooser and Dummer (1930b), as long as the rickettsiae are multiplying within an infected cell the surrounding tissue remains quiescent. As soon as the infected cell disintegrates and liberates the organisms, a sudden acute inflammatory reaction occurs in the vicinity of the disintegrating cell. The rickettsiae which do not invade new endothelial cells or find their way into the blood stream are taken up and digested by the polynuclear leucocytes which in turn are taken up by the large mononuclear leucocytes. These gather concentrically around the smaller leucocytes, forming a characteristic nodule which is not an early lesion but which, according to Mooser and Dummer, represents a healing stage of a typhus lesion and does not contain demonstrable rickettsiae.

The usual laboratory animals, rats, mice, guinea pigs, rabbits, and monkeys, are all susceptible to infection by the rickettsia of murine typhus. The eastern cotton rat has also been found to be highly susceptible.

In guinea pigs *R. prowazeki* var. *typhi* produces a febrile reaction sooner than does *R. prowazeki* var. *prowazeki*, usually within three to five days after inoculation. A scrotal involvement is almost always present with the murine virus. Usually the scrotum is swollen and reddened, but there is no subsequent sloughing. According to some observers, the scrotal swelling in guinea pigs infected with endemic typhus may be differentiated from that occurring in spotted fever by its location. That is, in spotted fever the swelling occurs more toward the dorsal end of the scrotum, whereas in endemic typhus the swelling makes more of a complete encirclement, occurring also in a more ventral position and toward the hind legs. Coincident with the fever there occurs an edema of the underlying tissues, and the scrotum becomes tightly stretched; the testicles cannot be pushed through the rings into the abdominal cavity as they can be in the normal animal. Fever and scrotal swelling usually lasts five or six days. The temperature may be irregular, usually going not much higher than 40° C. though occasionally it may reach as high as 40.6° C. Nearly all guinea pigs survive the infection. On autopsy the spleen may be seen to be enlarged 1 to 2 times; it is fairly light in color and the surface may be rough or "pebbly." Testes and tunica are injected and adherent with exudate present.

In 1934 (b) Dyer found the woodchuck (*Marmota monax monax*), the house mouse (*Mus musculus musculus*), the meadow mouse (*Microtus pennsylvanicus pennsylvanicus*), and the white-footed mouse (*Peromyscus leukopus noveboracensis*) to be susceptible to the rickettsia of murine typhus. Brigham (1936, 1937, 1938a,b) has determined experimentally the susceptibility of numerous animals native to the southeastern United States. These include the following species: the opossum (*Didelphis virginiana*), old-field mouse (*Peromyscus polionotus polionotus*), cotton mouse (*Peromyscus gossypinus gossypinus*), golden mouse (*Peromyscus nuttalli aureolus*), wood rat (*Neotoma floridana rubida*), cotton rat (*Sigmodon hispidus hispidus*), rice rat (*Oryzomys palustris palustris*), flying squirrel (*Glaucomys volans saturatus*), gray squirrel (*Sciurus carolinensis*), fox squirrel (*Sciurus niger niger*), cottontail rabbit (*Sylvilagus floridanus mallurus*), swamp rabbit (*Sylvilagus aquaticus aquaticus*), chipmunk (*Tamias striatus striatus*), and skunk (*Mephitis elongata*). A raccoon and a gray fox were found not susceptible. The possible role these animals may have in the epidemiology of murine typhus fever remains to be determined (see Brigham and Dyer, 1938). A strain of the murine typhus rickettsia has recently been isolated from the brain of a wild rat (*Rattus rattus* var. *alexandrinus*) in California.

Castañeda and Vargas-Curiel (1938) found horses to develop a febrile reaction of short duration when inoculated with the murine virus. They were unable to recover the virus from the blood of the animals, but immune bodies were produced in the animals' serum. European as well as Mexican investigators have reported the isolation of typhus rickettsiae from cats.

Philip and Parker (1938) found the virus of murine typhus to persist in white rats and white mice more consistently and for longer periods than in the same tissues of guinea pigs. The brain appeared to be the most favorable tissue for this persistence.

Immunologic Aspects. Human cases of murine typhus fever may be diagnosed by employing the Weil-Felix serologic test just as with European typhus. It is not practically possible, however, to differentiate between the two forms of typhus by this method. Other techniques may also be used to aid in diagnosis such as the microscopic agglutination test (Hudson, 1940) and the complement fixation test (Bengston, 1941b, 1944). With the latter test one may differentiate between European and murine typhus if the antigens are properly prepared.

In 1934 Castañeda reported the isolation from Mexican typhus rickettsiae of a soluble specific substance which gave, with *Proteus* X-19 antiserum and typhus human serum, the same precipitation reactions as the polysaccharides extracted from *Proteus* OX-19. Castañeda concluded that in the antigenic composition of both *Proteus* X-19 and typhus rickettsiae there was a common soluble specific factor which was responsible for the Weil-Felix reaction. Later (1935) this worker isolated polysaccharide substances from the soluble extracts of *Proteus* X-19; one is alkali-labile and the other is alkali-stable. The latter substance is the common antigenic factor in *Proteus* X-19 and typhus rickettsiae.

Antigenic differences between the European and murine types are indicated by the apparent failure of the vaccine of one type adequately or completely to protect against the virus of the other. In guinea pigs, vaccines of either type give a solid immunity against the homologous strain but usually do not completely protect against the heterologous strain.

Vaccines made from tissues of the embryonic chick and from the lungs of rats and guinea pigs have proved successful in vaccinating humans against the disease.

Both varieties of *R. prowazeki* produce a toxin or toxic substance when grown in yolk sac cultures and in the lungs of mice. In either case it appears that the amount of toxic substance produced is roughly proportional to the number of rickettsiae in the suspensions of these tissues. German workers (Otto and Bickhardt, 1941) believe it to be of the nature of an endotoxin, though it differs from most endotoxins of bacteria in being easily destroyed by heat and chemicals. Treatment with formol and phenol also destroys the toxic substance. One of the easiest ways to detect the toxin is to inoculate the suspension into the tail veins of mice. Lethal doses will usually kill the mice in from three to eight hours. The so-called toxin is neutralized by the specific immune sera of animals or man.

RICKETTSIA WOLHYNICA Jungmann (1917)

Synonym: *Rickettsia quintana* Schminke (1917). *Fossilis quintana* (suggested as a possible subspecies named by Megaw, 1943, "if necessary"). Probable synonym: *Rickettsia pediculi* Munk and Da Rocha-Lima (1917).

Disease Caused. During the World War of 1914–1918 there occurred among the several armies in Europe a disease characterized by recurrent fever, headache, vertigo, sweating, polyuria, severe muscular pains, and a slight leucocytosis at the height of the fever. The disease first appeared in 1915 among members of the British Expeditionary Force, and, according to Strong (1940), it caused more sickness in the British army than any other disease. It spread to the French, Italian, German, and Austrian armies and become known as "trench fever," "Wolhynian fever,"[1] "shinbone fever," and "five-day fever."

The blood from diseased individuals was infectious when inoculated into heathy men. Such blood was infective on the first day of the disease and remained so until at least the fifty-first day. Typical attacks of the disease have been initiated with filtrates prepared from sediments of infectious urines of patients and from excrement of infected lice. In 1918 the Trench Fever Commission of the American Red Cross Research Committee stated in its conclusions that the organism causing trench fever is a "resistant filterable virus." The Commission's report also stated that "at least one stage of development of the virus of trench fever is filterable and ultramicroscopic."

---

[1] This name relates to one of the centers of trench fever outbreaks, Wolhynia, a part of the western Ukraine.

In spite of this report other workers (Töpfer, 1916a; Jungmann, 1917; Munk and Da Rocha-Lima, 1917; Arkwright, Bacot, and Duncan, 1919; Byam, 1919; and others) obtained evidence indicating that trench fever is caused by a rickettsia. Töpfer described minute extracellular organisms in the intestines of lice and suggested their etiologic significance. Munk and Da Rocha-Lima observed in presumably normal lice similar extracellular organisms which they named *Rickettsia pediculi*.

According to Wolbach, Todd, and Palfrey (1922), while Bacot was studying typhus fever in Poland, he contracted trench fever at a time when he was feeding upon his person a stock of uninfected and rickettsia-free lice. Shortly after his attack, extracellular rickettsia began to appear in these stock lice. By feeding them on his person he was able to infect clean stock lice with rickettsia for four months after his trench fever attack. Wolbach, Todd, and Palfrey felt that these experiences constituted strong evidence for the identity of *Rickettsia pediculi* and *Rickettsia wolhynica* and for the etiologic relationship to trench fever. They wisely point out, however, that such a belief naturally implies that the mass population in central Europe is immune to the disease. It should be remembered that several investigators have demonstrated rickettsiae in normal lice fed on healthy people in trench-fever-free areas, and such lice fed on other healthy people produced no disease. Furthermore, rickettsiae have never been found in persons ill with this disease.

Since the first World War little as to its true nature has been observed or written concerning trench fever. Werner (1939) has reported the occurrence of the disease in individuals on whom supposedly normal lice were being fed in preparing Weigl's typhus vaccine. In 1942 (Arneth, 1942; Jacobi, 1942) and 1943 (J. Am. Med. Assoc., 1943, **122**, p. 239) it was reported that trench fever, with somewhat varying symptoms, was again affecting the armies of Europe. Various theaters of war were concerned: Poland, Russia, Rumania, Italy, France, and Belgium. Some German physicians have reported the successful treatment of the disease with atebrin and plasmochin, well known antimalarials. Arneth states, however, that treatment by sulphonamides, quinine, atebrin, and salvarsan appears to be useless, but that the intravenous administration of electrocollargol was useful in cases with fever of the relapsing type. (See also Hurst, 1942, and Sylla, 1942.) It will probably be some years, pending the study of data gathered by German and Russian workers, before a true picture of the occurrence of trench fever during the recent war can be had.

Insects Concerned and Biologic Relationships Involved. All indications point to the body louse, *Pediculus humanus corporis,* as the vector of trench fever. When fed on trench fever patients, the lice do not become infective for about five to nine days. Once they become infective they remain so for at least four months. Individuals suffering from trench fever may retain the power of infecting lice fed on them up to at least 443 days after the onset of the disease. The infectivity of louse excreta is very high. The louse may transmit the disease by bite or by the introduction of its feces into the scarified skin.

In the louse the rickettsiae are present mainly in the region of the epithelial lining of the gut, where they occur extracellularly. As they multiply they gradually fill the lumen, extending backwards down the intestine, and eventually are voided with the feces (Hindle, 1921). Pinkerton (1942) has suggested that since the organism concentrates along the borders of the lining cells, it may depend for its growth on some product of the metabolism of these cells.

Some observers have reported rickettsiae (*Rickettsia pediculi*) in the ova of lice, but others have reported that such transmission does not occur.

Morphology and Staining. *Rickettsia wolhynica* is more plump, tending to be oval, and is less pleomorphic than *Rickettsia prowazeki*. Most workers have found it to stain more deeply with Giemsa's and other stains than do most rickettsiae. Töpfer (1916a) described the organisms he found in lice taken from trench fever patients as small short rods, often arranged in pairs and frequently showing bipolar staining. When present in lice they usually occur in large numbers and are often arranged in heaps. Hindle (1921) gives their size as approximately 0.3 microns in diameter by 0.3 to 0.5 microns in length.

Properties of R. wolhynica. The rickettsia of trench fever resists a temperature of 60° C. moist heat for thirty minutes and dry heat at 80° C. for twenty minutes, but 70° C. moist heat for thirty minutes kills it. It resists desiccation in sunlight for four months.

The agent of trench fever is filterable under certain conditions but not when in plasma or serum. It is present in filtrates made from the sediments of infected urines and from the excrement of infected lice.

R. wolhynica has not been cultivated in any type of cell-free media or in tissue cultures.

Pathogenicity. So far as is definitely known, only man is susceptible to trench fever. Some workers have reported low-grade or inapparent infections in guinea pigs, but this has been unconfirmed. At any rate a strain has never been definitely established in any experimental animal. Mice and rats inoculated with infective blood were refractory.

Immunologic Aspects. One attack of the disease apparently produces only partial and limited immunity. Relapses may occur as late as two years after onset. The immunity is probably one of tolerance and is very slow in development. Second attacks have been produced experimentally by the inoculation of infectious material four or five months after the first attack.

RICKETTSIA TSUTSUGAMUSHI (Hayashi, 1920) [1]

Synonyms: *Theileria tsutsugamushi* Hayashi (1920); *Rickettsia nipponica* Sellards (1923; not valid since Sellards was unknowingly concerned with another organism); *Rickettsia tsutsugamushi* Ogata (1930, 1931); *Rickettsia orientalis* Nagayo, Tamiya, Mitamura, and Sato (1930); *Rickettsia akamushi* Kawamura and Imagawa (1931a); *Rickettsia orientalis* var. *schüffneri* Do Amaral and Monteiro (1932); *Rickettsia megawi* Do Amaral and Monteiro (1932); *Rickettsia tsutsugamushi-orientalis* Kawamura (1934); *Rickettsia pseudotyphi* Vervoort (1938); *Rickettsia sumatrana* Kouweenaar and Wolff (1939, 1942); *Dermacentroxenus orientalis* (Nagayo, et al.) Moshkovsky (1945). Probable synonyms: *Rickettsia megawi* var. *breinli* Do Amaral and Monteiro (1932); *Rickettsia megawi* var. *fletcheri* Do Amaral and Monteiro (1932).

Disease Caused. *Rickettsia tsutsugamushi* is the causative agent of the tsutsugamushi disease of Japan, which is the type disease of the tsutsugamushi fever group. The disease is sometimes known as "Japanese River fever," "flood fever," and "Kedani fever." It is also found in other regions of the Orient such as in Formosa, Sumatra, and the Malay States, and is probably widely distributed throughout the islands of the southwest Pacific. For a time such diseases as the rural form of tropical typhus (scrub typhus) and tsutsugamushi disease were thought to be separate diseases because of the presence

---

[1] Concerning this nomenclature, we quote from an editorial in the *United States Naval Medical Bulletin*, 1945, 44, 179: "Hayashi in his original description of Rickettsia tsutsugamushi emphasized the pleomorphic character of the organism. Ten years later Nagayo, reinvestigating the causative agent of tsutsugamushi, isolated a rickettsia which was monomorphic and which he called R. orientalis. He contended moreover that Hayashi's studies centered around a group of organisms composed of a variety of rickettsiae erroneously considered pleomorphic strains of a single species.

"Despite these assertions, pleomorphism in R. tsutsugamushi enjoys an almost general acceptance. . . . Restriction of the nomenclature of the disease to tsutsugamushi disease, and that of the causative agent to Rickettsia tsutsugamushi, consequently is not only timely but necessary if uniformity in reporting vital statistics is to be furthered."

On the other hand, Blake, Maxcy, Sadusk, Kohls, and Bell (1945), working under the auspices of the U.S. Typhus Commission, have preferred to accept the name *Rickettsia orientalis*. They did not accept Hayashi's (1920) specific name *tsutsugamushi* because they felt that from his descriptions and plates it was impossible to say whether or not this author had actually seen the real etiologic agent of tsutsugamushi disease. They supported their belief with the opinions of Philip (1943) and those of Dr. S. B. Wolbach.

Since it cannot be proved that Hayashi did not see the agent of tsutsugamushi disease, it would seem that ignoring the specific name he proposed is not entirely justified.

of a primary lesion and accompanying bubo in the latter (Lewthwaite, 1930). Lewthwaite and Savoor (1940) subsequently concluded that the two diseases were identical and suggested that the terms "rural typhus" and "scrub typhus" be eliminated. Similarly, Australian typhus or coastal fever and tick typhus or sporadic typhus of India very probably belong to the tsutsugamushi fever group. For a summary of the geographical distribution of this disease, the reader is referred to a review by Corson and Wilcocks (1944).

Tsutsugamushi disease is characterized by an abrupt onset, fever, and a severe headache. About the fifth day of the disease a macular to papular rash appears. Usually an initial lesion (eschar) is apparent at the site of the arachnid's bite, followed by swelling and tenderness of the local lymphatics. In general, the clinical manifestations resemble typhus and spotted fever, but there are marked differences epidemiologically. In Malaya, for example, the disease is not seasonal since cases occur every month of the year. The case mortality of tsutsugamushi disease in Japan is about 30 per cent; in Malaya it is about 15 per cent. The second World War found the disease to be a serious problem among troops in various parts of the southwest Pacific (e.g., see Ahlm and Lipshutz, 1944; and Blake, Maxcy, Sadusk, Kohls, and Bell, 1945).

The etiologic agent of this disease has been described independently by several workers (Ogata, 1931; Kawamura and Imagawa, 1931a; and others). Although he did not name the organism, Anigstein (1933) was among the first to demonstrate clearly its rickettsial nature. He made his observations in 1929 and 1930.

Arachnids and Insects Concerned. The word "tsutsugamushi" means "dangerous bug" and refers to the arthropod host of *Rickettsia tsutsugamushi*. In Japan and Formosa this agent is transmitted by the larval stage of the mite *Trombicula* (or *Trombidium*) *akamushi*, and in Malaya, Assam, Burma, and Sumatra by *Trombicula deliensis*. These are the only species of mites which have so far been proved to be vectors by animal experiments. On epidemiologic and other bases, *Schöngastia* (*Trombicula*) *schüffneri* in Sumatra and *Trombicula buloloensis* (*T. hirsti* or *T. minor*), *Trombicula fletcheri*, and probably *Trombicula walchi* in New Guinea, have been suggested as vectors. Isolations of *R. tsutsugamushi* have been made from the two last-named mites.[1] The natural hosts of the mites concerned in the transmission of tsutsugamushi disease rickettsiae have been discussed by Farner and Katsampes (1944), Womersley and Heaslip (1943), Williams (1944), and Kohls, Armbrust, Irons, and Philip (1945).

[1] In 1946, in the United States, it was discovered that another species of mite (*Allodermanyssus sanguineus*) is responsible for the transmission of a disease known as rickettsialpox, caused by *Rickettsia akari*.

Lewthwaite and Savoor (1940) mentioned experiments in which they tested the ability of ticks (*Dermacentor andersoni* and *Rhipicephalus sanguineus*) and fleas (*Xenopsylla cheopis*) to transmit the rickettsia experimentally. Most of the attempts failed, but there was some evidence that the rat fleas may possibly have a feeble power to transmit the organism. Other ticks (*Dermacentor auratus, Hyalomma aegyptium, Amblyomma* sp.) have been suggested as possible vectors, but in most of these cases confirmatory proof is lacking. Argasid ticks (e.g., *Ornithodoros moubata*) have not been found capable of transmitting the rickettsia. Anigstein (1933) observed that the human body louse, *Pediculus humanus,* does not play any part in the transmission of the rickettsia, but the organisms will multiply in and kill the insect.

Fig. 106. Larva of *Trombicula akamushi,* one of the vectors of *Rickettsia tsutsugamushi*. (From Hirst, 1915.)

Very little work has been done on the biologic relationships between the rickettsiae and the mites. In the nymphal and adult stages the mites concerned are entirely vegetable feeders and hence are not dangerous. It is believed that the rickettsiae pass from the larvae on through the nymphs to the adults. There is also evidence that the rickettsiae are transmitted to the larvae via the eggs. The importance of this has been pointed out by Farner and Katsampes (1944): the trombiculid mite takes only a single "blood meal" from a single host during its life cycle. Therefore, following the infection of a larva by feeding on an infected host, the rickettsiae are not transmitted to another vertebrate host until the life cycle of the infected larva has been completed and a new generation of larvae has been produced, and then only if these larvae have received the organisms from the female through the eggs.

Kawamura and Imagawa (1931b) have reported the presence of *R. tsutsugamushi* in the salivary glands of mites taken from infected field mice.

Morphology and Staining. *Rickettsia tsutsugamushi* is usually shorter and thicker than *Rickettsia prowazeki* and *Dermacentroxenus rickettsi*. It usually appears as a diplococcus or short diplobacillus with bipolar staining. On occasions it may appear vacuolated, and some strains are fusiform in shape. The size varies but on the average is limited to from 0.8 to 2.0 microns in length and from 0.3 to 0.5 microns in width. Some observers have reported monococcoid and lanceolate forms and short chains. With Giemsa's method they stain deep blue to purple. Syverton and Thomas (1944) found that proliminary treatment of infected yolk sacs with lipid solvents, such as chloroform, enhanced the permeability of the microorganisms to dyes of the Giemsa stain. Like other pathogenic rickettsiae they are nonmotile.

In tissue sections the organisms have not been seen to distend the cells, and instead of being compactly clustered they are usually scattered or very loosely clustered.

Growth and Cultivation. Limited studies on the growth and cultivation of *Rickettsia tsutsugamushi* indicate that in these respects the organism is similar to other rickettsiae. Cultivation on artificial media has not been attained. In 1923 Sellards reported the cultivation of the agent, but there has been considerable doubt as to whether he was actually working with *R. tsutsugamushi*.

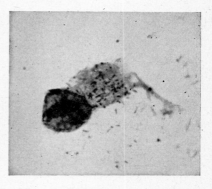

Fig. 107. *Rickettsia tsutsugamushi* in the smear of the spleen of an infected mouse.

Tissue cultures have been successful, and the rickettsia grows well in the cytoplasm of the interstitial cells of rabbit testes and in the endothelial cells overlying Descemet's membrane. Without transplanting it Yoshida (1935) kept the organism alive in rabbit testicular tissue cultures for more than four weeks. In such cultures no rickettsiae were observed until about the eighth day after inoculation, and in about two weeks the culture was at its maximum both in the number of organisms present and in their virulence. In older cultures the organisms became polymorphic, taking on various shapes not usually seen in the natural state. To perpetuate the cultures the medium should be replenished every two or three weeks.

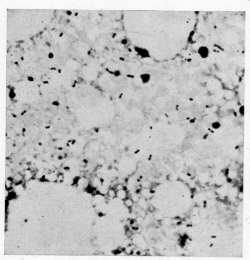

Fig. 108. *Rickettsia tsutsugamushi* growing in the yolk-sac membrane of a developing chick. (Photo by N. J. Kramis.)

Certain strains of *R. tsutsugamushi* have been grown with ease in the yolk sac of developing chick embryos. Duck eggs have also proved very satisfactory for this work, the chorio-allantois being especially suitable for their propagation.

The organism appears to be an obligate intracellular parasite apparently limited to the cytoplasm of the infected cell.

Pathogenicity. The pathogenicity of *Rickettsia tsutsugamushi* for man has already been mentioned.

In laboratory animals the disease is sometimes difficult to establish, although guinea pigs have been rendered more susceptible by a vitamin-deficient diet. After eight or ten passages normal animals may be used. The incubation period in vitamin-deficient guinea pigs varies from 9 to 18 days, with a febrile reaction of from 5 to 6 days' duration. The mortality in these animals is about 90 per cent. Lewthwaite and Savoor (1940) carried such a strain in guinea pigs for at least eight years without loss of virulence. For passage virus brain, spleen, heart blood, or ascitic fluid was used. On autopsy ascites and enlargement of the spleen were the most constant signs. Rickettsiae were found abundant in the ascitic fluid and in the fibrinous deposit on the spleen. Recovered guinea pigs were immune to subsequent inoculations of the agent. It should be pointed out, however, that many strains of so-called "scrub typhus" isolated in recent years are readily initiated and maintained in normal guinea pigs and white mice.

Rabbits may be infected by inoculating infectious material interocularly. In a few days an intense iridocylitis develops, and the rickettsiae may be demonstrated in the cells of the endothelium covering Descemet's membrane at the back of the cornea. According to Lewthwaite and Savoor, fever is fitful or absent and the animals recover and are immune. Agglutinins of the OX-K type develop in about half the rabbits, as they do when the animals are inoculated intraperitoneally. The rickettsia may also be maintained by means of passage in rabbit testicles. In white rats the infection is usually of the afebrile type with an enlarged spleen, and often ascites is seen on autopsy. Most of the infected rats die in about 15 days. Monkeys and gibbons are also susceptible, showing fever, leucopenia, dermal lesions at the site of inoculation, and a positive Weil-Felix reaction. The Syrian hamster has also been shown to be susceptible.

A vole, *Microtus montebelli,* is thought to be a reservoir of the disease agent in Japan. Evidence suggests that a similar role is played by the common house rat, *Mus rattus* var. *rufescens,* in Formosa; and in Sumatra the house rat, *Mus concolor,* and the field rat, *Mus diardii,* are believed to be so concerned. For discussions of these and other possibilities, the reader is referred to papers by Zinsser (1937), Lewthwaite and Savoor (1940), and Farner and Katsampes (1944). In nature the mite vector, *Trombicula akamushi,* has been found on

Rattus losea, Apodermus agrarius, and other Muridae. In Australia rats and bandicoots are apparently the main reservoir (Burnet, 1942). In New Guinea, *R. tsutsugamushi* has been isolated from the brain of the wild rat, *Rattus concolor browni.*

Immunologic Aspects. The immunity which results from an attack of tsutsugamushi disease does not appear to be as solid as in the case of spotted fever and in typhus, since second and third attacks frequently occur. These subsequent attacks, however, are usually quite mild.

As in typhus and spotted fever, the Weil-Felix reaction is used in the diagnosis of tsutsugamushi disease. In this case the immune serum agglutinates *Proteus vulgaris* OX-K in relatively high titers. *Proteus vulgaris* OX-19 is usually not agglutinated.

RICKETTSIA (COXIELLA Philip, 1943) BURNETI Derrick (1937)

Synonym (or subspecies): *Rickettsia diaporica* Cox (1939).

Disease Caused. In 1937 Derrick reported the occurrence of a hitherto-unrecognized disease in Australia, designated as "Q" (Queensland) fever. Most of the cases were in abattoir workers. The incubation period was 15 days or less, the duration of illness 7 to 24 days, and the mortality nil. In the same year Burnet and Freeman (1937) reported the finding of a rickettsia in the tissues of infected animals. The next year Davis and Cox (1938) and Cox (1938b) reported the recovery (in 1935) of a similar organism from ticks (*Dermacentor andersoni*) collected near Nine Mile Creek [1] in western Montana. This infectious agent was first established in guinea pigs before it was known to cause human illness. The first known human infection with this American disease occurred as a laboratory infection and was reported by Dyer (1938) who pointed out its relationship to Australian Q fever.

In January, 1939, Derrick named the Australian Q fever organism *Rickettsia burneti;* in October, 1939, Cox named the agent of the American disease *Rickettsia diaporica.* Since then it has become quite evident that the two organisms are very similar if not identical.

In man the two diseases appear to be clinically similar and are now generally considered to be essentially the same. The disease is characterized by a rather acute onset with chills, prostration, and fever. Most cases suffer severe headache. The fever is continuous and lasts from a few days to two to three weeks. In 1940, 15 cases (1 fatality) of Q fever pneumonitis occurred in an institutional outbreak in Washington, D.C. From three of these cases the rickettsiae were recovered. The avenue of infection was not determined.

---

[1] It was for this reason that the term "Nine Mile fever" was used in some of the earlier papers on this disease.

There was no evidence of an arthropod vector being involved. As reported by Hornibrook and Nelson (1940), two types of onset predominated. One was coryzalike and the other was manifested by headache, chilly sensations, and general malaise. Chills, fever, sweats, and generalized body aches and pains were present in the majority of cases. Pulmonary lesions were evidenced by roentgen-ray examinations. Similar cases have occurred in Europe.

Ticks and Insects Concerned. When *Rickettsia burneti* was so named by the Australian workers, no arthropod host was known, but the organism being rickettsial in nature, the existence of one was anticipated. Burnet and Freeman (1939) had suspected a mite, *Liponyssus bacoti,* of being a vector in one outbreak on purely circumstantial evidence. In 1940 Smith and Derrick isolated six strains of *R. burneti* from naturally infected ticks of the species *Haemaphysalis humerosa* collected from bandicoots in Australia. Since this tick does not readily attack man, it is doubtful whether it plays any direct role in the transmission of Q fever to man. Nevertheless, it is no doubt an important factor in the maintenance of *R. burneti* in the native animal population.

Fig. 109. Female of the tick *Amblyomma americanum*. (Courtesy Dr. R. A. Cooley.)

As we have already indicated, in the case of the American strains of *R. burneti* (*R. diaporica*), the arthropod host was discovered before the disease was observed in man. The original isolations were made from the tick *Dermacentor andersoni*. Parker and Davis (1938) also found the organism to be transmitted experimentally by this tick. Subsequent isolations were made by Davis (1939b) from *D. andersoni* collected in Wyoming. According to Cox (1940), the organism has been isolated from *Dermacentor occidentalis* from Oregon and California, and from *Amblyomma americanum* from Texas. The last-named tick was incriminated on the evidence obtained by Parker and Kohls (1943), who actually isolated the rickettsia from *A. americanum* in 1937. What appear to be strains of *R. burneti* (*R. diaporica*) have also been isolated from specimens of this tick collected in Texas in 1943 and 1944.

Other ticks which have been shown to transmit experimentally the rickettsia of Q fever include *Rhipicephalus sanguineus, Haemaphysalis bispinosa, Ixodes holocyclus, Ornithodoros moubata,* and *Ornithodoros hermsi.* Other ticks are able to ingest and retain the rickettsia but are unable to transmit it by feeding on susceptible animals. These include *Ornithodoros turicata, Ornithodoros parkeri, Ornithodoros gurneyi,* and *Boöphilus annulatus microplus.* (See papers by Smith, 1940a, 1941, 1942a, b, and Davis, 1940b, 1943a.)

Smith (1940b) found *Rickettsia burneti* incapable of survival or multiplication within the body of the cat flea, *Ctenocephalides felis.*

Biologic Relationships. Although the first arthropod host found to be naturally infected with the rickettsia of Q fever was *Dermacentor andersoni,* still more study of the actual biologic relationships involved is necessary. Parker and Davis (1938) found the organism to survive in, and be transmitted by, nymphal and adult *D. andersoni* that had ingested it as larvae. They also found it to survive through the eggs of infected females and to be transmitted by the progeny.

In the case of *Haemaphysalis humerosa, R. burneti* infects slightly more than half the larval, nymphal, and adult ticks fed upon infected guinea pigs. The immature stages pass the organism to the adults, which only occasionally transfer it to the next generation by transovarial passage (Smith, 1940a). Infected nymphal and adult ticks infect about one quarter of the host guinea pigs upon which they feed.

Upon histopathologic examination of infected ticks Smith found the relationship between the rickettsia and the host to be similar to that described by Cowdry (1925b,c) with regard to the tick *Amblyomma hebraeum* infected with *Rickettsia ruminantium.* In *H. humerosa* the Q fever rickettsia appears to be confined to the intestinal epithelial cells and the lumen of the gut. Smith never observed invasion of any other organs including the salivary glands. As with *Rickettsia prowazeki,* the cytoplasm of the infected cells are often packed and distended with *R. burneti,* which is never seen in the nucleus as is *Dermacentroxenus rickettsi. R. burneti* probably also undergoes active extracellular multiplication within the intestinal lumen, since in the more distal portions of the ceca of partially engorged ticks Smith found irregularly shaped vacuolated areas bordered by a zone of tightly packed rickettsiae. In the intestine anterior to the midgut rickettsiae have never been observed. The situation is very much the same with *Haemaphysalis bispinosa.*

Triturated tissues of *H. humerosa* are infectious to guinea pigs through the unbroken skin. Similarly, the feces of infected ticks are capable of infecting guinea pigs when applied to either abraded or unabraded skin. Davis (1940b) found the excrement of *Ornithodoros turicata* to be infective when inoculated into guinea pigs. He also found the organism to survive in this tick for at least 1,001 days, although transmission through feeding did not occur.

According to Smith (1941, 1942a), *Rhipicephalus sanguineus* and *Ixodes holocyclus* pass the rickettsiae from larvae to nymphs and from nymphs to adults, but not from the latter to their progeny. As with *Haemaphysalis humerosa,* rickettsiae are found only in the intestinal epithelial lining and in the lumen of the gut. They occur only in the cytoplasms, never in the nuclei of the cells. The feces of *R. sanguineus* are highly infective.

The dried feces of *Boöphilus annulatus* may retain their infectivity for months. This has been thought to explain some of the cases in Australian abattoir workers. That is, *Ixodes holocyclus* after feeding on infected bandicoots transfers the rickettsia to cattle, where it is picked up by the cattle ticks (*Boöphilus*) and later excreted in their feces. When the cattle are sent to the abattoir, the workers contact the infectious fecal dust then liberated from the hides. Derrick (1944) has designated a "basic" and a possible "secondary" cycle in the natural history of Q fever. The basic cycle is that in which the bandicoot acts as a reservoir and *Haemaphysalis humerosa* and probably *Ixodes holocyclus* are vectors. If a bush worker interrupts this cycle, he may acquire the Q fever rickettsia from *Ixodes holocyclus*. As we have already indicated, cattle may become infected from *Ixodes holocyclus,* and the source of human infections may be the feces of ticks (*B. annulatus* and *H. bispinosa*) which feed on the cattle. The possible secondary cycle may be: cattle—*Haemaphysalis bispinosa*—cattle.

Morphology and Staining. *Rickettsia burneti* is described by Derrick (1939) as being a rod-shaped, gram-negative organism, staining well by Giemsa's and Castañeda's methods. The usual size is about 1.0 micron long by 0.3 microns wide, but forms up to 2.0 or 3.0 microns long are occasionally seen.

Earlier Burnet and Freeman (1937) stated that with Castañeda's stain most of the rickettsiae stain blue, and in many clumps a large proportion of them are pink. With Giemsa's stain they take a reddish-purple color. Macchiavello's method also stains them well, the rickettsiae appearing bright red against a blue background. Ordinary bacterial stains color them poorly in tissue preparations. They are not acid-fast.

In mouse spleen sections the rickettsiae occur as intracellular microcolonies of close-packed individuals, nearly always sharply circumscribed within an oval or circular outline or vacuole. The accumulations range in diameter from about 3 to 12 microns. With hematoxylin-eosin the rickettsial microcolonies stain a rather pale blue. With Mann's stain they are a light reddish-purple.

In ticks the organisms appear as minute cocci when packed together in heavily infected cells. In the lumen of the gut, where they are likely to be scattered, they appear as short rods.

The rickettsiae of American Q fever have been described by Cox (1938b). Organisms free from tissue cells appeared as small lanceolate rods (0.25 by

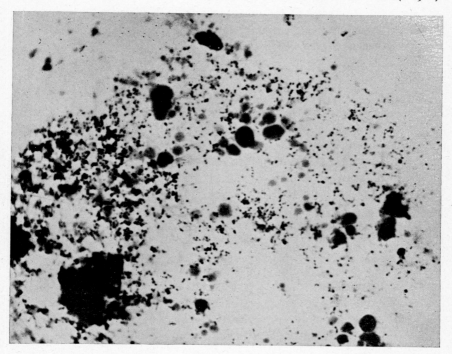

Fig. 110. *Rickettsia burneti* in a smear of the yolk sac of an infected chick embryo. (Photo by N. J. Kramis.)

0.5 microns), bipolar rods (0.25 microns by 1.0 micron), diplobacillary forms (0.25 by 1.5 microns), and occasionally segmented, filamentous forms. Chains of three to six or more minute rod or coccus forms were frequently seen. In the cytoplasm of infected cells individual bipolar or diplobacillary forms, as well as small spherical clusters (3 to 12 microns in diameter), were observed. Some of the intracytoplasmic clusters appeared to be made up of coccoid or granular organisms. It is generally agreed that the Q fever rickettsia does not occur intranuclearly, hence it is difficult to interpret Cox's observation that this part of the cell is occasionally invaded. In slides of splenic exudate he observed the nuclei of a number of cells to be vacuolated, and "sharply stained forms indistinguishable from the bipolar forms commonly observed, could be seen in the vacuoles." In general, *Rickettsia burneti* resembles *Bartonella bacilliformis* in the type of cell infection which occurs.

Growth and Cultivation. *Rickettsia burneti* has not been grown with certainty on lifeless bacteriologic media. In such preparations as leptospira medium it dies out after a few passages (Cox, 1939). In 1938 both Burnet (1938) and Cox (1938b) successfully grew the organism in tissue cultures. Burnet grew the rickettsia for 12 generations in Maitland tissue cultures of chick embryo viscera, and Cox maintained a strain through 29 passages in a similar medium. Burnet found that growth occurred principally during the second week of incubation, which is characteristic of most rickettsiae as contrasted to typical viruses which yield a maximum titer in 48 to 72 hours.

Using various types of tissue cultures, Cox and Bell (1939) obtained the best results with rubber-stoppered flasks containing minced yolk sac tissue suspended in filtered human ascitic fluid, transfers being made every 8 to 12 days. These workers were also successful in maintaining the rickettsia in serial passage in incubating fertile eggs for over 50 transfers. The greatest number of organisms were found in the yolk sac.

Singer (1941) claims to have propagated *R. burneti* using tyrode extracts from guinea pig spleen prepared by repeated grinding, freezing, and thawing, and changed every week. He found that propagation of the organisms was possible to the third week using fresh extracts, or extracts aerated for several hours in the cold to oxidize the tissue glutathione.

Other Properties of R. burneti. Both the Australian and the American workers found the Q fever rickettsia to be filterable through Berkefeld N and W filters, which ordinarily retain the typhus and spotted fever rickettsiae. It can also pass through gradocol membranes with an average pore diameter of 0.7 microns. It was because of this property of filterability that Cox (1939) named the organism isolated in America *Rickettsia diaporica*.

Other properties of the Q fever rickettsia include its resistance to glycerin and its inseparability from blood cells even after repeated washings.

Pathogenicity. The following bush animals are susceptible to *Rickettsia burneti*: the rodents—*Isoodon torosus* (the bandicoot), *Rattus assimilis, Rattus conatus, Rattus culorum culorum, Rattus culorum youngi, Rattus lutreolus, Hydromys chrysogaster, Melomys littoralis,* and *Thetomys gracilicaudatus;* and the marsupials—*Trichosurus vulpecula* and *Aepyprymnus rufescens*. The infection in most of the above animals is mild and frequently inapparent. Of these, the bandicoot is probably the most important in Australia from an epidemiologic standpoint since they have repeatedly been found naturally infected.

Of the experimental animals, the mouse and guinea pig are the most susceptible; the monkey, dog, albino rat, and rabbit are mildly susceptible.

In guinea pigs the incubation period varies from two to ten days after which ensues a febrile reaction lasting from two to eight days. Inapparent infections occur occasionally. When inoculated subcutaneously or intradermally, guinea pigs usually show a marked inflammatory thickening of the skin at the site of inoculation. On autopsy the inguinal and mesenteric nodes are usually enlarged but not injected with blood. The spleen is enlarged from 2 to 12 times by weight, and is smooth and engorged with blood. The polar fat of the testes is slightly icteric. The lungs and adrenals are frequently injected. The majority of the guinea pigs survive the infection though some die as long as two to three weeks after the temperature has returned to normal. In surviving animals Parker and Steinhaus (1943b) found *R. burneti* to be present in the materials tested for at least the following periods after defervescence: kidney and spleen, 110 days (the duration of the experiment); liver, 50 days; seminal vesicles and testes, 90 days; brain, 5 days; lungs, 20 days; and in the urine for 100 days.

In general, it appears that for most laboratory animals the American strains of Q fever are more virulent than are the Australian strains. With the American strains there is usually a shorter incubation period, more fibrinous exudate on the spleen, and more deaths.

The susceptibility of calves is interesting in the light of various epidemiologic factors observed in Australia. The association of many patients with cattle, the experimental infection of calves, the finding of dairy cows recovered from naturally acquired infection, and the demonstration that cattle ticks may be infected through feeding upon infected cattle caused Derrick, Smith, and Brown (1942) to suggest that the cow plays an important part in the transmission of Australian Q fever to man.

Immunologic Aspects. Dyer (1938) was apparently the first to suspect the relationship between Australian Q fever and the American Nine Mile fever caused by the rickettsia isolated from ticks in Montana. Cross immunity tests between the virus from a patient having the Montana disease and Rocky Mountain spotted fever, and epidemic and endemic typhus were negative, while guinea pigs recovered from Australian Q fever were immune to the American virus. Later Dyer (1939) showed that there was complete cross immunity between the Australian and American strains. In 1939 Burnet and Freeman also made a comparative study of the two strains and found them to be immunologically indistinguishable. Bengtson (1941a) found that agglutination and agglutinin absorption tests gave evidence of the identity of the rickettsiae of the Australian and the American diseases. There has been no cross immunity demonstrated between Q fever and any of the other rickettsial infections.

The preparation of vaccines which completely protect guinea pigs against the disease has been described by Cox (1940).

Perhaps the most noteworthy immunologic difference between Q fever and other rickettsial diseases is the failure of the Weil-Felix reaction in sera from individuals or animals infected with *R. burneti*. This reaction, therefore, is of no value in diagnosing the disease. However, satisfactory diagnosis can be made by means of agglutination tests, using purified suspensions of *R. burneti* for antigen. Bengtson (1941c, 1944) developed a specific complement-fixing test for Q fever. For antigen she used infected yolk sacs as the source for a purified suspension of rickettsiae.

RICKETTSIA RUMINANTIUM Cowdry (1925)

Synonym: *Rickettsia cowdria ruminantium* (Cowdry) Moshkovsky (1945).

Disease Caused. *Rickettsia ruminantium* is the causal agent of heartwater, a highly fatal disease of sheep, goats, and cattle in South Africa. The name of the disease originates from the fact that in infected small ruminants, though not so much in cattle, serous fluid accumulates in the pericardial cavity.

Following the bites of infected ticks, the incubation period of the disease varies from 11 to 18 days. It varies from 7 to 14 days after the injection of virulent blood. The resulting disease may be of a peracute, acute, chronic, or abortive form. The abortive or mild form is sometimes known as heartwater *fever*. Most animals contract the acute form in which the first indication of infection is a marked rise in temperature, which maintains itself a few days and drops rapidly just before death. The animal appears depressed and listless, and as the disease progresses nervous symptoms develop until it falls to the ground in a state of convulsions and death results. The mortality may run from 50 to 90 per cent and is usually highest in goats, lower in sheep, and lowest in cattle (Henning, 1932).

Heartwater was probably first observed in 1858 near the coast of South Africa along the borders of King Williams Town and Peddie. Later the disease was reported from various regions of South Africa and other parts of Africa and from Madagascar.

Many theories were advanced as to the cause of heartwater, but it was not until about 1900 that its infectious nature was proved. At first it was thought to be a filterable virus because no visible organisms could be found associated with the condition. In 1925, however, Cowdry (1925b,c) observed a rickettsia in the kidney and brain of sheep dying of the disease. To this organism he gave the name *Rickettsia ruminantium* and studied its relationship to ticks.

Ticks Concerned and Biologic Relationships Involved. Soon after heartwater was first observed, its relation to the bont tick, *Amblyomma hebraeum*, was noted. In 1900 Lounsbury showed that this tick could transmit the disease. When the tick is infected during the larval stage, it can transmit the infection in the nymphal and adult stages. Infected females, however, do not pass the rickettsiae through the egg to the larvae of the next generation. During his studies of *R. ruminantium* in the tick, Cowdry (1925c) confirmed Lounsbury's findings by demonstrating the organism in ticks which had fed on infected animals. This worker observed the rickettsia within the cells lining the alimentary tracts of the ticks. Intranuclear and filamentous forms were not observed.

In East Africa *R. ruminantium* has been shown by Daubney (1930) to be transmitted by *Amblyomma variegatum*.

Morphology and Cultivation. In size *Rickettsia ruminantium* is approximately 0.2 to 0.5 microns in diameter, and it is not filterable. Morphologically

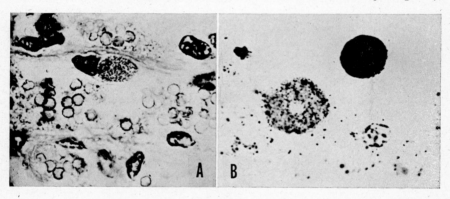

Fig. 111. *Rickettsia ruminantium*. *A.* A group of rickettsiae within an endothelial cell of a blood vessel of the kidney of an infected sheep. *B.* Rickettsiae in the tissues of the tick, *Amblyomma hebraeum*. (Courtesy Dr. E. V. Cowdry.)

it differs markedly from the rickettsiae of spotted fever and typhus. Rod-shaped forms are observed only occasionally, and most of the organisms are spherical or elliptical in shape. Once in a while horseshoe-shaped rickettsiae and other pleomorphic forms are seen. Considered from a morphologic standpoint, Alexander (1944) believes that the organism does not properly belong in the genus *Rickettsia*.

According to Cowdry (1925b), the rickettsia stains a deep, clear blue by Giemsa's method and is easily stained by Löffler's methylene blue and other basic aniline dyes. It is gram-negative.

R. ruminantium is exclusively intracytoplasmic. It does not grow in cell-free media, and apparently no attempts have been made to grow the organism in tissue cultures.

Pathogenicity. At first heartwater was thought to be a disease only of sheep and goats, but Lounsbury showed that bovines, too, were susceptible. The disease as its affects these animals has already been discussed. There were some early unconfirmed reports that antelopes were susceptible.

Small laboratory animals apparently are not susceptible to heartwater.

Immunologic Aspects. After recovery from heartwater cattle, sheep, and goats usually develop a certain amount of resistance against a second attack, though absolute immunity is obtained only after repeated attacks. Since rickettsiae may be found in the tissues of recovered animals long after recovery, the immunity appears to be one of tolerance. Some workers believe that various strains of *R. ruminantium* exist. These are thought to give fairly good homologous immunity, whereas heterologous strains give poor cross immunities.

Satisfactory vaccines and immune sera have not been produced.

RICKETTSIA SUIS Donatien and Gayot (1942b)

This rickettsia was reported by Donatien and Gayot (1942b) as causing a disease of swine, the pathology of which resembles heartwater of ruminants.

RICKETTSIA CONJUNCTIVAE Coles (1931)

Synonym: *Chlamydozoon conjunctivae* (Coles) Moshkovsky (1945).

In 1931 Coles reported on an acute purulent conjunctivitis (infectious or specific ophthalmia) occurring among sheep in South Africa. The disease appears to be the same as that occurring in Australia, New Zealand, Algeria, Germany, and probably other regions of the world. In scrapings of the epithelium lining the inner surface of the eyelid, Coles observed a rickettsialike organism which he named *Rickettsia conjunctivae*. This rickettsial disease is peculiar in that so far no arthropod vector has been discovered. Mechanical distribution by flies (*Musca domestica* and *Stomoxys calcitrans*) has been claimed by some authors (Mitscherlich, 1941, 1943), although transmission is thought to take place mainly through direct contact.

A similar, if not identical, disease has been observed in France by Lévi (1940).

Related Varieties. In 1935 Coles found an organism similar to *R. conjunctivae* in the conjunctival epithelium of goats and associated with infectious ophthalmia in these animals. In some of these, and in separate cases, an unknown intracellular organism was also found. Coles thought the latter organism might be associated with a very mild form of conjunctival catarrh. Unidentified and apparently harmless rickettsialike organisms have also been found in the conjunctival epithelium of sheep.

To a rickettsia he found associated with specific ophthalmia in cattle, Coles (1936) gave the name *Rickettsia conjunctivae bovis*.[1] (See also Rossi and Detroyat, 1938.) Later (1940), to one associated with conjunctivitis (ocular roup) in fowl, he gave the name *Rickettsia conjunctivae galli*. Convincing evidence is not available as to the relationship of these to *Rickettsia conjunctivae* of sheep. If Coles meant to designate these strains as varieties of *R. conjunctivae*, results of cross immunity and pathogenicity tests are needed showing this close relationship. If they are distinct species, they perhaps should be designated as *R. conjunctivae-bovis* and *R. conjunctivae-galli* in the interest of nomenclatorially preferred binomials.

Coles (1936), while in the United States, observed intracellular microorganisms in the conjunctival epithelium of pigs. These microbes were not identified.

Morphology and Staining. *Rickettsia conjunctivae* may occur extracellularly and intracellularly. In either location it may appear in masses or clumps made up of organisms of uniform size and shape. Individual rickettsiae appear as minute rods or granules 0.3 to 1.4 by 0.2 to 1.4 microns in size, the average being about 0.8 to 1.0 by 0.5 to 0.8 microns. They may be very pleomorphic, appearing triangular and ring-shaped. Most of the rod-shaped forms are bipolar, a few being solid in appearance. According to Coles, the chromatoid material in the larger forms is very dense in spots and these rickettsiae are inclined to bulge giving the impression of budding.

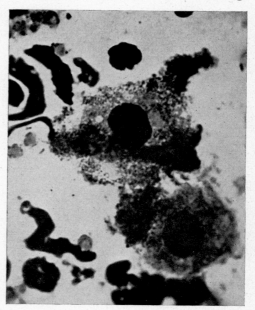

Fig. 112. *Rickettsia conjunctivae* in conjunctival scrapings of a sheep. (Photo by N. J. Kramis.)

With Giemsa's stain the organisms take on various shades of purplish-red. The minute bipolar forms stain blue. The rickettsia is gram-negative, stains

---

[1] Not to be confused with *Rickettsia bovis* Donatien and Lestoquard (1936b).

well but not as intensely as do bacteria with fuchsin, eosin, methylene blue, and carbolthioine blue. When used as a supravital stain, brilliant cresyl blue colors it a light purple.

Pathogenicity. Infectious or specific ophthalmia, the disease caused by *R. conjunctivae*, begins as an acute purulent conjunctivitis and often rapidly develops into an acute keratitis. The infection usually lasts from 2 to 20 days. According to Coles (1931), corneal ulceration, panophthalmia, and subsequent phthisis bulbi occur occasionally though in most cases the eyes apparently return to normal. Frequently only one eye may be affected with the other eye showing symptoms after two or three days.

One attack does not give the animal a solid immunity against the disease. Subsequent attacks, however, are usually milder and of shorter duration.

Information is lacking as to the infectivity of *R. conjunctivae* for the small laboratory animals.

RICKETTSIA LESTOQUARDI Donatien and Gayot (1942a)

In 1942 (a) Donatien and Gayot proposed the name *Rickettsia lestoquardi* for a rickettsia causing a benign conjunctivitis in swine similar to that which occurs in ruminants.

RICKETTSIA CANIS Donatien and Lestoquard (1935)

Synonym: *Ehrlichia canis* (Donatien and Lestoquard) Moshkovsky (1945).

Disease Caused. In 1935 Donatien and Lestoquard discovered a rickettsiosis occurring in dogs used for experimental purposes in Algeria. These dogs were kept where they were exposed to infestation by *Rhipicephalus sanguineus*. While examining the blood from the axillary artery of a dead dog, the workers observed a rickettsia in the cytoplasm of the monocytes. Donatien and Lestoquard (1935) named this organism *Rickettsia canis*.

Malbrant (1939b) has reported a disease of dogs in the French Congo apparently caused by a rickettsia similar in some respects to *R. canis*. However, on the basis of its pathogenicity to guinea pigs and its serologic and clinical manifestations he believes the disease to be a different one.

Tick Concerned. *Rickettsia canis* appears to be transmitted naturally by the dog tick *Rhipicephalus sanguineus*. All active stages of the tick may transmit the organism, and from the female it passes transovarially to the larvae of the next generation.

Morphology and Staining. *Rickettsia canis* usually occurs in small aggregates in the cytoplasm of the mobile cells of the reticulo-endothelial system.

The rickettsiae are from 0.2 to 0.4 microns in diameter and stain light purple with Giemsa's solution.

Donatien and Lestoquard (1938) observed that smears of liver and blood taken from the small ducts of the *pia mater* show not only granular inclusions

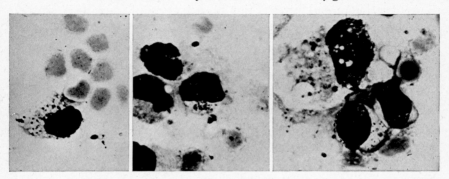

Fig. 113. Three fields showing *Rickettsiae canis* in the cytoplasm of monocytes. (Photos by N. J. Kramis.)

but also some larger forms which stain very deeply, almost black, with Giemsa's stain. They measure 3 to 5 microns in diameter and are generally spherical in shape, frequently situated in a notch of the nucleus of a monocyte. Both forms of the rickettsia may be found either in the aggregate or the dispersed state. The large, deeply stained forms are usually seen during the first days of the fever. After a few days the light purple, granular inclusions appear.

According to Neitz and Thomas (1938), a parasitized cell may contain from 1 to 12 colonies of rickettsiae varying from 2 to 10 microns in size, with the individual granules varying from 0.5 to 1.5 microns in diameter. These workers state that "some colonies consist of small granules, and then again large and small granules may be found in one colony."

According to Donatien (see Donatien and Lestoquard, 1938), *R. canis* is not stained by the Castañeda staining technique.

Pathogenicity. Dogs and monkeys (particularly *Macacus inuus*) are susceptible to *Rickettsia canis*. All breeds of dogs are apparently equally susceptible. Inapparent infections have been initiated in the jackal, *Thos mesomelas*.

In dogs infected through tick bites, death may occur as early as the fifteenth day. Following artificial inoculation, Neitz and Thomas (1938) found the period of incubation to vary from 8 to 16 days. The first symptom noticed is fever, which may be continuous for four to eight days, or remittent. The temperature may rise to 41.6° C. but usually does not exceed 40° C. After the primary reaction there may be febrile exacerbations and remissions at intervals of two to three days, or complete absence of fever until death. Infected dogs become listless and move only if forced to do so. The eyes are sunken, respiration is slow, the pulse is weak and slightly accelerated, and anemia is present. A rash may also appear. Two or three days of coma may precede death.

Rickettsiae can be found in blood smears three to five days after the initial rise in temperature, and they may thus be observed until death, which usually takes place four or five weeks after inoculation.

Recovered dogs harbor a latent infection, and splenectomy in such dogs produces a relapse.

Since the dog is also susceptible to Australian Q fever, Derrick, Johnson, Smith, and Brown (1938) compared this disease with that caused by *R. canis* and found the following differences:

The illness caused in dogs by *R. canis* is serious, often fatal; Q fever is mild. Guinea pigs do not react significantly with *R. canis;* they are readily susceptible to Q fever. *R. canis* is usually round in shape, and can be seen in the circulating monocytes; the Q rickettsia is bacilliform and has not so far been seen in blood cells. The two diseases are therefore distinct.

Similarly, Donatien and Lestoquard (1936a) showed *R. canis* and *Dermacentroxenus conori* to be distinct species.

Like guinea pigs, other small laboratory animals are not known to be susceptible to *R. canis*.

Carmichael and Fiennes (1942) have reported that the sulfonamides are very effective in the treatment of canine rickettsioses.

RICKETTSIA BOVIS Donatien and Lestoquard (1936b)

Synonym: *Ehrlichia bovis* (Donatien and Lestoquard) Moshkovsky (1945).

Disease Caused. This rickettsia is responsible for a cattle rickettsiosis observed in 1935 by Donatien and Lestoquard. These workers in Algeria received a number of *Theileria*-infected ticks (*Hyalomma* sp.) from Delpy, their colleague in Persia. The ticks were then fed on bulls. The blood of one was found to be infectious for normal bulls ten months later. Microscopic examinations of smears of the bull's liver revealed the presence of a minute granular inclusion in the monocytes which they considered to be a rickettsia, naming it *Rickettsia bovis*.

Tick Concerned. An unidentified tick of the genus *Hyalomma* is the only arthropod with which Donatien and Lestoquard were concerned as regards this rickettsia. This tick, which was originally found to transmit *R. bovis*, also transmitted *Theileria dispar* at the same time.

Morphology and Staining. *Rickettsia bovis* occurs in circular, elliptical, or round-angled polygonal masses of varying dimensions, the largest being 6 by 11 microns; the smallest from 1 to 2 microns in diameter. These masses consist of a large number of tightly pressed, minute, spherical granulations (the rickettsiae). Sometimes the rickettsiae are slightly larger and less regularly round, at which times they are more or less separated. However, their individual diameters are always considerably less than 1 micron.

The masses of rickettsiae are situated in the cytoplasm of the various monocytes. Sometimes the cytoplasm is completely filled by a single or by several joined inclusions. More frequently the cytoplasm contains a smaller mass, often situated in a notch of the nucleus.

When stained with Giemsa's stain, they take a purplish or deep mauve

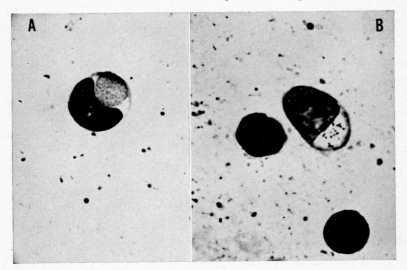

Fig. 114. *Rickettsia bovis.* A. Colony of rickettsiae in cytoplasm of monocyte of cow. B. Separated rickettsiae in cytoplasm. (Photos by N. J. Kramis.)

color. They are easily differentiated from azurophile granulations which are stained bright red and which, contrary to the rickettsiae, are always more or less dispersed. They are not characteristically stained by the Castañeda technique.

Pathogenicity. *R. bovis* causes a relatively light febrile disease in cattle. Donatien and Lestoquard's (1936b) original experiments were conducted using bulls as the experimental animals. About 14 or 15 days (sometimes a week) after the subcutaneous inoculation of infective blood, the animals begin to show febrile temperatures (40–41° C.) which continue for several days. Rickettsiae can usually be found in liver smears from the ninth to sixteenth day after inoculation. Prostration and emaciation is sometimes observed in weakened animals though experimentally the disease is rarely if ever fatal.

In order of abundance the rickettsiae are found to be most numerous in the lungs, then the kidneys, the blood of the meningial capillaries, the liver, the spleen, the blood of the peripheral circulation, the ganglia, the surrenal capsule, the myocardium, and the dermis. Rickettsiae are not found in the bone marrow, brain, or testicles.

Sheep appear to be susceptible but react only in the form of an inapparent or latent infection. In monkeys fever and depression result.

In defibrinated blood the rickettsia may remain virulent for at least 26 hours at room temperatures.

Immunologic Aspects. One attack apparently confers an immunity for an undetermined length of time. The virus persists in infected animals for long periods (at least ten months) after an acute attack. It does not confer an immunity in sheep against *Rickettsia ruminantium*.

RICKETTSIA OVINA Lestoquard and Donatien (1936)

Synonym: *Ehrlichia ovina* (Lestoquard and Donatien) Moshkovsky (1945).

In 1936 Lestoquard and Donatien described *Rickettsia ovina* in the blood of diseased sheep from Turkey and Algeria. The organisms were present only in the monocytes and never in endothelial cells. They occurred as minute coccoid granules, 5 to 30 in number, grouped in masses 2 to 8 microns in size. With Giemsa's solution they stained uniformly dark red, but with the Castañeda technique they did not stain.

From infected sheep Lestoquard and Donatien collected engorging *Rhipicephalus bursa*, which when injected into healthy sheep produced illness and fever. *R. ovina* appeared in the peripheral blood of these sheep. This suggested to these workers that *Rhipicephalus bursa* was the vector.

Schulz (1939) reported cases of this disease occurring in sheep in South Africa and differentiated it from heartwater.

RICKETTSIA AVIUM Carpano (1936)

In 1936 Carpano observed minute bodies in leucocytes and tissue cells of a bullfinch, *Pyrrhula europea*. He named the organism *Rickettsia avium*. The bird was brought to Egypt from Germany and kept for some time in the Gîza Zoological Gardens. Donatien and Lestoquard (1937) have suggested that *R. avium* and *"R. psittaci"* may be the same.

Not related to *R. avium* is another unnamed rickettsia found in birds and reported by Canham (1943) as occurring in the blood of pigeons (*Columba livia?*). Apparently no symptoms or pathologic lesions were associated with these inclusions found chiefly in the cytoplasm of the monocytes and lymphocytes. All attempts to transmit the organism failed and no vector was discovered.

RICKETTSIA PISCES Mohamed (1939)

Mohamed (1939) has observed an organism, which he named *Rickettsia pisces*, in the monocytes and plasma of the blood of a fish (*Tetraodon fahaka*) brought to him for pathologic examination. Pathologically the fish showed necrotic ulcers on either side of its body and head. The heart was congested, the liver showed advanced fatty infiltration, and the intestines were highly inflamed.

The organisms were minute coccoid forms varying from 0.2 to 0.4 microns in diameter. They frequently occurred in pairs resembling diplococci and stained pink with Giemsa's solution.

Mohamed observed the organisms in great numbers in the cytoplasm of the mononuclear leucocytes but in no other cells. Organisms present in the plasma apparently came from ruptured monocytes. He was unable to cultivate them on artificial media.

RICKETTSIA WEIGLI Mosing (1936)

In 1934 an epidemic disease broke out among employees engaged in feeding supposedly uninfected lice on their persons for purposes of typhus vaccine production in the Institute of Biology in Lwów. Mosing (1936) reported the outbreak, which affected 19 persons, and described the etiologic agent as a rickettsia which he named *Rickettsia weigli* after Prof. Weigl, director of the Institute. Mosing, among others, suggests the possibility that this rickettsia may be an extreme mutant of *Rickettsia pediculi,* and that the infection may have resulted because the feeding of such large numbers of lice at once on their persons brought them in contact with abnormal numbers of this rickettsia. Until proved otherwise, however, he is inclined to regard the causative agent as a new species. The resemblance of *R. weigli* to *R. wolhynica* is also noteworthy.

Characteristics of R. weigli. This rickettsia is a small coccoid to rod-shaped organism staining well with Giemsa's solution. It is usually slightly longer than *R. prowazeki*. In lice it is best cultivated at temperatures between 30 to 36° C. and will not develop at ordinary room temperatures. A temperature of 55° kills the organism, but it may be preserved for at least several weeks at refrigeration temperatures. It is not cultivable on ordinary bacteriologic media.

Relationship to the Louse. In the louse, *Pediculus humanus,* the rickettsiae occur extracellularly in the intestinal lumen. They multiply and form a layer covering the surface of the epithelial lining. Large numbers are eliminated with the excrement, in which the organisms may be found four or five days after being infected.

R. weigli is not pathogenic for the louse as is *R. prowazeki* or *R. rochalimae,* and the growth of the organism does not appear to lower its vitality.

The rickettsia does not pass from infected male to female or vice versa, and apparently the offspring of infected females are not infected. It may be passed from louse to louse by the intrarectal inoculation of infectious material such as hemolymph. Lice fed on infected persons may acquire the organism.

Pathogenicity. The outbreak in humans reported by Mosing as caused by *R. weigli* had the following general picture: After incubation periods of ten days to six weeks the onset was sudden with chills, sweats, and temperatures up to 104° F. and even to 105.8° F., which persisted 24 to 48 hours, after which they returned to normal. After a day or two the fever recurred, with the relapses occurring three to five times as in trench fever. Also characteristic of the disease were headaches, muscular aches and pains, slight leucocytosis during attacks, and a slight enlargement of the spleen. The blood remained infectious long after clinical recovery. The urine was also infectious.

Chodzko (1935) has designated this disease as "Rickettsiaemia Weigli."

R. weigli was agglutinated by the convalescent sera but not by sera from typhus patients. Convalescent sera gave no positive Weil-Felix reactions. The disease apparently immunizes against a reinfection. A vaccine prepared from lice dejecta protected workers against infection via infected lice.

RICKETTSIA MELOPHAGI Nöller (1917a)

In 1917, while working on the flagellates of the sheep ked, *Melophagus ovinus* (Fig. 115), Nöller discovered and named *Rickettsia melophagi,* which he found upon and in the cuticular layer covering the epithelium of the midintestine of this insect. Subsequent study has revealed the presence of this rickettsia in practically all sheep keds examined.

Location in Insect. The exact part of the sheep ked's alimentary tract which harbors the rickettsiae is not agreed upon by all workers. Sikora (1918) observed them occupying the two middle quarters of the mid-intestine. Jungmann (1918) claimed that they extended as a continuous layer from the midintestine to the anus. Hertig and Wolbach (1924) found that this layer is not always a continuous one, and in some cases only isolated patches can be found. They also observed clumps of organisms in the intestinal lumen and in the feces. The mouth parts and fore-intestine apparently never harbor the organisms.

In sections made of the sheep ked's mid-intestinal epithelium the rickettsiae are to be seen arranged in closely packed rows perpendicular to the epithelial surface. According to Sikora (1918), this arrangement results from occupying corresponding vertical cavities in the cuticular border. After first thinking he had observed the rickettsiae in an intracellular position, Nöller (1917b) later decided this to be an error and believed that they occur extracellularly only. Hertig and Wolbach (1924) also found only extracellular forms. Jungmann (1918) observed what he believed to be intracellular forms in the anterior portion of the mid-intestine. Katić (1940) also

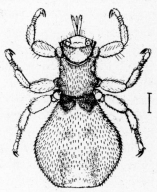

Fig. 115. A female sheep ked, *Melophagus ovinus*. (From Hoare, 1923.)

Fig. 116. Drawing of a section of the fore part of the midgut of *Melophagus ovinus* showing the rows of *Rickettsia melophagi* attached to the epithelial lining. (From Anigstein, 1927.)

reports their intracellular occurrence. According to Anigstein (1927), the arrangement of *R. melophagi* in the insect's gut is connected with the development of a lining (*Bürstenbesatzes*) which works against the penetration of the rickettsiae into the epithelial cells of that particular part of the gut. He considered the microorganisms in the secretory intestinal cells to be intracellular *R. melophagi*. (See also chapter IV.) At any rate, the great majority of rickettsiae are extracellular, and, as pointed out by Hertig and Wolbach, in tangentially cut sections the groups of rickettsiae in the cuticular border may simulate intracellular organisms.

Rickettsia melophagi appears to be transmitted transovarially from generation to generation. This would be indicated by the almost universal presence of the organism in *Melophagus ovinus*. Larval keds, as well as those freshly emerged from the pupal stage, contain rickettsiae, and some investigators have observed them in developing eggs. Different specimens vary considerably in the number of rickettsiae they contain; young insects usually harbor few, whereas older ones harbor many organisms.

Morphology and Staining. In the intestine of the sheep ked *Rickettsia melophagi* occurs characteristically in pairs and is of fairly uniform size. Generally

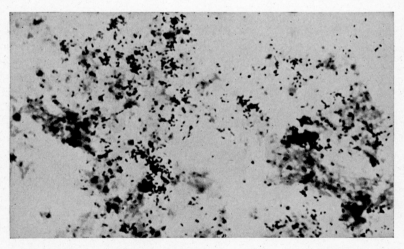

Fig. 117. *Rickettsia melophagi* growing in the tissues of an embryonic chick. (Photo by N. J. Kramis.)

it is coccoid in shape, averaging 0.4 to 0.6 microns in diameter. Minute rods forms are also seen and average about 0.3 to 0.4 by 0.5 to 0.6 microns in size. According to Hertig and Wolbach, the organisms are gram-negative and stain fairly well with carbol-fuchsin and gentian violet, but safranine and methylene blue color them very faintly. With Giemsa's method they stain a deep purple, and Macchiavello's method stains them a bright red as it does other rickettsiae. In sheep-ked eggs the organisms are rod-shaped, tend to be pleomorphic, and do not stain as intensely as they do in the insect's intestine.

In cultures on nonliving media the size of *R. melophagi* averages 0.3 to 0.35 by 0.4 to 1.0 micron. In old cultures of five to six weeks the organisms form swollen and globular forms containing deeply staining granules and disintegrating to amorphous granular material.

Of historic interest only is the fact that Woodcock (1923) refused to accept the organismal nature of *R. melophagi,* maintaining that it merely represented cytoplasmic granules of disintegrated flagellates.

Cultivation and Growth. Nöller (1917a), Hertig and Wolbach (1924), and possibly others have succeeded in growing *R. melophagi* on nonliving bacteriologic media. A glucose-blood-bouillon-agar medium was used in both cases. After 35 to 40 days' growth, Nöller obtained colonies as large as 0.4 to 0.6 mm., but in Hertig's and Wolbach's cultures the colonies remained at a size just visible to the naked eye. The morphology of the organisms in such cultures has already been described. In general, they are slightly longer and more rodlike than in the sheep ked.

The writer has several times initiated and maintained cultures of *R. melophagi* in fertile hens' eggs. One strain was maintained by serial transfer of infected embryonic fluid for more than 20 passages, when it was voluntarily abandoned.

Pathogenicity. The ability of *Rickettsia melophagi* to infect sheep, the natural host of *Melophagus ovinus,* has been the subject of contradictory claims. No ill effects which could be ascribed to the feeding of the keds upon them have ever with certainty been observed in sheep. Jungmann did not think that the organisms ever entered the sheep's blood. On the other hand, Nöller and Kuchling (1923) and others claim to have cultivated the rickettsia from that source. Gordan (1933) observed chromatin-staining bodies in lymphocytes of sheep's blood and believed them to represent *R. melophagi.* The bloods of 10 out of 49 sheep were found by Katić (1940) to contain *R. melophagi.*

The small laboratory animals (rabbits, mice, rats, guinea pigs, monkeys) all seem to be insusceptible. The writer has tried to infect vitamin-deficient guinea pigs with large doses of rickettsiae grown in the hen's egg, without observable effects.

Guinea pigs which have received massive doses of *R. melophagi* are not subsequently immune to inoculations of the rickettsiae of spotted fever, typhus, Q fever, or *maculatum* infection (Steinhaus, 1941b).

RICKETTSIA LECTULARIA Arkwright, Atkin, and Bacot (1921)

In 1921 Arkwright, Atkin, and Bacot observed in smears made from the gut of the bedbug, *Cimex lectularius,* filamentous and rodlike microorganisms which they considered as belonging to the rickettsial group, and which they named *Rickettsia lectularia*. In the same year Buchner (1921b) reported on a "new symbiotic organ" (the mycetome) of the bedbug. He later (1923) described this structure in considerable detail. Buchner observed the mycetome to contain bacteriumlike microorganisms, but he did not mention the presence of any organisms in the intestines of the insect as did Arkwright, Atkin, and Bacot. On the other hand, the latter workers did not report the presence of a mycetome containing "bacteria." Hertig and Wolbach (1924) believe that both reports were concerned with the same organism. Pfeiffer (1931) is of the opinion that both a rickettsia and a true bacterium are present in the mycetome and tissues of the bedbug.

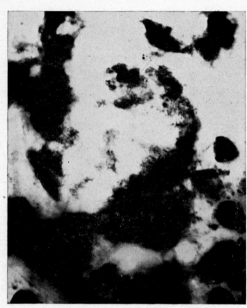

Fig. 118. A smear made from the mycetome of a bedbug showing a solid mass of filamentous *Rickettsia lectularia*.

Biologic Relationship with the Bedbug. It seems very probable that every bedbug harbors the organism concerned. Frequently the gut may have very few of them, but they seem to be invariably present in the mycetome. It or a similar microorganism is apparently present in other species of *Cimex*. Patton and Cragg (1913) diagram an organ which they call an "accessory lobe," and which in all probability is a mycetome, in *Cimex rotundatus*. Arkwright, Atkin, and Bacot (1921) observed a bacillary organism in the tissues of *Cimex hirundinis*. It is probably the same organism to which Cowdry (1923) referred as *Rickettsia hirundinis*. This name appears to be a *nomina nuda*.

Fig. 119. The bedbug, *Cimex lectularius*. (From Comstock, Comstock, and Herrick, 1938.)

As we have already indicated, the most important part of the insect from the standpoint of the localization of the rickettsia is the mycetome. This structure in insects has been described elsewhere (p. 196). The bedbug has two of these structures, which are slightly opaque and oval or pear-shaped, and which lie on either side of the body in about the third abdominal segment, near the gonads. The mycetome is supplied by a tracheal branch from the fourth abdominal spiracle. In addition to the mycetome the organisms may also be found in the alimentary tract, ovaries, testes, Malpighian tubes, and Berlese's organ.

Rickettsia lectularia is also found in the developing egg of the bedbug, which is evidence of the transovarial transmission of this organism. *R. lectularia* may also be readily observed in the embryo and in the newly hatched bedbug.

The location of the organism is essentially intracellular, and there is no evidence of multiplication within the lumen of the gut. According to Arkwright, Atkin, and Bacot, multiplication of the rickettsiae within the cells of the Malpighian tubes becomes so great that individual cells are swollen to more than twice their normal diameter. This frequently causes the tubes themselves to become swollen at these points to two or three times their normal diameter. In fresh preparations the enlarged cells are more transparent than normal ones. In sections, they have a "bird's nest" appearance, due to the intertwining of the long rickettsial thread forms.

Morphology and Staining. Arkwright, Atkin, and Bacot (1921) originally described *Rickettsia lectularia* as a nonmotile, gram-negative, coccoid, rod- or threadlike organism. They considered the typical form to be that of minute coccoid and diplococcoid bodies which stained deep purple with Giemsa. The bacillary, lanceolate, and thread forms stained more red than purple with Giemsa's solution. The latter forms appeared to have an outer sheath which took the eosin lightly and interior granules which stained more deeply and were of a purplish color. The threads were observed to liberate the granules when observed under dark-field illumination.

Arkwright, Atkin, and Bacot suggested a tentative scheme for the developmental cycle of the organism: The eggs in the ovary of the bedbug are presumably invaded with the small rickettsia forms at the time of fertilization. Some of these develop into the bacillary and thread forms while others continue to multiply in the original form. Practically every organ of the developing embryo becomes invaded, but only in a few are the conditions suitable for intracellular multiplication. In such tissues clusters of the minute granular forms develop, changing rapidly into the red-staining lanceolate forms through the development of an outer covering or envelope. These in turn develop into the long bacillary and thread forms containing the darkly staining minute granules, which are released and appear to be the same as the original minute rickettsia forms.

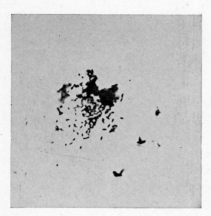

Fig. 120. *Rickettsia lectularia* in a smear of a bedbug mycetome, showing the small rod-shaped forms.

Buchner (1921b, 1923) and Hertig and Wolbach (1924) gave similar descriptions of the organisms observed in their specimens. The latter workers found the morphology of the microorganisms to vary with the season of the year. During the summer the short rods and filaments predominated over the small coccoid forms. In the winter the coccoid forms were dominant, though rods and filaments could always be found. Hertig and Wolbach found the short rods to measure 0.2 to 0.3 by 0.4 to 0.5 microns in size. The filaments (those forms over 3 or 4 microns in length) are nearly always irregularly curved and may be very slender, measuring 0.25 to 0.3 by 3.0 to 8.0 microns in size. Hertig and Wolbach are not convinced that the apparent granules contained in some of these filamentous or threadlike forms are actually the small rickettsial granules. Their findings, as well as those of Buchner's, also differ from those of Arkwright, Atkin, and Bacot in that the latter workers observed no motility in their preparations. Hertig and Wolbach found a certain proportion of the filaments to be motile. The filaments were fairly flexible, apparently possessing a terminal flagellum. No motility appears to be possessed by the short rods.

As has already been indicated, one of the outstanding characteristics of *Rickettsia lectularia* is its great pleomorphism. Not only are various shapes and sizes seen in different specimens and different tissues of the same specimen, but they may be seen in the same field of any one tissue. The typical mycetome picture shows a large proportion of curved and bent forms with occasional filaments and varying numbers of coccoid forms. The organisms in the ovaries are morphologically similar to those in the mycetomes though ovarian smears usually show more rods and filaments. The granular forms may be circular or oval; they may have a dense periphery and a light center or may appear homogeneous. Ring-shaped or C-shaped organisms may also be seen, and disk-shaped granules have been observed with dark-field illumination. Buchner, as well as Hertig and Wolbach, observed rather large,

RICKETTSIAE

strongly refractive bodies in small numbers in the mycetome. These spheroid bodies (2 to 3 microns in diameter) usually accompany groups of filaments.

Cultivation. Attempts to cultivate *R. lectularia* on a variety of media were made by Arkwright, Atkin, and Bacot but without success. Other investigators have met with similar negative results.

In eight out of ten attempts to culture on a special semisolid medium the intracellular organisms found in the bedbug, Steinhaus (1941a) isolated a diphtheroid which he named *Corynebacterium paurometabolum*. Whether or not this organism represents some of the forms just described is difficult to judge. If, as Pfeiffer (1931) believed, a bacterial symbiote and a true rickettsia occur together in the bedbug, then it is possible that the symbiote and not the rickettsia was cultivated. This work was repeated later, using the developing chick as a medium, and 16 strains of this same organism were isolated from 24 specimens of *Cimex lectularius*.

Pathogenicity. When triturated bedbugs containing *R. lectularia* are inoculated into guinea pigs, rabbits, and mice, no apparent infection results. Arkwright, Atkin, and Bacot (1921) inoculated two human volunteers with the emulsified guts of two such bugs, but no sign of infection or illness followed.

RICKETTSIA ROCHA-LIMAE Weigl (1921)

Weigl (1921, 1924) has described a rickettsia which occurs in lice (*Pediculus humanus*) and which apparently is nonpathogenic either to the lice or to vertebrates. On occasions its presence in lice has created trouble and confusion in the preparation of typhus (*R. prowazeki*) vaccine by the Weigl method. *Rickettsia rocha-limae* may occur spontaneously in stocks of lice (Sparrow, 1939) and spread among them by contact (Weigl, 1924; Mariani, 1940).

Differences between *R. rocha-limae* and *R. prowazeki* have been worked out by Weigl and others. Characterizations of *R.*

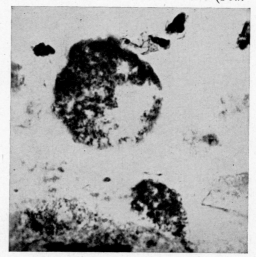

Fig. 121. *Rickettsia rocha-limae* in a section of the tissues of the louse *Pediculus humanus*.

rocha-limae may be stated as follows: It is larger and more pleomorphic than *R. prowazeki*. In smears or sections of the guts of lice *R. rocha-limae* occurs in agglomerated masses arranged in a manner similar to that of staphylococci and appears as if it were agglutinated. Frequently the individual organisms in the mass cannot be differentiated. They occur both intracellularly and extracellularly and stain more deeply and easily than do the typhus rickettsiae. *R. rocha-limae* is much more resistant to adverse conditions than is *R. prowazeki*. Weigl (1924) states: "I had succeeded in culturing it anaerobically."

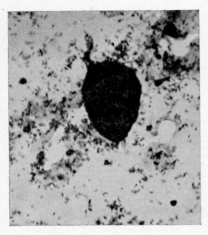

Fig. 122. *Rickettsia rocha-limae* in a smear of the gut of a louse.

In 1923 Arkwright and Bacot observed a rickettsia (sometimes called *Rickettsia cairo*) similar to *R. rocha-limae* in their stocks of lice in Egypt. However, some differences were noted, particularly in their staining and infective properties. The differentiating characters were not so definite as to exclude the possibility that the Cairo rickettsia was a nonvirulent variant of *R. prowazeki*.

R. rocha-limae is not pathogenic for guinea pigs, rabbits, rats, mice, or monkeys. Similarly, all attempts to infect man have failed (Durand and Sparrow, 1939).

RICKETTSIA CTENOCEPHALI Sikora (1918)

Of 100 cat fleas (presumably *Ctenocephalides felis*), Sikora (1918) found 5 in which rickettsiae (*Rickettsia ctenocephali*) were very numerous and 20 in which they were rather scant. Sikora believed the site of the rickettsiae to be on the surface of the organs in the body cavity and was able to demonstrate them in coelomic fluid drawn from a severed leg of the flea. Occasionally the fluid contains no organisms, but they may be found on the surface of such organs as the ovaries.

Sikora observed both large and small forms, which she suggests may be two species. The larger form resembles *Rickettsia pediculi,* and the smaller form is similar to the coccoid *Rickettsia melophagi*. Hertig and Wolbach (1924) observed *R. ctenocephali* to vary in size and shape from minute cocci (0.3 to 0.4 microns in diameter), to rather large, swollen, curved rods (0.3

by 1.5 to 2.0 microns). The organisms stained reddish with Giemsa, and short, red-stained rods with deeply stained poles were common in their preparations.

First attempts by Sikora (1918) to infect *Pediculus* with the cat-flea rickettsia failed, although later (1920) she reported that they multiplied with ease in the coelomic cavity of the louse.

RICKETTSIA TRICHODECTAE Hindle (1921)

In 7 to 8 per cent of lice (*Trichodectes pilosus*) examined, Hindle (1921) found small rickettsialike microorganisms which he named *Rickettsia trichodectae*. *T. pilosus* is a species of Mallophaga, the biting lice, and may be found on horses as were Hindle's specimens. Since this louse does not suck blood, it is assumed that the microorganisms are peculiar to the insect and not the horse.

R. trichodectae occurs extracellularly in the alimentary canal of the louse and, according to Hindle, closely resembles *Rickettsia melophagi* in its morphology. The average size is approximately 0.3 to 0.5 by 0.5 to 0.9 microns. Longer cells may be observed occasionally. The rickettsiae apparently multiply in the intestinal tract and pass out in the feces, by which means, Hindle believes, they pass from one insect to another. Transovarial transmission apparently was not considered.

In 1923 Arkwright found a similar organism in the alimentary tract of *Trichodectes caprae*.

RICKETTSIA LINOGNATHI Hindle (1921)

While examining smears from the alimentary tract of the goat louse, *Linognathus stenopsis,* Hindle (1921) found an organism he named *Rickettsia linognathi* in 2 out of 57 specimens. Morphologically it was very similar to *Rickettsia trichodectae* and apparently occurs only extracellularly in the lumen of the gut.

RICKETTSIA CULICIS Brumpt (1938)

In 1938 Brumpt reported the discovery of a rickettsia, which he named *Rickettsia culicis,* in the stomach epithelium of mosquitoes (*Culex fatigans*) 12 days after they had been fed on a patient carrying *Microfilaria bancrofti*. Brumpt believed the rickettsia to be pathogenic for the mosquito since a large number of the infected stomach cells were destroyed except in the anterior part. Whether it is essentially a parasite of the mosquito or of man is difficult to decide from the meager data available, although Brumpt is inclined to favor the latter possibility.

Rickettsia culicis occurs in the form of small granules (about 0.6 microns in diameter) and, more often, as small bipolar rods not longer than 1 micron in

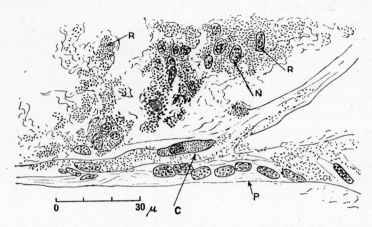

Fig. 123. *Rickettsia culicis* as seen in a saggital section of the stomach of *Culex fatigans*. C, cells invaded by the rickettsiae; N, nucleus of stomach cells; P, stomach wall; R, colonies of rickettsiae. (From Brumpt, 1938.)

length. It was first seen in sections stained with haemalum, but erythrosine-orange and toluidine blue also stain it well. It is gram-negative.

RICKETTSIA DERMACENTROPHILA Steinhaus (1942a)

Early investigators (Wolbach, 1919; Ricketts, 1911; Parker and Spencer, 1926; and Pinkerton and Hass, 1937), studying the rickettsia of Rocky Mountain spotted fever, reported the presence of microorganisms in the wood tick, *Dermacentor andersoni*, which simulated *Dermacentroxenus rickettsi*, yet were apparently nonpathogenic for laboratory animals. In 1942 (a) Steinhaus described a rickettsialike organism which occurs in apparently normal *Dermacentor andersoni* and which is possibly the microorganism observed by these early workers. It was tentatively assigned to the genus *Rickettsia* and given the specific name *dermacentrophila*.

Occurrence in D. andersoni. The ticks were from a stock laboratory strain maintained for about seven years and reared through several generations on rabbits. Frequent tests during this time failed to reveal any evidence that the ticks were carrying a disease agent of any kind.

R. dermacentrophila was found in every stage of the tick, including the egg, and was invariably present in each of several hundred adults. It usually was most abundant in the epithelial cells of the intestinal diverticula but in some specimens was found throughout the various tissues of the tick. In some preparations, perhaps as a result of the techniques used, it appeared extracellularly, but in isolated cellular elements it was seen within the cytoplasm of the cells. It was never observed with certainty in the nucleus.

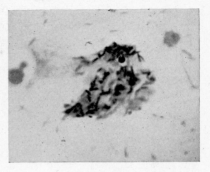

Fig. 124. *Rickettsia dermacentrophila* in a smear of the gut wall of *Dermacentor andersoni*.

Morphology and Staining. The organism is gram-negative, and like other rickettsiae it stains red by the Macchiavello method. With Giemsa it stains bluish-purple, slightly darker than the Rocky Mountain spotted fever rickettsia. This and the ordinary bacterial stains do not color it quite as deeply or as distinctly as they do most bacteria. The organism is not acid-fast.

R. dermacentrophila is usually somewhat larger than the spotted fever rickettsia though they frequently cannot be differentiated microscopically. In size it ranges from 0.3 to 0.8 by 0.5 to 2.8 and sometimes up to 4.5 microns. It frequently occurs in short chains of two or three closely joined members, and occasionally as filaments.

Cultivation. The organism has not been grown on artificial media. Fourteen such media were tried aerobically and anaerobically and at different incubation temperatures, but all without definite success. On the other hand, several strains were established in fertile chicken eggs incubated at 39° C. for five or six days and then inoculated with triturated tick viscera. The eggs were then incubated at 32° to 34° C. until the embryos died, usually between the second and sixth days. The rickettsia appeared to grow chiefly in the egg fluids, of which 0.5 cc. were used as the inoculum for serial passage, and on or in the yolk sac membrane. After five or six initial transfers, which seemed to be necessary definitely to establish the growth of the organism, it was carried through the thirtieth passage and probably could have been carried on indefinitely.

In the egg the organism appeared identical with that in ticks, though it was sometimes slightly longer.

Pathogenicity. The following animals were tested but found not to be susceptible to infection by *R. dermacentrophila:* guinea pigs (normal and

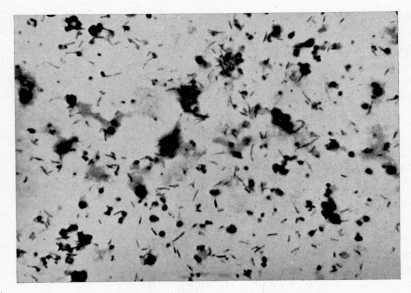

Fig. 125. *Rickettsia dermacentrophila* growing in chick embryonic membranes. (Photo by N. J. Kramis.)

vitamin-deficient), rabbits, white mice, monkeys, bushy-tailed wood rats, Columbian ground squirrels, pine squirrels, flying squirrels, chipmunks, and side-striped ground squirrels. The last six animals are natural hosts for *Dermacentor andersoni*.

Guinea pigs inoculated with massive doses of the organism were subsequently not immune to Rocky Mountain spotted fever, Q fever, epidemic or endemic typhus, boutonneuse fever, and *maculatum* disease.

Agglutination tests using the hanging drop method and the tube method were run with spotted fever immune sera. Although the spotted fever sera did agglutinate *R. dermacentrophila* in dilutions slightly higher than did normal sera, no definite complete agglutination occurred in any of the immune sera except in low dilutions (Steinhaus, unpublished experiments).

EHRLICHIA (RICKETTSIA) KURLOVI
Mochkovski [Moshkovsky] (1937)

In 1937 Moshkovsky described what he thought to be a new species of rickettsia causing a chronic infection of the monocytes of guinea pigs. He named it *Ehrlichia (Rickettsia) kurlovi*, after Ehrlich and Kurloff, thereby erecting a new subgenus. Moshkovsky suggested that this subgenus include all rickettsiae which primarily parasitize monocytes and that *Rickettsia canis* Donatien and Lestoquard (1935) become the type species.

According to Moshkovsky, in 1889 Kurloff, working in Ehrlich's laboratory, discovered certain inclusions in the mononuclear blood cells of guinea pigs. These inclusions became known as "Kurloff bodies," and similar bodies have been observed in bats, dogs, rabbits, frogs, and, very rarely, man. Moshkovsky believes the Kurloff bodies in guinea pigs are in reality rickettsiae and that probably those in other animals are too. Most authorities consider these bodies to be nonliving inclusions in guinea pig monocytes.

Morphology and Staining. Moshkovsky found the best method of staining *Ehrlichia kurlovi* to be that of Pappenheim or Giemsa. Castañeda's was not very satisfactory.

The size and shape of the organisms vary according to the length of time they have infected the cell. In early infections they appear as granules in tightly packed masses. In older infections the parasites are more coccobacillary in shape and occur in heaps which often surpass the size of the nucleus. Besides the granular and coccobacillary shapes, true bacillary, dumbbell, and filamentous forms also occur. The granules average from 0.2 to 0.3 microns in size and sometimes are arranged in minute chains of few members each.

Pathogenicity. The true pathogenic properties of *Ehrlichia kurlovi*, if it has any, remain to be investigated. The supposed infection in guinea pigs is apparently a chronic one, being initiated some time after the birth of the animal, since the fetus and newborn are free of the parasite. Moshkovsky believes that it is reasonable to assume that it is transmitted by an arthropod vector, but this assumption remains to be proved.

Systematic Status. The Kurloff (or Foà-Kurloff) bodies have often been taken as microorganisms parasitic in guinea pigs. However, recent workers (see Ledingham, 1940) have shown that the parasitic theory has no experimental evidence to support it. It seems likely, therefore, that Moshkovsky's *Ehrlichia kurlovi* will not be accepted as a bona fide member of the rickettsial group.

WOLBACHIA PIPIENTIS Hertig (1936)

In a postscript discussion of a paper by Sikora (1920) Nöller tells of the discovery of a rickettsia in the sucking stomach (esophageal diverticula) of *Culex pipiens*. This organism was found in association with a yeast described earlier by Schaudinn (1904). In 1924 Hertig and Wolbach reported the occurrence of rickettsiae in the ovaries or testes of 25 specimens of *Culex pipiens*. Nöller's organism in the esophageal diverticula was not found except possibly in one specimen. Hertig and Wolbach's organism was a minute rodlike or coccoid, gram-negative rickettsia possessing considerable pleomorphism. Hertig (1936) later published a detailed account of the organism and named it *Wolbachia pipientis,* thereby erecting a new genus as well as a new species. He gave the generic characteristics as follows:

Micro-organisms having the general characteristics of the rickettsiae, and exhibiting not only minute, bacterium-like forms appearing with darkfield illumination as luminous rods and points, but also enlarged forms within the body of which are contained one to several smaller individuals. Type and only known species *Wolbachia pipientis.*

The following discussion is based largely on Hertig's account.

Morphology and Staining. One of the outstanding morphologic characteristics of *Wolbachia pipientis* is its great pleomorphism. Microscopically the usual field observed consists of a mixture of forms of various shapes and sizes.

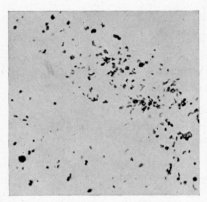

Fig. 126. *Wolbachia pipientis* showing rods and coccoid forms. (From Hertig, 1936.)

Rarely does an infected cell contain only a single form. Although it is difficult to designate any one form as typical, tiny coccoids and short straight rods are always present and hence may be considered typical. The coccoid forms measure 0.25 to 0.5 microns in diameter; the rods are of the same diameter and from 0.5 to 1.3 microns in length. With Giemsa's solution elongate and dividing rods stain relatively densely at each pole, while the intermediate portion, which may or may not be constricted, is pale by contrast. The ends of the long cells are typically rounded though one or both ends of the short rods may be drawn out to more or less of a point. As is typical with most rickettsiae, *Wolbachia pipientis* exhibits a poorly defined outline. Both rods and coccoids are irregular in shape; the coccoids are seldom perfectly spherical, and the rods may have parallel sides or may be oval in shape.

Together with the forms just described are enlarged coccoids of 0.6 to 0.9 microns, and at times of 1.0 to 1.8 microns, in diameter. These large forms stain densely, have a sharp outline, and are usually somewhat irregular in shape. All gradations from the typical small coccoids to the large ones may be found. Hertig believes that the large coccoids are the forerunners of sharply curved V-shaped organisms found in the same smears. The originally sharp outlines of the enlarged coccoids appear to become less distinct as the densely stained substance is transformed into a faintly staining matrix containing "daughter" forms, which, as the "parent" matrix disappears, become more definite in shape and finally become entirely free. According to Hertig, the sharply curved and bent forms would be those which have freed themselves from the parent matrix and are in the process of straightening out. The process appears to be a rearrangement of the protoplasm of the organism into one or more individuals embedded in a matrix which gradually disappears. Apparently the large coccoids are derived from the small coccoids.

In fresh preparations of dissected, uninjured gonads, the germ cells usually are of a glassy refractivity with no internal structures apparent. On allowing the preparation to stand or on injuring the gonad wall, a sudden change in the refractivity occurs, and the nuclei and rickettsiae immediately become visible. The rickettsiae in such preparations may show Brownian movement, but true motility has not been observed.

Relationship to Culex pipiens. *Wolbachia pipientis* may be found throughout all stages of the mosquito's development. The only organs consistently involved are the gonads of both sexes. Occasionally in larvae a few cells of a Malpighian tube may be found infected. The rickettsiae apparently do not invade the nuclei of the cells and may completely fill the cytoplasm.

The larval undifferentiated gonads lie in the sixth abdominal segment and consist of thin-walled, oval or fusiform sacs filled with germ cells. The thin membrane which constitutes the gonad wall apparently does not contain rickettsiae. On the other hand, the thin cytoplasmic shell surrounding the germ cell nucleus always contains rickettsiae.

The sex of the gonad may usually be determined late in the fourth larval instar. The testis is an ellipsoidal structure consisting of a series of transverse chambers filled with germ cells in various stages of spermatogenesis. The spermatogonia and spermatocytes are infected with rickettsiae, though in some cases with very few. The spermatozoa themselves have never been seen to contain the organisms, and Hertig assumes that they are separated with the fragments of cytoplasm and residual protoplasm normally sloughed during the latter stages of spermatogenesis.

The ovary consists of a large central duct with developing follicles radiating from it. The young follicles or several large cells (oögonia) surround the follicular epithelium, which consists of a layer of small cells. With the development of the follicle the oögonia increase in number, and a group consisting of seven nurse cells and a single oöcyte is divided off to form the primary follicle. According to Hertig, the oögonia, nurse cells, and oöcytes are invariably infected with rickettsiae. However, no rickettsiae have been observed either in the follicular epithelium or in other tissues of the ovary.

By means of smears Hertig demonstrated that rickettsiae are always present in eggs and embryos.

In a small proportion of the germ cells of the undifferentiated gonads and particularly in the testis, the rickettsiae encroach upon the area normally occupied by the nucleus and completely fill the cell except for a small mass of chromatin. Hertig has designated these as "pathological cells." Such cells had been observed earlier by other investigators who, however, were unaware of the associated rickettsiae. There is some variation in the types of pathologic cells such as certain peculiarly enlarged cells and those with pycnotic nuclei. For a description of these the reader is referred to Hertig's paper (1936).

In referring to the relationship of the rickettsia to *Culex pipiens,* Hertig makes the following general statement:

> The role of this rickettsia is that of a harmless parasite, which, however, approaches the status achieved by the many intracellular symbionts of insects, in that it infects all individuals of a given species and is invariably transmitted via the egg, but falls short of that status in that it apparently causes to a variable degree the degeneration of certain host cells.

NR Bodies. Hertig (1936) made the interesting observation that when freshly dissected gonads of *Culex pipiens* are placed in saline to which a little neutral red has been added, there become apparent in the gonad wall certain brilliantly stained, spheroid bodies which take the stain. Cresyl blue gives comparable results but Janus green, Niagara blue BX, toluidine blue, and dilute Lugol's solution do not stain the bodies. The true nature of these NR (neutral red) bodies is not known. In some ways they resemble the inclusion bodies of certain viruses, and at times they simulate masses of rickettsiae, but Hertig believes they are distinct from *Wolbachia pipientis*. Hertig was unable to find similar bodies in *Culex territans* but did observe similar bodies in specimens of *Cimex lectularius* and *Cimex pipistrelli,* and in *Phlebotomus vexator,* which he found had rickettsiae invariably present in its gonads.

Cultivation and Pathogenicity. Hertig attempted to cultivate *Wolbachia pipientis* and the NR bodies on various culture media (chiefly serum agar, NNN medium) and on the chorio-allantoic membranes of chick embryos. All such trials yielded only negative results.

The pathologic effects of the rickettsia on certain of the mosquito gonad cells has already been discussed. Laboratory animals are apparently not susceptible. This is indicated by Hertig's failure to infect mice with suspensions of mosquito ovary.

DERMACENTROXENUS RICKETTSI Wolbach (1919)

Synonyms: *Rickettsia rickettsi* Brumpt (1927); *Rickettsia dermacentroxenus* (this name, although widely used, is a corruption of *Dermacentroxenus rickettsi* and has no bona fide taxonomic standing); *Rickettsia brasiliensis* Monteiro (1931); *Rickettsia typhi* Do Amaral and Monteiro, 1932 (not Wolbach and Todd, 1920).

Disease Caused. *Dermacentroxenus rickettsi* is the etiologic agent of Rocky Mountain spotted fever, São Paulo typhus or Brazilian spotted fever, and Tobia fever or Colombian spotted fever. Some writers also include *fièvre boutonneuse,* Kenya typhus, and South African tick-bite fever as variants of this group (see *Dermacentroxenus conori*). There is some evidence that sporadic typhus of India is related to Rocky Mountain spotted fever (Banerji, 1942; Topping, Heilig, and Naidu, 1943).

As early as 1873 settlers along the Snake River in Idaho and in the Bitter Root Valley in Montana frequently in the spring of the year were attacked by a disease characterized by a continuous, moderately high fever, headache, severe arthritic and muscular pains, and occasionally vomiting, nosebleed, and delirium. The outstanding characteristic was the appearance of a profuse petechial or purpural eruption in the skin, first on the wrists and ankles but rapidly spreading to the back and to all parts of the body (fig. 127). The spotted appearance of the rash was the characteristic which caused the disease to be called "spotted" fever, and because it was thought to occur only in the northwest Rocky Mountain region, "Rocky Mountain spotted fever." It is now known that the disease occurs in most parts of the United States as well as in South America and probably in certain other parts of the world.

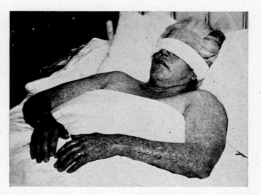

Fig. 127. A case of Rocky Mountain spotted fever showing the spotted appearance caused by the rash. (U.S.P.H.S. Photo.)

Unfortunately in almost every locality that the disease appeared a new name was attached to it; this was particularly true when slight variations from the original Rocky Mountain spotted fever were observed. Thus we have Eastern spotted fever, Brazilian spotted fever (São Paulo typhus and Minas Geraes typhus), Colombian spotted fever (Tobia fever), and Choix fever in Mexico. However, different species of ticks are commonly involved in the transmission in the different countries or sections of countries concerned.

Ticks and Insects Concerned. *Dermacentroxenus rickettsi* is known to be transmitted in nature only by ticks. In the United States five species of ticks have been shown to be naturally infected with the rickettsia: the Rocky Mountain wood tick, *Dermacentor andersoni;* [1] the American dog tick, *Dermacentor variabilis;* the rabbit tick, *Haemaphysalis leporis-palustris;* the lone-star tick, *Amblyomma americanum;* and *Ixodes dentatus* (Parker, unpublished experiments). In addition, several species of ticks have been shown experimentally to be capable of carrying the rickettsia of the United States strains of spotted fever, but they have not as yet been found naturally infected in this country. These include the Cayenne tick, *Amblyomma cajennense,* and *Amblyomma striatum,* which does not occur naturally in the United States and which is known to transmit the disease in Brazil; the Pacific Coast tick, *Dermacentor occidentalis;* the rabbit dermacentor, *Dermacentor parumapertus;* the brown dog tick, *Rhipicephalus sanguineus* (which also may possibly have been found naturally infected—Anigstein and Bader, 1943a); and the argasid ticks, *Ornithodoros parkeri, Ornithodoros hermsi,* and *Ornithodoros nicollei* (Mexico). Observations on the latter three were made by Davis (1939a, 1943b,c, and unpublished experiments), who also found that *Ornithodoros turicata* and *Ornithodoros rudis,* when infected on guinea pigs,

Fig. 128. *Dermacentor andersoni,* female. This species of tick is one of the principal vectors of *Dermacentroxenus rickettsi.* (Courtesy Dr. R. A. Cooley.)

---

[1] In this country usage has favored the name *Dermacentor andersoni* Stiles though strict adherence to the rules of nomenclature would probably favor *Dermacentor venustus* Banks. In his work on spotted fever Ricketts (1906) originally referred to the wood tick as *Dermacentor occidentalis* on the advice of Stiles who identified it as such. Later, however, Stiles recognized the Montana tick as a new species which he named *Dermacentor andersoni.* In the meantime Banks had described this tick under the name *Dermacentor venustus* Marx. Ricketts referred to the ticks from Idaho as *Dermacentor modestus,* but this also was the result of an erroneous identification, since they were identical with the Montana spotted fever or wood tick. For further discussions of these taxonomic confusions, see Wolbach (1919) and Cooley (1938).

failed to transmit the infection by feeding, although the infectious agent was present in the ticks as shown by injecting them into guinea pigs. In Brazil the ticks *Amblyomma ovale* and *Amblyomma brasiliensis* have been reported naturally infected.

In Colombia the spotted fever rickettsia is transmitted in nature by *Amblyomma cajennense*. Patiño-Camargo (1941) states that other ticks found capable of transmitting the Colombian strains are *Otocentor nitens, Dermacentor andersoni, Ornithodoros parkeri, Ornithodoros rudis,* and *Ornithodoros turicata*. The transmitting ability of the last two, however, has been questioned.

As to insects, none are known to transmit or carry the rickettsia in nature. In 1937 Dias and Martins found that the rickettsia of Brazilian spotted fever did not persist in the gut of *Panstrongylus megistus* for longer than 48 hours. Philip and Dias (1938) found the following triatomids unable to transmit the rickettsia of Rocky Mountain spotted fever to guinea pigs: *Eutriatoma uhleri, Triatoma protracta, Triatoma infestans,* and *Rhodnius prolixus*. In Brazil *Cimex rotundatus* and *Cimex lectularius* have been found naturally infected, but it is doubtful if they play much of a role in the transmission of the disease. Experimentally in *Cimex lectularius* the rickettsiae frequently do not survive for longer than 24 hours (however, see Kuczynski, 1927).

Biologic Relationships. Compared with most arthropod-borne disease agents the biologic relationships between *Dermacentroxenus rickettsi* and its tick vectors have received considerable study. As so often happens, the more information gained the more there appears yet to be known.

Dermacentroxenus rickettsi has been found to occur in nearly every tissue of the tick *Dermacentor andersoni* including the muscles, brain, salivary glands, intestinal epithelium, reproductive organs, and Malpighian tubes, and in the feces. This generalized infection of the tick apparently does it no harm. The rickettsiae occur intracellularly and may be seen either in the cytoplasm or, less frequently, within the nucleus. It was on the basis of the latter characteristic that Wolbach (1919) differentiated it generically (*Dermacentroxenus*) from other rickettsia, which so far are not known to occur within the nucleus of their host cells.

Early literature on experimental spotted fever used the term "tick virus" to differentiate between it and "serum virus" or "blood virus." It is with the properties of this "tick virus" and its relationship to its host that we are particularly concerned here. In 1923 Spencer and Parker found that the recovery of the spotted fever rickettsiae from ticks collected in nature could best be accomplished by first feeding them on normal animals and then inoculating the fed ticks into guinea pigs. When infected unfed ticks were inoculated into animals, no frank infection would result, but many of these animals would subsequently be immune to spotted fever. In other words, the ingestion of fresh blood is necessary to "reactivate" the virus in unfed infected ticks. As pointed out by Pinkerton (1942), this "reactivation" phenomenon has never been satisfactorily explained, and whether it represents an actual change from a nonvirulent immunity-producing phase to a virulent disease-producing one or merely a greater concentration of viable rickettsiae is not known. This phenomenon also explains why ticks do not infect unless they have been attached and feeding for several hours. A similar observation was made by Ricketts (1907b), who was able to produce spotted fever in animals by the injection of recently laid eggs from an infected female. When the eggs had been dried for four months, immunity but no frank infection developed. Later Spencer and Parker (1924b) found that when unfed hibernating nymphs and adults infected as larvae were incubated for 24 hours at 37° C. and subsequently inoculated into guinea pigs, a higher percentage of positive infections resulted than when similar ticks not incubated were inoculated. Blood or tissue virus showed no changes comparable to those just described for the tick virus.

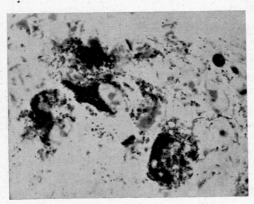

Fig. 129. *Dermacentroxenus rickettsi* in the tissues of the tick *Dermacentor andersoni*.

Spencer and Parker (1930) finally described in detail three functional phases of the virus as conceived by them:

1. A noninfectious phase in hibernating fasting ticks which is not capable of producing the typical disease unless stimulated by blood or heat but which will frequently immunize when injected into guinea pigs.

2. A highly infectious and virulent phase in nymphs and adults follows feeding. This phase in adult ticks which have been infected in an earlier stage is usually accompanied by a marked increase in rickettsiae throughout the tick tissues, but in the same phase in adults recently infected as adults, rickettsiae are not numerous and may be absent despite a high concentration of live virus. It is also this phase of the virus in fed adults with or without rickettsiae from which a protective vaccine can best be prepared.

3. A mammalian blood or tissue-virus phase which lacks the penetrating power or invasiveness of tick virus and which rarely possesses protective quality when treated with phenol.

With respect to the invisible form referred to in the second phase, Pinkerton (1942) believes this to be more apparent than real and that it may be explained by the fact that the smear method, used by Spencer and Parker, does not demonstrate intranuclear forms. Pinkerton has found that in paraffin sections

cytoplasmic rickettsiae may be absent while thousands of nuclei contain many rickettsia. In tissue cultures, infectivity was never found in the absence of demonstrable rickettsiae. It is the author's opinion that all the observed phenomena can be explained without assuming the existence of the etiological agent of spotted fever in an invisible form.

Other biologic relationships existing between *Dermacentroxenus rickettsi* and its arachnid hosts may be summarized as follows:

(1) Normal adult ticks infected by feeding on an infected guinea pig are not capable of transmitting the infection by feeding for at least the first nine days, although during this period the presence of virulent rickettsiae in the tick may usually be demonstrated by injecting it into susceptible guinea pigs.

(2) *Dermacentroxenus rickettsi* may be found in all stages of *Dermacentor andersoni, Dermacentor variabilis, Haemaphysalis leporis-palustris,* and *Amblyomma cajennense,* including the egg. The rickettsiae are passed from stage to stage and from generation to generation.

(3) In the cold-blooded tick the rickettsiae are required to withstand a far greater change in temperature than in mammalian hosts.

(4) The infecting dose of tick virus may be as low as 1:10,000 of a recently fed tick.

(5) The rickettsiae may be transmitted from infected ticks of one sex to normal individuals of the opposite sex during copulation (Philip and Parker, 1933).

(6) There is no cellular reaction to the rickettsia in the infected cells of the tick (Wolbach, 1919).

(7) The rabbit tick, *Haemaphysalis leporis-palustris,* consistently carries an extremely mild type of spotted fever virus (Parker, 1938).

There is very little information available concerning the biologic relationships between *D. rickettsi* and other species of ticks. The argasid ticks, though not as notorious vectors as the ixodid ticks, are nevertheless capable of harboring the spotted fever rickettsia for long periods of time. For example, Brumpt and Desportes (1941) have found the tissues of *Ornithodoros turicata* infectious for guinea pigs more than four and a half years after the tick had had an infective feeding. However, this tick apparently is unable to transmit the rickettsia, as pointed out by Davis (1939a). The data suggests that there is some biologic barrier in *O. turicata* which, though not opposed to the survival or possible multiplication of the rickettsiae, nevertheless operates effectively to prevent its transmission.

Morphology and Staining. According to Wolbach (1919), whose descriptions we shall follow, the morphology and distribution of *Dermacentroxenus rickettsi* in mammalian tissues are identical in man, monkey, rabbit, and guinea pig. In sections of these tissues, Wolbach found that the rickettsia has the form of a minute paired organism frequently surrounded by a narrow, clear zone or halo. The distal ends of the pair may be tapered resembling minute pneumococci. This form is found in the endothelial cells of the vascular lesions and especially in the smooth muscle cells of the media of blood vessels with the lesions. They are also found in certain other cells. In size the rickettsia averages about 1 micron long and from 0.2 to 0.3 microns wide. In smears stained with Giemsa's solution one may also find slender, pale-blue, rod-shaped forms in addition to the lanceolate forms. These rod-shaped forms sometimes possess polar granules which stain purplish or reddish. Also frequently seen in such preparations are tiny, pale-blue staining, rounded or granular forms.

In the tick Wolbach observed three morphologic forms: (1) Pale-blue bacillary forms curved and club-shaped and without chromatoid granules. These forms are first seen on and after the fifth day after infection and are found with constancy only in the gut of nymphs. Their presence in the nymph precedes the occurrence of form (2) which is a smaller, delicately staining bluish rod with deeply staining chromatoid granules. This form is most abundant on and after the fifteenth day. A later form (3) may be seen in smear preparations of the gut and of other organs. This is the more deeply staining, purplish, lanceolate form and in smear preparations of adult tissues is found only in the Malpighian tubes and salivary glands. In sections of tick tissues the larger lanceolate form, universally distributed in all tissues, and a much more minute form, in the nuclei of cells and packed in the muscle fibers of the intestinal

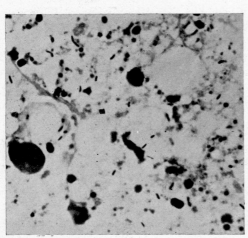

Fig. 130. *Dermacentroxenus rickettsi* in a smear of the yolk sac of an embryonic chick in which the rickettsiae are readily cultivable. (Photo by N. J. Kramis.)

tract, may be seen. The minute intranuclear forms may completely fill and even distend the nucleus, appearing as a tightly packed mass difficult to resolve into its individual members. Wolbach regarded this intranuclear form as the most characteristic one in infected ticks.

In smear preparations of eggs from infected ticks the larger lanceolate form predominates. In observing such preparations as well as those made from the tick itself, one should be careful not to confuse the spotted fever rickettsia (*Dermacentroxenus rickettsi*) with the nonpathogenic rickettsialike organism (*Rickettsia dermacentrophila*) found in normal ticks and eggs. Both are similar in appearance but may be differentiated by careful observation (Pinkerton and Hass, 1937; Steinhaus, 1942a).

Pinkerton and Hass (1932) and Pinkerton (1934) have studied the rickettsia in tissue cultures and concur with Wolbach that the morphologic range of *Dermacentroxenus rickettsi* is much greater than that of *Rickettsia prowazeki*. These workers found that the large lanceolate forms at times reach the size of pneumococci. They observed that frequently, in the same cell, clusters of very minute paired granules are contained in the nucleus, while diplobacilli of average size as well as a few of the very large lanceolate forms appear in the cytoplasm. This large lanceolate form is characteristic of *D. rickettsi* and is never seen with *R. prowazeki*. Occasionally long chains or filaments are seen in spotted fever tissue cultures similar to those observed in typhus.

As has already been indicated, one of the best stains for *D. rickettsi* is Giemsa's. With smears equally good results may be had by the proper use of Macchiavello's method with which the rickettsiae stain red against a blue background. Goodpasture's method (see Mallory, 1938) also works fairly well. Like other rickettsiae, *D. rickettsi* is gram-negative.

Growth and Cultivation. As with *Rickettsia prowazeki*, all attempts to cultivate *Dermacentroxenus rickettsi* on artificial media have been unsuccessful. Of the early investigators, Fricks (1916) claimed to have obtained cultures of the parasite grown under partial anaerobic conditions in human blood serum diluted with normal salt solution to which was added a piece of fresh guinea-pig kidney at the time of inoculation. Wolbach (1919) was unable to confirm Fricks's results, and after exhaustive attempts to grow the rickettsia on various media under a wide variety of conditions, he was unable to find any medium in which it would survive as long as in defibrinated blood.

Successful cultivation of the spotted fever rickettsia has been reported in tissue cultures and in the tissues of the developing chick embryo. In 1923 Wolbach and Schlesinger reported the cultivation of *D. rickettsi* in tissue plasma culture. Pinkerton and Hass (1932) initiated tissue cultures of the rickettsia in which the same infected cells were propagated indefinitely. They found the optimum temperature for growth to be approximately 32° C. In 1937 Bengtson grew *D. rickettsi* in modified Maitland media, using as tissue the chorio-allantoic membrane of chick embryos and guinea-pig tunica. She maintained the strain through 15 passages without any loss of virulence. Earlier Bengtson and Dyer (1935) cultivated the spotted fever rickettsia in the developing chick embryo and maintained it through 20 passages without diminution in virulence for either the embryo or the guinea pigs. In 1938 (a) Cox described a technique in which the yolk sac of the developing chick embryo is used for the cultivation of spotted fever and other rickettsiae. By use of this "egg technique" the effect of various substances on the growth of rickettsiae has been studied. Thus, it has been found that para-aminobenzoic acid (paba) exerts a striking inhibitory action on *D. rickettsi* and other members of the spotted fever group (Hamilton, 1945).

Using the agar slant tissue culture method devised by Zinsser, Wei, and Fitzpatrick (1937), Fitzpatrick (1939) grew the spotted fever rickettsia and made a vaccine with the yield obtained.

Properties and Longevity of Dermacentroxenus rickettsi. During his early work on the virus of spotted fever Ricketts (1907a,b) observed certain properties which still characterize the rickettsia. One of these properties was the nonfilterability of *D. rickettsi* in the blood of infected animals and in the eggs of infected ticks. Spencer and Parker (1924c) found that the rickettsia in blood and in emulsions of infected tick viscera was held back by Berkefeld N and V filters, nor did the filtrates of these produce immunity to spotted fever in guinea pigs. Later (1926) these same workers observed that when serum from guinea pigs infected with Rocky Mountain spotted fever was centrifuged at 8,800 revolutions per minute, the top portion retained infectivity. Red and white blood cells from infected guinea pigs retained their infectivity even after repeated washings. Normal red and white blood cells to which serum virus had been added likewise retained their infectivity after repeated washings.

Ricketts (1907b) found that Rocky Mountain spotted fever blood virus lost its pathogenicity in 24 to 48 hours when subjected to complete desiccation, and that when kept in an ice chest it retained its infectiousness for 16 days, though amounts as large as 5 ml. were required to infect guinea pigs. Wolbach (1919) found the rickettsia to survive in defibrinated blood for at least 5 days at room temperature and for at least 12 days when kept in the cold room at 7 to 10° C. Citrated blood held at 10° C. failed to infect at the end of 28 days. Blood virus remained infectious in ox bile for 3 hours but not for 23 hours. Complete desiccation destroyed it in 10 to 15 hours, and it withstood freezing longer than 4 and less than 9 days. Whereas Ricketts found the organism to be destroyed at 50° C. for 25 minutes but not at 45° C. for 30 minutes, Wolbach observed the virus to resist 45° C. for 15 minutes and 49° C. for 5 minutes. It was destroyed at 49° C. in 10 minutes and at 50° C. in 5 minutes.

Travassos and Biocca (1942) found the spotted fever rickettsia in peritoneal and vaginal washings of guinea pigs to remain virulent after 30 minutes' contact with electrolyzed silver.

In preserved tissues Wolbach found the virus to be viable for five days but not for one month in guinea-pig testes, liver, spleen, and kidney held in 25 per cent and 50 per cent glycerin at 7 to 10° C. Spencer and Parker (1924a) found that the rickettsia in guinea-pig tissues survive in glycerin for more than 10 months when kept at minus 10° C., the testicle being a more favorable tissue for preservation than spleen or liver. Tissue virus dried in vacuo survived two months when held at minus 10° C.

In living animals the rickettsiae usually do not remain viable for more than a month (Philip and Parker, 1938). According to Pinkerton (1942), *D. rickettsi* survives only for brief periods when set free by disintegrating cells.

Pathogenicity. Man is only an incidental victim to Rocky Mountain spotted fever and usually acquires the infection only when he accidentally becomes the host of an infected tick. He is in no way responsible for the maintenance of the infection in nature. This role is played largely by wild rodents and by the ticks themselves. Animals serving to transfer *Dermacentroxenus rickettsi* from infected to noninfected ticks are thought to include species of ground squirrels, tree squirrels, chipmunks, cottontail rabbits,[1] jack rabbits, snowshoe rabbits, marmots, wood rats, weasels, meadow mice, and deer mice. Various workers have found these animals to be susceptible in varying degrees (see Jellison, 1934), but as yet no animal has been found *naturally* infected. A probable exception to this is the possible isolation of *D. rickettsi* from a pocket gopher by Hassler, Sizemore, and Robinson (manuscript). The strain isolated by these workers was lost before it could be identified with certainty. In most cases, however, the animals' susceptibility consists of inapparent infections without diagnostic gross lesions or distinctive febrile reactions, and they seldom die. In Brazil the opossum, rabbit, and cavy have been found naturally infected, and the Brazilian plains dog, capybara, coati, and certain bats are also susceptible. The bats (*Histiotus velatus* and *Hemiderma perspicillatum*) die with typical lesions.

---

[1] What may prove to be a very significant observation was made recently by Jellison (1945). This worker noted that a close geographical association exists between spotted fever and one species of cottontail, *Sylvilagus nuttalli,* in the western United States. In 12 western states, 99.58 per cent of the spotted fever cases occur within the range of this rabbit, which is present in 55.88 per cent of the counties.

Most large domestic animals are insusceptible to the spotted fever rickettsia. Badger (1933b) found dogs and sheep to be mildly susceptible, isolating the virus on the sixth to eighth day after inoculation from the dogs and on the fourth to tenth day after inoculation from the sheep. Magalhães and Rocha (1942) have not only found dogs susceptible to the artificial inoculation of the rickettsiae of Brazilian spotted fever but have found dogs naturally infected.

Regarding the susceptibility of laboratory animals, the guinea pig is the most suitable for purposes of experimentation. In this animal the incubation period varies from 3 to 8 days. With well-established strains there is usually an incubation period of 2 or 3 days. The animals may become listless and lose weight. The febrile period lasts from 5 to 14 days, with the temperature ranging between 40° and 41° C. With some of the more virulent western strains the temperature may go slightly above this with death occurring in about 70 to 90 per cent of the animals during the second week. In male guinea pigs a scrotal reaction and swelling develops on about the third or fourth day of fever. This reaction begins on the skin of the scrotum as a macular rash, which later becomes petechial, the spots finally coalescing and becoming necrotic. This is followed by sloughing and subsequent healing and scar formation. Necrosis and sloughing of the foot pads and the tips of the ears also are frequent occurrences. Most of these symptoms are less marked with some of the less virulent strains which kill only about 25 per cent of the infected animals.

When guinea pigs are autopsied at the height of the infection, various manifestations may be observed. The spleen is enlarged two to five times its normal size and is smooth and dark red. Occasionally there is a very thin, translucent layer of fibrin upon its surface. In virulent strains the swollen testes are injected with blood as is the tunica, which is also very adherent to the testes and to the cremasteric muscles. Brain lesions may occur and there is a thrombonecrosis of the small blood vessels.

Other laboratory animals vary in their susceptibility to the disease. The infection is rarely fatal to rabbits. They do, however, develop a fever and may show ear and scrotal reactions similar to those seen in guinea pigs. They develop antibodies detectable by the Weil-Felix reaction. Monkeys are also susceptible, and the course of the disease in these animals may be rapid terminating in death in the case of the more virulent strains. They frequently develop a rash on the face, over the lower back, and on the outside of the thighs. Necrosis of the ears may occur. Like rabbits, monkeys develop antibodies for the Weil-Felix reaction. White mice are relatively insusceptible. White rats may not show frank infections but they are at least moderately susceptible.

Immunologic Aspects. One attack of spotted fever is generally considered to leave a more or less permanent immunity. Vaccines made from infected ticks or from infected yolk sacs confer a substantial degree of protection against the disease, nearly always reducing the severity of the infection when given annually. The tick-tissue vaccines retain their potency for long periods of time. A hyperimmune rabbit serum has been available for limited passive immunization, but convalescent serums and transfusions are usually without much beneficial effect.

Anigstein, Bader, and Young (1943) were able to protect guinea pigs against infection by spotted fever rickettsia inoculated into the skin by using minute doses of specific rabbit hyperimmune serum. The rickettsiae were presumed to be absorbed by the highly potent antiserum and to act as a sensitized vaccine, since these animals were immune to subsequent reinfection.

The antigenic components of *D. rickettsi* are similar but by no means identical with those of *R. prowazeki*. The Weil-Felix reaction is usually positive in spotted fever, titers of more than 1:160 usually being significant. The Weil-Felix test, however, is of very little assistance in differentiating spotted fever from typhus fever. Reliable specific agglutination tests are also possible, using concentrated rickettsial suspensions. Some workers have also found the complement-fixation test to be of considerable diagnostic value.

There is some evidence that *D. rickettsi* produces a toxin or toxic substance similar to that produced by *R. prowazeki*. The latter toxin appears to be the more potent of the two by present means of testing.

DERMACENTROXENUS CONORI (Brumpt, 1932)

Synonyms: *Rickettsia conori* Brumpt (1932); *Rickettsia blanci* Caminopétros (1932);[1] *Rickettsia megawi* var. *pijperi* Do Amaral and Monteiro (1932); *Dermacentroxenus rickettsi* var. *pijperi* Mason and Alexander (1939); *Dermacentroxenus rickettsi* var. *conori* Mason and Alexander (1939).

Disease Caused. Recent evidence tends to show a specific relationship between the agents of *fièvre boutonneuse,* Kenya typhus, and South African tick-bite fever. Until distinct differences, if they exist, are discovered, we shall consider these diseases to be the same, though epidemiologic differences are

---

[1] Both Caminopétros and Brumpt described the organism of *fièvre boutonneuse* and Brumpt named it *Rickettsia conori* in 1932. Later the same year at the First International Congress of Mediterranean Hygiene held at Marseille in September 1932 (Rapports et Comptes Rendus, Premier Congrès International, Hygiène Méditerranéenne, 1933c, Vol. 2, 202–212; see page 210), Caminopétros stated that the organism described and named by Brumpt did not correspond to the organism he described as the agent of *fièvre boutonneuse*. For this reason he proposed the name *Rickettsia blanci*. However, since most authorities believe they were probably both working with the same organism, Brumpt's *Rickettsia conori* has priority.

present. Differences in the rickettsiae concerned may show these to be caused by different varieties or subspecies of *Dermacentroxenus conori*.

Fièvre boutonneuse (Marseilles fever, *fièvre escharonodulaire*, eruptive fever) was first described by Conor and Bruch (1910) in Tunisia. Since then the disease has been reported from numerous regions of the Mediterranean littoral. The so-called "South African tick-bite fever" occurs in the general region of the Cape to Northern Rhodesia, and in Mozambique and Lorenzo Marqués. Clinically, *fièvre boutonneuse* is comparatively mild and of short duration, with few deaths. It is characterized by fever, headache, muscular and joint pains, an exanthema, and a lesion at the site of the tick bite. The rash spreads rapidly over the entire body including the face, hands, and feet. About the eleventh or twelfth day the temperature declines by lysis. Most cases occur during the warm summer months.

The generic relationship of *Dermacentroxenus conori* of *fièvre boutonneuse* to *Dermacentroxenus rickettsi* of spotted fever has been shown immunologically by Badger (1933a) and by Hass and Pinkerton (1936). Specific differences, however, do exist. One of the most noteworthy clinical differences is that *fièvre boutonneuse* exhibits distinctive lesions (*taches noires*) at the sites of the tick bites and an inflammatory reaction in the regional lymph nodes. Davis and Parker (1934) have shown that spotted fever vaccine made from the tissues of infected *Dermacentor andersoni*, which affords complete protection against the rickettsia of spotted fever in guinea pigs, gives no protection against the rickettsia of *fièvre boutonneuse*. This observation has been confirmed by Hass and Pinkerton (1936). Furthermore, *fièvre boutonneuse* may be differentiated from spotted fever by the complement fixation reaction (Plotz, Reagan, and Wertman, 1944).

Fig. 131. Female of the dog tick, *Rhipicephalus sanguineus*. (Courtesy Dr. R. A. Cooley.)

Ticks Concerned. The brown dog tick, *Rhipicephalus sanguineus* (fig. 131), appears to be the principal vector of *Dermacentroxenus conori* in the Mediterranean region and Kenya. In South Africa the larvae of *Amblyomma hebraeum* and all active stages of *Haemaphysalis leachi* are largely responsible for spreading the "tick-bite fever."

Prior to the proof by Durand and Conseil (1930), Blanc and Caminopétros (1932), and others that *Rhipicephalus sanguineus* carried the organism, Olmer (1930) suggested the possibility of *Ixodes ricinus* being involved. This, apparently, was not confirmed. Experimental data on other ticks capable of transmitting the infection are meager. In 1942 Davis (unpublished experiments) found *D. conori* to survive for at least 15 months in the tissues of *Ornithodoros erraticus* although this tick would not transmit the disease while feeding on susceptible animals.

In addition to *Amblyomma hebraeum* and *Haemaphysalis leachi,* the South African strains have been isolated from *Hyalomma aegyptium* and *Rhipicephalus appendiculatus. Boöphilus decoloratus* and *Rhipicephalus simus* have also been cited as vectors (Brumpt, 1927); experimental proof of this appears to be lacking, however. Neitz, Alexander, and Mason (1941) showed that the tick-bite virus could be transmitted by *Rhipicephalus sanguineus.*

Biologic Relationships. All stages of the brown dog tick, *Rhipicephalus sanguineus,* may be infected with *Dermacentroxenus conori,* and the organisms may pass from the female through the eggs to the larvae. Brumpt (1932) found the rickettsia to live for more than a year and a half in this tick.

Hass and Pinkerton (1936) made a study of *D. conori* as it occurs in the dog tick and found the rickettsiae to be present in nearly all tissues in some ticks. Usually the organisms are fairly well localized and sparsely distributed. They are found most frequently in the cells of the gut, hypoderm, and ovaries, and appear to occur more abundantly shortly after feeding. Frequently the nucleus of the cell is infected, but more often the cytoplasm is the site of their localization. When the nucleus is involved the rickettsiae occur in clusters in the center or are uniformly distributed throughout this structure, in which case they frequently cause great distension of the nuclear membrane.

The South African strains are similar in their relationships to their arthropod hosts to those just described. In the case of *Haemaphysalis leachi,* the rickettsiae may be found in any stage and are passed transovarially from the infected female to the next generation larvae. Gear and De Meillon (1941) observed that the nymphs were capable of transmitting the infection for as long as five months and that the rickettsiae do not kill the tick or seem to have any deleterious effect on it. In *Rhipicephalus sanguineus* the South African strains of *D. conori* apparently behave in a manner identical to the Mediterranean strains. All stages may be infected and the rickettsia passed to the next generation through the egg.

That *Ornithodoros moubata* may harbor the South African tick-bite organism for a considerable length of time is indicated by the finding of Parker (1942) who used the tick to transport, from Africa to the United States, the rickettsia which he recovered after 36 days.

Morphology and Staining. Morphologically and tinctorially *Dermacentroxenus conori* appears identical to *Dermacentroxenus rickettsi*. In the tick diplococcoid and diplobacillary forms frequently predominate, although when the rickettsiae occur in compact masses, they usually are smaller and more coccoid. Occasionally large lanceolate diplococci and diplobacilli are also seen.

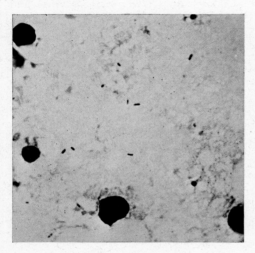

Fig. 132. *Dermacentroxenus conori* in a smear of the yolk sac of an infected hen's egg. (Photo by N. J. Kramis.)

In tissue cultures Hass and Pinkerton (1936) found the nuclei of approximately 20 per cent of the cells to be partially or completely filled with rickettsiae. Frequently those nuclei undergoing mitosis contained the microorganisms; there is no evidence that such parasitism is detrimental to the cell. Hass and Pinkerton encountered the characteristic lanceolate, diplococcoid, and diplobacillary forms of rickettsiae most often. Isolated organisms showed a tendency to be pleomorphic, and in many cells long chains were found. There is considerable variation in the size of individual rickettsiae, but in general they are similar in this respect to the spotted fever rickettsiae, averaging 0.2 microns wide by 0.9 microns, or slightly more, in length.

In animal tissues and smears of guinea-pig scrotal sac exudate, the rickettsiae are very similar in appearance to those just described. In such preparations they may be found in intracellular and extracellular locations though the majority of them are in the cytoplasm of macrophages and serosal cells. The organisms occur characteristically in pairs, and the lanceolate form is typical. Again, there is great variation in size, some being fairly large and others being extremely minute.

Giemsa's, Castañeda's, and Macchiavello's methods are the most satisfactory for staining this rickettsia.

Cultivation. The cultivation of *Dermacentroxenus conori* on nonliving media has not been successful. As with most rickettsiae, tissue cultures have been used successfully. Characteristically good media are those used by Hass and Pinkerton (1936), who applied the plasma clot tissue culture method and the Maitland method. Scrotal sac exudate and infectious splenic tissue were the basic tissues used.

Use of the embyronic tissues of developing chicks in the "egg method" is also practicable (Alexander and Mason, 1939; Cox, 1941).

Pathogenicity. As in man, the pathogenicity of *Dermacentroxenus conori* in animals is comparatively low. The evidence pointing to the dog as a reservoir of the rickettsia is convincing, although only young dogs appear to be susceptible to the extent of showing definite infection. Durand (1932) inoculated four dogs, of ages varying from two to eight months, with infected tick material. No fever or other symptom developed, but blood from these animals was infectious when injected into two paralytics. Durand believed that previous efforts to infect dogs failed because animals were used which had already been exposed to the bites of infected ticks.

With regard to laboratory animals the guinea pig has proved to be most practical for experimentation purposes. When inoculated intraperitoneally there is an incubation period of two to six days, then a rise in temperature to as high as 40.5 to 41° C. for four or five days, after which the fever gradually subsides. Scrotal swelling develops but there is no sloughing. There is practically no mortality. When the animal is inoculated subcutaneously, the fever does not rise very high and there is no scrotal reaction. On autopsy the spleen may be enlarged from one to two times; the testes are injected with blood and the tunicae are adherent.

Other animals reported as being susceptible in varying degrees include horses, spermophiles, monkeys, rabbits (including wild rabbits), and gerbils. In South Africa the last-named animal has been used to prepare rickettsial vaccines.

According to Blanc and Caminopétros (1931) the pig, goat, sheep, and pigeon are not susceptible, though the white mouse is partially susceptible and the rodent *Citellus citellus* considerably more so.

On the whole, animal susceptibilities to *Dermacentroxenus conori* have not been well worked out.

Immunologic Aspects. As has already been pointed out, Badger (1933a), Hass and Pinkerton (1936), and others have demonstrated an immunologic relationship between spotted fever and *fièvre boutonneuse*. That they are not absolutely identical, immunologically, is indicated by the fact that spotted fever tick-tissue vaccine does not protect against the Mediterranean or South

African strains of *fièvre boutonneuse*. This phenomenon is in need of further explanation, particularly since the diseases themselves do cross-immunize. Complement-fixation tests have also shown the existence of immunological differences between the two infections.

On convalescence or after the tenth day of the disease the Weil-Felix reaction becomes positive. *Proteus* OX-19 and *Proteus* OX-2 are of about equal importance, but *Proteus* OX-K is inconstant and agglutinates only in low titers.

Caminopétros and Contos (1933) showed that convalescence serum of *fièvre boutonneuse* has a neutralizing action on the virulence of the rickettsia. This property of the serum appears about the ninth day of fever and reaches a maximum on the fifteenth day. Earlier Blanc and Caminopétros (1931) concluded that serum of men recovered from the disease apparently has no protective value. They also determined that one attack of *fièvre boutonneuse* usually leaves a definite immunity.

UNNAMED PATHOGENIC RICKETTSIAE

Several unnamed rickettsiae have been described as having pathogenic properties or as being the etiologic agents of diseases of man or animals. In most cases these rickettsiae have been inadequately studied, and the biologic relationships existing between them and their arthropod hosts are practically unknown.

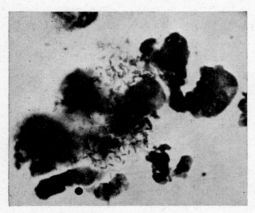

Fig. 133. Rickettsia of *"maculatum* disease" in a smear of the *tunica vaginalis* of an infected guinea pig.

One of these rickettsiae was isolated by Parker, Kohls, Cox, and Davis (1939) in 1937 from *Amblyomma maculatum* ticks collected in Texas. It is pathogenic for guinea pigs and in the literature the disease is sometimes referred to as *"maculatum* disease." In infected male guinea pigs a characteristic temperature curve usually occurs along with an edema and reddening of the scrotum. Although there is complete cross immunity between this infection and both Rocky Mountain spotted fever and *fièvre boutonneuse*, it agrees with none of these diseases in all particulars.

In 1940 Webb isolated a rickettsialike organism from the reduviid bug

Triatoma rubrofasciata. This microorganism produced pathologic lesions in laboratory animals and was maintained in guinea pigs for at least five passages. In the reduviid bug the rickettsiae may be transmitted to the next generation through the egg and could be found in the unhatched embryos and larvae.

Anigstein and Bader (1943a) isolated a spotted fever type of rickettsia from *Rhipicephalus sanguineus* taken from normal dogs. This rickettsia was pathogenic for rabbits and guinea pigs. These two workers also reported the isolation of a rickettsia from *Amblyomma americanum* collected in Texas. They believed this rickettsia to be the cause of Bullis fever (see Woodland, McDowell, and Richards, 1943), a human disease occurring in the area of Camp Bullis, Texas (Anigstein and Bader, 1943b,c).

A rickettsial infection in bison has been reported by Enigk (1942), who observed this unnamed rickettsia in the leucocytes of an infected calf. The original bison had been imported into Germany from Canada. No arthropod vector was associated with this rickettsia.

In 1944 Tatlock reported the isolation of a rickettsialike agent pathogenic for guinea pigs. The agent was isolated from guinea pigs which had been injected with blood from a patient with pretibial fever, a disease of unknown etiology. No arthropod vector was indicated.

THE BARTONELLA GROUP

There is a tendency in much of the current literature to classify the members of the genus *Bartonella* with the rickettsiae. Both groups of organisms consist of small, gram-negative, pleomorphic organisms which occur intracellularly in their vertebrate hosts and which are transmitted by arthropods. The true taxonomic status of the bartonellas is still in doubt. The best known of these organisms is *Bartonella bacilliformis,* the cause of Carrión's disease (*verruga peruana,* Oroya fever) which occurs in South America, principally in Peru, Colombia, and Ecuador. *Bartonella muris* occurs in mice, and similar organisms have been reported from rats, monkeys, dogs, and other animals.

Carrión's disease occurs in two main clinical forms: a rapid, severe anemia (Oroya fever), and a more benign type characterized by a cutaneous and sometimes subcutaneous eruption (*verruga peruana*). The Oroya fever or malignant type is approximately 40 or 50 per cent fatal; the verruga type has a very low mortality rate. For the historical, clinical, and epidemiologic aspects of the disease, the reader is referred to publications by Odriozola (1898), Strong *et al.* (1915), Rebagliati (1940), Mera (1943), Fox (1944), and Weinman (1944).

The causative agent of Carrión's disease in man, *Bartonella bacilliformis,* is found principally in the red blood cells. It usually appears as a small, gram-negative, dumbbell-shaped rod 0.2 to 0.5 by 1.0 to 2.0 microns in size or as small coccoid bodies 0.3 microns to 1.0 micron in diameter. Under dark-field illumination it has been seen to be motile. *B. bacilliformis* may be cultivated on nonliving media such as Noguchi's semi-solid leptospira media and blood agar.

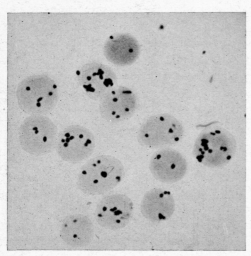

Fig. 134. *Bartonella bacilliformis* in the blood of a man suffering from Carrión's disease.

Insects Concerned. Although insects were suspected as being vectors of *B. bacilliformis* as early as 1889, it was not until Townsend (1913, 1915) completed his investigations that the sandfly, *Phlebotomus verrucarum,* was definitely incriminated. Shannon (1929) confirmed these observations and added *Phlebotomus noguchii* as a possible vector. The genus *Phlebotomus* (*Flebotomus*) belongs to the family Psychodidae of the Diptera. Only the females suck blood.

In 1942 Hertig reported on his extensive studies in Peru concerning the relation of *Phlebotomus* to Carrión's disease. This worker conducted several experiments in which wild sandflies (*P. verrucarum*) transmitted *B. bacilliformis* to monkeys by their bites. He believes that *P. noguchii,* the chief hosts of which appear to be field mice, plays very little, if any, part in the transmission of verruga to man. In microscopic studies of *P. verrucarum* fed on patients with Carrión's disease, Hertig observed the insects to develop a marked infection of the mid-intestine with a microorganism, presumably *B. bacilliformis*. Most of the organisms were eliminated with the feces, though usually some remained adhered to the surface of the gut epithelium. Hertig has also observed massive infections of the sandfly proboscis with an unidentified coccoid or rodlike microorganism which he was able to isolate in pure culture. Frequently as high as 40 or 50 per cent of the sandflies were infected. In many cases both the tip of the proboscis and pharynx of the insects were infected. What relationship, if any, this organism may have to the bartonellas is not clear. Hertig has cultured true *B. bacilliformis* directly from sandflies.

One very interesting aspect of the epidemiology of bartonella infections is the possibility that certain lactescent plants may act as reservoirs for the organism. These plants, which are found almost exclusively in endemic regions, probably also serve as food for the insect vectors. Mackehenie and Coronado (1933) have been able to culture bartonellalike organisms from the latex of these plants.

❖ CHAPTER VI ❖

YEASTS AND INSECTS

THE TERM "yeast" actually has no taxonomic significance, and the position of these organisms in relation to other fungi has not been agreed upon generally. Because of the production, by some groups of them under certain conditions, of asci and ascospores, they are usually included with the Ascomycetes (see chapter VII). In general, the yeasts are spherical, oval, or rod-shaped unicellular organisms, considerably larger than bacteria, and reproduce mostly by budding but also by fission and sporulation. The true yeasts may be divided into at least two groups: (1) those with ascospores and (2) those without ascospores. In our consideration of them as associated with insects and ticks we should keep in mind that some yeastlike fungi occur in both unicellular and mycelial arrangements. Some organisms in tissues look like yeasts yet grow out as filamentous molds.

The yeasts and yeastlike organisms associated with insects may be separated as to whether they occur extracellularly or intracellularly with respect to their host's tissues.

EXTRACELLULAR YEASTS

Yeast Flora of Insects. The yeasts associated extracellularly with insects and ticks have been very meagerly studied. Most of such studies have concerned the role of yeasts in the nutrition of insects. In addition to the general lack of interest in the group, one reason for this dearth of information may be the fact that the yeasts appear to be associated extracellularly with insects in much smaller numbers than are the bacteria or the protozoa. This may seem strange when we consider the relatively large number of yeasts and yeastlike fungi which occur intracellularly in insects.

Limited studies have shown that yeasts may be found both in and on insects taken from nature. According to Guilliermond (1920), in 1883 Boutroux showed that insects (mosquitoes, gnats, wasps, bees, and ants) distribute in nature such yeasts as *Saccharomyces cerevisiae*, *Saccharomyces ellipsoideus*, and *Pastorianus*. Berlese observed the presence of yeasts in the intestinal tracts of various insects and believed certain Diptera to be the normal habitat for certain yeasts. In the hornet, *Vespa crabro*, he observed the same yeast (*Saccharomyces apiculatus*) which he found in the nectar of flowers visited by this insect. In 1903 Parker, Beyer, and Pothier gave to what they thought was a protozoan, which they found in *Aëdes aëgypti*, the name *Myxococcidium stegomyiae*. Subsequent workers (see Castellani and Chalmers, 1919) have shown that this organism is a yeast found in normal mosquitoes. Hecht (1928) found a yeast in the esophageal diverticula of mosquitoes. He believed that its presence in these insects was incidental to the imbibing of saps which contain the yeasts. Holst (1936) found a new species of yeast, *Zygosaccharomyces pini*, associated with bark beetles on pine, and the same yeast was later isolated by Rumbold (1941) from seven *Dendroctonus ponderosae* beetles. The latter worker also found anascosporous yeasts (*Candida*) associated with the beetles *Dendroctonus monticolae* and *Dendroctonus ponderosae*. Lysaght (1936) observed an unidentified fungus which he thought to be related to *Saccharomyces* in the body cavity of the thrip *Aptinothrips rufus*. Steinhaus (1941a) isolated an unidentified yeast from the walkingstick, *Diapheromera femorata*, and one from the German cockroach *Blattella germanica*. Later (1942b) he isolated two species of yeasts from the Rocky Mountain wood tick, *Dermacentor andersoni*. Yeasts have also occasionally been isolated from houseflies and have been used to determine the dispersion of flies (Peppler, 1944). *Drosophila* flies have been found carrying the yeasts to grapes in French vineyards. These yeasts cause a fermentation of the grapes which provides optimum conditions for the developing larvae.

Yeasts as Food for Insects. In 1913 Guyénot reported, in one of a series of notes, that bacteria-free larvae of *Drosophila ampelophila* may breed entirely on yeast. Under natural conditions the larvae feed principally on yeasts and other microorganisms; and the absence of microorganisms renders certain foods unsuitable. (See also Guyénot, 1917). Northrop (1917) observed that the number of flies may be increased by the addition of banana, casein, or sugar to the yeast. Loeb and Northrop (1917a) went a step further and showed that while the larvae of *Drosophila* cannot grow on glucose agar unless yeast is added, the imago can live well on glucose agar alone. Baumberger (1917) maintained that the insect depends on yeast for its protein.

Later, in a very thorough report on a nutritional study of insects with special reference to microorganisms and their substrates, Baumberger (1919) clarified the situation with respect to *Drosophila ampelophila*. Sterile larvae grow rapidly on sterile food but die before pupating. Decaying fruit is not the food for *Drosophila* but merely a substrate for yeast cells, although the fruit also has some additional nutritive value. Further, the larvae grow on dead as well as on living yeast. Other microorganisms (bacteria and molds) are also suitable food, but yeast is a more complete food. In general, the use of microorganisms as food is widespread among insects. According to Baumberger, the feeding habits of insects may be grouped into three classes, as follows:

1. Ingestion of microorganisms with substrate, e.g., *Drosophila, Musca, Sciara,* worker termites.
2. Feeding directly on microorganisms, e.g., tree crickets, many adult Diptera.
3. Preparation by insects, of a substrate for the development of microorganisms, e.g., leafcutting ants, termites, ambrosia beetles.

Buddington (1941), in an investigation of the nutrition of mosquito larvae, found that the yeast *Saccharomyces cerevisiae* and the bacterium *Bacillus subtilis* supported the growth of larvae to maturity.

Disease-producing Yeasts Transmitted by Insects. Some instances of the dissemination of disease-producing yeasts by insects have been reported, but in many cases confirmatory proof is lacking.

Wingard (1925) claimed that *Nematospora phaseoli*, the cause of "yeast spot" of lima beans and cowpeas, was spread by the green stinkbug, *Nezara hilaris*. Fawcett (1929) demonstrated the *Nematospora coryli*, the cause of dry rot of citrus fruits, could be transmitted from pomegranates to citrus fruits by the plant bug *Leptoglossus zonatus*. A kernel spot of pecans also caused by *Nematospora coryli* has been described by Weber (1933), who associated it with the feeding punctures of the stinkbug *Nezara viridula*. Similarly in 1932 Wallace described a disease of the coffee bean caused by *Nematospora gossypii* and *N. coryli* which were transmitted by *Antestia lineaticollis*. As Leach (1940) points out, the widespread association of insects with yeasts of the genus *Nematospora* suggests that the relationship is more than a casual one. This author lists 11 species of insects as vectors of various species of *Nematospora*. In addition to those already mentioned, they are *Dysdercus cingulatus, Dysdercus fasciatus, Dysdercus intermedius, Dysdercus nigrofasciatus, Dysdercus superstitiosus, Dysdercus* spp., *Leptoglossus balteatus,* and *Phthia picta*.

According to Leach (1940), insects (*Drosophila ampelophila* and *Carpophilus hemipterus*) are responsible for the dissemination of the two or more species of yeasts which cause souring of figs. In 1942 Mrak and Phaff (unpublished experiments) isolated 123 strains of yeasts from the fig wasp, *Blastophaga psenes,* and the dried fruit beetle, *Carpophilus hemipterus,* taken from figs. Of these, 100 strains were imperfect yeasts and 23 sporogenous yeasts. The latter included 4 *Saccharomyces* spp., 6 *Hansenula* spp., and 13 *Hanseniaspora* spp.

In a few instances yeasts have been known to be the cause of diseases in insects themselves. In 1879 Hagen advocated the use of beer mash or diluted yeast to control grasshoppers, potato bugs, and the like. A yeast, which they called *Mycoderma clayi*, was found by Metalnikov, Ellinger, and Chorine (1928) in European corn borer larvae from Canada. After infection the insect develops a septicemia and dies in two to five days. On artificial media the yeast develops mycelium, a characteristic, incidentally, not typical of *Mycoderma*. Burnside (1930) fed yeasts (*Saccharomyces ellipsoideus* and *S. cerevisiae*) to honeybees, causing no more than a slight dysentery. However, when they were inoculated into the blood, death resulted. Evlakhova (1939) described a yeastlike fungus, *Blastodendrion pseudococci*, which he found to be pathogenic to the mealybug, *Pseudococcus citri*. This worker was able to increase the virulence of the organism by repeated passage through mealybugs. In experiments testing its pathogenicity it caused a mortality up to 50, and in some cases even 100, per cent of the insects. The fungus enters the body through the alimentary tract and soon parasitizes the fat body and the muscular system, but it does not penetrate the integument. Sweetman (1936) states that the pathogenicity of many yeasts for insects has been indicated but never proved, though it may be considerable when the yeasts gain entrance to the body tissues.

In 1920 Keilin changed the name of the genus *Monospora* to *Monosporella* and added the species *Monosporella unicuspidata*, which he found parasitizing the body cavity of *Dasyhelea obscura* larvae. The parasitized dipterous larvae could be recognized by the milky appearance of the body. Although the larvae are finally killed by the infection, they nevertheless are able to move about freely even when heavily parasitized, the fat body being the only organ completely destroyed before death.

INTRACELLULAR YEASTS

The majority of workers in this field have considered the intracellular symbiotes we are about to discuss as being yeasts or yeastlike fungi. For the time being, at least, the writer is inclined to follow this lead since no indisputable proof has been forthcoming that most of them are anything else in particular. Paillot (1929, 1933) in his work on the symbiotes of aphids has been one of the leaders in the belief that the intracellular organisms in some of these insects are not yeasts but bacteria. He asserts that their multiplication is by fission rather than budding and that they originate from bacteria which normally occur in the aphids. In certain other insects, what appear to be yeasts some workers believe are actually forms of molds. These statements are made so that whatever we have to say in the following pages the reader will consider accordingly. Certainly in many cases the true nature of the symbiotes is in doubt.

In chapter IV the reader was introduced to the nature of the mycetocytes and mycetomes which harbor many of these intracellular microorganisms. Hence, we shall proceed without any further elucidation of these structures and their general anatomic relationships to their hosts.

Nature of Yeastlike Symbiotes. As has just been pointed out, some workers believe the forms under discussion to be pleomorphic forms of bacteria. Others are just as convinced that they are yeasts or yeastlike fungi, and some believe them to be forms of molds. At any rate, most of them are larger than most bacteria, many have definite nuclei, and usually their protoplasm is granular and vacuolated. Some have been observed to multiply by budding, some by fission, and a few by sporulation. In a great many cases the symbiotes within the insect have been observed in the process of budding. Frequently the buds adhere together, thus forming chains of buds of three or four members. Sometimes these chains give the appearance of filaments or tubes which can be taken for other types of fungous growth.

Some of the yeast and yeastlike symbiotes have been cultivated on artificial media but the majority have not. Schwartz (1924) tried many procedures and techniques but found great difficulty in maintaining cultures of the symbiotes. When they do grow on artificial media they frequently give rise to both buds and mycelia. The mycelium may give off conidia, which in turn reproduce by budding. Such observations tend to give weight to the belief that the symbiotes are not true yeasts or, in some cases, may even be fungi of no immediate relation to the yeasts. At least it appears that they have lost the property of assuming a saprophytic existence when the host dies, since in most cases when this happens the symbiotes have been observed to die also.

Mahdihassan (1941) has cultivated certain pigmented bacteria from tissues of *Cicadella viridis* containing what appear to be yeastlike organisms. He maintains that the latter are not living units but are plasmatic pieces or debris of tissue cells attacked by bacteria, and that they only superficially resemble yeasts. These studies have not been confirmed.

Classification. Early attempts to classify certain of the intracellular yeasts were made by Moniez (1887), Lindner (1895), Šulc (1906–1910), and Buchner (1912). One of the most recent systems of classification for this group of organisms is that by Brain (1923). His system is based on the morphologic characteristics of type species as obtained from smears or sections. Unfortunately this is perhaps necessitated by the fact that very few of these intracellular organisms have as yet been cultivated outside of their hosts. Any system of classification which is not based also on cultural characteristics is probably due for considerable revision as soon as it becomes possible to study the organisms in cultures. This is undoubtedly the principal reason why any classification so far proposed for these intracellular yeastlike organisms is entirely inadequate.

Brain used certain suffixes on generic names to indicate the anatomical relationship of the symbiote to the host. Thus the suffix *-cola* on the generic name refers to those forms living free in the hemolymph or connective fat tissue, the suffix *-myces* refers to the symbiotes inhabiting an "obligatory mycetocyte" of a definite mycetome, e.g., *Lecaniocola, Cicadomyces*.

On the following pages we shall briefly consider a few of the distinguishing characteristics of some of the species of the various genera. It should be understood, however, that Brain's classification is used only in a temporary sense, since as we have already indicated, it contains several inconsistencies, is incomplete, and is subject to much future revision. Furthermore it should be remembered that this entire group of symbiotes has not received the attention it deserves from mycologists or specialists on yeasts and yeastlike organisms, and that for the most part the names used to designate these organisms are based on rather flimsy foundations. In no sense does the writer wish to perpetuate the taxonomic anomalies which are undoubtedly included in the systematics that have been employed. Their use in this chapter may be justified only because there is nothing better to offer at present. It is hoped that the very apparent inadequacies will serve as a challenge and inspiration to taxonomists, so that they may evolve some acceptable system of classification for the group. It is possible that many of these intracellular forms actually belong to groups already having bona fide nomenclature. Some could logically be placed in the genus *Trichosporon*. Others are perhaps not yeasts at all and belong to an entirely unrelated group of fungi. Especially to be criticized is the practice of placing a yeastlike symbiote in one of the "perfect" (sexual) genera prior to actual observation of the ascospores. Similarly symbiotes studied only microscopically should not be placed in a genus like *Saccharomyces* without their fermentative characteristics having been determined. With the great majority of these symbiotes much more needs to be known before they can be given a definite systematic status. The presentation which follows is for the purpose of orientation only, and the writer disclaims any attempt to advocate the adoption of this type of classification.

The question has frequently arisen as to whether certain groups of symbiotes associated with various species of insects are the same species. Some workers have assumed that each species of insect has its own species of symbiote. In any case, this question should be approached with caution since it is definitely known that the morphology of the symbiotes is dependent upon the metabolism of the host. Cultivation on artificial media and subsequent study of the cultures will doubtlessly clear up such problems.

LECANIOCOLA Brain

In this genus the symbiotes do not occur in a definite mycetome but are found free in the hemolymph or connective fat tissue. Although these symbiotes are not strictly intracellular in habitat, they are discussed here for the sake of order and because they comprise a large group in Brain's classification. They are spherical, pear-shaped, or elongate in shape, often with one or both ends somewhat pointed. Multiplication is by budding. In some ways this genus resembles *Saccharomyces*, but, according to Brain (1923), it is separated by the fact that all its members occur in coccids, particularly those of the subfamily Lecaniinae and closely related forms. As far as is known, no endospores are produced.

Lecaniocola parasitica (Lindner). Synonyms: *Lecaniascus polymorphus* Moniez (pro parte) and *Saccharomyces apiculatus* var. *parasiticus* Lindner. This species was one of the first of the yeast symbiotes observed. In 1854 Leydig described free-living lanceolate bodies in the hemolymph of *Coccus hesperidum* (*Lecanium hesperidum*), considering them to be parasites. He found them never to occur in cells; they were about 4 microns in length and reproduced by terminal budding. Moniez (1887) and Lindner (1895) studied this species further. After 20 hours in a hanging drop of wort the average size is 10 microns long and 3 microns wide. Most of the symbiotes have one pointed end, and the protoplasm possesses a few small vacuoles. The insect ova become infected by a few organisms which enter at the anterior poles.

Lecaniocola rosae (Buchner). Synonym: *Coccidomyces rosae* Buchner. This organism was observed by Buchner (1912) in *Lecanium corni*. It is elongate, usually with one end pointed and one end broadly rounded, though sometimes both ends are pointed. It averages about 8.5 microns in length and reproduces by budding at the pointed end. The protoplasm of this symbiote usually contains diffused granules.

Lecaniocola proteae Brain. The host insect for this species is *Lecanium proteae*. Brain (1923) describes *Lecaniocola proteae* as a relatively small organism, 4 to 5 microns long and almost as broad, being almost spherical with one end attenuated. The protoplasm usually contains one large granule and one small vacuole. The buds, which arise terminally, persist until fully grown. Brain observed the presence of a second organism in this insect, which appeared as a deeply staining rod. He suggests that this may represent a disymbiotic condition in which both forms live free in the hemolymph or connective fat tissue.

Lecaniocola lecanii viridi Brain. Königsberger and Zimmermann (1901) described this organism which is found in *Lecanium viride*.

The symbiotes of other species of *Lecanium* have been studied, including the irregular, knotty forms in *Lecanium longulum* as well as those in *Lecanium persicae* (see Teodoro, 1918).

Lecaniocola contei Brain. This symbiote is associated with *Lecanium hemisphericum* (now *Saissetia hemisphaerica*) and was described by Conté and Faucheron (1907). In the insect it is ovoid in form, usually pointed at one or sometimes at both ends. Budding occurs near the apical extremities and no mycelium has been observed. In the insect the size of the organism averages about 26 microns long and 13 microns wide. In cultures the dimensions are less, usually being about 8 microns long and 4 microns wide. No sporulation has been noted, and in old cultures the yeasts appear to be enclosed in a thick membrane.

Transverse sections of *S. hemisphaerica* show the yeasts to be present in large numbers in all the connective tissue, which fills the general body cavity and which corresponds with a fat body of other insects. The organisms occur intracellularly and in stained sections appear to be surrounded by clear halos. Transmission to the next generation takes place through the eggs.

Conté and Faucheron also observed these yeastlike organisms "always in very large numbers in *Lecanium oleae, Lecanium hesperidum, Pulvinaria floccifera*, etc. These present different aspects from those in *Lecanium hemisphericum*."

Lecaniocola saissetiae Granovsky. In all stages of *Saissetia oleae*, Granovsky (1929) found an elongated, spindle, and sometimes egg-shaped, yeastlike organism which varied considerably in size and shape. It was usually pointed at one end, though sometimes both ends were pointed. The protoplasm was coarse, very granular, and vacuolated. Multiplication was by terminal budding at the pointed end, though sometimes buds formed at both ends and were attached by long necks. As a rule, no more than two or three buds are ever found in chain formation. Granovsky gives the average size of the mature individuals as about 3.5 microns wide and 13 to 14 microns long. The newly separated forms are 3 to 3.5 microns wide and 6.5 to 7 microns long. In the insect the symbiotes are found free in the hemolymph and in the connective fat tissue. Transmission to the next generation occurs via the egg, the organisms entering the ovum soon after its differentiation from the nurse cells and the follicular epithelial cells.

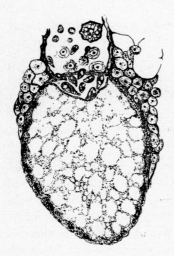

Fig. 135. *Lecaniocola saissetiae* in *Saissetia oleae*, showing penetration of ovum by means of mass attack of the yeastlike organisms. (From Granovsky, 1929.)

Lecaniocola ceroplastidis pallidi Brain. This yeastlike organism is found in *Ceroplastes pallidus* and is approximately 6 to 8 microns long. The protoplasm is coarse with one or two large granules and a few vacuoles. Reproduction occurs by budding, the buds sometimes adhering together in a chain of two or three members.

Lecaniocola egbarum fulleri Brain. *Ceroplastes egbarum fulleri* is the host of this symbiote, which is similar in appearance to *Lecaniocola ceroplastidis pallidi* though it is broader and has a coarser protoplasm. It reproduces by budding, the buds attached by long, thin necks.

Lecaniocola saccardiana (Berlese). Synonym: *Oöspora saccardiana* Berlese (pro parte). Berlese in 1906 described this species in *Ceroplastes rusci* but was probably concerned with a mixed "infection" of the symbiote and a fungus. Actually the organism does not form a mycelium in the insect. Instead, free yeastlike forms with granular protoplasm

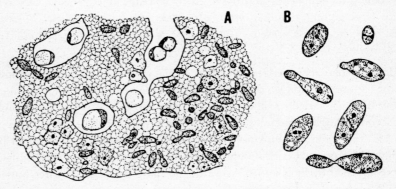

Fig. 136. The symbiote of *Pulvinaria innumerabilis*. *A*, section of fat body showing symbiotes in overwintered nymphal insects; *B*, isolated symbiotes. (From Brues and Glaser, 1921.)

occur which are long and egg-shaped. Frequently lemon-shaped forms occur with both ends pointed. They measure 6 to 7 by 2 to 2.5 microns in size and reproduce by terminal budding.

Lecaniocola putnami Brain. Putnam (1880), in a detailed study of *Pulvinaria innumerabilis,* observed in these insects small oval bodies 3 to 5 microns in width and 10 microns long. This worker was not certain of their nature. Brues and Glaser (1921) investigated this symbiote further and concluded it was not a yeast but a fungus, probably related to the genus *Dematium.* They arrived at this opinion through a study of their potato agar cultures in which during the first few days yeastlike budding cells were observed. After further incubation, however, a distinct mycelium occurred, at first white but later with black pigmented spots. The mycelium was branched, the larger hyphae measuring from 6 to 15 microns in diameter.

On the other hand, in the insect the symbiote appeared yeastlike, though extremely variable in size and shape. The average dimensions were 10 to 16.7 by 5 to 6.5 microns. Budding forms occurred frequently, the buds being of the same general size and shape as the mother cell. The internal structure of the symbiote varied but usually showed one or several deeply stained portions which resembled nuclei. Vacuoles were also present.

Although it seems likely that the species concerned is a *Dematium,* further study is necessary before this point can be settled.

Lecaniocola pulvinariae Brain. *Pulvinaria mesembryanthemi* is the insect host of this yeastlike organism which averages 4 or 5 microns in length and is pear-shaped. Its protoplasm is more or less homogeneous, being devoid of granules or vacuoles. It reproduces by budding at the pointed end, the buds sometimes forming a chain of two or three members.

Lecaniocola protopulvinariae Brain. This species occurs in *Protopulvinaria pyriformis.* It is a 6-to-11-microns-long, spindle-shaped organism with finely granular protoplasm. Buds occur terminally attached by a very thin, tapering neck.

Lecaniocola filippiae Brain. The host insect for this species is *Filippia chilianthi.* Brain (1923) found this symbiote to be rather variable in size and form. The average size is from 6 to 8 microns long and 3 microns wide. However, division forms, which frequently consist of several buds in a chain, may measure 25 to 30 microns in length. The protoplasm is finely granular in nature and seldom contains vacuoles.

Lecaniocola inglisiae Brain. This symbiote is found in *Inglisia geranii.* It is an elongated organism often with parallel sides and averages 11 microns in length. The protoplasm is coarsely granular and frequently contains one or more large vacuoles. Reproduction occurs by terminal budding, the buds often remaining attached by a long neck until they are almost as large as the mother cells.

Lecaniocola conomeli limbati (Šulc). As in the case of the species just listed, Brain (1923) was in doubt about placing this species in the genus *Lecaniocola*. In 1910 (a) Šulc had placed it with *Saccharomyces*. It is found in *Conomelus limbatus*. The symbiote is elliptical or egg-shaped. Its protoplasm is coarse and its nucleus small. Budding is terminal and slightly lateral. The buds are at first spherical, then egg-shaped, and they have not been observed to occur in chains.

Lecaniocola macropsidis lanionis (Šulc). Brain (1923) questionably places this species in the genus *Lecaniocola* since Šulc (1910a) had placed it in the genus *Saccharomyces*. It was found by the latter worker in *Macropsis lanio*. It is elongated and may be egg-shaped or with both ends pointed (fig. 137). The average size is 3 microns long and 1 micron wide. The protoplasm is coarse and the nucleus is distinct, round, and centrally located. Budding is terminal, the buds being first elliptical, later egg-shaped, with the broad end toward the mother cell. They are vacuolated but apparently do not occur in chains.

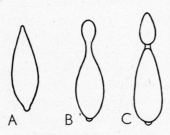

Fig. 137. *A. Lecaniocola (Saccharomyces) macropsidis lanionis. B, C.* Reproductive forms. (From Šulc, 1910a.)

KERMINCOLA (Šulc)

The microorganisms in this genus do not occur in a definite mycetome. They are very long and narrow and have parallel sides. One end is usually pointed. The nucleus is large and distinct, sometimes in long individuals there appear to be two to four nuclei. The average length of the symbiotes is about 20 microns though some individuals are 40 to 50 microns long. The outer membrane is dense and hyaline. Multiplication is by budding.

Kermincola kermesina Šulc. This organism was found in *Kermes quercus* by Šulc (1906). It is an elongated organism (4 by 20 microns in size) with parallel sides and one or both ends frequently pointed. The nucleus is very large and distinct, the protoplasm granular, and the periphery hyaline. Occasionally forms 40 to 60 microns in length are seen which have two to four nuclei present. Multiplication is by terminal budding. (See also Vejdovsky, 1906.)

Kermincola tenuis (Buchner). Synonym: *Psyllidomyces tenuis* Buchner. In 1912 Buchner observed in the fat cells and hemolymph of a psyllid an organism very similar to *Kermincola kermesina*.

PHYSOKERMINCOLA Brain

This genus so far contains only one known species, *Physokermincola physokermina* (Šulc). Šulc (1906) had originally placed it in the genus *Kermincola*. It is found in *Physokermes abietis*.

The symbiote is not harbored in a definite mycetome. It is pear-shaped or pointed at both ends and about 3 microns wide and 10 microns long. The protoplasm is dense and the periphery is not hyaline. The nucleus is very large. Multiplication is by terminal budding.

CICADOCOLA Brain

This genus also contains a single species, *Cicadocola cicadarum* (Šulc), which is found in *Cicada orni* (Šulc, 1910b). Šulc originally named it *Saccharomyces cicadarum*. The symbiote does not occur in a definite mycetome but in the fat tissues and hemolymph. It is narrow and long (2 to 3 by 10 to 12 microns), with one end attenuated but not sharply pointed. The protoplasm is finely reticulate, containing one or more small nuclei. Usually present is a single vacuole. Multiplication is by terminal budding, the buds often united in a chain of four or more members (fig. 138).

CICADOMYCES Šulc

The organisms in this genus are enclosed in a definite mycetome. They are spherical, bean-shaped, or polygonal in shape and from 3 to 10 microns in diameter. The protoplasm usually contains several large granules and vacuoles. Multiplication is by budding and fission, the buds often occurring in long chainlike groups.

Cicadomyces ptyeli lineati Šulc. In introducing this genus and species in his original paper, Šulc (1910b) spelled the name *Cycadomyces ptyeli lineati*. This was probably a misspelling of the generic name since on other occasions in the same manuscript he spelled it *Cicadomyces*.

This organism is the type species of the genus and was observed by Šulc (1910a) in *Ptyelus lineatus*. The symbiote is round, bean-shaped, or roundly polygonal. The average size is from 6 to 10 microns long. The cell membrane is very fine, but the protoplasm is granular and contains large vacuoles. Multi-

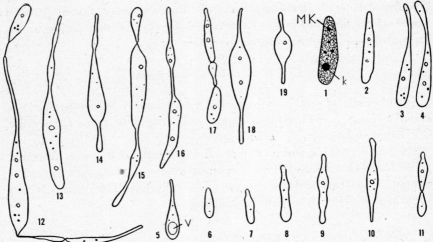

Fig. 138. *Cicadocola cicadarum*. *1*, a single, mature, nongerminating yeast cell (*k*, nucleus; *MK*, metachromatic granule); *2, 3, 4, 6, 7, 8, 9, 10, 11, 18, 19*, different stages of growth; *12-17*, different forms of cell division; *5*, a cell with a large vacuole (*v*). (From Šulc, 1910b.)

plication is by budding and fission. The buds are frequently united in long chains (fig. 139).

Cicadomyces minor Buchner. Šulc (1910b) considered this to be another form of *Cicadomyces ptyeli lineati* occurring in the same insect, *Ptyelus lineatus*. According to Brain (1923), this organism is found in the small yellow mycetome of the insect (figs. 140 and 141). The symbiote itself is much smaller than *C. ptyeli lineati*, averaging about 3 microns in diameter, and its general appearance is different and more distinct.

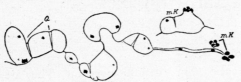

Fig. 139. *Cicadomyces ptyeli lineati* (a germination band). *mk*, metachromatic granules; *Q*, the beginning of cross division. (From Šulc, 1910a.)

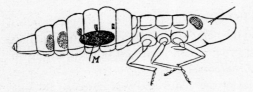

Fig. 140. *Ptyelus lineatus*. Side view of a young larva showing position of mycetome (*M*). (From Šulc, 1910a.)

Cicadomyces aphrophorae alni Šulc. This species is without an adequate description. Šulc (1910b) observed it in the mycetome of *Aphrophora alni*.

Cicadomyces aphrophorae salicis Šulc. This species inhabits the inner portion of the mycetome of *Aphrophora salicis*. It is elongated and often curved and has parallel sides.

Cicadomyces rubricinctus Buchner. The host insect of this species is also *Aphrophora salicis,* which has a second mycetome (the orange-red portion described by Šulc, 1910b). It is similar to *C. aphrophorae salicis* but somewhat thicker.

Cicadomyces aphalarae calthae Šulc. This symbiote is enclosed in a definite mycetome of *Aphalara calthae*. It is similar in appearance to *Cicadomyces ptyeli lineati* but more spindle-shaped. It contains numerous small metachromatic granules and averages about 10 microns in size. Šulc (1910a) also observed a smaller form, which he calls form II in the same insect.

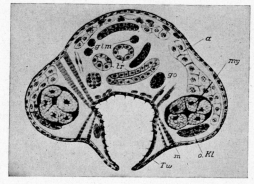

Fig. 141. A cross section through the fifth abdominal segment of a young larva of *Ptyelus lineatus* showing the mycetomes (*my*). (From Šulc, 1910a.)

Cicadomyces dubius Buchner. According to Brain (1923), this is an undescribed species found in the central portion of the mycetome of a psyllid. It has been figured by Buchner (1912).

Fig. 142. Larva of *Cicada orni* showing the position of the mycetome (*M*). (From Šulc, 1910b.)

Cicadomyces cicadarum Šulc. This organism is very pleomorphic and contains colored granules. It is found in the mycetome of *Cicada orni* (fig. 142). The cells are elongated and elliptical (fig. 143), averaging 6 microns wide and 15 microns long. The cell membrane is very fine and the plasma is finely reticulated.

Cicadomyces liberiae Buchner. Buchner (1912) found this symbiote in the peripheral portion of the mycetome of a cicada from Liberia. The organism is round, oval, or elongate in shape.

Cicadomyces minimus Buchner. This symbiote inhabits the inner portion of the mycetome of the unnamed cicada just mentioned. It is very small, being only 1 to 3 microns in diameter.

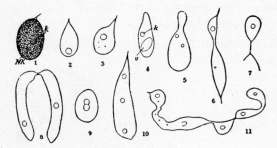

Fig. 143. *Cicadomyces cicadarum* from *Cicada orni*. *1*, a single, mature, nongerminating yeast cell (*k*, nucleus; *MK*, metachromatic granules); *2*, *3*, tear-shaped yeast cells; *4*, *5*, *6*, formation of new cells (*k*, nucleus; *v*, vacuole); *7*, two thread-connected individuals; *8*, two cells with tapering ends; *9*, beginning of the direct division of the nucleus; *10*, a sack-shaped individual with three nuclei; *11*, a germination band or cluster. (From Šulc, 1910b.)

CISSOCOCCOMYCES Brain

This genus contains only one named species, *Cissococcomyces natalensis* Brain, found in *Cissococcus fulleri*. It inhabits mycetocytes which resemble large oenocytes. The symbiotes are spherical, oval, or elongate in shape and stain uniformly.

COCCIDOMYCES Buchner

Coccidomyces dactylopii Buchner is the single named species of this genus and is found in *Pseudococcus citri*. The specific name of the microorganism is taken from the old generic name of the insect host, *Dactylopius*.

In 1893 Berlese described an unpaired *corpo ovale* or *corpo giallo* in *Pseudococcus citri*, lying below the intestine and about a third of the length of the host. In the female the structure is larger than in the male. Pierantoni (1910a, 1913) showed that this yellowish-orange structure was a mycetome, and he described its structure in more detail (see fig. 62). On dissecting out the mycetome one notices an abundance of tracheae. According to Pierantoni, a trachea leads to each mycetocyte and there becomes intimately associated with the symbiotes. Each mycetocyte has a large nucleus rich in chromatin and about 10 to 12 round or oblong, colorless, cell-like balls in the cytoplasm. These cell-like spheres contain the microorganisms. *Pseudococcus diminutus* and

Pseudococcus medanieli are similar in their symbiotic arrangements to *Pseudococcus citri*.

According to Buchner (1930), the symbiotes are spherical, oval, or sausage-shaped and always have blunt, rounded ends, though Pierantoni observed some sickle-shaped forms, i.e., forms pointed at both ends. The protoplasm is delicately honeycombed and contains small granules. Rarely, some organisms possess lateral branches or processes occurring at right angles. Sometimes spherical or oval-shaped forms predominate. Pierantoni found the spherical or round forms to predominate in the spring and elongated forms in the winter. In some of the round forms there appear to be evidences of division by fission. Embedded between these forms small, dense, and apparently degenerate forms frequently occur. Not infrequently a peculiar circumscribed accumulation of nearly round or oval forms is found among the symbiotes within one of the cell-like balls or spheres containing the symbiotes. These seem to be contained within membranes of their own and appear to include budding forms. The nature or purpose of these islandlike accumulations is not definitely known, though Buchner feels they may have something to do with the transmission arrangements. This may be indicated by the fact that before the symbiotes leave the mycetocytes for transmission to the next generation they go through a pleomorphic stage and take on a new characteristic shape.

As has been brought out in chapter IV, the yeast or yeastlike nature of this organism is not generally agreed upon. Buchner (1930) comes to the conclusion that the symbiotes of *Pseudococcus citri*, like those of *Pseudococcus adonidum*, are bacteria. He feels that they are similar to other "swollen" bacteria associated with insects. The solution to this problem awaits the study of pure cultures.

Transmission to the next generation takes place through the egg. The cell-like spheres of the mycetocytes move through the follicle at the base of the nurse cells and occupy a position around the nutritive cord. They stay here until the oöcyte is mature and the nutrient cells atrophy, after which they are taken up by a depression at the anterior pole of the egg.

Buchner (1930) studied symbiotic arrangements of *Phenacoccus aceris* and *Phenacoccus piceae* and in general found them to be similar to the *Pseudococcus* type. Both of these two species possess unpaired mycetomes with large mycetocytes surrounded by flat epithelium. The symbiotes occur singly and are rodlike, or swollen and roundish in form. Carter (1935) observed large, bizarre, pleomorphic, yeastlike forms in the mycetome of *Pseudococcus brevipes*, but their relation to bacteriumlike forms found in the same mycetocytes is not clear.

ICERYMYCES Brain

The single species of this genus was originally named *Coccidomyces pierantonii* by Buchner (1912). Brain (1923), however, places it in a new genus with the name *Icerymyces pierantonii*. This symbiote was originally found by Pierantoni (1910b, 1912, 1913) in *Icerya purchasi*. It differs from the genus *Coccidomyces* in that invasion of the ovum takes place by the entry of single organisms rather than by the cell-like spheres of them. The symbiote is round, oval, or pear-shaped and usually multiplies by fission. In artificial cultures, however, Pierantoni observed budding.

The mycetomes lie on each side of the mid-intestine in the region of the first to seventh segments. Each mycetome is divided into seven small parts which lie one behind the other. Each part consists of a few large mycetocytes, five or more in number according to the age of the host.

ALEURODOMYCES Buchner

This genus consists of the single species *Aleurodomyces signoretti* Buchner, found in an unidentified *Aleurodes*. The symbiote occurs in paired mycetomes, the mycetocytes of which are from 8 to 13 microns in diameter. The individual organisms are from 2 to 5 microns in diameter, round, oval, or pear-shaped. The nucleus is small, the cytoplasm coarse. Multiplication is by budding.

APHIDOMYCES Brain

Brain (1923) established this genus to include those small forms of more or less spherical organisms which occur as symbiotes in aphids and related insects. In some respects such as in their method of multiplication (fission) they resemble *Schizosaccharomyces* Buchner. Spore formation occurs in some forms.

Aphidomyces aphidis (Šulc). Synonym: *Saccharomyces aphidis* Šulc. The insect host of this species is *Aphis amenticola*. The symbiotes inhabit a mycetome in this insect. They are spherical, about 4 microns in diameter, and multiply by fission. Spore formation occurs.

Aphidomyces drepanosiphi (Buchner). This species was described by Buchner (1912) as occurring in a *Drepanosiphum*.

Aphidomyces aphalarae calthae (Šulc). Synonym: *Saccharomyces aphalarae calthae* Šulc. This species was found by Šulc (1910a) in the mycetome of

Aphalara calthae in association with two other fungi. Šulc observed the formation of spores in this species, usually three being formed in each cell.

Aphidomyces psyllae forsteri (Šulc). Synonym: *Schizosaccharomyces psyllae forsteri* Šulc. The insect host of this species is *Psylla forsteri,* in which it inhabits a mycetome. The symbiote is oval or egg-shaped.

Aphidomyces sulcii (Buchner). Synonyms: *Cicadomyces sulcii* Buchner (1911); *Schizosaccharomyces sulcii* Buchner (1912). Brain (1923) has this to say about the organism: "This is a very distinct form described and figured by Buchner from the fat-body of a Japanese cicada. It possesses several striking and unusual characteristics and does not appear to belong, properly, to any of the known genera. I allow it to stand, for the present, in the hope of discovering other forms showing similar characters."

CHERMOMYCES Brain

The members of this genus are similar to those in *Aphidomyces* except that they are elongate, with ends frequently pointed. The nucleus is distinct and the cytoplasm contains vacuoles. Multiplication is by fission.

Chermomyces chermetis strobilobii (Šulc). Synonym: *Schizosaccharomyces chermetis strobilobii* Šulc. This is the type species of the genus; it inhabits a mycetome in *Chermes strobilobius*. It is 1 to 2 microns in length with ends frequently pointed. Multiplication is by transverse fission.

Chermomyces chermetis abietis (Šulc). Synonym: *Schizosaccharomyces chermetis abietes* Šulc. This species lives in a mycetome in *Chermes abietes*. It is long, oval, and broad, and its sides are more nearly parallel than those of *C. chermetis strobilobii*.

COCCIDIASCUS Chatton

Coccidiascus legeri Chatton. In 1913 (b) Chatton described and figured this species which he found in the cells of the mid-intestine of the muscid *Drosophila funebris*. In the inhabited cell a vacuole is formed about the yeast, which multiplies by ordinary budding thereby enlarging the vacuole. Multiplication also takes place by the formation of asci, each of which contains eight fusiform ascospores. A mycelium is not produced, but as yet this organism has not been cultivated on artificial media. The formation of the banana-shaped asci apparently is preceded by an isogamic copulation. Guilliermond (1920) places this genus and species in the family Saccharomycetaceae.

TORULOPSIS Berlese
(*Torula* sensu Turpin)

Torula lecanii corni (Benedek and Specht). This species was found by Benedek and Specht (1933) to be the chief symbiote in the eggs and in the young and adult forms of *Lecanium corni* at all times of the year, and regardless of the insect's host plant. Each insect contained on the average 300 to 500 symbiotes, which appeared as budding conidia in the living insect. After the insect's death mycelium appeared. Benedek and Specht claimed to be able to cultivate the organisms in artificial media, and the limited growth could be reproduced in drop cultures independently of the host. Hence, they concluded that the conidial appearance of the microorganism in the host was caused by the limited amount of nutritive fluid in the insect. It is questionable whether this organism can rightly be considered a yeast. In about 50 per cent of the insects a second symbiote was observed which appeared to be closely related to *Bacillus megatherium*.

SACCHAROMYCES (Meyen) Reess

The yeasts in this genus are short-oval to elongated cells, sometimes in small clusters of buds. Reproduction is by budding and by the formation of endospores. The spores may be round, oval, kidney-shaped, or hat-shaped, smooth, and one to four per ascus. The species we are about to discuss has not been observed to produce endospores, but Brain (1923) retained the organism in this genus because of lack of sufficient information to place it elsewhere. True *Saccharomyces* yeasts always ferment dextrose, levulose, and maltose, and frequently other sugars. Since the symbiote about to be described has never been cultivated with certainty, it seems to have been premature to place it originally in the genus *Saccharomyces* which is characterized by its spores and fermentative characteristics.

Saccharomyces anobii Buchner. In 1899 Karawaiew described peculiar cecumlike structures attached to the beginning of the mid-intestine of the drug-store weevil, *Stegobium paniceum* (=*Sitodrepa panicea*). He observed that the cells of these structures were filled with what he thought were flagellates. The next year Escherich (1900) asserted that they were not parasites but represented some type of mutualism. He recognized them as yeastlike fungi and claimed success in cultivating them in artificial media. Subsequent investigations by Buchner (1912, 1913), Heitz (1927), and Breitsprecher (1928) have yielded considerable additional knowledge concerning this relationship.

Buchner named the organism *Saccharomyces anobii* (after the insect's family name, Anobiidae).

At the beginning of the mid-intestine of *Stegobium paniceum* larvae are

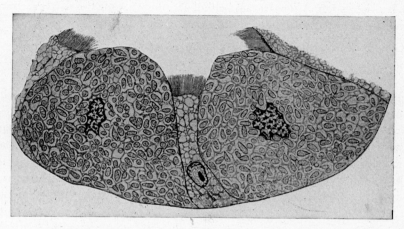

Fig. 144. Mycetocytes from the wall of the intestinal diverticulum of *Stegobium paniceum,* showing large numbers of the intracellular yeast, *Saccharomyces anobii.* (From Buchner, 1930.)

four peculiar protrusions, somewhat botryoidal in character, which more or less cover the point at which the fore- and mid-intestine join. The epithelium of these structures consists of two different types of cells: (1) typical, slender, epithelial cells with rounded nuclei and with the edge of the cell possessing a well-developed brush border and (2) large mycetocytes with larger and jagged-edged nuclei and no brush border. The latter cells harbor the symbiotes, around each one of which is a vacuolated area. Rod-shaped mitochondria are present in both types of cells.

According to Buchner (1930), larvae of *Ernobius abietis* and *Ernobius mollis* have essentially the same arrangement though the mycetocytes are somewhat smaller, the nuclei large, and the number of symbiotes less than in *Stegobium paniceum*. In larvae of *Anobium striatum* the cecumlike structures are very much larger, and the mycetocytes are extremely full of symbiotes. *Xestobium rufovillosum* and *Trypopitys carpini* are somewhat similar to the *Anobium* type. Similar arrangements are also present in *Oligomerus brunneus* and *Lasioderma redtenbacherei*. However, all these types may vary with respect to their development and histologic detail.

In the adults of *Stegobium paniceum* we find similar symbiote-containing protrusions at the corresponding anatomical location, though they do not appear as well developed as those in the larvae. Histologically the mycetocytes extend into the lumen of the ceca forcing the regular epithelial cells further back. Buchner, as well as Breitsprecher, has shown that in old larvae ready to pupate the symbiotes migrate into the intestinal lumen to be eliminated with the last defecation. A few remain behind and these invade the newly forming mycetocytes of the shrunken ceca in the adult.

Generation-to-Generation Transmission. The transmission of the symbiotes from generation to generation in *Stegobium paniceum* and related insects is especially interesting. Briefly what happens is this: Under the wrinkled vagina of the female are two long, symbiote-filled, chitinous pockets or pouches shut off from the outside by two overlapping chitin plates. These vaginal pockets and the vagina are entirely independent of each other, though finally toward the outside opening they unite. Presumably the vaginal pockets are filled during the first defecation of the young adult when the symbiotes, which have left the mycetocytes, pass from the intestine through the anus to the narrow space between the intersegmental membrane and ovipositor where they have the opportunity of entering the transmission organs. This takes place before the beetle leaves the puparium.

Fig. 145. *Stegobium paniceum*. The emerging larvae eating parts of the egg shells to which the symbiotes are attached. (From Buchner, 1930.)

Now when an egg is oviposited the symbiotes are pressed from the vaginal pockets and are "glued" to the outside of the shell. When the larva hatches it leaves the eggshell head first. It eats off the edges of the egg opening until about one half of the shell is consumed (fig. 145), after which it seeks other food. In this manner the symbiotes enter the hitherto sterile larva with the pieces of eggshell. The yeasts soon invade the epithelium of the mid-intestine at the site of the future cecumlike structures. Thus the symbiotes have been safely implanted in a new generation. (See also Nolte, 1938.)

The Nature of the Symbiotes. The symbiotes in *Stegobium paniceum* are generally considered to be yeasts (*Saccharomyces anobii* Buchner). They are more or less pear- or tear-shaped cells about 4.5 microns long and 3.5 microns wide. One end is usually broadly rounded and the other pointed. Multiplication is by budding, with the buds occurring either terminally or slightly to one side of the pointed end. Each cell has a large vacuole and generally a highly refractive nucleus. Spore formation has never been observed.

In an artificial medium, such as 1 per cent glucose broth, Escherich (1900) found the cells to continue to increase by budding until after eight days when chainlike unions or a kind of mycelium appeared. Heitz (Buchner, 1930) was unable to reproduce the luxuriant growth obtained by Escherich and thought the latter worker might perhaps have worked with a different fungus. Breitsprecher (1928) likewise was unable to cultivate the organism on an artificial medium.

The symbiotes of other Anobiidae are in general fairly similar though somewhat modified in particulars. Those of *Ernobius mollis* are rounded, oval, or lemon-shaped; those of *Anobium striatum* are elongated and have rounded ends; those of *Xestobium rufovillosum* are pear- and lemon-shaped; and those of *Ernobius abietis* are similar to those of *E. mollis* and apparently can be cultivated on artificial media with ease.

The symbiotes in the various insect hosts mentioned, except those of *Stegobium paniceum,* have apparently never been named. It is perhaps just as well since the generic allocation of this group has not yet been made certain.

Role of the Symbiotes. As in the cases of intracellular symbiosis discussed in chapter IV, the entire and exact role of the symbiotes in the life processes of *Stegobium paniceum* has not been determined. Most of the present evidence indicates that the symbiotes are in some way concerned with the insect's nutrition, perhaps serving as a source of vitamins or growth factors. Some experimental work has been accomplished which supports this belief.

Koch (1933a), after obtaining symbiote-free *Stegobium paniceum* larvae, found that the insects would not properly develop unless yeast was added to otherwise satisfactory food (*Erbswurst*). Working with symbiote-free larvae of *Stegobium paniceum* and *Lasioderma serricorne,* Blewett and Fraenkel (1944) concluded that the intracellular symbiotes of these insects supply vitamins of the B group. On food deficient in vitamins of the B group normal larvae grew much better than symbiote-free larvae. As Koch observed, the addition of yeast to the food eliminated this difference in the growth rate.

An interesting observation was made by Heitz (1927) concerning the effect of temperature on the symbiotes in *Ernobius mollis*. He found that in January, when the temperature was about zero, only a few mycetocytes were present in the cecumlike structures and that most of these were symbiote-free. Eight days at room temperature brought about a marked increase in symbiotes. Conversely, in larvae with normally filled mycetocytes most of the symbiotes disappeared after ten days in the refrigerator. Starvation was found to have a similar influence.

Yeast Symbiotes in Other Coleoptera. In addition to *Stegobium paniceum* (family Anobiidae) other Coleoptera have been found to harbor yeasts or

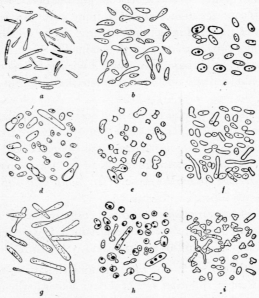

Fig. 146. The symbiotes of nine different cerambycids. *a, Rhamnusium bicolor; b, Oxymirus cursor; c, Cerambyx scopolii; d, Rhagium bifasciatum; e, Tetropium castaneum; f, Strangalia maculata; g, Necydalis major; h, Leptura rubra; i, Leptura cerambyciformis.* (From Buchner, 1930.)

yeastlike symbiotes. The Cerambycidae in particular have been studied from this standpoint. In 1927 Heitz observed intracellular yeastlike organisms in *Rhagium inquisitor, Rhagium bifasciatum,* and *Leptura rubra.* This worker claimed to have cultivated the organisms from these cerambycids on artificial media. Ekblom in 1931 also made a study of the symbiotes of *Rhagium inquisitor.* He observed that the yeastlike symbiotes live in cells of the intestinal wall and are of variable sizes and shapes. Both gram-positive and gram-negative strains appear to be present. Ekblom also claimed to have cultivated the organism.

One of the most thorough investigations of the symbiotes of the cerambycids is that of Schomann (1937). According to this worker, all genera of the following tribes of the Cerambycidae harbor symbiotes: Spondylini, Asemini,

Saphanini, Necydaliini, Trichomesiini, and Tillomorphini. All genera and species of the tribe Lepturini examined have symbiotes except *Toxotus schaumi, Toxotus vittiger, Toxotus vestitus, Pidonia lurida, Akimerus schäfferi, Stenocorus meridianus, Stenocorus quercus,* and *Vesperus* sp. Of the Cerambycini, symbiotes are harbored only by *Dialeges pauper*. In general, those cerambycid larvae which live in the fresh wood of deciduous trees appear to be without symbiotes while those that live in either living or dead coniferous trees contain symbiotes.

Schomann made a detailed study of the mycetomes and symbiotes of the larva and adults of several cerambycids. The larval mycetome is made up of small tissue masses which circle the midgut in one or two girdles. Each tissue mass is a sacciform evagination of the midgut wall whose cells are filled with symbiotes. In some species (*Rhagium inquisitor, Rhagium bifasciatum, Tetropium castaneum, Criocephalus* sp.) the mycetome undergoes considerable variation in size, being large and acinous during the summer, and smaller in the winter. The mycetocytes of Spondylini, Asemini, and Saphanini larvae continually degenerate and are cast off into the gut lumen while new embryonic cells appear in the margin of the mycetome, are infected, and become mycetocytes. In species of *Rhagium* the mycetocytes themselves are retained, and only cytoplasmic masses containing the symbiotes are cast off. When the insect pupates the mycetome has become much smaller, most of its cells having been shed into the lumen. Enough mycetocytes remain to infect the adults. The latter have no gut mycetomes, but the symbiotes are found in the intersegmental pouches of the ovipositor. These pouches vary in size and shape from a short pocket to an elongated tube. They are lined with a glandular epithelium and contain symbiotes in numbers directly proportional to the size of the pouches. The symbiotes enter the pouches from the gut feces during the first ten days of imaginal life. At the time of egg laying, other integumental folds in the region of the ovipositor become "infected." The principal ones of these are known as "vaginal pockets" (already described in the case of *Stegobium paniceum*) and may occur ventrally or dorsally. These

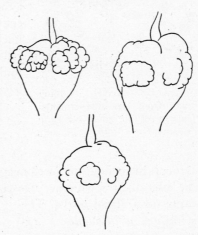

Fig. 147. *Leptura rubra*. Mycetomes (shrunken) on the intestines of larvae about to pupate. (From Schomann, 1937.)

various integumental pockets apparently do not function primarily as abodes for the symbiotes since they also occur in species which harbor no such organisms. Schomann concludes that these structures also furnish glandular secretions in connection with egg laying.

Transmission of the symbiotes from one generation of beetles to the next takes place during the laying of the egg. At this time the symbiotes from the vaginal pockets are smeared over the outside of the egg. When hatching, the larva eats part of the eggshell and thus the symbiotes gain entrance into the insect. The newly hatched larva does not possess a mycetome as such, although certain large cells of the midgut epithelium characterized by especially large nuclei eventually become mycetocytes. These cells take up the symbiotes, which grow and multiply causing the mycetocytes to pouch out from the gut wall.

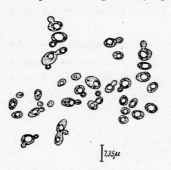

Fig. 148. Actively budding yeasts from the intersegmental pouches of an adult *Rhagium bifasciatum*. (From Schomann, 1937.)

The symbiotes themselves are yeasts or yeastlike fungi. Those of the Spondylini, Asemini, and Saphanini go through a cycle of morphologic types from the fusiform through filamentous to an elongated dumbbell shape. Sporulation occurs in the latter, and asci with four ascospores are formed. Schomann cultivated these easily on artificial media. The symbiotes of *Rhagium*, *Leptura*, and *Oxymirus* apparently do not go through a morphologic cycle in the insects, nor do they form spores. In cultures, those of *Rhagium* show some form changes, but those of *Leptura* do not. The *Oxymirus* symbiotes have not been grown outside their specific hosts.

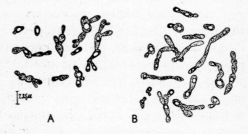

Fig. 149. *A*. Yeasts from the intestinal tissues and the neighborhood of the "fungus organ" of *Leptura rubra*. *B*. Symbiotes from the "fungus organ" of a 14-hour dead larva of *Rhagium inquisitor*. In both cases the yeasts are actively budding. (From Schomann, 1937.)

The Aphididae (Plant Lice). As we have noted, Brain (1923) included a few of the symbiotes of aphids in his classification of the intracellular yeasts. Since then much study has been undertaken with respect to the symbiotes of aphids, and two schools of thought have arisen as to their nature. Many workers still consider the symbiotes to be yeasts or closely related organisms. Others led by Paillot (1923) believe that the yeastlike organisms originate from bacteria which are normally associated with aphids. Although those supporting the latter belief may be in the minority, their theory deserves careful consideration. Paillot was a careful and experienced worker in the field, and his many observations supporting this view are rather convincing. Here again, however, we must wait for the cultivation of these forms on artificial media and a thorough study of the cultures.

The reader is referred to the preceding chapter in which the aphid was used as an exemplary insect to show the nature of mycetome development. The details of this will not be repeated here. To recapitulate a bit, however, it may be repeated that the mycetome of the newly born aphid is a longitudinally bipartite structure, each lobe occupying the cavity on either side of the alimentary tract. As the aphid develops the mycetome becomes reduced to isolated groups of two, three, or more mycetocytes. Transmission of the symbiotes to the following generation takes place via the egg. The organisms invade the follicular epithelium from which place they contact the yolk of the developing egg, thus bringing about their rapid multiplication and subsequent organization compatible with the embryologic development of the aphid.

The symbiotes themselves vary in appearance from small granular or swollen bacteriumlike bodies to typical yeast or yeastlike organisms. Frequently the various forms appear together in the same aphid. What appears as a nucleus to some workers, Paillot claims to be a pseudonucleus of deeper staining material. With iron-hematoxylin stain the center of the large symbiotes stain more deeply than the periphery, but this does not represent a nucleus. The internal structure appears homogeneous by some stains, but Giemsa's stain shows chromatin and basophilic granules. Paillot compares these larger, more rounded forms with those of the large, rounded forms of ordinary entomophytic bacteria found in the general cavity of larvae immune to these bacteria. He believes that the two forms may be the same.

As the writer has indicated, there is perhaps as yet too little information available as to the true nature—bacterial or yeast—of these organisms. They have already been arbitrarily discussed in the preceding chapter along with the intracellular bacteria. The reader is referred to such works as those by Buchner (1930) and Paillot (1933) for comprehensive treatments of this controversial subject.

Yeast Symbiotes in Other Homoptera. As has already been indicated, Brain's (1923) classification did not include all the families or subfamilies of Homoptera that harbor symbiotes. An example is the family Diaspidinae in which the symbiotes are found intracellularly in definite mycetocytes located throughout the fat tissue. Šulc originally observed this relationship in a species of *Lepidosaphes*. Buchner (1921c, 1930), Rickter (1928), and others studied this group further, adding the following species to the list of those harboring yeastlike symbiotes: *Aspidiotus latiniae, Aspidiotus hederae, Aspidiotus cyanophylli, Aspidiotus pini, Aspidiotus nerii, Chrysomphalus dictyospermi, Pseudoparlatoria parlatorioides, Lepidosaphes ulmi, Lepidosaphes gloverii, Parlatorea olea,* and *Chionaspis salicis.* The symbiotes themselves are round, oval, somewhat elongate, and sometimes lobate and probably reproduce by budding. The mycetocytes also contain colored or colorless, refractive granules and fat droplets, depending on the insect species. Transmission to the next generation takes place via the egg. In most cases the symbiotes leave the mycetocytes and apparently are carried through the blood stream to the ovarioles.

Shinji (1919) and Rickter (1928) have touched on the symbiotes of some of the Asterolecaniae. In adult *Asterolecanium aureum* the mycetocytes contain nucleated, round to oval, yeastlike organisms of different sizes. In larvae the mycetocytes occur in clusters and the symbiotes are smaller, of uniform size and are dumbbell-shaped. Transmission to the next generation occurs transovarially. According to Rickter, although *Asterolecanium variolosum* contains no easily recognized mycetocytes, it harbors rod-shaped bacteria in large homologous cells in addition to the small round or oval inclusions. This association, however, needs further clarification. Rickter has reported the presence of symbiotes in the following species of Homoptera: *Mogannia hebes, Scieroptera splendidula, Rihana ocrea, Platypleura kaempferi, Clovia bipunctata, Plinia ampla, Pilagra fusiformis, Tettigoniella ferruginea, Tettigoniella viridis, Thompsoniella porresta, Oliarus horisanus, Euricania ocellus, Sabimamorpha* sp., *Tartessus* sp., *Ormensis* sp., *Tambina* sp., and *Lecaniodiaspis pruinosa.*

In the Margarodinae we have an interesting instance of marked difference in the symbiotic arrangements of two species of *Margarodes*. The female of one of these species (unidentified), was found by Šulc (Buchner, 1930) to have paired mycetomes which not only join the inner side of the oviduct but grow through its wall so that the mycetocytes form the only boundary of part of its lumen. Transmission to the next generation takes place via the eggs through the use of peculiar "plugs" associated with the oöcytes in the oviduct. Invasion of the symbiotes occurs at the posterior pole of the egg by way of canallike lumen usually running through the middle of the plug.

On the other hand, *Margarodes polonicus* differs markedly from the species just discussed. In the female, the two mycetomes are elongated, slightly bean-shaped bodies lying along both sides of the intestine. They are situated in an area the length of which is covered by the first five abdominal segments, but nowhere does it come into the immediate vicinity of the oviduct. At the time the female is ready for oviposition, the mycetocytes become free and migrate into the fatty tissue in the neighborhood of the ovaries, eventually invading the developing eggs. In the male the mycetomes are shorter and somewhat more slender than in the females, but histologically they are much alike.

Mahdihassen (1924–1935) has published several very interesting accounts of his studies on the symbiotes of scale insects, particularly the shellac-producing insects. One of the best known of these species is *Laccifer lacca*. This species, according to Tschirch (1922–1924), harbors symbiotes which appear as rather short, pear-, tear-, or tube-shaped organisms which he named *Laccomyces symbioticus*. *Laccifer albizziae* has symbiotes suggesting organisms similar to *Monilia candida* (*Candida albicans?*). Though the symbiotes of *Lakshadia communis* are yeastlike organisms, some of the other symbiotes of this genus are difficult to classify. Those of the Indian *Lakshadia* species seem closely related to *Monilia nigra*. Those of *Metatachardia conchiferata* have been considered by some workers to be similar to *Actinomyces*. Mahdihassen cultivated what he considered a *Torula* species from *Tachardina lobata*. *Tachardina sylvestrii* apparently harbors two different species of symbiotes. One of these organisms appears as uniformly thick chains of actinomycete-like members. The other is rod-shaped and quite different in appearance. Both symbiotes are transmitted to the next generation via the eggs.

An interesting sidelight to such studies as these is the assertion by Mahdihassen that he has been able indirectly to differentiate species of coccids by examining blood smears containing their symbiotic microorganisms, which show morphologically distinct forms dependent on the species of insects harboring them.

❖ CHAPTER VII ❖

FUNGI AND INSECTS

FOR the most part, the known relationships of fungi to insects are parasitic or semiparasitic. This is indicated by the numerous epidemics of fungous diseases of insects which occur in nature. The muscardine diseases and those caused by the various species of *Empusa* are well-known examples. Some fungi, such as the Laboulbeniaceae, live on the chitin of living insects, the death of which means the death of the attached fungus. In other cases, as with *Septobasidium* fungi and scale insects, the fungi and the insects set up a mutual relationship and together parasitize a third organism such as a tree. Insects may employ fungi for their own use, as do the wood-boring beetles, certain ants, and termites which take pains to cultivate fungous gardens upon which they feed. Some insects (certain Diptera) do not keep fungous gardens but parasitize and feed on fungi, especially putrescent fungi. Not to be forgotten is the fact that many saprophytic fungi play an important role in the final disintegration of dead insects.

Although various authors use the term "fungi" to designate different groups of thallophytes, in this chapter we shall limit it to that group of non-chlorophyll-containing organisms usually possessing the filamentous vegetative structure known as mycelium. Because of the nature of the subject matter, the yeasts (Saccharomycetales) have already been considered in another chapter. Most mycologists divide the fungi into four large classes: Phycomycetes, Ascomycetes, Basidiomycetes, and Deuteromycetes or Fungi Imperfecti. The Phycomycetes are frequently spoken of as the "lower fungi" and the Ascomycetes and Basidiomycetes are referred to as the "higher fungi." Entomogenous fungi are contained in each of the four classes.

Considering the almost incredible number of fungi associated with insects, it is apparent that they cannot all be discussed or even mentioned here. As pointed out by Petch (1925a), there are at least 130 species of entomogenous fungi known in Ceylon alone, and Seymour (1929) has listed approximately

750 North American insect and arachnid hosts of fungi. Especially is the reader's attention directed to the list of North American entomogenous fungi and their hosts compiled by Charles (1941b). In the following pages we shall treat briefly the predominant features of the biologic relationships existing between the various groups of fungi and their insect hosts and, where practical, list or discuss a few of the better known examples of entomogenous fungi.

PHYCOMYCETES

Phycomycetes, which include many of the fungi associated with insects, are a very large and diverse group containing both aquatic and terrestrial members. The hyphae which make up the mycelium are usually plurinucleate and nonseptate except where the reproductive organs are delimited. Sexual organs are commonly present, and as a result of their function, spores (oöspores or zygospores) are produced. As a whole the most characteristic structures of the group are the sporangium, coenocytic mycelium, and zoöspores (Fitzpatrick, 1930).

The Phycomycetes frequently are divided into the following eight orders: Chytridiales, Lagenidiales, Blastocladiales, Monoblepharidales, Saprolegniales, Peronosporales, Mucorales, and Entomophthorales. Of these, as the name would imply, the order Entomophthorales is the most important from the standpoint of having insect hosts, though the Mucorales and Chytridiales also contain many entomogenous species.

ENTOMOPHTHORALES

The Entomophthorales are generally placed in a single family, the Entomophthoraceae, or as designated by some workers, Empusaceae, which may be divided into five and possibly six genera, two of which are not associated with insects. The genera *Empusa, Entomophthora,* and *Massospora* include numerous species, nearly all of which are entomogenous. A fourth genus, *Basidiobolus,* does not contain strictly entomogenous species. However, with one species, *Basidiobolus ranarum,* beetles ingest the sporangia, and in turn are eaten by frogs where the sporangia undergo further development.

Considerable confusion still prevails with respect to the use of the generic names *Empusa* and *Entomophthora.* In 1855 Cohn erected the genus *Empusa* with the housefly parasite *Empusa muscae* as the type species. Earlier the name *Empusa* had been used for a genus of orchids, and for this reason Fresenius (1856) proposed the name *Entomophthora* to take the place of

Empusa for the fungus. Subsequently both names were used simultaneously by Brefeld (1877) and Nowakowski (1884), two genera thereby being recognized. *Entomophthora* was characterized by branched conidiophores, the presence of rhizoids and cystidia, and the formation in some species of zygospores. *Empusa* was without rhizoids and cystidia, had unbranched conidiophores, and produced azygospores. Nowakowski added a third genus, *Lamia,* which he considered intermediate and which was similar to *Empusa* except that it possessed cystidia.

This arrangement has not been entirely accepted, and some authorities prefer to incorporate all the species in the single genus *Empusa.* This has been done mainly on the basis of the work of Thaxter (1888), who did not accept Nowakowski's taxonomic treatment of the group and who used the single genus *Empusa.* It appears likely, however, that the two genera, *Empusa* and *Entomophthora,* will continue to be used separately, though it is acknowledged that a great deal of cytologic study is necessary before the separation of all the species concerned in these two genera can be finally accomplished.

Empusa and Entomophthora. All the members of these genera are entomogenous, having been found associated with many species of insects, including those of the orders Lepidoptera, Coleoptera, Hemiptera, Diptera, and Orthoptera. According to Thaxter (1888), the Diptera are the greatest sufferers from infection by *Empusa,* with the Hemiptera coming next, followed by the Lepidoptera and Coleoptera. Larvae, pupae, and adults are all affected. Occasionally two species of fungi may develop on one host at the same time. Until recent years these fungi have been considered to be obligate parasites. However, indications are that with the employment of proper techniques and conditions they can be made to grow saprophytically. This is borne out by the successful attempts of Sheldon (1903), Speare (1912), and Sawyer (1929), who cultured a few of the species on artificial media.

In nature, infection of the insect host takes place with the penetration of the cuticulum by the conidial germ tube which enters the body cavity. Apparently, infection rarely occurs as the result of the ingestion of the spores. In the insect a profusely branched mycelium does not develop, but the filamentous infection threads soon become segmented and break down into their component cells, which are then termed hyphal bodies. The disease usually progresses very rapidly, and frequently within 48 to 72 hours all of the host's internal structures become liquefied, absorbed, and replaced by the fungus. Eventually, or under certain environmental conditions, the vegetative stage may be followed by a resting spore stage (zygospores if formed by conjugation and azygospores if formed asexually). Usually the internal hyphal bodies produce conidiophores, which penetrate the insect's outer covering, forming conidia which are forcefully projected into the air in large numbers. On the return to favorable conditions the sexual spores germinate, also forming conidiophores and conidia.

The conidia of the various species vary considerably in size and are spherical to ellipsoidal in shape. The adhesive material on the smooth conidial walls aids in fastening the conidium to the object on which it happens to fall. If it should fail to land on an insect, it might put out a germ tube which would form a secondary conidium. This in turn would be discharged and might land on a suitable host, or recur until the protoplasm became exhausted. Other variations in form and manner of dissemination also occur.

Many fungus-infected insects before death crawl upward, and the favorite place sought appears to be the underside of leaves about the woods, or in houses. Here they may be found dead, often in considerable numbers.

Empusa muscae (fig. 150) causes epidemics in the fall of the year among houseflies, blowflies, and other flies. Perhaps because of lack of extensive research, it is one of the few *Empusa* reported south of the equator. *E. muscae* usually occurs about houses and in great abundance. The flies affected with the fungus attach themselves to the walls and ceilings in a lifelike position. The spores cling among the hairs on the upper portion of the fly's body, and the germinating hypha or germ tube penetrates to the interior through the breathing pores or through the thinner parts of the covering. The hyphal bodies multiply and are carried to all parts of the body by the hemolymph. In a few days the flies are plugged with fungi and die. The germinated hyphal bodies within the insect produce germ tubes which pierce the softer portions of the body wall, especially the back, where the spores from the conidiophores are discharged in a circle. After the fly has been dead for a few days, a halo of spores around the insect may be observed (fig. 151A).

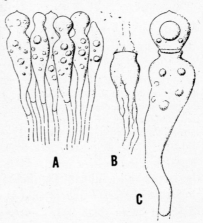

Fig. 150. *Empusa muscae. A.* A group of conidiophores showing conidia in several stages of development. *B.* Basidium after the discharge of the conidium. *C.* Basidium bearing conidium before discharge. (From Thaxter, 1888.)

Empusa grylli, another fairly well known fungus, is parasitic on grasshoppers and crickets. It parasitizes grasshoppers in a manner very similar to that by which *E. musca* does flies. The tissues are interlaced with mycelial threads, and after the death of the grasshopper the germ tubes emerge and form spores. The spores are sticky, and direct contact as well as cannibalism seems to be important in the spread of the fungus.

Other examples of species of *Empusa* are *Empusa culicis*, which affects the adults of a species of *Culex; E. aphidis*, which affects aphids; *E. conglomerata*, which parasitizes

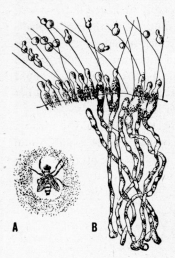

Fig. 151. *A.* A fly parasitized by *Empusa muscae*, the cast off spores forming an aureole about the dead insect. *B.* Mycelium and conidiophores of *E. muscae*. (From Paillot, 1933; after Brefeld.)

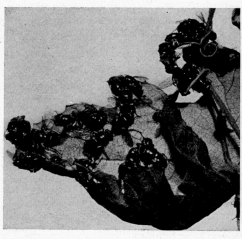

Fig. 152. Muscid flies killed by *Empusa muscae*. The flies are clinging to a grapevine leaf. (From Sweetman, 1936.)

the larvae and adults of Tipulidae; *E. radicans* (fig. 153), which parasitizes the larvae of *Pieris brassicae;* and *E. sciarae*, which affects the larvae and adults of *Sciara* sp. (a small fly), *E. delpiniana, Polyete lardaria,* and other insects.

Entomophthora sphaerosperma produces epidemics of considerable proportions among all insect orders except Orthoptera. The germ tube rapidly penetrates the host and forms mycelium with a few septa, which consume the fat bodies especially. When the hyphae reach the body cavity, they divide into hyphal bodies, which multiply and consume the whole insect, leaving it mummified.

Altogether there are about 40 species of *Empusa* and *Entomophthora* known in the United States.

Massospora. The only well-studied species of this genus was described by Peck (1879) and named *Massospora cicadina*. It is a parasite mainly of the adult males of the seventeen-year cicada, *Magicicada septendecim*. One of the outstanding characteristics of this fungus is that, unlike *Empusa,* it produces conidia within its host instead of on its surface.

Growth of the fungus takes place in the posterior end of the body of the insect. The intersegmental membranes of the abdomen become weakened and eventually as much as a quarter to a half of the body drops off. The cicada may continue to fly and crawl about, thus widely disseminating the internally borne conidia or the later-forming resting spores which resemble the typical azygospores of the *Empusa.* Dissemination does not take place by the forcible ejection of conidia but rather by the wind and through contact with other individuals. How the fungus survives the sixteen and three-quarters years of the host's immature and subterranean existence has not been determined. It is probable that infection takes place while it is underground.

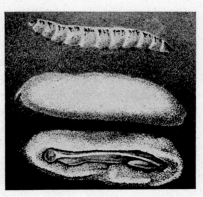

Fig. 153. Larvae of *Pieris brassicae* killed by *Empusa radicans* and embedded in the fungus. The lower view shows the remains of a dead larva surrounded with spores. (From Lohde, 1872.)

CHYTRIDIALES

In this order, regarded by some as the lowest of all fungi, have tentatively been placed several species of interesting fungi parasitic in insects. We refer especially to those of the genus *Coelomomyces.* The true systematic position of this genus, however, is not clear, and when more is known concerning the life history of its members, it probably may have to be placed elsewhere.[1] Some authorities (Sparrow, 1944) do not believe *Coelomomyces* to have the proper characteristics to warrant its inclusion in the Chytridiales, and further study is necessary to decide the question of its taxonomic position. A similar enigma prevails concerning an organism (*Sporomyxa scauri*) described by Léger (1908) and found in the fat body, reproductive organs, and blood of the beetle *Scaurus tristis.*

Coelomomyces stegomyiae was found by Keilin (1921c) to be a parasite in the body cavity of an *Aëdes albopictus* (=*Stegomyia scutellaris*) larva. Portions of the parasitized larva showed oval bodies 37.5 to 57 microns long and

[1] Since this was written, Couch (1945, *Jour. Elisha Mitchell Sci. Soc.,* **61**, 124-136) has revised the genus, described several new species, and placed these fungi in a new family, Coelomomycetaceae in the Blastocladiales.

20 to 30 microns in diameter, surrounded by a thick yellowish wall. In the neighborhood of the viscera, especially the midgut and the five anterior intestinal ceca, the mycelium was well developed. Except for the insect's fat body, which had completely disappeared, the internal organs were apparently healthy.

In 1922 Bogoyavlensky described a parasitic fungus under the name of *Zografia notonectae* which he had found in the body cavity of *Notonecta* (Hemiptera) collected in Russia. Keilin (1927) showed *Zografia* to be synonymous with *Coelomomyces,* thereby giving this species the name *Coelomomyces notonectae.* Bogoyavlensky found this parasite present in the bugs from the end of May until September. In the adult *Notonecta* spores were readily found, and mycelium could be observed in the early stages of the insect's development. Like *C. stegomyiae* in the mosquito larva, *C. notonectae* lives in the body cavity of its host. The mycelium may lie freely between the lobes of the fat body but apparently does not penetrate this tissue.

Two additional species were found by Iyengar (1935) in India parasitizing larvae of the following species of *Anopheles: A. barbirostris, A. hyrcanus* var. *nigerrimus, A. subpictus, A. vagus, A. annularis, A. jamesi, A. ramsayi, A. varuna,* and *A. aconitus.* To the two fungi he found in these mosquito larvae he gave the names *Coelomomyces indiana* and *Coelomomyces anophelesica* (fig. 154). In the case of either fungus, development is completed during the mosquito's larval stage. The progress of the infection is rapid, setting in during the first larval instars and having fully developed sporangia in the third and fourth instars. According to Iyengar, the infection begins in the thoracic region and spreads posteriorly into the abdominal segments traveling along the adipose tissue on which the fungus lives. The fat body shrinks and eventually disappears, together with all fat tissues. Infected larvae rarely develop into adult mosquitoes and nearly always die before pupating.

De Meillon and Muspratt (1943) found a *Coelomomyces* infecting mosquito larvae (*Mucidus* sp.) in Northern Rhodesia. Of particular interest is the fact that these investigators observed the germination of the sporangia, an observation apparently not made by earlier workers.

Chytridiaceous fungi have been observed in abundance on submerged and empty exuviae, or castoff integuments, of the larval, pupal, and nymphal stages of certain insects which pass their immature stages in fresh water.

ASCOMYCETES

Some of the Ascomycetes are saprophytic, such as the blue and green molds, the yeasts, and the truffles, and many are parasitic causing plant and animal

FUNGI AND INSECTS

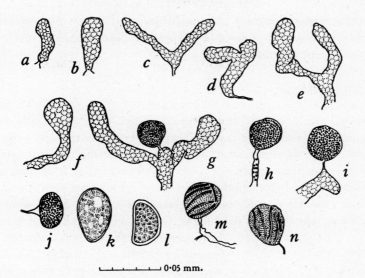

Fig. 154. *Coelomomyces anophelesica.* *a* and *b*, young mycelia; *c–f*, older mycelia; *g* and *i*, formation of sporangium; *h*, formation of sporangial wall and withdrawing of contents of mycelium; *j*, young sporangium with attached strand; *k*, vacuolated stage of sporangium; *l*, older sporangium with dense contents and thick wall; *m* and *n*, mature sporangia with partially developed ribs and connecting strand. (From Iyengar, 1935.)

diseases. A large number of the fungi associated with insects belong to this class.

The outstanding characteristic of the group is the formation, somewhere in the life history, of a unicellular sac (ascus) containing a definite number of endogenous spores (ascospores), usually eight. In addition to ascospores many of the Ascomycetes also produce conidia. In contrast to the Phycomycetes, the hyphae are septate and each cell thus formed may contain one or more nuclei. While there are exceptions, in most Ascomycetes the asci are produced in numbers in special, large, fruit bodies or ascocarps. The ascocarp is either an apothecium or a perithecium, depending on whether it opens widely to a cup or saucer-shaped structure, or remains closed as a more or less spherical body enclosing the asci. The perithecium usually possesses a small apical opening (ostiolum) through which the ascospores escape at maturity.

In this discussion we shall follow the taxonomic arrangement used by Clements and Shear (1931), who divide the class Ascomycetes into the following orders: Laboulbeniales, Gymnoascales, Perisporiales, Sphaeriales, Dothideales, Microthyriales, Phacidiales, Pezizales, Agyriales, and Tuberales. The yeasts (Saccharomycetaceae) are usually included in the Ascomycetes but since they have been considered in a separate chapter we shall exclude them from the present discussion.

LABOULBENIALES

The exclusively entomophytic Laboulbeniales occur principally on the chitinous cuticula of living insects. They are generally thought to be commensals, though they produce what may be considered contagious cutaneous diseases. These, however, do not reach the proportions of fatal epidemics as in the case of the Entomophthoraceae. They are true parasites in the sense that their existence depends on the continued life of the host, the death of which means the death of the fungus. On the other hand, there is so little inconvenience to the insect that the life of the fungus is relatively secure and long. A detailed treatment of this group, which includes over 600 known species, has been made by Thaxter (1896, 1908, *et seq*.) and by Colla (1934).

Fig. 155. The fungus *Cantharosphaeria chilensis* on a cucujid beetle showing distribution of perithecia. (From Thaxter, 1920.)

These organisms are small, frequently minute, fungi and appear as scattered or densely crowded bristles or bushy hairs, which, on certain areas of the host's integument, form a furry or velvety patch. They usually develop on the exterior of their host (fig. 155), but in rare instances, with certain soft-bodied insects, they penetrate the interior with extensive rhizoidal processes of the basal cell. In some members of the group the fungus may penetrate the integument over a pore canal, thus obtaining nourishment from the blood of the insect host. This penetrating structure, known as the haustorium, may also secure the fungus' attachment to the host, since it is highly branched and ramifies through the fatty tissue. In some species the haustorium penetrates only a little way into the chitin of the insect and never reaches the tissues. It has been suggested that the fungi may decompose the chitin using it for food.

The external fungus is usually attached to the body of the host by means of a blackened base or foot. It is interesting that the site of this attachment on different host species is limited to definite regions. As pointed out by Gäumann and Dodge (1928), "not only will the distance from the apex of the elytron of the area occupied by a given species always be about the same, but its relation to either margin will be more or less definitely fixed. Of species inhabiting the left elytron, none will be found even in a corresponding position on the right." Furthermore, each species of the Laboulbeniales appears to be limited to a certain genus of Hexopoda. Most of them are found on members of the Coleoptera.

Transmission from one insect to another occurs chiefly by direct contact such as during copulation. At such times the ascospores are discharged or forced out of the asci and perithecia and adhere, because of their mucilaginous matrix, to the body of the uninfected insect.

On the basis of the relative development of the male sexual apparatus the order Laboulbeniales has been divided into three families: Ceratomycetaceae, Laboulbeniaceae, and Peyritschiellaceae.

Ceratomycetaceae. This family consists of several genera, the principal ones of which are *Caenomyces, Zodiomyces, Euzodiomyces, Coreomyces, Autoecomyces, Ceratomyces, Hydrophilomyces,* and *Rynchophoromyces.* Most of these occur on aquatic insects (Hydrophilidae). The Ceratomycetaceae are characterized by the fact that their antherids or male sex organs are more or less undifferentiated cells of the appendages or their branches.

Laboulbeniaceae.[1] This family is the largest and perhaps the best studied of the order Laboulbeniales and is characterized by the fact that the antherids are differentiated single cells with free efferent tubes. For the most part its members live on insects of the orders Coleoptera, Diptera, and Neuroptera. There have been some reports of the finding of these fungi on specimens of Orthoptera, Isoptera, and possibly Hymenoptera; Thaxter (1896) also reports their occurrence among the Arachnida. Of the Coleoptera, the Carabidae or ground beetles have been found to be the most frequent hosts, especially of the genus *Laboulbenia*. The fungus also occurs on Staphylinidae, Gyrinidae, Cicindelidae, and on certain flies, ants, and mites. According to

---

[1] The family Laboulbeniaceae includes the following better-known genera, all of which are associated with insects:

Amorphomyces	Distichomyces	Rhadinomyces
Arthrorhynchus	Ectinomyces	Rhizomyces
Ceraiomyces	Eucorethromyces	Rickia
Chaetomyces	Herpomyces	Smeringomyces
Clematomyces	Idiomyces	Sphaleromyces
Compsomyces	Laboulbenia	Stigmatomyces
Corethromyces	Moschomyces	Symplectromyces
Dioecomyces	Rhachomyces	Teratomyces
Diplomyces		

Thaxter (1908), in some instances the distribution of the specific fungi corresponds to the distribution of the insect genera. *Laboulbenia cristata,* for example, occurs in all continents, and the members of the large insect genus

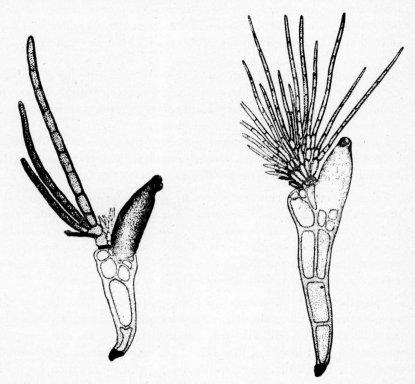

Fig. 156. *Laboulbenia cristata* (left) and *Laboulbenia variabilis.* Greatly enlarged. (From Thaxter, 1895.)

Paederus which it infests are widespread and numerous. *Laboulbenia pheropsophi* is coextensive with the distribution of its hosts (*Pheropsophus* spp.) in five continents. Others, such as *Laboulbenia variabilis,* occur on a variety of hosts but are not found outside the American continents.

Peyritschiellaceae.[1] The antheridial cells of this family are endogenous and

---

[1] The better-known genera of Peyritschiellaceae include the following:

Acallomyces	Dimoromyces	Hydraeomyces
Acompsomyces	Dimorphomyces	Limnaeomyces
Camptomyces	Enarthromyces	Monoecomyces
Cantharomyces	Eucantharomyces	Peyritschiella
Chitonomyces	Euhaplomyces	Polyascomyces
Clidiomyces	Eumonoicomyces	Stichomyces
Dichomyces	Haplomyces	

are united in a specialized organ. They extrude their spermatia into a common chamber from which these are liberated. Hence, they are known as compound antheridia. Some of the members are fairly large, especially those occurring in the tropics.

OTHER ORDERS

Besides the Laboulbeniales, several other orders of the class Ascomycetes contain entomogenous members. Chief of these is Sphaeriales, although others, such as Dothidiales and the family Myriangiaceae, are also concerned.

The genera of Sphaeriales containing entomogenous species include *Berkelella, Hypocrella, Nectria, Ophionectria, Sphaerostilbe* (all of which have been observed to attack species of Homoptera), and *Cordyceps*.

The genus *Cordyceps* contains about two hundred species, the majority of which parasitize insects, especially species of Lepidoptera, Coleoptera, Homoptera, Diptera, Hymenoptera, and Orthoptera. The imperfect (asexual) forms of this genus of fungi as they appear on infected insects have, in the past, been referred to by some authors as "muscardine." The fungus permeates almost the whole insect body, where it increases by yeastlike budding, while aerial hyphae and conidia form on the exterior surface. According to the character of the conidiophore the fungus may be of the *Isaria, Verticillium,* or *Penicillium* type. The fungus survives the winter within the insect body, where it is transformed into a hard, dark-colored mass known as the "sclerotium," which roughly keeps the normal shape of the host. In the spring a stalklike stroma, with the perithecia at the distal end, grows from the sclerotium. The stromata are usually pale or bright-colored, the well-known British species, *Cordyceps militaris,* being red. Others are flesh-colored, lemon-yellow, brown, or purple. Eventually the ascospores are liberated into the air with the chance of encountering another insect host.

Other examples of *Cordyceps* which may be mentioned include *Cordyceps barnesii,* which occurs on cockchafer grubs and, according to Petch (1924a), was apparently fairly common in Ceylon in the days when coffee was grown there extensively. *Cordyceps coccinea* and *Cordyceps flacata* have been collected from coleopterous larvae in decaying wood. *Blattella germanica* on two occasions has yielded cultures of *Cordyceps blattae,* whose brown mycelium overruns the insect and fastens it to the leaf to which it is attached. *Cordyceps dipterigena* usually occurs on flies of the genus *Mydaea*. The dying flies attach themselves to the underside of living leaves. The Brazilian *Cordyceps rhynchoticola* is found on dead leaf-lice. The fungus spreads out in a loose hyphal felt and clings firmly to the substrate. Many of the species of *Cordy-*

ceps fungi are divided into a sterile base and fertile top, with the fertile parts often assuming bizarre forms. Most of the United States species of *Cordyceps* (approximately 30) have been studied and listed by Mains (1934, 1939a,b).

Fig. 157. *Cordyceps thyrsoides* on a dead fly. (From Möller, 1901.)

Several examples of *Torrubiella* (similar to *Cordyceps* but without a clava) may be cited. *Torrubiella aranicida* as well as *Cordyceps gonylepticida* occurs on spiders; *Torrubiella rostrata, T. ochracea, T. sericicola,* and *Cordyceps cristata,* on species of Lepidoptera; and *Torrubiella tomentosa, T. ruba, T. luteorostrata, T. lecana, T. tenuis,* and *T. barda,* on scale insects. The mycelium often appears as pink or red spots (Petch, 1923b).

The Myriangiaceae are parasitic on leaves, bark, and insects. There are two generic types, *Kusanoöpis* and *Myriangium,* divided according to the kind of asci formation. The majority of the *Myriangium* type are parasitic on insects, especially plant lice. In a comprehensive study of entomogenous fungi Petch (1924b) states that the following species of *Myriangium* are known to be parasitic on scale insects: *M. acaciae, M. duriaei, M. montagnei, M. curtisii, M. (Phymatosphaeria) brasiliensis, M. dolichosporum,* and *M. philippinense.* The mycelium permeates the body of the insect, and the stroma is formed over the insect or at the side of it on the plant host. *Myriangium duriaei* is widely distributed.

Other members of the Ascomycetes will be discussed in the section on insects and fungous diseases.

BASIDIOMYCETES

The Basidiomycetes have a special kind of sporophore called a basidium which produces spores (basidiospores) at the end of several (usually 4) slender, pointed protuberances (sterigmata). The ascus and basidium are homologous, the latter bearing its spores exogenously at the ends of the sterigmata. Varying numbers of orders are recognized in this class. Sometimes, for convenience, the class is simply divided into three groups: the smut fungi, the rust fungi, and the fleshy and woody fungi.

Very few associations between Basidiomycetes and insects are known to exist. Where they do occur they are often of a doubtful relationship. Some of the fungi cultivated by termites have been considered Basidiomycetes by some workers and placed in the genus *Volvaria*. Other Basidiomycetes (*Pholiota, Lentinus*) have been recorded as occurring in the fungus gardens of ants.

Septobasidium. The most noteworthy group of Basidiomycetes to be associated with insects is that of the genus *Septobasidium* of the family Auriculariaceae, order Tremellales. The reader will find one of the best accounts of this genus in a treatise by Couch (1938), who describes the biologic relationships concerned and presents the known species in detail.

All species of *Septobasidium* are found living on plants in association with scale insects. This association was first reported in 1907 by Von Höhnel and Litschauer who observed that beneath the stroma of species of *Septobasidium* numerous scale insects were present. Other investigators soon confirmed these observations. In 1929 and 1931 Couch worked out in considerable detail the relationship between *Septobasidium burtii* and the scale insect *Aspidiotus osborni*.

Fig. 158. *Cordyceps militaris* on a dead caterpillar. (From Lohde, 1872.)

Septobasidium burtii is perennial, growing actively between April and November. This fungus normally reproduces by the formation of spores which develop only after rains while the fungus is damp. According to Couch (1938), the fungous body consists of oblong or circular resupinate patches up to several centimeters in diameter. Each patch is made up of varying numbers of irregular, concentric rings of growth, a ring being formed each year (fig. 160). The fungus consists of top and bottom layers between which are numerous tunnels and chambers, many of which are in direct communication with the outside. These chambers contain scale insects, usually one to a chamber but sometimes two or three. The shape of each chamber is adapted to the shape of the insect's body but is considerably larger. (See fig. 159.)

Whereas earlier workers had assumed that the relationship between the fungus and the scale insects is one of parasitism, Couch (1931) maintains that the fungus and insects live symbiotically at the expense of the host plant. The fungus furnishes a home and protection for the insects. In return, the insects serve as a source of food for the fungus. The latter enters the circulatory system of living insects, and there develops numerous coiled haustoria, which the insects digest. Some of the insects may be killed by the fungus. Unless they are infected when young, the insects remain free of the fungus, for the older insects appear to be immune. The parasitized insects may be firmly imbedded in the fungus with sucking tubes extended into the bark of the tree (fig. 161).

Fig. 159. Insect house of *Septobasidium apiculatum* with half the roof cut away exposing a scale insect giving birth to young, one of which is crawling out through the door. (Courtesy Dr. J. N. Couch.)

Couch (1938) lists 217 species of *Septobasidium* known in various countries of the world. Of these, 36 are found in the United States, mainly in the southeastern part. Doubtless numerous other species exist, especially in the tropical jungles. In the United States the most common species appears to be

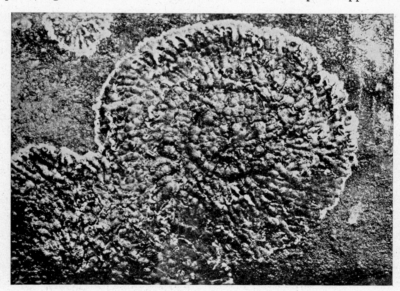

Fig. 160. *Septobasidium burtii* growing on the bark of *Quercus palustris*. Surface view showing radiating ridges and openings to tunnels. If the top layer were to be removed numerous scale insects would be exposed. (Courtesy Dr. J. N. Couch.)

Septobasidium curtisii. Other common ones are *S. pseudopedicellatum, S. sinuosum, S. castaneum,* and *S. alni.* Most of these live on several different species of trees, but *S. canescens, S. grandisporum,* and *S. sabalis* live on only one. Couch has found 76 species of trees to be subject to attack by this genus of fungus.

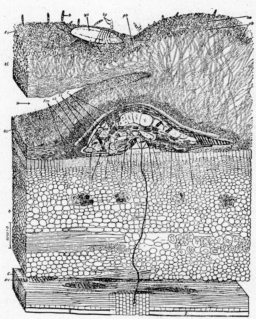

Fig. 161. A diagrammatic sectional view of the symbiotic association between the fungus *Septobasidium burtii* and the scale insect *Aspidiotus osborni.* The sucking tube of the insect may be seen extending down through the bark of the tree into the cambium region. A young scale insect may be seen crawling over the fruiting surface of the fungus. (From Couch's *The Genus Septobasidium,* 1938, University of North Carolina Press, Chapel Hill.)

In the United States about 20 species of scale insects have been found associated with *Septobasidium.* These include species in the genera *Aspidiotus, Cerecoccus, Chermes, Chionaspis, Chrysomphalus,* and *Lepidosaphes.* The majority of the species are of the genus *Aspidiotus* (*A. ancylus, A. juglansregiae, A. osborni, A. forbesi,* and others). Some species of *Septobasidium* are associated with only one species of scale insect; others, perhaps most of them, are associated with several species. Thus, *Septobasidium alni* and its variety *squamosum* are found with at least seven different species of scale insects.

Furthermore, in some cases, such as that of *Septobasidium apiculatum,* as many as three different species of scale insects have been found under the same specimen.

Transmission or distribution of *Septobasidium* apparently occurs in two ways. In the case of *Septobasidium burtii,* when the young scale insect crawls out, it is frequently infected. When it settles down to start a new colony, patches of *Septobasidium* growth develop. However, the dissemination of the wingless young is limited, and perhaps the transportation of *Septobasidium* for great distances occurs by a second means, that of the transplanting of infected plants.

Trees on which *Septobasidium* grow may be damaged severely or practically not at all. The damage is done by the combination of the fungus and the activities of the scale insect.

Fig. 162. *Hirsutella saussurei* on the hornet *Vespa cincta*. This fungus occurs on hornets throughout the tropics. (From Petch, 1924a.)

Agaricales. In the order Agaricales the genus *Hirsutella* contains several interesting entomophytic species. *Hirsutella saussurei* occurs throughout the tropics on hornets. According to Petch (1923a, 1924a), in the specimens observed the clavae consist of long, black, rigid hairs radiating in all directions from the body (fig. 162). These hairs may be as long as 5 cm. and as many as 60 or more arise from one insect. They are composed of longitudinally parallel hyphae, 2 to 3 microns in diameter. When first observed they were thought to be part of the insect. The host insect settles on some plant to which it becomes attached by brown mycelium, and if one of the hairs comes in contact with a leaf or stem, it becomes connected by a brown pad of mycelium from which new black hairs may arise. Other species of *Hirsutella* observed by Petch (1924a) include *H. arachnophila* from spiders; *H. floccosa* from a leafhopper; *H. citriformis* from a pentatomid; and *H. entomophila* from an undetermined coleopteron. According to Charles (1941b), *H. subulata* causes a disease among codling moths.

Cartwright (1938) has isolated another Agaricales, *Stereum sanguineolen-*

tum, from *Sirex gigas.* A similar fungus from *Sirex cyaneus* was also thought to be a form of the same species.

Ustilaginales. One genus of Ustilaginales contains a species (*Ustilago violaceae*) which is the cause of anther smut of pinks. The spores of this fungus are formed in the anthers of the infected flowers. When one of several species of sphinx moths visits the plants in search of nectar, the spores of the fungus adhere to its body and are transmitted to the stigmatic surface of a healthy flower.

DEUTEROMYCETES
(Fungi Imperfecti)

This class, Deuteromycetes or the Fungi Imperfecti, was established for the accommodation of a large number of fungi known only in the asexual or imperfect stage. In other words, they are fungi with incomplete or incompletely known life cycles. In most cases they are believed to be imperfect (asexual) stages of Ascomycetes though some eventually may be found to belong with the Phycomycetes or Basidiomycetes.

Scattered through the Fungi Imperfecti are numerous fungi which have been found associated with insects. Some are extremely pathogenic to arthropods, but others are saprophytic and merely present adventitiously on those animals. Occasionally they may be found fortuitously in the intestinal tracts of insects (Steinhaus, 1941a). Thaxter (1920) has described others which appear to be more or less harmless exterior parasites of living insects (fig. 163).

Fig. 163. The fungus *Termitaria coronata* on the termite *Eutermes morio.* (From Thaxter, 1920.)

The parasitic Fungi Imperfecti are largely of the order Moniliales and include the genera *Acrostalagmus, Aegerita, Aspergillus, Beauveria, Botrytis, Cephalosporium, Cladosporium, Fusarium, Isaria, Oculium, Oöspora, Penicillium, Peziotrichum, Phymatotrichum, Spicaria, Sporotrichum, Trichoderma,* and *Verticillium.* Most of the entomogenous members of these genera are parasitic on species of Lepidoptera, Coleoptera, and Homoptera.

Metalnikov and Toumanoff (1928) and Toumanoff (1928) experimentally infected the larvae of the European corn borer, *Pyrausta nubilalis,* with several fungi. Of these the most virulent was *Aspergillus flavus. Beauveria bassiana* [1]

---

[1] Clements and Shear (1931) use the generic name *Phymatotrichum* in place of *Beauveria.* However, since most authors still appear to prefer *Beauveria* we shall use it here.

and *Spicaria farinosa* were also very virulent and *Aspergillus niger* less so. The *Pyrausta* larvae were much more sensitive to these fungi than were *Galleria mellonella* larvae.

Charles (1938) found *Spicaria heliothis* causing an infection among pupae of *Heliothis armigera*, the corn earworm. The bodies of the insects were completely ensheathed in the white mold and filled with a mass of tightly packed mycelium, the alimentary canal being the only organ not invaded. A similar reaction of *Spicaria verticillioides* on *Conchylis ambiguella* has been observed. *Spicaria rileyi* attacks last instar caterpillars of *Thermesia gemmatilis*. Petch (1925a) isolated *Spicaria javanica* from specimens of a *Ceroplastes*, caterpillars of *Euproctis flava*, and egg masses of *Homona coffearia*. On the insects it formed loose white masses, covering the insect and spreading to the leaves, while in culture a violet-gray color was formed. *Spicaria prasina* has been isolated from larvae of *Pionea forficalis*, larvae of *Spodoptera*, and larvae of *Anticarsia gemmatilis; Spicaria arneae*, from spiders collected in Ceylon; *Spicaria aleyrodis*, from *Aleyrodes variabilis;* and *Spicaria cossus*, from larvae of *Cossus* by Petch (1925b).

Beauveria bassiana, already mentioned as being pathogenic to the larvae of the corn borer, also parasitizes numerous other insects. In North America it has been reported on approximately 30 species of insects. In Europe and other parts of the world where sericulture is practiced, this fungus (at one time known as *Botrytis bassiana*) is notorious as the cause of a muscardine disease of silkworms (*Bombyx mori*). A detailed account of this disease may be found in a publication of Paillot's (1930b) on the maladies of the silkworm. *Beauveria globulifera* has been recorded as occurring on numerous species of insects including more than 70 in North America alone (see Charles, 1941b). In the United States it is well known as a parasite of the chinch bug *Blissus leucopterus*.

Beauveria stephanoderis was observed by Petch on *Stephanoderes hampei*. According to Petch (1926), with *Beauveria densa*, originally isolated from cockchafers, Giard infected the larvae of *Tenebrio molitor, Anomala frischii,* and *Polyphylla fullo;* and caterpillars of *Acherontia atropos, Sphinx ligustri, Mamestra brassicae, Brotolomia meticulosa,* and *Bombyx mori*. He failed to kill *Schistocerca peregrina, Decticus verrucivorus, Locusta viridissima,* and several species of *Stenobothrus*. From these results he concluded that *Beauveria densa* could attack Coleoptera and Lepidoptera but not Orthoptera. Other workers have infected larvae of *Cetonia aurata* and *Rhizotrogus solstitialis,* caterpillars of *Nygmia phaeorrhoea, Anthonomus pomorum,* and *Cheimotobia brumata* with *Beauveria densa*. The species of *Beauveria* found on insects have hyaline hyphae, white or pale yellow in mass, though pinkish or reddish in patches in some cases. The mycelium may appear either chalky or cottony. The conidia are borne in loose, globose heads, either on the main hyphae or on short lateral branches. The morphologic differences between the species are very slight, but there is a marked difference in the habits of these fungi in nature and in culture, and these have been used to separate the species.

A number of species of *Aspergillus* were found by Burnside (1930) to be pathogenic for the honeybee. He (1927) also isolated from bees several species of *Penicillium*, none of which were pathogenic.

The order Phomales also has a few entomogenous fungi such as those included in the genus *Aschersonia* which parasitize species of Lepidoptera and Homoptera. Another entomophytic species belonging to Phomales is *Macrophoma* (now *Sphaeropsis*) *coronillae* associated with the dipteran *Asphondylia coronillae* which produces galls on the plant *Coronilla emerus*. The fungus is probably in a mutualistic association with the insect and is disseminated by the female gnat, being deposited at the time of oviposition.

Another interesting fungus, which Speare (1920) placed in the old order Hyphomycetes, is *Sporosporella uvella*. The first description of the fungus was made by Krassilstschik in 1886, who discovered it in the body of *Cleonus punctiventris*. Speare found that unlike most species of entomogenous fungi *Sporosporella uvella* completes its entire development within the walls of its host, producing no growth externally.

The bodies of infected cutworms occasionally show red-colored patches a few hours before death. When the insect is broken open, a brick-red, powdery mass of spores escapes (fig. 164). Upon exposure to moist air, the spores germinate. The organs of the insects are completely destroyed by the fungus. Experimentally the only external growth obtained from unbroken larvae was found to be confined to those spores which lay immediately beneath the integument. *Sporosporella uvella* also affects *Agrotis ypsilon*, *Euxoa segetum*, *Euxoa tessellata*, *Feltia subgothica*, and *F. jaculifera*.

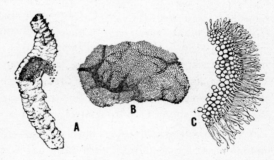

Fig. 164. *Sorosporella uvella*. *A*, infected cutworm torn open, exposing the resting-spore aggregations; *B*, a single resting-spore aggregation; *C*, a portion of a resting-spore aggregation germinating in water, showing promyceliallike germination and conidia. (From Speare, 1920.)

Fungi are also known to parasitize certain arachnids. For example, in Florida the fungus *Rhinotrichum depauperatum* is found on the spider mite *Paratetranychus yothersi*.

FUNGOUS DISEASES OF INSECTS

Although we have just mentioned several specific examples of fungi causing diseases of insects, it is perhaps desirable, at this point, to discuss in further detail some of the more general aspects of this subject.

One of the first to observe a disease of insects caused by a fungus was Bassi de Lodi, who in 1835 found a parasitic fungus attacking silkworms. Some years later Metchnikoff demonstrated the infectiousness of the fungus *Metarrhizium anisopliae* for the beetles *Anisoplia austriaca* and *Cleonus punctiventris*. Forbes (1895, 1896) found *Beauveria globulifera* to be an enemy of the chinch bug and other insects. With these beginnings the possible role of fungi in the biologic control of insects was soon realized.

Fig. 165. A dead moth killed by a fungus and attached to the bark of a tree. (From Sweetman, 1936.)

Though the early expected efficiency of fungi in controlling destructive insects has not as yet been forthcoming, it is apparent that this group of microorganisms may be an important natural factor in this regard. No doubt when more effective means of dispersing the microorganisms and of controlling the conditions under which they are used are known, greater success will be achieved. Sweetman (1936) lists five conditions or precepts which should be considered in any attempt to control insect pests with microorganisms.

(1) The receptivity of the insect during its various stages of development; (2) the optimum environmental conditions favoring the attack of the disease-producing organism; (3) the conditions which influence the virulence of the microorganisms; (4) the exact way in which the parasite attacks the host; (5) the necessity that the optimum activity of the disease coincide with the favorable abundance and a stage of the host which can be attacked.

These conditions are especially applicable to the fungous diseases.

Symptoms of Fungous Diseases in Insects. The syndrome of each fungous disease varies from that of the others. In general, however, insects parasitized by pathogenic fungi have a characteristic appearance. The dead hosts, particularly larvae, may appear apparently normal or soft, though after they have been dead awhile, they usually have a dry, powdery appearance (fig. 166), especially as compared to the blackened, limp, soft appearance of larvae infected with bacteria. Frequently, the fungus may be seen on the exterior of the mummylike insect.

In most cases no external symptoms are noticeable the first few days after the germinating spore penetrates the insect's cuticle. Later one may observe small yellow, brown, or gray spots on various parts of the body. In the absence of these, the first symptoms are frequently those associated with disturbances of the nervous system, though this type of syndrome usually occurs in the later stages of the disease. The insect becomes sluggish in movement and in irritability, eventually losing its ability to retain its equilibrium.

Symptoms usually depend on the manner in which the fungus infects the insect. If the fungus merely lives on the integument of the host or saprophytically within its gut, very little inconvenience is given the arthropod. Untoward reactions are observed, however, if a fungus in the gut gives off toxic substances absorbed by the tissues, though some investigators feel this sort of thing to be infrequent. Actual interference with the mechanics of the insect is noted when the fungus penetrates the body wall and destroys the underlying tissues. Suffocation results when the branching mycelium fills the tracheae.

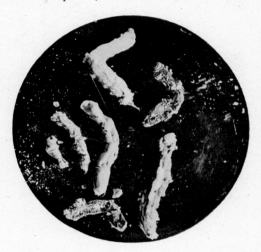

Fig. 166. A group of silkworms dead from green muscardine, *Metarrhizium anisopliae*. (From Glaser, 1926.)

Optimum Conditions for Fungous Diseases. Conditions prerequisite for the optimum development of fungi pathogenic for insects usually concern the factors of temperature and moisture.

Although spores may survive for long periods in an extremely dry environment, a relatively high humidity is necessary for their germination and rapid development. On the other hand, it has been observed (Metalnikov and Toumanoff, 1928) that in the absence of moisture European corn borer larvae may still be readily infected with several species of fungous spores. Wet and dry spores inoculated into cornstalks were found to be equally infective for this same insect by Wallengren and Johansson (1929). In 1928 Toumanoff observed that relative humidities between 40 and 70 per cent had little or no influence on infection of insects with *Beauveria bassiana, Spicaria farinosa,* and *Aspergillus flavus.* However, a thoroughly saturated atmosphere did seem to accelerate the infection. The formation of spores on insects dead of a fungous disease is retarded or does not take place when the conditions are too dry.

Though the optimum temperature varies with the species of fungus con-

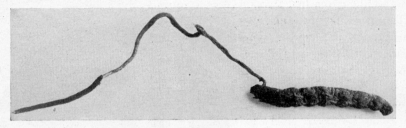

Fig 167. A fatal infection of a scarabeid grub by *Cordyceps robertsi.* The spores are held at the distal end of the fungous stalk. (From Sweetman, 1936.)

cerned, most develop best at temperatures between 20° and 30° C. The extreme limits are usually between 5° and 45° C.

Another important condition conducive to the development and spread of fungous diseases is physical crowding of the insects. In fact, most epidemics break out with the simultaneous occurrence of warm temperatures, a humid atmosphere, and crowded conditions of the insects.

The amount of light may constitute another factor since most fungi appear to be favored by shade although some need light for spore production.

Practical Use of Fungous Diseases. As we have already indicated, the early belief that most of our insect pests could be controlled by the proper use of pathogenic fungi has not as yet been substantiated. Most of the natural epidemics which occur from time to time seem to be largely transitory in effect. They are apparently one of Nature's ways of biologically controlling the numbers of a given species of insect. Man's efforts in this direction do not seem to have yielded much more permanent results. However, it should not be concluded that the field holds no promise of practical results. As frequently happens, initial reports about the use of fungi as a means of biological control were sometimes overenthusiastic, and after a few failures the early optimism was replaced by undue pessimism. It is entirely possible that most of the failures in the past have been due to our lack of understanding of the basic factors concerned. It should be remembered that we are still far from knowing very much about the fundamental biologic relationships between insects and fungi. Although microbic diseases may never constitute a universal means for the control of destructive insects, it is certainly too early to abandon all hope as to their practical use in a moderate but important degree.

Some of the considerations as to whether the efficiency of parasitism can be increased by artificial means have been discussed by Fawcett (1944) as follows:

(a) If the conditions for natural distribution of a fungus are such that there is the maximum number of spores capable of infecting the maximum possible number of insects under prevailing environmental conditions, then no added results could be expected from artificial distribution without at the same time changing the conditions. A relationship of this kind may be known as the "saturation point" for insect and fungus. . . . A saturation point might be expected most of the time from the abundance of wind-borne spores which . . . [some fungi are] capable of producing.

(b) When conditions are such that natural distribution is inefficient, i.e., when this distribution is retarded or lags behind that necessary for maximum infection under prevailing conditions, then artificial distribution would be expected to increase the degree of infection. Whether artificial distribution would constitute an economical measure of control would depend, as it does with entomophagous insects, on what degree of increase could be effected by this means and its cost. It is believed that the natural distribution of a number of these fungi often lags far behind the possible maximum degree of infection. . . . Artificial distribution preceding migration would appear to be suggested.

(c) As with insect parasites on insects, so with the fungus parasites, where the saturation point has been reached for natural infection under a given set of conditions, there is still some possibility of artificially manipulating or changing the natural conditions to shift the saturation point and to increase infection. Changes may be made in humidity by overhead irrigation to increase the moisture on the surface of leaves or fruits for a short time, or by other devices, as the growing of intercrops. . . . With fungus diseases on plants it is known that such changes in conditions (rapid weather changes, etc.) do shift the saturation point and greatly increase infection.

(d) There is also the possible increase or decrease of susceptibility of an insect to infection because of changed nutritional conditions influenced by the nutrition of the host plant.

In general, it has been found that where conditions of humidity and temperature are optimum, success may be had in eradicating destructive insects by the use of entomogenous fungi. As stated earlier, if the fungi are already present in a given area, it is difficult artificially to increase their infectiveness; however, in some instances they may be helped toward attaining epidemic proportions. If they are absent, success may be had in introducing them, though their absence may indicate unfavorable conditions. Further study of the various ecologic aspects and habits of both the fungi and the insects will greatly assist in the practical control of insects by fungi.

One of the earliest attempts to control an insect by the introduction of a parasitic fungus was made in this country in Illinois from 1888 to 1896.

Beauveria globulifera, known to be naturally pathogenic for the chinch bug, *Blissus leucopterus,* was used artificially against this insect by Snow (1890) and Forbes (1895, 1896). Though partially successful in some instances the attempt was eventually abandoned. This same fungus was introduced into Algeria in 1892 where it is supposed to have caused epidemics among flea beetles (*Haltica*). It was claimed by Vaney and Conté (1903) that *Haltica* larvae also were very susceptible to the fungus, *Beauveria bassiana,* cultivated upon silkworms.

One very interesting and important observation on the relation of fungous diseases to the numbers of insects has been made by Watson (1912) and Hill, Yothers, and Miller (1934). They found the number of scale insects infesting citrus trees to increase markedly when the trees were sprayed with Bordeaux mixture. The spray appeared to destroy the entomogenous fungi parasitizing these insects. Recent work by Holloway and Young (1943), however, may force a somewhat different conclusion from that originally made. These investigators believe that the most important factor causing abnormal purple scale increase following the use of fungicidal sprays is the inert granular residue content of the sprays and that any abnormal scale increase is not associated with the fungicidal properties of the sprays.

The fungi most frequently used against white flies and scale insects on citrus trees include the red fungus, *Aschersonia aleyrodis;* the yellow fungus, *Aschersonia goldiana;* the brown fungus, *Aegerita webberi;* the red-headed fungus, *Sphaerostilbe aurantiicola;* and the white-headed fungus, *Podonectria coccicola* (fig. 168), one of the first entomogenous fungi noted in Florida. The first three are usually applied by spraying an aqueous solution of spores on the trees. The last two species are usually disseminated by attaching pieces of fungus-bearing material to the trees. The results appear to have been sufficiently satisfactory to have warranted the distribution of these fungi by the Florida State Plant Board some years ago.

Other instances of the practical use of entomogenous fungi against destructive insects include *Spicaria farinosa* against vine moth pupae (*Polychrosis* and *Clysia*) in France; *Metarrhizium anisopliae* (fig. 169) against the European corn borer, *Pyrausta nubilalis;* and *Acrostalagmus aphidium* against the aphids *Ropalosiphum persicae* and *Aphis gossypii.*

Frequently epidemics break out of their own accord in nature and in insectaries killing insect pests in large numbers. An example of this is a fungous disease of the Comstock mealybug, *Pseudococcus comstocki,* in Virginia caused by a fungus named *Endosclerotium pseudococci* by Harrar and McKelvey (1942). According to Cox, Bobb, and Hough (1943), it has also been identified as the *Isaria* stage of *Cordyceps clavulata. Phoma (Isaria) stenobothri* has been observed affecting grasshoppers in Europe and the United States. Insectary locusts and grasshoppers have been found similarly infected in Argentina by *Aspergillus parasiticus* which in Hawaii and Puerto

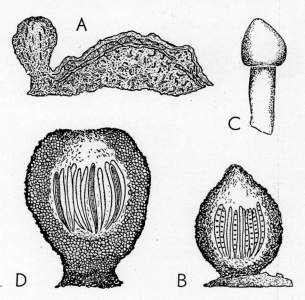

Fig. 168. Two fungi which infect scale insects. *A*. Cross section of a scale insect with a pustule of the red-headed fungus *Sphaerostilbe aurantiicola*. *B*. Single perithecium of the red-headed fungus showing the spores. *C*. Conidia-bearing head of the white-headed fungus *Podonectria coccicola*. *D*. Single perithecium of the white-headed fungus showing the spores. (From Sweetman, 1936; after Rolfs and Fawcett.)

Rico has been shown to be highly parasitic on sugar-cane mealybugs. Entomogenous fungi are also common on aphids. *Empusa fresenii* was reported to be an important factor in the control of the citrus aphid, *Aphis spiraecola*, in Florida.

Fungi also attack useful insects such as the honeybee, *Apis mellifera*. Burnside (1930) found that *Aspergillus flavus* and *A. fumigatus* were very virulent for adult honeybees. Infection may result either from ingesting the pathogenic fungi or by puncturing the exoskeleton. The tissues are penetrated by the developing mycelium and are digested by the fungous enzymes. One strain of *A. flavus* was shown to produce a toxic substance fatal to the bees. The pathogenicity of a fungus appears to be determined by the ability of the spores and mycelium to resist the action of the intestinal fluids of the bees. *Aspergillus nidulans*, *Aspergillus niger*, *Aspergillus glaucus*, and *Aspergillus ochra-*

ceus have also been shown to attack bees in nature. Experimentally the brood as well as the adults are affected by fungous diseases. In nature the brood is rarely attacked, probably because of the fact that the larval food does not contain enough viable spores to cause infection.

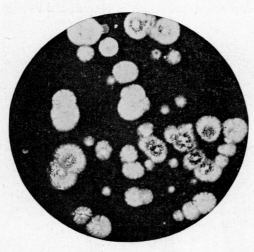

Fig. 169. Colonies of *Metarrhizium anisopliae* after eight days of growth on potato agar. Fruiting and nonfruiting areas may be seen. (From Glaser, 1926.)

Mucor hiemalis (Burnside, 1935) is pathogenic for young adult bees when they are exposed to a temperature of 20° C. or thereabout. That not many bees are killed by this fungus may be explained by the fact that the young bees always remain in the brood combs where the temperature is several degrees above that at which the fungus can attack. When the bees are old enough to leave the hive, they are no longer susceptible.

Bees often take in with their food spores of nonpathogenic fungi which do not germinate in the alimentary tract of normal bees, but remain viable and germinate after the death of the bee.

AMBROSIA FUNGI

An outstanding example of ectosymbiosis is that existing between certain wood-boring beetles and the fungi which they cultivate in their galleries and upon which they feed. Not knowing the true nature of a glistening white substance upon which he observed the beetle larvae to feed, Schmidberger in 1836 called it "ambrosia." Ratsburg (1839) thought the ambrosia might be the result of a mixture of insect spittle and plant sap. A few years later, however, Hartig (1844) discovered this material to be of a fungous nature. To the fungus associated with *Xyleborus dispar* he gave the name *Monilia candida*.

These and other early investigators failed to grasp the full significance of the interesting biologic relationships between the ambrosia beetles and the ambrosia fungi. Over fifty years later Hubbard (1897) studied in considerable detail the ambrosia beetles of the United States, but his account placed emphasis on the beetles rather than on the fungi. Even today a comprehensive investigation of the ambrosia fungi in general is wanting. The works of Neger (1908, 1909, 1910, 1911) and Schneider-Orelli (1911, 1913) are steps in the right direction.

Galleries and Breeding Habits. The true ambrosia beetles belong to the family Scolytidae (Ipidae) and are sometimes called timber beetles because they burrow in the solid wood of trees. As a rule they attack weakened, sickly, or sometimes dying trees, boring deep into the sapwood and forming galleries characteristic of each species (fig. 170c). Generally there is a main gallery which may be branched and which extends deep into the solid wood, and numerous short galleries or chambers, called cradles, extend from its sides. An egg is laid and a larva develops in each cradle. Another group of ambrosia beetles rears the young in large communal galleries where young and old live together. In case a member of the colony dies it is sealed off in a special death chamber.

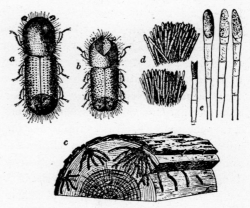

Fig. 170. Ambrosia beetle, *Xyleborus celsus* of the hickory (after Hubbard). *a*, female beetle; *b*, male; *c*, piece of hickory, showing burrows of *X. celsus* in the sap-wood; *d*, ambrosia grown by *X. celsus* on the walls of the burrows; *e*, same, more enlarged. (From Wheeler's *Social Life among the Insects*, 1923, Harcourt, Brace and Co., New York.)

In boring the galleries the adults (usually females, though in some species the males may help) enter the bark of the tree leaving small "shot-hole" entrances similar to those made by bark beetles. If conditions are favorable, excavation of the tunnels may proceed during two or three generations. The galleries of the ambrosia beetles may be differentiated from those of other wood-boring insects by the uniform size of the various ramifications and the absence of refuse material and wood dust. Furthermore, their walls are always stained a dark color, usually brown or black, by the ambrosia fungus.

In that group of beetles in which the larvae are reared in individual cradles the mother constantly attends to her young during their development. A mass of the ambrosia fungus is used to plug the mouth of each cradle. This plug of fungus is used as food by the larvae, which from time to time perforate it to clean their cells of excrement pellets and other refuse. The mother beetle cleans this away and constantly renews the supply of fresh fungus. It has been assumed that in the case of the larvae which live in separate chambers and which eat wood, the fungi serve as a source of nitrogenous material supplementing the wood. The larvae of those species which rear their young in a communal chamber apparently depend solely on the fungus for all the nutritive elements.

The ambrosia fungi are carefully cultivated by the beetles in a characteristic bed or garden. Each species of ambrosia beetle cultivates only one species of fungus, and only the most closely allied species of beetles cultivate the same fungous species. All extraneous or contaminating fungi are somehow suppressed, though such secondary fungi overgrow the ambrosia fungi as soon as the galleries are deserted. The mother beetle prepares a carefully packed bed or layer of chips upon which she deposits some conidia, thus starting her fungous garden. New beds are started using the excrement of the larvae as a substratum. Some workers believe the mother inoculates the bed by regurgitating the spores from her crop where they had been stored to insure transmission to the next generation. Others think the spores are distributed from the body through the fecal pellets. Schneider-Orelli found that the spores of fungi taken directly from the galleries of *Xyleborus dispar* did not germinate, while those taken from the crops of female beetles would do so readily. Leach (1940) points out, however, that this is not true for all ambrosia fungi. A third means of transporting the fungous spores has been suggested by Strohmeyer (1911). This worker has observed ambrosia beetles from the tropics the females of which possess chitinous bristles on the front part of the head. Spores and mycelium of ambrosia fungi are always found in these and it is supposed that they are used to seed new fungi beds.

The most common ambrosia beetles in this country include *Platypus* (*compositus, flavicornis, quadridentatus*), *Xyleborus* (*pubescens, fuscatus, dispar, tachygraphus, obesus, celsus, xylographus*), *Corthylus* (*punctatissimus, columbianus*), *Monarthrum* (*fasciatum, mali, scutellare, dentiger*), *Trypodendron* (*bivittatum, politus, retusum, scabricollis*), and *Gnathotrichus* (*materiarius, asperulus, retusus*).

The Fungi. The ambrosia fungi themselves do not invade the wood more than a short distance from the galleries. Although the gallery walls are deeply stained, they apparently are uninjured by the activities of the fungus which lives on the contents of the sapwood cells but does not destroy the cell walls.

Almost no study has been made of the ambrosia fungi from the taxonomic standpoint. As we have mentioned elsewhere, in 1844 Hartig gave the name *Monilia candida* to the species he found in the galleries of *Xyleborus dispar*. Neger (1919) believed the fungus associated with this beetle was an Endomycete. Trotter (1934) also studied the fungus associated with a species of *Xyleborus,* and to the pleomorphic fungus he described he gave the new generic and specific name *Ambrosiamyces zeylanicus.* Leach, Hodson, Chilton, and Christensen (1940) studied the two ambrosia beetles, *Trypodendron*

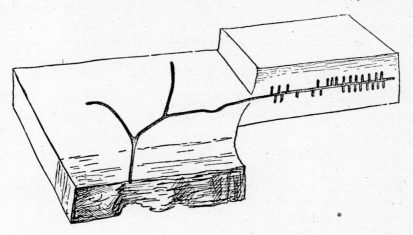

Fig. 171. Gallery of the ambrosia beetle, *Monarthrum mali,* in maple. (From Comstock, 1936; after Hubbard.)

retusum and *Trypodendron betulae,* and their respective fungi, which they were able to grow in pure culture. These seemingly are Fungi Imperfecti and of the genus *Monilia,* being somewhat similar to *Monilia candida.* Obviously here is a field awaiting the attention of the systematic mycologists.

It is interesting to note that under controlled conditions the ambrosia beetle *Xyleborus germanus* can transmit the Dutch elm disease fungus, *Ceratostomella ulmi,* to nursery elm trees. Larvae removed from brood chambers in elm or red pine developed into adult beetles after they had been fed for over a month on one of the following fungi: *Ceratostomella ulmi, Ceratostomella pluriannulata, Pestalozzia* sp., and one unidentified species (Buchanan, 1942).

In 1897 Hubbard distinguished two principal types among the ambrosia fungi: "(1) Those with erect stems, having at the terminations of the stems, or their branches, swollen cells (conidia). (2) Those which form tangled chains of cells, resembling the piled-up beads of a broken necklace." Those beetles whose larvae are reared free in communal galleries (*Platypus* and *Xyleborus*), are associated with the erect type. Those whose larvae live in separate cradles (*Corthylus, Monarthrum,* etc.) are found among the bead-like type.

The growing parts of the fungus are juicy and very tender. Young larvae nip off the tender conidial tips, but the older larvae and adult beetles eat the entire fungus down to its base from which it rapidly grows up again. Hubbard compares its growth to that of asparagus which remains tender and edible only when continually cropped but is no longer desirable as food when permitted to go to seed. Similarly if the ambrosia is allowed to ripen, it can no longer serve as food for the insects, and in fact may be a source of danger to them by choking the galleries and suffocating the inhabitants.

ANTS AND FUNGI

Another outstanding example of ectosymbiosis between insects and fungi is that existing between ants and the fungi that they cultivate. The work of Wheeler (1907, 1923, 1937), Weber (1937, 1938), and others has revealed extremely interesting and involved associations between these two forms of life.

The fungus-growing and the fungus-feeding ants belong to the tribe Attini of the subfamily Myrmicinae. Some investigators believe that certain ants in other groups also have these habits (see Bailey, 1920). The Attini are exclusively American ants, being confined to the tropical and subtropical climates, though at least one species ranges as far north as New York. Over a hundred species of attine ants have been described and all are fungivorous, apparently a different fungus being maintained by each species or group of closely related species.

Taxonomically very little has been accomplished concerning the fungi associated with these ants. Möller (1893) gave the name *Rozites* (now *Pholiota*) *gongylophora* to the fungus cultivated by *Acromyrmex disciger*. Spegazzini (1899) described the fungus *Xylaria micrura* which is associated with the nest of *Acromyrmex lundi*. The yeastlike fungus cultivated by *Cyphomyrmex comalensis* Wheeler (1907) is called *Tryidiomyces formicarum*. *Lentinus atticolus* is the fungus associated with *Atta cephalotes* (Weber, 1938).

The ants under discussion are sometimes known as the leaf-cutting or the parasol ants. As these names indicate, the ants cut pieces from the leaves of trees and carry them like parasols to their nests. Here the pieces are cut into smaller fragments from which they construct the spongelike masses which form the substrata of their fungus gardens. These spongelike masses are "seeded" by the female and are soon covered with a white mycelium. In some unknown manner the workers treat this fungous growth so that the hyphae produce large numbers of small, spherical swellings known as bromatia (fig. 172). The bromatia are used as food by the ants and are fed to the larvae. When grown on artificial media the bromatia are not formed, which indicates that they are probably induced by the ants. The gardens are carefully and continuously tended and all contaminating growths are suppressed.

When the virgin queen is about to leave the parental nest for the period of swarming, she takes a mass of hyphae and leaf tissue and retains it as a pellet in her infrabuccal pouch (fig. 173). This pouch is a spherical sac below the floor of the mouth but opening into the mouth; most of the time it acts as a repository for particles of solid material which are not ingested by the adult ant. When she has mated and discarded her wings, she makes a small chamber for herself in the soil. Here she uses the saved pellet as spawn to start a new fungus garden. As the garden begins to grow, she fertilizes it with her feces or occasionally even by breaking up an egg and adding it to the garden. The garden rapidly develops and soon it is large enough to serve as a nest for the first-generation eggs. The resulting larvae eat of the fungus, pupate, and develop into small workers which go out and bring in more pieces of leaves to add to the garden. From this time on the queen devotes her time to laying eggs and the workers take over the duties connected with the fungus garden.

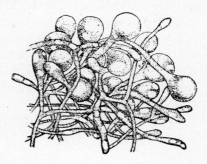

Fig. 172. Modified mycelium (bromatium) of fungus cultivated by the ant *Moellerius heyeri*. These globular swellings of the hyphae are produced and eaten by the ants. (From Wheeler's *Social Life among the Insects*, 1923, Harcourt, Brace and Co., New York. After Bruch.)

 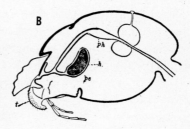

Fig. 173. Sagittal sections through the heads of ants showing the infrabuccal pouch and pellet. *A*, the head of a queen of *Lasius niger* with the mouth open; *B*, queen of *Camponotus brutus* with the mouth closed. *t*, tongue; *o*, oral orifice; *ph*, pharynx; *h*, infrabuccal pouch; *pe*, pellet made up of solid particles of food refuse and strigil sweepings. When the ants swarm to establish a new colony, the female retains a mass of fungus "spawn" in her infrabuccal pouch. With this she is able to start a new fungus garden. (From Wheeler's *Social Life among the Insects*, 1923, Harcourt, Brace and Co., New York.)

Of more than a hundred species of fungus-growing ants only a few have been the subject of detailed studies as to the biologic relationships between them and their fungi. These include such species as *Atta sexdens, Atta texana, Atta vollenweideri, Acromyrmex lundi, Acromyrmex discigera, Cyphomyrmex rimosus, Moellerius heyeri, Poroniopsis bruchi, Trachymyrmex septentrionalis,* and others. For a more complete list see Wheeler (1907).

Ants have been blamed for transmitting certain fungous diseases of plants such as the mealy-pod disease of cocoa caused by *Trachysphaera fructigena*. Examination of the contents of their infrabuccal pouch (Bailey, 1920; Leach and Dosdall, 1938) has revealed the presence of large numbers of fungous spores even in ants which do not normally cultivate fungi. Unknown relationships with other fungi and probably fungous diseases are therefore indicated.

TERMITES AND FUNGI

The termites or "white ants" occur principally in the tropics and to a lesser extent in the temperate zones. They differ from ants in the pale-yellow or white color of many of them and in the form of the abdomen, which is broadly joined to the thorax instead of being a slender petiole as in ants. Like the ants, however, they are social insects and each species contains several distinct castes. They may be divided into two large groups according to their food habits. One group feeds largely on wood and the associated fungi, the other feeds principally on fungi cultivated in special "gardens" which they tend with great care.

Termites of this latter group generally live in large nests called "termitaria" which they may build under or above ground in large mud-covered structures. In special compartments of the termitarium the fungi are cultivated on the excreta of the termites. This fecal substrate is inoculated automatically when the spores of the fungus are ingested by the workers and pass uninjured through the insects' intestinal tracts. The spores germinate giving rise to a thick growth of fungus, which is then used as food for the royal or first reproductive castes and the young. The workers and soldiers feed on other plant material and do not use the cultivated fungus.

Most of the termites with which we in the United States are familiar are the wood-eating type. Frequently, the wood penetrated by these termites show indications of rot produced by fungi. The termites may carry spores or hyphae on their bodies into their burrows, which soon give rise to patches of fungi which invade the walls of the galleries.

Working with the termites *Kalotermes minor, Reticulitermes hesperus,* and *Zoötermopsis angusticollis,* Hendee (1933) found fungi associated with each of them. She isolated fungi from the exteriors as well as the guts of the insects and reported that conditions in their burrows are favorable to the growth of fungi which are present in the walls of the burrows and in the fecal pellets which the termites eat. Later Hendee (1935) stated that fungi offer a source of proteins in the diet of *Zoötermopsis angusticollis* and that they probably supply vitamins which are essential to the normal growth and development of termites. She also suggests that through the secretion of extracellular enzymes the fungi may render the wood itself nutritively more available.

Taxonomically very little is known concerning the fungi associated with termites. According to Leach (1940), the fruiting bodies of *Xylaria* and *Volvaria* have been found in the termites' gardens and are thought to be identical with the fungi making up the gardens. In the case of the wood-feeding termites discussed in the preceding paragraph, Hendee (1933) isolated representatives of 33 known genera and 20 unidentified fungi from their colonies. Most frequently isolated were the genera *Penicillium* and *Trichoderma*.

INSECTS AS VECTORS OF FUNGI PATHOGENIC FOR ANIMALS AND PLANTS

It seems strange that while there are many fungous diseases of plants which are spread by insects, there are remarkably few such diseases of man and animals which have an arthropod vector. One searches the literature almost in vain for references to arthropod-transmitted fungous diseases of man, finding more discussion of a hypothetical nature than of fact. Some diseases for example, coccidioidomycosis caused by *Coccidioides immitis,* have been suspected of being transmitted by the bites of insects but most such cases lack definite proof (Jacobson, 1932).

Because of this dearth of information relating to insect-borne fungous diseases of animals, our discussion will be concerned largely with diseases affecting plants. In the light of the splendid treatment of the subject of insects and fungous diseases by Leach (1940) only a brief outline of the field will be given here, and for much of this we have drawn heavily from this author.

According to Leach, fungi are responsible for more plant diseases than any other group of microorganisms. While most of these diseases are disseminated by wind, a considerable number depend on wounds, such as those made by insects, for their transmission. By considering a few representative examples, we may be able to point out some of the biologic relationships involved.

Dutch Elm Disease. This disease, first discovered in Holland in 1919, was recognized in the United States in 1930. Nearly all species of elms are susceptible to this very destructive disease which is caused by the fungus, *Ceratostomella ulmi*. (It has also been placed in the genera *Craphium* and *Ophiostoma*.) The following insects (bark beetles) have been shown to be vectors of the disease agent: *Scolytus scolytus, Scolytus multistriatus, Scolytus sulcifrons, Scolytus affinis, Scolytus pygamaens,* and *Hylurgopinus rufipes*. Other insects have been suspected of acting as vectors but convincing proof is lacking; these include: *Ptelobius vittatus, Ptelobius kroatzi, Saperda punctata, Saperda tridentata,* and *Magdalis armigera*. Mites (*Monieziella arborea, Histiogaster fungivorax, Megninietta ulmi,* and *Pseudotarsonemoides innumerabilis*) have also been suspected of being potential vectors of the fungus (Jacot, 1934, 1936; Fransen, 1939). There may yet be undiscovered vectors, but arthropods appear to be obligatory vectors of the disease in nature.

Infection of the trees takes place mechanically when the contaminated beetles inoculate the feeding wounds or when they bore into the bark to establish brood tunnels. These types of transmission are facilitated by the fact that the elm-disease fungus fruits in the old brood and larval tunnels and that when the beetles leave these infected trees they are usually heavily contaminated both externally and internally with fungous spores.

For more details of the relationships between *Ceratostomella ulmi* and its insect vectors the reader is referred to the works of Welch, Herrick, and Curtis (1934), Readio (1935), Clinton and McCormick (1936), and McKenzie and Becker (1937).

Blue Stain of Conifers. The market value of many species of coniferous trees may be greatly reduced when the sapwood of such trees is attacked by fungi which stain it dark blue. The fungi do not decay the wood, but the discoloration they cause renders it useless for most purposes. Usually the blue stain is found in felled timber, but when living trees are attacked the trees may be seen dying from the top downward.

There are several fungi concerned in such destructive processes; these include *Ceratostomella ips, Ceratostomella pseudotsugae, Ceratostomella piceaperda,* and *Tuberculariella ips*. It should be mentioned that Von Höhnel (1918) divided the genus *Ceratostomella* into two groups; to those similar to *Ceratostomella pilifera* he gave the generic name *Linostoma*. This name, however, had already been used to designate a genus of plants, hence Sydow and Sydow (1919) replaced it with the name *Ophiostoma*. In 1936 Goidanich created the genus *Grosmannia*, using *Ceratostomella penicillata* Grosmann as the type species. Most writers still prefer to use the generic name *Ceratostomella*.

The insect vectors include the beetles *Dendroctonus pseudotsugae, Dendroctonus piceaperda, Glischrochilus pini, Glischrochilus grandicollis, Glis-*

chrochilus emarginatus, *Glischrochilus integer,* and *Glischrochilus oregoni.* When bark beetles emerge from infected trees, they are contaminated both externally and internally with the fungi. Thus when the beetles bore their new brood tunnels the sapwood and inner bark become infected. The fungi fructify and the sticky matrix of the conidia and ascospores makes it certain that the new brood of beetles is adequately contaminated.

According to Leach (1940), the fungus apparently plays no important role in the insect's nutrition, but probably makes the living tree a more suitable environment for beetle development by weakening the tree and reducing its water content. On the other hand, the fungus depends upon the insect for its dissemination and penetration into the trees.

Brown-staining fungi have been found associated with several insects. *Trichosporium symbioticum* was found on *Scolytus ventralis* and *Spicaria anomala* on *Scolytus praeceps* and *Scolytus subscaber* (Wright, 1935, 1938).

Fungous Diseases of Figs. Of the California "fruit spoilage diseases," those affecting figs are among the most destructive. The three most commonly affecting figs are smut, endosepsis, and souring.

The smut disease of figs is not a true smut (Hemibasidiomycete). It is caused by the mold, *Aspergillus niger.* The spores of this fungus gain entrance into the figs through the activities of the dried-fruit beetle, *Carpophilus hemipterus,* and to some extent through those of *Drosophila ampelophila.* Thrips and mites at times enter the young fruit and they have been accused of carrying *Aspergillus niger* in with them.

Endosepsis is a firm dry rot affecting the caprificated fig fruit and is most easily detected from the outside after the fig begins to ripen. The causative agent of the disease is the fungus *Fusarium moniliforme* var. *fici.* It is disseminated by the fig wasp, *Blastophaga psenes.* The fungus is associated only with the adult wasp, which becomes contaminated after emerging from the galls in which it overwinters.

Souring is caused by yeasts. Later bacteria and molds aid in the decomposition of the ripe fruit. The molds as well as the yeasts are introduced into the figs by *Carpophilus hemipterus* and *Drosophila ampelophila.*

Other Plant Disease Fungi Spread by Insects. The list of plant diseases of which insects are the proved or suspected vectors is a long one. Those we have just discussed are merely representative. The mention here of a few others may be pertinent.

Claviceps purpurea, the cause of ergot which attacks many cereals and grasses, is thought to be transmitted by many species of flies including the fungous gnat, *Sciara thomae.* A list of 40 or more different insects which have been observed feeding on the sugary secretion of ergot is given by Atanasoff (1920).

Perennial canker of apple trees is believed to occur through wounds made by the woolly aphis. These wounds become infected with the fungus *Gloeosporium perennans*, which lives as a saprophyte in the dead bark of old cankers. European canker is caused by *Nectria galligena* and a New Zealand canker by *Diplodia griffoni*—infections by both apparently occurring through wounds made by the woolly aphis. There is some question, however, as to just how direct a relationship exists between the fungi and the insects. Bees have been indicted for carrying several species of fungi pathogenic to plants. Blossom blight of red clover, caused by *Botrytis anthophila,* and downy mildew of lima beans, caused by *Phytophthora phaseoli*, are examples. Tree crickets (*Oecanthus niveus* and *Oecanthus angustipennis*) transmit the fungus, *Leptosphaeria coniothyrium,* which causes tree-cricket canker of apple trees. The following grasshoppers have been found acting as vectors of *Fusarium vasinfectum,* the cause of cotton wilt: *Encoptolophus texensis, Melanoplus differentialis, Melanoplus femur-rubrum, Melanoplus mexicanus, Tomonatus aztecus, Spharagemon cristatum, Chortophaga viridifasciata* var. *austratior, Trimentropis citrina, Dissosteira carolina, Schistocerca obscura,* and *Schistocerca americana*. The method by which species of *Dysdercus* transmit *Nematospora gossypii* and allied fungi, the cause of internal boll disease of cotton, has been greatly clarified by Frazer (1944). She found that the fungous material (spores or mycelium) is carried as an external contaminant of the mouth parts of these insects, although it is located in the deep stylet pouches which serve both for protection of the fungus and for spore germination.

In other cases insects are indirectly responsible for infection of plants by fungi. They wound the plants, permitting the entrance of the pathogens. For example, the cabbage maggot, *Hylemya brassicae,* injures cabbage roots allowing the fungus *Phoma lingam,* the cause of cabbage blackleg, to enter. *Lasiodiplodia triflorae,* the cause of plum wilt, enters the plum trees through wounds made by the peach-tree borer, *Aegeria exitiosa*. Several aphids and scale insects, especially the white fly, *Aleyrodes citri,* secrete honeydew, which on the surface of leaves serves as a rich medium for sooty mold fungi, such as *Capnodium citri,* to grow luxuriantly. These saprophytic fungi do no direct harm to the plant on which they are found but damage it indirectly by interfering with photosynthesis. The activities of *Monochamus scutellatus* and *Monochamus notatus* facilitate the invasion of the heartwood of red-pine logs by the fungus *Peniophora gigantia,* which causes a reddish-brown decay.

❖ CHAPTER VIII ❖

VIRUSES AND INSECTS

THE KNOWN relationships existing between insects and viruses are concerned with (1) virus diseases of insects, (2) vectors of viruses causing diseases of man, animals, and plants, and (3) bacteriophage. A possible fourth group, relating to nonpathogenic viruses of insects, may be added to this list in the future when techniques are so improved as to afford a method of their detection.

I. VIRUS DISEASES OF INSECTS

The first description of a disease which characterizes this group was made by Cornalia in 1859. It was not until 1909 to 1913, however, that Wahl, Von Prowazek, and Escherich by independent experiments demonstrated the virus nature of these maladies. Up to this time the virus diseases of insects had been confused etiologically with the bacterial and protozoan diseases. White, in 1913, demonstrated sacbrood of bees as being due to a filterable virus. In the same year Glaser and Chapman discovered a virus disease in gypsy-moth larvae. Thus was clarified the confusion which had enveloped the cause of this group of diseases in insects, permitting the virologists to include one more form of life in the list of those experiencing virus diseases.

Since that time, considerably more work has been done on this type of disease and more insects have been found to be natural hosts of pathogenic viruses. The following, adopted from Chapman and Glaser (1915) and Sweetman (1936), with additions, is an incomplete but representative list of the names of the insects reported to be susceptible to viruses:

Coleoptera
    Dermestidae
        Dermestes lardarius
        Anthrenus museorum

Lepidoptera
    Oecophoridae
        Chimabache fagella

Tortricidae
- *Acleris variana*
- *Conchylis ambiguella*

Saturniidae
- *Hemileuca maia*
- *Hemileuca oliviae*
- *Antherea pernyi*
- *Antherea yamamaï*
- *Antherea mylitta*
- *Philosamia cynthia*
- *Saturnia pavonia major*

Liparidae
- *Porthetria dispar*
- *Lymantria monacha*
- *Hemerocampa leucostigma*
- *Notolophus antiqua*

Noctuidae
- *Cirphis unipuncta*
- *Laphygma frugiperda*
- *Prodenia litura*
- *Alabama argillacea*
- *Heliothis armigera*
- *Autographa gamma californica*
- *Autographa brassicae*
- *Noctua clandestina*
- *Euxoa segetum*
- *Nephelodes emmedonia*

Bombycidae
- *Bombyx mori*

Lasiocampidae
- *Malacosoma americana*
- *Malacosoma disstria*

Geometridae
- *Bupalus piniarius*
- *Ellopia fiscellaria lugubrosa*

Dioptidae
- *Phryganidia californica*

Sphingidae
- *Celerio* sp.
- *Smerinthus atlanticus*

Notodontidae
- *Harpyia bifida*
- *Heterocampa guttivitta*

Pieridae
- *Colias (Eurymus) philodice*
- *Colias (Eurymus) eurytheme*
- *Pieris brassicae*
- *Pieris rapae*

Nymphalidae
- *Vanessa urticae*

Tineidae
- *Tineola biselliella*

Arctiidae
- *Hyphantria cunea*
- *Callarctia virgo*

Diptera

Calliphoridae
- *Calliphora vomitoria*

Hymenoptera

Diprionidae
- *Diprion rufus*
- *Gilpinia hercyniae*

Apidae
- *Apis mellifera*

Most authorities now divide the virus diseases of insects into two groups: (1) the polyhedral diseases and (2) the nonpolyhedral diseases. The principal ones of these may be grouped as follows:

(1) **Polyhedral diseases**
   (a) Jaundice (e.g., of the silkworm, *Bombyx mori*).
   (b) Wilt (e.g., of the larva of the gypsy moth *Porthetria dispar*).
   (c) *Wipfelkrankheit*.
   (d) Disease of the cutworm *Euxoa segetum*.
   (e) Diseases of the larva of *Vanessa urticae*.

(2) **Nonpolyhedral diseases**
   (a) Sacbrood of the honeybee (*Apis mellifera*) larva.
   (b) Pseudojaundice No. 1 of the cutworm *Euxoa segetum*.
   (c) Pseudojaundice No. 2 of the cutworm *Euxoa segetum*.

There are probably other names of virus diseases of insects which may be added to this list. A case in point is the virus which may be the primary cause of *gattine* and flacherie, diseases of the silkworm *Bombyx mori*. Bacteria, *Streptococcus bombycis* and *Bacillus bombycis,* have been described as the causative agents of these diseases. Paillot (1930b) believes, however, that an ultramicroscopic virus is the primary cause which predisposes the silkworms to the disastrous effects of the secondary invasion by *Streptococcus bombycis* in the case of *gattine,* or by *Bacillus bombycis* in the case of flacherie.

Another recently discovered virus disease of insects is a paralysis which afflicts the honeybee *Apis mellifera* (Burnside, 1945). Symptoms include sprawled legs and wings and a general trembling of the whole insect.

THE POLYHEDRAL DISEASES

It is a curious fact that a majority of the polyhedral diseases of insects occur in members of the order Lepidoptera. As we have just indicated, those which have elicited the most study are jaundice or *grasserie* of the silkworm (*Bombyx mori*), "wilt" disease of the larva of the gypsy moth (*Porthetria dispar*), "*Wipfelkrankheit*" of the nun-moth (*Lymantria monacha*) larva, and the polyhedral diseases of the cutworm, *Euxoa segetum,* and of the larva of *Vanessa urticae*. The following discussion is based largely on these typical polyhedral diseases.

Polyhedra. A few hours after larvae die of a polyhedral disease their internal body tissues become liquefied. If this fluid is examined microscopically, large numbers of polyhedral bodies may be seen among the disintegrating tissues. Individual polyhedra may vary considerably in size though they usually range anywhere from 0.5 to 15 microns in diameter. According to Glaser (1928), the average size of the polyhedron affecting the silkworm is 3 to 5 microns in diameter; that of the gypsy moth, 3.4 microns; that of the tent caterpillar, 2.6 microns; that of the nun moth, 2.65 microns; that of *Phryganidia californica*, 1.6 microns; and that of the tussock moth larvae, 1.5 microns in diameter. These bodies never occur in the shape of spheres and are usually very refractive and crystallike in appearance, having a relatively large number of faces. They may occur singly or in pairs. Onionlike concentric layers are frequently observed within the bodies, suggesting that they "grow" by accretion. In glass slide preparations, when pressure is applied the bodies may be seen to crack and fragment. Whereas some observers believe that the polyhedra contain no particulate material, others have described their contents as consisting of minute virus granules.

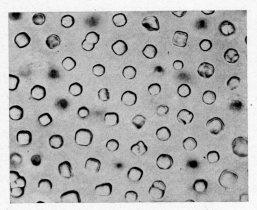

Fig. 174. Polyhedral bodies of silkworm jaundice (*grasserie*). (Courtesy Dr. R. W. Glaser.)

It is difficult to stain polyhedra with ordinary stains using ordinary techniques. They resist all manner of staining unless mordants or steaming methods are used. When they are stained they either take the stain uniformly or show structures similar to refractive granules and striations. Excellent observation of polyhedra may be gained by dark-field examination.

Very little can be said as to their chemical structure. That they contain protein material is indicated by the fact that they stain yellow with picric acid. They apparently contain no fat since they do not blacken with osmic acid and are not stained by Sudan III. It is known that polyhedra contain iron. They are not dissolved by alcohol, benzol, hydrogen peroxide, glycerol, benzine, carbon bisulfide, or by hot or cold water. They are soluble in weak alkalies and are precipitated by acids. In the light of these reactions, most authorities believe the polyhedra to be nucleoprotein in composition and crystallike degeneration products of the diseased tissue with which they are associated.

Working with the virus and polyhedral bodies of silkworm jaundice, Glaser and Stanley (1943) concluded that the view that polyhedra represent crystallized virus was untenable. These workers found that although suspensions of polyhedral bodies were usually infectious, no great concentration of virus activity was found within them.[1]

---

[1] Contrasting results were obtained in 1939 by Paillot and Gratia, and in 1943 by Bergold, but war conditions have made their publications, along with certain others, unobtainable to the author.

VIRUSES AND INSECTS

Symptoms of Polyhedral Diseases. Some species of insects affected by polyhedral disease exhibit their own peculiar symptoms. In some cases the appearance of the disease is responsible, in certain parts of the world, for its name. Thus, the silkworm polyhedral disease is known as "jaundice" in America, *grasserie* in France, and *Gelbsucht* or *Fettsucht* in Germany, since a few days before the affected caterpillars die they are covered with bright yellow blotches. Before death they cease to eat and become inactive, and their skin is shiny and opaque and takes on a yellowish color. Such larvae are called *vacca* (milk cows) by the Italians and *Fettraupen* by the Germans because of the swollen appearance of the larvae.

In larvae infected with other polyhedral diseases the symptoms are not distinctive. The nun moth loses its appetite and migrates to the tree tops, where it hangs by its prolegs and dies. The polyhedral disease of gypsy moths and tent caterpillars is known in America as "wilt." Besides loss of appetite, very few symptoms are apparent. Shortly after death, however, the larvae are flaccid and shiny, and give off an offensive odor.

JAUNDICE (*GRASSERIE* OR *GELBSUCHT*) IN SILKWORMS

Early History. As was common in the study of many virus diseases, the first supposed etiologic agent of jaundice in the silkworm, *Bombyx mori,* was a bacterium (*Micrococcus lardarius* Krassilstschik, 1896). Later Bolle (1898, 1908) observed polyhedral bodies in infected worms and considered them to be a sporozoan which he named *Microsporidium polyedricum*. Marzocchi (1908) concurred with Bolle, and Von Prowazek (1907) also considered the

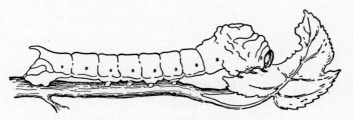

Fig. 175. The silkworm larva, *Bombyx mori.* (From Comstock, 1936.)

etiologic agent to be a protozoan which he named *Chlamydozoa bombycis*. The last worker considered the polyhedra to be a by-product of the disease, since he found diseased material to be infective after the removal of the polyhedral bodies. Wolff (1910), working with a polyhedral disease of *Bupalus piniarius,* supported Von Prowazek's views, naming his supposed protozoan *Chlamydozoa prowazeki*.

In 1912 Von Prowazek apparently began to doubt his chlamydozoa theory when he found jaundice material still infectious after being passed through an ultrafilter made of a Berkefeld candle covered with agar. In the same year Hayashi and Sako (1913) concluded that the polyhedra carry the virus or infectious agent. Acqua (1918–19) showed that the cause of jaundice is a filterable agent too small to be seen through an ordinary microscope and capable of passing through fine Berkefeld filters. The association of the virus with the polyhedra was maintained by Aoki and Chigasaki (1921), who stated that silkworms were just as susceptible to infection by polyhedral bodies shaken in physiologic salt solution and centrifuged more than ten times as by unwashed polyhedra.

A more recent theory as to the nature of the causative agent of jaundice is that proposed by Pospelov and Noreiko (1929). These workers suppose that the virus of jaundice is an ultramicroscopic involution form of the yeast *Debaryomyces tyrocola*. They arrived at this conclusion after typical jaundice resulted in caterpillars to which they had fed the yeast. No one has yet provided further evidence supporting this theory, and for the present, at least, it must be considered unproved.

The Virus of Jaundice. If the blood of diseased silkworms is examined under the dark-field microscope, two types of granules may be seen. One kind appears as a very brilliant body which normally occurs in the blood of silkworms. The other kind is less brilliant, smaller (being 0.1 micron in size), and invisible by ordinary illumination. According to Paillot (1933), this last type occurs in all infected caterpillars. Twenty-four hours after inoculation with jaundice virus, only a few of these bodies appear in the blood, although after 48 hours they are very numerous and have reached their greatest numbers. They are retained by L_4 and L_5 Chamberland filters, the filtrates thus obtained not being infectious. They remain in the infectious supernatant fluid after centrifuging. These bodies may be present in the absence of polyhedra. Paillot found that he could transmit the disease by inoculating silkworms with blood of individuals infected two days previously and in which no polyhedra were present. Because infectiousness appeared to be directly associated with the presence of these bodies, Paillot believed them to be the virus and gave them the name *Borrellina bombycis*. Glaser and Cowdry (1928) in their study of the visibility of polyhedral viruses came to the conclusion that no observable qualitative difference exists between the particles found in the blood of normal and diseased larvae.

In the case of jaundice, as with the other polyhedral diseases, most authorities are of the opinion that the polyhedra are not, in themselves, the causative agents of the disease. Instead, they are generally regarded as "products of a specific disturbed metabolism; of diagnostic value, but of no etiological importance" (Glaser and Lacaillade, 1934). The two workers just cited showed that the polyhedra are apparently carriers of the virus, which, however, may

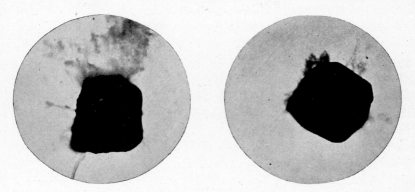

Fig. 176. Washed polyhedral bodies obtained from jaundice-diseased silkworms. Magnification 7,600 (electron microscope). (From Glaser and Stanley, 1943.)

be separated from these inclusions. When fresh blood containing polyhedra was immediately centrifuged, the supernatant fluid containing no visible inclusions was highly infectious. However, if the blood were allowed to stand long enough for the polyhedra to settle, no virus could be detected in the upper layer of the fluid. In agreement with Aoki and Chigasaki (1921), Glaser and Lacaillade found that though the polyhedra continuously lost virus when washed in water, they could not be freed of the virus. When the virus was rendered inactive by heat, the polyhedra were able to reabsorb fresh virus brought in contact with them.

Glaser and Lacaillade (1934) report that virus containing blood held in ampules at 0° to 4° C. for one year loses its infectivity only to a slight degree and that full activity may be restored after a single passage through the silkworm. According to Glaser (1928), it remains infectious for at least two years when dried at room temperature. It may be kept for at least six months in 89 per cent glycerol without losing its virulence, although this is lost when the material is held at 60° C. for 15 to 20 minutes. The virus is inactivated when placed in 80 per cent alcohol for 15 minutes.

Biochemical studies on jaundice virus and inclusion bodies by Glaser and Stanley (1943) showed that whereas the virus itself is stable only between pH 5 and about pH 9, the polyhedral bodies retain their virus activity after exposures to a hydrogen ion concentration as high as pH 2. This indicates that the polyhedral bodies protect the virus held within them. A similar protective action may be noted against other reagents. A purified suspension from diseased caterpillars was found to consist essentially of a nucleoprotein component having a sedimentation constant of 17S, a particle diameter of 10 millimicrons, and a molecular weight of about 300,000. By use of the electron microscope the blood of diseased silkworms was found to contain particles which may be the virus (fig. 177).

Pathology. Silkworms infected with jaundice usually become inactive and fail to eat, and their external covering possesses characteristic lemon yellow blotches. Before death the integument becomes opaque and assumes a shiny, yellowish appearance. Usually they take on the jaundiced appearance in from 4 to 7 days and die in from 10 to 14, or as long as 18 days after infection. Shortly before death most of the tissues disintegrate.

The histopathology of the disease has been studied both in diseased insects and in tissue cultures. Referring to the tiny granular bodies he believes to be the virus, Paillot (1933) states that at the beginning of the infection the bodies

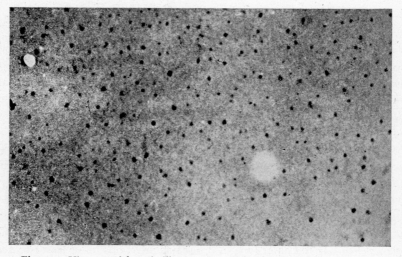

Fig. 177. Virus particles of silkworm jaundice taken at a magnification of 39,700 with the electron microscope. This purified material was obtained from the blood of jaundice-diseased silkworms by differential centrifugation. (From Glaser and Stanley, 1943.)

are found in the cytoplasm of infected cells (hypodermal, fat, tracheal matrix, and certain blood cells). After the second day, however, they are found in the nucleus where the bodies form a ring around the condensed chromatin mass. The nucleus contains a liquid, formed by liquefaction of chromatin material by the virus, in which the polyhedra are in suspension and in which the tiny virus granules are formed. The nucleus becomes hypertrophied (fig. 178) and two kinds of bodies may be found in the nuclear fluid. One has a slow vibratory movement and is the regular polyhedral body; the other is smaller, vibrates more rapidly, and presumably is the virus.

Trager (1935c) developed a tissue culture medium in which certain cells from the gonads of female silkworms survived and multiplied for periods of two to three weeks. In such cultures he was able to maintain successive passages of the jaundice virus. Typical intranuclear polyhedra were always present in at least some of the infected tissue culture cells. The most rapid formation of polyhedra occurred in the healthiest cells in which the inclusions frequently began to appear within 24 hours after inoculation. Within 48 hours most of the infected cells showed polyhedra, although they usually increased in size and number during and on following days. Figure 179 shows the difference in appearance between noninfected cells and infected cells. The latter are hypertrophied, and nearly all contain polyhedra varying in size and number. When the cells die the polyhedra are frequently liberated and occur freely in the degenerated culture.

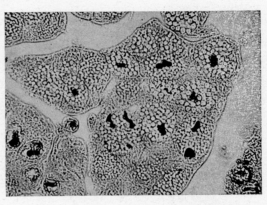

Fig. 178. Silkworm jaundice polyhedral bodies within the nuclei of fat cells; also chromatin remains; X540. (Courtesy Dr. R. W. Glaser.)

THE WILT DISEASES

One of the first and best known of the wilt diseases [1] is that affecting the gypsy moth larva, *Porthetria dispar*. Approximately 35 other species of insects have been reported to be affected by the same or very similar diseases (see Chapman and Glaser, 1915, and the list on page 414).

There seems to have been no record of wilt diseases of the gypsy moth in America prior to 1900. It was first studied in this country by Reiff (1909, 1911) and Jones (1910). In 1912 Glaser and Chapman described as the etiologic agent of the disease a small, motile coccus (*Gyrococcus flaccidifex*). In 1913,

---

[1] The term "wilt disease" is more and more being used as a general term to designate any virus-caused disease of insects.

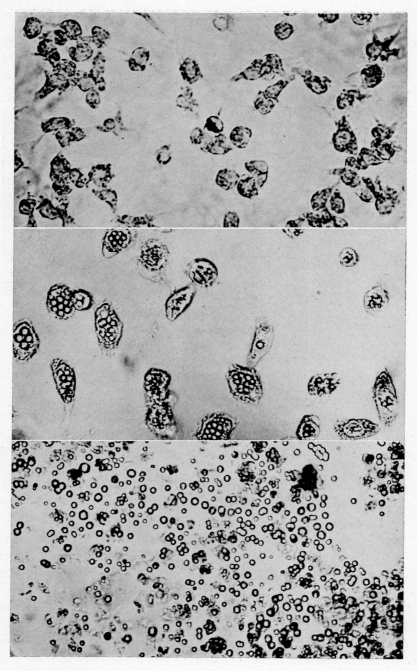

Fig. 179. The cultivation of the virus of jaundice in silkworm-tissue cultures. (Top) Normal 6-day old culture. (Center) A culture 5 days after infection with the polyhedral virus. (Bottom) A culture 8 days after infection with the polyhedral virus. (From Trager, 1935c.)

however, these workers corrected their mistake by proving that the disease was caused by a filterable virus. Since this discovery they have made extensive observations and contributions concerning the wilt diseases.

Symptoms and Other Characteristics. An insect suffering from wilt disease soon stops eating and becomes sluggish in its activity, usually crawling upward toward the top of a plant or other object where it remains motionless. After several hours a dark, foul-smelling liquid oozes from its mouth and anus. The caterpillar becomes extremely flaccid, the legs losing their grip one after the other until finally, with only one or two of its false feet or with its anal claspers, it hangs limp and dead. The black skin may now be easily ruptured and a dark, thin liquid having an extremely offensive odor runs out.

Age but not sex seems to have some influence on the susceptibility of the insect to wilt diseases. The virus usually kills older larvae more rapidly than younger ones, although all larval instars are susceptible, as are young pupae. Adults, however, appear to be immune. Thus far the eggs have not been shown to be destroyed by the virus.

Factors predisposing to the wilt diseases include starvation, poor food, high temperatures, and other environmental conditions.

The incubation period varies, the time from inoculation by feeding to death being from 13 to 29 days, an average of 21 days.

The symptoms of wilt disease in the larvae of other insects (tent caterpillar, armyworm, tussock moth caterpillar, oak caterpillar, range caterpillar, etc.) appear similar to, if not identical with, those just described for the gypsy moth caterpillar.

Pathology. According to Chapman and Glaser (1916), the pathology of caterpillars dead or dying of wilt does not vary with the age of the insect. In general, the tissues of caterpillars dead of the infection are completely disintegrated and polyhedra may be found throughout. These polyhedral bodies originate in the nuclei of the tracheal matrix, hypodermal, fat, and certain blood cells; the nuclei of the blood cells and tracheal matrix appear to be the first affected. Some of the blood cells may rupture, liberating the polyhedra into the blood. The alimentary canal is usually the last organ in the body to disintegrate.

Microscopically, the nuclei of infected cells contain many minute, violently dancing granules.

The Virus. In 1913 Glaser and Chapman were able to prove that the exciting cause of the wilt disease in gypsy moths is a filterable virus. As with most of the other virus diseases of insects, these workers considered the polyhedra to represent reaction products of the disease and not the etiologic agent.

The virus passes through a Berkefeld N candle and with the dark-field microscope tiny dancing granules may be observed in the filtrate which are not present in filtrates of uninfected caterpillars. These granules appear identical with those seen in infected tissues.

Virus-containing blood loses its virulence when held at 60° C. for 15 to 20 minutes. When dried at room temperature it remains infectious for at least two years and may be kept for at least six months in 98 per cent glycerol without losing its virulence. The virus apparently resists putrefaction for an indefinite time but is inactivated by nine hours' exposure to sunlight or when placed in 80 per cent alcohol for 15 minutes. When in 5 per cent phenol it takes about three weeks for it to lose its infectiousness.

WIPFELKRANKHEIT

Wipfelkrankheit, considered as a wilt disease by most American workers, has also gone under the names of "flacherie" and *Schlaffsucht.* The German name *Wipfelkrankheit* is taken from the fact that the diseased caterpillars of the nun moth (*Lymantria monacha*) migrate to the tops (*Wipfeln*) of the trees or branches on which they happen to be located. Here they gather and die in small or large blackened masses.

The Disease. In nearly all respects *Wipfelkrankheit* is very similar to, if not identical with, the wilt diseases just described. The symptoms, in general, are the same. Infected nun moths show a loss of appetite, and as has already been indicated, tend to migrate to the tops of the trees where they hang by their prolegs and die.

The Virus. During *Wipfelkrankheit* epidemics in nun moths in 1889 and 1892, both Hofmann (1891) and Von Tubeuf (1892a,b) isolated bacteria (*Bacillus "B"* and *Bacterium monachae,* respectively) which they considered as the etiologic agents. Eckstein (1894) considered the organisms of Hofmann and of Von Tubeuf to be identical, although in 1911 Von Tubeuf changed his opinion and postulated that the disease occurs when a certain variety of intestinal bacteria is present in large numbers.

Polyhedral bodies were seen by Hofmann and by Von Tubeuf, who considered them to be reaction products. That they originate within tissue cells was observed by Wachtl and Kornauth (1893).

However, it remained for Wahl (1909 to 1912) to demonstrate that the disease was caused by a virus not bacterial in nature. This worker observed the seat of polyhedra formation to be the nuclei of certain cells, especially those of fat tissue and blood. Escherich and Miyajima (1911) concerned themselves particularly with the polyhedra, which they thought carried the virus. Knoche (1912) attached etiologic significance to the polyhedra, considering them a microsporidian protozoan. That the polyhedra are not in themselves the agents of the disease was concluded by Komárek and Breindl (1924), who maintained that they merely contained the virus. Occasionally the virus occurred free, in which case these workers were able to pass the virus through Berkefeld filters.

In a dried state the virus of *Wipfelkrankheit* may retain its virulence in glycerol for at least five days. The virus also withstands putrefaction.

Practical Value of Wilt Diseases. Each year the New England states of our country lose millions of dollars through the destructive habits of the gypsy moth, *Porthetria dispar,* which is a serious pest of shade trees. Since this imported insect is so far confined principally to the New England states, great effort is continuously being made by our entomologists to keep it from spreading farther. In addition to the insect parasites and sprays, much of the credit for its control must go to epidemics of wilt disease which take a tremendous annual toll.

Experiments have been conducted to determine the effect of virus artificially introduced into given areas, but the results, in general, have not been revealing. Apparently, the polyhedral disease is present wherever there is a large number of the insects and these natural epidemics seem to be just as effective as those which are supposedly assisted by the virus artificially introduced.

Natural epidemics of wilt disease have been observed in other destructive insects, including the eastern tent caterpillar (*Malacosoma americana*), forest tent caterpillar (*Malacosoma disstria*), armyworm (*Cirphis unipuncta*), tussock moth (*Hemerocampa leucostigma*), oak caterpillar (*Phryganidia californica*), and the range caterpillar (*Hemileuca oliviae*). Most of these outbreaks have occurred in the northeastern states of this country. In 1945, however, the writer observed typical polyhedral virus diseases affecting caterpillars of the alfalfa butterfly, *Colias eurytheme,* collected in California by R. F. Smith, and larvae of the hemlock looper, *Ellopia fiscellaria lugubrosa,* collected by R. L. Furniss in the state of Washington. The diagnoses were confirmed by laboratory examinations. The collectors of these specimens had noticed the high mortality in these species in nature for several years prior to these findings. No doubt unrecorded outbreaks of virus diseases in lepidopterous larvae have been seen in other parts of the country. Several outbreaks in other insects (Sphingidae, Saturnidae, Noctuidae, etc.) have been observed in Europe.

OTHER POLYHEDRAL DISEASES

In 1936 Paillot reported the natural occurrence of two polyhedral diseases in the larvae of *Vanessa urticae* and *Euxoa segetum,* respectively. In the case of the first insect the epidemic was severe, the larvae being unable to become chrysalids. In the case of *Euxoa segetum* no true epidemic was observed, only a small number of individual insects having been found diseased.

Infected *Vanessa urticae* larvae are difficult to distinguish from healthy larvae in the early stages of the disease. Just before death their blood is turbid and milky in appearance. A microscopic examination of the blood reveals numerous characteristic small bodies. When the larvae die they liquefy very rapidly. According to Paillot, the virus multiplies only in the hypodermis, peritrachea, blood, adipose tissues, and sometimes in the genital capsule. The virus infects chiefly the nucleus and destroys the chromatic substance, the nucleolus, and the nuclear fluid, and polyhedra appear in their places. The disease is very contagious and may be transmitted by the oral route or by the direct inoculation of infected blood. The incubation period is from four to five days.

The infected larvae of the cutworm, *Euxoa segetum,* are also very difficult to distinguish from healthy ones. The blood is not very turbid and contains few polyhedra. The virus multiples in the adipose, hypodermal, and tracheal cells, having the greatest affinity for the last site where apparently the first lesions of the disease appear. The polyhedral bodies are larger than those in the *Vanessa* disease, being 3 to 5 microns in diameter. They are usually triangular in shape. There is practically no mortality from the disease and the morbidity, very low compared with other polyhedral diseases, is less than 1 per cent in regions where it has been found.

THE NONPOLYHEDRAL DISEASES

SACBROOD

The only nonpolyhedral virus disease of insects that has been studied to any extent is sacbrood of honeybees (*Apis mellifera*). Although the disease was probably recognized by beekeepers as early as 1857, any discussion of the characteristics of the disease and its virus will necessarily have to concern itself with the relatively recent investigations of White (1913, 1917). This worker was the first to make a careful scientific study of the disease, which he named "sacbrood," as suggested by the saclike appearance of the dead larvae at the time the beekeeper is most likely to see them. Very little additional information has been forthcoming since White published his observations, which will briefly be reviewed here.

Symptoms. Sacbrood is an infectious disease which does not attack adult bees, but rather the larvae and less frequently the pupae. In general, it is a rather benign and insidious disease somewhat transient in character. Rarely does it kill off entire colonies, but the aggregate loss of individual bees is great. The death of individual bees weakens the colony and entails considerable economic loss.

Most of the larvae die during the last two days of the 4-day prepupal period, although many deaths occur during the first two days of the period as well. Since the larvae are motionless during this period, it is difficult to detect sick individuals. Usually, the first symptom observed is the presence of obviously dead brood.

The beekeeper's attention is usually drawn to the irregular appearance of the brood nest. Almost invariably the infected brood dies in capped cells, but frequently the adult bees remove the caps from the cells containing the dead individuals and sometimes the caps are punctured. The dead larvae are usually found lying extended lengthwise with the dorsal side on the floor of the cell.

Post-Mortem Appearance. Larvae dying of sacbrood change in color from bluish-white to yellow and then to brown a few days after death. As the process of decay continues, the shade deepens until in some instances it appears almost black. During this period of decay the body wall toughens, per-

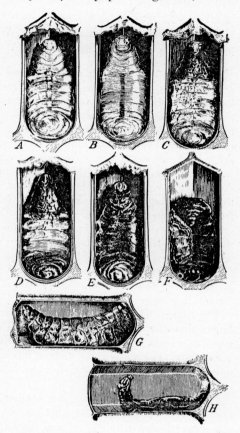

Fig. 180. Symptoms of sacbrood. *A*, ventral view of a healthy larva at the age when death usually occurs from sacbrood; *B–F*, stages in the decay and drying of larvae dead of sacbrood, ventral views; *G*, lateral view of larva recently dead of sacbrood; *H*, lateral view of scale. (From Burnside and Sturtevant, 1936; U.S.D.A. Circular 392.)

mitting it to be removed intact from the cell. The contents of the decaying larvae are watery and granular in appearance. However, if the dead larva is allowed to remain in its cell, the watery contents evaporate, distorting it into a dry, wrinkled form which has been termed "scale." It should be pointed out that all larvae dead of sacbrood do not undergo the same changes in appearance and the description which would fit them one day may not do so the next. For convenience, however, White divided the continual change in appearance into five arbitrary stages.

First stage: Larva is a light yellowish color. The posterior and lateral margins are still deeply notched and often appear transparent which is due to the watery-looking fluid beneath the cuticular portion of the body-wall. The cuticle is less easily broken at this time than in the healthy larva. When the body wall is broken the fluid tissue mass flows out and is less milky in appearance than that from a normal larva. The granular contents of the sac, which is marked in later stages of decay is already noticeable. This granular appearance is due chiefly to fat cells.

As determined by feeding the remains, emulsified in syrup, to healthy larvae, the dead larva is particularly infectious at this stage. The importance of this fact is that it is during this stage that the dead larva is frequently removed piecemeal from the cell by the workers.

Second stage: The color has changed from yellow to one of a brownish tint, though some traces of yellow remain. The ridges and furrows, representing the segments, are less pronounced. The subcuticular fluid at the lateral and posterior margins has increased in quantity. The decaying contents consist of a brownish, granular mass suspended in a watery fluid. The remains of the larvae at this stage are still infectious in some instances.

Third stage: The color of the larva is brown, the anterior third being a deeper brown than that of the other two thirds. The cuticular sac is very tough and its granular content is brownish in color and is suspended in a small quantity of clear watery fluid. The status of the virus in this stage has not been definitely settled but tests indicate that it is probably dead and the larvae in this condition are not infectious.

Fourth stage: This stage is characterized by the marked evidence of drying present. The subcuticular fluid is no longer present and the color of the larva is a still deeper brown with the anterior third the darkest. The decaying tissue mass still appears granular and the contents are paste-like in consistency. The larval remains are not infectious.

Fifth stage: In this last stage also the larva is noninfectious. All of its moisture has been lost by evaporation, leaving the dry, mummy-like remains or "scale" which can easily be removed intact from the cell (fig. 180H). Out of their cells the scales vary markedly in appearance. Usually the anterior third is a deeper shade of brown than the other two thirds. The dorsal side of the middle and posterior thirds is shaped to conform to the floor of the cell, being in general convex, with a surface that is smooth and polished. The margin is thin and wavy. The anterior third and the lateral sides of the middle and posterior thirds being turned upward, the ventral surface being concave, and the posterior side being convex, the scale in general represents a boat-like appearance and could be styled "gondola-shaped."

Histopathology. The greater portion of the body of a larva dead of sacbrood consists of fat tissue made up of relatively large fat cells which are irregular in outline and have irregular-shaped nuclei. The cells contain more or less spherical black bodies which Glaser (1928) says probably represent the so-called protein bodies found within the fat cells of insects prior to and during metamorphosis. According to White, these fat cells are the chief cause for the granular appearance of the contents of larvae dead of sacbrood. The granular aspect is enhanced, however, by other cellular elements such as oenocytes, and various tissue cells.

Between the molt skin, which is at a considerable distance from the hypodermis, and another cuticula lying near the hypodermis is an area filled with a watery fluid. The fact that this fluid coagulates in the preparation of histologic sections indicates that it is not pure water. The molt skin is the sac which encloses the decaying larval mass, although the cuticula mass assists in this capacity somewhat.

Thus, one of the chief diagnostic signs of sacbrood, the granular mass suspended in a watery fluid, can be more clearly understood.

The Virus. The activities of the virus of sacbrood may be influenced by certain predisposing causes. As we have already stated, adult bees do not appear to be susceptible to the disease and pupae are rarely infected. The sex of the larva seems to make little difference in susceptibility. Although it has not been proved that the queen larva is susceptible, both worker and drone larvae are. No complete immunity against the disease is known to exist in any race of bees commonly kept in the United States. Sacbrood appears to be geographically widespread (Germany, England, Switzerland, Denmark, Australia) and hence climate may not be an important predisposing factor. The spring of the year seems to be the optimal season for sacbrood, although it may appear at any season of the year at which brood is being reared. Neither the quality nor quantity of food predisposes the insect to the disease.

In 1913 White showed the exciting cause of sacbrood to be a filterable virus. He also found that a single larva recently dead of the disease was sufficient to infect and kill at least 3,000 healthy larvae within one week. On the other hand, the virus is also rather easily killed even in nature. Suspended in water, the virus is destroyed in ten minutes at a temperature of 59° C. In honey or glycerol it is destroyed in ten minutes at 70° to 73° C. It withstands room temperature for approximately three weeks. Dried virus is destroyed by the direct rays of the sun in from four to seven hours; when suspended in water the sun kills it in from four to six hours and in honey, from five to six hours. When shielded from direct sunlight, the virus suspended in honey at room temperature during the summer remains virulent for slightly less than one month.

In the presence of fermentative processes taking place in a 10 per cent cane sugar solution at room temperature, the virus is destroyed in about five days—the same period of survival as in a 20 per cent honey solution at outdoor temperatures. The virus remained virulent for about ten days in the presence of putrefactive processes. The virus withstands 0.5, 1.0, and 2.0 per cent aqueous solutions of phenol for more than three weeks.

Transmission. White (1917) has shown that sacbrood may be induced experimentally in larvae by the virus being added directly to their food or by its being mixed with syrup fed to them. This would indicate that whenever the food or water used by the bees contains the active virus, the larvae are likely to become infected.

The tendency of adult bees to remove piecemeal diseased or dead larvae from the cells would appear to be a likely means of disseminating the virus. If the removed fragments were fed to healthy young larvae within a week, sacbrood would very probably result. It is not known for certain, however, if feeding these fragments to young larvae is frequently done. If this were the case, it would seem that the disease would increase more rapidly than it usually does.

Other possible means of transmission exist, but none of them have been conclusively proved. It is possible, for instance, that if infectious material reaches the bees' water supply, some of the virus might be returned to the hive and given to the healthy larvae. Stray bees drifting from infected colonies to healthy ones might be considered as possible transmitters, but White considers this method of transmission unlikely.

OTHER NONPOLYHEDRAL DISEASES

The natural occurrence of two virus diseases in the larvae of the cutworm *Euxoa segetum* was reported by Paillot in 1936. He termed them "pseudograsserie 1" and "pseudo-grasserie 2."

Pseudograsserie 1, or as we shall designate it, pseudojaundice 1, was first observed by Paillot near Lyons, France. He was able to distinguish the sick larvae from the healthy ones by the white color of the ventral region of the former's body. This is due to the porcelain-white color of the adipose tissue which in normal larvae is more or less transparent. The blood of diseased insects is usually turbid because of small coccoid granules. These microscopic granules, which Paillot believes represent the virus, are also present in the cells of the adipose tissue from which they are liberated into the surrounding fluids. Stained with fuchsin they appear light red in color and are from 0.2 to 0.3 microns in diameter. Paillot found the Fontana-Tribondeu spirochete stain best for staining the granules; Giemsa stain was of little value.

The granular elements seem to multiply first in the cytoplasm of the fat cells. Soon the nucleus hypertrophies, the chromatin loses its granular aspect and forms masses and clumps, and eventually the fat cell resembles a homogeneous structure or one composed of very fine granules according to the method of fixation used. In diseased larvae the adipose tissue proliferates more than it does in normal larvae. The cells which are not attacked frequently divide rapidly. At various places in the adipose tissue follicular nodules are formed resembling tubercules in mammalian tuberculosis, although the structure is entirely different.

Paillot was unable to reproduce the disease by introducing diseased blood into the general cavity of healthy larvae or by feeding fluid rich in virus bodies. He therefore concludes that the disease is not very contagious and that it probably is transmitted from one generation to the next through the egg.

Pseudojaundice 2 was observed by Paillot (1936) in Saint-Genis-Laval, France. The symptoms in *Euxoa segetum* are very similar to those of pseudojaundice 1 except that in disease No. 2 the body wall is opaque, unlike that in disease No. 1. Pseudojaundice 2 is a disease of the adipose body and the tracheal and hypodermal cells. The histopathology is similar to that in disease No. 1, and most of the nuclear area in the adipose, as well as in the tracheal and hypodermal cells, is filled with the virus granules.

Although there is considerable proliferation of the cells in pseudojaundice 2, there is no formation of follicular nodules as in disease No. 1. In the adipose tissue the parasitized cells proliferate as do those which are not parasitized. The latter, however, return to the young embryonic state, especially when the disease progresses slowly and is very much advanced. Paillot believes that this phenomenon resembles the neoplasms or malignant tumors in the higher vertebrates, although the proliferation of the cells does not reach the extent seen in neoplasms.

The disease is not very contagious and healthy larvae placed with diseased larvae do not contract the disease. Actual epidemics do not occur. However, the disease is more contagious than pseudojaundice 1. It may be transmitted by the inoculation of infected blood and occasionally by the ingestion of fluids containing the virus.

Paillot has observed pseudojaundice 2 in association with the polyhedral disease of *Euxoa segetum* mentioned in the preceding section. The association of these two viruses does not modify the respective affinity of each for the cells. The presence of the polyhedra seems to hypertrophy the nucleus of the new-formed cells, and the course of the pseudojaundice is the same as though it were alone.

II. BIOLOGIC RELATIONSHIPS BETWEEN ARTHROPODS AND VIRUSES CAUSING DISEASES OF ANIMALS AND PLANTS

A. VIRUS DISEASES OF MAN AND ANIMALS

Compared to the number of plant disease viruses transmitted by insects, surprisingly few animal disease viruses have arthropod vectors. However, some cases, such as the mosquito transmission of yellow fever virus, are classic examples of the relation of insects to the health and well-being of man and animals. We shall discuss briefly some of these.

Yellow Fever. Josiah Nott in 1848 expressed the belief that mosquitoes were in some way responsible for the occurrence of yellow fever. In 1854 the French physician, Louis Daniel Beauperthuy, although a believer in the telluric origin of yellow fever, claimed that it was transmitted by mosquitoes. Almost forty years later, just before the turn of the century, Carlos Finlay advanced the hypothesis that the disease was carried by the mosquito now known as *Aëdes aegypti*. This fact was proved beyond doubt by the illustrious investigation of the American Army Yellow Fever Commission composed of Reed, Carroll, Lazear, and Agramonte.

According to many authorities (e.g., Hargett, 1944) two epidemiologic types of yellow fever are now recognized: (1) that transmitted by *Aëdes aegypti* and (2) that transmitted by mosquitoes other than *Aëdes aegypti*. The first is commonly referred to as the "domestic type" in which man is the reservoir and *Aëdes aegypti* the vector. This type may be subdivided into urban and rural forms depending on the place where it occurs. The second type is divided into the "jungle" and the "Nubian" forms. Jungle yellow fever has been defined by Soper (1936) as "yellow fever occurring in rural, jungle and fluvial zones in the absence of *Aëdes aegypti*." The maintenance of jungle yellow fever is independent of man and relies on certain jungle-dwelling vertebrates (reservoirs) and mosquitoes (vectors). Man is infected when he happens to be fed upon by these mosquitoes. Nubian yellow fever is that form which has been observed in the Nuba Mountains of Africa and in which man is the vertebrate host and certain aëdines other than *Aëdes aegypti* appear to be the main vectors.

Although it was believed at first that only one species of mosquito (*Aëdes aegypti*) carried the yellow fever virus, today approximately twenty species are known to be capable of transmitting the agent, at least on the basis of experimental results. In addition to *Aëdes aegypti*, the virus has been recovered from the following mosquitoes caught wild: *Aëdes leucocelaenus, Aëdes simpsoni, Haemagogus capricorni,* a pool of *Sabethini,* and possibly from *Psorophora ferox.*

Directly after the capture of *Aëdes leucocelaenus* and *Haemagogus capricorni,* rhesus monkeys have been infected with yellow fever by their bites.

The following mosquitoes have been found capable of transmitting the yellow fever virus experimentally by bite:

Aëdes aegypti
Aëdes africanus
Aëdes albopictus
Aëdes fluviatalis
Aëdes geniculatus
Aëdes luteocephalus
Aëdes metallicus
Aëdes scapularis

Aëdes simpsoni
Aëdes stokesi
Aëdes taeniorhynchus
Aëdes taylori
Aëdes triseriatus
Aëdes vittatus
Culex thalassius
Taeniorhynchus africanus

The association of the name *Stegomyia fasciata* with the yellow fever mosquito is a strong one. The insect has also been known by the names *Aëdes calopus* and *Aëdes argenteus*. However, the correct scientific name apparently is *Aëdes aegypti* (fig. 181). This mosquito is distributed within the limits of 40° N. and 40° S. latitude, does not thrive in dry hot climates, is diurnal, lives in the vicinity of man's habitations, and breeds in rain barrels, water-filled tin cans, and other artificial containers. Eggs are laid in masses of 30 to 50 at intervals of several days each and hatch in a minimum of two days. The larvae prefer moderately clean water and feed principally on bacteria. From egg, through larva and pupa, to adult, the entire cycle usually takes place in 10 to 15 days. Males rarely live longer than 50 days, but females have been observed to live for more than four months in the laboratory.

The blood of a yellow fever patient is infectious to mosquitoes only during the first three days of illness. Adult mosquitoes are able to transmit the virus of yellow fever any time from 10 or 12 days after infection to the end of their lives. This depends upon the temperature, however. At 37° C. the mosquito may become infective in 4 days; at 21° C. 18 days may be necessary.

Since we are here primarily concerned with the biologic relationships between insects and viruses, we must be aware of the dearth of information on such relationships between the virus of yellow fever and its vectors. Very little study has been made of the effects of the virus upon the mosquito but apparently there are none, since none of the insect's life processes nor its longevity appears to be adversely affected. It appears that most of the mosquito's tissues are infected and retain the virus the rest of the insect's life. Davis and Shannon (1930) demonstrated the presence of the yellow fever virus in the head, thorax, and abdomen of mosquitoes before the bites were infective. The legs, ovaries, salivary glands, midgut, and hindgut of *Aëdes aegypti* also contained virus, although the mouth parts and hemocoelic fluid were negative. Occasionally the dejecta of infected mosquitoes were infectious.

Fig. 181. The yellow fever mosquito *Aëdes aegypti*. (From Hegner, Root, Augustine, and Huff's *Parasitology*, 1938, D. Appleton-Century Co., New York. After Howard.)

The immediate fate of the ingested virus in the insect's body is a point for argument. Some workers (Gay and Sellards, 1927; Davis, 1932; and Whitman, 1937) maintained that multiplication of the virus does occur. Later, however, Davis apparently reversed his opinion when he, Frobisher, and Lloyd (1933), and Davis (1932) were unable to recover as much virus from the mosquitoes as was present shortly after the insects were infected. They explained the incubation period as the time required for the mechanical transportation of the virus to the salivary glands. Davis, Frobisher, and Lloyd estimated that the average mosquito, immediately after engorging on highly infectious blood with a titer of one billion lethal doses of virus per cubic centimeter, contained between one and two million lethal doses. During the two weeks after the infectious blood meal the titer was reduced to not more than 1 per cent of that in the freshly fed mosquitoes. At later periods the titer was somewhat higher but these workers believed that this increase did not represent multiplication of the virus but merely "an increase of extracellular virus and of that easily freed by grinding to a titratable form." Whitman (1937) expressed the belief that the virus is capable of multiplying in the body of *Aëdes aegypti* and that more virus can be recovered from the mosquitoes following an incubation period than immediately after feeding. Following the ingestion of infected blood the virus content falls for several days, reaching a minimum during the first week. It then rapidly increases until quantities of virus greater than those previously present can be detected. Whitman felt that the maximal titer obtainable following incubation might be lower than the artificially produced high titer apparent after the ingestion of the fully virulent monkey blood. Beyond a certain point the growth requirements of the virus might surpass the supporting ability of the mosquito's cells. Hence there would be a much greater supply than demand.

A life cycle of the yellow fever virus in mosquitoes does not seem to be present. It is ruled out, in part, by the fact that injection of mosquitoes into monkeys at any time or interval after an infectious blood meal produces yellow fever. A supply of virus ample to produce the disease appears to be continuously present. Davis (1934) fed infected mosquitoes on mice which were immediately killed and extracts titrated. The results indicated that each mosquito during the act of feeding injected at least 100 infective doses of virus. Probably about 1 per cent of the mosquito's total virus content was injected at a single biting.

Transovarial transmission of the virus is not known to occur (Philip, 1929, and others). Only rarely does transference of the virus from infected female to normal male mosquitoes occur. It is interesting, however, that stage-to-stage transmission may take place in *Aëdes aegypti*. Whitman and Antunes (1938) were able to infect this mosquito in the larval stage, and the resultant male and female adults were infected. Earlier, Davis and Shannon (1930) were unable to transmit the virus by adults bred from larvae which had consumed large numbers of infected mosquitoes.

As it exists in *Aëdes aegypti,* the virus of yellow fever is filterable through Berkefeld N filters when suspended in normal monkey serum (Sawyer and Frobisher, 1929). Apparently the virus, when suspended in physiologic saline, will not go through the filters.

Besides mosquitoes, other arthropods have been used in transmission experiments. Davis (1933a), in a search for possible accessory vectors of the yellow fever virus in Brazil, found *Triatoma megista* to take up the virus during an infective blood meal, but except after an interrupted blood meal it was unable to transmit yellow fever by biting. The virus appears to survive in this insect about a week. Philip (1930) was unable to produce infection in monkeys by the bites of bedbugs (*Cimex lectularius*) previously given an infective blood meal or by inoculating triturated specimens directly. Working with *Cimex hemipterus,* Kumm and Frobisher (1932) found the virus of yellow fever to die off so rapidly in the insect that the disease could not be transmitted by injection of the triturated bugs later than the second day after their infecting meal. Davis (1933b) observed that the yellow fever virus may remain alive in ticks for considerable periods of time. Injection into monkeys of the following ticks produced yellow fever the indicated number of days after infective blood meals: *Argas persicus*, 6 days; *Amblyomma cajennense*, 15 days; *Rhipicephalus sanguineus*, 23 days; *Boöphilus microplus* (larvae), 10 days. There was no indication that the virus was transmitted from one generation to the next via the egg. The ticks did not transmit the virus by biting, and it would seem that in order to transmit the disease the ticks would have to be crushed on the skin of a susceptible person.

While studying the epidemiology of jungle yellow fever in eastern Colombia, Roca-Garcia (1944) isolated three neurotropic viruses immunologically distinct from yellow fever. Two of the viruses were isolated from lots of the mosquito *Anopheles boliviensis* and the third from the mosquito *Wyeomyia melanocephala;* all were wild caught mosquitoes. These viruses also seemed to be immunologically distinct from the encephalitis viruses and the distemper virus.

Dengue. Dengue, or breakbone fever, may occur in practically every part of the tropical and subtropical world, although occasionally it is also found in temperate regions. Between 500,000 and 600,000 cases of the disease occurred during an epidemic in Texas in 1922. The mortality from dengue is very low, and death usually occurs only when complications set in. Because of its debilitating effects, however, the disease is of considerable importance, especially economically. It is caused by a filterable virus which is present in the blood stream during the first three days of fever.

The virus of dengue appears to be transmitted principally by mosquitoes. This was first demonstrated by Graham in 1902. In 1923 Chandler and Rice were able experimentally to transmit the infection with mosquitoes (*Aëdes aegypti*) fed on patients. Although these workers report transmission in 24 to 96 hours after feeding, others (Shule, 1928; Siler, Hall, and Hitchens, 1926) found 8 and 11 days to be the necessary incubation periods. Besides *Aëdes aegypti,* other mosquitoes have been incriminated as possible vectors of dengue. These include *Aëdes albopictus, Aëdes scutellaris hebrideus, Culex fatigans* (or *C. quinquefasciatus*), *Desvoidea obturbans,* and probably others.

The biologic relationships between the virus and its insect vectors have been little studied. As we have already indicated, a period of incubation apparently is necessary before the insect can transmit the agent. There is no indication, however, that this represents cyclic development. According to Simmons, St. John, and Reynolds (1931), a possibility exists that the virulence of the virus may lessen upon serial passages through mosquitoes, accomplished by feeding normal *Aëdes aegypti* on infected mosquitoes triturated and suspended in normal blood. These workers also found that dengue virus remains alive in its living mosquito and animal hosts for considerably longer periods than in the dead tissues of these hosts. Some evidence suggests that the virus may multiply in its insect hosts. Blanc and Caminopétros (1929) found that naturally or experimentally infected stegomyia mosquitoes can live at least 200 days under favorable conditions. They remain infective as long as the temperature is above 18° C. but lose infectivity when the temperature drops below this point. Virulence is regained when the temperature returns to above 18° C. These workers also observed that mosquitoes can transmit dengue for at least 174 days after being infected, which indicates that they may carry the virus from one year to the next. Dengue virus is not known to be transmitted through the egg.

The Encephalitides. For an epidemiologic designation of this group of viruses, Hammon, Reeves, and Gray (1943) have suggested the term "the arthropod-borne virus encephalitides." These include the St. Louis virus, western equine virus (and probably the eastern equine virus), Japanese B virus, and the Russian spring-summer encephalitis virus. With the exception of the last, which is tick-borne (*Ixodes*), most of the vectors of this group of viruses are mosquitoes. A Venezuelan type of virus has been described, but its arthropod vector has not been determined with certainty.

In 1933 Kelser demonstrated the ability of *Aëdes aegypti* to transmit the virus of equine encephalomyelitis from infected to normal guinea pigs and to horses. The mosquitoes were able to bring about the disease as early as 6 days after an infectious feeding and they remained infective for at least 36 days. Later (1938) this worker found *Aëdes taeniorhynchus* capable of transmitting the western type of the virus experimentally. Kelser states that "transmission is not mechanical but occurs after multiplication, maturation or, less probably, cyclic changes of the virus within the mosquito." Other workers have found other species of *Aëdes* mosquitoes capable of transmitting the virus of equine encephalomyelitis, including *Aëdes albopictus, A. sollicitans, A. dorsalis, A. lateralis, A. geniculatus,* and *A. vexans.* Merrill, Lacaillade, and Ten Broeck (1934) found a 1,000- to 10,000-fold increase of the virus within the bodies of both *Aëdes aegypti* and *Aëdes sollicitans.* Merrill and Ten Broeck (1934, 1935) showed that the virus could be carried by serial passage through *Aëdes aegypti,* and such treatment does not appear to alter the serologic characteristics or the virulence of the virus. They found the virus to be distributed throughout the body of the mosquito including the legs, head, thorax, abdomen, and body fluid. The mosquito does not in any way seem to be adversely affected by infection with the virus. Larvae do not take up the virus when it is added to water in which they are living, and eggs from infected females apparently do not contain the virus. There is no transmission from male to female or from female to male. It is interesting that *Aëdes aegypti* appears to be unable to transmit the eastern type of equine encephalomyelitis virus experimentally. Merrill and Ten Broeck (1935) suggest that this is probably due to the inability of the virus to penetrate the mosquito's intestinal mucosa since, when the virus is inoculated into the body cavity, it persists and transmission experiments are positive. These workers demonstrated by inoculation that the virus is present in the bodies of infected mosquitoes for the duration of life. However, they were able to transmit the virus for a period of only slightly more than two months.

Syverton and Berry (1941) found that under experimental conditions the wood tick, *Dermacentor andersoni,* can act as a transmitting agent and as a reservoir host of the western type of equine encephalomyelitis virus. They were thus able to transmit the virus to guinea pigs and gophers by any active stage (larva, nymph, adult) of the tick. The virus may pass through the egg from generation to generation in the tick as well as from stage to stage. Very little has been worked out as to the biologic relationships between the virus and its arachnid host. Syverton and Berry believed that infected ticks may be stimulated to complete their life cycles in a shorter period of time than do normal ticks. Blattner and Heys (1941) reported the experimental transmission of St. Louis encephalitis by the tick *Dermacentor variabilis.* They also observed (1944) that this virus may be transmitted transovarially to the third generation of ticks and that all stages of the tick, including the eggs, remained infective after being stored for a period of ten months. Possibly of considerable epidemiologic importance is the isolation of the virus of St. Louis encephalitis by Smith, Blattner, and Heys (1944) from chicken mites (*Dermanyssus gallinae*) collected in nature in the St. Louis area during a nonepidemic period. This suggests that the mite may be an important interfowl vector. The virus of western equine encephalomyelitis has also been isolated from chicken mites (Sulkin, 1945).

Hammon and his co-workers have shown the important role played by *Culex tarsalis* in the transmission of certain types of encephalitis. Hammon, Reeves, and Gray (1943) present evidence indicating that in at least one human epidemic area both St. Louis and western encephalitis were mosquito-borne and that the source of the mosquito infection was a large inapparent reservoir among vertebrates, particularly domestic fowl. Both viruses have been found in naturally infected *Culex tarsalis* and in *Culex pipiens.* The latter, however, apparently does not transmit the western equine virus. *Culex pipiens* was probably a vector during the great 1933 St. Louis epidemic. *Culiseta inornata* has been found naturally infected with western equine virus and is capable of transmitting it. *Culex coronator* and *Theobaldia incidens* are also capable, at least experimentally, of transmitting the virus. In the laboratory Hammon and Reeves (1943) transmitted the St. Louis virus by the following mosquitoes: *Culex tarsalis, Culex pipiens, Culex coronator, Culex quinquefasciatus* (see their footnote 5), *Aëdes lateralis, Aëdes taeniorhynchus, Aëdes vexans, Aëdes nigromaculis, Theobaldia incidens,* and *Culiseta inornata.* Although the following did not transmit the virus, they were capable of harboring it more than a few days: *Culex stigmatosoma, Psorophora ciliata,* and *Anopheles maculipennis freeborni.*

Kitselman and Grundmann (1940) isolated a western strain of the virus of equine encephalomyelitis from *Triatoma sanguisuga* collected in pastures in Kansas. This reduviid is a large bloodsucking bug known to feed upon horses. One of its natural hosts is the wood rat. Insects tested but found unable to transmit this virus include the horn fly, *Haematobia serrata,* a horsefly, *Tabanus punctifer,* the stablefly, *Stomoxys calcitrans,* and the mosquitoes, *Aëdes dorsalis, Anopheles maculipennis,* and *Culex quinquefasciatus.* In a subsequent investigation Grundmann, Kitselman, Roderick, and Smith (1943) transmitted a western strain of the virus experimentally by *Triatoma sanguisuga* from guinea pig to guinea pig but not to horses. However, the mechanism of the infection was not clear. Experimentally they infected the host by methods simulating contamination of the skin by the insect's feces. These same workers could not transmit the virus experimentally with the tick *Dermacentor variabilis.* By inoculation of the virus into the hemocoeles of the ticks, the latter became infected but would not transmit the infection by feeding. The virus could not be recovered from the progeny of such ticks.

Lépine, Mathis, and Sauter (1941) found *Triatoma infestans* capable of ingesting the virus from infected guinea pigs but unable to transmit it by bite or by dejecta of the bugs. By animal inoculation it was shown that the virus disappeared from the insects after 17 days.

The Russian spring-summer or verno-aestival encephalitis is transmitted to humans through the bite of the tick *Ixodes persulcatus,* which has been proved to be naturally infected at all stages of its development (see Chumakov and Gladkikh, 1939, and Chumakov and Seitlenok, 1940). The distribution of the virus within the tick has been shown by Pavlovsky and Soloviev (1940) to be extensive, penetrating into all organs. The virus appears to remain undiminished during at least 25 days in the gut of the tick. In the genital organs the virus content at first decreases but later increases again. High concentrations of the virus occur in the salivary glands and some is also found in the brain, although the integument of the tick is free of the agent. Transovarial transmission of the virus takes place by the viruses penetrating the walls of the ovary and thence entering the developing ovum. It appears that the virus is preserved over winter in the hibernating ticks. Experiments with *Haemaphysalis concinna* indicate that it is not as well adapted to harboring the virus as is *Ixodes persulcatus*. Levkovich and Skrinnik (1940–41) observed that when infected ticks (*I. persulcatus*) were fed on white mice a subclinical infection and an immunity to subsequent lethal doses of the virus resulted. The antibody content of the blood and the animal's resistance were proportional to the number of tick bites and length of feeding.

Lymphocytic Choriomeningitis. Although the method of the natural transmission of the virus of lymphocytic choriomeningitis is not known, it has been suggested that it may be transmitted by arthropod vectors among certain rodent reservoirs. In 1939 Shaugnessy and Milzer reported the experimental infection and transmission of the virus by the tick *Dermacentor andersoni*. They were able to demonstrate stage-to-stage and generation-to-generation transmission of the virus in the tick, but attempts to transmit it by feeding infected adults on normal animals were unsuccessful. The virus was transmitted, however, by feeding nymphs which, as larvae, had previously fed on infected guinea pigs. The tick's feces were also infectious.

In 1939 Coggeshall reported the experimental transmission of the virus of lymphocytic choriomeningitis to guinea pigs by the bite of infected *Aëdes aegypti*.

In a later investigation Milzer (1942) was able to transmit the virus by the bite of mosquitoes of this species when they were incubated at various constant temperatures ranging from 26 to 34° C. No virus was detected in mosquitoes incubated at 37° C., or at 25° C. and lower. *Culex pipiens* and *Aëdes albopictus,* incubated at 22 to 25° C., failed to transmit the virus. Bedbugs, *Cimex lectularius,* incubated at 22 to 25° C., transmitted the virus from infected to normal guinea pigs in 11 of 18 attempts at intervals ranging from 10 minutes to 85 days. Both male and female adults and young bedbugs were capable of being transmitters, although the evidence indicated that none of the bugs were able to transmit virus by bite alone. Infection occurred only when the bugs were allowed to defecate on the bitten area. In one instance transovarial transmission occurred. Neither the monkey louse *Eupedicinus longiceps* nor the mite *Atricholaelaps glasgowi* was able to transmit the virus. However, both of these arthropods were capable of retaining the virus— the louse for at least 24 hours and the mite for at least 25 days. Infection of mice and guinea pigs could be brought about by feeding these animals living infected mites in gelatin capsules.

Humphreys, Helmer, and Gibbons (1944) reported the probable isolation of a virus from *Dermacentor andersoni* ticks collected in nature in Canada. Subsequent work on this agent indicates that it is similar to, if not identical with, the virus of lymphocytic choriomeningitis. The Canadian virus on two occasions was isolated from cockroaches (*Blattella germanica*) in a laboratory in which the disease was spreading among guinea pigs (Steinhaus, 1944).

Poliomyelitis. Since before the first World War there has been a seesaw of opinion back and forth regarding the actual role of insects in the spread of the virus of poliomyelitis. Until recently very little was known concerning the biologic relationships between the virus and the insects concerned, and there is still much to be learned in order to determine definitely whether or not insects may act as more than mechanical carriers of the virus.

In 1911 Flexner and Clark showed that flies fed on infectious material retained the virus either on the surface of their bodies or in the alimentary tract for at least 48 hours. Similar results with flies were obtained by Howard and Clark (1912), who also experimented with lice (*Pediculus humanus*) and bedbugs (*Cimex lectularius*) fed on human beings or monkeys ill with poliomyelitis. With the exception of one bedbug, none of the latter insects were found to harbor the virus. Fleas collected from poliomyelitis patients have also been tested but with negative results. In 1912 Rosenau and Brues claimed the experimental transmission of the virus to monkeys by the stablefly *Stomoxys calcitrans*. Although at first these observations were confirmed by Anderson and Frost (1912), these same workers later (1913) obtained nothing but negative results, as did several other groups of investigators. This series of negative experiments more or less discouraged further research on the problem until 1937 when Rosenow, South, and McCormack observed that poliomyelitis developed in one out of three monkeys inoculated with filtrates of flies collected in nature. Since then the work of Toomey, Takacs, and Tischer (1941), Paul, Trask, Bishop, Melnick, and Casey (1941), Sabin and Ward (1941, 1942), Trask, Paul, and Melnick (1943), and others leaves no doubt that flies (Muscidae, Calliphoridae, etc.) may harbor the virus in nature. Repeated isolations of the virus from wild-caught flies have demonstrated this. The true significance or correct interpretation of these findings, however, has yet to be made.

Using Theiler's mouse strain and the Lansing mouse-adapted strain of human poliomyelitis, Bang and Glaser (1943) found that these viruses may be recovered from adult flies only when the adult itself acquires them by feeding. Attempts to recover the viruses from flies which had developed from infected larvae were unsuccessful. Theiler's virus was recovered from *Musca domestica* as long as 12 days after infective feeding, and the Lansing strain survived only 2 days in this fly. They were unable to recover the latter strain from the four other species tested: *Lucilia caesar, Muscina stabulans, Calliphora erythrocephala,* and *Sarcophaga haemorhoidalis*. The results with the

Theiler strain in these species were essentially the same as those obtained with *Musca domestica*. Rendtorff and Francis (1943) obtained results with the housefly similar to those of Bang and Glaser with the Lansing strain. The former workers observed that the virus appears to persist primarily in the abdomen of the insect. In from two to seven hours after ingestion there is a sharp decrease in the amount of detectable virus in the fly's adbomen. Although some of the virus may die off in the fly, they believed that its disappearance is probably due to the rapid excretion by the insect. Active virus was recovered from fecal and vomit spots.

Other Virus Diseases. Most of the cases in which viruses causing human and animal diseases are carried by arthropods have been little studied from the standpoint of the biologic relationships existing between the two forms of life.

Although some significant information is available, nevertheless, much scarcity of knowledge surrounds the relationships between the virus of pappataci fever and the sandfly *Phlebotomus papatasii;* Rift Valley fever and *Mansonia fuscopennata, Mansonia versicolor, Mansonia microannulata,* and the tick *Rhipicephalus appendiculatus;* fowl pox and *Anopheles maculipennis, Theobaldia annulata, Aëdes aegypti, Aëdes stimulans, Aëdes vexans, Culex pipiens,* and *Stomoxys calcitrans;* Nairobi disease of sheep and *Amblyomma variegatum* and *Rhipicephalus appendiculatus;* catarrhal fever of sheep and *Aëdes caballus* and *Aëdes lineatopennis;* louping ill of sheep and *Ixodes ricinus* and *Rhipicephalus appendiculatus;* bluetongue of sheep and *Culicoides* spp.; rabbit papilloma (Shope) and *Haemaphysalis leporis-palustris;* South African horse sickness and mosquitoes (*Aëdes, Culex, Anopheles*), *Culicoides,* and flies (*Stomoxys, Tabanus, Culicoides, Haematobia*); Semliki Forest virus and *Aëdes abnormalis;* Colorado tick fever and *Dermacentor andersoni;* and the virus of equine infectious anemia and mosquitoes (*Psorophora columbiae = P. confinnis*).

In 1926 (b) Noguchi reported the isolation from ticks (*Dermacentor andersoni*) of a virus pathogenic for guinea pigs. The virus was filterable and was maintained in special semisolid culture media for at least seven generations. Noguchi was able to transmit the virus from infected guinea pigs to ticks and in one instance was able to transmit it by tick feeding from an infected tick to a guinea pig. The true identity of this virus was never determined with certainty. Davis and Cox (1938) have suggested the possibility that it is the same filterable agent now known as *Rickettsia burneti* (*R. diaporica*), the cause of Q fever (see chapter V).

A peculiar instance of virus transmission by insects has been reported by Donisthorpe (1945). Outbreaks of smallpox among patients in a Middle East hospital were explained by the observation that ants were carrying bits of "scab" and skin from the floors of smallpox isolation wards (tents) to patients in other wards of the hospital. The ants concerned in this curious situation were not identified.

B. VIRUS DISEASES OF PLANTS

Several excellent treatises have been written on the insect transmission of viruses causing diseases of plants (e.g., those by Cook, 1935, 1936, and Leach, 1940). The subject matter concerned is immense, and it shall be our purpose here to limit the discussion to those salient points having to do with the biologic relationships existing between the plant viruses and their insect vectors. There are in the neighborhood of 150 known virus diseases of plants; insects are concerned in the transmission of most of these.

Insects Concerned. Leach (1940) has effectively discussed the virus diseases of plants according to the order of their insect vectors. In general, we shall follow the same method of treatment.

The great majority of the insect vectors of virus diseases of plants belong to five orders: Orthoptera, Thysanoptera, Homoptera, Hemiptera, and Coleoptera.

METHODS OF TRANSMISSION

As in the case of other infectious agents, plant disease viruses may be transmitted by insects in three ways: (1) external transmission, in which the virus is carried as a contaminant on external parts of the insect, (2) mechanical internal transmission, in which the virus passes through the insect's body but in so doing undergoes no development changes or multiplication, and (3) biologic internal transmission, in which the virus multiplies or develops in the insect's body (Rand and Pierce, 1920).

Another classification of the methods by which insects transmit viruses has been suggested by Storey (1939) who divides the vectors into two large groups: (1) those vectors which fail to infect any of the first few series of plants to which they are exposed but thereafter will begin and continue to infect their hosts and (2) those which infect the first series of plants but few or none of the succeeding hosts. While most of the study has been concerned with the first type of transmission, it is probable that the second type occurs more frequently in nature.

One of the most convenient methods of arranging the virus diseases of plants according to the methods of transmission is that used by Leach (1940). This author separates the diseases into the following eight groups:

1. Transmission by mechanical sap inoculation; insect transmission purely mechanical.
2. Transmission by aphids, showing group specificity; insect transmission not entirely mechanical.
3. Transmission by leaf hoppers; biological and highly specific.
4. Transmission by thrips; biological and specific.
5. Transmission by lace bugs; biological and specific.
6. Transmission by white flies; biological and specific.
7. Transmission by mites.
8. Nature of transmission obscure.

Following Leach's method, examples of each of these groups will be given on the following pages.

Examples of Insect-transmitted Virus Diseases of Plants

1. Transmission by Mechanical Sap Inoculation; Insect Transmission Purely Mechanical. Potato spindle tuber is an example of a virus disease of plants which may be transmitted mechanically and nonspecifically by insects. Smith (1937) classified the agent as Solanum virus 12. It is distributed principally in the potato-growing areas of the United States. Infected potatoes become elongated and spindle-shaped, tapering particularly at one end, and the yield is greatly reduced from the normal. Besides other means of transmission, the agent of this highly infectious disease is also transmitted by a considerable number of potato-frequenting insects. Schultz and Folsom (1923) first showed that insects could act as vectors when they demonstrated the role of aphids (*Myzus persicae* and *Macrosiphum gei*) in this capacity. Goss (1931) reported the following vectors: the tarnished plant bug, *Lygus pratensis,* the flea beetles *Epitrix cucumeris* and *Systena taeniata,* the leaf beetle *Disonycha triangularis,* grasshoppers of the genus *Melanoplus,* and larvae of the Colorado potato beetle, *Leptinotarsa decemlineata*. It is probable that in all cases transmission takes place mechanically.

Yellow dwarf of onions, caused by Allium virus 1, has been transmitted by artificial sap inoculation by more than fifty different species of aphids. The disease causes yellowing, crinkling, and dwarfing of the leaves, and the diseased plants yield underdeveloped bulbs of little commercial value. According to Leach (1940), the following species of aphids are most commonly found on onions and are probably of greatest importance in the transmission of the virus: *Macrosiphum pisi, Aphis rumicis, Aphis helianthi, Hyalopterus atriplicis,* and *Rhopalosiphum prunifoliae*. Other insects have also been reported as vectors of this agent.

Other virus diseases transmitted by mechanical sap inoculation include:

cucumber mosaic (Cucumis virus), which may be transmitted by insects such as *Aphis gossypii, Myzus persicae, Macrosiphum gei, Diabrotica vittata,* and others; tobacco mosaic (Nicotiana virus 1), which may be transmitted to some extent by aphids, especially from tomato plant to tobacco plant; and Western celery mosaic (Aphium virus 1), which can also be transmitted mechanically and nonspecifically by aphids.

2. **Transmission by Aphids, Showing Group Specificity; Insect Transmission Not Entirely Mechanical.** In this group are considered those virus diseases which may be exemplified by sugar-cane mosaic (Saccharum virus 1), and potato leaf roll (Solanum virus 14).

Sugar-cane mosaic manifests iteself by chlorotic streaks on the leaves, stunted plants, and reduced yields. The virus also affects a variety of other plants including sorghum, corn, millet, and several wild grasses. Although sugar cane is not the preferred host of *Aphis maidis,* this insect is an effective vector of the mosaic virus. The aphids breed on the wild grasses in the vicinity of the sugar cane. When these wild grasses are cut, the insects migrate to the sugar cane, carrying the virus with them. Brandes (1923) demonstrated that when the aphids insert their mouth parts into the plant, they seek out the phloem and there is an abundant flow of saliva into the tissues. It is believed that infection of the plant takes place during this process. Other insects capable of transmitting the virus include the rusty plum aphid *Hysteroneura setariae, Carolinaia cyperi,* and possibly *Toxoptera graminum.*

Potato leaf roll virus includes the following in its list of vectors: *Aphis rumicis, Aphis rhamni, Myzus persicae* (fig. 182), *Myzus pseudosolani, Myzus circumflexus, Macrosiphum solanifolii, Calocoris bipunctatus, Typhlocyba ulmi, Eupterix auratus, Psylliodes affinis,* and larvae of *Tipula paludosa.*

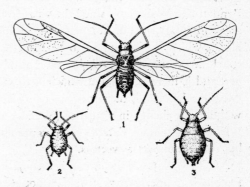

Fig. 182. *Myzus persicae,* the vector of leaf roll and more than a dozen other virus diseases. *1,* adult winged female; *2,* immature female; *3,* adult wingless female. (From Leach's *Insect Transmission of Plant Diseases,* 1940, McGraw-Hill Book Co., New York.)

Myzus persicae has also been found capable of transmitting the virus of peach mosaic disease in Colorado.

3. **Transmission by Leafhoppers; Biologic and Highly Specific.** Of this group, one of the best known virus diseases is that causing curly top of sugar

beets and other plants (Beta virus 1). The disease in the United States occurs only in the western part of the country. So far as is known at present, the virus of this disease is transmitted only by the beet leafhopper, *Eutettix tenellus,* which apparently transmits no other disease agent. In Argentina another species of leafhopper, *Agallia sticticollis,* is known to be a vector of a similar if not identical disease of sugar beets.

In *Eutettix tenellus* the virus undergoes an incubation period varying from 4 or more hours in a single individual to 20 minutes when 100 or more infected insects are used on one plant. The insects retain their ability to transmit the virus throughout their nymphal and adult life, although this ability decreases with age and is lost by overwintering females in an average time of about 84 days.

The relation of curly top virus to *Eutettix tenellus* has been studied by several investigators, particularly Bennett and Wallace (1938). These workers found that the leafhoppers acquired virus from diseased beet plants in one minute of feeding time and that viruliferous leafhoppers were able to infect healthy plants in a feeding time of one minute. Nonviruliferous leafhoppers placed on diseased sugar beets in two days picked up enough virus to give them their maximum ability to infect seedling beets. To enable them to acquire their maximal charge of virus a longer feeding period was necessary. Bennett and Wallace obtained the virus from the blood, salivary glands, feces, and alimentary tract of *E. tenellus*. They found it to be most abundant in the blood, which they assumed to be the chief virus reservoir. These workers concluded that there is probably no multiplication of the virus in the leafhopper, and if there is, it is not sufficient to maintain the original virus content. On the other hand, the virus can overwinter in the body of the leafhopper without any change in virulence (Wallace and Murphy, 1938).

Insects other than *Eutettix tenellus* are able to acquire the virus by feeding, but they do not act as vectors.

Another disease of this group is aster yellows (Callistephus virus 1 and 1a). Despite its name, aster yellows does not affect only asters but has a host range of about two hundred species of plants. It occurs throughout North America and is rare or absent in Europe.

Aster yellows is transmitted chiefly by one species of leafhopper, the identity of which has had considerable argument. Some have designated it as *Cicadula sexnotata,* which others maintain to be a European species only. Most authorities consider the insect to be *Macrosteles divisus,* which is widely distributed throughout North America. Both nymphs and adults may transmit the virus after an incubation period of ten days or more. The presence of the virus causes no apparent change in the insect, which may remain infective the rest of its life or for only a short time. It is possible that warm summer months may cause the leafhoppers to lose their virus, since it has been found that viruliferous insects held at temperatures above 31° C. for several days become noninfective (Kunkel, 1937). Freezing viruliferous insects for 20 minutes at minus 10° C. destroys most of the virus.

Streak of corn (Zea virus 2) is a serious virus disease of corn (*Zea mays*) in Africa, where it is transmitted by leafhoppers of the genus *Cicadulina* (*C. mbila, C. zeae,* and *C. storeyi*). The minimal incubation period of the virus in the leafhoppers varies from 6 to 12 hours at 30° C. to 84 hours at 16° C. The nymph transmits the virus more readily than the adult, but both are efficient vectors and retain the virus the rest of their lives.

According to Storey (1933), the streak of corn virus may be detected in the contents of the rectum of infected insects (*C. mbila*), but is not found in naturally voided feces. Apparently all viruliferous individuals contain the virus in their blood, and Storey was able to demonstrate its presence there before expiration of the normal incubation period. Storey concluded that normally the virus enters the intestinal tract through the mouth, and passes through the intestinal wall into the blood and thence to the salivary glands from whence it is introduced into the plant during the feeding process. In certain individuals the virus is unable to penetrate the wall of the intestine, and hence such insects are unable to transmit the virus.

Potato yellow dwarf is a North American disease of potatoes, the agent (Solanum virus 16 or *Aureogenus vastans*) of which is transmitted by the clover leafhopper, *Aceratagallia sanguinolenta*. Other insects have been reported as vectors, but most of these instances lack confirmation. Differences between the New York and the New Jersey strains of potato yellow dwarf virus in relation to their specific vectors has been pointed out by Black (1944).

Dwarf disease of rice was the first plant virus disease shown to be transmitted by insects (*Nephotettix apicalis* var. *cincticepts*). A number of Japanese workers independently contributed to this revelation between 1901 and 1909. Later a second species of leafhopper (*Deltocephalus dorsalis*) was incriminated. Transmission is biologic, and of particular interest is the fact that the virus (Oryza virus 1) is transmitted through the egg from one generation to the next—the only known instance of transovarial transmission of a plant virus. This occurs only when the females harbor the virus; infected males do not transmit the agent to the progeny of noninfected females.

Peach yellows is a North American disease transmitted by the plum leafhopper, *Macropsis trimaculata,* but not by the large number of other common sucking insects found on peaches (Kunkel, 1933). Both nymphal and adult *M. trimaculata* insects are capable of acquiring and transmitting the virus

(Prunus virus 1). The incubation period varies from 8 to 26 days. Hartzell (1937) has described inclusion bodies in the cells of the intestinal walls and salivary glands of infected insects. The plum may harbor both the virus and the insect and is therefore a source of infection for many peach orchards.

Other plant virus diseases which may be assigned to this group include the Fiji disease of sugar cane (Saccharum virus 2) transmitted by the leafhoppers *Perkinsiella vastatrix* and *Perkinsiella saccharicida,* and false blossom of cranberry (Vaccinium virus 1) transmitted by the blunt-nosed leafhopper, *Euscelis striatulus.*

4. **Transmission by Thrips; Biologic and Specific.** The first virus disease proved to be transmitted by thrips was spotted wilt of tomatoes and other plants. The disease is caused by Lycopersicum virus 3 and occurs principally in Australia, New Zealand, and in a few localities in Canada and the United States. The species of thrips involved include *Thrips tabaci, Frankliniella lycopersici, Frankliniella occidentalis,* and *Frankliniella moultoni.*

The incubation period of the virus in the thrips requires a minimum of five to seven days. Only the larval thrips acquire the virus from the plants. The larvae may transmit the virus to uninfected plants or the agent may survive metamorphosis and be transmitted by the adults. Transovarial transmission does not occur.

Yellow spot of pineapple is also transmitted biologically by thrips, *Thrips tabaci* being the only known vector of the virus (Ananas virus 1). In the insect the virus requires an incubation period of about ten days before it can be transmitted to fresh plants. As with spotted wilt of tomatoes, the insect must acquire the virus in the larval stage. The virus may persist through metamorphosis to the adult, in which stage it may also be transmitted.

5. **Transmission by Lace Bugs; Biologic and Specific.** In this group Leach (1940) has placed the two virus diseases leaf curl (*Kräuselkrankheit*) of sugar beet and "savoy" of beets, caused respectively by Beta virus 3 and Beta virus 5.

Leaf curl is transmitted by the lace bug *Piesma quadrata.* The nymphs of this insect apparently may take up the virus, but they are unable to transmit it. As stated by Leach, the effectiveness of the transmission and the severity of the symptoms caused apparently depend on the number of insects and the feeding time. For example, only mild symptoms result when one insect feeds for 24 hours or ten insects feed for 1 hour. Ten lace bugs that feed for 2 hours are as effective as three lace bugs that feed for 24 hours. The virus may overwinter in the adult bugs or in infected plants.

"Savoy" of sugar and garden beets is transmitted by *Piesma cinerea,* its only known insect vector.

6. Transmission by White Flies; Biologic and Specific. The two best known plant virus diseases spread by white flies are leaf curl of cotton (Gossypium virus 1) and cassava mosaic (Manihot virus 1). The former occurs chiefly in the Sudan and Nigeria and is transmitted by *Bemisia gossypiperda*. The incubation period of the virus in this white fly is not longer than 30 minutes. Insects infected as larvae are viruliferous as adults and remain so during their lifetime.

Mosaic disease of cassava is known in Africa where it causes considerable economic loss. The virus is transmitted by white flies. *Bemisia gossypiperda* var. *mosaicivectura* and *Bemisia nigeriensis* have been incriminated.

7. Transmission by Mites. Although definite, clear-cut proof is lacking, there is a great deal of evidence pointing to the transmission by a mite of the virus agent (Ribes virus 1) causing "reversion" of black currants in England and western Europe. The mite concerned is *Eriophyes ribis*. So far as is known, this is the only virus disease of plants transmitted by mites.

8. Nature of Transmission Obscure. In this group Leach (1940) places two diseases of which the nature of transmission is obscure, although insects may possibly play some role in the dissemination of the viruses. These diseases are wheat mosaic and latent (X) virus of potato, and others could be added. Since the possible relationships between the viruses of these diseases and insects is unknown, they will not be discussed here.

III. BACTERIOPHAGE

The housefly, *Musca domestica,* was the first arthropod found to harbor bacteriophage (Shope, 1927; Glaser, 1938b). In a salt solution extract of houseflies, Shope found a bacteriophage active against four species of bacteria (*Eberthella typhosa, Salmonella paratyphi, Escherichia coli,* and *Staphylococcus muscae*). An attempt to obtain a streptococcus bacteriophage by feeding the organism to the flies was unsuccessful.

Glaser (1938b) conducted experiments on the origin of bacteriophage by establishing in the alimentary tract of sterile houseflies a nonlysogenic staphylococcus known to be susceptible to lysis. No bacteriophage was formed in these flies. When given to sterile flies in the absence of the staphylococcus, the bacteriophage survived for one generation. In the presence of the bacterium it persisted for at least eight generations of flies and very probably longer. According to Glaser, these results do not support the theory that bacteriophage is the result of the interaction of host and bacteria. It is interesting to note that Glaser found houseflies caught in nature or bred in the contaminated state to harbor bacteriophage invariably.

Girard (1935) reports that he was able to isolate from a triturated suspension of fleas (*Xenopsylla cheopis*) a bacteriophage against the plague bacillus, *Pasteurella pestis*.

Fig. 183. Photograph showing "moth-eaten" appearance of the colonies of a gram-negative small rod (*Proteus*) infected with a strain of bacteriophage isolated from *Dermacentor albipictus*. (Photo by N. J. Kramis.)

Bacteriophage has also been recovered from ticks (Steinhaus, 1942b). From *Dermacentor andersoni* a strain was isolated in association with a strain of *Micrococcus*, and on two different occasions from *Dermacentor albipictus* a gram-negative small rod was isolated which was infected with bacteriophage (fig. 183).

CHAPTER IX

SPIROCHETES ASSOCIATED

WITH INSECTS AND TICKS

ACCORDING to the sixth edition [1] of *Bergey's Manual,* the order Spirochaetales is divided into two families. The first (Spirochaetaceae) contains the three genera [2] *Spirochaeta, Saprospira,* and *Cristospira.* The second (Treponemaceae) contains *Borrelia, Treponema,* and *Leptospira.* Members of the family Spirochaetaceae are larger in size and less susceptible to the action of bile salts than are those of the family Treponemaceae. Numerous other classifications have been proposed and various generic names have been used interchangeably. Since throughout this book the systematics of *Bergey's Manual* have been followed, and since we wish to be consistent, we shall arbitrarily follow it with respect to the spirochetes.

Compared with the true bacteria and true protozoa, the number of spirochetes known to be associated with insects and ticks is not large. Nevertheless, many of those instances in which such associations do occur are very important from the standpoint of health and economics. Most important in this regard is the relapsing fever group of spirochetes. In addition there are some entomogenous spirochetes which are nonpathogenic for man and animals, and some which are found in only a fortuitous association with arthropods.

---

[1] As this chapter is written, the sixth edition of *Bergey's Manual* has not yet appeared. However, through the courtesy of Dr. R. S. Breed, the writer has been privileged to see the outline and key to families and genera as they are to be presented in the new edition.

[2] Members of the bacterial genus *Spirillum* should not be confused with spirochetes. The genus *Spirillum* contains organisms which, unlike spirochetes, have rigid preformed spirals and terminal flagella. For example, the agent which is thought by many to be a cause of rat-bite fever was called *Spirochaeta morsus-muris* by Japanese workers but is now known to be a bacterium, *Spirillum minus.*

THE RELAPSING FEVER GROUP OF SPIROCHETES

Relapsing fever is a disease characterized by a sudden onset of chills and fever which continue for several days when they end by a crisis. After a week or more the fever recurs and several such recurrent attacks may take place. Usually the mortality is low, but it varies in different epidemics and sometimes may reach as high as 75 per cent. Outbreaks occur on every continent in the world, except possibly Australia, being most severe in Africa and India and less so in eastern Europe and the Americas.

Species of Spirochetes Concerned. Apparently several different relapsing fevers occur in man, caused by closely related species of morphologically similar spirochetes. The various types of the disease differ in the species of arthropods which transmit the spirochetes concerned, and in their geographic distribution.

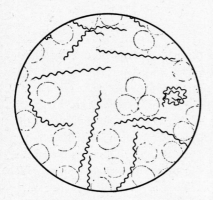

Fig. 184. *Borrelia recurrentis*. (From Wenyon's *Protozoology*, 1926, Wm. Wood and Co., New York.)

Much has been written on the species unity or plurality of relapsing fever spirochetes. The final word is still to be spoken on this question, but evidence of different species or types is indicated by the fact that immunity to one type does not protect against another type or even against variations of the same strain.

In 1873 Obermeier discovered the causative agent of louse-borne European relapsing fever, and the next year Lebert named it *Spirochaeta* (now *Borrelia*) *recurrentis*. Specific names given to other strains include the following:

    B. duttoni Novy and Knapp (1906) West Africa
    B. novyi Schellack (1907) America
    B. kochi Novy (1907) East Africa
    B. carteri Manson (1907) India
    B. rossi Nuttall (1908) East Africa
    B. berberum Sergent and Foley (1910) North Africa
    B. persica Dschunkowsky (1913) Persia
    B. aegypticum Mühlens (1913) Egypt
    B. venezuelense Brumpt (1921b) Venezuela and Colombia

B. neotropicalis Bates and St. John (1922) Panama
B. hispanicum De Buen (1926) Spain
B. turicatae Brumpt (1933b) America
B. hermsi Davis (1942) America
B. parkeri Davis (1942) America

These strains with certainty cannot be distinguished morphologically and most criteria advanced as means of differentiating the species concern the type

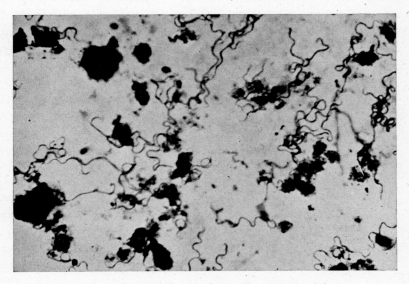

Fig. 185. *Borrelia hermsi.* (Photo by N. J. Kramis.)

of vector, the geographic locations, and the pathogenicity for experimental animals. With respect to the strains transmitted by ticks, an interesting possible means of differentiation has been suggested by Davis (1942). In studies on three United States species of ticks Davis suggests that "(1) each species of *Ornithodoros* that is a relapsing fever vector carries a spirochete that is tick host-specific and (2) this host-specific relationship offers a more accurate approach to the differentiation of spirochetes than any of the several criteria heretofore used."

Other spirochetes have been described which may cause various types of relapsing fevers or which may have a distant relationship to some of the better-known forms. An example of one of these is *Borrelia* (= *Spirochaeta*) *sogdianum* found by Nicolle and Anderson (1928) in *Ornithodoros tholozani* (= *O. papillipes*). This spirochete was very pathogenic for guinea pigs, but mice and rats were not very susceptible to it. Experimentally it could be transmitted by *Ornithodoros moubata* and *Ornithodoros normandi*. Another of these spirochetes is *Borrelia* (= *Spirochaeta*) *cobayae*, isolated from a guinea pig by Knowles and Basu (1935). In *Argas persicus* it behaves similarly to *Borrelia anserina* though it is not transmitted by the bite of this tick.

Insects and Ticks Concerned. The first arthropod proved to be capable of transmitting the spirochetes of relapsing fever was *Ornithodoros moubata* in 1904 and 1905. In the latter year *Ornithodoros savignyi* was also suspected, and since that time many species of *Ornithodoros* have been incriminated in various parts of the world. These include *O. talaje, O. rudis* (= *O. venezuelensis*), *O. turicata, O. hermsi, O. parkeri, O. erraticus* (= *O. marocanus*), *O. tholozani* (= *O. papillipes*), *O. tartakovskyi, O. nereensis,* and possibly *O. lahorensis* and *O. normandi*. Other ticks possibly concerned include *Rhipicephalus sanguineus* in Algeria and *Dermacentor andersoni*, which has been suspected of transmitting relapsing fever spirochetes in British Columbia. These two possibilities, particularly the latter, lack confirmation and are based largely on circumstantial evidence. *Borrelia hispanicum* has been transmitted to man and guinea pigs experimentally by *Rhipicephalus sanguineus* (Seargent, 1938).

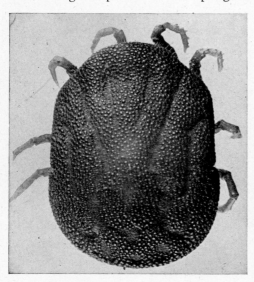

Fig. 186. *Ornithodoros moubata,* one of the vectors of *Borrelia recurrentis*. (Courtesy Dr. R. A. Cooley.)

In 1907 observations in India and in Algeria led to the conclusion that the spirochetes could be conveyed by lice (*Pediculus humanus* var. *corporis*). It is thought that the louse-borne strains probably originated from tick-borne strains through the agency of man's infection. Adler and Ashbel (1942) recovered a strain of *Borrelia persica* from lice that was also transmissible by ticks (*O. tholozani*). Louse-borne epidemics arise principally in Europe, India, and China. Francis (1938) demonstrated spirochetes in monkey lice (*Pedicinus longiceps*) after 17 days of subsistence on spirochete-positive monkey blood. Such lice, however, failed to infect monkeys on which they subsequently fed.

Bedbugs (*Cimex lectularius*) have been suspected of being vectors of the spirochetes since they may become infected by feeding and they retain the spirochetes in their bodies for long periods of time (20 to 60 days). However, they are not able to transmit the organisms by their bites except mechanically and are infective only when their triturated bodies are inoculated into animals or when the animals eat infected bedbugs. (See Rosenholz, 1927; Francis, 1938.)

Biologic Relationships Involved. Of the two groups of transmitting agents, lice and ticks, only the ticks transmit the relapsing fever spirochetes by their bites. In the case of lice, man usually becomes infected through crushing or otherwise damaging the lice on the skin. Since the coelomic fluid contains spirochetes, infection may occur when this fluid escapes through the broken legs of this delicate insect. Wounds made by the lice or scratching may then afford a portal of entry for the organisms.

During the first day after feeding the spirochetes disappear entirely from the stomach of the louse. At first most of them are seen to become immobile and then to disintegrate, while others penetrate the cytoplasms of the epithelial lining cells, after which their behavior has not been traced. The apparent absence of spirochetes anywhere in the louse persists for about six days. From this time on spirochetes commence to appear in the body cavity fluids and gradually spread to all parts of the body, though they never reappear in the stomach lumen or lining cells. Nothing is known concerning their behavior during the interval between their disappearance from the stomach and their reappearance in the coelomic cavity, where they have been observed to persist for at least 25 days.

In one investigation Adler and Ashbel (1942) studied five strains of *Borrelia persica* derived from human cases and found them to vary considerably in their behavior in lice. The types of behavior observed were as follows:

(a) When ingested by lice in large quantities, the spirochaetes all disappeared, and inoculation of the lice at various intervals into susceptible animals gave negative results. When inoculated into the coelom of the louse, the spirochaetes rarely survived up to 24 hours.

(b) When ingested by lice in large quantities, the spirochaetes all disappeared, and inoculation of the lice into susceptible animals gave negative results. When inoculated into the coelom of lice, the spirochaetes survived up to three days.

(c) When ingested by lice in large quantities, the spirochaetes all disappeared, and inoculation of the lice into susceptible animals gave negative results. When inoculated into the coelom, they survived up to seven days.

(d) When ingested by lice in large quantities, the spirochaetes survived up to 10 days, as proved by inoculation into susceptible animals. (This finding is in accordance with the original experiment of Nicolle and Anderson (1928) in the case of *Sp. hispanica*.) When injected into the coelom of the louse, they survived up to seven days.

In no case did we find evidence of multiplication in the coelom, whereas *Sp. obermeieri* multiplies and is found in enormous numbers in the coelom of infected lice.

Adler and Ashbel also observed that *B. persica* was rapidly destroyed in the monkey louse, *Pedicinus eurygaster*.

Exactly what happens to the spirochetes when they are ingested by ticks is not known with certainty. That it is different from what takes place in lice is agreed upon. Within an hour after an infectious feeding ticks are capable of transmitting the infection and may continue to do so for from one to six and a half years afterwards (Francis, 1938) and perhaps even longer. Furthermore, no tick-transmitted strains of spirochetes which are able to multiply in the coelom of the louse have been discovered.

Working with *Ornithodoros hermsi,* Wheeler (1942) observed that within three hours after the ingestion of infected blood, the spirochetes begin to migrate and to penetrate the gut wall. By the third day they have invaded the coelomic cavity, and from the tenth day spirochetes may be found in various muscles of the tick's body as well as in the coelomic fluid. Wheeler found spirochetes, recognizable as such, present in the gut contents of ticks up to at least 38 days after an infective blood meal. The organisms continue to undergo transverse division in the gut contents several days following their ingestion by the tick.

In 1907 Dutton and Todd found that *Borrelia duttoni* broke up into granules in the Malpighian tubes of *Ornithodoros moubata* and suggested that this may possibly represent a method of reproduction. Leishman (1918, 1920) made similar observations and reported that the granules are able to reproduce by simple or multiple division. These granules, which often appear in clumps enclosed by a membrane, are spherical, rod-, and comma-shaped and occur principally in the Malpighian tubes and ovaries. It has been postulated that this is a method of multiplication which occurs at low temperatures, for when ticks containing these granules are exposed to temperatures of 34° to 37° C. for several days, regular spirochetes reappear.

There has been no general acceptance of any particular theory as to the fate of the spirochetes in ticks. Many agree that the granules are degenerated spirochetes but do not believe in their regeneration. Some have claimed to find similar granules in ticks free from spirochetal infection. Others think that the granules may be other microorganisms such as rickettsiae. Though it is now known that some species of ticks infect man through their bites, it is also thought that the coxal fluid and feces may spread the disease by contaminating skin wounds. Davis (1941) found that *Ornithodoros turicata* males may transmit spirochetes at each feeding throughout life, or they may transmit them irregularly or after several successive transmissions they may thereafter fail to transmit them.

The relapsing fever spirochetes in ticks are transmitted transovarially to succeeding generations. This was established simultaneously by Koch in East Africa and by Dutton and Todd in the Belgian Congo in 1905. Zasukhin (1936b) has described some interesting unpublished experiments of Troitsky, who found that eggs laid by infected females of *Ornithodoros tholozani* (= *O. papillipes*) were unable to cause infection when large numbers of them were inoculated into guinea pigs. Larvae from such eggs were also incapable of infecting guinea pigs, although when the nymphs were fed on the animals infection resulted. This last fact has recently been frequently observed by other workers in other species of ticks (*Ornithodoros hermsi* and *Ornithodoros turicata*).

In 1911 Hindle observed that about 30 per cent of the *Ornithodoros moubata* ticks from Uganda were immune to infection with *B. duttoni*. This worker also noted that when infected ticks were held at temperatures of about 21° C., the Malpighian tubes, the sexual organs, the gut plus its contents, and the feces were infectious, but the salivary glands and the coxal fluid were noninfectious. On the other hand, when infected ticks were held at a temperature of about 35° C. for two or three days, all the organs of the body were infectious including the salivary glands. No spirochetes as such could be detected in any of the ticks' organs when kept at 21° C. When warmed to a temperature of 35° C. for two or three days, the spirochetes reappeared in the gut lumen, in the coelomic fluid, and in all the organs of the tick. From his experiments Hindle concluded that infection following the bite of an infected tick does not result from the inoculation of infective material from the salivary glands, but rather from the entrance of infective material, excreted by the tick during the feeding, into the wound caused by the tick's bite.

OTHER SPIROCHETES ASSOCIATED WITH TICKS AND INSECTS AND CAUSING DISEASES OF VERTEBRATES

Borrelia anserina (B. gallinarum). Most authorities consider these two spirochetes causing septicemia in chickens and geese, respectively, to be the same species. It may also infect other birds including turkeys, ducks, and canaries.

The following ticks are known to transmit the spirochetes: *Argas persicus* (= *A. miniatus*), *Argas reflexus,* and *Ornithodoros moubata*. The mite *Dermanyssus avium* has also been implicated.

The situation regarding the development of *B. anserina* in the tick is similar to that which prevails concerning the relapsing fever spirochetes. Various investigators, particularly Hindle (1912), have reported the breaking up of the spirochetes into granules. Hindle described what he considered to be the life cycle of the spirochete in *Argas persicus* and in the fowl (fig. 188). According to his conception, the ticks introduce the granules into the fowl where they develop into spirochetes, reproducing by fission. These may also break up into granules and become spirochetes again. When the spirochetes are taken up by the tick, they invade the cells of the body (particularly those of the Malpighian tubes) and the lumen of gut, in which locations they break down into granules which continue to multiply. Some of these granules are supposed to enter the eggs, thus ensuring transmission to the next generation. The same relation to high and low temperatures as in the case of the supposed granules of the relapsing fever spirochetes was observed. Absence of proof of this theory has, in recent years, given rise to the growing belief that whereas some of the granules in ticks may be derived from spirochetes, there is no conclusive evidence that they can again develop into spirochetes. Indications are that the spirochete may remain in the tick in its spirochete form and that for some reason it is difficult to demonstrate except when occurring in relatively large numbers, but this has by no means been proved.

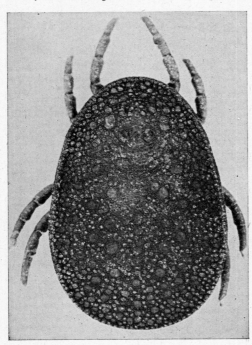

Fig. 187. The fowl tick, *Argas persicus*. (Courtesy Dr. R. A. Cooley.)

Nieschulz and Bos (1940) showed that females of the mosquitoes *Culex pipiens, Aëdes aegypti,* and *Theobaldia annulata* could take up the spirochetes from infected fowls. However, the organisms lost their virulence in one to two days and the authors concluded that mosquitoes were of no practical importance in the transmission of the spirochete.

Borrelia theileri. This spirochete closely resembles the relapsing fever spirochetes of man and is found in the blood of cattle. Natural transmission takes place by means of the ticks *Margaropus decoloratus* and *Rhipicephalus evertsi*. Experimentally transmission has been effected with *Margaropus australis*. In the first two ticks, at least, the spirochete is transmitted to the next generation through the egg.

Treponema pertenue. Several investigators have attributed the spread of *T. pertenue*, the cause of yaws, to insects. Although the housefly, *Musca domestica*, has been incriminated, especially indicted are the *Hippelates* flies, such as *Hippelates pallipes* and *H. flavipes*. Large numbers of these flies frequent the cutaneous lesions of persons with yaws and may actually penetrate under the scabs to feed. Kumm, Turner, and Peat (1935) observed that when the flies are fed on highly infectious lesions, they will take in large numbers of spirochetes, some of which may later be regurgitated in drops of vomit. These workers found that most of the spirochetes ingested by *Hippelates pallipes* remained actively motile in the esophageal diverticulum for about seven hours. On the other hand, they lose their motility very rapidly in the fly's stomach. The proboscis of the fly retains very few spirochetes and these quickly lose their motility. In a subsequent paper Kumm and Turner (1936) report the transmission of *T. pertenue* from man to rabbit by *Hippelates pallipes*. Transmission appears to be purely mechanical. It should be emphasized, however, that the actual role of insects in the natural transmission of the yaws spirochete has not been ascertained. Transmission by direct contact is also a factor.

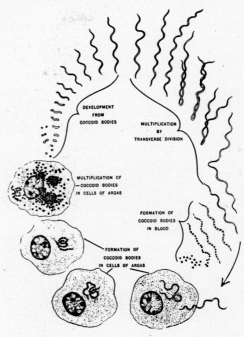

Fig. 188. A diagrammatic representation of the life cycle of *Borrelia anserina* (*B. gallinarum*). (From Hindle, 1911.)

SPIROCHETES OBSERVED ONLY IN INSECTS

Many workers have found spirochetes associated with insects and other arthropods but have not been able to associate them with any vertebrate hosts.

No doubt many of these have no vertebrate host and are peculiar to insects. They are usually found in the intestines of the insect, but occasionally other parts of the body are also involved.

One of the earliest such findings was the observation of *Borrelia glossinae* in the intestines of *Glossina palpalis* by Novy and Knapp (1906), who originally named it *Spirochaeta glossinae*. In the larva of an unidentified *Culex* Jaffé (1907) described a form he called *Spirochaeta culicis*. *Spirochaeta melophagi* was observed by Porter (1910) in the sheep ked, *Melophagus ovinus*. To two forms seen in the intestines of cockroaches Dobell (1912) gave the names *Treponema stylopygae* and *Treponema parvum*. *Spirochaeta blattae* has been described from the cockroach *Blabera atropus* (see Tejera, 1926). Patton (1912) found *Spirochaeta ctenocephali* in the flea *Ctenocephalides felis*, and *Spirochaeta drosophilae* has been recorded from *Drosophila confusa*.

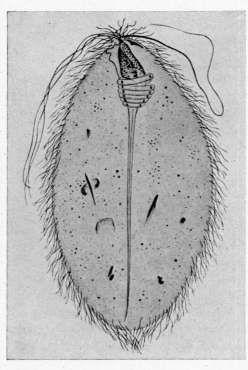

Fig. 189. Spirochetes attached to the surface of the flagellate *Caduceia bugnioni* from the termite *Neotermes greeni*. See also fig. 12C. (From Kirby, 1942a.)

In the intestine of *Phlebotomus perniciosus* Pringault (1921) observed a new species he designated *Spirochaeta phlebotomi*. In 1925 Zuelzer gave the name *Spirochaeta noelleri* to a spirochete he found in the intestine of *Simulium noelleri*. Jírovec (1930) reported the same spirochete from *Simulium reptans*. In the intestines of certain termites, *Termes lucifugus* of Japan, Von Prowazek (1910) found a spirochete he named *Spirochaeta minei*. Similarly Dobell (1911) gave the name *Spirochaeta termitis* to organisms he saw in termites of Ceylon, and Doflein (1911) called the forms he saw in termites of Italy, *Spirochaeta grassii*. Spiral organisms, probably spirochetes, were seen by Damon (1926) in all of ten species of termites he examined.

Mention should be made of the fact that attached to the surface of certain flagellates occurring in insects, particularly in termites, spirochetes are invariably present and are frequently mistaken for flagella or cilia. The majority of these spirochetes are unnamed and for the most part little studied. They range in length from 2 up to 20 or 40 microns, depending upon the species of protozoa to which they are attached. (See Kirby, 1941.) The spirochetes may be separated from the protozoa by feeding the termites cellulose moistened in a 5 per cent aqueous solution of acid fuchsin. The protozoa are unharmed by this treatment but they lose their spirochetes. Similar spirochetes have been observed in the wood-eating roach *Cryptocercus punctulatus* (Cleveland, Hall, Sanders, and Collier, 1934).

Paillot (1940) has described an interesting spirochete, *Spirochaeta pieridis,* which he found causing a septicemia in larvae of the cabbage butterfly, *Pieris brassicae*. Most spirochetes occurring in insects are harmless commensals, but this organism causes a disease in *Pieris brassicae* that gives the blood the same diseased appearance found in cases of bacterial septicemia. However, there are no external symptoms which distinguish sick caterpillars from healthy ones. The spirochetes are found principally in the blood, though some hypodermal cells are affected and spirochetes may occasionally be seen in the intercellular spaces of the fat tissue. Phagocytes take up considerable numbers of them. The infection cannot be initiated by the digestive routes but can be through the skin, though this does not always succeed. There is often a secondary bacterial septicemia present also. The silkworm, *Bombyx mori,* is not susceptible to *Spirochaeta pieridis*.

Spirochaeta culicis has been reported from mosquitoes by several investigators. Sinton and Shute (1939) found large numbers of spirochetes, morphologically indistinguishable from *S. culicis,* in the salivary glands of *Anopheles maculipennis* var. *atroparvus*.

Unnamed spirochetes have been observed in the following insects: *Anopheles maculipennis* (by Sergent and Sergent, 1906), *Ptychoptera contaminata* (Léger and Duboscq, 1909), *Culex pipiens* (Schaudinn, 1904), *Culex nebulosus* (Taylor, 1930), *Aëdes aegypti* (Stévenel, 1913, and Noc, 1920), *Trichoptera* (Mackinnon, 1910), *Chironomus plumosus* (Léger, 1902b).

CHAPTER X

PROTOZOA AND INSECTS

(EXCEPT TERMITES)

THE NUMBER of species of protozoa known to be associated with insects and ticks is surprisingly large. Whereas most of these arthropods harbor varying numbers of bacteria readily cultivable on artificial media, a microscopic search usually has to be made to detect the protozoa. Nevertheless, the protozoologists have probably investigated the protozoa associated with insects more intensively than the bacteriologists have investigated the bacteria—at least the protozoa appear to have been named and classified more satisfactorily.

As with the bacteria, very little detailed study has been made of the biologic relationships existing between the protozoa and their insect hosts. Such studies have in most cases been limited to investigations of the life cycle or that part of the life cycle which the protozoan spends in its arthropod host. Except in the case of the protozoa of the termite and wood-eating roach, very little study has been made on the physiologic processes involved.

The Arthropod Hosts. Protozoa have been found associated with nearly all orders of insects and with many species of Acarina as well. Most of the protozoa found in the latter arthropods are also parasites of other animals. In the case of the Hexapoda, however, most of the protozoa are either commensals or parasites of the insects themselves.

Very few studies of the protozoa associated with insects have been approached from the standpoint of any particular group of insects as a whole. A perusal of the protozoa associated with each order of insects indicates that they are about equally distributed throughout the insect world. One exception to this is the great abundance of Mastigophora (flagellates) in the order Isoptera (termites). Because of the large numbers of protozoa in these insects and the great amount of work done on the associations involved, the protozoa of termites will be discussed in another chapter (chap. XI).

An example of what was found in a survey of one order is provided by the work of Semans (1936, 1939, 1941), who studied the protozoa of Ohio Orthoptera exclusive of the domestic cockroaches and the wood-eating roach *Cryptocercus punctulatus*. This worker examined a total of 1,287 individual orthopteran specimens, representing 6 families and 99 species. Of these, 33 species and 192 individuals were found to contain gregarines in their alimentary tracts, and 34 species and 244 individuals harbored other types of protozoa; in all, 34 per cent of the species and 19 per cent of the individuals harbored protozoa. In addition to the gregarines, flagellates and a ciliate were observed. The gregarines were found in the adult Orthoptera about twice as frequently as in the nymphs and in females more often than in males. Whereas the flagellates and ciliates were found particularly in the colonic region of the hindgut, the gregarines were usually located in the extreme anterior end of the midgut at the base of the enteric ceca. Sometimes the enteric ceca also harbored them.

The Protozoa. The phylum Protozoa is usually divided into two subphyla separated on the basis of their means of locomotion: Plasmodroma, which have pseudopodia or flagella, and Ciliphora, which have cilia. The subphylum Plasmodroma may be divided into three classes: Mastigophora, Sarcodina, and Sporozoa. The subphylum Ciliophora is divided into two classes: Ciliata and Suctoria. Certain members of each of these five classes have been found associated with insects or ticks in some way or other.

PLASMODROMA

MASTIGOPHORA

(The Flagellates)

An entirely satisfactory classification of the flagellates is lacking. Generally, however, the Mastigophora are divided into two subclasses: Phytomastigina and Zoomastigina, each of which are subdivided into several orders. Of the two subclasses, Zoomastigina includes certain species which are associated with insects or ticks and which are found in each of the four orders: Protomonadida, Rhizomastigida, Polymastigida, and Hypermastigida.

Most of the Mastigophora associated with insects are limited to the intestinal tracts of these arthropods. Becker (1930) has separated this flagellate fauna of insects into four groups: (1) those which live in the intestines of certain termites; (2) those belonging to the family Trypanosomidae which spend a part of their life cycle in the intestines of insects and the remainder in the blood stream or tissues of vertebrates; (3) those belonging to the family Trypanosomidae but which are exclusively insect parasites apparently having no definite vertebrate or plant host; and (4) all intestinal flagellates of insects belonging to families other than the Trypanosomidae. This grouping is satisfactory for a general discussion of the intestinal flagellates of insects and will be more or less followed here with the exception that the first group —those associated with termites—will be considered in the next chapter.

PROTOMONADIDA

Trypanosomidae. The family Trypanosomidae is sometimes divided into six genera: *Trypanosoma, Crithidia, Herpetomonas, Leishmania, Leptomonas,* and *Phytomonas,* although some protozoologists recognize fewer. In considering the relation of these genera to each other, one is less likely to become confused if he also thinks of these protozoa in terms of morphologic types, i.e., the trypanosome type, the crithidia type, the leptomonas type, and the leishmania type.

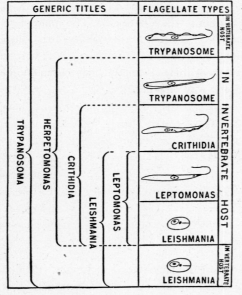

Fig. 190. Diagram of the classification and types of flagellates. (From Wenyon's *Protozoology,* 1926, Wm. Wood and Co., New York.)

Wenyon (1926) has grouped the Trypanosomidae according to whether they are limited to an invertebrate host, whether they have both a vertebrate and invertebrate host, or whether, in their highest stage of development, they reach the leptomonas, crithidia, or trypanosome form (see fig. 190). On this basis the various types may be grouped into the following six genera:

1. The genus *Leptomonas* includes those flagellates which never develop beyond the leptomonas stage and which during their life history show only

the leishmania and leptomonas forms. They are limited to invertebrate hosts and pass from one to another by means of cysts voided with the dejecta. The leptomonas form is generally considered to be the simplest type and all other flagellates of the family Trypanosomidae are regarded as having originated from the leptomonas form.

2. Members of the genus *Crithidia* during their life history assume leishmania, leptomonas, and crithidia forms. They are confined to invertebrate hosts, to which they are transmitted in the form of cysts.

3. In the course of the development of the flagellates of the genus *Herpetomonas* all four types are found: leishmania, leptomonas, crithidia, and trypanosome. They are limited to invertebrate hosts and are passed from one to another in the encysted state.

4. The genus *Phytomonas* resembles *Leptomonas* but has both an invertebrate and a plant host.

5. Flagellates of the genus *Leishmania* are very similar to those of the genus *Leptomonas* since they also have only the leishmania and leptomonas forms. The members of the two genera differ, however, in that those of the genus *Leishmania* have both a vertebrate and an invertebrate host. Evidence with respect to the manner of transmission indicates that it passes from invertebrate to vertebrate or from vertebrate to invertebrate.

6. The members of the genus *Trypanosoma* are similar to those of the genus *Herpetomonas* in that they show all four types during the course of their development. They have both a vertebrate and an invertebrate host, however, and transmission takes place from invertebrate to vertebrate or from vertebrate to invertebrate. For a discussion of the evolution of the morphology in the genus *Trypanosoma* the reader is referred to a paper by Lavier (1942-43).

Those Trypanosomidae which have both the vertebrate and invertebrate hosts are included in the genera *Leishmania* and *Trypanosoma*. These are frequently known as the hemoflagellates and any or all of the various types may be observed in insects, but only the leishmania and trypanosome forms are found in the blood of vertebrates. Members of the genus *Phytomonas* bear a similar relationship to the vascular system of plants. Chandler (1936) expresses the belief that since both the hemoflagellates and the flagellates parasitizing plants undergo cycles of development in the intestines of insects, one may assume that they were originally and primitively parasites of insects. By adapting themselves to the ingested blood or plant juices in the intestines of insects, they may possibly, upon their inoculation into the bodies of the animals or plants on which the insects feed, have become adapted to the hosts of the insects. Those Trypanosomidae which appear to occur only in arthropods and other invertebrates and not in vertebrates have been placed in the genera *Leptomonas, Crithidia,* and *Herpetomonas.*

Entomogenous Trypanosomidae: Leptomonas. Much discussion has taken place concerning the validity of the generic names *Leptomonas* and *Herpetomonas.* Some writers consider them to be synonymous, and more study of the *Leptomonas* type species, *Leptomonas bütschlii* from the nematode *Trilobus gracilis,* is necessary in order to decide whether it and *Herpetomonas muscarum* from the housefly, *Musca domestica,* are actually cogeneric. As we have already indicated, the present method is to consider those flagellates which have only the leptomonad and leishmania forms as *Leptomonas* and those which in addition have the crithidial and trypanosome forms as *Herpetomonas.*

The best-known species of *Leptomonas* is *Leptomonas ctenocephali,* first seen by Basile (1910) in the intestine of the dog flea, *Ctenocephalides canis.* Basile and other early investigators mistook it for developmental forms of *Leishmania donovani.* In 1912 Fantham named the flagellate *Herpetomonas ctenocephali,* and the next year Brumpt (1913) called it *Herpetomonas pseudoleishmania.* Morphologically similar forms have been found in other fleas, and in some cases, at least, they may be identical with *L. ctenocephali.* These include *Herpetomonas ctenophthalmi* in *Ctenophthalmus agyrtes, Herpetomonas pattoni* in species of *Pulex* and *Ceratophyllus, Herpetomonas debreuli* in *Ceratophyllus sciurorum,* and *Herpetomonas ctenopsyllae* in *Leptophylla segnis* (= *Ctenopsylla musculi*). According to Wenyon (1926) these belong to the genus *Leptomonas.*

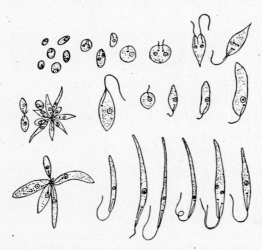

Fig. 191. *Leptomonas ctenocephali* from the intestines of the dog flea *Ctenocephalides canis.* (From Wenyon's *Protozoology,* 1926, Wm. Wood and Co., New York.)

Leptomonas ctenocephali (fig. 191) is limited to the intestinal tract and Malpighian tubes of the dog flea and only exceptionally occurs in the stomach, the infection stopping at the pyloric opening. Most of the flagellates attach themselves by their anterior ends to the epithelial lining which may become thoroughly covered with a mosaic of short, stubby forms and characteristic rosettes. Others remain free and may pass with the insect's dejecta, which consists of digested, partly digested, or whole blood. Also found in the flea's feces are the small, ovoid, leishmania forms which occur in the posterior end of the gut. These leishmania forms appear to possess a cyst wall, and it has been supposed that in this form they are protected against desiccation and are passed from one insect to another in the feces. The larva of the dog flea feeds upon the feces of the adult and in so doing acquires the flagellates, which also survive through the pupal stage. Like the flagellates in other fleas, *L. ctenocephali* may be maintained in cultures on N.N.N. (Novy, MacNeal, and Nicolle) medium. It is generally considered to be nonpathogenic, though some investigators (Laveran and Franchini, 1919, 1920) have claimed ability to infect mice and guinea pigs as well as *Euphorbia* plants with this flagellate.

Fig. 192. *Leptomonas pulicis* of the human flea, *Pulex irritans*. (From Wenyon's *Protozoology*, 1926, Wm. Wood and Co., New York.)

Leptomonas pulicis (fig. 192) is found in the human flea, *Pulex irritans*, and has a life history similar to that of *L. ctenocephali*. Both flagellates have

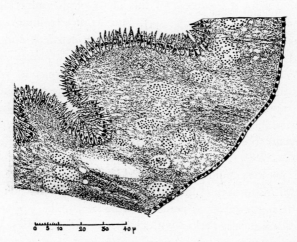

Fig. 193. *Leptomonas pyraustae* attached to the midgut of the European corn borer *Pyrausta nubilalis*. (From Sweetman, 1936; after Paillot.)

been found capable of living in the gut of bedbugs (*Cimex lectularius*) for several weeks. *Leptomonas jaculum* is found in *Nepa cinerea* and has been observed penetrating the eggs of its host. Other members of the genus *Leptomonas* include *L. vespae* found in *Vespa crabo*, *L. culicis* in *Culex quinquefasciatus* (= *Culex fatigans*) and probably other mosquitoes, *L. familiaris* in *Lygaeus familiaris*, *L. agilis* in *Rhinocoris iracundus*, *L. foveati* in *Notocyrtus foveatus*, *L. naucoridis* in *Naucoris maculatus*, *L. pangoniae* in *Pangonia neavei*, *L. pyrausta* in *Pyrausta nubilalis*, and *L. mirabilis* in *Sarcophaga melanura*. According to Wenyon (1926), Tejera reported an unnamed *Leptomonas* in an unidentified species of the tick genus *Amblyomma*. Tejera (1926) has also described *L. blaberae* from *Blabera* sp., and Semans (1939) has described a similar leptomonad from the midgut of *Parcoblatta virginica*, *Parcoblatta lata*, and *Parcoblatta pennsylvanica*.

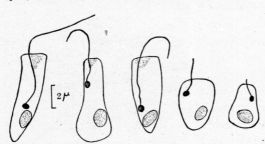

Fig. 194. Various forms of *Crithidia fasciculata*. (From Wallace, 1943.)

Crithidia. The genus *Crithidia* was created by Léger (1902a) for *Crithidia fasciculata*, which he discovered in the gut of *Anopheles maculipennis*. Considerable confusion has prevailed concerning the correct nomenclature for this species, and for a treatment of this phase of the subject the reader is referred to a paper by Wallace (1943). This worker maintains that *C. fasciculata* has no undulating membrane (a fact often debated) and that Patton's (1908, 1909) life cycle of *Herpetomonas culicis* (= *C. fasciculata*) involving preflagellate, flagellate, and postflagellate stages is erroneous. Wallace found further that the organisms are normally attached to the mosquito's gut lining in a small rounded form but become free when disturbed. In larval mosquitoes he was able to produce transitory infections by feeding them cultures of *C. fasciculata*, but the parasites did not persist into the adult stage. Adult mosquitoes (*Culex pipiens* and *Aëdes aegypti*) could be easily infected by feeding them cultures, and the parasitized insects (*Culex*) were able to transfer the organisms to normal individuals. In parasitized mosquitoes the entire lining of the hindgut, rectal pouch, and rectal papillae are covered like a carpet, and the lumen is packed with clusters of the flagellates. Wallace gives the following names as synonyms of *Crithidia fasciculata*: *Herpetomonas culicis* (as used by certain authors), *H. culicidarum*, *Leptomonas fasciculata*,

Leptomonas michiganensis, Leptomonas (Strigomonas) fasciculata, and *Crithidia anophelis.*

In 1908 Patton discovered the flagellate which he named *Crithidia gerridis.* This protozoan is an intestinal inhabitant of the water bugs *Gerris fossarum* and *Microvelia* sp., and also of a water bug related to *Perittopus.* The encysted forms of *Crithidia gerridis* may be found in the crop of the nymphs *Microvelia* which acquire them with ingested water. These forms soon produce flagella and begin to multiply by binary fission, forming characteristic rosettes which attach themselves to the epithelial lining. Gradually this grouping breaks up, and the elongated crithidial forms swim away and subsequently multiply again. Transmission from bug to bug occurs by way of feces and perhaps through the insects' cannibalistic habits. Becker (1923a,b), who has reported the occurrence of trypanosome forms in its life history, found *Crithidia gerridis* in *Gerris remiges, Gerris marginatus, Gerris rufoscutellatus,* and *Microvelia americana.* It has also

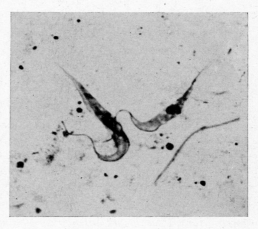

Fig. 195. *Crithidia (Herpetomonas?) hyalommae.* (Photo by N. J. Kramis.)

been reported in *Gerris paludum* in which the parasite occurs throughout the alimentary tract, in the ovaries, and in the feces.

In 1913 O'Farrell reported the occurrence of *Crithidia hyalommae* in the body fluid or hemocele of the tick *Hyalomma aegyptium.* This flagellate (fig. 195) is apparently limited to the fluids of the body cavity and does not occur in the intestine, though rarely the salivary gland may be infected with non-flagellate forms. O'Farrell also reports its transmission to the next generation via the infected egg, where the flagellate occurs in the leishmania form. By cutting off one of the tick's legs and examining the exuded fluid, the greatest number of protozoa may be found present just before the tick oviposits. After oviposition, round leishmania forms may appear in the fluid just before the death of the tick. Some authors place this organism in the genus *Herpetomonas.*

An interesting *Crithidia* was described by McCulloch in 1917 as occurring in the gut of the bug *Euryophthalmus convivus,* which feeds on the plant *Lupinus arboreus.* This flagellate, *Crithidia euryophthalmus,* takes on various forms in different parts of the insect's alimentary tract. McCulloch also describes two new phases of development: multiple segmentation and internal budding. The flagellate goes through its life history in the anterior half of the gut rather than in the hindgut as in most such cases.

Other species of *Crithidia* include *C. cleti* found in *Cletus varius*, *C. oxyca-*

Fig. 196. *Crithidia ortheae*, showing the various types of flagellates and aflagellate elements occurring in the crop of *Orthea bilobata*. (From Uribe, 1926.)

reni in *Oxycareni lavaterae*, *C. nalipi* in *Apiomerus nalipa*, *C. liturae* in *Leogorrus litura*, *C. vacuolata* in *Rhinocoris albopilosus*, *C. simuliae* in *Simulium columbaczense*, *C. tennuis* in *Haematopota duttoni* and *Haematopota vandenbrandeni*, *C. nycteribiae* in *Cyclopodia sykesi*, *C. ctenophthalmi* in *Ctenophthalmus agyrtes*, *C. cleopatrae* (*C. xenopysllae*) in *Xenopsylla cleopatrae*, *C. campanulata* in *Chironomus plumosus*, *Ptychoptera contaminata*, and certain Trichoptera larvae, *C. minuta* in *Tabanus tergestinus*, *C. subulata* in *Tabanus glaucopis* and *Haematopota italica*, *C. tabani* in *Tabanus hilaris*, *C. tullochi* in *Glossina palpalis*, *C. haematopini* in *Haematopinus spinulosus*, *C. hystrichopsyllae* in *Hystrichopsyllae talpae*, *C. pulicis* in *Loemopsylla cleopatrae* (*C. pulicis* was also the name given to a flagellate in *Pulex irritans*). *C. ortheae* was found in the alimentary tract of two species of reduvids, *Orthaea bilobata* and *Orthaea vincta*, from the crop to the rectum. A variety of forms of this flagellate were found occurring simultaneously in the gut of these insects (fig. 196). *C. haemaphysalidis* is the name of a flagellate found in the tick *Haemaphysalis flava*, and *C. christophersi* was found in *Rhipicephalus sanguineus*. Unnamed species of *Crithidia* have been observed in *Glossina fusca*, *Glossina morsitans*, *Xenopsylla cleopatrae*, and *Ctenocephalides felis*.

Herpetomonas. The best known species of this genus is *Herpetomonas* *muscarum* (fig. 197), a flagellate first observed in the gut of the housefly, *Musca domestica*. The first reference to this flagellate was made by Burnett (1851, 1852), who considered it under the genus *Bodo* but did not give it a specific name. It was next mentioned but not described by Leidy (1856), who referred to it as *Bodo muscarum* which, following the rules of priority, would become the correct specific name. Becker (1923c), however, considers the flagellate observed by Burnett and by Leidy as a *nomen nudum* and prefers to accept the specific name first used by Stein (1878), *Cercomonas muscae domesticae*. Later Kent (1880–1882) placed it in his newly erected genus *Herpetomonas* with the name *H. muscae domesticae*.

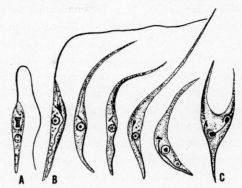

Fig. 197. *Herpetomonas muscarum* from the alimentary tract of *Calliphora erythrocephala*. *A*, typical leptomonas form; *B*, leptomonas with rhizostyle; *C*, dividing form. (From Wenyon's *Protozoology*, 1926, Wm. Wood and Co., New York.)

Other proposed names have been *Schedoacercomonas muscae domesticae* and *Monomita muscarum* (Grassi, 1879, 1882). For the present discussion we shall follow the suggestion of Hoare (1924) and use the name *Herpetomonas muscarum*.

The host of *H. muscarum* is generally listed as the housefly, *Musca domestica*, because the flagellate is very common in the intestinal tract of this insect in most parts of the world and especially in the tropics. Apparently other insects may also serve as hosts to this protozoan. It is probable that some of the flagellates regarded as *H. muscarum* are in reality other species. Among the various hosts in which *H. muscarum* is supposed to have been found are *Musca nebulo*, *Musca humilis*, *Fannia canicularis*, *Lucilia argyricephala*, *Lucilia craggi*, *Scatophaga lutaria*, *Neuroctena anilis*, and species of *Homalomyia*, *Borborus*, and *Drosophila*. By cross-infection experiments Becker (1923d) found *H. muscarum* to infect *Sarcophaga bullata*, *Cochliomyia macellaria*, *Phormia regina*, *Lucilia sericata*, and *Calliphora erythrocephala*. The flagellates (*Herpetomonas luciliae*, *Herpetomonas calliphorae*, *Herpetomonas sarcophagae*, and those in *P. regina*, and *C. macellaria*) which occur naturally in these insects Becker believes to be the same as *H. muscarum*. Similar identity is thought to exist between *H. muscarum* and the flagellate occurring in *Lucilia caesar* (Drbohlav, 1925).

In the housefly *H. muscarum* occurs in any part of the intestinal tract up to the proventriculus and is usually seen in the leptomonas form, which has a slender body, 30 by 2 to 3 microns. One often sees many of them in a state of active division just prior to which time they may appear as biflagellate organisms. The flagellate passes from insect to insect in the form of cysts eliminated with the feces of infected flies. It is possible, however, for flies to become infected through the ingestion of adult flagellates or the pre-encysting forms.

Several workers have reported the cultivation of *H. muscarum* on artificial media. Patton (1921) cultured it on N.N.N. medium from the peritrophic membrane of *Lucilia argyricephala*. Glaser (1922) cultivated the flagellate and was able to infect grasshoppers via the body cavity. A strain obtained in 1930 from Glaser was maintained by Packchanian (1944) on "N.N." medium by monthly or bimonthly transfers.

The five species of flagellates described by Chatton and his colleagues as occurring in *Drosophila confusa* have all been placed in the genus *Herpetomonas* by Wenyon (1926). Those given specific names are *Herpetomonas drosophilae* and *Herpetomonas roubaudi*. Other drosophila flies have also been found harboring flagellates; thus we have *Herpetomonas ampelophilae* from *Drosophila ampelophila* and *Herpetomonas rubrostriatae* from *Drosophila rubrostriata*.

The genus *Herpetomonas* contains numerous species not well enough known to be discussed in detail here. Among these are *H. pyraustae* which occurs occasionally in the European corn borer *Pyrausta nubilalis*, *H. ctenophthalmi* described from the flea *Ctenophthalmus agyrtes*, *H. jaculum* in the water scorpion *Nepa jaculum*, *H. vespae* in the hornet *Vespa crabro*, *H. aretocorixae* in the gut of the water boatman *Arctocorixa interrupta*, and *H. algeriense*, *H. culicis*, *H. myzomyiae*, and *H. pessoai* in mosquitoes. Some, such as *H. mirabilis*, *H. mesnili*, and *H. caulleryi* were at one time placed in the genus *Cercoplasma* erected by Roubaud (1911), though this genus has not been generally accepted. Other such genera are *Cystotrypanosoma*, which included *Herpetomonas grayi* (*Trypanosoma grayi*) from species of *Glossina*, and *Rhynchoidomonas*, which included *Herpetomonas siphunculinae* found in the eye fly *Siphunculina funicola*.

Phytomonas. The members of this genus, which have only the leptomonas and leishmania forms, all have a plant host, otherwise they are similar in all respects to members of the genus *Leptomonas*. The name *Phytomonas* was suggested by Donovan (1909).

According to their plant hosts the three best-known species of this genus are *Phytomonas davidi* found in the latex of the plants *Euphorbia pilulifera*, *Euphorbia thymifolia*, and *Euphorbia hypericifolia*; *Phytomonas elmassiani* found in *Araujia angustifolia,* and *Phytomonas bordasi* found in *Morreira odorata.* Wenyon (1926) considers these species as probably synonymous. The latex of infected plants becomes clear and is devoid of starch and other granules normally present. The chlorophyll gradually diminishes, and the plant eventually withers and dies.

Several species of insects may be responsible for the natural distribution of these flagellates. The bug *Stenocephalus agilis* has been incriminated in Portugal and *Chariesterus cuspidatus* in Central America. The relation of the flagellates to *S. agilis* has been studied in some detail by França (1919, 1920), and to *C. cuspidatus* by Strong (1924). Experimentally these protozoa have been transmitted by the bugs *Nysius euphorbiae* and *Dieuches humilis.* Flagellates have been found in the salivary glands of milkweed bugs, *Oncopeltus fasciatus,* which were found on infected milkweeds. The exact role of this insect in the spread of *Phytomonas elmassiani* is still unknown. *Lygaeus kalmii,* a hemipterous insect related to *Oncopeltus,* was found to contain a flagellate which Noguchi and Tilden (1926) named *Herpetomonas (Phytomonas) lygaeorum.* It is very probable that *Phytomonas leptavasorum,* the cause of phloem necrosis of the coffee plant, is spread by insects—most likely *Lincus spathuliger.*

Leishmania. We come now to the first of the two genera of Trypanosomidae which have both a vertebrate and an invertebrate host, the genus *Leishmania.* The members of this genus are very similar to those of the genus *Leptomonas* in that they occur in only the leishmania and leptomonas forms, but, as we have seen, members of the latter genus are found only in invertebrate hosts.

The genus Leishmania was named by Ross (1903) in honor of Leishman, who, just a few weeks prior to Donovan in 1903, described the small round parasites in the spleen of a victim of kala azar or dumdum fever. Kala azar is caused by *Leishmania donovani* and is an important disease in India, China, and parts of Africa. *Leishmania infantum* or *L. donovani* var. *infantum* are the names at one time given to the agent of infantile kala azar. It is now thought to be synonymous with *L. donovani.* Other species pathogenic for man are *Leishmania tropica,* the cause of Oriental sore, and *Leishmania brasiliensis,* the cause of espundia.

The insect transmisison of these diseases is a subject still not thoroughly understood. As Chandler (1936) has said of kala azar, and the same applies to Oriental sore and espundia, "Few problems in parasitology have caused more fruitless effort, more blasted hopes, more false conclusions, or more unfounded speculation than the transmission of kala azar, and the end is not yet." One of the first insects to be suspected was the bedbug (*Cimex hemipterus*), but proof of its guilt was never found and in 1925 the Indian

Kala-Azar Commission definitely exonerated this insect. Fleas were next suspected, but were likewise dismissed as were lice, mosquitoes, *Triatoma, Culicoides,* ticks, and others.

In recent years the sand flies (*Phlebotomus*) have been incriminated in each of the three diseases, and much supporting evidence has been accumulated. Those suspected of transmitting *L. donovani* include *Phlebotomus argentipes, Phlebotomus perniciosus, Phlebotomus major, Phlebotomus longicuspis,* and *Phlebotomus major* var. *chinensis.* Transmission is thought to take place in some cases through the bite of the insects and in others by the crushing of them on the skin. *L. tropica* is thought to be transmitted by *Phlebotomus papatasii* and *Phlebotomus sergenti. Phlebotomus lutzi* and *Phlebotomus intermedius* are thought to be concerned in the spread of *L. brasiliensis,* the flagellate causing espundia.

Leishmania tarentolae, a parasite of the gecko (*Tarentola mauritanica*) is believed to be transmitted by *Phlebotomus minutus* which feeds upon this lizard. *L. tarentolae* was at one time thought to be identical with *L. tropica,* and the gecko was considered the reservoir. The correctness of this idea still remains to be decided.

Trypanosoma. The types of biologic relationships existing between insects and trypanosomes are varied and extremely interesting. Limitation of space prevents us from discussing all these in detail. We shall, however, mention some of the more outstanding examples with the realization that no doubt numerous others remain yet to be discovered.

Because of the immediate practical value to man, a great deal of the study on the biologic associations between trypanosomes and insects has been made with the trypanosomes pathogenic to man. The three most important of these are *Trypanosoma gambiense,* the cause of central and west African sleeping sickness; *Trypanosoma rhodesiense* (a race of *Trypanosoma brucei*), the cause of east and south African sleeping sickness; and *Trypanosoma cruzi,* the etiologic agent of Chagas' disease in South America.

In addition, trypanosomes pathogenic for animals and nonpathogenic trypanosomes are also carried by various arthropods.

Trypanosoma gambiense. The tsetse fly, *Glossina palpalis,* transmits *T. gambiense* from man to man. This fact was demonstrated by Bruce and Nabarro (1903) and the development of the trypanosome in this insect was described by Kleine (1909). In some parts of western Africa *Glossina tachinoides* transmits the organism, and some investigators believe it to be a more important vector in these areas than is *G. palpalis.* Experimental evidence indicates that other species of *Glossina* (*G. morsitans, G. fusca, G. pallidipes, G. submorsitans*) may also transmit *T. gambiense.* Insects such as mosquitoes

(*Mansonia uniformis*) and other biting flies have been found capable of transmitting the parasite experimentally and in a mechanical manner. The chief vertebrate host appears to be man, but numerous wild and domestic animals are also thought to be natural hosts.

Wenyon (1926) lists six alternatives any one of which may occur when a tsetse fly first ingests blood and trypanosomes from an infected animal:

1. The trypanosomes may be destroyed and disappear in the fifty to seventy-two hours during which the blood is being digested.
2. Trypanosomes may not entirely disappear with digestion of the first feed of blood.
3. They may survive and multiply in the gut, although a second feed of blood has been superadded.
4. They may survive and multiply in the crop for as long as twelve days, provided the crop has never been entirely emptied of blood. In such cases the gut may be entirely free from trypanosomes. Those in the crop are unable to survive a complete emptying of the organ, and no permanent infection of the fly results if this takes place.
5. The trypanosomes may persist in greater or less numbers both in the gut and in the crop of the same fly.
6. The whole of the partially digested blood which survives from the first feed may be displaced by the fresh blood of the second feed without the trypanosomes which are present in the stomach disappearing. The crop in these cases may be either empty or filled with new blood.

The last of these alternatives seems to be the one which will bring about a permanent infection of the fly. The development cycle completes itself in only about 10 per cent of the flies, and in these active and progressive multiplication of the trypanosomes begins in the mid- and hind-intestines soon after the infective feeding.

The development cycle in the tsetse fly covers a period of about 20 to 30 days. The multiplication in the mid- and hindgut takes place until by the tenth day considerable numbers of broad forms are present. From the tenth to the fifteenth day these gradually develop into long, slender forms which move forward in the digestive tract to the proventriculus. Further development takes place when the flagellates migrate through the esophagus, hypopharynx, and salivary ducts to the salivary glands where they attach themselves by their flagella and multiply rapidly. After 18 to 30 days they assume the crithidial form, followed 2 to 5 days later by the free-swimming trypanosome forms. The latter forms are capable of infecting a vertebrate host when the fly bites its victim.

Trypanosoma rhodesiense. Considerable controversy has taken place concerning the relationship between *Trypanosoma rhodesiense* (fig. 198) and

Trypanosoma brucei. The latter causes nagana, a disease in domestic animals, but is apparently nonpathogenic to man and most wild game. Most authorities believe that *T. rhodesiense* is a strain of *T. brucei* which has adapted itself to infecting the human host.

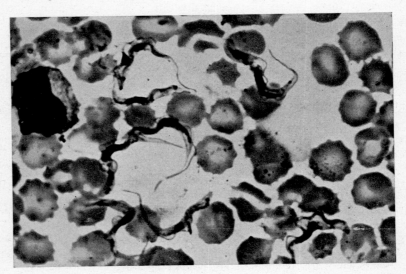

Fig. 198. *Trypanosoma rhodesiense* in the blood of a man. (Photo by N. J. Kramis.)

T. rhodesiense is transmitted by *Glossina morsitans* and in some areas by *Glossina swynnertoni*. Experimentally *Glossina palpalis* and *Glossina brevipalpis* also transmit it. Some strains of *T. brucei* after prolonged maintenance by direct passage in mammalian hosts or old laboratory strains are not transmissible by *Glossina*.

The life cycle of *T. rhodesiense* in its insect hosts appear to be identical with that of *T. gambiense*.

Trypanosoma cruzi. In South America and particularly in Brazil *T. cruzi* (fig. 199) causes a trypanosome infection, Chagas' disease, named after its discoverer. Its natural hosts include armadillos, opposums, and rodents.

In Brazil the insect vector of *T. cruzi* is the reduviid, *Triatoma megista*. In Venezuela and Colombia *Rhodnius prolixus* transmits the parasite to man. It is probable that nearly all of the bloodsucking reduviids may serve as intermediate hosts. *Triatoma infestans, Triatoma chagasi, Triatoma sordida,* and *Rhodnius pictipes* have been shown to become very easily infected. The following species of *Triatoma* are among those found infected in nature: *T. megista, T. infestans, T. sordida, T. dimidiata, T. chagasi, T. geniculata,*

T. hegneri, T. vitticeps, T. sanguisuga, T. longipes, T. rubida, T. gerstaekeri, T. heidemanni (the last four in the U.S.A.), *Rhodnius prolixus, R. pictipes,* and *Eratyrus cuspidatus. Apiomerus pilipes,* which will feed on *Rhodnius prolixus* (but not on man), may also serve as a host to the flagellates. *T. cruzi* may survive in such ticks as *Ornithodoros furcosus, Ornithodoros parkeri* and *Ornithodoros amblus* for many weeks or months. For a list of other vectors of this trypanosome the reader is referred to a bulletin by Usinger (1944).

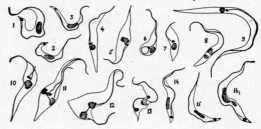

Fig. 199. *Trypanosoma cruzi.* Development in gut of *Rhodnius prolixus. 1–3,* trypanosomes of the blood type which are ingested by the bug; *4–9,* various crithidia forms; *10–12,* dividing crithidia forms; *13–16,* metacyclic trypanosomes which escape from the feces of the bug. (From Wenyon's *Protozoology,* 1926, Wm. Wood and Co., New York.)

When reduviid bugs ingest infected blood, the trypanosomes quickly change into stumpy crithidial forms which undergo rapid multiplication. The progeny develop into the long crithidial forms in the posterior end of the midgut. The flagellates (noninfective) eventually pass into the rectum where they become smaller and give rise to metacyclic (infective) trypanosomes. The entire life cycle of the trypanosome in the insect takes from 6 days in larval bugs to about 15 days in adults. With the appearance of the trypanosomes in the bug's gut, its feces become infective in varying numbers up to 3,500 per cubic millimeter. Although under certain conditions infection may result from the insect's bite, in most cases transmission occurs by rubbing the feces into the wound, into skin abrasions, or into the mucous membranes of the eyes and mouth. Normally the salivary glands of the insect do not become infective.

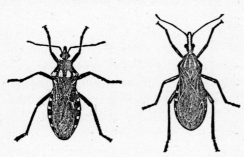

Fig. 200. Adult males of *Triatoma megista* (left) and *Rhodnius prolixus* (right). (From Hegner, Root, Augustine, and Huff's *Parasitology,* 1938, D. Appleton-Century Co., New York. After Brumpt.)

Of interest is an observation by Wood (1942) who was able to detect living *T. cruzi* in *Triatoma protracta, Triatoma rubida,* and *Triatoma longipes* which had been dead for 15 days. Since the rodents which harbor *T. cruzi* are entomophagous and if they can become infected orally, the ingestion of these insects may explain part of the maintenance of the infection in the burrows under natural conditions.

Trypanosoma lewisi. This trypanosome is commonly found in rats, particularly *Rattus rattus* and *Rattus decumanus,* in nearly all parts of the world. Four to six days after the intraperitoneal inoculation of infected blood into a rat, the trypanosomes may be found in the peripheral blood. They usually disappear within two or three months after infection, but may occasionally persist for a longer period of time. Although under certain conditions the virulence of *T. lewisi* may be enhanced, it is generally considered that this trypanosome causes no marked pathologic changes in and is not fatal to rats. Once freed of the parasite rats show a definite immunity and usually cannot be reinfected with the same strain.

Transmission of *T. lewisi* takes place through the agency of fleas (*Nosopsyllus fasciatus*). The infected feces of these insects may be swallowed by the rat when it licks its fur, as may the fleas themselves; or wounds made by the bite of the insect may become contaminated with the dejecta. Transmission does not take place directly through the insect's bite, and its salivary glands are free of the parasite. The fleas do not become infective until after about six days from the time of their feed in infected blood. During this interval a definite cycle of development occurs in the flea's intestine. Transmission of *T. lewisi* has been demonstrated with other fleas including *Ceratophyllus lucifer, Ceratophyllus hirundinis, Pulex brasiliensis, Pulex irritans, Xenopsylla cheopis, Ctenocephalides canis, Leptopsylla segnis,* and *Ctenopthalmus agyrtes.*

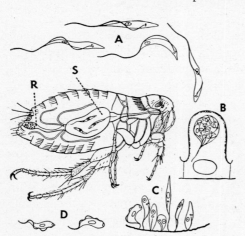

Fig. 201. *Trypanosoma lewisi* in the blood of the rat and in the flea *Nosopsyllus fasciatus.* *A*, trypanosomes as seen in the rat at a late phase of infection; *B*, intracellular phase of development in stomach of flea; *C*, attached flagellates in rectum; *D*, free metacyclic trypanosomes which bring about infection when ingested by rat; *R*, rectal phase; *S*, trypanosomes in stomach of the flea. (From Wenyon's *Protozoology,* 1926, Wm. Wood and Co., New York.)

The rat louse, *Haematopinus spinulosus,* may mechanically bring about infection of rats, but apparently no established infection is produced in this insect as in the case of the flea.

The cycle of *T. lewisi* in the flea has been studied by numerous workers, the study of Minchin and Thomson (1915) probably being the most detailed. During the first six hours after the trypanosomes enter the flea's stomach they undergo a physiologic change evidenced by the fact that at this time they will not cause infection when inoculated into rats. After this period they invade the cells lining the stomach. In the cell the trypanosome forms a pear-shaped body within a vacuole. This pear-shaped body increases in size, and the kinetoplast and nucleus multiply by repeated divisions. Each of the resulting spherical flagellated bodies divides into a number of trypanosomes similar in appearance to those found in the blood of the rat. The stomach cell eventually ruptures, and the trypanosomes are released into the flea's stomach. This intracellular phase, which may occur in any part of the insect's stomach, may end as early as 18 hours or as long as five days after the infective feeding. Trypanosomes which are released from the lining cells may enter other cells and repeat the process.

From the stomach the flagellates migrate back to the hindgut and rectum where they change in structure to crithidial forms. These crithidial forms multiply by fission, and the ensuing rectal phase is characterized by many forms including herpetomonad, crithidial, and trypanosome types. These forms also occur in the feces, but most workers believe that only the small trypanosome forms will give rise to infection.

Other trypanosomes found infecting small animals and associated with insects include *T. duttoni* (found in mice), which is able to go through a complete cycle of development in the swallow flea, *Ceratophyllus hirundinis.* *T. blanchardi* (in the dormouse) has been experimentally transmitted by the flea *Ceratophyllus laverani.* The vector of *T. conorrhini* (in rats) was discovered to be *Triatoma rubrofasciata.* *T. rabinowitschi* (in the hamster) has been found in the rectums of the fleas *Ctenophthalmus assimilis, Nosopsyllus fasciatus,* and *Ctenocephalides canis,* which probably are able to transmit the infection. *T. nabiasi* (in rabbits) is transmitted by and undergoes its development in the rabbit flea *Spilopsyllus cuniculi.* *T. talpae* (in the mole) has been observed in the mole flea *Palaeopsylla gracilis.* *T. vespertilionis* and other trypanosomes of bats are thought to be transmitted by mites (*Liponyssus arcuatus, Leiognathus laverani,* etc.) and possibly *Cimex pipistrelli.*

Among the larger animals there are several which harbor apparently harmless trypanosomes in their blood streams. *Trypanosoma theileri,* or its varieties, is an example of an apparently harmless trypanosome being found in the blood of cattle in many parts of the world. Tabanid flies (e.g., *Tabanus glaucopis*) are considered the vectors of this flagellate.

Trypanosoma melophagium likewise is not pathogenic for its vertebrate host, the sheep. This organism was first observed in the sheep ked, *Melophagus ovinus,* and later in the blood of sheep. Infection of the sheep does not take place through the bite of the ked but probably through contamination of the bite or other wounds with infected feces or through the ingestion of the ked or its feces.

Though it has been shown that *T. melophagium* produces metacyclic trypanosomes in the hindgut of the ked, the early stages of its development in this insect have not been thoroughly elucidated. A crithidia seems to be the predominating type and appears to be confined to the stomach (figs. 202 and 203). The protozoa multiply rapidly and attach themselves to the wall of the hindgut, especially near the pyloric opening.

Some workers have observed what might be forms of *T. melophagium* in ticks (*Ixodes ricinus*) taken from sheep.

Trypanosoma congolense is a small trypanosome found in cattle, horses, and sheep. Wild game constitutes the main reservoir of *T. congolense* which is conveyed by a number of species of *Glossina*. Among those known to be associated with *T. congolense* or its varieties are *Glossina morsitans, G. longipalpis, G. tachinoides, G. palpalis, G. brevipalpis*. Some species, e.g., *Glossina austeni,* are suspected of being natural vectors on the basis of laboratory transmission experiments, but naturally infected flies have not yet been found. Experimentally *T. congolense* has also been transmitted by *Aëdes aegypti* and *Stomoxys calcitrans*.

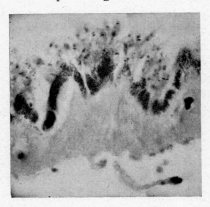

Fig. 202. Cross section of the gut of a sheep ked showing the flagellate *Trypanosoma melophagium*. The black-appearing border consists of *Rickettsia melophagi* in closely packed rows perpendicular to the epithelial surface. (Courtesy Dr. S. B. Wolbach.)

The life cycle of *T. congolense* in the tsetse fly begins with its development in the intestine, after which the trypanosomes migrate forward to the hypopharynx. Here they change into attached crithidia forms and then into true trypanosomes. There is no invasion of the salivary glands, instead the inner surface of the labrum appears to be the most heavily infected part of the insect. The exact mode of transmission has not been determined though the infected insect's feces as well as infected mouth parts may be important in this regard.

Trypanosoma simiae is a trypanosome similar to *T. congolense* in morphology though somewhat larger in size. It is pathogenic for monkeys and domestic pigs, but small laboratory animals do not appear to be susceptible to infection by it. *T. simiae* may be transmitted by *Glossina morsitans* and *G. brevipalpis*. The cycle of development in these flies is similar to that of *T. congolense*, commencing in the stomach and eventually infecting the labial cavity and the hypopharynx. The cycle takes about 20 days for completion.

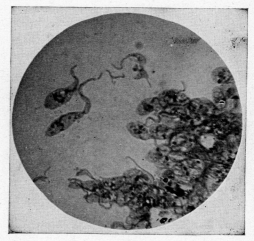

Fig. 203. Crithidia form of *Trypanosoma melophagium*. (From Hagan, 1943.)

Trypanosoma vivax occurs in Africa and is found most frequently in cattle, sheep, goats, and horses. Like those species just discussed, it is transmitted by tsetse flies: *Glossina palpalis, G. tachinoides, G. longipalpis, G. morsitans,* and possibly others.

In the fly the trypanosome multiplies in the proboscis only, with crithidial forms being produced in the labial cavity, where they attach themselves to its walls. The infective metacyclic trypanosomes are finally produced in the hypopharynx.

Similar to *T. vivax* are *Trypanosoma caprae* and *Trypanosoma uniforme*. *T. caprae* is transmitted by *Glossina morsitans* and *G. brevipalpis*. *T. uniforme* is known to be carried by *Glossina palpalis*. Some workers believe these two trypanosomes are merely varieties of *T. vivax*.

Trypanosoma evansi causing surra, a disease of horses, is transmitted from animal to animal by various bloodsucking flies, chiefly of the genus *Tabanus*. Transmission is believed to take place through purely mechanical means. Ticks (*Ornithodoros tholozani* and *O. lahorensis*) have also been known to transmit the flagellate, at least experimentally, and there is some indication that a developmental cycle may occur in these arthropods. In *Glossina* flies *T. evansi* dies and is digested soon after being ingested. This probably indicates that morphologically similar *T. brucei* and *T. evansi* are really separate species with distinct phylogenies.

Several trypanosomes of birds have been described as being possibly transmitted by mosquitoes and fleas. Though many of the trypanosomes of reptiles are transmitted by leeches, some are carried by insects, such as *Trypanosoma grayi* (= *T. kochi*) of crocodiles of which the vectors are tsetse flies (*Glossina palpalis*). Hoare (1931) has studied *T. grayi* in considerable detail and has worked out its life cycle in the tsetse fly and in the crocodile. The chronologic order of the flagellate's development in the fly is shown in figure 204.

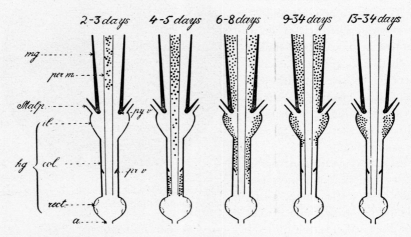

Fig. 204. Diagram showing the distribution of *Trypanosoma grayi* (indicated by dots) in the intestine of *Glossina palpalis* in chronological order of its development. The time interval covered is from 2 to 34 days after the fly becomes infected. *a*, anus; *col*, colon; *hg*, hindgut; *il*, ileum; *Malp*, Malpighian tubes; *mg*, midgut; *per m*, peritrophic membrane; *pr v*, prorectal valve; *py v*, pyloric valve; *rect*, rectum. (From Hoare, 1931.)

Other Protomonadida. Besides the Trypanosomidae several other Protomonadida have been found associated with insects. The genus *Retortamonas* contains at least four species found in Orthoptera: *R. gryllotalpae* and *R. wenrichi* found in the mole cricket *Gryllotalpa gryllotalpa*, *R. blattae* found in the hindgut of the cockroach *Blatta orientalis*, and *Retortamonas orthopterorum* found in the gut of *Blatta orientalis, Ectobia lapponica, Gryllotalpa gryllotalpa*, and the dipteran *Tipula abdominalis*.

Retortamonas belostomae was isolated from a water bug, *Belostoma* sp. (Hemiptera).

Several Coleoptera have been reported as harboring species of *Retortamonas. R. phyllophagae* was found in the larva of a beetle, *Phyllophaga* sp., and in the larva of the Japanese beetle *Popillia japonica*. In aquatic coleopteran larvae belonging to the family Gyrinidae, a protozoan under the name *Retortamonas caudacus* was described.

Retortamonas agilis was the name given to a flagellate found in Trichoptera larvae of the genera *Limnophilus* and *Anabolia*. *Embadomonas alexeieff* was also discovered in trichopterous larvae.

RHIZOMASTIGIDA

Very few members of this order are associated with insects. In 1913 Mackinnon reported finding *Rhizomastix gracilis* in the intestinal tract of tipulid larvae. This organism was originally observed in an axolotl.

POLYMASTIGIDA

The majority of the flagellates in the order *Polymastigida* associated with insects are found in termites and will be discussed in the next chapter. However, there are a few of these flagellates not associated with termites.

Polymastix melolonthae was originally isolated from the intestine of the cockchafer *Melolontha melolontha*. Similar forms have been described from larvae of *Cetonia* sp., *Tipula* sp., *Phyllophaga* spp., *Oryctes nasicornis*, *Popillia japonica*, and *Anomala orientalis*. *Polymastix phyllophagae* has been observed in the rectum of several species of larvae of *Phyllophaga*.

Tetratrichomastix blattidarum has been found in the intestine of the roaches *Blattella germanica*, *Blatta orientalis*, and *Periplaneta americana*, and *Tetratrichomastix mackinnoni* in the Japanese beetle, *Popillia japonica*.

Mackinnon (1910) described *Eutrichomastix trichopterorum*, a flagellate occurring in members of several genera of trichopterids or caddis-fly larvae. Those of the genus *Limnophilus* (*L. rhombicus* and *L. flavicornis*) contained the greater number of the protozoa, and those of the genera *Stenophylax* and *Sericostoma* contained fewer. No trace of the flagellate was found in the pupae or adults. In the gut of the larvae it occupies a definite position, just below the point at which the Malpighian tubes are attached. It may also occur in small numbers in the lower part of the intestine and in the rectum, but it has not been observed in the midgut. Travis and Becker (1931) observed *Eutrichomastix passali* and *E. phyllophagae* in the rectum of several species of *Phyllophaga*.

Travis (1932) described *Monocercomonoides ligyrodis* from larvae of *Ligyrodes relectus*.

Monocercomonas mackinnoni and *M. melolonthae* have been found in larvae of the Japanese beetle *Popillia japonica*.

Hexamastix batrachorum var. *tipulae* has been observed by Geiman (1932) in the larvae of the crane fly, *Tipula abdominalis*.

Hexamita periplanetae is found in the gut of certain cockroaches.

There is some evidence that cysts of *Giardia lamblia* may pass unharmed through the alimentary tract of cockroaches. Young (1937) has observed that the *Giardia* cysts when eaten by cockroaches may reach the colon of *Periplaneta brunnea, Eurycotis floridana,* and in some individuals of *Periplaneta americana* in two hours. The cysts were present in the colon of all three species after three hours, and in both species of *Periplaneta* they remained there for as long as 12 days.

HYPERMASTIGIDA

All species of Hypermastigida are found in the alimentary tracts of insects, most of them being inhabitants of the guts of termites (see next chapter) just as are the Polymastigida.

Two of the best-known among those Hypermastigida associated with insects other than termites are *Lophomonas blattarum* and *Lophomonas striata,* both found in the colon of cockroaches and in *Gryllotalpa* sp.

Prolophomonas tocopola and *Trichonympha grandis* are found in the gut of the roach *Cryptocercus punctulatus*. *Trichonympha grandis* undergoes encystment only once a year at the time of molting of the host roach. *Rhynchonympha tarda, Urinympha talea, Idionympha perissa, Barbulanympha ufalula, Eucomonympha imla, Leptospironympha europa,* and *Macrospironympha xylopletha* also occur in *Cryptocercus punctulatus*. A more complete list of the flagellates associated with wood-eating roaches may be found in chapter XI.

SARCODINA

The Sarcodina may be divided into two subclasses: Rhizopoda and Actinopoda, which in turn are separated into several orders. The most important from the standpoint of insect associations is the order Amoebida, of which the families Endamoebidae and Amoebidae are particularly significant. The members of Endamoebidae are exclusively endoparasitic, and generic differentiation is based upon the morphologic characteristics of the nucleus (Kudo, 1939).

In the family Endamoebidae the type species of the genus *Endamoeba*[1] is *Endamoeba blattae,* found in the colon of the cockroach, *Blatta orientalis,* by Leidy in 1879. It also occurs in *Periplaneta americana. Entamoeba thomsoni* occurs in the same roaches and also in *Blattella germanica.* Other *Endamoeba* species occur in the gut of termites.

The genus *Entamoeba* likewise contains several species whose habitat appears to be the gut of insects: *Entamoeba minchini* found in the gut of tipulid larvae, *Entamoeba apis* in *Apis mellifera,* an unidentified *Entamoeba* in *Limnophilus rhombicus,* and *Entamoeba nana* in *Blaberus atropos.*

In 1917 Keilin observed in larvae of the winter gnat, *Trichocera hiemalis,* an amoeba which he named *Entamoeba mesnili.* Bishop and Tate (1939) restudied this protozoan and placed it in a newly created genus as *Dobellina mesnili.* In Cambridge, England, the *Trichocera* larvae did not harbor *D. mesnili* earlier than January and the infection rate appeared to be highest (92 per cent) toward the end of January and in February. By May the amoebae were no longer to be found.

Frye and Meleney (1936) fed cysts of *Entamoeba histolytica* to cockroaches (*Periplaneta americana*) and observed them to pass through the insect's alimentary tract within 16 to 20 hours after a single feeding. By microscopic examination cysts were observed in the hindgut 72 hours after feeding. The authors conclude that it is possible that cockroaches which have access to human feces containing cysts of *E. histolytica* may be capable of contaminating food with their droppings, and thus act as passive transmitting agents of this organism. A similar situation apparently is possible with the housefly, *Musca domestica.* In this insect *E. histolytica,* particularly the cysts, pass unaltered through the alimentary tract and are viable in the feces. In fact, viable cysts may be passed by the flies as early as one minute after feeding (Sieyro, 1942).

The smallest amoeba to have been found in cockroaches is *Endolimax blattae,* which Lucas (1927) found in the hindgut and rectum of *Blatta orientalis* and *Periplaneta americana.* It has also been reported from the colon of *Periplaneta australasiae.*

Unidentified species of *Entamoeba* and *Endolimax* have been observed in larvae of the Japanese beetle, *Popillia japonica.*

The family Amoebidae contains several genera whose members are found in insects. The protozoa of two genera, *Malpighiella* and *Malpighamoeba,* parasitize the Malpighian tubes of their hosts. *Malpighiella refringens* is found in the rat flea *Nosopsyllus fasciatus,* and *Malpighamoeba* (also called *Vahl-*

---

[1] Concerning the nomenclature of the genera *Endamoeba* and *Entamoeba,* see a paper on this subject by H. Kirby (*J. Parasitol.,* **31**, 177–184, 1945).

kampfia) *mellificae* occurs in the excretory organs of adult bees (*Apis mellifera*). Several species of *Vahlkampfia* have been described from larvae of *Phyllophaga* spp.

Malameba (formerly *Malpighamoeba*) *locustae* was found to occur in the epithelial cells of the midgut and gastric ceca of laboratory-reared grasshoppers (*Melanoplus differentialis, M. mexicanus,* and *M. femur-rubrum*). In addition to these, Taylor and King (1937) were able to parasitize 37 species of grasshoppers with this protozoan. These workers collected in nature 633 specimens of grasshoppers representing 51 species and found only 2 species (*Trachyrhachis k. kiowa* and *Hadrotettix trifasciatus*) to harbor *M. locustae*.

Hartmannella blattae occurs in the oriental cockroach.

In 1909 Porter described *Amoeba chironomi* from the gut of *Chironomus* larvae (probably *C. dorsalis* or *C. plumosus*). The amoebae were found throughout the gut of the larvae except near the mouth. Very few moving forms occurred in the rectum though cysts were present in this region. Cysts were also found in the insect's feces.

The genus *Mycetosporidium* contains two species found in weevils. *M. talpa* was described from *Otiorhynchus fuscipes* and *M. jacksonae* from a species of *Sitona*. In the case of the latter the parasite is found in the tissues of the gut and in the Malpighian tubes of the weevils (fig. 205). As Tate (1940) points

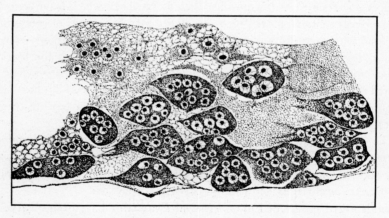

Fig. 205. *Mycetosporidium jacksonae*. Tangential section of a Malpighian tube of *Sitona* (weevils) showing a large, much vacuolated, ramifying plasmodium within which islands of dense cytoplasm are being formed. These islands later become massive, multinucleate schizontlike bodies. (From Tate, 1940.)

out, the systematic position of *Mycetosporidium* is obscure and the species mentioned may possibly belong to the order Haplosporidia (Sporozoa).

SPOROZOA

Members of the class Sporozoa are parasitic in habitat, bear spores in some stage of their development, and may have two phases in their life cycles, one asexual and one sexual. Their hosts are in every animal phylum including Arthropoda.

Asexual reproduction, or schizogony, takes place by repeated binary or multiple fission or budding of intracellular trophozoites or schizonts. Sexual reproduction occurs by isogamous or anisogamous fusion or autogamy. The zygote or oöcyst (or oökinete, if motile) usually again undergoes multiple fission, called sporogony, producing either naked sporozoites or walled spores. Many Sporozoa have life cycles in which the sexual and asexual generations alternate, i.e., there is an alteration of schizogony and sporogony. Frequently this occurs in relation to a change of hosts.

Many protozoologists divide the Sporozoa into two subclasses, the Telosporidia and Neosporidia. The first includes the gregarines, in most of which asexual multiplication does not occur and in which the two uniting sex cells are the same in size. The second subclass includes the coccidia-*like* forms, in which both sexual and asexual multiplication does occur, and in which the sex cells usually vary in size much like eggs and sperms. Other authors divide the Sporozoa into three subclasses: Telosporidia, Acnidosporidia, and Cnidosporidia. However, this variation in the manner of subdividing the Sporozoa need not greatly trouble one interested in the entomophilic members of the class since the great majority of such Sporozoa are included in the subclass Telosporidia, which is common to both systems of classification.

The subclass Telosporidia may be separated into three orders: Gregarinida, Coccidia, and Haemosporidia.

GREGARINIDA

The members of this order, the gregarines, are of special interest to us because of the large number of them which are parasitic in digestive tracts of insects. They also are found in annelids, tunicates, mollusks, and other invertebrates.

When the parasites are young (sporozoites or merozoites) they are usually intracellular, but as they grow they leave or protrude from the host cell, moving freely about in the body of the host, or remain attached externally to the digestive or coelomic lining. The grown, elongated, wormlike trophozoites vary from 10 to 16,000 microns (16 mm.) in length. They move in various amoeboid, gliding, or wormlike fashions. These protozoa are frequently considerably differentiated: their ectoplasm sometimes consists of several layers including a zone of myonemes (the myocyte); their anterior ends may possess spines, filaments, hooks, and the like which are used in adhesion; the endoplasm may contain a variety of inclusions, granules, spheres, and other bodies.

The Gregarinida fall into two suborders, the Eugregarinaria which do not multiply asexually but undergo a sexual reproduction by sporogony, and the Schizogregarinaria, which undergo sporogony and schizogony or asexual multiplication, producing merozoites that enter other of the host's epithelial cells and again grow into adults.

EUGREGARINARIA

When the spore of a eugregarine is ingested by a suitable host, it soon germinates, releasing the sporozoites which enter the epithelial cells of the digestive tract. In this location they grow at the expense of the host cells until they leave or protrude from the cells, frequently remaining attached for a while by various means of attachment. The detached forms (this state is known as sporadin) move about in the gut lumen until they encyst in pairs and become gametocytes. Within this cyst the nucleus of each individual divides repeatedly until a large number of small nuclei are formed. By the process of budding the nuclei form gametes, which may be isogamous or anisogamus. Each of the gametes of one gametocyte unites with one of the gametes formed in another gametocyte, thus forming a large number of zygotes. The contents of each zygote develop into sporozoites (usually eight), and with the formation of a resistant membrane surrounding it a spore is formed. When ingested by a suitable host the spore germinates and the cycle is repeated. Since asexual reproduction does not occur in this group, it is apparent that there is no actual multiplication in the host, and hence the host does not become intensely infected as in the case of other protozoan parasites.

In the lumen of the host gut the gregarine is usually rather large and vermiform and its body covered by a pellicle and its cytoplasm clearly differentiated into ectoplasm and endoplasm.

The suborder Eugregarinaria is usually divided into two groups or tribes. In one, Acephalina, the protozoan body consists of a single compartment. In the other group, Cephalina, the body consists of two compartments separated by an ectoplasmic septum.

Acephalina. As stated above, the bodies of the members of this tribe consist of a single compartment. One of the best-known species is *Lankestria culicis* (fig. 206) found in the gut and Malpighian tubes of the mosquitoes *Aëdes* *aegypti* and *Aëdes albopictus*. However, some workers (Ray, 1933) maintain that *L. culicis* is a cephaline gregarine and should not be placed with the Acephalina.

Lankestria culicis was first observed by Ross (1895) in India. It has since

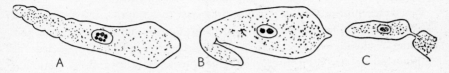

Fig. 206. *Lankesteria culicis* of *Aëdes aegypti*. *A* and *B*, gregarines free in the gut of larval mosquitoes. *C*, gregarine attached to epithelial cell. (From Wenyon, 1911.)

been found in nearly all parts of the world where the mosquito hosts live. According to Wenyon (1926), the infection begins when the mosquito larvae ingest the oöcysts or spores which adult mosquitoes have deposited in the water. In the digestive tract of the larva the spore germinates, allowing eight sporozoites to escape and enter the epithelial cells of the stomach (fig. 207).

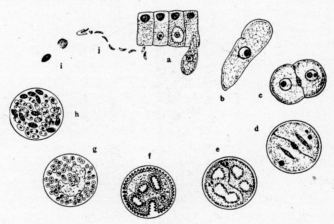

Fig. 207. The life cycle of *Lankestria culicis*. *a*, entrance of sporozoites into the epithelial cell and growth stages of trophozoites; *b*, mature trophozoite; *c*, association of two trophozoites; *d–f*, gamete formation; *g*, gametogony; *h*, development of spores from zygotes; *i*, a spore; *j*, germination of spore in host gut. (From Kudo's *Protozoology*, 1939, C. C. Thomas, Baltimore.)

Here each one becomes a more or less spherical intracellular parasite, which grows until it is an elongate body with a large central nucleus. During this time the host cell is destroyed. The gregarine remains attached at one point to the degenerating cytoplasm of the cell by means of an organella (epimerite). The average size of fully grown *L. culicis* is about 50 microns in length, though considerably larger forms have been seen. After remaining attached to the gut wall for some time, the parasite drops into the gut lumen and moves about in various types of characteristic movements.

When the mosquito larva pupates, the gregarines migrate from the gut to the Malpighian tubes where they soon begin to associate in pairs. Each pair becomes enclosed in a spherical gametocyst within which the nucleus of each gregarine divides forming daughter nuclei which undergo repeated mitotic divisions. The successive divisions of the nuclei in each gregarine take place simultaneously. When this period of multiple division finally ends, each gregarine (i.e., gametocyte) contains several hundreds of nuclei, which pass to the surface periphery where they form small protoplasmic buds each containing one nucleus. Each of these small elevations or buds then separates off becoming a gamete. As explained by Wenyon, when there are a large number of nuclei there may not be enough surface area to accommodate them all. Hence, a number of large vacuoles appear in the cytoplasm and the nuclei assume positions upon the surface of these vacuoles. After gamete formation the residual protoplasm disintegrates.

In *L. culicis* the gametes produced are equal in size though the size of their nuclei differ—those with large nuclei being female gametes and those with small nuclei being male. Thus the gregarine yielding gametes with large nuclei becomes the female gametocyte while that yielding gametes with small nuclei becomes the male gametocyte. The male and female gametes in each gametocyst conjugate in pairs, thus forming a slightly elongated zygote which develops a cyst wall (oöcyst). The zygote splits longitudinally into eight sporozoites lying around a central mass of residual cytoplasm.

About the time the adult mosquito emerges from the pupal case the gametocyst ruptures, and the oöcysts or spores are liberated into the lumen of the Malpighian tubes and then into the mosquito's intestine, from which they are ejected into the water with the insect's feces. Mosquito larvae living in this water may then ingest the oöcyst and the cycle is repeated.

Other Acephalina parasitic in insects include:

Enterocystis ensis in the gut of the May fly, *Caenis* sp.

Diplocystis schneideri found in the general body cavity of the roaches *Periplaneta americana* and *Blatta orientalis*. The young stages occur in the gut epithelium. *Diplocystis major* and *Diplocystis minor* are found in *Gryllus domesticus*.

Allantocystis dasyhelei occurs in the midgut of *Dasyhelea obscura* larvae. All stages of this gregarine are located in a space between the intestinal epithelium and the peritrophic tube. They always lie with their main axis parallel to that of the larva.

Cephalina. The bodies of cephaline gregarines are divided into two compartments. The anterior part, which is smaller, is called the "protomerite," and the larger posterior part is called the "deutomerite." The latter part usually contains the nucleus, although in the genus *Pileocephalus* it is claimed that the nucleus is in the protomerite. Two of the best-known Cephalina are probably *Gregarina blattarum* (fig. 208) of the cockroach *Blatta orientalis* and *Gregarina cuneata* (fig. 209) of the mealworm *Tenebrio molitor*. Both of these gregarines have well-developed epimerites by which they attach themselves to the gut cells of the host. The sporadins of both species are frequently seen in syzygy, a phenomenon in which several individuals are grouped in an endwise association of two or more sporadins. Upon encystment they become globular, attaching themselves to each other by the protomerite. The thick wall of the gametocyst is perforated in various places, and to each one of these openings is attached a tube which becomes everted outward from the cyst wall after which the oöcysts are forced through the tubes. *Gregarina blattarum* also occurs in *Parcoblatta pennsylvanica, Blattella germanica,* and *Periplaneta americana. Gregarina legeri* is another gregarine of the American roach.

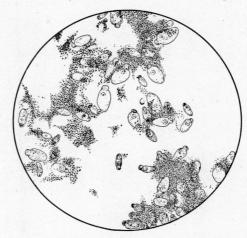

Fig. 208. *Gregarina blattarum* of cockroaches. Smear of intestinal contents. (From Wenyon's *Protozoology*, 1926, Wm. Wood and Co., New York.)

According to Semans (1936, 1943), *Gregarina rigida rigida* is known to occur in the alimentary canal of the following hosts: *Chorthippus curtipennis curtipennis, Arphia sulphurea, Encoptolophus sordidus, Dissosteira carolina, Spharagemon bolli, Brachystola magna, Schistocerca americana americana, Hesperotettix viridis pratensis, Melonoplus obovatipennis, Melanoplus differentialis, Melanoplus bivittatus, Melanoplus femur-rubrum femur-rubrum, Melanoplus mexicanus mexicanus, Melanoplus keeleri luridus, Melanoplus angustipennis.*

Other species of *Gregarina* include *G. locustae* in the gut of *Dissosteira carolina,* and *G. oviceps* in the crickets *Gryllus abbreviatus* and *Gryllus ameri-*

canus. *Gregarina nigra* occurs in *Encoptolophus sordidus, Zonocerus elegans, Cyrtacanthacris ruficornis, Melanoplus differentialis,* and *M. femur-rubrum femur-rubrum*. *Gregarina kingi* inhabits *Gryllus assimilis* as does *Gregarina galliveri,* which has also been found in *Hapithus agitator agitator*. From vari-

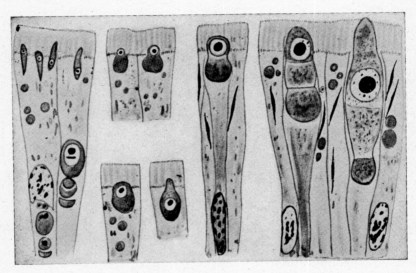

Fig. 209. *Gregarina cuneata* in the intestinal epithelium of the larva of *Tenebrio molitor*. (From Sweetman, 1936; after Léger and Duboscq.)

ous Orthoptera, Semans (1939) reported the following *Gregarina: G. ohioensis* from *Parcoblatta virginica; G. thomasi* from *Parcoblatta pen(n)sylvanica; G. parcoblattae* from *Parcoblatta uhleriana* and *P. pen(n)sylvanica; G. indianensis* from *Chorthippus longicornis, Melanoplus differentialis, Melanoplus bivittatus,* and *Melanoplus mexicanus mexicanus; G. hadenoeci* from *Hadenoecus puteanus; G. proteocephala* from *Ceuthophilus gracilipes; G. ceuthophili* from *Ceuthophilus gracilipes, Ceuthophilus brevipes,* and *Ceuthophilus divergens; G. prima* from *Ceuthophilus uhleri; G. acrydiinarum* from *Acrydium arenosum angustum* and *Paratettix cucullatus cucullatus;* and *G. rigida columna* from *Arphia sulphurea, Chortophaga viridifasciata, Encoptolophus sordidus, Pardalophora apiculata, Dissosteira carolina, Spharagemon bolli, Spharagemon collare collare, Melanoplus obovatipennis, Melanoplus scudderi scudderi, Melanoplus differentialis, Melanoplus bivittatus, Melanoplus mexicanus mexicanus,* and *Melanoplus keeleri luridus*. *Gregarina segmentata* is an inhabitant of the fungus-eating beetle *Cis bidentatus,* and restricted to the posterior region of the midgut. *Gregarina mystacidarum* has been observed in certain Trichoptera larvae. *Gregarina hyalocephala* is found in several species of *Tridactylus;* *G. conica* in Gryllidae spp.; *G. oblonga* in *Locusta migratoria* and *Acheta campestris; G. paranensis* in *Schistocerca paranensis; G. acridiorum* in *Tryxalis* sp., *Oedipoda coerulescens, Sphingonotus* sp., *Pamphagus* sp., and *Calliptamus italicus; G. consobrina* in *Ceuthophilus uniformis; G. neglecta* in *Ceuthophilus nigricans; G. udeopsylla* in *Udeopsylla robusta; G. ovata* in *Acheta campestris* and *Forficula auricularia; G. longiducta* in the alimentary canal of *Ceuthophilus pallidipes* and *Ceuthophilus meridionalis; G. stygia* in the alimentary canal of *Ceuthophilus gracilipes; G. macrocephala* in *Acheta campestris, Gryllus domesticus,* and *Nemobius sylvestris; G. davini* in *Gryllomorpha dalmatina; G. sphaerulosa* in *Gryllotalpa* spp. and *Oedipoda* spp.; *G. platydema* in the intestine of *Platydema excavatum;* and *G. diabrotica* in the intestine of *Diabrotica vittata.*

Gregarina chagasi, according to Semans, is found in *Neoconocephalus fratellus,* and is the only protozoan parasite described for any of the Tettigoniidae.

Besides *Gregarina blattarum,* the oriental cockroach, *Blatta orientalis,* also harbors *Protomagalhaensia serpentula,* a gregarine occurring in the gut and coelom of its host. The roach *Blatta lapponica* is known to support the gregarine *Gamocystis tenax* in its gut.

In addition to *Gregarina cuneata* mentioned above, larvae of the mealworm, *Tenebrio molitor,* also harbor *Gamocystis polymorpha, Gamocystis steini,* and occasionally *Steinina ovalis.*

Other cephaline gregarines include:

Tettigonospora stenopelmati in ceca and mid-intestine of *Stenopelmatus pictus* and *Stenopelmatus fuscus,* and in excrement of *Stenopelmatus intermedius.*

Hyalospora affinis in the gut of *Machilis cylindrica.*

Hirmocystis harpali in the gut of *Harpalus pennsylvanicus erythropus. Hirmocystis gryllotalpae* in the alimentary tract of a gryllid.

Anisolobus dacnecola in the midgut of *Dacne rufifrons.*

Leidyanna erratica has been found in the intestines of the crickets *Gryllus abbreviatus* and *Gryllus assimilis.* It has also been reported from other orthopteran hosts including *Nemobius fasciatus fasciatus, N. fasciatus socius, Anaxipha exigua,* and *Hapithus agitator agitator.* A species very similar to *Leidyanna erratica* is *Leidyanna gryllorum* found in *Gryllus assimilis, Gryllus domesticus, Nemobius sylvestris,* and *Nemobius fasciatus fasciatus.* Other species of *Leidyanna* found in insects include *L. xylocopae* of the carpenter bee, *Xylocopa aestuans,* which was the first gregarine observed in a hymenopteran host. This gregarine occurs throughout the length of the alimentary tract, tightly packing the gut without apparent damage to its host. *Leidyanna tinei* was described from the midgut of lepidopterous larvae of *Endrosis fenestrella*.

Sphaerocystis simplex in *Cyphon pallidulus*.

Lepismatophila thermobiae in the ventriculus of the firebrat, *Thermobia domestica*, which also harbors a larger gregarine, *Colepismatophila watsonae*.

Menospora polyacantha in the gut of *Agrion puella*.

Dendrorhynchus systeni in the midgut of larvae of *Systenus* sp. (probably *S. scholtzii*), a dolichopodid fly found in the decomposed sap of the elm tree.

Stylocephalus giganteus in the Coleoptera, *Asida opaca*, *Asida* sp., *Eleodes* sp., and *Eusattus* sp.

Bulbocephalus elongatus in the gut of a cucujid larva.

Cystocephalus algerianus in the gut of *Pimelia* sp.

Lophocephalus insignis in the gut of *Helops striatus*.

Acanthospora polymorpha in the gut of *Hydrous caraboides*.

Corycella armata in the gut of larvae of *Gyrinus natator*.

Prismatospora evansi in the intestines of *Tramea lacerata* and *Sympetrum rubicundulum*.

Ancyrophora gracilis in the intestines of larvae and adults of *Carabus auratus*, *Carabus violaceus*, and *Carabus* sp., and in the larvae of *Silpha thoracica*. *Ancyrophora uncinata* has been reported from Trichoptera larvae.

Cometoides capitatus in the gut of larvae of *Hydrous* sp.

Actinocephalus acutispora in the gut of *Silpha laevigata;* an unidentified *Actinocephalus* in larvae of the Japanese beetle, *Popillia japonica;* and another unidentified species in the midgut of *Pardalophora apiculata*. *Actinocephalus elongatus* occurs in the alimentary tracts of *Dichromorpha viridis*, *Arphia sulphurea*, *Chortophaga viridifasciata*, and probably inhabits the intestines of other Orthoptera; *Actinocephalus giganteus* in a "majority of the grasshopper species"; *A. amphoriformis* in "practically all the species of grasshoppers"; and *A. fimbriatus* in *Dissosteira carolina* (see Bush, 1928).

Asterophora philica in the gut of *Nyctobates pennsylvanica*, and *Asterophora elegans* in certain Trichoptera.

Steinina rotunda in the gut of *Amara augustata*.

Pileocephalus striatus in the gut of larvae of *Ptychoptera contaminata*. *Pileocephalus chinensis* and *Pileocephalus heerii* have been observed in certain Trichoptera. Semans (1939) gave the name *Pileocephalus tachycines* to a gregarine he found in the midgut of the orthopteran *Tachycines asynamorus*. *Pileocephalus blaberae* occurs in *Blaptica dubia* and related species.

Discorhynchus truncatus in the gut of larvae of *Sericostoma* sp.

Pyxinia bulbifera occurs in the gut of adult *Dermestes lardarius*. *Pyxinia rubecula* has been found in the adult of *Dermestes vulpinus* and in the larva of *Dermestes lardarius*. Both the larvae and adults of *Dermestes peruvianus* and *D. vulpinus* harbor *Pyxinia crystalligera*. *Pyxinia frenzeli* occurs in the larva of *Attagenus pellio* and *Pyxinia möbuszi* in the larva of *Anthrenus verbasci*. In both the larva and adult of *Stegobium paniceum* was found a gregarine described as *Pyxinia anobii* (fig. 210).

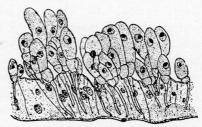

Fig. 210. Section through part of the mesenteron of *Stegobium paniceum* showing cephalonts of *Pyxinia anobii* attached to the epithelial cells. (From Vincent, 1922.)

Schneideria mucronata in the intestinal ceca of *Bibio marci* larvae.

Beloides firmus in the gut of larvae of *Dermestes lardarius*.

Taeniocystis mira in the gut of *Ceratopogon solstitialis* larvae.

Stictospora provincialis in the gut of larvae of *Melolontha* sp. and *Rhizotrogus* sp.

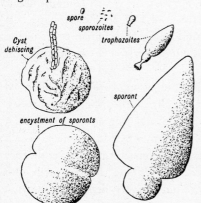

Fig. 211. *Agrippina bona*, showing the relative sizes of the various stages in the life history. Each stage drawn to the same scale. (From Strickland, 1912.)

Bothriopsis histrio in the gut of *Hydaticus* sp.

Coleorhynchus heros in the gut of *Nepa cinerea*.

Legeria agilis in the gut of larvae of *Colymbetes* sp.

Phialoides ornata in the gut of *Hydrophilus piceus* larvae.

Geneiorhynchus aeschanae in *Aeschna constricta*.

Caulocephalus crenata in the beetle *Aulacophora foveicollis*.

Agrippina bona (fig. 211) in *Nosopsyllus fasciatus*. Strickland (1912) found this gregarine in nearly every flea larva he examined, but he did not find it in the adult fleas. The gregarine was also present in the larval excreta. As far as Strickland was able to determine, the protozoan was in no way injurious to its host.

SCHIZOGREGARINARIA

The suborder Schizogregarinaria is a smaller group than the Eugregarinaria. Like the latter, the schizogregarines parasitize arthropods, annelids, and similar forms of life. They vary greatly in size including both the smallest and the largest gregarines known. Both sporogony and schizogony occur in this group. Sporogony is of the type already described for the eugregarines, but schizogony may take place outside or within the host cell. Schizogony may take place by binary fission, multiple fission, or by budding. As with the eugregarines, the fully grown trophozoites become paired, encyst, undergo sexual reproduction followed by sporogony, and produce gametes. The zygote is formed by the fusion of two gametes forming spores which contain from one to eight sporozoites.

The suborder Schizogregarinaria is sometimes divided into two families: Ophryocystidae, in which one spore is formed from two gametocytes, and Schizocystidae, in which two or more spores result from the union of two gametocytes.

Ophryocystidae. The principal genus in this family which includes gregarines associated with insects is the genus *Ophryocystis,* which contains about ten species occurring in the Malpighian tubes of Coleoptera. Perhaps best known of these is *Ophryocystis mesnili* found in the Malpighian tubes of the yellow mealworm, *Tenebrio molitor,* and *Ophryocystis hessei,* parasitic in *Omophlus brevicollis.*

When a spore is ingested by a suitable host, eight sporozoites are liberated and attach themselves to the surface of the cells of the Malpighian tubes. Here each sporozoite grows and becomes a multinucleate adult, which then segments into a number of merozoites. These merozoites, in turn, attach themselves to the cells and grow into adults. After repeating this several times schizonts containing several nuclei are produced. These then segment into forms, not merozoites, which will come together in pairs, being gametocytes, as in the case of the eugregarines. Within each of the encysted gametocytes three nuclei are produced, only one of which goes to form the nucleus of a single gamete. Thus, each of the two gametocytes produces only one gamete making two solitary gametes in the gametocyst. These two gametes conjugate and are encysted in a spindle-shaped spore or oöcyst within which are formed eight sporozoites.

Schizocystidae. In this family two or more spores are formed in a pair of gametocytes.

Some of the best known species are as follows:

Schizocystis gregarinoides is found in the gut of larvae of the midge *Ceratopogon solstitialis*. In this insect the sporozoites attach themselves to the gut epithelium and there develop into elongated, vermiform bodies which may reach the length of 150 microns. In each individual the nucleus multiplies producing, by schizogony, a number of merozoites which again attach themselves to the gut and become schizonts. After this process repeats itself several times, the merozoites develop into gametocytes which pair up and form gametocysts. The oöcysts finally formed each contain eight sporozoites. A closely related species, *Schizocystis legeri,* found in the intestines of *Systenus adpropinquans* and *Systenus scholtzii,* has its schizonts separated into a small protomerite and deutomerite.

Syncystis mirabilis occurs in the coelomic fluid and fat bodies of *Nepa cinerea*.

Mattesia dispora is found in the adipose tissue cells of larvae of the flour moth, *Ephestia kühniella,* and in the larva, pupa, and adult of the Indian meal moth *Plodia interpunctella*. According to some workers, the protozoan is highly pathogenic for its hosts.

Caulleryella pipientis occurs in the larval gut of *Culex pipiens; Caulleryella anophelis* in the gut of *Anopheles claviger; Caulleryella annulatae* in *Theobaldia annulata; Caulleryella maligna* in *Cellia allopha*. All of these hosts are mosquitoes. The genus was originally established by Keilin (1914) for *Caulleryella aphiochaetae* found in the larval gut of the Diptera *Aphiochaeta rufipes*. *Caulleryella* adults are more or less globular in shape. Otherwise the life history is very similar to that of *Schizocystis*. It is interesting to note that when the larval mosquitoes pupate, the infection dies out leaving the adult mosquitoes uninfected.

Wenyon (1926) lists two species in the genus *Lipotropha* (*L. macrospora* and *L. microspora*) which are found in the dipterous genus *Systenus*. The fat body of the infected larvae of these insects usually contains parasites in all stages of development. Interestingly enough, most of the life cycle takes place in the cells of the fat body. The merozoites are released in the hemocele fluid, where the gametocysts are formed eventually yielding sixteen oöcysts.

Machadoella triatomae occurs in the Malpighian tubes of the reduviid *Triatoma dimidiata*.

COCCIDIA

The Coccidia parasitize the tissues of vertebrates and invertebrates and attack chiefly the epithelium of the digestive tract and its associated glands. In the majority of cases the life cycle is passed in a single host. When the oöcysts are ingested by a suitable host, each of the liberated sporozoites enters an epithelial cell and becomes a trophozoite. The trophozoite increases in size until it becomes a spherical schizont nearly filling the cell. The schizont nucleus undergoes multiple division, each daughter nucleus eventually forming a small merozoite. When the epithelial cell ruptures, the motile merozoites enter new cells and the asexual cycle is repeated.

Instead of growing into schizonts some merozoites enter the cells and develop into male and female gametocytes. After a period of growth the male gametocyte divides into a number of microgametes while the female gametocyte forms a large oval macrogamete, which is penetrated by and unites with a microgamete. The zygote produced secretes a cyst wall about itself and becomes an oöcyst. The oöcyst nucleus divides into daughter nuclei which, surrounded by cytoplasm, form sporoblasts. Each sporoblast secretes an enveloping membrane (sporocyst) transforming itself into a spore containing the infecting sporozoites. If the oöcyst is passed in the feces the spore is formed outside of the body.

The order Coccidia may be divided into the suborders Eimeridea, in which the gametocytes are similar, and Adeleidea, in which the gametocytes are dissimilar.

The suborder Eimeridea contains very few members associated with insects. *Barrouxia ornata* is found in the gut of *Nepa cinerea*.

Some of the species in the suborder Adeleidea are entomophilic.

A dozen or more species of *Adelina* have been discovered in insects. Some of these (e.g., *A. mesnili* in *Tineola biselliella*) were previously in the genus *Adelea*. Examples of other *Adelina* are *A. tribolii* in *Tribolium castaneum* and *A. cryptocerci* in *Cryptocercus punctulatus*.

Legerella hydropori (fig. 212) is found in the epithelium of the Malpighian tubes of *Hydroporus palustris*. *Legerella parva* is the name given to a similar parasite in the Malpighian tubes of the chicken flea, *Ceratophyllus gallinae,* and the pigeon flea *Ceratophyllus columbae*. Another species, *Legerella grassii,* has been observed in the Malpighian tubes of *Nosopsyllus fasciatus*.

Chagasella hartmanni is found in the gut of the bug *Dysdercus ruficollis,* as is *Chagasella alydi*. The latter organism occurs in the gonads of its host as well as in the intestine.

Haemogregarina triatomae, the vertebrate host of which is a South American lizard (*Tupinambis teguixin*), has for its invertebrate host the reduviid *Triatoma rubrovaria*.

Hepatozoon muris is found in common rats in nearly all parts of the world

It was first observed in the brown rat, *Rattus norvegicus,* and since in several other species including white rats. The schizogony cycle takes place in the liver of the rat, and the young gametocytes invade the monocytes and resemble hemogregarines. When the infected blood is ingested by the mite, *Laelaps echidninus,* the parasites are liberated from the monocytes and associate in pairs; the union of the two gametes results in a vermiform organism. It then penetrates the intestinal epithelium and goes through the intestinal wall to the surrounding tissues where it becomes spherical and grows. It finally becomes a cyst, the contents of which break up into sporoblasts and then into spores. Each spore contains several sporozoites which infest the rat when the mite is ingested.

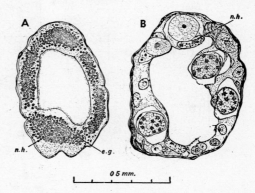

Fig. 212. *Legerella hydropori.* *A.* Section of a normal Malpighian tube of *Hydroporus palustris* showing the large cells charged with excretory granules. *B.* Section of Malpighian tube heavily parasitized by *Legerella hydropori.* *e.g.,* excretory granules; *n.h.,* nucleus of host cell. (From Vincent, 1927.)

Hepatozoon canis (fig. 213) is similar to *H. muris* and occurs in dogs in various parts of the world. The gametocytes undergo development in the dog tick *Rhipicephalus sanguineus.* Dogs are probably infected when they devour infected ticks.

Hepatozoon criceti is a parasite of the hamster (*Cricetus frumentarius*). Its invertebrate host is apparently the mite *Liponyssus arcuatus. Hepatozoon balfouri* parasitizes jerboas, and the invertebrate host is probably a mite of the genus *Dermanyssus* or the flea *Pulex cleopatrae. Haematopinus stephensi* is apparently the arthropod host of *Hepatozoon gerbilli* of the gerbil (*Gerbillus indicus*).

Unidentified species of *Hepatozoon* have been observed in *Phlebotomus papatasii* and probably in *Glossina palpalis.*

Karyolysus lacertarum parasitizes the lizard *Lacerta muralis.* Sexual reproduction of the protozoan takes place in the female mite *Liponyssus saurarum,* in which the sporokinetes enter the ova and mature. During the tissue differentiation of the mite embryo the sporocysts occupy cells which become the gut epithelium. After the mite's first blood meal the spores are cast out of the body, and when ingested by a lizard this animal becomes infected.

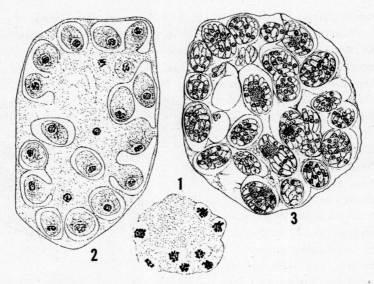

Fig. 213. *Hepatozoon canis.* Developmental stages in the tissues of the tick *Rhipicephalus sanguineus.* *1*, zygote (sporont) which has increased considerably in size and has developed a number of nuclei; *2*, section of fully grown sporont in which sporoblast formation is taking place; *3*, section of fully-developed oöcyst in which the sporocysts, each containing about sixteen sporozoites, have formed. (From Wenyon's *Protozoology,* 1926, Wm. Wood and Co., New York.)

HAEMOSPORIDIA

The Haemosporidia undergo schizogony in the blood of vertebrates and sporozoite formation in the alimentary tracts of invertebrates, always remaining within the body of one or two hosts. For this reason the sporozoites do not possess a protective envelope.

The order Haemosporidia is generally separated into three families: Plasmodiidae, Haemoproteidae, and Babesiidae.

Plasmodiidae. Included in this family are the four species of *Plasmodium* which cause malaria in man. These are *Plasmodium vivax* (tertian malaria), *P. falciparum* (malignant or aestivo-autumnal), *P. malariae* (quartan), and *P. ovale* (similar to tertian type). Other species occur in mammals, birds, and lizards.

That the malaria parasite is transmitted by mosquitoes was suspected by several early well-known observers as well as by the natives in certain of the malaria-infested regions of the world. This history makes extremely interesting reading, especially so the reports of such men as King (1882), Laveran (1891), Manson (1896, 1898), Ross (1897, 1898), Grassi (1898), and Nuttall (1898). Now, fifty years later, the true role of the mosquito in the spread of malaria is better understood but is still one of the greatest problems of medical entomology. The second World War brought the fight against this mosquito-borne disease to a new crescendo of research activity and importance (see Coggeshall, 1943a,b; Faust, 1944).

The clinical and therapeutic aspects of malaria are discussed in the numerous books on the subject. For the parasitologic aspects of malaria the reader is referred to such books as those by Chandler (1940), Belding (1942), Culbertson (1942), and others. It will be our chief purpose here to summarize briefly the biologic relationships existing between the parasites and their arthropod hosts.

In brief, the developmental cycle of *Plasmodium vivax* is as follows (fig. 214): The infected anopheline mosquito introduces the sporozoites (fig. 215) into the blood stream of a man. Here, after a preliminary exoerythrocytic stage, they invade the erythrocytes, grow, and undergo schizogony, producing a number of merozoites. These merozoites are liberated into the blood stream, those escaping phagocytosis enter other red blood cells and the schizogonic cycle is repeated.

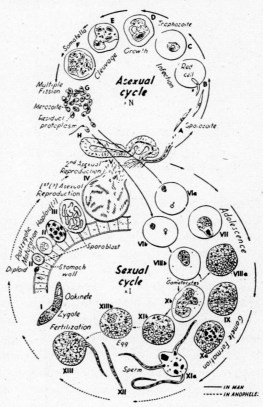

Fig. 214. Diagram of the life cycle of *Plasmodium vivax*, parasitic in *Anopheles* mosquitoes and in the red blood cells of man. (From Kofoid, in Calkins and Summers' *Protozoa in Biological Research*, 1941, Columbia University Press, New York.)

After several generations some of the merozoites appear as macrogametocytes (female) and microgametocytes (male). When a suitable species of mosquito

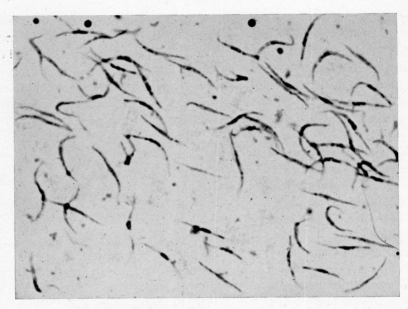

Fig. 215. Free sporozoites from the salivary gland of a mosquito. (Photo by N. J. Kramis.)

feeds on the infected blood, some of these sexual forms are taken into its stomach where they develop into macrogametes and microgametes. These unite in pairs forming zygotes called "oökinetes." The latter elongate and worm themselves through the stomach wall, lodging between the inner and outer linings of the stomach. Here each one grows rapidly while a cyst wall develops formed in part by the parasite and in part by the elastic membrane of the mosquito's stomach. The cysts (fig. 216) may be seen on the outer surface of the stomach protruding like tiny tumors or warts (40 to 60 microns). Inside, the nuclei rapidly undergo repeated divisions forming a large number of sporozoites. With the rupture of the cyst wall these sporozoites are released and make their way to the salivary glands (fig. 217) where they wait to be inoculated into the blood stream of another human.

The developmental cycle of *Plasmodium malariae, Plasmodium falciparum,* and *Plasmodium ovale* are essentially the same as that just described for *P. vivax.* There are minor and diagnostic differences, however. *P. malariae* does not cause the invaded erythrocyte to enlarge and does not have active movements as does *P. vivax,* which grows very large causing the blood cells to enlarge and which is extremely active during the early stages of its development in the erythrocytes. Unlike the other malarial parasites which sporulate in 48 hours, *P. malariae* sporulates once in 72 hours. Morphologic differences in the parasites also exist. Whereas the gametocytes of *P. vivax* and *P. malariae* are rounded, those of *P. falciparum* are crescentric. The shape and size of

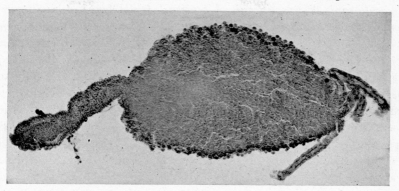

Fig. 216. The outer surface of the stomach of a mosquito infected with *Plasmodium falciparum*, showing the numerous protruding oöcysts.

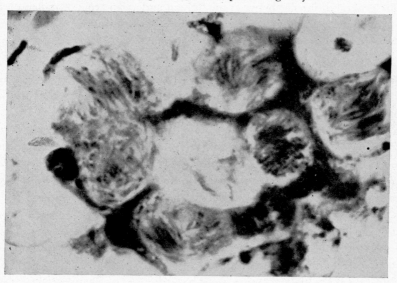

Fig. 217. Sporozoites grouped in the salivary gland of a mosquito. (Photo by N. J. Kramis.)

the schizonts also vary. The four species show very few differences in their development in the mosquito.

Species of Mosquitoes Concerned. The names of the mosquitoes concerned in transmission of malaria in North America have been listed by various writers. Simmons (1941) discusses the roles played by 7 nearctic and 23 neotropical anopheline species. Of these, the following are considered or suspected of being effective vectors of malaria: *Anopheles quadrimaculatus, A. maculipennis, A. punctipennis, A. crucians, A. albimanus, A. pseudopunctipennis, A. tarsimaculatus, A. argyritarsis, A. darlingi, A. albitarsis, A. punctimacula, A. hectoris,* and *A. bellator.* As Simmons points out, however, there is very little information available to indicate the relative importance of these anophelines as vectors. Some species have received a great deal of study and others have had very little. Numerous other species have been infected with parasites, but their true relationship to malaria has not been adequately determined. In Europe several varieties of *Anopheles maculipennis* are the most important vectors. In Asia *A. culcifacies, A. elutus, A. hyrcanus, A. ludlowii, A. minimus, A. stephensi,* and *A. superpictus* are important. *A. albimanus, A. albitarsis, A. argyritarsis,* and *A. darlingi* are the chief vectors in South America. In Africa *A. gambiae* and *A. funestus* are among the most effective carriers (Coggeshall, 1943b; Ross and Roberts, 1943).

One general difference between the human and avian types of malaria is in the genus of the mosquito vector concerned. The plasmodia of the human types are carried by *Anopheles,* whereas those of the avian types are transmitted chiefly by *Culex* mosquitoes.

Biologic relationships. The biologic relationships existing between the malarial parasites and their mosquito hosts have not been as well studied as one might suppose.

Although it varies with the temperature, the time usually considered necessary for mosquitoes to become infective is 10 or 12 days after an infective blood meal. The lower the temperature the longer the time necessary. *Plasmodium vivax* and *P. falciparum* may produce sporozoites in seven or eight days at 85° to 90° F. though this temperature will kill many of the mosquitoes. At 65° to 75° F., *P. vivax* becomes infective in 15 to 17 days and *P. falciparum* in about 19 days. The latter parasite is unable to produce sporozoites below 65° F. though *P. vivax* will slowly continue to do so down to 60° F. (See also Stratman-Thomas, 1940.)

Huff (1941) has made mention of the hazards faced by the malarial parasites from the time that the gametocytes are ingested by a mosquito until the sporozoites are delivered to a new host. At such critical times as when the oökinetes penetrate the stomach wall "it is known that far from all oökinetes succeed in getting through." Huff believes that the most important factors operating against such penetration are intrinsic, being connected with the physiology and chemistry of the stomach wall. Both extrinsic and intrinsic factors may be involved in the location of the oöcyst on the outside of the mosquito's stomach. Even after the liberated sporozoites make their way through the hemocele to the salivary glands, certain as yet unknown factors operate for or against the preservation of the sporozoite in a viable condition.

Just what determines whether a given species of mosquito is capable of transmitting the malarial parasites is far from being understood. Huff (1941) points out that, assuming all other factors as conducive to transmission, then susceptibility, or the lack of it, may play the "all-important part in determining whether a given mosquito may act as a vector." While different species may vary greatly in their susceptibilities to the parasites, individuals within a species may likewise vary in their capacities for becoming infected. In mass selection experiments with *Culex pipiens* and *Plasmodium cathemerium*, Huff (1929) was able to raise or lower the percentage of susceptible mosquitoes in a given stock by selecting from infected or uninfected individuals respectively. He (1931) found the characters responsible for susceptibility to be inherited in the 1:3 ratio and to be recessive.

Although it is known that the sporozoites may lose their infectiousness with age, the factors involved have not been determined. Humidity apparently exerts little or no effect on the sporozoites or their development. Other factors such as food, elevation, etc., have had little study. Temperature and the inherent make-up of the individual mosquito appear to be the greatest factors concerned in the infection by and the development of the malarial parasites in anopheline mosquitoes.

Other Plasmodiidae, presumably transmitted by mosquitoes, have been found in a variety of mammals, birds, and reptiles. Representative examples include *Plasmodium reichenowi*, *P. kochi*, *P. inui*, and others of monkeys; *P. bubalis* of a buffalo; *P. murinum* and *P. melanipherum* of bats; *P. brodeni* of a jumping rat; *P. pteropi* of a flying fox; *P. vassali* of a squirrel; and *P. agamae*, *P. mabuiae*, *P. tropiduri*, *P. minasence*, and others in lizards.

A considerable amount of experimental work has concerned the malarial parasites of birds, most of them transmitted by mosquitoes of the genera *Culex*, *Theobaldia*, and *Aëdes*. In the United States the better known species are *Plasmodium praecox*, *P. cathemerium*, *P. elongatum*, *P. vaughni*, *P. nucleophilum*, *P. polare*, and *P. circumflexum*.

Haemoproteidae. In this family, unlike the Plasmodiidae, schizogony does not take place in the peripheral blood of vertebrates, but in the endothelial cells. The merozoites, after entering the circulating blood cells, develop into gametocytes which, when ingested by certain bloodsucking insects, develop into gametes. The gametes unite forming zygotes, which undergo changes similar to those of the Plasmodiidae. Members of the Haemoproteidae occur in birds and reptiles. Representative species of the family's two genera include:

Haemoproteus columbae found in the pigeon *Columba livia*. It is transmitted by the flies *Lynchia maura, Lynchia brunea, Lynchia lividicolor, Lynchia capensis,* and *Microlynchia pusilla*. *Haemoproteus lophortyx* found in California Valley quail (*Lophortyx* spp.). Sexual reproduction takes place in the fly *Lynchia hirsuta*. *Lynchia fusca* may also act as a vector of this protozoan. The incubation period in the latter insect appears to be from 9 to 13 days.

A number of *Haemoproteus* species pathogenic for birds have been reported to pass part of their life cycle in both *Culex* and *Theobaldia* species of mosquitoes.

Of the various species of *Haemoproteus* occurring in reptiles very little or nothing is known of their invertebrate hosts.

Leucocytozoon anatis occurs in wild and domestic ducks, and its sexual cycle takes place in the black fly *Simulium venustum*.

Babesiidae. Members of this family are minute, nonpigmented parasites of the red blood cells of mammals and are transmitted by ticks. Kudo (1939) recognizes three genera: *Babesia (Piroplasma), Theileria,* and *Dactylosoma*.

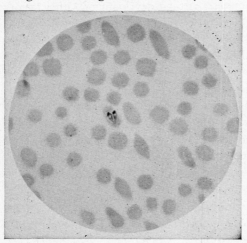

Fig. 218. *Babesia bigemina* in the blood of a cow suffering from Texas fever. The one parasitized erythrocyte located near the center of the field contains two parasites. (From Hagan, 1943.)

The outstanding species of the genus *Babesia* is *B. bigemina,* the cause of Texas fever or red-water fever of cattle (fig. 218). It is transmitted by the tick *Boöphilus annulatus*. This fact was discovered by Smith and Kilbourne in 1893 and is of great historic as well as practical importance since it was the very first demonstration that an arthropod was capable of transmitting a protozoan. Likewise startling was the finding that the parasite was transmitted to the next generation through the egg. This was evident by the fact that the tick remains on the one host through all stages of development and only drops off as an adult to lay its eggs.

Dennis (1931, 1932) has presented in considerable detail the life history of *Babesia bigemina* in the tick (*Boöphilus annulatus*). According to his version, not accepted by all authorities, after the parasites are ingested by the tick the isogametes fuse and develop into motile, club-shaped oökinetes. These bodies then penetrate the gut wall, grow, and form sporonts, or they may pass completely through the gut wall, enter the ovary, and invade the ova. Here the oökinete also forms a sporont which divides to form multinucleated, amoeboid sporokinetes. These migrate throughout the embryonic tissue cells of the tick including those of the anterior embryonic cell mass, part of which are destined to develop into the salivary glands. Here the sporokinetes undergo multiple fission and form sporozoites—usually before the tick hatches though some may wait until after it has hatched. According to Dennis, the sporozoites are inoculated into the bovine blood stream in the saliva of the feeding larval tick.

Working with *Boöphilus microplus,* Regendanz (1936) failed to detect sporonts and sporoblasts of *Babesia bigemina* as described by Dennis in *B. annulatus*. In fact, the parasite did not develop at all well in this tick, and Regendanz could find no signs whatever of invasion of the salivary glands. (See also Reichenow, 1935.)

It is of interest to note that more than 99 per cent of the territory in the United States originally infested with this tick has been freed of it and of the disease it carried.

Babesia bovis causes a disease in cattle very similar to Texas fever but it occurs chiefly in Europe. It is transmitted by *Ixodes ricinus* and occasionally other ticks. *B. berbera* occurs in zebus and in young cattle in Madagascar.

Species of *Babesia* also occur in sheep, goats, horses, dogs, felines, pigs, and other animals. There are probably three species which attack sheep: *Babesia motasi* transmitted by *Rhipicephalus bursa; B. ovis* and *B. sergenti* of which the transmitting agents are unknown. *B. caballi* and *B. equi* of horses are transmitted by the ticks *Dermacentor reticulatus* and *Rhipicephalus evertsi*, respectively. *B. canis,* the cause of canine babesiosis or malignant jaundice in dogs, is transmitted by *Rhipicephalus sanguineus* in the United States and India, by *Dermacentor reticulatus* in Europe, and by *Haemaphysalis leachi* in South Africa. Like *B. bigemina, B. canis* passes through the eggs of the ticks to the next generation. *B. gibsoni* also may occur in dogs. *B. felis,* found in pumas, may be transmitted by *Rhipicephalus simus.*

The genus *Theileria* includes *T. parva* (*T. kochi*), the agent of an important disease of cattle in Africa known as "East Coast fever" (fig. 219). (Neitz, 1943, has suggested the possibility that the disease is caused by the combination of a "virus" and *T. parva.*) It is transmitted most commonly by the tick *Rhipicephalus appendiculatus* (*R. nitens*), although *R. evertsi, R. simus,* and

R. capensis are also capable of transmitting it. Experimental transmission has been successful with *Hyalomma impressum* near *planum*. In the case of *R. appendiculatus*, development of *T. parva* was found by Lewis and Fotheringham (1941) not to be retarded when the engorged tick larvae were exposed to a temperature of 4° to 6° C. for 3 days at intervals of 2, 4, 6, 8, and 9 days after repletion, indicating that climatic conditions do not kill or weaken the parasite as long as the tick survives. The protozoa tend to disappear from "hungry" ticks kept under laboratory conditions for about a year or more. These workers also noted that ticks do not become infected with *T. parva* from the blood of a bovine for the first four days of the reaction period.

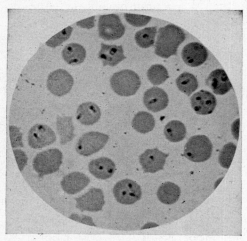

Fig. 219. *Theileria parva* in the erythrocytes of the blood of a cow during the febrile period of East Coast fever. (From Hagan, 1943.)

All of the details of the life cycle of *Theileria parva* are not as yet known with certainty. One of the first reports to describe the significant stages of the parasite's life cycle in the tick *R. appendiculatus* was that of Cowdry and Ham (1932). An outline of their findings is given in figure 221.

Reichenow (1940) has also described the life cycle of *Theileria parva*. According to this worker, the organisms multiply in the lymphocytes of the lymphatic system of cattle, eventually entering the blood stream where they may enter the erythrocytes in small numbers but do not multiply in them. When the blood is ingested by ticks (*R. appendiculatus*), it is digested intracellularly in the intestinal cells. Some of the parasites get into the hemocele and eventu-

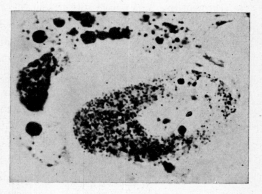

Fig. 220. *Theileria parva* (sporozoites) in the salivary glands of the tick *Rhipicephalus appendiculatus*. (Courtesy Dr. E. V. Cowdry.)

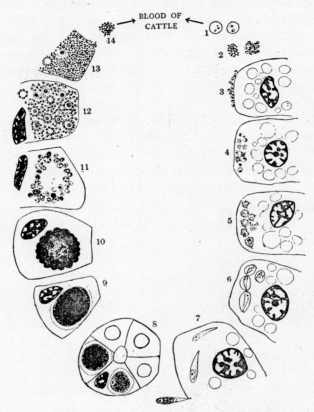

Fig. 221. Diagrammatic representation of the life cycle of *Theileria parva* in *Rhipicephalus appendiculatus* according to Cowdry and Ham (1932). *1.* After the tick has had an infective feeding the ingested red blood cells may be seen to contain parasites of varying size and shape. Some are distinctly small and some distinctly larger. 2. In the gut of the tick the large and small types of parasites escape from the erythrocytes and (*3*) locate themselves on the surface of the epithelial cells lining the gut. *4.* The parasites enter these cells and the small forms rapidly disappear. The large forms grow and give rise to a zygote without distinct nuclei. 5. These zygotes increase in size with nuclei reappearing in them. 6. Large, elongated forms ("oökinetes") develop. *7.* These escape into the cytoplasm of the gut epithelial cells and then make their way into the body cavity. 8. The oökinetes enter the cells of the salivary glands where they become rounded and surrounded by a halo. The nucleus soon disappears as further growth takes place (*9*). *10.* "Buds" (sporoblasts) appear about the periphery of the mass. *11.* The sporoblasts become more distinct and develop rapidly. *12.* The peripheral substance gives rise to sporozoites which subsequently become detached from the parent sporoblasts. *13.* The sporozoites become reduced in size and consist of a slightly elongated mass of cytoplasm, in one extremity of which is situated a nucleus. *14.* The sporozoites are discharged into the lumen of the salivary gland acinus from whence they are passed into the animal upon which they are feeding.

ally reach the salivary glands where they remain dormant until the next time the tick feeds, when they multiply in large numbers within the cells of the salivary glands. Reichenow observed some cells to contain as many as 30,000 parasites. When these cells break down the organisms are liberated with the saliva into a new vertebrate host. No schizogony or gametogony occurs; multiplication is by fission only. There is apparently no transovarial transmission of the parasites in the ticks.

Other species of *Theileria* (*T. dispar, T. mutans,* and *T. annulata*) have been reported as causing diseases in cattle. These organisms differ biologically but very little morphologically. For a description of the developmental cycle of *T. dispar* in the tick *Hyalomma mauritanicum,* the reader is referred to a paper by Sergent, Donatien, Parrot, and Lestoquard (1936).

Anaplasmata. At this point may be mentioned a group of organisms of doubtful nature which are transmitted by arthropods. In the course of their work on Texas cattle fever Smith and Kilbourne (1893) observed small coccuslike bodies or granules on the margins of the red blood cells of cattle. Though these workers thought them to be related to the Texas fever parasite, it has since been thought that they were dealing with two diseases simultaneously—Texas fever and an anaplasmosis. Many believed these bodies to be artifacts or granules similar to the "Jolly bodies" of young and anemic animals. In recent years evidence has been accumulated which supports their parasitic nature, though whether they are protozoa or bacteria is still in some doubt (see Cowdry and Rees, 1935). The name *Anaplasma marginale* (fig. 222) has been given the organism, the generic name referring to the apparent absence of cytoplasm about the parasite. Almost 20 species of ticks have been proved capable of transmitting the agent of an anaplasmosis. Though most of these are probably mechanical carriers, some are biological carriers, e.g., *Boöphilus annulates, Dermacentor occidentalis, Dermacentor andersoni,* and *Ixodes ricinus.* Sev-

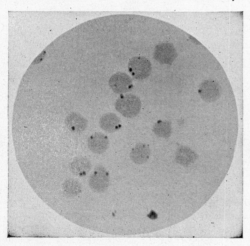

Fig. 222. *Anaplasma marginale* in the blood of a cow suffering from acute anaplasmosis. (From Hagan, 1943.)

eral species of mosquitoes and horseflies (Tabanidae) have been suspected or shown to be capable of being at least mechanical carriers.

Practically nothing is known concerning the life cycle of *Anaplasma marginale* in its arthropod hosts. It is known, however, that in the ticks named above, with the possible exception of *D. andersoni,* the agent passes through the egg to the next generation.

Anaplasma centrale is the name given to a similar organism characteristically located in the central part of the cattle's erythrocytes instead of in the periphery. The agent apparently is also transmitted by ticks such as *Boöphilus annulatus* and others.

ACNIDOSPORIDIA

This inadequately known group of protozoa has spores of rather simple structure and produces a characteristic cycle of development. The subclass Acnidosporidia is usually divided into two orders: Sarcosporidia, occurring in the muscles of higher vertebrates, and Haplosporidia, found in fish and invertebrates. The latter group contains the only entomophilic species of note, namely, *Coelosporidium periplanetae* and *Coelosporidium blattellae* which commonly occur in the Malpighian tubes of cockroaches. Some workers consider these organisms to be microsporidia.

CNIDOSPORIDIA

MICROSPORIDIA

Of the subclass Cnidosporidia the order Microsporidia is one of the most important from the standpoint of the parasitism of insects. In fact, the greater part of the protozoa known to be pathogenic for insects belongs to the Microsporidia, almost all of which invade and destroy the cells of their hosts. According to Kudo (1924), externally and internally the infected insect may show a number of changes due to infection with these protozoa. If the insect is transparent, it may become opaque, and coloration different from normal often appears. In silkworms infected with *Nosema bombycis,* brown spots are observed (fig. 223). Red colorations have been noticed in a few cases in other insects. The activity of the host decreases and often the insect is smaller than normal. In silkworms the infected insects are noticeably smaller. When the infection is very intensive, a deformity of the host is noted, tumor masses causing distention of the infected parts. *Simulium* larvae infected with microsporidia never pass through the pupal stage to the adult. The most conspicuous change is that the host cell undergoes striking hypertrophy of its nucleus, which becomes greatly enlarged and may increase in number or may show various ways of division. The cytoplasm also is greatly enlarged. Kudo says he has observed enormous enlargement of the cell body and nucleus of the infected fat body of mosquito larvae and of *Baëtis* sp. due to microsporidian infection.

Fig. 223. The middle segments of a *Bombyx mori* larva infected with *Nosema bombycis*. Note characteristic dark spots. (From Stempell, 1909).

The Microsporidia apparently have no intermediate host.

There seems to be a specific relation between the host and its Microsporidia. In a great number of cases the parasite invades only a specific tissue of the insect. A few of the Microsporidia are specific for a certain species or certain genus of insect, as *Thelohania legeri, Thelohania obesa,* and *Nosema anophelis* seem to be exclusive parasites of *Anopheles* mosquitoes found in United States and in France, while *T. opacita, T. rotunda, T. minuta,* and *Stempellia magna* are exclusive parasites of the *Culex* genus (Kudo, 1924).

The spores of microsporidian parasites are usually rather small (3 to 6 microns in length) and may resemble yeasts or bacteria in appearance. The spore membrane surrounds the sporoplasm and the polar filament which is a long, fine filament that can be extruded under pressure or by the aid of certain chemicals. This structure may be coiled in the spore or encased in the polar capsule. When the spores are ingested by a suitable host, the polar filaments are extruded and perhaps anchor the spores to the gut epithelium (Kudo, 1939). After the filaments become detached the sporoplasms emerge through the opening and by amoeboid movements they make their way through the gut epithelium entering the blood stream or body cavity. From here they find their way to the particular part or parts of the body which they specifically parasitize. After entering the host cells they undergo schizogony, the schizonts becoming sporonts, and the latter producing characteristic spores. When these spores reach a new host the cycle is repeated, although in many cases the spores are capable of germinating in the same host body, thus increasing the number of infected cells.

The Microsporidia may be divided into two suborders: Monocnidea, in which the spore has a single polar filament, and Dicnidea, in which the spore has two polar filaments. The suborder Monocnidea consists of three families: Nosematidae, Coccosporidae, and Mrazekiidae.

Nosematidae. Nosema. The most noteworthy genus of this family is *Nosema* in which each sporont develops into a single spore. Of the Hexapoda, the Lepidoptera seem to be the most frequently infected with *Nosema,* the Diptera next frequently so.

Nosema bombycis is the cause of the well-known silkworm disease, pébrine, which often assumes epidemic proportions. If it had not been for the illustrious work of Pasteur with this disease, the entire silk industry of France would probably have been ruined. The parasite invades practically all the tissues of the embryo, larva, pupa, and adult silkworm (*Bombyx mori*). The larvae become undernourished, atrophy, and finally die. The silk glands are usually heavily invaded, which hinders the formation of silk and hence the larvae cannot spin cocoons. Even when cocoons are spun, the silk threads are not uniform, and are easily broken. Infected moths are generally weak and often cannot copulate. The females lay fewer and abnormal eggs. Often the embryo dies within the egg. On the posterior ventral side of the host body, dark brown or black irregular spots occur. The organs of heavily infected individuals exhibit a milky-white appearance because of the presence of large numbers of spores.

Nosema bombycis may be acquired either through infected eggs or by ingestion of the spore with food. After the microsporidian spore is ingested, a small amoeboid body (sporoplasm) escapes. It multiplies in the intestine of the insect until some of the parasites penetrate the intestinal epithelium and pass into the hemocele space. From here they reach the various tissues of the body including the ovary. When they eventually enter the cytoplasm of the cells, they multiply by schizogony, gemmation, or by binary fission at the expense of the host cell. Spores containing four nuclei each finally develop. Two of these nuclei, together with

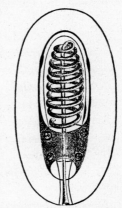

Fig. 224. A schematic sketch of the spore of *Nosema bombycis* greatly enlarged. (From Stempell, 1909.)

some of the cytoplasm, go to make up the spore capsule; a third helps form the terminal polar capsule, and the fourth becomes the nucleus of the infective sporoplasm. In size the spores (fig. 224) are 3 to 4 microns long by 1.5 to 2 microns wide. The polar filament is between 57 to 72 microns long. Trager (1937), who was able to bring about the partial development of the protozoan in tissue cultures, was able to prove the presence of minute extracellular forms, the "planonts" of Stempell, in the blood of silkworms which had recently ingested spores.

Besides the silkworm, other insects have been experimentally infected with *Nosema bombycis*. These include the larvae of *Arctia caja, Margaronia pyloalis, Chilo simplex,* and *Hyphantria cunea*. Kudo and De Coursey (1940) studied the parasite in the last-named insect, which they found to be susceptible when the spores were ingested. Schizogony and sporogony were similar to those which occur in the silkworm, and the symptoms were characteristic of pébrine.

The following species of *Nosema*, as well as *N. bombycis*, occur in the Lepidoptera:

Nosema	*Host*
N. auriflammae	Scea auriflamma
N. junonis	Dione juno, Papilio pompejus
N. micrattaci	Micrattacus nanus
N. lysimniae	Mechanites lysimnia
N. lophocampae	Lophocampa flavostica
N. hydriae	Hydria sp.
N. heliotidis	Heliothis armigera
N. halesidotidis	Halesidotis (Halsidota) sp.
N. ephialtis	Ephialtes angulosa
N. vanillae	Dione vanillae
N. erippi	Danais erippus, D. gilippus
N. eubules	Catopsilia eubule
N. caeculiae	Caeculia spp.
N. astyrae	Brassolis astyra
N. sabaunae	Bombycidae (larva)

Nosema schneideri has been found in the gut epithelium of *Ephemera vulgata; Nosema ephemerae* in the intestine of *Ephemera* sp.; and *Nosema baëtis* in the fat body of *Baëtis* sp.

Nosema bombi has been observed to parasitize *Bombus agrorum, Bombus hortorum, Bombus latreillelus, Bombus lapidarius, Bombus sylvarum, Bombus terrestris, Apis mellifera,* and *Apis florea*. The parasite may be found in the Malpighian tubes and alimentary tract. There is no evidence of hereditary infection.

Nosema apis causes a disease of the adult honeybee, *Apis mellifera,* attacking the midgut epithelium and occasionally the Malpighian tubes. It does not seem to multiply in the blood. The spores enter the body of the adult bee with water or food. They germinate in the stomach and attack the tissues lining the stomach or midgut, which swell and become discolored. The bees often have a dysentery, and inability to fly any distance is observed. Many bees will be seen crawling about on the ground or on the hive. The legs seem to be paralyzed, and the rear wings are held at an abnormal angle. Adult workers, drones, and queens are affected by the disease. Fantham and Porter (1913) produced experimental infections among the following insects: *Bombus terrestris, Bombus lapidarius, Bombus hortorum, Bombus venustus, Bombus latreillelus,* Mason bees, *Vespa germanica, Pieris brassicae, Callimorpha jacobeae, Abraxas grossulariata, Calliphora erythrocephala, Tipula oleracea,* and *Melophagus ovinus.*

Several species of *Nosema* attack mosquitoes. *Nosema anophelis* attacks the epithelial cells of the gastric pouch and midgut of *Anopheles quadrimaculatus; Nosema stegomyiae,* the adults and larvae of *Aëdes aegypti* (=*Stegomyia fasciata*); *Nosema culicis,* the larvae of *Culex pipiens;* and an unnamed species of *Nosema,* the larvae of *Aëdes nemorosus* and *Aëdes cantans.*

Nosema ctenocephali was found in the salivary glands, Malpighian tubes, and fat body of the dog flea, *Ctenocephalides canis* by Korke (1916) in India. Nöller (1912) had earlier isolated a microsporidium from the dog flea in Germany and named it *Nosema pulicis.*

Nosema adiei is supposed to occur in the salivary glands and other tissues of bedbugs in India.

Nosema chironomi occurs in the larvae of *Chironomus* sp.

Nosema longifilum is found in the fat body of *Otiorhynchus fuscipes.*

Perezia. In this genus there are three species which attack the larva of the cabbage butterfly *Pieris brassicae,* and one that parasitizes the corn borer *Pyrausta nubilalis.* The former are *Perezia mesnili, P. legeri,* and *P. pieris;* the latter is *Perezia pyraustae. Perezia mesnili* and *P. pieris* attack chiefly the silk gland and the Malpighian tubes. *Perezia aeschnae* has been reported from the nymph of the dragonfly, *Aeschna grandis.*

Gurleya. *Gurleya francottei* occurs in the epithelial cells of the midgut of the larva of *Ptychoptera contaminata;* and *G. legeri,* in the fat body, muscle, and connective tissue of nymphs of *Ephemerella ignita* and in the fat body of the larval caddis fly (Trichoptera).

Thelohania. There are large numbers of the *Thelohania* which parasitize insects, especially the Diptera. Mosquitoes seem especially susceptible. *Thelohania legeri* has been isolated from the following species of *Anopheles: A. claviger, A. crucians, A. maculipennis, A. punctipennis, A. quadrimaculatus, A. barbirostris, A. hyrcanus* var. *nigerrimus, A. annularis, A. ramsayi, A.*

gambiae, A. varuna, A. subpictus, and *A. vagus. Thelohania indica* and *T. anomola* have been reported from *Anopheles ramsayi; T. indica,* from *A. hyrcanus* var. *nigerrimus; T. pyriformis,* from the fat body of larva of *Anopheles crucians* or *A. quadrimaculatus; T. obesa,* from the fat body of an anopheline larva, perhaps *A. quadrimaculatus;* and *T. obscura,* from *A. varuna.*

Thelohania rotunda and *T. minuta* have been found in the fat body of larva of *Culex leprincei;* a *Thelohania* sp. in *Culex pipiens; Thelohania opacita,* in *Culex restuans, Culex testaceus,* and *Culex* sp. A species of *Thelohania* has also been isolated from *Aëdes nemorosus.* For a list of protozoan and other parasites of mosquitoes, the reader is referred to a compendium of such by Speer (1927).

Various species of *Simulium* larvae are also attacked by *Thelohania;* the fat bodies of *Simulium bracteatum, S. hirtipes, S. maculatum, S. ochraceum,* and *S. venustum* by *Thelohania bracteata* and *T. fibrata; S. bracteatum, S. maculatum,* and *S. vittatum* by *T. multispora;* and *S. ornatum* and *S. reptans* by *T. varians.*

Other Diptera affected are *Tanypus varius* by *Thelohania pinguis; Fannia scalaris* by *T. ovata; Limnophilus rhombicus* by *T. janus; Corethra plumicornis* by *T. corethrae;* and *Corethra* sp. by *T. brasiliensis.*

Among the Hexapoda other than Diptera, it has been found that *Thelohania ephestiae* causes a disease in *Ephestia kühniella* and *T. mesnili* in the corn borer, *Pyrausta nubilalis. T. mutabilis* was found in the adipose tissue of the nymph of *Ameletus ludeus, T. cepedei* in the Malpighian tubes of *Omophlus brevicollis,* and *T. baetica* in the adipose tissue of *Baëtis pygmaea.*

Stempellia. *Stempellia mutabilis* parasitizes the cells of the fat body of the nymphal May fly, *Ephemera vulgata,* in France. *S. magna* (fig. 225) occurs in the fat body of the larva of *Culex pipiens* and *C. restuans.*

Trichoduboscqia. This genus contains the single species *Trichoduboscqia epori* which infects the fat bodies of nymphs of *Phithrogena semicolorata* and *Epeorus torrentium.*

Plistophora. This species develops as many as 16 spores. *Plistophora simulii,* of which there are several variants, occurs in the larvae of *Simulium venustum, Simulium ochraceum,* and *Simulium maculatum.* From the epithelial cells of the ceca and midgut of *Blatta orientalis, P. kudoi* has been isolated. *Plistophora vayssierei* was noted as causing an infection among larvae of *Baëtis rhodani. P. stegomyiae* parasitizes both larvae and adults of *Aëdes aegypti* (=*Stegomyia fasciata*). In the adult the seats of infection are the stomach, esophagus, airsac, coelom, the Malpighian tubes, ovary, ova, thoracic

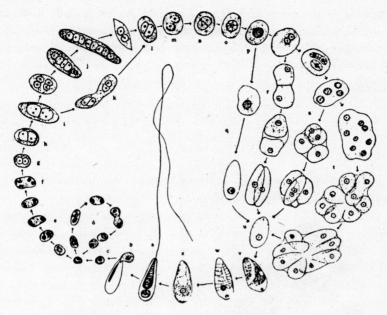

Fig. 225. The life cycle of *Stempellia magna*. *a, b,* germination of spore in the midgut of culicine larva; *c–k,* schizogony; *l–p,* sporont formation; *q–t,* formation of 1, 2, 4, and 8 sporoblasts; *u,* sporoblast; *v–x,* development of sporoblast into spore. (From Kudo's *Protozoology,* 1939, C. C. Thomas, Baltimore.)

muscles, the large ganglion, and the tracheal epithelium. The larvae are not so generally infected. The most frequent method of new infection seems to be through the eggs, though infection through the mouth also occurs. *Plistophora periplanetae* occurs in the Malpighian tubules of *Blattella germanica* and *Blatta orientalis,* and in the alimentary tract of *Periplaneta americana.*

Coccosporidae. This family consists of a single genus, *Coccospora,* earlier known as *Cocconema.* The spores of this genus are spherical, resembling cocci. *Coccospora micrococcus* parasitizes the fat body of the larva of *Tanypus setiger,* the infected larvae appearing milky-white and swollen. The spore is very small having a diameter of 1.8 to 2 microns. *C. polyspora,* the spores of which are from 2 to 3.2 microns in diameter, occurs in the adipose tissue of the larvae of *Tanypus* sp. *Coccospora octospora* was found in the gut epithelium of the larva of a species of *Tanytarsus.* The spore diameter in this case is slightly over 2 microns.

Mrazekiidae. Kudo (1939) recognizes six genera in this family of which at least four contain members which parasitize insects.

Mrazekia. In the genus *Mrazekia* we have *M. brevicauda* which occurs in the fat body of larvae of *Chironomus plumosus*. *M. bacilliformis* was found in the fat body of larvae of *Orthocladius* sp. and *M. tetraspora* in the fat body of larvae of *Tanypus* sp.

Octosporea. The best-known species of the genus *Octosporea* is *O. muscae-domesticae* first seen in the gut and germ cells of the housefly *Musca domestica*. Soon afterward it was observed in *Drosophila confusa* and *Drosophila plurilineata*. Parasitic in the same species of *Drosophila* and in *Fannia scalaris* is *Octosporea monospora*.

Spiroglugea and Toxoglugea. In the fat body of *Ceratopogon* sp. have been seen the species *Spiroglugea octospora* and *Toxoglugea vibrio*.

Dicnidea. The suborder Dicnidea (in which the spores have two polar filaments) consists of the single genus *Telomyxa* with only one species, *Telomyxa glugeiformis*, which parasitizes the cells of the fat body of *Ephemera vulgata* larvae. The infected larvae are chalky-white, and the parasite usually kills the insect.

HELICOSPORIDIA

This order was erected by Kudo to include an unusual organism, *Helicosporidium parasiticum*, discovered and described by Keilin (1921a). It has been placed in the Cnidosporidia more or less tentatively. Its hosts were found to be the Diptera *Dasyhelea obscura* and *Mycetobia pallipes* and the mite *Hericia hericia*. In *Dasyhelea obscura* heavy infection is found during the larval stage, though pupae also occasionally harbor the parasite. The adults apparently are free of the infection.

Fig. 226. Living larva of *Dasyhelea obscura* heavily parasitized by *Helicosporidium parasiticum*. The larva has been pressed between a slide and a cover slip so as to rupture the posterior end of its body, liberating a large mass of spores. (From Keilin, 1921a.)

When a larva becomes infected, small, round trophozoites invade the host tissues or body cavity. These multiply by schizogony. The schizonts form a small morula consisting of four or eight merozoites which become free. Spores are eventually formed and these consist of four cells surrounded by a thin wall or sporocyst. Three of the four cells are amoeboid and form the real sporozoites. The fourth cell develops into a peripheral spiral filament surrounding the central cells. The sporozoites are liberated when the unrolling of the spiral filament opens the spores inside the dead body of their host. When entirely unrolled the spiral filament is 60 to 65 microns long, and 1 micron thick at its widest part, being pointed at both ends. This is the only species of this genus known.

CILIOPHORA

As noted at the beginning of this chapter, the phylum Protozoa may be divided into two large subphyla: Plasmodroma and Ciliophora. We come now to the latter of these two subdivisions in which we find considerably fewer protozoa associated with insects than are in the Plasmodroma.

The Ciliophora are characterized by the possession of cilia, which are used as a means of locomotion and for capturing food. Most Ciliophora contain two kinds of nuclei, a large macronucleus which has to do with the organism's metabolic activities, and a small micronucleus which controls the reproductive processes. This group of protozoa reproduce asexually by binary fission or by budding, and sexually chiefly by conjugation.

The subphylum Ciliophora is usually divided into two classes:

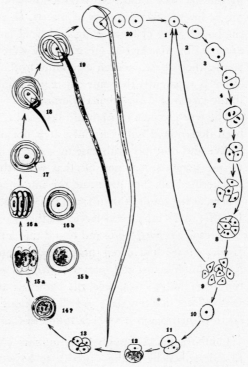

Fig. 227. The life cycle of *Helicosporidium parasiticum*. Stages 1–16 occur in the living larva of *Dasyhelea*, but stages 17–20 are found only in the dead body of the host. (From Keilin, 1921a.)

Ciliata, in which the organisms have cilia present throughout their trophic life, and Suctoria, in which cilia are present only while the organisms are young, the matured adults having tentacles. Some authors describe a third class, Opalinata, most species of which are found in the intestines of Amphibia.

CILIATA

Of the class Ciliata only two of the orders are known to contain species associated in some capacity or other with insects: Holotricha and Spirotricha.

HOLOTRICHA

Glaucoma. In this genus has been placed an interesting species, *Glaucoma pyriformis,* which has been reported inhabiting naturally and experimentally the bodies of several species of insects. *G. pyriformis* was first discovered in the body cavity of both living and dead larvae of the mosquito *Theobaldia annulata* by MacArthur (1922). They were found in the head region of the larvae, their presence even extending to the antennae and other appendages. What is apparently the same ciliate was found by Treillard and Lwoff (1924) in the larvae of *Chironomus plumosus.*

The true biologic relationship between *G. pyriformis* and its hosts has not been determined with certainty. It is not obligately parasitic since it is able to live and multiply with ease in water or other suitable media. As Wenyon (1926) suggests, it is possible that when the ciliates are ingested by certain sick individuals they penetrate the intestinal wall and enter the body cavity where, as the result of their rapid multiplication, they hasten the larva's death. Lwoff (1924) inoculated larvae of the wax moth, *Galleria mellonella,* with cultures of *G. pyriformis* and found that a fatal infection resulted. Janda and Jírovec (1937) inoculated this protozoan into annelids, mollusks, crustaceans, fishes, amphibians, and insects (*Dixippus morosus* and other species). Only the insects were susceptible, the hemocele and adipose tissue becoming filled with ciliates in a few days and all infections terminating fatally. Ciliates similar to *G. pyriformis* have been found in *Chironomus plumosus.*

Lambornella. Although some authors (Wenyon, 1926) claim this genus, first described by Keilin (1921b), to be without definition, some protozoologists recognize it. The genus was erected to include a ciliate (*Lambornella stegomyiae*) found by Lamborn (1921) in the hemocele of larvae of *Aëdes scutellaris.* Lamborn observed that none of the infected larvae reach maturity and that the ciliates could be found in all parts of the body cavity including the siphon and gills (fig. 228).

Espejoia. According to Kudo (1939), *Espejoia mucicola* is found in the gelatinous envelope of the eggs of certain insects and mollusks.

Ophryoglena. *Ophryoglena collini* occurs in various tissues including the generative organs of ephemerid larvae (*Baetis*).

Isotricha. *Isotricha caulleryi* is found in the alimentary tract of the American roach, *Periplaneta americana*.

SPIROTRICHA

In this order the genera *Balantidium* and *Nyctotherus* are apparently the only ones containing members associated at some time or other with insects.

Balantidium praenucleatum is a rather large ciliate found in the colon of the cockroach *Blatta orientalis*. *Balantidium blattarum* and *B. ovatum* are two ciliates found in the cockroach *Periplaneta americana*.

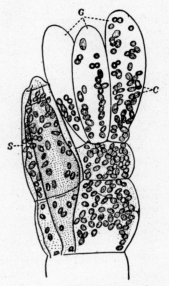

Fig. 228. The posterior end of a larva of *Aëdes* (*Stegomyia*) *scutellaris*, siphon (*S*) and gills (*G*) parasitized by the ciliate *Lambornella stegomyiae* (*C*). (From Keilin, 1921b.)

Balantidium orchestium has been described as occurring in the sand fleas *Orchestia agilis* and *Talorchestia longicornis*. *Balantidium knowlesi* has been found in the coelomic cavity of *Culicoides peregrinus* in India.

It is possible that *Balantidium coli*, the cause of balantidial dysentery in man, may occasionally be carried in or on flies from infective feces to man and his food.

Keilin (1921b) cites Bütschli as stating that a *Nyctotherus* occurs in the alimentary tract of adults of *Blatta*, *Gryllotalpa*, and *Hydrophilus*. This probably referred to *Nyctotherus ovalis*, which is found in *Periplaneta americana* and *Periplaneta australasiae* as well. It has been observed (Armer, 1944) that in *P. americana* the persistence of *N. ovalis* is apparently not influenced by the diet of the host although starvation does decrease the size and number of the protozoa. According to Semans (1939), this ciliate also lives in the alimentary canal of *Blattella germanica*, *Blatta orientalis*, *Gryllotalpa gryllotalpa*, and *Parcoblatta pen(n)sylvanica*. *Nyctotherus buissoni* has also been found

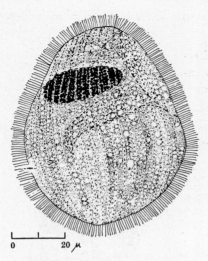

Fig. 229. *Nyctotherus osmodermae*. (From Zeliff, 1933.)

in cockroaches. There are a number of other *Nyctotherus* species which are found in insects. Among them are *Nyctotherus hormeticae* found in *Hormetica laevigata; N. uichancoi,* in *Panesthia javanica; N. neocurtillae,* in the mole cricket *Neocurtilla hexadactyla; N. panesthiae,* in *Panesthia angustipennis; N. pintoi,* in *Stethotorax ater* var. *brasiliensis;* and *N. osmodermae* (fig. 229), in the intestine of *Osmoderma scabra.*

SUCTORIA

Members of the class Suctoria are rarely found associated with insects. *Rhynchophrya palpans* has been found on the water beetle *Hydrophilus piceus.*

PROTOZOAN DISEASES OF INSECTS

Although the various species of protozoa which are known to cause diseases in insects have already been given in the foregoing parts of this chapter, there are a few aspects of this relationship which remain to be discussed.

The earliest protozoan disease to be investigated thoroughly was pébrine, a microsporidian disease of silkworms caused by *Nosema bombycis*. At first the minute protozoa were believed to be merely a symptom and not the inciting agent of the disease. Even Pasteur, who discovered the mode of infection, originally considered these bodies to be similar to the granulations of cancer cells or of pulmonary tubercles. Their protozoan nature was first recognized by Balbiani in 1866.

Since then numerous other protozoan diseases of insects have been observed and studied. Their number and applications, however, have not been as great as have those concerned with the bacterial and fungous diseases. For a discussion of all the types of diseases of insects the reader is referred to the book by Sweetman (1936) on the biologic control of insects.

GROUPS OF PROTOZOA CONCERNED

Sweetman (1936) groups the chief forms of protozoa parasitic in insects as follows:

    Sporozoa
      Microsporidia
        Nosematidae
        Cocconematidae
        Mrazekiidae
        Telomyxidae
      Gregarinida
        Eugregarinaria
        Schizogregarinaria
        Aggregata
    Mastigophora
      Trypanosomidae

It is apparent that of the several classes of Protozoa, the Sporozoa include most of the parasitic species and most of those pathogenic to insects belong to the order Microsporidia. Such a group as the gregarines have very little if any deleterious effect upon the life of their hosts and are for the most part commensal in their relationship.

SYMPTOMS AND PATHOLOGY

Most insects infected with a protozoan have very few visible symptoms until shortly before death. At that time the insect may become sluggish in movement, lose its appetite, and show some change in color. Many of the symptoms will, of course, depend on the particular tissues affected. With some species, e.g., *Nosema bombycis,* most of the infected insect's tissues are involved. In other cases only certain tissues show signs of infection. Thus, *Thelohania mesnili* infects only the adipose tissue of the European corn borer; *Perezia pieris* and *P. mesnili,* the Malpighian tubes and silk glands of the cabbage-butterfly larvae, etc.

Though there is frequently some gross pathology connected with protozoan diseases in insects, most of the changes are histopathological. In some cases the tissue and cellular alterations are not very characteristic, but in other cases they are fairly uniform and diagnostic. In cases where there is little cellular degeneration the death of the insect usually results from the impregnation of vital tissues and organs with so many of the parasites that they simply cease to function. Many times the infected larval insect is made unable to pupate, and hence eventually dies as a larva. Sometimes the insect is able to pupate but is incapable of becoming an adult.

Some protozoa, especially certain of the microsporidia, bring about a marked hypertrophy of the cells. Some of the microsporidia also cause the nucleus to enlarge abnormally. In such cases the infected cell soon ceases to function normally, if at all, since occasionally the entire contents of the cell are destroyed.

TRANSMISSION

The two most common methods of transmission of the protozoan diseases of insects are by way of the mouth, and from the female through the egg to the next generation. Infection through ingestion is especially likely to occur in the case of those species which multiply in the cells of the gut epithelium. On the other hand, some protozoa, such as *Thelohania mesnili* of the corn borer, are unable to infect caterpillars when their spores are administered with the insects' food.

Transmission through the egg occurs with many species (e.g., *Perezia legeri, Perezia mesnili, Nosema bombycis*), and in many of these it is the principal method of transfer. In other species (*Perezia pyraustae*) transmission can take place very readily through the contamination of the hosts' food as well as through the eggs.

Several instances have been reported in which insects other than the ones affected may serve to disseminate the protozoan spores. Certain flies including a tachinid have been incriminated in the case of *Nosema bombycis* and the silkworm. *Thelohania ephestiae,* which causes a disease in the larvae of the meal moth, *Ephestia kühniella,* can be transmitted from host to host by *Microbracon hebetor,* a hymenopterous parasite. According to Payne (1933), the first focus of infection is in the ganglion pierced by the wasp during oviposition. Subsequent to this the protozoa may be found in both the nervous and the storage tissues. A similar mode of transmission may occasionally occur with *Thelohania mesnili* in *Pieris brassicae* by the parasite *Apanteles glomeratus.*

PRACTICAL UTILIZATION OF PROTOZOAN DISEASES

Although Paillot (1928a) contends that protozoan diseases are more destructive to insects than bacterial diseases, most authorities believe otherwise. It is perhaps true that protozoan spores are less susceptible to changing environmental conditions and are therefore more likely to maintain constant contact with susceptible hosts than are bacteria. As pointed out by Sweetman (1936), however, such protozoa as *Perezia pyraustae* commonly occur in the European corn borer yet do not appear to be of much consequence in the natural destruction of this insect. Similarly, *Perezia mesnili* in Europe is common in the larvae of the white cabbage butterfly, *Pieris brassicae,* and yet never seems to be responsible for any significant decrease in the number of these insects. According to Sweetman, Krassilstschik in 1886 described a deadly epidemic among the larvae of the beet webworm, *Toxostege sticticalis,* caused by *Microklossia*. The cabbage moth of Russia, *Mamestra oleraceae,* is also attacked by this parasite.

Unlike the bacterial and fungous diseases, no significant attempts have been made to disseminate the protozoan diseases artificially in order to eradicate pests.

❖ CHAPTER XI ❖

PROTOZOA IN TERMITES

IN RECENT YEARS the work of numerous investigators has opened a field of research the rewards of which are both fascinating and important. We refer to the habitation of protozoa in the intestinal tracts of termites. The relationship of these two forms of life was at first considered to be merely an interesting association—one for academic study only. It is now beginning to be realized that a knowledge of the biologic relationships between these two forms of life may be a key to the understanding of many general host-parasite relationships including certain aspects concerned with evolution and genetics. We shall have the opportunity to discuss some of these relationships later in this chapter.

No present-day discussion of the protozoa which occur in termites could be written without relying heavily upon the writings of Cleveland and those of Kirby. Much of what will be said in this chapter will be based on the investigations and reports of these two men. To be sure, numerous other workers have contributed to the subject and their work will also be referred to when appropriate.

THE TERMITES

Termites belong to the order Isoptera and are sometimes known popularly as "white ants." They are social insects and live in colonies like ants but differ from the latter chiefly in their habits and structure. In general, the termite can be differentiated from the true ant by the fact that the abdomen of the termite is broadly joined to the thorax, whereas with the ant the abdomen is pedunculate. Unlike ants, most termites are of a pale color. The cuticula of termites is very delicate, and when exposed to the open air the insects shrivel quickly and die. For this reason they build long tubes with earth and excrement which they use for passageways.

Each species of termite consists of a number of castes, usually four, each of which includes both male and female individuals. The following castes are known to occur: (1) first reproductive caste, with dark bodies and long wings, known as kings and queens; (2) second reproductive caste, with short wings, known as substitute kings and queens; (3) third reproductive caste,

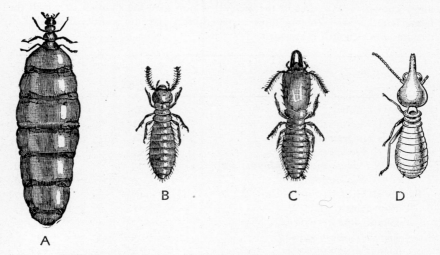

Fig. 230. Termite castes. *A*, queen; *B*, worker; *C*, soldier; *D*, nasutus. (From Comstock, 1936.)

with no wings, known as ergatoid kings and queens; (4) workers; (5) soldiers; and (6) nasuti. The reproductive organs of the workers and soldiers are undeveloped. The second and third reproductive castes can reproduce their own kind, and workers and soldiers, but not the long-winged kings and queens. The winged kings and queens are usually the only termites that normally appear in open air. With few exceptions, the other castes remain in the nest and are usually blind.

The order Isoptera consists of five families: Mastotermitidae, Hodotermitidae, Kalotermitidae, Rhinotermitidae, and Termitidae. Every species of the first four of these, the so-called lower families, contains flagellates in the hindgut. According to Kirby (1937a), these protozoa are present in all stages of the insect's life history except the eggs, the youngest instars, and preceding or immediately following a molt. Certain specialized members of the fifth family, Termitidae, contain very few protozoa. In 1937 (a) Kirby listed in tabular form the approximate number of termite species known and the number which have been examined for protozoa. This data, which is of value in giving a general idea of the type of information available, is reproduced in table II and has kindly been revised for the writer by Dr. Kirby.

TABLE II

CLASSIFICATION OF TERMITES *

(The figures in the first column give the approximate number of species of termites known. The figures in the second column give the number in which the fauna is rather well known; those in parentheses give the additional number of termites in which little is known.)

Classification	Number of species	Number examined
Four lower families	429	146 (38)
Family I. Mastotermitidae	1	1
Genus *Mastotermes*	1	1
Family II. Hodotermitidae	21	4 (5)
Genera *Hodotermes* and *Microhodotermes*	15	2 (2)
Genus *Anacanthotermes*	6	2 (3)
Family III. Kalotermitidae	266	112 (27)
Lower genera	15	9 (1)
Genus *Archotermopsis*	1	1
Genus *Zoötermopsis*	3	3
Genus *Hodotermopsis*	2	(1)
Genus *Stolotermes*	6	3
Genus *Porotermes*	3	2
Kalotermes group		
Genus *Kalotermes*	49	20 (5)
Genus *Neotermes*	75	12 (11)
Genus *Paraneotermes*	1	1
Genus *Rugitermes*	10	2
Genus *Cryptotermes*	40+	30 (2)
Genus *Procryptotermes*	11	9
Genus *Eucryptotermes*	1	0
Genus *Calcaritermes*	10	6
Genus *Glyptotermes*	49	24 (8)
Unassigned to genus	5	5
Family IV. Rhinotermitidae	141	29 (7)
Genus *Psammotermes*	6	3
Genus *Leucotermes*	26	4 (1)
Genus *Reticulitermes*	14	6
Genus *Stylotermes*	1	0
Genus *Coptotermes*	46	11 (2)
Genus *Prorhinotermes*	13	3
Genus *Termitogeton*	2	(1)

* After Kirby (1937a), with additions.

Classification	Number of species	Number examined
Genus *Parrhinotermes*	5	(2)
Genus *Rhinotermes*	27	2 (1)
Subg. *Rhinotermes* (s. str.)	9	(1)
Subg. *Schedorhinotermes*	18	2
Genus *Serritermes*	1	0
Family V. Termitidae	about 1200	about 40 (51)

Host-Parasite Specificity. There is a tendency among many of the workers in insect microbiology to consider a microorganism as a separate and distinct species simply because it has been found in a host different from that in which it was originally observed. The fallacy of this tendency is clearly brought out in the varying degrees of host-parasite specificity maintained by the protozoa in termites.

To be sure, many species of protozoa are known to occur in only one host, but certain species are found in several hosts. Kirby (1937a) tells of fifteen species of *Foaina* which he distinguished on the basis of morphologic characteristics, and of these flagellates thirty-three termites harbored one or two species, one termite harbored three, and another was found to contain four species. On the other hand, eight of the species of *Foaina* were found in one host each, three were found in two different hosts, and one species of the flagellate was found inhabiting the guts of fifteen different species of termites from different parts of the world. Kirby also notes that four of thirteen species of *Devescovina* were each found in one host only, while one species, *D. lemniscata*, was found in seven different species of termites from India, Australia, Fanning Island, and Central America. In at least two instances probably all the species of each of the genera of termites examined have been found to harbor a single species: *Tricercomitus divergens* in *Kalotermes* and *Trichonympha agilis* in *Reticulitermes*. In general, most termites do not harbor more than three species of flagellates, frequently only two, and usually only one. When ten or more species are present, it is very unusual.

In most cases, all the termites of a given species have identical faunas. There are exceptions to this, however, and it cannot be accepted as a definite or invariable rule. The general constancy of the natural occurrence of certain species of flagellates in certain species of termites has made it possible in many cases to use this association as an aid in identifying the termites. In the case of the Kalotermitidae practically all species that have been examined have a characteristic flagellate fauna, which is especially helpful in identifying nymphs when no other castes are present. In the case of *Zoötermopsis neva-*

densis and *Zoötermopsis angusticollis* this is not possible for they have similar faunas. In some instances specific flagellates may not be present in certain individuals or they may be absent from all the members of an entire colony. The latter is possible when a pair of termites which start a new colony are both devoid of the flagellate. However, the transmission of the flagellates among members of a colony is usually very efficient and successful. Too much reliance should not be placed in the method of separating species of termites by their faunas. Instead, it should be used as a helpful support in the study of other taxonomic characteristics. When the faunas of two termites are markedly different, the indication is strong that one is dealing with different species.

Two different opinions exist as to whether or not there is any resistance to cross infection among the termites to protozoa different from those found constantly present. On the assumption that the internal environment of the intestine of one wood-feeding termite probably does not differ greatly from that of most others, Kirby states: "There seems to be no resistance to cross infection." This statement is supported by the experimental work of Light and Sanford (1927, 1928), and Cleveland *et al.* (1934). They showed that by injecting via the anus one defaunated termite (*Zoötermopsis*) with the protozoa of another species (*Kalotermes hubbardi*) that the flagellates survived for at least 100 days. Cleveland was able to transfer the fauna of the roach *Cryptocercus* to *Zoötermopsis*, and *Zoötermopsis* to *Cryptocercus* with ease. On the other hand, Dropkin (1941) using species of *Zoötermopsis, Reticulitermes,* and *Kalotermes,* claims that there is a resistance to cross infection to protozoa from one species to another widely different species of termite. He concludes that the protozoa are physiologically as well as morphologically distinct.

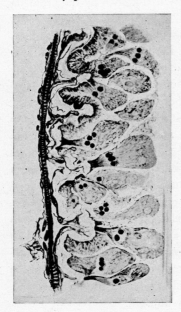

Fig. 231. Section of intestine of *Cryptotermes dudleyi,* showing flagellates (*Proboscidiella kofoidi*) attached by rostella to wall. (From Kirby, 1928.)

Kirby believes it not likely that cross infection among species of termites has been an important factor in spreading the flagellates, at least not in the four lower families. Since protozoa do not survive for any length of time outside of termites, and since colonies of termites are isolated from one another, it is improbable that one species of termite may acquire an infestation from another.

Evolutionary Development. To the student of evolution the protozoa of termites offer a valuable storehouse of interesting facts and possibilities. According to Kirby (1941), many species of termite flagellates have a present host distribution which indicates greater stability in their characteristics than has existed during the same period of time in the insects. Speciation has occurred in the termites without having taken place in certain of their protozoa. In other words, it seems probable that many of the flagellate types known today existed in the termite ancestors. It is possible that because of continuous evolutionary changes, in addition to occasional losses of certain members of the faunas, the composition of the flagellate inhabitants of termites is as we find it today. Kirby (1937a) also suggests that since members of the genus *Trichonympha* are found both in termites and in the wood-eating roach *Cryptocercus punctulatus*, it may be supposed that the flagellates passed into both of these insects from ancestral protoblattids in the late Paleozoic or Mesozoic ages. Furthermore, a close study comparing the protozoan faunas of termites and roaches lends support to Cleveland's hypothesis that evolutionary development of certain groups of the flagellates took place in the ancestral insects from which termites and roaches developed. Hence, we find them in both groups of insects. As Cleveland points out, regardless of how widely distributed a species of termite may be, the protozoa in the individual members are identical. Therefore, it is probably safe to assume that the evolution pertaining to species differentiation in the protozoa occurred at a time prior to the termite's distribution.

The faunas of the more primitive termites are not necessarily of a primitive character. Some of the most advanced hypermastigotes occur in *Mastotermes* and the more primitive Kalotermitidae. The hypermastigotes predominate also in the faunas of *Cryptocercus* which is even more primitive than *Mastotermes*. It appears that certain of the polymastigote flagellates have undergone their evolution in special groups of termites. The Pyrsonymphinae are known to occur only in *Reticulitermes*. The Oxymonadidae, Descovininae, and Calonymphidae are, as far as is known, found only in Kalotermitidae with two exceptions. On the other hand, species of *Trichonympha* have a wide and uneven distribution, being found in both termites and wood-eating roaches. Most of the hypermastigotes apparently underwent much of their evolution before termites and roaches evolved from their ancestral blattoid stem.

In regard to the family Termitidae, the members of which have few flagellates, it is interesting to speculate (as did Cleveland, 1926) as to whether these termites are losing their protozoa or are just beginning to harbor them.

Transmission. Details of the mechanisms whereby the protozoa are transmitted from one termite to another are not entirely understood. Since true encystation does not take place [1] and since living protozoa are rarely present for any length of time in the fecal pellets of termites, it has been assumed that transmission of the flagellates takes place in the active state. This assumption is in part supported by experimental observations. It is known that termites can be infected by contaminating the mouth parts with the flagellates. It is also known that defaunated termites, when allowed to remain in contact with normally faunated individuals, become refaunated. When freshly hatched young termites are left to themselves, they never acquire protozoa and die in a few days. When placed with faunated termites, however, the young quickly obtain protozoa. Since it appears likely that, except for cannibalism, the flagellates can be obtained only from the anal opening of another termite, acquisition of the protozoa must be acquired in some manner such as proctodaeal feeding, which is a common habit among termites (Kirby, 1941). Goetsch (1936) has described just such a mode of transmission in the case of *Kalotermes flavicollis* (fig. 232). Cleveland (1926) has suggested that perhaps the faunated members of a colony give off extra liquid material with the pellets from the anus from which the young may acquire the flagellates. Such a provision, however, lacks experimental confirmation.

Fig. 232. A one-day-old nymph of *Kalotermes flavicollis*, receiving proctodaeal food from the female termite. In this manner the transfer of the flagellates occurs. (From Goetsch, 1936.)

It should be remembered that in most species of termites the flagellates disappear prior to each molt except the last. Hence, refaunation must take place following each molt as well as at the beginning of the insect's life. In any case, intimate contact with faunated individuals appears necessary for the termite to regain its fauna.

---

[1] Encystment may take place with some of the small polymastigotes (see Trager, 1934, and Duboscq and Grassé, 1934).

PHYSIOLOGIC ASPECTS

Defaunation. In order to study the physiologic relationships existing between them, methods of freeing the termites from their protozoa had to be developed. Cleveland (1926) has described several ways of doing this, including incubation, starvation, and oxygenation of the termites.

Incubation of the termites at 36° C. for 24 hours frequently frees them of their protozoa. Some of the insects may be injured by this treatment, but this is minimum when the proper moisture conditions are maintained.

All species of protozoa are not killed in the same length of time at a given temperature, nor is the time-temperature relation constant for all the protozoa of certain termites. With *Kalotermes tabogae* at about 35.5° C., Cleveland found that the trichonymphids are all killed in two days whereas it takes from eight to ten days before all the calonymphids are killed. At a higher temperature, 36.5° C., the trichonymphids die off within 24 hours, while it takes three days to kill the calonymphids.

After the termites are freed of their protozoa, they usually die in the neighborhood of three weeks, though this depends on the kind of food eaten. When fed wood, the more it is decayed the longer they live. Humus or fungus-digested paper allows them to live much longer, and under optimum conditions they could probably live a normal lifetime.

Starvation of the termites affords another method of freeing these insects of their protozoa since starvation diet causes most of the flagellates to die before the termites do. In nearly every case the large protozoa die more rapidly than do the small ones. Thus, in the case of *Zoötermopsis nevadensis,* its largest protozoan, *Trichonympha campanula,* perishes first (within six days). At the end of eight days the next largest protozoan, *Trichonympha sphaerica,* dies. With continued starvation the numbers of the remaining protozoa, *Trichomonas termopsidis* and *Streblomastix strix,* are appreciably reduced, but it is difficult to remove all of them before the termite's starvation limit of approximately 25 days is reached.

Of considerable interest are the experiments of Cleveland in which he found that if, after the six days' starvation which removes the flagellate *Trichonympha campanula,* the termites are allowed to return to their normal diet of wood, they will live indefinitely. In the absence of this protozoan, another, *Trichonympha sphaerica,* undergoes rapid multiplication, taking the place of *T. campanula,* and becomes the main symbiote. When both of these protozoa are removed, the termites are able to live 60 or 70 days when returned to their normal diet. However, if all four species, including the remaining *Tricho-*

monas termopsidis and *Streblomastix strix* are removed, the termites live only three or four weeks. This indicates that this termite is dependent upon at least two of its symbiotes for its survival.

Oxygenation is a third method of removing the protozoa from their termite hosts. This method is more satisfactory than either incubation or starvation because it does not injure the insects. Oxygenation may be accomplished by placing the termites in oxygen under pressure. As with the other two methods of defaunation, the various species of protozoa are affected differently by the various percentages of oxygen.

An atmosphere of air consists of approximately 20 per cent oxygen. Some termites lose their protozoa merely by moving them from this percentage to one containing 95 to 98 per cent oxygen. With others it is necessary to increase the oxygen to more than that in one atmosphere. As the oxygen pressure is increased, the protozoa die more rapidly until three and a half to four atmospheres have been reached. Above this pressure, however, defaunation is not as rapid. Cleveland decided that the best oxygen pressure for separating the termites from their protozoa is between three and four atmospheres. The time required for the oxygen to kill the protozoa depends markedly on the temperature. For example, Yamasaki (1931) found the protozoa of *Reticulitermes speratus* under one atmosphere of oxygen to die in 9 to 10 days at 25° C., in 1 to 6 days at 10° C., and in 24 to 30 hours at 0° C., but at 20° C. the termites would not lose more than three of their seven species of protozoa (see also Cleveland *et al.*, 1934).

As in the case of starvation, oxygenation removes some of the termite protozoa but not others. Thus, when *Zoötermopsis nevadensis* is held at one atmosphere of oxygen, it loses its *Trichomonas* within 24 hours but retains the other protozoa for three days. At one and a half atmospheres for seven hours *Streblomastix* is lost, leaving *Trichonympha* and *Leidyopsis,* the first two species to be lost under conditions of starvation. Hence, one can readily see the possibilities for selective studies when various combinations of the defaunating methods are used.

Diet. In nature most termites feed exclusively on wood. The protozoa, in turn, take into their bodies the particles of wood and digest them, the termites gaining thereby. Wood always consists of at least 50 per cent cellulose, which is the chief fraction utilized by the termites. The lignin derivatives have no food value for these insects. When fed a diet of pure cellulose, the termites (*Zoötermopsis* and *Reticulitermes*) suffer no apparent lack and they behave exactly like normal controls (Cleveland, 1925b). Hungate has shown that the fermentation of cellulose carried on by the protozoa is anaerobic in nature.

In further experiments Cleveland (1925d) found that when *Zoötermopsis* is fed sucrose the hypermastigotes (three species of *Trichonympha*) disappear in 6 to 10 days and the termites survive 30 to 50 days. With pure rice starch, the hypermastigotes are gone in 12 days or sooner. On this diet the termites may live at least seven months, but when returned to a wood diet they die within six weeks.

In regard to the various termite castes it is a general rule that protozoa are present at every stage in the life cycle where wood is eaten. The protozoa are never present when wood is not eaten. According to Cleveland (1925c) in the second and third forms of young *Reticulitermes flavipes* the protozoa disappear as the host loses its ability to eat wood. In all castes the protozoa are lost during the period of molting and regained shortly afterwards except in the case of the final molt of the second and third forms, which are unable to eat wood because of the degeneration of their jaw muscles. These individuals are then fed salivary secretions. Cleveland suggests that it is possible these forms are fed so much salivary secretion that they will no longer feed on wood, and because of this their jaw muscles disintegrate through disuse and the termites completely lose their ability to feed on wood. The first form and the worker always eat wood except in the postadult stage of the first form when it can no longer eat wood and depends on other members of the colony to feed it.

Because of their large mandibles the adult soldiers cannot chew wood, hence they obtain it from the ani of the xylophagous termites. Soldier nymphs can themselves chew wood.

The manner in which the various species of termite protozoa ingest their food has not been adequately studied. Swezy (1923) reported that protozoa of the genus *Trichonympha* ingest their food by a "pseudopodial method of feeding." In the case of *Trichonympha campanula* Cleveland (1925a) found the food to be taken in through the protozoan's posterior end, which, by invaginating, forms a temporary cytostome. This cavity then slowly closes by the flowing together of the undifferentiated portion of the protozoan's body. Emik (1941) maintains that *Trichonympha* ingests food according to the pseudopodial method of Swezy rather than the invagination method of Cleveland. Emik finds the pseudopodial method to function as two distinct types —one type for the large particles and the other type for the fine particles of food.

Protozoa in Cryptocercus. A discussion of the protozoa of termites would not be practical unless it also included mention of the fauna of the wood-eating roach *Cryptocercus punctulatus*. The fact that the protozoa of this orthopteran are closely related to those of termites is perhaps not as surprising as one might at first suppose since, according to some authorities (Holmgren, 1909; Imms, 1919; and others), termites and roaches are both descendants of the same primitive group, the Protoblattoidae. Even today there are many resemblances between Isoptera and Blattidae. In any case, the protozoan fauna of *Cryptocercus punctulatus* is of sufficient similarity to that of termites to necessitate simultaneous consideration.

The genus *Cryptocercus* is usually placed in the subfamily Panesthiinae of the family Blattidae. *Cryptocercus punctulatus,* the only known species of Panesthiinae in the Western Hemisphere, has been reported from the eastern and far western parts of the United States. The principal habitat of this insect is dead logs covered with leaves and humus in heavily forested areas.

The relationships between the roach and its protozoa have been described in detail by Cleveland and his associates (1934) in a monograph which the reader is urged to consult since only the very briefest mention can be made of them here. According to these workers, *Cryptocercus punctulatus* harbors 12 genera and 25 species of flagellate protozoa, as follows:

Hypermastigida	Polymastigida
Barbulanympha coahoma	*Hexamita cryptocerci*
Barbulanympha estaboga	*Monocercomonoides globus*
Barbulanympha laurabuda	*Saccinobaculus ambloaxostylus*
Barbulanympha ufalula	*Saccinobaculus doroaxostylus*
Eucomonympha imla	*Saccinobaculus minor*
Idionympha perissa	
Leptospironympha eupora	
Leptospironympha rudis	
Leptospironympha wachula	
Macrospironympha xylopletha	
Prolophomonas tocopola	
Rhynchonympha tarda	
Trichonympha acuta	
Trichonympha algoa	
Trichonympha chula	
Trichonympha grandis	
Trichonympha lata	
Trichonympha okolona	
Trichonympha parva	
Urinympha talea	

In general, there is a close relationship between the flagellates of the roach and those of termites. Some of the roach protozoa are species belonging to the same genera or families as those of the termites. For example, members of the genus *Trichonympha* occur both in termites and the roach. Experiments indicate that some of the *Cryptocercus* protozoa may survive in the gut of termites when transferred artificially. As with termites, the protozoa of the roach are essential for its existence in nature since they digest the insect's food. The protozoa contain the enzymes cellulase and cellobiase which the roach is unable to liberate. Essentially the nutrition of the roach is dependent upon certain constituents, of which one of the most important is dextrose, supplied it by the action of the protozoa. As with termites, roaches can be effectively defaunated by oxygenation, starvation, and incubation.

One interesting difference between the protozoa of *C. punctulatus* and those of termites is in the encystment of some of the former. At the time of the ecdysis of the roach, encystation occurs in nine species of protozoa, and the flagella and other extranuclear organelles disappear in seven species.

Two other wood-feeding roaches, *Panesthia javanica* and *Panesthia spadica*, also harbor protozoa. The flagellates, *Monocercomonoides panesthiae* and *Hexamita cryptocerci*, occur in the hindgut of both species and are thought by Kidder (1937, 1938) to be commensals. *Endamoeba philippinensis*, found rarely in *P. javanica*, is a bacteria feeder, and *Endamoeba javanica*, found in both *P. javanica* and *P. spadica*, is xylophagous. *Nyctotherus uichancoi* was found in 90 per cent of all the specimens of *P. javanica* examined, while *P. spadica* was very lightly infected. Kidder also isolated a number of new ciliates from these two roaches. He found *Paraclevelandia panesthiae, Paraclevelandia contorta, Paraclevelandia constricta, Paraclevelandia brevis,* and *Paraclevelandia simplex* present in both roaches, while *Paraclevelandia elongata, Paraclevelandia parapanesthiae, Paraclevelandia hastula* were present only in *Panesthia javanica*. *Paraclevelandia nipponensis* was present only in *Panesthia spadica*.

* * *

Most of the protozoa found in termites are flagellates (Mastigophora). However, members of the classes Sarcodina, Sporozoa, and Ciliata are also represented.

Approximately 250 species of flagellates from termites are recognized. Most of these are among the most highly developed Mastigophora, and except for a few found in certain roaches are found almost exclusively in termites. They occur in these insects in large numbers and, according to Cleveland (1926), weigh almost as much as the termites themselves, though by actual experiment Katzin and Kirby (1939) found the protozoa in *Zoötermopsis* to represent from 16 to 36 per cent of the termite's weight.

Perhaps the first to observe an abundant fauna of protozoa in a termite (*Reticulitermes lucifugus*) was the French worker Lespès in 1856. Later Leidy reported on the protozoa in *Reticulitermes flavipes* of the eastern United States, and Grassi described the flagellates which he found in *Reticulitermes lucifugus* and *Kalotermes flavicollis* of Italy. In late years many investigators have described numerous species from these insects, revealing their interesting morphologic and physiologic characteristics.

MASTIGOPHORA

Nearly all of the flagellates found in termites are included in the two highest orders—Polymastigida and Hypermastigida—of the class Mastigophora. In 1937 (a) Kirby recognized in termites 30 genera comprising 133 species of Polymastigida and 18 genera with 63 species of Hypermastigida.

Most of the studies on the new termite flagellates are concerned chiefly with their morphology and taxonomy. For this reason a discussion of the biologic relationships involved in each case is out of the question at this time. Very likely, however, the reader will wish to gain some idea as to the distribution of these protozoa, their names, and their hosts. For this information the following listing is presented.[1] The list is not complete, and with further taxonomic studies is probably due for considerable revising.

For a list of some of the protozoa arranged according to their termite host, the reader is referred to Kirby (1926).

Order: PROTOMONADIDA

Retortamonas termitis
   Amitermes beaumonti

Order: POLYMASTIGIDA

Chilomastix minuta
   Amitermes beaumonti

Tricercomitus divergens
   Neotermes holmgreni
   Kalotermes tabogae

---

[1] The author is indebted to Prof. Harold Kirby for numerous helpful suggestions concerning this list.

Kalotermes snyderi
Kalotermes clevelandi
Kalotermes emersoni
Rugitermes kirbyi
Rugitermes panamae
Glyptotermes angustus (= G. barbouri)
Glyptotermes contracticornis
Cryptotermes longicollis
Cryptotermes dudleyi

Monocercomonas termitis (= *Eutrichomastix termitis*)
    Anacanthotermes murgabicus

Monocercomonas axostylis (= *Eutrichomastix axostylis*)
    Subulitermes kirbyi

Hexamastix termopsidis (fig. 233, 3)
    Zoötermopsis angusticollis
    Zoötermopsis nevadensis

Hexamastix termitis
    Kalotermes flavicollis (probably Reticulitermes lucifugus)

Hexamastix claviger
    Kalotermes perezi (= K. marginipennis)
    Cryptotermes dudleyi
    Calcaritermes brevicollis

Hexamastix conclaviger
    Cryptotermes piceatus

Hexamastix disclaviger
    Cryptotermes brevis
    Lobitermes longicollis

Hexamastix laticeps (fig. 233, 1)
    Zoötermopsis laticeps

Trichomonas linearis
    Orthognathotermes wheeleri

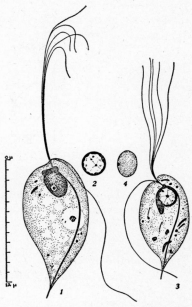

Fig. 233. 1. *Hexamastix laticeps* from *Zoötermopsis laticeps*. 2. More heavily stained nucleus. 3. *Hexamastix termopsidis* from *Zoötermopsis angusticollis*. 4. Nucleus much destained. (From Kirby, 1930.)

Trichomonas termitis
    Archotermopsis wroughtoni

Trichomonas macrostoma (fig. 234, B)
    Hodotermes mossambicus

Trichomonas trypanoides
    Reticulitermes lucifugus

Trichomonas termopsidis (fig. 234, A) (= *Trichomitus termitidis*)
    Zoötermopsis angusticollis
    Zoötermopsis laticeps
    Zoötermopsis nevadensis

Trichomonas vermiformis
    Anacanthotermes murgabicus

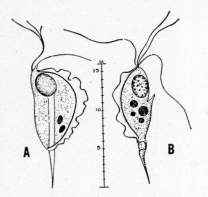

Fig. 234. *A. Trichomonas termopsidis.* *B. Trichomonas macrostoma.* (From Kirby, 1931.)

Trichomonas barbouri
    Glyptotermes angustus (= G. barbouri)

Trichomonas cartagoensis
    Glyptotermes contracticornis

Trichomonas labelli
    Mirotermes hispaniolae

Trichomonas lighti
    Amitermes emersoni
    Amitermes minimus
    Amitermes wheeleri
    Amitermes beaumonti
    Amitermes coachellae
    Amitermes silvestrianus

Trichomonas brevicollis
    Calcaritermes brevicollis

Tritrichomonas holmgreni
    Neotermes holmgreni

Pseudotrypanosoma giganteum
    Porotermes adamsoni (= P. grandis)

Pseudotrypanosoma minimum
    Porotermes adamsoni

Pentatrichomonoides scroa
    Cryptotermes dudleyi
    Cryptotermes longicollis

Devescovina elongata
    Anacanthotermes murgabicus
    Anacanthotermes ochraceus

Devescovina cometoides
    Kalotermes sp., Damaum, Portuguese India
    Cryptotermes dudleyi
    Cryptotermes sp., Java (possibly C. dudleyi)
    Cryptotermes sp., Java
    Cryptotermes sp., Madagascar, Mauritius
    Cryptotermes sp., Tanganyika Territory
    Glyptotermes dilatatus
    Glyptotermes sp., Java
    Kalotermes repandus

Devescovina fissa
    Procryptotermes sp., Reunion

Devescovina tendicula
    Kalotermes jeannelanus

Devescovina glabra
    Cryptotermes havilandi
    Cryptotermes spp., Tanganyika Territory
    Cryptotermes sp., Kenya Colony
    Cryptotermes sp., Zanzibar
    Cryptotermes sp., Portuguese East Africa, South Africa
    Cryptotermes sp., Madagascar
    Cryptotermes sp., Java
    Glyptotermes caudomunitis
    Glyptotermes sp., Sumatra
    Kalotermes sp., Madagascar
    Neotermes zuluensis

Neotermes meruensis
Neotermes sp., Madagascar

Devescovina coghilli
Cryptotermes merwei

Devescovina insolita
Neotermes sp., Tanganyika Territory
Kalotermes sp., Transvaal, South Africa

Devescovina robusta
Neotermes erythraeus

Devescovina parasoma
Neotermes tectonae
Neotermes dalbergiae
Neotermes sonneratiae
Cryptotermes sp., Java
Cryptotermes sp. (possibly *C. cynocephalus*)
Cryptotermes sp., Kenya Colony

Devescovina vestita
Glyptotermes niger

Devescovina exilis
Neotermes connexus

Devescovina lepida
Calcaritermes brevicollis
Calcaritermes emarginicollis
Calcaritermes nearcticus
Calcaritermes parvinotus
Cryptotermes longicollis
Neotermes castaneus
Neotermes holmgreni

Devescovina arta
Glyptotermes angustus
Glyptotermes minutus
Neotermes castaneus

Devescovina vittata
Kalotermes sp., Tanganyika Territory

Devescovina striata
Cryptotermes grassii
Cryptotermes brevis
Cryptotermes piceatus
Cryptotermes darwini
Cryptotermes sp., South Africa

Devescovina lemniscata
Cryptotermes hermsi
Cryptotermes breviarticulatus
Cryptotermes dudleyi
Cryptotermes domesticus
Cryptotermes fatulus
Cryptotermes queenslandis
Cryptotermes sp., Tanganyika Territory
Cryptotermes sp., Kenya Colony
Cryptotermes sp., Mauritius, Madagascar
Cryptotermes sp., Java
Cryptotermes sp., Ecuador
Kalotermes sp., Portuguese India
Glyptotermes tuberculatus
Glyptotermes sp., Java
Kalotermes [1] (s.l.) *castaneiceps* (= *N. castaneiceps*)
Kalotermes (s.l.) sp., Madagascar
Neotermes insularis
Neotermes larseni
Kalotermes [2] (s.l.) *longus*

Devescovina uniflexura
Kalotermes perezi

Devescovina hawaiensis
Neotermes connexus

Devescovina transita
Glyptotermes sp., Java
Glyptotermes dilatatus

[1,2] Used by Kirby (1944) for a new but undesignated genus of Kalotermitinae recognized by A. E. Emerson.

Devescovina transita (*Cont.*)
   Cryptotermes cynocephalus
   Kalotermes repandus

Devescovina similis
   Cryptotermes sp., Madagascar

Caduceia bugnioni
   Neotermes greeni

Caduceia theobromae
   Neotermes gestri
   Neotermes aburiensis
   Kalotermes (s.l.) *jeannelanus*
   Kalotermes sp., South Africa

Caduceia monile
   Neotermes tectonae

Caduceia kalshoveni
   Neotermes dalbergiae

Caduceia pruvoti
   Kalotermes sp., Lifou, Loyalty Islands

Caduceia nova
   Neotermes erythraeus

Macrotrichomonas pulchra (fig. 235)
   Glyptotermes parvulus
   Glyptotermes ceylonicus
   Glyptotermes contracticornis
   Glyptotermes brevicornis (= *G. dubius*; *G. perangustus*)
   Glyptotermes iridipennis
   Glyptotermes montanus
   Glyptotermes neotuberculatus
   Glyptotermes taveuniensis
   Glyptotermes sp., Uganda
   Glyptotermes sp., Belgian Congo
   Glyptotermes sp., Ruanda-Urundi
   Glyptotermes sp., Philippines

Macrotrichomonas unguis
   Glyptotermes caudomunitis
   Glyptotermes sp., Java
   Glyptotermes minutus

Macrotrichomonas ramosa
   Glyptotermes brevicaudatus

Macrotrichomonas restis
   Kalotermes jouteli

Macrotrichomonas virgosa
   Procryptotermes sp., Madagascar

Macrotrichomonas lighti
   Paraneotermes simplicicornis

Macrotrichomonas procera
   Calcaritermes brevicollis
   Calcaritermes emarginicollis

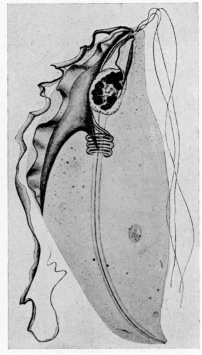

Fig. 235. *Macrotrichomonas pulchra* from a termite of the genus *Glyptotermes*. (From Kirby, 1942.)

Calcaritermes nearcticus
Calcaritermes parvinotus

Foaina gracilis
   Neotermes connexus

Foaina grassii
   Kalotermes flavicollis

Foaina solita
   Cryptotermes dudleyi
   Cryptotermes longicollis
   Cryptotermes queenslandis
   Cryptotermes sp., Madagascar, Mauritius
   Glyptotermes angustus
   Glyptotermes dilatatus
   Glyptotermes parvulus
   Glyptotermes tuberculatus
   Glyptotermes sp., Uganda
   Glyptotermes sp., Belgian Congo
   Kalotermes (s.l.) *jeannelanus*
   Kalotermes repandus
   Kalotermes sp., Madagascar
   Neotermes castaneus
   Neotermes connexus
   Neotermes greeni
   Neotermes holmgreni
   Neoteremes sonneratiae
   Neotermes tectonae
   Procryptotermes sp., Reunion, South Africa
   Rugitermes kirbyi
   Rugitermes panamae

Foaina parvula
   Cryptotermes sp., Madagascar
   Glyptotermes contracticornis
   Glyptotermes montanus
   Glyptotermes neotuberculatus
   Glyptotermes taveuniensis
   Kalotermes (s.l.) *castaneiceps*
   Kalotermes sp., Tanganyika Territory
   Kalotermes sp., Transvaal
   Kalotermes (s.l.) *longus*
   Neotermes erythraeus
   Neotermes sp., Tanganyika Territory
   Neotermes sp., Madagascar

Foaina humilis
   Cryptotermes brevis
   Cryptotermes grassii
   Cryptotermes piceatus

Foaina exempta
   Neotermes insularis

Foaina ovata
   Neotermes howa var. *mauritiana*
   Neotermes desneuxi
   Neotermes gracilidens
   Neotermes howa

Foaina taeniola
   Paraneotermes simplicicornis

Foaina nana
   Cryptotermes hermsi
   Cryptotermes cynocephalus
   Cryptotermes darwini
   Cryptotermes domesticus
   Cryptotermes dudleyi
   Cryptotermes sp., Java
   Cryptotermes fatulus
   Cryptotermes sp., Tanganyika Territory, Kenya Colony, Madagascar, and Java
   Kalotermes sp., Portuguese India
   Kalotermes condonensis var. *chryseus*
   Kalotermes repandus
   Glyptotermes dilatatus
   Procryptotermes sp., Reunion
   Neotermes connexus
   Neotermes dalbergiae
   Neotermes greeni
   Neotermes larseni
   Neotermes sonneratiae
   Neotermes tectonae
   Rugitermes kirbyi
   Rugitermes panamae

Foaina hilli
    Glyptotermes iridipennis
    Glyptotermes brevicornis (= *G. dubius; G. perangustus*)

Foaina dogieli
    Kalotermes flavicollis

Foaina hamata
    Cryptotermes sp., Madagascar
    Glyptotermes sp., Uganda, Belgian Congo
    Glyptotermes parvulus
    Glyptotermes caudomunitis

Foaina minuscula
    Procryptotermes sp., Madagascar

Foaina delicata
    Glyptotermes iridipennis
    Glyptotermes brevicornis (= *G. dubius; G. perangustus*)
    Glyptotermes neotuberculatus

Foaina acontophora
    Cryptotermes sp., Zanzibar
    Cryptotermes sp., Madagascar
    Kalotermes (s.l.) *castaneiceps*
    Kalotermes (s.l.) *longus*

Foaina signata
    Kalotermes sp., Tanganyika Territory
    Neotermes meruensis
    Neotermes zuluensis
    Neotermes europae

Foaina falcifera
    Glyptotermes contracticornis
    Glyptotermes montanus

Foaina funifera
    Glyptotermes tuberculatus
    Cryptotermes queenslandis

Foaina decipiens
    Neotermes aburiensis

Foaina appendicula
    Neotermes minutus

Foaina reflexa
    Kalotermes (s.l.) *perezi*
    Kalotermes (s.l.) *jeannelanus*
    Kalotermes sp., Tanganyika Territory
    Kalotermes sp., Kenya Colony
    Calcaritermes brevicollis
    Calcaritermes emarginicollis
    Calcaritermes nearcticus
    Calcaritermes parvinotus
    Cryptotermes breviarticulatus
    Cryptotermes cavifrons
    Cryptotermes longicollis
    Cryptotermes havilandi
    Neotermes castaneus
    Neotermes erythraeus
    Neotermes holmgreni
    Neotermes meruensis
    Neotermes zuluensis
    Neotermes europae
    Neotermes sp., Tanganyika Territory

Foaina duo
    Glyptotermes niger

Foaina nucleoflexa
    Cryptotermes merwei
    Cryptotermes sp., Tanganyika Territory

Foaina ramulosa
    Cryptotermes sp., Madagascar

Foaina pectinata
    Anacanthotermes ochraceus

Parajoenia grassii
    Neotermes connexus

Parajoenia decipiens
    Neotermes aburiensis

Metadevescovina polyspira
   Kalotermes occidentis

Metadevescovina mediocris
   Glyptotermes sp., Belgian Congo
   Glyptotermes sp., Ruanda-Urundi

Metadevescovina extranea
   Mastotermes darwiniensis

Metadevescovina modica
   Kalotermes marginipennis

Metadevescovina modesta
   Glyptotermes angustus Snyder

Metadevescovina stereociliata
   Glyptotermes parvulus
   Glyptotermes sp., Uganda

Metadevescovina cuspidata
   Kalotermes minor

Metadevescovina cristata
   Procryptotermes sp., Tanganyika Territory

Metadevescovina patula
   Kalotermes condonensis

Metadevescovina turbula
   Kalotermes jouteli

Metadevescovina nudula
   Kalotermes jouteli

Metadevescovina carina
   Kalotermes madagascariensis

Metadevescovina fulleri
   Kalotermes durbanensis
   Kalotermes sp., South Africa
   Kalotermes sp., Madagascar, near Tulear

   Kalotermes sp., Madagascar, near Maevetanana

Metadevescovina debilis
   Kalotermes hubbardi

Metadevescovina nitida
   Procryptotermes paradoxus
   Procryptotermes sp., Madagascar

Metadevescovina magna
   Kalotermes marginipennis

Pseudodevescovina uniflagellata
   Neotermes insularis

Pseudodevescovina ramosa
   Kalotermes (s.l.) sp., Tanganyika

Pseudodevescovina punctata
   Neotermes aburiensis

Pseudodevescovina brevirostris
   Neotermes aburiensis

Bullanympha silvestrii
   Neotermes erythraeus

Coronympha clevelandi
   Kalotermes clevelandi
   Kalotermes immigrans

Coronympha octonaria
   Kalotermes emersoni
   Kalotermes pacificus
   Kalotermes tabogae
   Kalotermes platycephalus
   Kalotermes lighti

Metacoronympha senta (fig. 236)
   Kalotermes emersoni
   Kalotermes tabogae
   Kalotermes pacificus
   Kalotermes lighti
   Kalotermes platycephalus

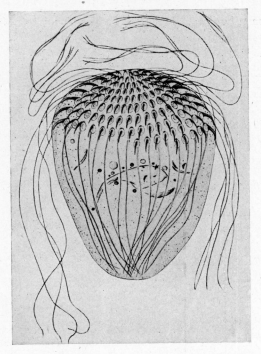

Fig. 236. *Metacoronympha senta* from *Kalotermes tobogae*. (From Kirby, 1939.)

Stephanonympha nelumbium
  Cryptotermes hermsi

Stephanonympha silvestrii
  Neotermes connexus

Stephanonympha silvestrii var. *cryptotermitis havilandi*
  Cryptotermes havilandi

Stephanonympha silvestrii var. *neotermitis erythraei*
  Neotermes erythraeus

Diplonympha parvulus
  Glyptotermes parvulus

Calonympha grassii
  Cryptotermes grassii

Calonympha sp.
  Calcaritermes brevicollis

Snyderella tabogae
  Cryptotermes longicollis

Monocercomonoides globus
  Cryptocercus punctulatus

Oxymonas projector
  Kalotermes perparvus

Oxymonas dimorpha
  Paraneotermes simplicicornis

Oxymonas granulosa
  Neotermes connexus

Oxymonas parvula
  Cryptotermes hermsi

Oxymonas pediculosa
    Kalotermes nigriceps

Oxymonas gracilis
    Kalotermes magninotus

Oxymonas ovata
    Calcaritermes brevicollis

Oxymonas minor
    Kalotermes minor

Oxymonas clevelandi
    Kalotermes clevelandi
    Kalotermes tabogae

Oxymonas panamae
    Kalotermes snyderi

Oxymonas kirbyi
    Rugitermes kirbyi
    Rugitermes panamae

Oxymonas barbouri
    Glyptotermes angustus
    (= *Kalotermes barbouri*)

Oxymonas grandis
    Neotermes dalbergiae
    Neotermes tectonae

Oxymonas hubbardi
    Kalotermes sp.

Oxymonas janicki
    Kalotermes sp.

Oxymonas brevis
    Kalotermes sp.

Oxymonas snyderi
    Kalotermes sp.

Oxymonas jouteli
    Kalotermes sp.

Oxymonas spp.
    Glyptotermes contracticornis
    Neotermes holmgreni

Streblomastix strix
    Zoötermopsis angusticollis

Microrhopalodina enflata
    Kalotermes flavicollis

Microrhopalodina multinucleata
    Cryptotermes dudleyi
    (= *Kalotermes nocens*)

Microrhopalodina occidentis
    Kalotermes occidentis

Microrhopalodina kofoidi
    Cryptotermes dudleyi

Barroella zeteki (= *Kirbyella zeteki*)
    Calcaritermes brevicollis

Pyrsonympha vertens
    Reticulitermes flavipes

Pyrsonympha grandis
    Reticulitermes speratus
    Reticulitermes flaviceps

Pyrsonympha modesta
    Reticulitermes speratus

Pyrsonympha affinis
    "*Coptotermes* from Japan" = *Reticulitermes*?

Pyrsonympha vacuolata
    Reticulitermes lucifugus

Pyrsonympha elongata
    Reticulitermes lucifugus

Pyrsonympha minor
    Reticulitermes hesperus

Pyrsonympha granulata
    Reticulitermes hesperus

Pyrsonympha major
    Reticulitermes hesperus

Dinenympha exilis
    Reticulitermes speratus
    Reticulitermes flaviceps

Dinenympha fimbriata
    Reticulitermes hesperus

Dinenympha gracilis
    Reticulitermes flavipes
    Reticulitermes lucifugus

Dinenympha leidyi
    Reticulitermes speratus

Reticulitermes flaviceps

Dinenympha nobilis
    Reticulitermes speratus

Dinenympha parva
    Reticulitermes flaviceps
    Reticulitermes speratus

Dinenympha porteri
    Reticulitermes speratus
    Reticulitermes flaviceps

Dinenympha rugosa
    Reticulitermes speratus

Dinenympha sp.
    Kalotermes lucifugus

Order: HYPERMASTIGIDA

Prolophomonas kalotermitis
    Kalotermes flavicollis

Joenia annectens
    Kalotermes flavicollis

Joenia intermedia
    Hodotermes mossambicus

Joenina pulchella
    Porotermes adamsoni

Joenopsis polytricha
    Archotermopsis wroughtoni

Joenopsis cephalotricha
    Archotermopsis wroughtoni

Kofoidia loriculata
    Paraneotermes simplicicornis

Hoplonympha natator
    Paraneotermes simplicicornis

Staurojoenina mirabilis
    Kalotermes aethiopicus

Staurojoenina assimilis
    Kalotermes minor

Staurojoenina sp.
    Kalotermes hubbardi

Trichonympha turkestanica
    Anacanthotermes murgabicus
    Anacanthotermes macrocephalus
    Anacanthotermes vagans
    Anacanthotermes viarum

Trichonympha fletcheri
    Anacanthotermes viarum

Trichonympha chattoni
    Glyptotermes iridipennis
    Glyptotermes brevicaudatus
    Glyptotermes brevicornis
    Glyptotermes ceylonicus
    Glyptotermes contracticornis
    Glyptotermes montanus
    Glyptotermes neotuberculatus
    Glyptotermes parvulus
    Glyptotermes taveuniensis
    Glyptotermes sp., Uganda
    Glyptotermes sp., Belgian Congo
    Glyptotermes sp., Ruanda-Urundi
    Glyptotermes sp., Philippine Is.
    Kalotermes milleri
    Kalotermes schwartzi

Trichonympha divexa
    Kalotermes sp., South Africa

Trichonympha zeylanica
    Neotermes militaris
    Kalotermes obscurus

Trichonympha tabogae
    Kalotermes tabogae

Trichonympha subquasilla
    Kalotermes clevelandi
    Kalotermes immigrans

Trichonympha lighti
    Kalotermes emersoni

Trichonympha saepicula
    Rugitermes kirbyi
    Rugitermes panamae

Trichonympha ampla
    Kalotermes occidentis

Trichonympha quasilla
    Kalotermes (s.l.) *perezi*

Trichonympha corbula
    Procryptotermes sp., Madagascar

    Kalotermes (s.l.) *castaneiceps*
    Kalotermes (s.l.) *longus*
    Kalotermes sp., Madagascar

Trichonympha teres
    Neotermes meruensis

Trichonympha peplophora
    Neotermes howa
    Neotermes desneuxi
    Neotermes gracilidens
    Neotermes amplus

Trichonympha agilis
    Reticulitermes flavipes
    Reticulitermes lucifugus
    Reticulitermes tibialis
    Reticulitermes hesperus
    Reticulitermes speratus
    Reticulitermes fukienensis

Trichonympha campanula (fig. 237, 3)
    Zoötermopsis angusticollis
    Zoötermopsis nevadensis
    Zoötermopsis laticeps

Trichonympha minor
    Reticulitermes lucifugus

Trichonympha magna
    Porotermes adamsoni

Trichonympha sphaerica (fig. 237, 2)
    Zoötermopsis angusticollis
    Zoötermopsis nevadensis

Trichonympha collaris (fig. 237, 1)
    Zoötermopsis angusticollis
    Zoötermopsis nevadensis

Trichonympha spp.
    Hodotermopsis japonicus

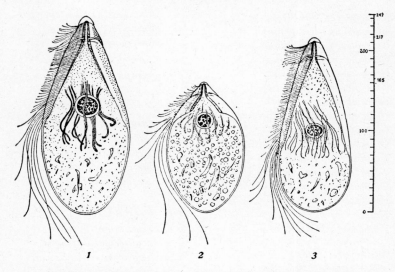

Fig. 237. Species of *Trichonympha* from Zoötermopsis. *1, T. collaris. 2. T, sphaerica. 3. T, campanula.* (From Kirby, 1932a.)

Trichonympha spp. (*Cont.*)
    Kalotermes obscurus
    Kalotermes rufinotum
    Kalotermes taylori
    Glyptotermes borneensis
    Glyptotermes nigrolabrum
    Glyptotermes satsumensis
    Glyptotermes trilineatus
    Cryptotermes dolei
    Glyptotermes contracticornis

Trichonympha serbica
    Reticulitermes lucifugus

Pseudotrichonympha hertwigi
    Coptotermes hartmanni
    Coptotermes lacteus
    Coptotermes sjöstedti

Pseudotrichonympha grassii
    Coptotermes formosanus

Pseudotrichonympha introflexibilis
    Schedorhinotermes putorius

Pseudotrichonympha magnipapillosa
    Schedorhinotermes putorius

Pseudotrichonympha parvipapillosa
    Schedorhinotermes intermedius

Pseudotrichonympha pristina
    Archotermopsis wroughtoni

Pseudotrichonympha sphaerophora
    Rhinotermes nasutus

Pseudotrichonympha spp.
    Leucotermes aureus
    Coptotermes corvignathus
    Coptotermes travians
    Coptotermes ceylonicus

Deltotrichonympha numidica
    Anacanthotermes ochraceus

Deltotrichonympha operculata
    Mastotermes darwiniensis

Mixotricha paradoxa
    Mastotermes darwiniensis

Holomastigotes elongatum
  Reticulitermes lucifugus
  Reticulitermes speratus
  Reticulitermes flaviceps
  Hodotermes mossambicus

Holomastigotes crassum
  Hodotermes mossambicus

Holomastigotoides hartmanni
  Coptotermes formosanus

Holomastigotoides hertwigi
  Coptotermes hartmanni

Holomastigotoides hemigymnum
  Coptotermes lacteus

Holomastigotoides mirabile
  Coptotermes sjöstedti

Rostronympha magna
  Anacanthotermes ochraceus

Spirotrichonympha africana
  Hodotermes mossambicus

Spirotrichonympha mirabilis
  Porotermes adamsoni

Spirotrichonympha elongata
  Schedorhinotermes intermedius

Spirotrichonympha flagellata
  Reticulitermes lucifugus

Spirotrichonympha flagellata var. *coptotermitis lactei*
  Coptotermes lacteus

Spirotrichonympha flagellata var. *schedorhinotermitis intermedii*
  Schedorhinotermes intermedius

Spirotrichonympha leidyi
  Coptotermes formosanus

Spirotrichonympha pulchella
  Reticulitermes hageni

Spirotrichonympha polygyra
  Paraneotermes simplicicornis

Spirotrichonympha bispira
  Paraneotermes simplicicornis

Spirotrichonympha gracilis
  Reticulitermes hageni

Spirotrichonympha grandis
  Porotermes adamsoni (= *P. grandis*)

Spirotrichonympha kofoidi
  Reticulitermes flavipes
  Reticulitermes hageni

Spirotrichonympha segmentata
  Reticulitermes lucifugus

Spirotrichonymphella pundibunda
  Porotermes adamsoni

Spironympha ovalis
  Reticulitermes hesperus

Spironympha porteri
  Leucotermes flaviceps

Spironympha elegans
  Leucotermes tenuis

Microjoenia hexamitoides
  Reticulitermes lucifugus

Microjoenia axostylis
  Archotermopsis wroughtoni

Microjoenia pyriformis
  Reticulitermes hageni

Microjoenia ratcliffei
  Reticulitermes hesperus

552   INSECT MICROBIOLOGY

Torquenympha octoplus
  Reticulitermes hesperus

Teratonympha mirabilis
  Reticulitermes speratus

Teratonympha mirabilis var. *formosana*
  Reticulitermes flaviceps

Spirotrichosoma capitata
  Stolotermes victoriensis

Spirotrichosoma obtusa
  Stolotermes victoriensis

Spirotrichosoma magna
  Stolotermes ruficeps

SPOROZOA

Only a few species of Sporozoa have been reported from termites. The first was *Hirmocystis* (now *Gregarina*) *termitis* described in 1881 by Leidy. It is found in *Reticulitermes flavipes, Zoötermopsis angusticollis, Zoötermopsis nevadensis,* and possibly others. Other species are as follows:

Duboscqia legeri (fig. 238) is from the body cavity of *Reticulitermes lucifugus* and *Reticulitermes flavipes.* This parasite produces white nodules up to 500 microns in diameter in the body cavity.

Nosema termitis invades and develops in the epithelial cells of the midintestine of *Reticulitermes flavipes.* Continued extrusion or sloughing of infected epithelial cells into the gut lumen probably is responsible for the presence of the light infection observed in most of the insects.

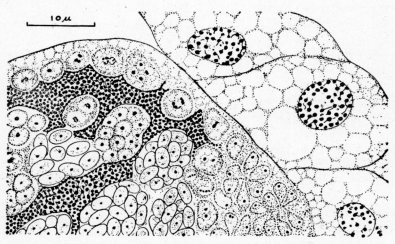

Fig. 238. *Duboscqia legeri.* Peripheral portion of infected and uninfected adipose tissue cells of *Reticulitermes flavipes,* showing the normal and hypertrophied nuclei of the host cells. (From Kudo, 1942.)

Kofoidina ovata is from the midgut of *Zoötermopsis angusticollis* and *Z. nevadensis*.

Stylorhynchus sp. lines the intestinal wall of *Neotermes castaneus*. There is some question as to the actual existence of this species since the termite's intestinal wall is known to be lined with an *Oxymonas*.

Gregarina mirotermitis is in *Mirotermes panamaensis*.

COCCIDIA

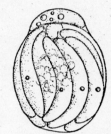

Fig. 239. An unidentified coccidian from *Amitermes minimus*. This stage contains sixteen merozoites. (From Kirby, 1932b.)

Kirby (1932b) has reported the presence of an unidentified coccidian in a specimen of *Amitermes minimus* (fig. 239). This is the only available record of the coccidian in a termite.

SARCODINA

The first termites in which amoebae have been found are of the family Termitidae (Kirby 1927, 1932b). Amoebae in termites other than Termitidae are very few. The amoebae which have been observed are listed below.

Order: AMOEBINA

Family: ENDAMOEBIDAE

Endamoeba disparata
  Mirotermes hispaniolae

Endamoeba majestas (fig. 240)
  Mirotermes hispaniolae

Endamoeba simulans
  Mirotermes panamaensis

Endamoeba sabulosa
  Mirotermes panamaensis

Endamoeba beaumonti
  Amitermes beaumonti
  Amitermes coachellae
  Amitermes minimus
  Amitermes wheeleri
  Amitermes medius

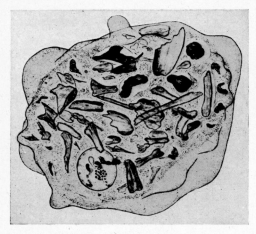

Endamoeba pellucida
　Cubitermes sp.

Endamoeba granosa
　Cubitermes sp.

Endamoeba lutea
　Cubitermes sp.

Endolimax suggrandis
　Cubitermes sp.

Endolimax termitis
　Mirotermes hispaniolae

Endolimax gohceni
　Coptotermes heimi

Fig. 240. *Endamoeba majestas* from *Mirotermes hispaniolae*. (From Kirby, 1927.)

CILIATA

The first member of the Ciliata described from termites was *Nyctotherus termitis* reported from *Neotermes militaris* by Dobell in 1910. Since then several other species have been described and are listed as follows:

Order: HETEROTRICHA

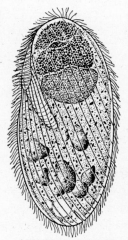

Nyctotherus termitis
　Neotermes militaris

Nyctotherus silvestrianus (fig. 241)
　Amitermes silvestrianus
　Amitermes minimus
　Amitermes emersoni

Nyctotherus fletcheri
　Coptotermes sp.
　Leucotermes indicola

Nyctotherus spp.
　Termitidae [1]

Fig. 241. *Nyctotherus silvestrianus* from *Amitermes minimus*. (From Kirby, 1932b.)

---

[1] In a personal communication (March, 1944) Kirby states that he has in his collection species of *Nyctotherus* which occur in Termitidae. These findings have not as yet been published.

❖ CHAPTER XII ❖

IMMUNITY IN INSECTS

THE FIRST serious consideration was given to the immunity principle in insects when Pasteur selected a race of silkworms naturally immune to the ravages of the microsporidian disease pébrine. It was not until after the turn of the century, however, that an active interest was taken in this field. In 1905 Metchnikoff published an account on immunity to infectious diseases in which he included some of his observations on immunity in insects. In the years just following this Metalnikov began an extensive series of experiments and publications on the immunity principles in insects. He and his associates have made some of the most informative contributions to the field, particularly as concerns cellular immunity. The French worker Paillot has also been a leader in this work and has published many interesting observations on mechanisms of the various types of immunity. Paillot, in contrast to Metalnikov, was a great supporter of the theories of humoral immunity. In recent years there has been a tendency to reconcile the two theories of immunity—cellular and humoral—in the case of insects as well as vertebrates.

The literature apparently contains no references to the study of immunity in ticks.

It perhaps is well to call the reader's attention to the fact that a study of the principles of immunity in invertebrates, particularly insects, is not without its applications to the study of vertebrate immunity. This will be pointed out in later paragraphs of the chapter. In the case of insects we have, as compared to vertebrates, much simpler anatomic and physiologic situations which bear upon their immunologic responses. The effects which various systems (e.g., nervous, digestive, etc.) have upon the production of immune bodies can be and has been easily exploited with insects.

The study of immunity in insects has had only a laggard start. Here lies an almost virgin field waiting to be investigated. The practical applications to such fields as medical entomology, insect pathology, and plant pathology would undoubtedly be great.

NATURAL IMMUNITY

Like normal men and animals, normal insects are known to possess a so-called natural immunity to many infectious agents. As with higher animals, the natural immunity is apparently an inherent quality or depends on mechanical and physiologic characteristics and may not be the result of the insect's previous contact with the agent, although sometimes this may take place unknowingly—in which case we are actually concerned with an acquired immunity. Natural immunity may be confined to the species, race, or possibly even individuals, but it is generally concerned with groups rather than with individuals.

As a group, insects in general have a relatively high natural immunity to most organisms pathogenic for higher animals. For example, most insects are insusceptible to most strains of *Pasteurella pestis, Corynebacterium diphtheriae, Mycobacterium tuberculosis, Mycobacterium leprae,* and various streptococci. Of 34 strains of bacteria tested, Metalnikov and Chorine (1929b) found *M. tuberculosis* to be the least virulent of all when given *per os* or inoculated into the corn borer *Pyrausta nubilalis*. Moreover, this bacterium was destroyed and completely digested in the body of the larvae. The larva of the wax moth (*Galleria mellonella*) possesses an even stronger immunity to animal pathogens than does the corn borer. Similar immunities are apparent in certain ticks to the rickettsiae of spotted fever (*Dermacentroxenus rickettsi*) and in most species of culicine and anopheline mosquitoes to the protozoa causing human and avian malaria.

On the other hand, *Rickettsia prowazeki,* the cause of typhus fever in man, is also pathogenic for the louse, its insect host. Other instances include the marked susceptibility of *Galleria mellonella* to the bacteria causing cholera and dysentery. *Salmonella enteritidis* has been observed to kill a certain number of ticks (*Dermacentor andersoni*) which ingest it. Piscine strains of the tubercle bacillus (*Mycobacterium piscium*) are infectious for larvae of *Achroia grisella*.

Of equal interest is the fact that many microorganisms, bacteria particularly, which are generally not pathogenic to man and higher animals, are extremely virulent for insects. These include such species as *Escherichia coli, Pseudomonas aeruginosa, Bacillus subtilis, Proteus vulgaris,* and *Serratia marcescens*. The latter organism has also been observed to kill a certain number of ticks (*Dermacentor andersoni*) which ingested it.

Huff (1940) has called attention to three groups into which the microorganisms could be divided showing varying degrees of pathogenicity for certain insects; specifically, the caterpillars of the wax moth *Galleria mellonella*. Group 1 included those to which the caterpillars showed complete immunity (*Mycobacterium tuberculosis, Corynebacterium diphtheriae,* streptococci, *Clostridium tetani, Pasteurella pestis,* and *Trypanosoma brucei*). In group 2 were placed those microorganisms to which the insects showed an incomplete immunity (*Staphylococcus aureus, Clostridium perfringens, Clostridium septicum, Pasteurella aviseptica, Pasteurella avicida,* and *Vibrio comma*). Group 3 contained those to which the caterpillars showed no immunity even to small doses (*Bacillus subtilis, Bacillus megatherium, Proteus vulgaris, Pseudomonas aeruginosa, Serratia marcescens,* and *Escherichia coli*). From this grouping it can be seen that what occurs with the caterpillars of the wax moth conforms more or less to the general rule expressed in the preceding paragraph.

While studying the processes of immunity in insects, Cameron (1934) obtained results which, in general, are in fairly close agreement with those of Metalnikov (1927). Cameron presents a tabular comparison:

Pathogenicity of Various Bacteria for *Galleria mellonella*
(O = nonpathogenic, + = pathogenic, V = variable)

Bacterium	Metalnikov	Cameron
Mycobacterium tuberculosis var. *hominis*	O	O
Mycobacterium tuberculosis var. *bovis*	O	O
Mycobacterium avium	O	O
Mycobacterium smegmatis (*M. lacticola*)	-	O
Vibrio comma	V	O
Eberthella typhosa	V	O
Clostridium tetani	O	O
Clostridium sporogenes	O	O
Diplococcus pneumoniae	O	O
Salmonella schottmülleri	O	O
Pasteurella pestis	V	O
Neisseria gonorrheae	V	O
Hemophilus influenzae	-	O
Shigella dysenteriae	V	O
Staphylococcus aureus	V	O
Corynebacterium pseudodiphthericum	-	O
Bacillus subtilis	+	V
Staphylococcus albus	O	V
Clostridium oedematiens	V	V
Clostridium septicum	V	V

Bacterium	Metalnikov	Cameron
Clostridium perfringens	–	V
Streptococcus haemolyticus (*S. pyogenes*)	O	+
Streptococcus faecalis	–●	+
Corynebacterium diphtheriae	O	+
Proteus vulgaris	+	+
Bacillus mycoides	+	+
Pseudomonas aeruginosa	+	+
Escherichia coli	+	+

Several examples of natural species immunity may be cited. One of the best known is the natural immunity silkworms (*Bombyx mori*) have to *Coccobacillus acridiorum*, which is very pathogenic for grasshoppers, ants, crickets, and certain caterpillars. Peculiarities in the transmission of certain disease agents by one insect species and not by others cause one to wonder concerning the possible role natural immunity may have in biologically determining the insect vector.

Striking examples of individual natural immunity are known among insects. Of great economic importance was the fact that Pasteur was able to breed a stock of silkworms (*Bombyx mori*) which were insusceptible to *Nosema bombycis*, the microsporidian which causes pébrine. He was able to do this because certain individual silkworms were immune to the disease and by a careful selection of these individuals he was able to initiate and maintain an immune strain or race. Similar phenomena occur among other insects exposed to other infectious agents. Huff (1940) has pointed out and discussed cases of individual immunity among species of insects which act as vectors of various diseases. *Culex pipiens* exemplifies this when a considerable number of individuals feed on avian blood containing thousands of infective forms of *Plasmodium cathemerium* or *Plasmodium relictum*. Some of them become heavily infected, others lightly infected, and some are not infected at all. These noninfected individuals remain so even after repeated infective feedings. By a selection of the progenies from susceptible and insusceptible females, Huff was able to maintain stocks with increasing and decreasing degrees of susceptibility. He observed that this character of insusceptibility behaved as a Mendelian dominant and concluded that the natural immunity barrier in this case appeared to be the intestinal wall, since in insusceptible as well as in susceptible mosquitoes the parasite developed as far as the oökinete stage.

Nature of Natural Immunity. The phenomenon of natural immunity is an inherent, genetically determined characteristic and very little about it is well understood. It can be decreased by such factors as a lowered temperature, fatigue, and malnutrition, but in the normal healthy animal it cannot be increased since it is not dependent upon specific antibodies. Among the mechanisms which are believed to be involved in some cases are the body temperature, the anatomical structure, or perhaps some physiologic manifestation.

According to Paillot (1933), the natural immunity among insects is of a dual nature, depending upon a complexity of humoral and cellular factors. Of greatest importance is the bactericidal action of the blood plasma. This action reduces the invading microorganisms to granules, which may then be taken up by the phagocytic blood cells. Paillot believes that the bactericidal action of insect blood manifests itself by a chemical or physiologic change in the microrganism's substance and cannot be attributed to substances formed in the blood or produced by the amoebocytes.

It is very possible that in some cases the natural immunity of adult insects may actually be an acquired immunity gained in an earlier stage. For example, the immunity of the larvae of *Galleria mellonella* against certain agents is carried to the adult stage.

ACQUIRED IMMUNITY

Like higher animals insects may acquire an immunity as the result of an attack by an infectious agent (naturally acquired immunity) or when such agents or vaccines are inoculated into the insect (artificially acquired immunity). In the case of insects also the acquired immunity may be active or passive.

Active Immunity. Early investigators (e.g., Metchnikoff, 1905) at first were unable to produce an active acquired immunity in insects. Later, however, when workers had a better idea of the relative virulence of microorganisms with respect to insects, they were able not only to immunize them successfully but to do it with ease. Various methods have been used to bring about such an immunization: inoculating the insects with (1) old attenuated cultures, (2) cultures heated to 60° C., and (3) very small doses of virulent organisms.

Perhaps the most startling thing concerned with the active immunization of insects is the fact that they can be immunized in so short a time—usually within 24 hours following a single injection of an old culture or vaccine! Here certainly is a field waiting to be investigated by the immunologist interested in the mechanisms of immunity. Another characteristic of acquired immunity in insects is that in most cases it is not very strong. An immunized insect can survive several lethal doses but not extra heavy doses. In fact, immunized insects given heavy doses frequently die sooner than unimmunized individuals given the same dosage.

One of the first investigators successful in bringing about an acquired immunity in insects was Paillot (1920), who had found that larvae of *Euxoa segetum* died rapidly of a septicemia when inoculated with a suspension of *Bacillus melolonthae non-liquefaciens*. This worker observed that if he first inoculated into the larvae a little of a culture of this bacterium three months old or older, they did not become infected and were immune to ordinary lethal doses. Regular examinations of the blood from these insects showed that the organisms were gradually transformed into granules.

Shortly after this other workers were able to produce acquired immunities in insects. Metalnikov and Gaschen (1921) thus immunized caterpillars of the wax moth, *Galleria mellonella*, against *Escherichia coli* and *Proteus vulgaris* by the injection of vaccines and found that small doses bring about an immunity more quickly than do large ones. Of equal interest is the fact that they found the results persisted into the adult stage. Ishimori (1924) immunized larvae and pupae of *Galleria mellonella* against *Vibrio comma* with a heat-killed vaccine as well as heated (120° C.) or unheated culture filtrates of this organism. Glaser (1925a) immunized silkworms (*Bombyx mori*) against a "flacherie"-like disease by inoculating them with killed cultures of the causative bacterium, although the vaccine administered orally failed to immunize the insects. Metalnikov and Chorine (1929b) showed that the European corn borer, *Pyrausta nubilalis*, can be easily immunized against moderately virulent bacteria (*Bacterium galleriae* No. 1 and *Salmonella enteritidis* var. Danysz) but not against more virulent species (*Bacterium galleriae* No. 2, *Bacillus thuringiensis*, *Bacillus canadensis*, *Vibrio leonardi*, and *Coccobacillus ellingeri*). Other insects which have been successfully immunized against bacteria pathogenic to them include *Pieris brassicae*, *Carausius morosus*, and *Apis mellifera*.

Attempts to immunize some insects against certain microorganisms have failed repeatedly. For example, the oriental cockroach *Blatta orientalis* could not be experimentally immunized against *Pseudomonas aeruginosa* and *Escherichia coli*. Acquired immunities to protozoa and fungi are also difficult to produce.

Very little is definitely known with regard to the specificity of acquired immunity. Some workers have claimed that it is as specific as in higher animals, but others have maintained that very little specificity exists. Sufficient results to decide this question are yet to be obtained. Ishimori and Metalnikov (1924) carried out an interesting series of experiments along these lines.

They were able to immunize *Galleria mellonella* against *Vibrio comma* by infecting them with such bacteria as *Bacillus anthracis* and *Escherichia coli,* and with Chinese ink. A stronger and more stable immunity, however, appeared to result from the use of the specific vaccine. On the other hand, the organisms *Micrococcus galleriae* No. 3 and *Bacterium galleriae* gave a stronger immunity against the cholera vibrio than did *Vibrio comma* itself. The work of other investigators using the same insect but different antigens and different test organisms in general support the nonspecificity of these reactions both *in vivo* and *in vitro*. Zernoff (1934) obtained similar results with *Carausius morosus*. In general it may be said that in most cases the specific immunity is stronger and of longer duration than the nonspecific immunity, though it does appear that a greater degree of nonspecific immunity occurs in insects than in mammals.

Passive Immunity. From the few studies made of passively acquired immunity in insects, it is evident that in certain cases at least immunity acquired in one insect may be transferred passively to another. This passively acquired immunity is usually of short duration and sometimes not very specific. For example, Zernoff (1928a) found that when he transferred Danysz bacillus (*Salmonella enteritidis*) immune blood from one caterpillar (*Galleria mellonella*) to another, the latter was more immune against *Coccobacillus acridiorum* than against the specific organism.

Passive immunity may be conferred by the transfer of either the blood plasma or the blood cells alone (Zernoff, 1928b).

Nature of Acquired Immunity. Very little is known concerning the mechanisms involved in the manifestations of acquired immunity in insects. Certainly the complicated systems and tissues thought to be the site of antibody production in mammals is lacking in these invertebrates.

Metalnikov (1927) and others have studied the possible effect of an insect's nervous system upon its ability to produce an immunity against an infectious agent. It is difficult to decide just what conclusions should be drawn from their experiments since in destroying parts of the nervous system other vital functions of the insect were somewhat disrupted. In brief, it was found that when the cerebral ganglia and the first and second thoracic ganglia of *Galleria mellonella* larvae were destroyed with a hot needle, the caterpillars were immunized against the cholera vibrio and the bacillus of Danysz as easily as were normal insects. Similar results were obtained when the larvae were decapitated along with the first two thoracic segments. However, when the third thoracic ganglion was destroyed, which is a serious operation for the insects, they could not be immunized. Furthermore, there occurred a drop in the natural resistance of the caterpillars to staphylococci. The number of leucocytes in the operated caterpillars was decreased, as was the phagocytic index, leaving the bacteria to multiply rapidly and soon kill the host. Equally interesting is the experiment in which ligatures were placed tightly around the caterpillars—in which condition they continued to live two or three weeks. It was found that each of the two portions could be immunized separately. When the anterior half was immunized this immunity was transmitted to the posterior half. However, immunization of the posterior half did not afford protection to the anterior half. Metalnikov explained this phenomenon on the basis that in the caterpillar the central nervous system is concerned with immunization and that the immunity is transmitted through the ventral chain of ganglia. It is possible, however, that the blood supplying the living nerve tracts is also involved in this transmission of immunity and that the process is not a purely nervous phenomenon.

Of the various types of antibodies possible, only two have with certainty been produced experimentally in insects. These are agglutinins (Glaser, 1918b) and lysins (Paillot, 1920; Metalnikov, 1920a, 1924; Zernoff, 1930, 1931). It is possible that in one or two cases antitoxins have also been produced. An immunity against diphtheria toxin has been produced in larvae of *Galleria mellonella* by the use of a toxoid (Chorine, 1929a,b, 1931).

CELLULAR IMMUNITY

During the latter part of the nineteenth century Metchnikoff and his students called attention to the role of certain cells in the defense mechanisms of animals. On the basis of his findings, Metchnikoff proposed his theory of cellular immunity. Most of his observations have since been confirmed and elaborated. To those blood cells which engulf or ingest foreign materials including microorganisms, Metchnikoff gave the name "phagocytes" ("devouring cells"). These he divided into two groups depending on whether they were "wandering" or "fixed" cells. The former are called "microphages" and the latter, which are larger, "macrophages." The macrophages belong to or are derived from what is known as the reticulo-endothelial system. Such a system as it is known in mammals does not exist in insects, but cells (nephrocytes) are present which may be considered as analogous to it (see Yeager, McGovran, Munson, and Mayer, 1942).

The Circulatory System. In considering the various aspects of either cellular or humoral immunity one should bear in mind the nature of an insect's circulatory system. Essentially, it consists of a *hemocele* containing the circulating blood which bathes all the organs and tissues directly. There is no complex system of veins and arteries as in higher animals. In most insects the hemocele is divided into sinuses by diaphragms or fibromuscular septa. The main organ of the circulatory system is the *dorsal vessel* which is divided into the heart (the posterior part) and the aorta (the anterior part). This structure is simply a narrow, continuous or chambered, tubelike affair the sides of which are perforated with small valvular openings called "ostia." Through the waves of contraction or pulsations of the heart, or in some cases of the entire dorsal vessel, the blood is kept in motion throughout the body cavity of the insect. Pulsatile organs may also occur in other parts of the body to assist in the thorough distribution of the blood. As in vertebrate animals the main function of the blood of an insect is to convey nutrient substances to the tissues and to carry waste materials to the excretory organs. On the other hand, it plays a very small part in the respiration of most insects except chironomids which have hemoglobin in their blood. In other insects the trachea takes care of the gaseous exchanges.

The Blood. As with most other animals the blood of insects consists of two parts. The fluid part or blood plasma is known as "hemolymph," the cellular parts or blood corpuscles are called "hemocytes."

The hemolymph is a more or less viscid liquid which may be clear or tinged with green, yellow, orange, or brown pigment. The color is generally characteristic of the species but has no relation to food or geographic location. In some lepidopterous larvae or pupae the color differs with the sex, usually being green in female caterpillars and yellow or colorless in males. The composition of the hemolymph varies with the species concerned. Most of the constituents are associated with products of the insect's metabolism. In the blood of some insects four main constituents have been distinguished: hemoxanthin, fibrin, lutein, and uranidin. Other analyses of the plasma have brought out a long list of substances (see Muttkowski, 1923).

The hemocytes, or cellular elements of the blood, occur in a variety of forms and sizes, and a general classification of their types is difficult. Different authors have erected different systems of classification. Hollande (1911) divided the hemocytes into four groups: *proleucocytes, phagocytes, granular leucocytes,* and *oenocytoids* (the latter are absent in Orthoptera). In the blood of some species of Coleoptera and Lepidoptera he observed certain spherule cells the cytoplasms of which are filled with characteristic spherules. The proleucocytes are young leucocytes from which the more specialized forms of hemocytes are derived. Of special importance from the standpoint of cellular immunity are the phagocytes which are characterized by their large size and by their phagocytic activities. The granular leucocytes also are sometimes phagocytic, but the oenocytoids (named for their resemblance to oenocytes) are not.

Muttkowski's (1924) classification distinguishes two principal types of cells: *chromophile leucocytes* and *amoebocytes*. Paillot (1933) recognizes four types of blood cells: *micronucleocytes, micronucleocytes with spherules, macronucleocytes,* and *oenocytoids*. Under this nomenclature only the micronucleocytes are phagocytic. One of the most recent classifications is that of Cameron (1934) who distinguishes the hemocytes as follows: *lymphocytes* (primitive cells, deeply staining nucleus, small amount of cytoplasm, actively amoeboid—comparable to proleucocytes, amoebocytes, macronucleocytes), *leucocytes* (larger, abundant cytoplasm, markedly phagocytic—comparable to large amoebocytes, micronucleocytes, phagocytes), and *spherule cells* (cytoplasm contains large acidophilic spherules eventually liberated from cell), and *oenocytes* (large cells with deeply staining nuclei, homogeneous cytoplasms and associated with the fat body). In addition there are the *pericardial cells* which are distributed along the diaphragm, supporting the heart (fig. 243). The pericardial cells vary from small, granular cells with deeply staining nuclei to huge cells which may contain two to four nuclei. For a further detailed account of the blood cells of certain insects, the reader is referred to a booklet by Rooseboom (1937) and a paper by Yeager (1945).

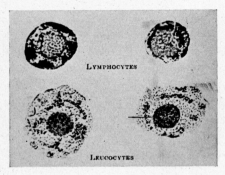

Fig. 242. Normal blood cells of *Galleria mellonella*. (From Cameron, 1934.)

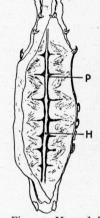

Fig. 243. Ventral dissection of *Galleria mellonella* showing location of heart (*H*) and pericardial cells (*P*). (From Cameron, 1934.)

For those interested in serologic techniques, a statement may be made relative to the clotting of blood. Yeager and Knight (1933) divided insects into three groups on the basis of the clotting properties of their blood: (1) species in which the blood does not clot, (2) species in which the clot is formed by agglutination of the hemocytes, and (3) species in which the clot is formed by coagulation of the plasma.

Phagocytosis. Some of the earlier investigators were of the opinion that the phagocytic cells of the blood were of greater importance in the immunity of an insect than were the humoral constituents. Since then most of the experimental data seems to support the opposite view. In the United States Glaser (1918b) was one of the first to point out the overemphasis placed on phago-

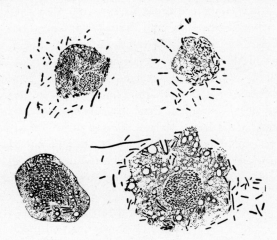

Fig. 244. Phagocytosis of *Bacillus melolonthae non-liquefaciens gamma* by the phagocytic blood cells of *Euproctis* (*Nygmia*) *chrysorrhoea* 24 hours after inoculation. (From Paillot, 1933.)

cytosis in insects. This worker found the blood cells of the grasshopper *Melanoplus femur-rubrum* to be rather passive; they did not phagocytose bacteria (*Bacillus poncei*) in an amoeboid fashion. Immunity resulted instead from the bactericidal properties of the hemolymph. Paillot (1933), as the result of an extensive study in this field, likewise maintains that phagocytosis is secondary to humoral reactions in protective ability. In cases where the bactericidal action of the blood has reduced the microorganisms to granules, the intensity of phagocytosis frequently increases at this point, thus aiding in the removal of the granules.

The most active role in the process of phagocytosis is generally assumed by the leucocytes, which absorb the bacteria and foreign substances such as carmine and India ink introduced into the insect's blood. The lymphocytes are next in importance, but these cells are not as active as the leucocytes. Occasionally the spherule cells are phagocytic but the oenocytes are not. With respect to the pericardial cells Cameron (1934) differs from most observers in attributing phagocytic properties to them.

Working with larvae of the European corn borer, *Pyrausta nubilalis,* Metalnikov and Chorine (1929b) found that phagocytosis of micrococci is most intense after three to five hours. The bacteria (e.g., *Coccobacillus ellingeri, Vibrio leonardi, Bacillus thuringiensis,* etc.) most virulent to the caterpillars do not provoke phagocytosis. On the other hand, when the dosage of less virulent bacteria is great enough to cause a mortal infection, there is frequently intense phagocytosis which lasts until death. Sometimes (e.g., with *Bacillus subtilis*) phagocytosis occurs only during the first stages of a mortal infection, then diminishes, until finally the bacteria multiply freely in the septicemic blood. Metalnikov and Chorine observed still another manner of phagocytosis. When *Proteus vulgaris* OX-19 is inoculated into the corn borer, no phagocytosis whatever appears at first; instead the bacteria multiply freely. Fifteen to twenty hours later the lymphocytes are found to have taken up bacteria. This reaction, however, occurs too late to save the insect's life and it dies a few hours later. Thus in the larvae of *Pyrausta nubilalis* at least four types of phagocytosis are known to occur depending on the kind of invading bacterium.

One of the more recent studies made of the phagocytic cells of insects is that of Cameron (1934). A summary of his work is very much in order. Most of Cameron's observations were based on the larva of the wax moth, *Galleria mellonella,*[1] though he extended his studies to other species of Lepidoptera as

---

[1] *Galleria mellonella,* the wax moth or bee moth, has to some extent become the "guinea pig" of insect microbiology, particularly as related to immunity studies. Of interest, therefore, is Cameron's account concerning the characteristics of this insect:

In the eighth and ninth books of the *Historia animalium,* Aristotle describes a moth found in beehives, which does great damage to the comb. He notes that it is never stung by a bee and can be got out of the hive only by fumigation. It engenders a caterpillar of a species nicknamed, so Aristotle says, the "borer," so that it seems likely that the wax moth was as well known in ancient Greece as it is in modern Europe, and indeed in most countries of the world.

Under natural conditions it completes its life cycle in 6–8 weeks (Réaumur, Fletcher). The eggs which are laid by the female in the uncapped cells of the honeycomb or along broken ridges of wax develop during 8–10 days. The larvae on hatching are barely visible to the naked eye but grow rapidly, eating voraciously and attaining their maximum size in about 25 days. The pupal stage lasts from 10 to 25 days. The optimum temperature for development is between 30° and 40° C. At lower temperatures growth of the larvae is retarded, and ceases at about 10° C. The food requirements of the larvae are simple, for old wax, no matter how dirty, serves admirably for the raising of large broods of wax moths. There is some evidence that wax from hives infested with foulbrood is avoided by the larvae (Phillips, 1907). Metalnikov has shown that purified beeswax, free from nitrogenous substances, will support life though it is not sufficient for the growth of the larvae. On the other hand the wax is necessary for the existence of the caterpillars. Because of the shortness of its life cycle and the ease with which it can be reared the wax moth is very suitable for experimental purposes. The larvae shortly before pupation reach a length of 2.5 cm. Although the body wall is somewhat delicate, with careful handling the caterpillar can be easily injected.

well. He concerned himself with the phagocytic response of the larvae to three types of substances: foreign particles, foreign cells, and bacteria.

Cameron found that the foreign particles (India ink, carmine, colloidal iron) were rapidly phagocytosed by the leucocytes and lymphocytes of the blood, the phagocytic cells of the fat body, and, in the case of colloidal iron and carmine, by the pericardial cells. Except when they occasionally took up India ink, the spherule cells did not appear to participate in the phagocytosis.

Fig. 245. *Galleria mellonella*. (From Comstock, 1936.)

When foreign cells (human and rabbit blood cells) are introduced into the body cavity of wax moth larvae, they are removed partly by hemolysis, partly by phagocytosis, and partly by clumping and subsequent encapsulation. Apparently phagocytosis plays only a very minor role in removing the foreign cells.

When inoculated with living bacteria Cameron observed that the *Galleria mellonella* larvae may respond with (1) active phagocytosis and complete destruction of the organisms (*Diplococcus pneumoniae, Staphylococcus aureus, Hemophilus influenzae*), (2) active phagocytosis, growth of the bacteria, and rapid death of the larvae (*Proteus vulgaris, Bacillus mycoides*), (3) active phagocytosis with survival of some organisms and their subsequent active growth, the larvae dying after several days (*Staphylococcus albus, Clostridium perfringens*), and (4) slight phagocytosis but destruction of the bacteria, presumably by other immunity principles (*Eberthella typhosa, Shigella dysenteriae, Vibrio comma*). Acid-fast bacteria (*Mycobacterium tuberculosis* and *Mycobacterium smegmatis* [*M. lacticola*]) may be inoculated into the larva without any apparent subsequent harm to the host. These bacteria tend to be segregated in the pericardial cells where they survive throughout the metamorphosis of the larva and can be isolated in a living virulent state from the adult moth. However, the acid-fast organisms also elicit a response from the blood phagocytes soon after inoculation. In 12 hours the lymphocytes become very numerous, composing up to 80 or 90 per cent of the hemocytes. After 24 hours encapsulation of clumps of bacilli occurs, forming small, rounded nodules which can frequently be seen with the naked eye. Cameron found these nodules to consist of several concentric layers of lymphocytes arranged about masses of tubercle bacilli, debris, and pigment. They may occur freely in the body cavity, attached to the layer of pericardial cells or enclosed in the lobules of the fat body.

Very little is known regarding the proportional numbers of the different cells occurring in the blood of insects. Very likely there is considerable variation according to species. Cameron found the relative proportion of hemocytes to be as follows: lymphocytes 38.3 per cent, leucocytes 57.5 per cent, spherule cells 2.9, oenocytes less than 1.0 per cent. Metalnikov and Chorine (1929b) found the differential count in normal larvae of the corn borer, *Pyrausta nubilalis,* to be: lymphocytes 27 to 45 per cent; leucocytes 30 to 69 per cent; and spheroidal cells 2 to 8 per cent. In the larvae infected with fatal doses of bacteria, the lymphocytes and spherules increased while the percentage of leucocytes decreased rapidly, marked changes in the blood picture usually being apparent within one or two hours. After 6 to 20 hours most of the blood cells had completely disintegrated. Apparently, there is very little change in the number of oenocytes. The increase in lymphocytes is thought to be due to hematopoietic stimulation. The fact that the cells are all in immediate contact with the stimulating substance and increase only by the division of existing blood cells may account for the rapidity of the increase of lymphocytes. This increase is frequently accompanied by the production of a large number of mitotic cells (Paillot's *karyokinetosis*) which may play a special role in the insect's immunity (Huff, 1940).

Other conditions favorable for phagocytosis have not been well studied. Temperature is known to exert some effect—low temperatures (18 to 15° C.) retard the beginning of phagocytosis, and below 10° C. no reaction occurs in most insects. The age and condition of the insect also affects the action of the phagocytes. The production of opsonins has not been demonstrated in the blood of insects though there very probably is a natural presence of this type of antibody.

The final manifestations of phagocytosis are concerned with the intracellular digestion and disposal of the engulfed organisms or foreign substances. In the case of an insect inoculated with *Mycobacterium tuberculosis,* for example, the phagocytes gradually dissolve the bacteria until their presence can no longer be detected by staining. Metalnikov and Chorine (1929b) have observed the presence of pigment in the bacteria during digestion and the gradual transformation to a brown-pigmented residue. Other bacteria may behave differently after being phagocytosed. Some become swollen and are then gradually absorbed; others are transformed into granules which are rapidly digested.

In most cases of phagocytosis the particular type of reaction is specific. That is, each bacterium or foreign substance elicits a typical and specific response.

Successful phagocytosis is occasionally thwarted by certain circumstances. The effect of adverse temperatures has already been mentioned. Sometimes phagocytes engulf so many organisms and are filled to such a point that they cannot digest the inclusions. In such a case the phagocytes themselves are destroyed. Occasionally certain bacteria liberate toxic substances which kill the leucocytes. At other times large vacuoles are formed in the phagocytes after ingesting the microorganisms. The blood cells become greatly swollen and are destroyed, and the invading bacteria eventually kill the insect.

Giant Cells and Other Formations. In case the phagocytes are not able to destroy the microorganisms with reasonable speed, they frequently come together and unite in masses to form *giant cells* (plasmodia). As the result of this formation two things are accomplished: (1) large numbers of bacteria are concentrated at one place and (2) there is an intensification of the intracellular digestion. Metalnikov and Chorine (1929b) inoculated corn-borer larvae with tubercle bacilli and found the fusion of the phagocytes to begin almost immediately. Three or four hours after injection large numbers of giant cells were seen forming

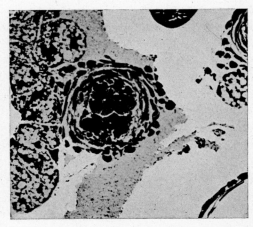

Fig. 246. Giant cell in the tissues of *Galleria mellonella;* cross section. (From Cameron, 1934.)

in the hemolymph. They became progressively more compact. As the bacteria were digested they were transformed to a brown-pigmented inclusion. After a day or two lymphocytes and leucocytes were seen to accumulate in concentric layers about the surface of the giant cell until a compact capsule or nodule was formed which completely imprisoned the bacteria. These same workers observed giant cells to form directly beneath the cuticle of *Galleria mellonella* into which they had inoculated large doses of slightly virulent bacteria. Black spots appeared on the epidermis at these points. Upon sectioning these areas, they found them to consist of aggregates of lymphocytes and leucocytes surrounding masses of half-digested bacteria. Eventually the epidermis and cuticle over these capsules burst, extruding the contents much the same as does an abscess. A new epidermis is then quickly formed underneath.

As has been pointed out by some writers, the term "giant cell" is perhaps not an ideal one since these formations in insects are histologically not the same as those known in vertebrates. These structures in insects are actually formed simply by layers of hemocytes around the phagocytosed material. The reaction is nonspecific since aseptic foreign bodies elicit the same response

(Huff, 1940). In some cases living bacteria have been isolated from these nodules during the remainder of the insect's life.

In addition to bacteria, nodules or similar formations have been seen in insects parasitized with sporozoa, nematodes, and fungi.

Specialized Phagocytosis. At this point may be mentioned a special function of phagocytosis which has been reported by Paillot (1930). Paillot believes that the intracellular symbiotic microorganisms of the aphid *Macrosiphum jaceae* gain entrance to the mycetome as follows: After the multiplication of specific microorganisms in the general body cavity, during which time they behave as parasites, phagocytosis takes place. At the same time there is an agglutination of the free microorganisms of the blood which are in contact with the mycetocytes. These microorganisms then penetrate the cytoplasm of the mycetocytes which also "fix" the free leucocytes filled with microorganisms. The leucocytes are later incorporated in the mass of the mycetome. Once securely in the mycetome the bacteriumlike microorganisms are transformed into giant rounded forms. The bacillary and giant forms are eventually passed into the egg where they form, in the yolk, a large clump at one of the poles. Paillot compares this clumping together with the agglutination phenomena seen in the blood of higher animals.

HUMORAL IMMUNITY

Most authorities agree that of the two types of immunity—cellular and humoral—the latter is of greater importance in the protection of the insect from agents of infection than is the former. In pointing out the importance of humoral immunity in insects Paillot (1923) stated three principles: (1) There may be immunity without phagocytosis. An example of this occurs in the case of *Porthetria dispar* caterpillars infected with *Bacillus melolonthae liquefaciens gamma,* which at first multiplies rapidly in the insect. After the ninth hour a crisis is reached and involution forms may be seen in the blood. These forms later become enlarged and rounded and eventually disappear as the result of a "leucolytic" process. At the same time that the organisms are disappearing the phagocytes and micronucleocytes are destroyed. Thus, Paillot concludes, the action is purely humoral. (2) There may be a humoral reaction without the intervention of leucocytic products secreted in the course of the infection. Paillot demonstrated this in *Agrotis* (*Euxoa*) larvae against *Bacillus melolonthae non-liquefaciens* and other organisms. (3) Phagocytosis in insects cannot be considered a consequence of the vital activity of the phagocytes. Most authorities are not prone to go as far as this in their consideration of the role of humoral immunity, but Paillot's experimental work does indicate that humoral immunity is extremely important in the insect's protection against infectious agents.

The lower limit of temperature at which humoral reactions will take place is about 12° C. (Paillot, 1933).

Antitoxins. Very little experimental work has been accomplished to demonstrate the presence or formation of antitoxins in insects. Chorine (1929a, 1931), however, has found diphtheria toxin to be toxic for larvae of *Galleria mellonella*. He was also able to bring about an immunity of the insects to the toxin through the use of a toxoid.

Bactericidins. Bactericidal substances in the blood of insects have been detected by several investigators. One of the first to demonstrate such killing power in the hemolymph was Glaser (1918b) who found the immune blood of the grasshopper *Melanoplus femur-rubrum* capable, *in vitro,* of killing *Bacillus poncei*.

According to Paillot (1933) the bactericidal action of insect blood manifests itself by bringing about a chemical or physiologic change in the bacterial substance and not by morphologic changes or disintegrations. Paillot believes that none of the bactericidal properties of the blood can be attributed to substances formed in the blood or produced by the hemocytes in the course of an infection.

Bacteriolysins. The most easily demonstrated antibodies found in the blood of insects are the bacteriolysins, which not only kill but also dissolve the invading bacteria. These antibodies may occur naturally or be produced artificially by vaccination or sublethal infection. However, unlike the bacteriolysins of vertebrates, those of insects cannot be separated into two portions by heating. Paillot (1933) found that the blood of *Euxoa segetum* does not lose its bacteriolytic power against *Bacillus melolonthae non-liquefaciens* even when heated at 70 to 72° C., though the latter temperature may decrease it slightly. When heated at temperatures over 75° C. the lytic properties disappear.

According to Paillot, immunized caterpillars lose their immunity as quickly as immune blood loses its bacteriolytic properties when held in a sealed tube.

Although Metalnikov supported the effects of cellular immunity over those of humoral immunity, he did acknowledge the presence of bacteriolysins in caterpillars of *Galleria mellonella* specific for cholera vibrio and dysentery bacilli.

Agglutinins and Precipitins. Metalnikov insisted that the larvae of *Galleria* *mellonella* were almost completely devoid of such antibodies as agglutinins and precipitins. Though most authorities prefer not to deny dogmatically their existence, there is very little in the literature to make one certain of their presence. Certainly the lack of adequate serologic microtechniques has a great deal to do with it.

One instance in which agglutinins were definitely claimed to have been demonstrated is that reported by Glaser (1918b). This worker added *Bacillus poncei* to the immune blood of *Melanoplus femur-rubrum* and observed the bacteria to agglutinate in large masses. After six days the clumped organisms were dead, whereas those in the control preparations were still viable.

Other Humoral Phenomena. The presence of opsonins in insect blood has been postulated, but definite proof of their presence is lacking. Complement also remains to be demonstrated in hemolymph. A substance known as arachnolysin occurs in certain spiders and in their eggs. A similar substance has been demonstrated in the eggs of certain ticks. The problems of hypersensitivity and anaphylaxis in insects has been practically untouched. Metalnikov and Gaschen (1921) have reported a hypersensitivity in *Galleria mellonella* and other insects after vaccination with *Vibrio comma*.

CHAPTER XIII

METHODS AND PROCEDURES

THE METHODS and procedures for investigating the various microbes associated with insects and ticks demand the same scientific bases as those used in any field of research. Each of the branches of microbiology concerned is dependent upon specialized tools, techniques, and manner of approach. It is obvious, since our object is the study of the microbiology of *insects and ticks,* that the tools and techniques of entomology are of prime importance. The other half of the unique combination with which we are dealing is the compounded science of microbiology, embracing the fields of bacteriology, protozoology, mycology, and their divisions and associated sciences. Frequently investigations are carried out by the collaboration of two or more workers in different fields. Occasionally, but all too infrequently, an investigator is trained both in entomology and in microbiology. Such a worker enjoys many advantages that are denied the pure entomologist or the pure microbiologist. Only recently have students become aware of the fact that instead of specializing in a single field it is not only possible but highly desirable to specialize in a compatible combination of fields. The practicability of the merger of microbiology and entomology on certain problems has already proved itself.

Therefore, in considering the methods and procedures involved in research in the field under discussion, it should be realized that the varied techniques of any or all the sciences concerned may be used. It is impossible to treat them in detail in a book of this kind. Many excellent works have been written concerning the methods and procedures involved and this chapter is not an attempt to substitute for them. It is the author's intention merely to point out the high spots and to furnish a general, elementary, indicative guide for the prospective research worker in this field. The treatment given the various sections of this chapter will perhaps appear inconsistent in the relative amounts of discussion given them. This arises from the desire to spare the reader detail that can easily be found in readily available reference books and yet to furnish some detail on those subjects concerning insect microbiology not treated elsewhere.

COLLECTING AND HANDLING INSECTS AND TICKS

It would not be appropriate to consider here in detail the numerous specialized methods developed for the handling of arthropod specimens. The interested reader is referred to such books on the subject as those by Peterson (1934, 1937) and others, in which equipment and methods are explained graphically and bibliographic references are furnished. For the benefit of the microbiologist, however, a few general comments may not be out of place here.

From the standpoint of their manner of collection there are at least two large groups of arthropods: those which are parasitic and suck blood, and those which occur freely in nature on vegetation or in water. The latter group may be collected through the proper use of nets, traps, and various other devices. Some of these methods may also be used for certain of the bloodsucking insects which for part of their lives live in nature away from their animal host. Ticks are an example of this; the unfed specimens of *Dermacentor andersoni*, for example, may be collected by dragging a flannel cloth over the vegetation. In areas where such ticks occur in any appreciable number, they cling to the flannel readily. (See MacLeod, 1932, and Boardman, 1944.)

Many arthropod parasites are collected directly from their hosts. Since such insects may be carrying disease agents acquired from the blood of their hosts, it is always wise to be careful not to crush the arthropod in one's hand and thus expose one's self to infection. Some arthropods, e.g., ticks, are attached so firmly to the body of their host that it is frequently necessary to use forceps to remove them intact.

If it is desired to study the microbes on the external surfaces of the arthropods, it is necessary to retain them in a sterile test tube or other container in order to prevent them from being contaminated before they reach the laboratory. If convenient, it is desirable to keep each specimen in a separate tube since where several are kept together they frequently become contaminated with each other's fecal deposits. To be sure, this is also likely to happen when a specimen is retained alone, since contamination may result through contact with its own fecal deposits on the walls of the tube. In all cases where a study is to be made of the external surfaces of the insects, the examination should be made as soon as possible after the collection of the specimens.

Shipping and Storing. The manner of sending living specimens over long distances varies with the species concerned. Many insects are too fragile and too short-lived to withstand more than a day in transit; others are very hardy and can be shipped across the country with ease. Since a discussion of the many methods employed for the various species of insects and ticks would necessitate many pages of diagrams and photographs, we shall merely list a few of the more pertinent articles which include descriptions of methods of shipping living specimens. The following are a few of the many which might be cited: Barraud (1929); Thompson (1930), Peterson (1934, 1937), Bradley (1934), Dowden (1934), Simmons (1935b), Bradley (1936), and Cooley (see Parker, 1942, p. 1964).

Once specimens are received at the laboratory it is sometimes necessary to store them before or while in use. The method used will vary with the species concerned. Temperature, humidity, and proper food supply are the most important factors to be considered. Most ticks may be held in desiccator jars, at room temperature, in which are kept saturated solutions of ammonium chloride. This set-up maintains a humidity of approximately 79 per cent. Many arthropods may be held relatively inactive at refrigeration temperatures.

Rearing. Since there are almost as many methods of rearing insects as there are insects which have been reared, it can only be suggested that the reader consult the entomologic literature on rearing methods. Such books are recommended as those by Peterson (1934, 1937) and that edited by Needham, Galtsoff, Lutz, and Welch (1937). Section II on the rearing of test insects in *Chemical Control of Insects,* edited by Campbell and Moulton (1943), is also valuable. Some insects may be reared very easily; for example, the milkweed bug (*Oncopeltus fasciatus*) will thrive in jars provided with milkweed pods and a source of water. Others may be reared in the laboratory only with extreme difficulty and under conditions simulating their natural habitat. In many cases field insectaries may be constructed so as to provide a suitable environment for the species being investigated.

The rearing of ticks and other bloodsucking parasites necessitates the use of animals as hosts. Man may use himself as a host in certain cases such as with lice where they may be placed in a feeding capsule and strapped to his leg. Ticks may be fed on rabbits, guinea pigs, mice, or birds, depending on the species. For detailed accounts of methods dealing with bloodsucking arthropods see Bacot (1914b), Jellison and Philip (1933), Kohls (1937), Schuhardt (1940), Yolles and Knigin (1943), and others.

Occasionally it is desirable to rear insects aseptically. This can usually be done if the eggs are obtained in a sterile condition. For accounts of methods for the sterile culturing of various insects see MacGregor (1929), Trager (1935), and Glaser (1938a).

576　　　　　　　　INSECT MICROBIOLOGY

DISSECTION OF INSECTS AND TICKS

The proper dissection of insects and ticks is necessary to make certain types of microbiologic examinations, to observe gross pathology in case of certain

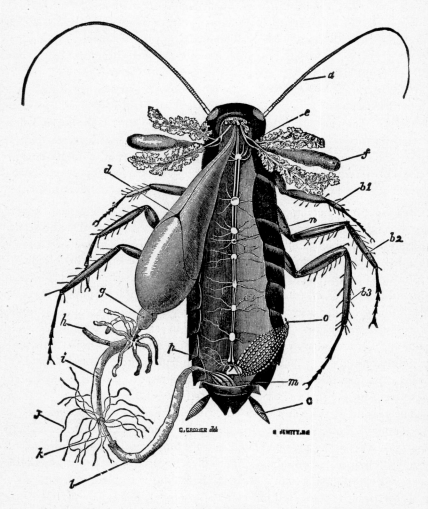

Fig. 247. The internal anatomy of an insect (*Blatta orientalis*). *a*, antennae; b_1, b_2, b_3, first, second, and third legs; *c*, cerci; *d*, ventricular ganglion; *e*, salivary duct; *f*, salivary bladder; *g*, gizzard or proventriculus; *h*, hepatic ceca; *i*, midgut; *j*, Malpighian tubes; *k*, small intestine; *l*, large intestine; *m*, rectum; *n*, first abdominal ganglion; *o*, ovary; *p*, sebaceous glands. (From Comstock, 1936; after Rolleston.)

insect diseases, to locate and study specialized organs such as mycetomes, and for other purposes. Needless to say, great care, diligence, and accuracy is necessary for this type of work. To the beginner the best words of advice are: Have patience and practice.

As applied to the microbiology of insects, most dissections are done on living or freshly killed specimens. If the insects are to be killed, this should be done immediately before dissecting. They may be killed with ether, chloroform, or cyanide. If the dissection is to be made under aseptic conditions, the disinfecting solution (alcohol, mercuric chloride, merthiolate) in which they are first immersed often is toxic enough to kill the specimens.

Large insects such as grasshoppers may be dissected in a glass dish or a small dissecting pan embedded with a mixture of paraffin wax and lamp black. The dissection should be made under a dissecting fluid which may consist of water or saline. If the material is to be preserved, the grade of alcohol used for washing or storage may be used. Small insects such as mosquitoes may be dissected on a slide placed over a dark background. The specimen should be moistened with saline or other dissecting fluid. Ticks may be prepared for dissection by placing them, ventral side down and legs extended, on the sticky side of a strip of adhesive tape tacked to a small block of wood.

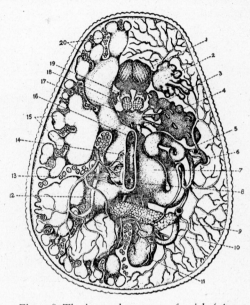

Fig. 248. The internal anatomy of a tick (*Argas persicus*). *1*, Gené's organ; *2*, glandular portion of Gené's organ; *3*, salivary gland; *4*, accessory genital gland; *5*, trachea; *6*, oviduct; *7*, uterus; *8*, rectum; *9*, ovary; *10*, Malpighian tube; *11*, rectal sac; *12*, Malpighian tube; *13*, heart; *14*, stomach; *15*, alimentary ceca; *16*, esophagus; *17*, brain; *18*, muscles of chelicera; *19*, chelicera; *20*, dorsoventral body muscles. (From Hegner, Root, Augustine, and Huff's *Parasitology*, 1938, D. Appleton-Century Co., New York. After Robinson and Davidson, 1913.)

For the actual dissecting operations three things are imperative: a good binocular dissecting microscope, a good source of brilliant light, and the proper dissecting instruments. The latter consist chiefly of dissecting forceps, needles, scalpels, probes, etc. For the proper use and construction of these

578   INSECT MICROBIOLOGY

instruments see the manual by Kennedy (1932). Iridectomy knives and various dental tools also make useful dissecting instruments.

The technique of dissection varies with the species of insect used. The conventional method is to open the insect from the dorsal surface. With some insects, however, dissection from the ventral surface is preferred. Furthermore, the location of certain organs, such as the heart, make them more easily dissected from the ventral position.

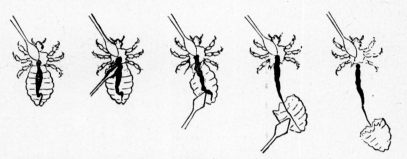

Fig. 249. The steps in a method for extracting the alimentary tract of a louse. (From Clavera and Gallardo, 1943.)

To begin the dissection, the wings and, if the insect is rather large, the legs are clipped off. A slit is then made along the middorsal or midventral line from the neck to the last abdominal segment. Since loops of intestine lie tightly against the body wall, care must be taken that only the integument is cut. While sharp-pointed scissors may be used to open the larger insects, iridectomy knives, razor-blade scalpels, or very fine needles are necessary for the small insects. If the insect is large, the cut edges of the body wall are then pinned out with small sizes of insect pins, the pins sloping outward, so their heads are as near the surface of the wax as possible. Small insects may be secured to the wax of the pan by forcing their legs into the wax just melted with a hot needle. In dissecting large numbers of small insects, it is frequently unnecessary to fasten them down. For example, the alimentary tract, attached to the body wall chiefly at the mouth and the anus, may be drawn out intact by first pulling off the head, then cutting off the end of the abdomen and gradually pulling this last part, thus drawing the tract out backwards. The binding tracheae which appear may be cut away. In the same fashion the salivary glands may be extracted along with the head. In nearly all dissecting operations the dissecting fluid should be used freely and the fluid kept clean by frequent changing. Nearly every worker develops his own dissecting techniques, and if a correct start is made, detailed instructions are as unnecessary as they are difficult to give.

It is frequently necessary to carry out dissections under aseptic conditions. In such cases it is necessary that the operator locate himself in a position with a minimum of air movements or other environmental sources of contamination. All instruments must be sterilized, and an adequate supply of each sterilized instrument should be at hand. The method of sterilization will vary with the type of instrument used. Though flaming is one of the best possible methods, it often is injurious to the metal of knives and other delicate tools. However, if inexpensive razor-blade scalpels and the like are being used, sterilization by flaming is practical. Other methods of sterilization include boiling, autoclaving, and the use of hot-air ovens. The use of disinfecting chemicals is not recommended since, apart from their corrosive action on metals, they may penetrate through delicate tissues, thus killing microbes which one may desire to isolate in a viable condition. For certain types of work the instruments can be disinfected in alcohol, which may be quickly burned off by passage through a flame. In addition to sterile instruments, it is necessary to disinfect the exterior surface of the insect, especially in the region where it is to be opened. The entire insect may be immersed in the disinfectant, or after the specimen is made ready for dissection it may be daubed with a cotton swab containing the disinfectant. Again, care must be taken to prevent the chemical from coming in contact with the inner organs of the insect. Various disinfectants have been used for this purpose, ranging from ether to mercuric chloride, and including alcohol, merthiolate, hexylresorcinol, and others. When dissecting fluid is used, this must also be sterile.

Since the methods of dissecting various species of insects and ticks differ greatly, the reader is referred to detailed accounts by the following workers: Ashworth (1929), Imms (1929), Patton and Evans (1929), Currie (1936), Douglas (1943), and others.

HISTOLOGIC METHODS

Histologic studies are frequently necessary to confirm observations made by simple smears. Of greater value, however, is the fact that through the use of histologic methods accurate information can be obtained with respect to the anatomical relationships between the microorganism and its arthropod host. These methods are particularly desirable in the study of intracellular microorganisms, or to microbes localized in some particular tissue.

Since there are many comprehensive books on histologic techniques, it is sufficient here that we merely acknowledge the value of these methods to the field of insect microbiology. The reader is urged to consult such books as those by Kingsbury and Johannsen (1927), Eltringham (1930), Kennedy (1932), Guyer (1936), McClung (1937), Lee (1937), and Cowdry (1943).

MICROBIOLOGIC EXAMINATION OF INSECTS AND TICKS

According to the techniques required for their study the groups of microorganisms associated with insects and ticks may be listed as follows:

Aerobic bacteria
Anaerobic bacteria
Acid-fast bacteria
Rickettsiae and rickettsialike microorganisms
Yeasts
Molds
Spirochetes
Protozoa
Viruses

Naturally there is some overlapping in the techniques used to examine arthropods for these various types of microbes. In general, however, special techniques are necessary in the case of each in order to accomplish a complete study. Certain of these will be discussed briefly in the following paragraphs.

Preparation of Specimens. Microbiologic examinations of arthropods may be carried out on whole, dissected, or triturated specimens. The first has to do with gaining a knowledge of the arthropod's external flora and fauna; the last two with the internal microorganisms.

To study the external flora, the living insects or ticks may be immersed in sterile saline or broth, shaken for five or ten minutes, and cultures made of the washings. With some insects such treatment may cause them to exude excreta, in which case the cultures would not represent only the external microorganisms. In some cases it is desirable to test the severed appendages or mouth parts separately. Here care must be taken that any organisms found are not coming from the hemolymph which may ooze from the cut appendage. Some knowledge of the external flora may often be gained simply by causing the insects to walk over sterile media on which may be left the microorganisms which had been clinging to their feet.

Although the nature of the internal flora and fauna of living insects may sometimes be ascertained from the examination of their excretory deposits, it is more frequently necessary, first either to dissect the arthropods or to triturate them in a suitable menstrum. Methods of dissection have been discussed in an earlier paragraph. It may be well to re-emphasize, however, the necessity of maintaining as strict asepsis as possible while dissecting. In the majority of cases it is surprising how easily this is done when ordinary precautions (sterile instruments, etc.) are taken. When various internal parts of the arthropod are to be examined separately, different specimens should be used for each part. Where this is not practicable it is wise to remove the divisions of the alimentary tract last, since with their disruption the remaining parts of the insect almost invariably become contaminated.

Aside from its usefulness in culturing the internal flora of arthropods, the trituration of the specimens is necessary in instances when they are to be inoculated into animals for the detection of pathogenic agents. Before triturating or comminuting the specimens, first thoroughly disinfect their exteriors with a suitable disinfectant. The reagent to use for this purpose will depend on the nature of the insect. Substances which penetrate into the interior of the insect should not be used. When the integument is thick, stronger solutions may be used. Suitable disinfecting solutions include 70 per cent alcohol, hexylresorcinol, a solution of 1:1000 mercuric chloride in 70 per cent alcohol, iodine, metaphen, merthiolate, and others. Depending on the thickness of the integument, the insects are held in the disinfectant from one minute to one hour. Ticks may be held thus for one to three hours. At the end of the disinfecting time the specimens must be thoroughly washed in sterile water or saline to remove all traces of the disinfectant. This usually requires from five to eight complete changes of the washing fluid. Special continuous washing devices may also be used.

After the arthropods have been satisfactorily washed, they may be placed in a sterile mortar or other comminuting device and triturated with the aid of sterile sand, alundum, or other abrasive. The diluting menstrum should be sterile saline or Ringer's solution. If the arthropods are being examined for the presence of viruses, it may be more desirable to use another diluent such as Tyrode's solution.

AEROBIC BACTERIA

In general, there are four groups into which bacteria may be divided on the basis of their oxygen requirements: (1) obligate aerobes, which grow in the presence of air or free oxygen, (2) obligate anaerobes, which will not grow in the presence of air or free oxygen, (3) facultative anaerobes, which grow in the presence or absence of air or free oxygen, and (4) microaerophilic organisms, which grow best in small amounts of free air or oxygen. The existence of these four groups must be kept in mind in attempting to isolate bacteria from insects and ticks.

Of the aerobic bacteria, certain groups demand highly specialized techniques and conditions for their handling. However, by far the largest majority of those associated with insects and ticks may be studied by the more common bacteriologic methods.

Staining. Some idea as to the type of bacterial flora found in arthropods may frequently be gained by making smears of the intestinal contents or tissues concerned. In some instances the tissues may be sectioned and then stained. When the insects are triturated, stained smears of the suspension may give one some idea of the types of bacteria to be expected on culturing, but of course this gives no idea as to their anatomical location. Several staining methods may be used to detect bacteria in insects, the two most important being Giemsa's and Gram's.[1]

Culturing. A variety of media may be used in culturing the aerobic bacteria in arthropods. Unless a certain bacterial species is sought demanding special nutrient requirements, it is wise to select a few media which will support the growth of the largest number of species. Several such combinations of media are available: (1) nutrient blood or chocolate-blood agar, and glucose agar; (2) North's gelatin chocolate-blood agar, and glucose agar. If a particular species or group of bacteria is to be isolated, specific media can be used. Thus, cystine heart agar may be used to cultivate *Pasteurella tularensis* and bismuth sulfite agar may be used to isolate *Eberthella typhosa* and certain other intestinal pathogens. (The formulae for these and other media may be found in such compilations as those by Fred and Waksman, 1928; Levine and Schoenlein, 1930; Eyre, 1930; and *Difco Manual,* 1943.)

Since insects are cold-blooded animals, it is desirable to incubate routinely the inoculated culture media at room temperature (22 to 25° C.) as well as at 37° C.

Leach (1933) has introduced an interesting method for the isolation or observation of microorganisms which occur at certain stages in insects in extremely small numbers. While studying the survival of bacteria in the puparia of the seed-corn maggot, he found that aid in tracing the organisms through metamorphosis could be gained by subjecting the insect to subfreezing temperature. The low temperature usually killed the insect but did not injure the bacteria, which began to grow luxuriantly when the dead insect was returned to a warmer temperature.

---

[1] Formulae for the more common stains and reagents mentioned in this chapter are omitted since these may readily be found in the numerous textbooks on microbiologic techniques. For the reader who may be unfamiliar with these sources the following are suggested: McClung (1937), Gradwohl (1938), Mallory (1938), Wadsworth (1939), Kolmer and Boerner (1941), Strong (1942), Craig (1942), Todd and Sanford (1943).

Identification of the bacteria isolated is carried out through a study of the organisms' morphologic, physiologic, and cultural characteristics. The latter two are studied by the use of differential media such as various carbohydrates, gelatin, milk, and the like. After this has been accomplished, an accepted standard of classification should be consulted and the bacteria properly identified. In the United States the most generally accepted system of bacterial classification is that presented in *Bergey's Manual of Determinative Bacteriology*.

ANAEROBIC BACTERIA

An essential difference between aerobic and anaerobic bacteria is in their methods of isolation and cultivation.

For the details of the various methods of cultivating anaerobes the reader is referred to any of the numerous textbooks of bacteriology and to such publications as those by Buchanan and Fulmer (1930), Fildes (1931), McCoy and McClung (1939), and McClung and McCoy (1941). Among the methods best suited for insect bacteriology are the following: (1) use of pyrogallic acid crystals and 10 per cent sodium hydroxide (Spray dish, Dack's method, inverted tube method); (2) use of anaerobe jars (Brown jar, Fildes jar, vacuum jar); (3) use of special media (thioglycollate broth and agar, shake cultures in deep semisolid agar, deep vaseline-covered broth containing chopped brain or kidney); and (4) use of animal inoculation methods.

ACID-FAST BACTERIA

It is sometimes necessary to rely largely on stained smears to detect the acid-fast bacteria which may be found associated with insects. Culture methods may be used, however, though the organisms of the tuberculosis group grow slowly and with difficulty. Media used for such purposes include Dorset's egg medium, Petroff's egg medium, Petragnani's medium, and others. Some species, such as the saprophytes and certain of those parasitic on cold-blooded animals, grow well on many ordinary media. (See Topley and Wilson, 1936; Todd and Sanford, 1943.)

In making routine examinations for acid-fast bacteria the most commonly used staining method is perhaps the Ziehl-Neelsen method using carbol fuchsin. Other methods are those of Gabbet, Pappenheim, and the fluorescence technique (Richards, 1941).

RICKETTSIAE AND RICKETTSIALIKE MICROORGANISMS

With few exceptions rickettsiae and rickettsialike microorganisms have not been cultivated on nonliving media. Because of this, most of our knowledge of this group of organisms has been gained through a study of their characteristics and reactions in stained preparations, tissue cultures, and experimental animals. As concerns the pathogenic rickettsiae (particularly those of typhus) one of the best treatments of laboratory methods and techniques is that by Clavera and Gallardo (1943).

Slide Preparations. In pure suspensions rickettsiae stain fairly well with the ordinary aniline dyes. In tissues, however, such stains are not adequate for accurate differentiation. One of the most dependable stains for this group of organisms is that of Giemsa. With this solution the rickettsiae stain reddish-purple (see chapter V). Other methods used in staining rickettsiae have been devised by Castañeda (1930), Fielding (1943), and Macchiavello (Zinsser, 1940). Macchiavello's method is one of the most practical and widely used of all for the staining of smears (it is not as good as the Giemsa method for staining rickettsiae in tissue sections). In this case the rickettsiae stain red against a blue background. For the formula for this stain the reader is referred to Zinsser (1940, p. 896).

Fluorescent microscopy has also been used to study rickettsiae (Gildemeister and Peter, 1943).

A technique of preparing slides directly from lice to determine the location of the rickettsia has been described by Nauck and Wayer (1941b).

Culture Methods. Isolation of the pathogenic rickettsiae from their arthropod hosts is usually best accomplished by the use of susceptible animals, as will be described shortly. From the infected animals tissue cultures may be commenced or fertile chicken eggs inoculated.

One of the very first methods used to cultivate pathogenic rickettsiae outside the animal body was that used by Wolbach, Pinkerton, and Schlesinger (1923) and Wolbach and Schlesinger (1923), who used brain tissue cultures Other methods have been used with varying degrees of success by Zilber and Dosser (1933), Zinsser and Macchiavello (1936), Zinsser, Wei, and Fitzpatrick (1937), Singer (1941), and Spaet (1944). Nonpathogenic forms have been cultivated by Nöller (1917b), Hertig and Wolbach (1924), Aschner and Kligler (1936), and others.

In 1934 Da Cunha found that rickettsiae were able to grow on the chorioallantoic membrane of the developing chick embryo. Later Cox (1938a, 1941) found that by injecting the yolk of six- to seven-day old chick embryos with infectious material, large numbers of rickettsiae could be found in three to six days growing in the yolk sac membrane. This is now one of the most widely used techniques in the cultivation of rickettsiae. Certain of the non-pathogenic rickettsiae or rickettsialike organisms have been observed to grow in the extraembryonic fluids of the egg (Steinhaus, 1942a).

Under the heading of culture methods may also be mentioned the propagation of rickettsiae in arthropods. This has been accomplished with lice in the case of *Rickettsia prowazeki* (Weigl, 1924; Pshenichnov, 1943; Snyder and Wheeler, 1945) and with ticks in the case of *Dermacentroxenus rickettsi* (Spencer and Parker, 1930). Such methods were used in propagating the organisms in numbers large enough to produce vaccines. Other possibilities along these lines are suggested by the work of Wei Hoi in China, who used silkworms (*Bombyx mori*) for the propagation of rickettsiae.

Experimental Animal Tests. One of the earliest and simplest methods of establishing a strain of Rocky Mountain spotted fever rickettsia was to inoculate triturated, recently fed, infected ticks into guinea pigs. Likewise, and with almost equal success, strains of other rickettsial infections have been established by the parenteral inoculation of their arthropod host into laboratory animals. Routine tests for arthropod-carried infectious agents are still, for all practical purposes, made by the introduction into animals of the comminuted suspension of the arthropod concerned and by feeding experiments (which have already been discussed).

In the case of Rocky Mountain spotted fever, Spencer and Parker (1923) found that it is unwise to rely upon either feeding or inoculation alone as an indication of the presence or absence of the rickettsia in unfed adult ticks. However, the inoculation method alone is probably the most reliable when testing recently fed ticks in all stages. The ingestion of fresh blood or holding the ticks at $37°$ C. for 24 hours is necessary, in the case of the spotted fever rickettsia, to "reactivate" the virus in unfed infected ticks.

After the ticks or insects are thoroughly triturated, 1 ml. of the resulting suspension is then inoculated intraperitoneally into two to ten guinea pigs or other laboratory animals depending on the nature of the experiment. The animals are then caged, temperatures recorded, and observations made. Care should always be taken in triturating engorged ticks or insects that the contents of the arthropods are not sprayed into the face or on the hands of the operator when they are first crushed. The writer has found that a petri dish cover with a hole cut in its center to permit the handle of the pestle to come through makes an excellent shield to guard against such a mishap. Many laboratories have found that the trituration of considerable numbers of arthropods can be easily accomplished by the use of one of several mixing machines on the market. One of the most suitable is the "Waring Blendor," [1] which in a few minutes comminutes the specimens as thoroughly as can be done by hand with a mortar and pestle.

Fig. 250. Photograph showing a method of feeding ticks in a screw-top capsule taped on a guinea pig. Ticks shown are attached and feeding. (From Jellison and Philip, 1933.)

Once a strain of most rickettsial infections is definitely established in laboratory animals, it is usually a simple matter to maintain it. The most widely used animal for this purpose is the guinea pig. Of course, as we explained in an earlier paragraph, fertile chicken eggs may also be used to maintain various strains of rickettsiae. Then too, infective animal or egg material may also be preserved by low-temperature freezing and by the dehydration or lyophile methods. With most strains of pathogenic rickettsiae, transfers to fresh animals may be made on the second to fourth day of fever. The material or tissue passed depends on the species of rickettsia concerned. Those of the spotted fever group are usually passed by blood, endemic typhus and boutonneuse fever by testicular washings, epidemic typhus by brain, the Q fever group by spleen, etc.

Identification of Rickettsiae. Identification of the nonpathogenic rickettsiae must be accomplished largely through morphologic studies and consideration of their sources. The identification of the pathogenic rickettsiae, on the other hand, necessitates the use of animals and serologic methods.

After a strain of the unknown rickettsia has been established in guinea

---

[1] Made by The Waring Corporation, 1697 Broadway, New York.

pigs, those animals which recover from the infection are held for about 10 to 20 days, after which they are given an "immunity test." This test is made by inoculating the recovered animals with strains of known rickettsiae. Rectal temperatures are taken for 15 days or longer. If, during this time, the animals inoculated with a certain strain remain afebrile, they are probably immune, and the original infection may be considered to be identical or closely related to the challenging strain. If the animals show a fever, the chances are they are not immune to the challenging strain and hence the rickettsiae are probably not related. When the animals appear to be immune, it is generally advisable to give them a second inoculation of the known strains about 15 days after the first inoculation. For some reason the first inoculation occasionally does not "take," and this fact is revealed by the second injection.

Frequently, in experimental work especially, there is occasion to run cross immunity tests. For example, let us say that we wished to make a comparison of a strain of *Rickettsia prowazeki* (typhus fever), and *Rickettsia burneti* (Q fever). One of the tests employed would be to inoculate guinea pigs which had recovered from a *R. prowazeki* infection with a strain of *R. burneti*. Conversely, guinea pigs which had recovered from a *R. burneti* infection would be inoculated with a strain of *R. prowazeki*. Similar tests could thus be made with other strains of rickettsiae in order to study strain differences.

Another animal test used in identifying rickettsiae is the neutralization or protection test. This test is particularly useful in identifying and diagnosing spotted fever infections. In brief, it consists of combining in varying amounts the serum in question and virulent spotted fever "serum virus." After standing about thirty minutes the mixture is inoculated into normal male guinea pigs. If the serum were immune serum and contained antibodies against the spotted fever rickettsiae, most of the inoculated animals would be protected. On the other hand, if the serum did not contain spotted fever antibodies, the guinea pigs would not be protected and would be susceptible to the virulent rickettsiae.

It should be pointed out that some idea as to the identity of the rickettsia in question may frequently be gained by its reaction in animals (i.e., type of fever produced, the pathology, etc.).

Serologic methods are also employed for purposes of identification. These have been discussed elsewhere, and for a detailed discussion of them the reader should consult the literature. Of practical use have been the Weil-Felix test, complement fixation tests, and agglutination tests (see Zinsser, 1937, 1940; Bengtson, 1941a,b,c; Hudson, 1940).

YEASTS

The methods used to isolate and study yeasts are similar to those used for bacteria. Many species of yeasts may be isolated readily on ordinary bacteriologic media. Special media may also be employed for this purpose, the most widely used of which is probably beer wort or beer wort agar, but this medium is not readily available to the average laboratory except in a dehydrated form. An agar containing 5 per cent glucose and 1 per cent peptone is a convenient medium to use, as is Sabouraud's maltose agar. The majority of the yeasts grow best in the neighborhood of 25° C.

The determination of both morphologic and cultural characters is required for the identification of yeasts. As stated by Henrici (1941):

Of the morphological characters it is necessary to determine the relative size and shape of the vegetative cells; whether they multiply by fission or by budding; if by budding whether buds are formed at any place on the cell (multipolar) or only at the ends of elongated cells (bipolar); whether the buds are separated from the mother cell by abstriction or by the formation of a cross-wall followed by fission. Further, it is necessary to determine whether the buds separate quickly after they are formed, giving rise to free cells or groups of only two or three, or whether they tend to remain attached forming clusters ("sprossverbände"); in the latter case the cells, if elongated, may give rise to a branched structure resembling closely true mycelium ("sprossmyzel," pseudomycelium); in some cases true mycelium with cross-walls is formed. With yeasts forming pseudo- or true mycelium, clusters of yeast-like cells may arise by budding, usually at the nodes ("sprosskonidien," "blastospores," "appareil sporifère").

Many of the characteristics of yeasts vary with the composition of the medium used for their culture. A variety of formulae have been suggested for the special study of various genera and species.

In addition to the morphologic characters mentioned above, the presence or absence of spores is important in the identification of yeasts. If spores are present attention must be paid to their number, shape, and manner of formation. Although in some cases ascospores may be formed in ordinary cultures, they sometimes require the use of special media or methods to force their development. Among the methods and media used to induce sporulation are plaster of Paris blocks, Gorodkowa's medium, potato and carrot plugs, raisins and raisin agar, Kufferath's medium, and others. For details of these methods and media the reader is referred to such publications as those by Guilliermond (1920, 1940), Henrici (1930, 1941), and Stelling-Dekker (1931).

Experienced workers may study the spores of yeasts by the microscopic examination of wet preparations. The average observer usually relies more on stained preparations. One of the best staining methods is that suggested by Henrici (1941): 5 per cent chromic acid solution for five minutes, Ziehl's carbol fuchsin steaming five minutes, decolorization for two minutes in 1 per cent aqueous sulfuric acid, counterstain with Loeffler's methylene blue one minute. The spores stain red, the asci and vegetative cells stain blue.

Other means of identifying yeasts depend on their reactions and fermentations in carbohydrates and their specific nutrient requirements.

MOLDS AND HIGHER FUNGI

The fungi popularly known as molds are handled in much the same manner as are bacteria. Like yeasts, they grow best at temperatures around 25° C.

Many of the common molds may be isolated on ordinary bacteriologic media. Others require special media, and still others, particularly many of the entomogenous fungi, have never been cultivated on an artificial medium. Among the media devised for the cultivation of fungi are Sabouraud's dextrose or maltose agar, malt extract broth, and Czapek's solution or agar. The optimum pH for most molds is on the acid side (pH 4.0 to 6.0). Solid media consisting of potatoes, carrots, or beans may also be used.

In the identification of the molds the dry, living cultures are first examined under the low-power objective of the microscope. In this manner the characteristics of the aerial mycelium, conidiophores, fruiting heads, spores, and other structures may be ascertained. Next, slide preparations should be made for examination under the high-power lenses. Care should be taken in removing the parts of the mold growth from the medium since they are very fragile and easily broken. They should be mounted in lactophenol or similar mounting fluid. Water should not be used since this causes a shrinkage of the hyphae (by osmosis) and the various parts adhere in a tangled mass. When desired, stains may be added to the fluid to color the various parts of the plant.

Another excellent way to study molds is by the preparation of slide cultures. These are prepared by inoculating the mold into a tube of liquefied and partially cooled malt agar. A few drops of this inoculated agar are placed on a sterile slide and covered with a sterile cover slip. The slides are incubated in a petri dish containing moist filter paper. In a few hours or days the mold will appear growing luxuriantly about the edge of the coverslip where it may be observed under the microscope.

For details of other techniques in the study of molds and higher fungi publications by the following are suggested: Lafar (1904–1914), Henrici (1930), Raistrick (1932), Rawlins (1933), Gwynne-Vaughan and Barnes (1937), and Smith (1938).

SPIROCHETES

The presence of spirochetes in insects and ticks may be detected by at least three methods: dark-field observation, staining, and feeding the arthropods on or inoculating them into animals.

The dark-field method is more satisfactory if the triturated suspension is dilute enough to prevent confusion with the large amount of foreign material which is usually found in such suspensions.

In general, one of the best methods of staining spirochetes is Giemsa's spirochete stain which may be purchased already prepared. The Fontana-Tribondeau silver method has also been recommended especially for the examination of tissues. Negative staining with India ink or collargol is also used. The staining of spirochetes in sections is usually more difficult than in smears or films. For this purpose, however, Dobell's modification of Levaditi's silver nitrate stain is one of the best. (See Wenyon, 1926; Knowles, 1928; and Hindle, 1931.)

As exemplified by the spirochetes of relapsing fever, feeding arthropods on animals such as white mice is one of the most satisfactory means of detecting pathogenic spirochetes in their invertebrate hosts. Ticks are allowed to engorge on the blood of a white mouse. After a varying number of days, usually five to ten, spirochetes may be found in the blood of the mouse. If the infected arthropods are inoculated into the mouse, the spirochetes are frequently found in the blood somewhat sooner—as early as the second day.

PROTOZOA

The proper method of examination of insects and ticks for protozoa depends entirely upon the type of protozoa for which one is looking. From the standpoint of the techniques involved the protozoa found in insects may be divided into two large groups: (1) extracellular protozoa (including such forms as are found among the free-living and the intestinal protozoa), and (2) intracellular protozoa (including parasitic forms and those stages spent by some protozoa within the cells of their invertebrate hosts). In many cases, of course, the methods involved in the study of these two groups overlap or are identical. In any event, when a particular species or group of protozoa is being sought, the available literature pertaining to it should be consulted for any special techniques which may be involved.

Extracellular Protozoa. For this group of protozoa, which may include forms found in the intestines of insects or free-living forms which have fortuitously gained entrance into the body of the insect, the simple expedient of a wet mount will often reveal their presence. Compensation for evaporation may be made by adding water at the edge of the cover slip. A hanging drop slide may also be used, but this is not necessary. This method, however, is convenient when one wishes to make a prolonged study of a small number of organisms. The movements of ciliates and flagellates may be slowed by thickening the suspension with 2 or 3 per cent gelatin or quince-seed jelly, or by adding a small piece of cotton wool to the drop on the slide. In many cases an anesthetizing agent (alcohol, ether, chloroform, menthol, chloretone, nicotine, etc.) added to the suspension will restrict the activities of actively motile protozoa.

For certain intestinal protozoa of insects, such as the flagellates of termites, the arthropod's intestinal contents should first be diluted with saline or 67 per cent Locke's solution.

The dark-field microscope is convenient for observing certain protozoan structures such as the flagella of flagellates.

Intravitam staining methods may also be used to great advantage in studying protozoa. Food vacuoles may be detected by placing finely divided carmine or carbon in the suspension or culture medium, where they are ingested by the protozoa. Cytoplasmic structures may be stained with Janus green or neutral red, and basic dyes in dilute concentrations may be used for the cytoplasm. For other methods of staining, killing, and preserving protozoa see Wenrich (1937).

In making a general survey of the protozoa present in any particular group of insects, generalized methods must be adopted. As an example of this we quote from a paper by Semans (1936), who made a survey study of the protozoan parasites of the Orthoptera of Ohio. Semans described the steps following the opening of the insect anteriorly on the midventral line with a small pair of scissors:

1. The slide mount was placed on the stage of a low-power (x10.5 to x30) binocular microscope. The alimentary tract was broken apart by means of two pairs of forceps—cleaning them after each operation—into fore-intestine, mid-intestine (removing the enteric ceca for separate consideration), and hind-intestine—or whatever divisions of the latter were present. Each section was placed in a separate drop of saline solution.

2. The alimentary tract and malpighian tubules were examined externally for

celomic parasites, and then, with two stout insect pins (cleaned after each operation), each section was torn into small shreds, the number and position of the larger Gregarinida being noted. The malpighian tubules were left more or less intact for separate examination.

3. The drops of saline solution impregnated with gut contents were spread over the slide to make them as thin as possible and were examined—usually without a cover-glass, the use of the latter making it difficult to obtain organisms for permanent staining—with the x60–100 combination of the microscope, stepping up the magnification as high as x880 for minute flagellates. The examination was facilitated by the use of a graduated mechanical stage.

4. In a thorough examination of the temporary mount, everything of apparent significance was recorded, measurements of the organisms being determined by means of a calibrated ocular micrometer. Drawings were made with or without the use of a camera lucida, depending upon the amount of movement in the organisms. Weak stains were used at the start but were abandoned in favor of the saline mount followed by the permanent staining of all representative parasites, including helminths.

5. Cysts of interest, if not permanently mounted, were put in a watch glass or the well of a concavity slide and kept in a moist chamber, but a technique for invariably bringing about cyst dehiscence was not perfected, as the number of cysts was too small.

6. Representative gregarinids (i.e., of the order Gregarinida) were removed, by means of a capillary pipette, to a watch glass for fixing and staining. It was found to be impracticable to make permanent mounts of flagellates from saline, as it was difficult to make them adhere to glass.

7. The fixing and staining of flagellates and ciliates (from step 3): A portion of the hind-intestine—usually the colon—of cockroaches was mixed with a small drop of sterile horse serum (adhering agent) on a clean cover-glass, spread over the entire surface of the latter, and plunged into a fixing agent (preferably Kahle's fixing fluid: 32%, 95% ethyl alcohol; 11%, 40% formaldehyde; 4% glacial acetic acid; 53% distilled water) in a syracuse watch glass for about two hours; then it was passed through 70% and 50% alcohol down to 30%, at fifteen-minute intervals, and stained ten to fifteen minutes by Ehrlich's hematoxylin (not longer, as in the case of other hematoxylin methods, otherwise overstaining results), and back up through the same alcohols to absolute alcohol and then xylol for the mounting in balsam.

8. The fixing and staining of Gregarinida (from step 6): Instead of using adhering agents, which created an undesirable background when stained (except for the smallest specimens), the saline was removed and the fixing agent (preferably Kahle's) applied by means of a capillary pipette—the method used throughout the entire fixing and staining procedure. Kahle's fluid was left about two hours, followed by two changes of 70% alcohol (two hours each), 95% alcohol (two minutes), the stain, Fast Green FCF (two minutes), 95% alcohol again (one minute), absolute alcohol (five minutes). The organism was placed by means of a capillary pipette directly into balsam (mixed with xylol) on a slide cleaned with xylol and a clean cover-glass was applied.

The Kahle's-Fast Green method is a slight modification of that suggested by

C. H. Kennedy (1932) for the staining of the cytoplasm of the insect cell, and was discovered to be effective for the Gregarinida through the staining of sections of the alimentary tract of *Melanoplus differentialis* (Thomas) containing sporonts of *Gregarina rigida* (Hall). No additional nuclear stain is necessary, as the refractive index is quite different from that of the cytoplasm. Several other staining methods were tried, but none appeared to be as satisfactory as that described above.

Intracellular Protozoa. Most of the protozoa which occur intracellularly in insects and ticks belong to the class Sporozoa. Most of these species may be observed at some stage in an extracellular position also.

In the intracellular stages in the arthropod, sectioning of the parasitized organs of the host is frequently necessary. This procedure also facilitates the observation of those intracellular forms which later become extracellular but remain attached to the host's tissues (e.g., gregarines). To be sure, in many of these cases ordinary smears should also be made. Sections may be stained with any of a variety of stains. Among those suited for a study of entomophytic protozoa are Giemsa's stain, Hansen's hematoxylin, Heidenhain's hematoxylin, picrofuchsin. Excellent counterstains are Bordeaux red, safranine, and orange G. Books on microscopical technique should be consulted for the many other stains which may also be used.

In certain of the Sporozoa it becomes desirable to identify the spores. In the case of Microsporidia this may be done by exerting pressure on the cover slip to cause the filaments to be extruded. This may also be brought about by the application of hydrogen peroxide. Another method of identifying microsporidian spores and at the same time differentiating them from yeast cells is to stain them with Ziehl's fuchsin and decolorize with a weak solution of sulfuric acid. With this treatment the yeast cells are decolorized while the microsporidian spores retain the red dye.

Culture Methods. The cultivation of protozoa from insects requires expert technique and much care and diligence. Relatively few protozoa have been successfully isolated and maintained in cultures. Among the most successful media used for the cultivation of protozoa are Boeck's Locke-egg-serum medium, Hogue's egg medium, N.N.N. medium, Noguchi's serum medium, and Ponselle's medium.

In cultivating flagellates from the intestines of insects it is frequently difficult to obtain them free from bacteria. Bacteria-free cultures have been obtained from some insects, however, by the use of several ingenious methods. For a discussion of some of these and other methods of cultivation the reader is referred to the book by Wenyon (1926). This author also gives a good discussion of the maintenance of protozoa in animals.

The culturing of protozoa directly from the intestines of insects may be aided by the use of such substances as quinanil, which is toxic to bacteria but has little harmful effect on protozoa.

VIRUSES

Viruses may be isolated from infected insects by allowing the latter to feed upon a susceptible plant or animal, or by inoculating the triturated insect into these hosts. Viruses pathogenic only to insects must be maintained in insects or, as shown by Trager (1935c), in tissue cultures. In triturating infected insects preparatory to inoculating them into animals or plants it is advisable to use a diluent other than saline, which exerts a deleterious effect on some viruses. Tyrode's solution appears to be a satisfactory diluent.

The techniques for handling viruses vary with the agent concerned. For this reason and because the methods are so involved, it is suggested that the interested reader refer to such works as those by Smith (1937), Levaditi and Lépine (1938), Sanders (1939), Storey (1939), Gildemeister, Haagen, and Waldmann (1939), Van Rooyen and Rhodes (1940).

KEEPING OF RECORDS

Most scientific investigators have their own methods of keeping records of their experimental data. This is usually done to one's own liking and it is not the writer's intention here to dictate how this should be done. On the other hand, since certain peculiarities in the field of insect microbiology make for peculiarities in the keeping of its records, a few suggestions in this direction may not be out of place.

The first requisite in making a microbiologic study of insects or ticks is the proper identification of the arthropod being examined. This may appear to be superfluous advice but one has only to consult the literature to find how all too often microorganisms have been reported as being isolated from "grasshoppers," "fleas," "flies," "mosquitoes," and the like. If the proper identification cannot be made at once, representative specimens of both sexes and of all stages available should be preserved until they can be identified by an expert.

From experience gained through the routine microbiologic examination of a large number of insects and ticks the writer has found the use of printed or mimeographed data sheets to be of tremendous aid in the keeping of the necessary records. These sheets may be kept in appropriate files or in looseleaf notebooks. Figures 251 and 252 show the front and back of the Preliminary Data and Examination Sheet. It is hoped that these forms are self-explanatory.

After the preliminary examinations are made, a more detailed study is made of each of the microorganisms isolated, and these records are likewise kept on data sheets. Such forms are shown for bacteria in figures 253 and 254, and for viruses and other pathogenic agents in figures 255 and 256. Similar forms may be used for protozoa, fungi, rickettsiae, and spirochetes.

PRELIMINARY DATA AND EXAMINATION SHEET

Date _____ Examination No. _____

Arthropod _____

_____

Where collected (originally) _____

By _____ Date _____

How collected _____

Host or habitat _____

Condition of Host _____

Accession No. _____ Collector's No. _____

No. of specimens: Dead _____ Alive _____ Stage _____

Sex _____ No. recently fed _____ Host _____

If known to be infected, with what? _____

Naturally or laboratory infected _____ Animal No. _____

At what stage last infected? _____ Date inf. _____

Method of examination _____

Exterior Sterilization. Disinfectant used _____ Time _____

No. of subsequent washings _____ Control: Pos. _____ Neg. _____

Dissected? _____ Triturated? _____ Menstrum _____

EXAMINED FOR:

1. AEROBIC BACTERIA. Positive _____ Negative _____ Not run _____

No. of strains isolated _____ Strain Nos. _____

Abundant _____ Moderate _____ Few _____

Check media used: Nutrient agar _____ Glucose agar _____ North's _____

gelatin choc. agar _____ Cystine-heart-agar _____ Blood agar _____

Other media _____

Incubation: 37° C. _____ Room temp. (20–30° C.) _____

(over)

Fig. 251 (Front)

Examination No. _____

2. ANAEROBIC BACTERIA. Positive _____ Negative _____ Not run _____

No. of strains isolated _____ Strain Nos. _____

Medium used: _____ Thioglycollate _____

Method used: _____ Incubation _____

3. ACID-FAST BACTERIA. Positive _____ Negative _____ Not run _____

Abundant _____ Moderate _____ Few _____

Description _____

4. YEASTS AND MOLDS. Positive _____ Negative _____ Not run _____

No. of strains isolated _____ Strain Nos. _____

Abundance _____ Media used _____

5. RICKETTSIAE or RICKETTSIALIKE ORGANISMS. Pos. _____ Neg. _____ Not run _____

Abundant _____ Moderate _____ Few _____

Intracellular _____ Extracellular _____ Intranuclear _____

Stain: Macchiavello _____ Giemsa _____ Other _____

6. SPIROCHETES. Positive _____ Negative _____ Not run _____

Abundant _____ Moderate _____ Few _____

Stain _____ Dark-field _____

Description _____

7. PROTOZOA. Positive _____ Negative _____ Not run _____

Abundant _____ Moderate _____ Few _____

Description _____

8. ALGAE. Positive _____ Negative _____ Not run _____

Description _____

9. VIRUSES. Positive _____ Negative _____ Not run _____

Method of examination _____

Animal Nos. _____

Fig. 252 (Back)

EXAMINATION OF BACTERIA

Date of Isolation _____ Strain No. _____

Arthropod _____ Examination No. _____

If dissected, list anatomic parts in which found _____
_____

Medium from which isolated _____

Temp. at which isolated _____ Incubation period _____

DESCRIPTION

Cellular morphology _____
_____

Size _____. microns. Gram reaction _____

Special staining characteristics _____
_____

Spores (size): _____ Position _____

Colony morphology _____
_____

Nutrient agar slant _____
_____

      agar slant _____

Potato slant _____

Blood agar (hemolysis) _____

Nutrient broth _____
_____

Temperature relations: Growth in refrigerator (°C.) _____

Growth at room temp. (°C.) _____ Growth at 37° C. _____

Growth at 50° C. _____ Relation to free oxygen _____

(over)

Fig. 253 (Front)

Strain No. _____

Litmus Milk _____

Gelatin liquefaction _____

Hydrogen sulfide _____

Motility _____	Glucose _____	Dextrin _____
Capsule _____	Galactose _____	Inulin _____
NO₃ to NO₂ _____	Mannose _____	Raffinose _____
Indole _____	Arabinose _____	Melezitose _____
M.R. _____	Xylose _____	Mannitol _____
V.P. _____	Fructose _____	Sorbitol _____
Citrate _____	Rhamnose _____	Dulcitol _____
Ammonia _____	Levulose _____	Glycerol _____
Starch hyd. _____	Sucrose _____	Adonitol _____
Uric acid _____	Maltose _____	Erythritol _____
_____	Lactose _____	Arabitol _____
_____	Cellobiose _____	Salicin _____
_____	Trehalose _____	Aesculin _____
Inositol _____	Melibiose _____	Coniferin _____

CLASSIFICATION

According to *Bergey's Manual of Determinative Bacteriology*, this organism

    Is identical with _____

    Is close enough to be _____

    May be _____

      Note exceptions:

Is probably a species belonging to _____

Is apparently a new species _____

Proposed name _____

Fig. 254 (Back)

EXAMINATION OF VIRUSES AND OTHER PATHOGENIC AGENTS

Date _____ Strain No. _____

Special Examination No. _____ Regular Examination No. _____

Arthropod _____

If dissected, list anatomic parts inoculated _____
_____
_____

Animals inoculated _____

Animal No. _____ Animal No. _____ Animal No. _____

Nature of material inoculated _____
_____

Diluent _____

Amt. inoculated _____ Route of inoculation _____

RESULTS

(See reverse side for temperature and autopsy recordings.)

Fever present? _____ No. days after inoculation _____

No. of days of fever temp. _____ Highest temperature _____

Other symptoms: _____
_____
_____

Transfers made to animals numbered _____

Organs used in transfer _____

Bacteriologic culture of _____ : _____
_____
_____

(over)

Fig. 255 (Front)

TEMPERATURES

Animal No. _____   Animal No. _____   Animal No. _____

Date	Temp. °C.		Date	Temp. °C.		Date	Temp. °C.

Autopsy   Autopsy   Autopsy

Fig. 256 (Back)

REFERENCES

Abe, N. 1907 Ueber den Nachweis von Typhusbacillen in Läusen Typhuskranker. Münch. Med. Wochschr., **54,** 1924.

Acqua, C. 1918–1919 Ricerche sulla malattia del giallume del baco da seta. Rend. Inst. Bact. Scuola Super. Agr. Portici, **3,** 243–256.

Adler, S., and Ashbel, R. 1942 The behavior of *Spirochaeta persica* in *Pediculus humanus*. Ann. Trop. Med. Paras., **36,** 83–96.

Adlerz, G. 1890 Om digestionssekretionen jemte några dermed sammanhängande fenomen hos insekter och myriopoder. Bihang Kgl. Svensk. Vet. Akad. Handlingar., **16,** 1–51.

Ahlm, C. E., and Lipshutz, J. 1944 Tsutsugamushi fever in the southwest Pacific theater. J. Am. Med. Assoc., **124,** 1095–1100.

Alexander, R. A. 1944 Personal communication.

Alexander, R. A., and Mason, J. H. 1939 Studies of the rickettsias of the typhus-Rocky-Mountain-spotted-fever group in South Africa. II. Morphology and cultivation. Onderstepoort J. Vet. Sci. Animal Ind., **13,** 25–39.

Allen, T. C., and Riker, A. J. 1932 A rot of apple fruit caused by *Phytomonas melophthora*, n. sp., following invasion of the apple maggot. Phytopathology, **22,** 557–571.

Amaral, A. do, and Monteiro, J. L. 1931 Histoire naturelle et classification des rickettsioses. Position systématique du "typhus exanthématique de São Paulo." Rev. Sud-Am. de Méd. et de Chir., **4,** 781–817.

Amaral, A. do, and Monteiro, J. L. 1932 Ensaio classificação das rickettsioses á luz dos nossos actuaes conhecimentos. Mem. Inst. Butantan, **7,** 345–376.

American Association for the Advancement of Science. 1941 A symposium on human malaria. Publication 15. Science Press, Lancaster, Pa. 398 pp.

Anderson, C. R. 1944 Survival of *Rickettsia prowazeki* in different diluents. J. Bact., **47,** 519–522.

Anderson, J. F., and Frost, W. H. 1912 Transmission of poliomyelitis by means of the stable fly (*Stomoxys calcitrans*). Pub. Health Repts., **27,** 1733–1735.

Anderson, J. F., and Frost, W. H. 1913 Poliomyelitis. Further attempts to transmit the disease through the agency of the stable fly (*Stomoxys calcitrans*). Pub. Health Repts., **28,** 833–837.

André, C. 1908 [On flies and the dissemination of tubercle bacilli.] (Quoted by Herms, 1939, p. 298.)

Anigstein, L. 1927 Untersuchungen über die Morphologie und Biologie der *Ricksettsia melophagi* Nöller. Arch. Protistenk., **57,** 209–246.

Anigstein, L. 1933 Researches on tropical typhus. Kyle, Palmer and Co., Kuala Lumpur. 184 pp.

Anigstein, L., and Amzel, R. 1927 Recherches sur l'étiologie du typhus exanthématique chez les cobayes infectés par les cultures du germe. Compt. Rend. Soc. Biol., 96, 1502.

Anigstein, L., and Amzel, R. 1930 Studien zur Aetiologie des Fleckfiebers. Zentr. Bakt. Parasitenk. Infekt., I Orig., 115, 149.

Anigstein, L., and Bader, M. N. 1943a Investigations on rickettsial diseases in Texas. I. Epidemiological role of ticks common to the Gulf Coast in relation to local spotted fever. Texas Repts. Biol. Med., 1, 105–115.

Anigstein, L., and Bader, M. N. 1943b Preliminary report on investigations of Bullis fever. Texas Repts. Biol. Med., 1, 298.

Anigstein, L., and Bader, M. N. 1943c Investigations on rickettsial diseases in Texas. Part 4. Experimental study of Bullis fever. Texas Repts. Biol. Med., 1, 389–409.

Anigstein, L., Bader, M. N., and Young, G. 1943 Protective effect of separate inoculation of spotted fever virus and immune serum by intradermal route. Science, 98, 285–286.

Aoki, K., and Chigasaki, Y. 1915a Ueber die Pathogenität der sog. Sotto-Bacillen (Ishiwata) bei Seidenraupen. Mitteil. der Med. Fakult. der Kaiser Univ. zu Tokyo, 13, 419–440.

Aoki, K., and Chigasaki, Y. 1915b Über das Toxin von sog. Sotto-Bacillen. Mitt. Med. Fakult. Univ. Tokyo, 14, heft 1, 59–80.

Aoki, K., and Chigasaki, Y. 1921 Immunisatorische Studien über die Polyederkörperchen bei Gelbsucht von Seidenraupen (Zelleinschluss). Zentr. Bakt. Parasitenk. Infekt., I Orig., 86, 481–485.

Ark, P. A., and Thomas, H. E. 1936 Persistence of *Erwinia amylovora* in certain insects. Phytopathology, 18, 459.

Arkwright, J. A. 1923 Rickettsia (laboratory demonstration). Trans. Roy. Soc. Trop. Med. Hyg., 17, 3–4.

Arkwright, J. A., Atkin, E. E., and Bacot, A. 1921 An hereditary rickettsia-like parasite of the bed bug (*Cimex lectularius*). Parasitology, 13, 27–36.

Arkwright, J. A., and Bacot, A. 1921 A bacillary infection of the copulatory apparatus of *Pediculus humanus*. Parasitology, 13, 25–26.

Arkwright, J. A., and Bacot, A. 1923 Observations on the morphology of *Rickettsia prowazeki* occurring in lice (*Pediculus humanus*) infected with the virus of typhus fever. Parasitology, 15, 43–48.

Arkwright, J. A., Bacot, A., and Duncan, J. T. 1919 The association of *Rickettsia* with trench fever. J. Hyg., 18, 76–94.

Armer, J. M. 1944 Influence of the diet of Blattidae on some of their intestinal protozoa. J. Parasitol., 30, 131–142.

Arneth, J. 1942 Periodisches Fieber (Wolhynisches Fieber, Funftagefieber) im Felde. Klin. Wochschr., 21, 998–999. (Abstract in Trop. Diseases Bull., 40, 301. 1943.)

Aschner, M. 1931 Die Bakterienflora der Pupiparen (Diptera). Eine Symbiosestudie an Blutsaugenden Insekten. Z. Morphol. Ökol. Tiere, 20, 368–442.

Aschner, M. 1932 Experimentelle Untersuchungen über die Symbiose der Kleiderlaus. Naturwissenschaften, **20**, 501–505.

Aschner, M. 1934 Studies on the symbiosis of the body louse. I. Elimination of the symbionts by centrifugalisation of the eggs. Parasitology, **26**, 309–314.

Aschner, M., and Kligler, I. J. 1936 Behavior of louse-borne (epidemic) and flea-borne (murine) strains of typhus rickettsia in tissue cultures. Brit. J. Exptl. Path., **17**, 173.

Aschner, M., and Ries, E. 1933 Das Verhalten der Kleiderlaus bei Ausschaltung ihrer Symbionten. Eine experimentalle Symbiosestudie. Z. Morphol. Ökol. Tiere (Abt. A. Z. Wiss. Biol.), **26**, 529–590.

Ashworth, J. H. 1929 [Dissection of the salivary glands of a mosquito.] (Referred to by Patton and Evans, 1929.)

Atanasoff, D. 1920 Ergot of grains and grasses. U.S. Dept. Agr., Bur. Plant Indus., Mimeo. Publ. 127 pp. (Cited by Leach, 1940.)

Atkin, E. E., and Bacot, A. 1917 The relation between the hatching of the eggs and the development of the larvae of *Stegomyia fasciata* (*Aëdes calopus*), and the presence of bacteria and yeasts. Parasitology, **9**, 482–536.

Atkin, E. E., and Bacot, A. 1922 Experiments on the infectivity of typhus virus contained in lice (*Pediculus humanus* and *Pedicinus longiceps*). Brit. J. Exptl. Path., **3**, 196–203.

Babers, F. H. 1938 A septicemia of the southern armyworm caused by *Bacillus cereus*. Ann. Entomol. Soc. Am., **31**, 371–373.

Bacot, A. 1911 On the persistence of bacilli in the gut of an insect during metamorphosis. Trans. Roy. Entomol. Soc. London, pt. II, 497–500.

Bacot, A. 1914a On the survival of bacteria in the alimentary canal of fleas during metamorphosis from larva to adult. J. Hyg., **13**, Plague Suppl. III, 655–664.

Bacot, A. 1914b A study of the bionomics of the common rat fleas and other species, etc. J. Hyg., **13**, Plague Suppl. III, 447.

Bacot, A. 1916 Reports on questions connected with the investigation of non-malarial fevers in West Africa. Yellow Fever Commission to West Africa, **3**, 1–191.

Bacot, A. 1917 The effect of the presence of bacteria or yeasts on the hatching of the eggs of *Stegomyia fasciata* (the yellow fever mosquito). J. Roy. Microscop. Soc., pt. I, 173–174.

Bacot, A., and Martin, C. J. 1914 Observations on the mechanism of the transmission of plague by fleas. J. Hyg., **13**, Plague Suppl. III, 423–439.

Badger, L. F. 1933a Rocky Mountain spotted fever and boutonneuse fever. A study of their immunological relationship. Pub. Health Repts., **48**, 507–511.

Badger, L. F. 1933b Rocky Mountain spotted fever: susceptibility of the dog and sheep to the virus. Pub. Health Repts., **48**, 791–795.

Baer, W. S. 1929 The use of a viable antiseptic in the treatment of osteomyelitis. Southern Med. J., **22**, 582–583.

Baer, W. S. 1931 The treatment of chronic osteomyelitis with the maggot (larva of the blow fly). J. Bone Joint Surg., **13**, 438.

Bahr, L. 1919 Paratyphus nos Honningbien. Skand. Vet. Tids., **9,** 25-40, 45-60.

Bahr, P. H., and Comb, H. 1914 A study of epidemic dysentery in Fiji Islands. Brit. Med. J., **1,** 294-296.

Bailey, I. W. 1920 Some relations between ants and fungi. Ecology, **1,** 174-189.

Baker, A. C. 1943 The typical epidemic series. Am. J. Trop. Med., **23,** 559-566.

Balbiani, E. G. 1866 Sur la reproduction et l'embryongénie des pucerons. Compt. Rend. Acad. Sci., **62,** 1231, 1285, 1390.

Balbiani, E. G. 1869 Mémoire sur la génération des aphides. I. Ann. Sci. Nat. Zool., **11,** 5-89.

Balbiani, E. G. 1870a Mémoire sur la génération des aphides. II. Ann. Sci. Nat. Zool., **14,** 1-39 (art. 2).

Balbiani, E. G. 1870b Mémoire sur la génération des aphides. III. Ann. Sci. Nat. Zool., **14,** 1-36 (art. 9).

Balbiani, E. G. 1871a Mémoire sur la génération des aphides. IV. Ann. Sci. Nat. Zool., **15,** 1-30 (art. 1).

Balbiani, E. G. 1871b Mémoire sur la génération des aphides. V. Ann. Sci. Nat Zool., **15,** 1-63 (art. 4).

Banerji, R. N. 1942 Tick typhus of India. Antiseptic, July, 1942. 3 pp.

Bang, F. B., and Glaser, R. W. 1943 The persistence of poliomyelitis virus in flies. Am. J. Hyg., **37,** 320-324.

Barber, M. A. 1912 The susceptibility of cockroaches to plague bacilli inoculated into the body cavity. Philippine J. Sci., Ser. B., **7,** 521-524.

Barber, M. A. 1914 Cockroaches and ants as carriers of the vibrios of Asiatic cholera. Philippine J. Sci., Ser. B., **9,** 1-4.

Barber, M. A. 1927 The food of anopheline larvae. Food organisms in pure culture. Pub. Health Repts., **42,** 1494-1510.

Barber, M. A. 1928 The food of culicine larvae. Food organisms in pure culture. Pub. Health Repts., **43,** 11-17.

Barber, M. A., and Jones, O. R. 1915 A test of *Coccobacillus acridiorum* D'Herelle on locusts in the Philippines. Philippine J. Sci., Ser. B., **10,** 163-176.

Barraud, P. J. 1929 A simple method for the carriage of living mosquitoes over long distances in the tropics. Indian J. Med. Research, **17,** 281-285.

Basile, C. 1910 Sulla Leishmaniose del cane e sull' ospite intermedio del Kala-Azar infantile. Atti Accad. Lincei, Rend., **19,** 523-527.

Bates, L. B., and St. John, J. H. 1922 Suggestion of *Spirochoeta neotropicalis* as name for spirochaete of relapsing fever found in Panama. J. Am. Med. Assoc., **79,** 575.

Baumberger, J. P. 1917 The food of *Drosophila melanogaster* Meigen. Proc. Natl. Acad. Sci. U.S., **3,** 122-126.

Baumberger, J. P. 1919 A nutritional study of insects, with special reference to microorganisms and their substrata. J. Exptl. Zool., **28,** 1-81.

Beard, R. L. 1944 Susceptibility of Japanese beetle larvae to *Bacillus popilliae*. J. Econ. Entomol., **37**, 702–708.

Becker, E. R. 1923a Life history studies of *Crithidia gerridis* in the water-strider. Am. J. Hyg., **3**, 202.

Becker, E. R. 1923b Observations on the morphology and life-cycle of *Crithidia gerridis* Patton in the water-strider, *Gerris remigis* Say. J. Parasitol., **9**, 141–152.

Becker, E. R. 1923c Observations on the morphology and life-history of *Herpetomonas muscae-domesticae* in North American muscoid flies. J. Parasitol., **9**, 199–213.

Becker, E. R. 1923d Transmission experiments on the specificity of *Herpetomonas muscae-domesticae* in muscoid flies. J. Parasitol., **10**, 25–34.

Becker, E. R. 1930 The intestinal flagellates of insects. *In* Hegner, R., and Andrews, J., Problems and methods of research in protozoology, chap. 28, pp. 248–256. The Macmillan Co., New York. 532 pp.

Begg, A. M., Fulton, F., and Van den Ende, M. 1944 Inclusion bodies in association with typhus rickettsiae. J. Path. Bact., **56**, 109–113.

Belding, D. L. 1942 Textbook of clinical parasitology. D. Appleton-Century Co., New York. 888 pp.

Bell, J. F. 1945 The infection of ticks (*Dermacentor variabilis*) with *Pasteurella tularensis*. J. Infectious Diseases, **76**, 83–95.

Bell, J. F., and Chalgren, W. S. 1943 Some wildlife diseases in the eastern United States. J. Wildlife Manag., **7**, 270–278.

Beltiukova, K. I., and Romanevich, B. V. 1940 On the bacterial diseases of the beet weevil and the application of excreted bacteria in combating the latter. Mikrobiologichniĭ Zhurnal, **7**, 121–134.

Benecke, W. 1905 Über *Bacillus chitinovorus*, einen Chitin zersetzenden Spaltpilz. Botan. Z., **63**, 227–242.

Benedek, T., and Specht, G. 1933 Mykologischbakteriologische Untersuchungen über Pilze und Bakterien als Symbionten in Kerbtieren. Zentr. Bakt. Parasitenk. Infekt., I Orig., **130**, 74–90.

Bengtson, I. A. 1922 Toxin producing anerobe. Pub. Health Repts., **37**, 164–170, 2252–2253.

Bengtson, I. A. 1937 Cultivation of the rickettsiae of Rocky Mountain spotted fever *in vitro*. Pub. Health Repts., **52**, 1329–1335.

Bengtson, I. A. 1941a Immunological relationships between the rickettsiae of Australian and American "Q" fever. Pub. Health Repts., **56**, 272–281.

Bengtson, I. A. 1941b Complement fixation in endemic typhus. Pub. Health Repts., **56**, 649–653.

Bengtson, I. A. 1941c Complement fixation in "Q" fever. Proc. Soc. Exptl. Biol. Med., **46**, 665–668.

Bengtson, I. A. 1944 Complement fixation in the rickettsial diseases—technique of the test. Pub. Health Repts., **59**, 402–405.

Bengtson, I. A., and Dyer, R. E. 1935 Cultivation of the virus of Rocky Mountain spotted fever in the developing chick embryo. Pub. Health Repts., **50**, 1489-1498.

Bennett, C. W., and Wallace, H. E. 1938 Relation of the curly top virus to the vector, *Eutettix tenellus.* J. Agr. Research, **56**, 31-51.

Bensley, R. R. 1942 Chemical structure of cytoplasm. Science, **96**, 389. (See also Biological Symposia, **10**. 334 pp.)

Bequaert, J. 1942 A monograph of the Melophaginae, or ked-flies of sheep, goats, deer and antelopes (Diptera, Hippoboscidae). Entomol. Americana, **22**, 1-210. (See p. 198.)

Bergey, D. H., Breed, R. S., Murray, E. G., and Hitchens, A. P. 1939 Bergey's manual of determinative bacteriology. 5th ed. Williams and Wilkins Co., Baltimore. 1032 pp.

Berlese, A. 1893 Le cocciniglie italiane, viventi sugli agrumi. I. Dactylopius. Riv. Path. Vegetale, **2**, 70-109, 129-193.

Berlese, A. 1906 Sopra una nuova specie mucidinea parassita del *Ceroplastes rusci.* Redia, **3**, 8-15.

Berliner, E. 1915 Über die Schlaffsucht der Mehlmottenraupe (*Ephestia kühniella,* Zell) und ihren Erreger, *Bacillus thuringiensis,* n. sp. Z. Angew: Entom., ii, no. 1, April, 29-56.

Bertarelle, E. 1910 Verbreitung des Typhus durch die Fliegen. Zentr. Bakt. Parasitenk. Infekt., I Orig., **53**, 486-495.

Beuperthuy, L. D. 1854 Transmission of yellow fever and other diseases by mosquito. Gaz. Oficial Cumanà, Año 4, no. 57, May 23. (Cited by Herms, 1939.)

Bilal, S. 1941 Kobaylarda tabii tularemi infeksiyonu ve tularemi suslarinin sekerlere te'siri. (Natural infection of tularemia in guinea pigs and the action of tularemic strains on some carbohydrates.) Poliklinik, **9**, 15-19.

Bishop, A., and Tate, P. 1939 The morphology and systematic position of *Dobellina mesnili* nov. gen. (*Entamoeba mesnili* Keilin, 1917). Parasitology, **31**, 501-510.

Black, L. M. 1939 Inhibition of virus activity by insect juices. Phytopathology, **29**, 321-336.

Black, L. M. 1944 Some viruses transmitted by agallian leafhoppers. Proc. Am. Phil. Soc., **88**, 132-144.

Blake, F. G., Maxcy, K. F., Sadusk, J. F., Kohls, G. M., and Bell, E. J. 1945 Studies on tsutsugamushi disease (scrub typhus, mite-borne typhus) in New Guinea and adjacent islands: epidemiology, clinical observations, and etiology in the Dobadura area. Am. J. Hyg., **41**, 374-396.

Blanc, G. R., and Baltazard, M. 1940 Comportement du virus du typhus épidémique chez la puce du rat *"Xenopsylla cheopis."* Bull. Acad. Med., **123**, 126-136.

Blanc, G. R., and Baltazard, M. 1941 Transmission du bacille de Whitmore par la puce du rat *Xenopsylla cheopis.* Compt. Rend. Acad. Sci., Paris, **213**, 541-543.

REFERENCES

Blanc, G. R., and Baltazard, M. 1942a Sur le mécanisme de la transmission de la peste par *Xenopsylla cheopis*. Compt. Rend. Soc. Biol., **136,** 646–647.

Blanc, G. R., and Baltazard, M. 1942b Transmission de l'infection à bacille de Whitmore par insectes piquers. I. Maladie expérimentale du Cobaye. Ann. Inst. Pasteur., **68,** 281–293.

Blanc, G. R., and Caminopétros, J. 1929 Durée de conservation du virus de la dengue chez les Stégomyas. L'influence de la saison froide sur le bouvoir infectant. Compt. Rend. Acad. Sci., Paris, **188,** 1273–1275.

Blanc, G. R., and Caminopétros, J. 1931 De la sensibilité du spermophile (*Citellus citellus*) au virus de la fièvre boutonneuse. Compt. Rend. Acad. Sci., Paris, **193,** 374–375.

Blanc, G. R., and Caminopétros, J. 1932 Epidemiological and experimental studies of boutonneuse fever conducted at the Pasteur Institute of Athens [trans. title]. Arch. Inst. Pasteur Tunis, **20,** 343–394.

Blanc, G. R., and Woodward, T. E. 1945 The infection of *Pedicinus albidus* Rudow, the maggot's louse, on typhus carrying monkeys (*Macacus sylvanus*). Am. J. Trop. Med., **25,** 33–34.

Blattner, R. J., and Heys, F. M. 1941 Experimental transmission of Saint Louis encephalitis to white Swiss mice by *Dermacentor variabilis*. Proc. Soc. Exptl. Biol. Med., **48,** 707–710.

Blattner, R. J., and Heys, F. M. 1944 Blood-sucking vectors of encephalitis: experimental transmission of St. Louis encephalitis (Hubbard strain) to white Swiss mice by the American dog tick, *Dermacentor variabilis* Say. J. Exptl. Med., **79,** 439–454.

Blewett, M., and Fraenkel, G. 1944 Intracellular symbiosis and vitamin requirements of two insects, *Lasioderma serricorne* and *Sitodrepa panicea*. Proc. Royal Soc., London, B., **132,** 212–221.

Blochmann, F. 1884 Ueber die Reifung der Eier bei Ameisen und Wespen. Festschr. Feier 500-jährigen Bestehens Ruperto-Carola, dargebracht von Naturh. Med. Verein Heidelberg, 143–172.

Blochmann, F. 1886 Ueber eine Metamorphose der Kerne in den Ovarialeiern und über den Beginn der Blastodermbildung bei den Ameisen. Verhandl. Naturh.-Med. Ver., **3,** 243–247.

Blochmann, F. 1887 Über die Richtungskörper bei Insekteneiern. Gegenbaurs Morph. Jahrb., **12,** 544–574.

Blochmann, F. 1888 Ueber das regelmässige Vorkommen von bakterienähnlichen Gebilden in den Geweben and Eiern verschiedener Insekten. Z. Biol., **24,** 1–15.

Blochmann, F. 1892 Ueber das Vorkommen von bakterienähnlichen Gebilden in den Geweben und Eiern verschiedener Insekten. Zentr. Bakt. Parasitenk. Infekt., **11,** 234–240.

Boardman, E. T. 1944 Methods for collecting ticks for study and delineation. J. Parasitol., **30,** 57–59.

Bode, H. 1936 Untersuchungen über die Symbiose von Tieren mit Pilzen und

Bakterien: die Bakteriensymbiose bei Blattiden und das Verhalten von Blattiden bei aseptischer Aufzucht. Arch. Mikrobiol., **7,** 391–403.

Bogdanow, E. A. 1906 Über das Züchten der Larven der gewöhnlichen Fleischfliege (*Calliphora vomitoria*) in sterilisierten Nährmitteln. Pflügers Arch. ges. Physiol., **113,** 97–105.

Bogdanow, E. A. 1908 Über die Abhängigkeit des Wachstums der Fliegenlarven von Bakterien und Fermenten und über Variabilität und Vererbung bei den Fleischfliegen. Arch. Anat. Physiol., Jahr. 1908, Suppl., 173–200.

Bogoyavlensky, N. 1922 *Zografia notonectae* n.g., n. sp. Arch. Russian Protistol. Soc., **1,** 113–119.

Bolle, J. 1898 Der Seidenbau in Japan, nebst einem Anhang: die Gelboder Fettsucht der Seidenraupe, eine parasitäre Krankheit. Hartlebens, Budapest, Wien, and Leipzig. 141 pp.

Bolle, J. 1908 Studien über die Gelbsucht der Seidenraupen. Ber. über die Tätigkeit der K. K. landw.-chem. Versuchsta. in Gorz im Jahre 1907. Z. Landwirt. Versuchsw. Deut.-Oesterr., **14,** 279.

Bollinger, O. 1874 Milzbrand. In Ziemssen's Handb. spec. Path. Therapie, **3,** 457, 482. (See Nuttall, 1899.)

Bonde, R. 1930a The cabbage maggot as a disseminating agent of bacterial rots in the Cruciferae. Phytopathology, **20,** 128.

Bonde, R. 1930b Some conditions determining potato-seed-piece decay and blackleg induced by maggots. Phytopathology, **20,** 128.

Bonde, R. 1939a Comparative studies of the bacteria associated with potato blackleg and seed-piece decay. Phytopathology, **29,** 831–851.

Bonde, R. 1939b The role of insects in the dissemination of potato blackleg and seed-piece decay. J. Agr. Research, **59,** 889–917.

Bonnamour, S. 1925 Elévages et nouvelle liste de diptères fongicoles. Ann. Soc. Linnéenne, Lyon, **72,** 85–93.

Borchert, A. 1926 Die seuchenhaften Krankheiten der Honigbiene. 2d ed. Richard Schoetz, Berlin. 98 pp., 45 fig.

Borchert, A. 1935a Über die Pathogenität einiger in der Bienen Pathologie bisher unbekannter Bazillenarten für die Bienenbrut. Berlin. Tierärztl. Wochschr., **51,** 673–675.

Borchert, A. 1935b Untersuchungen über die Lebenseigenschaften des bienenpathogenen *Bac. orpheus,* sowie Bemerkungen über Durchführung und Erfolgsaussichten von Ansteckungsversuchen bei Bienenvölkern. Z. Infektionskrankh. Parasit. Krankh. Hyg. Haustiere, **47,** 60–77.

Borzenkov, A., and Donskov, G. 1933 The experimental infection of the tick *Hyalomma volgense* P. Schulze & E. Schlottke, 1929, with plague. Rev. Microbiol. Epidemiol. et Parasitol., **12,** 30.

Bourne, B. A. 1921 Report of the Assistant Director of Agriculture on the entomological and mycological work carried out during the season under review. Rept. Dept. Agric., 1919–20, 322–323.

Bowers, R. E., and McCay, C. M. 1940 Insect life without vitamin A. Science. **92**, 291.

Bozhenko, V. P. 1935 The role of bed bugs (*Cimex lectularius* L.) in transmission and preservation of tularaemia virus. Rev. Microbiol., **14**, 436–440. (In Russian.)

Bozhenko, V. P. 1936 On the role of the mosquitoes, *Culex apicalis* Ad., as carriers and transmitters of tularaemia. Rev. Microbiol., **15**, 445–449. (In Russian.)

Bradley, G. H. 1934 A method of shipping mosquito eggs. *Culex quinquefaciatus* Say. J. Econ. Entomol., **27**, 289.

Bradley, W. G. 1936 A thermally insulated unit for the transportation of adult insect parasites. U.S. Dept. Agr., Bur. Entomol., Entomol. Tech., 77.

Brain, C. K. 1923 A preliminary report on the intracellular symbionts of South African Coccidae. Ann. Univ. Stellenbosch, **1**, 1–48.

Brandes, E. W. 1923 Mechanics of inoculation with sugar-cane mosaic by insect vectors. J. Agri. Research, **23**, 279–283.

Brecher, G., and Wigglesworth, V. B. 1944 The transmission of *Actinomyces rhodnii* Erikson in *Rhodnius prolixus* Stål (Hemiptera) and its influence on the growth of the host. Parasitology, **35**, 220–224.

Breed, R. S. 1929 Some additional stories about scientific names. Science, **70**, 480.

Breed, R. S. 1940 Personal communication.

Brefeld, O. 1877 Ueber die Entomophthoreen und ihre Verwandten. Botan. Z., **35**, 345–355, 368–372.

Breinl, F. 1924 Studies on typhus virus in the louse. J. Infectious Diseases, **34**, 1–12.

Breitsprecher, E. 1928 Beitrage zur Kenntnis der Anobiidensymbiose. Z. Morphol. Ökol. Tiere, **11**, 495–538.

Brigham, G. D. 1936 Susceptibility of the opossum (*Didelphis virginiana*) to the virus of endemic typhus fever. Pub. Health Repts., **51**, 333–337.

Brigham, G. D. 1937 Susceptibility of animals to endemic typhus fever. Pub. Health Repts., **52**, 660–662.

Brigham, G. D. 1938a Endemic typhus virus in mice. Pub. Health Repts., **53**, 1251–1256.

Brigham, G. D. 1938b Susceptibility of animals to endemic typhus virus. Second report. Pub. Health Repts., **53**, 2078–2079.

Brigham, G. D. 1941 Two strains of endemic typhus fever virus isolated from naturally infected chicken fleas (*Echidnophaga gallinacea*). Pub. Health Repts., **56**, 1803–1804.

Brigham, G. D., and Dyer, R. E. 1938 Endemic typhus fever in native rodents. J. Am. Med. Assoc., **110**, 180–183.

Brill, N. E. 1898 A study of seventeen cases of a disease clinically resembling typhoid fever, but without the Widal reaction; together with a short review of the present status of the sero-diagnosis of typhoid fever. N.Y. Med. J., **67**, 48-54, 77-82.

Brinley, F. J., and Steinhaus, E. A. 1942 Some relationships between bacteria and sewage-inhabiting insects. Unpublished observations.

Brown, F. M. 1927 Descriptions of new bacteria found in insects. Am. Museum Novitates, 251. 11 pp.

Brown, F. M. 1928 Enzymes and bacteria in the honey-bee. Am. Museum Novitates, 304. 5 pp.

Brown, J. H. 1944 The fleas (Siphonaptera) of Alberta, with a list of the known vectors of sylvatic plague. Ann. Entomol. Soc. Am., **37**, 207-213.

Bruce, D. 1895 Tsetse-fly disease or Nagana in Zuzuland. Preliminary report. Bennett and Davis, Field St., Durban, 28 pp.

Bruce, D., and Nabarro, D. 1903 Progress report on sleeping sickness in Uganda. Rept. Sleeping Sickness Comm. Roy. Soc., **1**, 11-88.

Brues, C. T., and Dunn, R. C. 1945 The effect of penicillin and certain sulfa drugs on the intracellular bacteroids of the cockroach. Science, **101**, 336-337.

Brues, C. T., and Glaser, R. W. 1921 A symbiotic fungus occurring in the fat-body of *Pulvinaria innumerabilis* Rath. Biol. Bull. Marine Biol. Lab., **40**, 299-324.

Brumpt, E. 1913 Evolution de *Trypanosoma lewisi, duttoni, nabiasi, blanchardi,* chez les puces et les punaises. Transmission par les déjections. Comparaison avec *T. cruzi*. Bull. Soc. Path. Exotique, **6**, 167-171.

Brumpt, E. 1921a Les spirochétoses. *In* P. Brouardel's *Nouveau Traité de Médecine,* vol. V. Masson et Cie., Paris. 530 pp.

Brumpt, E. 1921b (See Lavier, 1921.)

Brumpt, E. 1927 Précis de parasitologie. 4th ed. Masson et Cie., Paris. 1452 pp. (see p. 872).

Brumpt, E. 1932 Longévité du virus de la fièvre boutonneuse (*Rickettsia conori* n. sp.) chez la tique, *Rhipicephalus sanguineus*. Compt. Rend. Soc. Biol., **110**, 1199-1202.

Brumpt, E. 1933a Etude de la fièvre récurrente sporadiques des Etats-Unis, transmise dans la nature par *Ornithodorus turicata*. Compt. Rend. Soc. Biol., **113**, 1366-1369.

Brumpt, E. 1933b Etude du *Spirochaeta turicatae* n. sp., agent de la fièvre récurrente sporadique des Etats-Unis transmise par *Ornithodorus turicata*. Compt. Rend. Soc. Biol., **113**, 1369-1372.

Brumpt, E. 1933c Recherches expérimentales sur la fièvre boutonneuse. Rapports et Comptes Rendus, Premier Congrès International d'Hygiène Méditerranéenne, Marseille, Sept., 1932, **2**, 185-192.

Brumpt, E. 1938 Rickettsia intracellulaire stomacale (*Rickettsia culicis* n. sp.) de *Culex fatigans*. Ann. Parasitol. Humaine et Comparée, **16**, 153-158.

Brumpt, E., and Desportes, C. 1941 Grande longévité du virus de la fièvre pourprée des Montagnes Rocheuses et de celui du typhus de São Paulo chez *Ornithodorus turicata*. Ann. Parasitol. Humaine et Comparée, **18**, 145-153.

Brunchorst, J. 1885 Ueber die Knöllchen an den Leguminosenwurzeln. Ber. Deut. Botan. Ges., **3**, 241–257.

Buchanan, D. 1932 A bacterial disease of beans transmitted by *Heliothrips femoralis* Reut. J. Econ. Entomol., **25**, 49–53.

Buchanan, E. D., and Buchanan, R. E. 1938 Bacteriology. 4th ed. The Macmillan Company, New York. 548 pp.

Buchanan, R. E., and Fulmer, E. I. 1930 Physiology and biochemistry of bacteria. Vol. II. Effects of environment upon microorganisms. Williams and Wilkins Co., Baltimore. 709 pp.

Buchanan, W. D. 1942 Experiments with an ambrosia beetle *Xylosandrus germanus* (Blfd.). J. Econ. Entomol., **34**, 367–369.

Buchner, P. 1911 Über intrazellulare Symbionten bei zuckersaugenden Insekten und ihre Vererbung. Münch, Sitzber. Ges. Morphol. Physiol., **27**, 89–96.

Buchner, P. 1912 Studien an intracellularen Symbionten. I. Die intracellularen Symbionten der Hemipteren. Arch. Protistenk., **26**, 1–116.

Buchner, P. 1913 Neue Erfahrungen über intrazellulare Symbioten bei Insekten. Naturw. Wochschrift N.F., **28**, 401–406, 420–425.

Buchner, P. 1918 Vergleichende Eistudien. 1. Die akzessorischen Kerne des Hymenoptereneies. Arch. Mikroskop. Anat., **91**, 1–202.

Buchner, P. 1919 Zur Kenntnis der Symbiose niederer pflanzlicher Organismen mit Pedikuliden. Biol. Zentr., **39**, 535–540.

Buchner, P. 1921a III. Die Symbiose der Anobiinen mit Hefepilzen. Arch. Protistenk., **42**, 319–336.

Buchner, P. 1921b Über ein neues, symbiotisches Organ der Bettwanze. Biol. Zentr., **41**, 570–574.

Buchner, P. 1921c Tier und Pflanze in intracellularer Symbiose. Gebrüder Borntraeger, Berlin. 462 pp.

Buchner, P. 1922 Haemophagie und Symbiose. Naturwissenschaften, **33**, 1.

Buchner, P. 1923 Studien an intracellularen Symbionten. IV. Die Bakteriensymbiose der Bettwanze. Arch. Protistenk., **46**, 225–263.

Buchner, P. 1925 Studien an intracellularen Symbionten. V. Die symbiontischen Einrichtungen der Zikaden. Z. Morphol. Ökol. Tiere, **4**, 88–245.

Buchner, P. 1926 Studien an intracellularen Symbionten. VI. Zur Akarinensymbiose. Z. Morphol. Ökol. Tiere, **6**, 625–644.

Buchner, P. 1928 Holznahrung und Symbiose. Julius Springer, Berlin. 64 pp.

Buchner, P. 1930 Tier und Pflanze in Symbiose. Gebrüder Borntraeger, Berlin. 900 pp.

Buchner, P. 1933 Studien an intracellularen Symbionten. VII. Die symbiontischen Einrichtungen der Rüsselkäfer. Z. Morphol. Ökol. Tiere, **26**, 709–777.

Buddington, A. R. 1941 The nutrition of mosquito larvae. J. Econ. Entomol., **34**, 275–281.

Buen, S. de. 1926 Note préliminaire d'une fièvre récurrente à tiques, en Espagne. Bull. Acad. Méd., **95**, 294.

Bullamore, G. W. 1922 *Nosema apis* and *Acarapis* (*Tarsonemus*) *woodi* in relation to Isle of Wight bee disease. Parasitology, **14**, 53–62.

Burnet, F. M. 1938 Tissue culture of the rickettsia of Q fever. Australian J. Exptl. Biol. Med. Sci., **16**, 219–224.

Burnet, F. M. 1942 The rickettsial diseases in Australia. Med. J. Australia, **2** (29th yr.), 129–134.

Burnet, F. M., and Freeman, M. 1937 Experimental studies on the virus of "Q" fever. Med. J. Australia, Aug. 21, 299–305.

Burnet, F. M., and Freeman, M. 1939 Note on a series of laboratory infections with the rickettsia of "Q" fever. Med. J. Australia, Jan. 7, 11–12.

Burnett, W. I. 1851, 1852 The organic relations of some of the Infusoria, including investigations concerning the structure and nature of the genus *Bodo* (Ehr.). Proc. Boston Soc. Nat. Hist., **4** (1851–54), 1851, p. 124 (extract); Boston J. Nat. Hist., **6** (1850–57), 1852, p. 319.

Burnside, C. E. 1927 Saprophytic fungi associated with the honey bee. Mich. Acad. Sci., **8**, 59–86.

Burnside, C. E. 1930 Fungous diseases of the honey bee. U.S. Dept. Agr. Tech. Bull., 149. 43 pp.

Burnside, C. E. 1932 A newly discovered brood disease. Am. Bee J., **72**, 433.

Burnside, C. E. 1934 Studies on the bacteria associated with European foulbrood. J. Econ. Entomol., **27**, 656–668.

Burnside, C. E. 1935 A disease of young bees caused by a Mucor. Am. Bee J., **75**, 75–76.

Burnside, C. E. 1945 [News note.] *In* Science News, Feb. 16, suppl. to Science, **101**, 10.

Burnside, C. E., and Foster, R. E. 1935 Studies on the bacteria associated with parafoulbrood. J. Econ. Entomol., **28**, 578–584.

Burnside, C. E., and Sturtevant, A. P. 1936 Diagnosing bee diseases in the apiary. U.S. Dept. Agr. Bull., 392. 34 pp.

Burri, R. 1904 Bacteriologische Forschungen über die Faulbrut. Schweiz. Bienenzeitung, no. 10 (Oct.), 335–342, and no. 11 (Nov.), 360–365.

Burrill, T. J. 1881 Anthrax of fruit trees; or the so-called fire-blight of pear and twig blight of apple trees. Proc. Am. Assoc. Adv. Sci., **29**, 583–597.

Burrill, T. J. 1882 The bacteria; an account of their nature and effects together with a systematic description of the species. Rept. Illinois Ind. Univ., **11**, 126, 134.

Burrill, T. J. 1883 New species of micrococcus. Am. Naturalist, **17**, 319.

Burroughs, A. L. 1941 Bacterial flora of the alimentary tract of *Grylloblatta campodeiformis campodeiformis* Walker. Montana State College Master's Thesis. 47 pp.

Burroughs, A. L. 1944 The flea *Malaraeus telchinum* a vector of *P. pestis*. Proc. Soc. Exptl. Biol. Med., **55**, 10-11.

Busacca, A. 1935 Un germe aux caractères de rickettsies (*Rickettsia trachomae*) dans les tissus trachomateux. Arch. Ophthalmol. (Paris), **52**, 567.

Bush, S. F. 1928 A study of the gregarines of the grasshoppers of Pietermaritzburg, Natal. Ann. Natal Mus. Pietermaritzburg, **6**, 97–169.

Byam, W. 1919 Trench fever. *In* Lloyd, L., Lice and their menace to man. Oxford Univ. Press, London. (See pp. 120–130.)

Cahn, A. R., Wallace, G. I., and Thomas, L. J. 1932 A new disease of moose. III. A new bacterium. Science, **76**, 1974.

Calkins, G. N., and Summers, F. M. (eds.) 1941 Protozoa in biological research. Columbia Univ. Press, New York. 1148 pp.

Callen, E. McC. 1940 Hymenopterous parasites of willow insects. Bull. Entomol. Research, **31**, 35–44.

Cameron, G. R. 1934 Inflammation in the caterpillars of *Lepidoptera*. J. Path. Bact., **38**, 441–466.

Caminopétros, M. J. 1932 La fièvre boutonneuse en Grèce. Recherches épidémiologiques et expérimentales. Rapports et Comptes Rendus, Premier Congrès International d'Hygiène Méditerranéenne, Marseille, Sept., 1932, **2**, 202–212. (Published in 1933.)

Caminopétros, J., and Contos, B. 1933 Intradermoréaction à la fièvre boutonneuse. Compt. Rend. Acad. Sci., Paris, **196**, 967–969.

Campbell, F. L., and Moulton, F. R. (eds.) 1943 Chemical control of insects. Pub. Am. Acad. Advancement Sci., 20. 206 pp.

Campo, G. C., and Gallardo, F. P. 1943 Técnicas de laboratorio en el tifus exantemático. S. A. Serrano, Madrid. 185 pp.

Canham, A. S. 1943 A rickettsia-like organism found in the blood of pigeons. I. S. African Vet. Med. Assoc., **14**, 83–89.

Cantacuzène, J. 1923 Le problème de l'immunité chez les invertébrés. Célébration du 75me Anniversaire Society de Biologie, Paris, 48–119.

Cao, G. 1898 Sul passagio dei microorganismi a traverso l'intestino di alcuni insetti. Ufficiale Sanit. Riv. Igiene Med. Patrica, **11**, 337–348, 385–397.

Cao, G. 1906a Nuove osservazioni sul passagio dei microorganismi a traverso l'intestino di alcuni insetti. Ann. Igiene Sper., **16**, 339–368.

Cao, G. 1906b Sul passagio dei germi a traverso le larve di alcuni insetti. Ann. Igiene Sper., **16**, 645–664.

Carmichael, J., and Fiennes, R. N. T.-W. 1942 Rickettsia infection in dogs. Vet. Record, **54**, 3–4.

Carpano, M. 1936 L'infezione du *Rickettsia* negli uccelli (*Rickettsia avium*). Profilassi, Riv. Pat. Comp., Jan.–Feb., 1936, pp. 1–9.

Carter, W. 1935 The symbionts of *Pseudococcus brevipes* (Ckl.). Ann. Entomol. Soc. Am., **28**, 60–72.

Carter, W. 1936 The symbionts of *Pseudococcus brevipes* in relation to a phytotoxic secretion of the insect. Phytopathology, **26**, 176–183.

Cartwright, K. St. G. 1938 A further note on fugus association in the Siricidae. Ann. Applied Biol., **25**, 430–432.

Caspari, E. 1939 Kann die Bettwanze bakterielle Infektionserreger verbreiten? Tib. Fakültesi Mecm. Istanbul, **1**, 972–983.

Castañeda, M. R. 1930 A study of the relationship of the scrotal swelling and rickettsia bodies to Mexican typhus fever. J. Exptl. Med., **52**, 195–199.

Castañeda, M. R. 1934 The antigenic relationship between *Proteus* X-19 and typhus rickettsia. II. A study of the common antigenic factor. J. Exptl. Med., **60**, 119–125.

Castañeda, M. R. 1935 The antigenic relationship between *Bacillus proteus* X-19 and rickettsiae. III. A study of the antigenic composition of the extracts of *Bacillus proteus* X-19. J. Exptl. Med., **62**, 289–296.

Castañeda, M. R. 1939 Massive typhus infection in laboratory animals. Proc. 6th Pacific Sci. Congr. Univ. Calif. Press, Berkeley, 1942. pp. 613–618.

Castañeda, M. R., and Silva, G. R. 1939 Varieties of Mexican typhus strains. Pub. Health Repts., **54**, 1337–1345.

Castañeda, M. R., and Silva, G. R. 1941 Immunological relationship between spotted fever and exanthematic typhus. J. Immunol., **42**, 1–14.

Castañeda, M. R., and Vargas-Curiel, J. 1938 Response in the horse to Mexican typhus infection. J. Immunol., **35**, 47–54.

Castañeda, M. R., and Zinsser, H. 1930 Studies on typhus fever. III. Studies of lice and bedbugs (*Cimex lectularius*) with Mexican typhus fever. J. Exptl. Med., **52**, 661–668.

Castellani, A., and Chalmers, A. J. 1919 Manual of tropical medicine. 3d ed. William Wood and Co., New York. 2436 pp.

Cattani, G. 1886 Studj sul colera. Gazz. Ospedali Milano, **7**, 611.

Ceder, E. T., Dyer, R. E., Rumreich, A., and Badger, L. F. 1931 Typhus fever in feces of infected fleas (*Xenopsylla cheopis*) and duration of infectivity of fleas. Pub. Health Repts., **46**, 3103–3106.

Cellí, A. 1888 Trasmissibilità dei germi patogeni mediante le dejectione delle mosche. Bull. Soc. Lancisiana Ospedali Roma, 1.1. (Quoted by Nuttall and Jepson, 1909, p. 27.)

Chandler, A. C. 1936 Introduction to human parasitology. 5th ed. John Wiley and Sons, Inc., New York. 661 pp.

Chandler, A. C. 1940 Introduction to human parasitology. 6th ed. John Wiley and Sons, Inc., New York. 698 pp.

Chandler, A. C., and Rice, L. 1923 Observations on the etiology of dengue fever. Am. J. Trop. Med., **3**, 233–262.

Chapman, J. W., and Glaser, R. W. 1915 A preliminary list of insects which have wilt, with a comparative study of their polyhedra. J. Econ. Entomol., **8**, 140–150.

Chapman, J. W., and Glaser, R. W. 1916 Further studies on wilt of gipsy moth caterpillars. J. Econ. Entomol., **9**, 149–167.

Chapman, R. N. 1924 Nutritional studies on the confused flour beetle, *Tribolium confusum* Duval. J. Gen. Physiol., **6**, 565–585.

Charles, V. K. 1938 A new entomogenous fungus on the corn earworm, *Heliothis obsoleta*. Phytopathology, **28**, 893–897.

Charles, V. K. 1941a A fungous disease of codling moth larvae. Mycology, **33**, 344–349.

Charles, V. K. 1941b A preliminary check list of the entomogenous fungi of North America. Bur. Entomol., Insect Pest Survey Bull., Suppl. to no. 9, **21**, 707–785.

Chatton, E. 1913a Septicémies spontanées à coccobacilles chez le hanneton et le ver-à-soie. Compt. Rend. Acad. Sci., Paris, **156**, 1707–1709.

Chatton, E. 1913b *Coccidiascus legeri*, n. g., n. sp., levure ascosporée parasite des cellules intestinales de *Drosophila funebris* Fabr. Compt. Rend. Soc. Biol., **75**, 117–120.

Chatton, E., and Léger, A. 1911 Eutrypanosomes, leptomonas et leptotrypanosomes chez *Drosophila confusa* Staeger (muscide). Compt. Rend. Soc. Biol., **70**, 34–36.

Cheshire, F. R., and Cheyne, W. W. 1885 The pathogenic history and the history under cultivation of a new bacillus (*B. alvei*), the cause of a disease of the hive bee hitherto known as foul brood. J. Roy. Microscop. Soc., Ser. **2**, pt. II, **5**, 581–601.

Chittenden, F. H. 1926 The common cabbage worm and its control. U.S. Dept. Agr. Farmers' Bull., 1461. 13 pp.

Chodzko, W. 1935 Une nouvelle infection à rickettsia, *Rickettsiaemia weigli*. Bull. Mens. Office Intern. Hyg. Publ., **27**, 1–4.

Cholodkowsky, N. 1891 Die Embryonalentwicklung von *Phyllodromia* (*Blatta*) *germanica*. Mém. Acad. St. Petersbourg, VIIme Sér., **38**, 1–121.

Cholodkowsky, N. 1904 Zur Morphologie der Pedikuliden. Zoo. Anz., **27**, 120–125.

Chorine, V. 1929a New bacteria pathogenic to the larvae of *Pyrausta nubilalis* Hb. Internat. Corn Borer Invest., Sci. Repts., **2**, 39–53.

Chorine, V. 1929b Nouveaux microbes pathogènes pour les chenilles de la pyrale du maïs. Ann. Inst. Pasteur, **43**, 1657–1678.

Chorine, V. 1930a On the use of bacteria in the fight against the corn borer. Internat. Corn Borer Invest., Sci. Repts., **3**, 94–98.

Chorine, V. 1930b On the use of bacteria in the fight against the corn borer. Corn Borer Invest., Sci. Repts., **3**, 99–110.

Chorine, V. 1931 Contribution à l'étude de l'immunité chez les insectes. Bull. Biol. France Belg., **65**, 291–393.

Chow, C. Y. 1940 The common blue-bottle fly, *Chrysomyia megacephala*, as a carrier of pathogenic bacteria in Peiping, China. Chinese Med. J., **57**, 145–153.

Chumakov, M. P., and Gladkikh, S. 1939 On the role of Ixodidae in communicating spring and summer encephalitis. Bull. Biol. Méd. Exptl. U.R.S.S., **7**, 221–223.

Chumakov, M. P., and Seitlenok, N. A. 1940 Tick-borne human encephalitis in the European part of USSR and Siberia. Science, **92**, 263–264.

Ciferri, R. 1934 Associazione tra larve della Mosca delle frutta (*Ceratitis capitata*) en un coccobacillo (*Escherichia ellingeri*). Lavori Ist. Botan. Palermo, **4**, 168–200.

Clavera, G. C., and Gallardo, F. P. 1943 Técnicas de laboratorio en el tifus exantemático. Prensa Española, Madrid. 185 pp.

Clements, F. E., and Shear, C. L. 1931 The genera of fungi. H. W. Wilson Co., New York. 496 pp.

Cleveland, L. R. 1924 The physiological and symbiotic relationships between the intestinal protozoa of termites and their hosts, with special reference to *Reticulitermes flavipes* Kollar. Biol. Bull., **46**, 178–227.

Cleveland, L. R. 1925a The method by which *Trichonympha campanula*, a protozoön in the intestine of termites, ingests solid particles of wood for food. Biol. Bull., **48**, 282–287.

Cleveland, L. R. 1925b The ability of termites to live perhaps indefinitely on a diet of pure cellulose. Biol. Bull., **48**, 289–293.

Cleveland, L. R. 1925c The feeding habit of termite castes and its relation to their intestinal flagellates. Biol. Bull., **48**, 295–308.

Cleveland, L. R. 1925d The effects of oxygenation and starvation on the symbiosis between the termite, *Termopsis,* and its intestinal flagellates. Biol. Bull., **48**, 309–326.

Cleveland, L. R. 1926 Symbiosis among animals with special reference to termites and their intestinal flagellates. Quart. Rev. Biol., **1**, 51–60.

Cleveland, L. R. 1928 Further observations and experiments on the symbiosis between termites and their intestinal protozoa. Biol. Bull., **54**, 231–237.

Cleveland, L. R., Hall, S. R., Sanders, E. P., and Collier, J. 1934 The wood-feeding roach *Cryptocercus,* its protozoa, and the symbiosis between protozoa and roach. Mem. Am. Acad. Arts Sci., **17** (no. 2), 185–342.

Clinton, G. P., and McCormick, F. A. 1936 Dutch elm disease—*Graphium ulmi*. Conn. Agr. Exptl. Sta. Bull., 389.

Cochrane, E. W. W. 1912 A small epidemic of typhoid fever in connection with specifically infected flies. J. Roy. Army Med. Corps, **18**, 271–276.

Coggeshall, L. T. 1939 The transmission of lymphocytic choriomeningitis by mosquitoes. Science, **89**, 515–516.

Coggeshall, L. T. 1943a Malaria as a world menace. J. Am. Med. Assoc., **122**, 8–11.

Coggeshall, L. T. 1943b War malaria. Med. Clinics N. America, May, 1943, 617–631.

Cohn, F. 1855 *Empusa muscae* und die Krankheit der Stubenfliegen. Nova Acta K. Acad. Caes. Leop. Carol. Germ. Nat., **25**, 301–360.

Coles, J. D. W. A. 1931 A rickettsia-like organism in the conjunctiva of sheep. 17th Rept. Director Vet. Services Animal Ind., Union S. Afr., 175–186.

Coles, J. D. W. A. 1935 A rickettsia-like organism and an unknown intracellular organism of the conjunctival epithelium of goats. Onderstepoort J. Vet. Sci. Animal Ind., **4**, 389–395.

Coles, J. D. W. A. 1936 A rickettsia-like organism of the conjunctival epithelium of cattle. J. S. African Vet. Med. Assoc., **7**, 1–5.

Coles, J. D. W. A. 1940 Conjunctivitis of the domestic fowl and an associated rickettsia-like organism in the conjunctival epithelium. Onderstepoort J. Vet. Sci. Animal Ind., **14**, 469–478.

Colla, S. 1934 Laboulbeniales: Peyritsciellaceae, Dimorphomycetaceae, Laboulbeniaceae, Homothallicae, Ceratomycetaceae. Flora Italica Cryptogama (Soc. Bot. Italiana) Fasc., **16**, 1–157.

Comstock, J. H. 1936 An introduction to entomology. 8th ed. Comstock Publishing Co., Ithaca, N.Y. 1044 pp.

Comstock, J. H., Comstock, A. B., and Herrick, G. W. 1938 A manual for the study of insects. Comstock Publishing Co., Ithaca, N.Y. 401 pp.

Conor, A., and Bruch, A. 1910 Une fièvre éruptive observée en Tunisie. Bull. Soc. Path. Exotique, **3**, 492–496.

Conté, A., and Faucheron, L. 1907 Présence de levures dans le corps adipeux de divers coccides. Compt. Rend. Acad. Sci., Paris, **145**, 1223–1225.

Cook, C. E. 1944 Observations on the epidemiology of scrub typhus. Med. J. Australia, **2**, 539–543.

Cook, M. T. 1935 Index of the vectors of virus diseases of plants. J. Agr. Univ. Puerto Rico, **19**, 407–420.

Cook, M. T. 1936 First supplement to index vectors of virus diseases of plants. J. Agr. Univ. Puerto Rico, **20**, 729–739.

Cooley, R. A. 1938 The genera *Dermacentor* and *Otocentor* (Ixodidae) in the United States, with studies in variation. Nat. Inst. Health Bull., **171**, 1–89.

Cooley, R. A., and Kohls, G. M. 1944a The genus *Amblyomma* (Ixodidae) in the United States. J. Parasitol., **30**, 77–111.

Cooley, R. A., and Kohls, G. M. 1944b The Argasidae of North America, Central America and Cuba. Am. Midland Naturalist, Mono. no. 1. 152 pp.

Cornalia, E. 1859 Sui carratteri che presenta il seme sano dei bachi da seta, e come questo si possa distinguere dal seme infetto. Atti. Soc. Ital. Milano, **2**, 255–270.

Corson, J. F., and Wilcocks, C. 1944 The geographical distribution of mite-borne typhus fever. Trop. Diseases Bull., **41**, 431–439.

Couch, J. N. 1929 Monograph of *Septobasidium*. Part I. J. Elisha Mitchell Sci. Soc., **44**, 242–260.

Couch, J. N. 1931 The biological relationships between *Septobasidium retiforme* (B. & C.) Pat. and *Aspidiotus osborni* New. and Ckll. Quart. J. Microscop. Sci., **74**, 383–438.

Couch, J. N. 1938 The genus *Septobasidium*. Univer. of North Carolina Press, Chapel Hill. 480 pp.

Couvreur, E., and Chahovitch, X. 1921 Sur un mode de défense naturelle contre les infections microbiennes chez les invertébrés. Compt. Rend. Acad. Sci., Paris, **172**, 711–713.

Cowan, T. W. 1911 British bee-keeper's guide book. 20th ed. London. 226 pp.

Cowdry, E. V. 1923 The distribution of *Rickettsia* in the tissues of insects and arachnids. J. Exptl. Med., **37**, 431–456.

Cowdry, E. V. 1925a The occurrence of *Rickettsia*-like microorganisms in adult "locusts" (*Tibicen septendecim* Linn.). Biol. Bull., **48**, 15–18.

Cowdry, E. V. 1925b Studies on the etiology of heartwater. I: Observation of a rickettsia, *Rickettsia ruminantium* (n. sp.), in the tissues of infected animals. J. Exptl. Med., **42**, 231–252.

Cowdry, E. V. 1925c Studies on the etiology of heartwater. II: *Rickettsia ruminantium* (n. sp.) in the tissues of ticks transmitting the disease. J. Exptl. Med., **42**, 253–274.

Cowdry, E. V. 1925d A group of microorganisms transmitted hereditarily in ticks and apparently unassociated with diease. J. Exptl. Med., **41**, 817–830.

Cowdry, E. V. 1926 Rickettsiae and disease. Arch. Path. Lab. Med., **2**, 59–90.

Cowdry, E. V. 1943 Microscopic technique in biology and medicine. Williams and Wilkins Co., Baltimore. 206 pp.

Cowdry, E. V., and Ham, A. W. 1932 Studies on East Coast fever. I. The life cycle of the parasite in ticks. Parasitology, **24**, 1–49.

Cowdry, E. V., and Olitsky, P. K. 1922 Differences between mitochondria and bacteria. J. Exptl. Med., **36**, 521–533.

Cowdry, E. V., and Rees, C. W. 1935 An attempt to ascertain the behavior of *Anaplasma marginale* in ticks transmitting anaplasmosis. Am. J. Hyg., **21**, 94–100.

Cox, G. L., Lewis, F. C., and Glynn, E. E. 1912 The numbers and varieties of bacteria carried by the common house-fly in sanitary and insanitary city areas. J. Hyg., **12**, 230–319.

Cox, H. R. 1938a Use of yolk sac of developing chick embryo as medium for growing rickettsiae of Rocky Mountain spotted fever and typhus groups. Pub. Health Repts., **53**, 2241–2247.

Cox, H. R. 1938b A filter-passing infectious agent isolated from ticks. III. Description of organism and cultivation experiments. Pub. Health Repts., **53**, 2270–2276.

Cox, H. R. 1939 Studies of a filter-passing infectious agent isolated from ticks. V. Further attempts to cultivate in cell-free media. Suggested classification. Pub. Health Repts., **54**, 1822–1827.

Cox, H. R. 1940 *Rickettsia diaporica* and American Q fever. Am. J. Trop. Med., **20**, 463–469.

Cox, H. R. 1941 Cultivation of rickettsiae of the Rocky Mountain spotted fever, typhus and Q fever groups in the embryonic tissues of developing chicks. Science, **94**, 399–403.

Cox, H. R., and Bell, E. J. 1939 The cultivation of *Rickettsia diaporica* in tissue culture and in the tissues of developing chick embryos. Pub. Health Repts., **54**, 2171–2178.

Cox, J. A., Bobb, M. L., and Hough, W. S. 1943 A fungous disease of the Comstock mealybug. J. Econ. Entomol., **36**, 580–583.

Craig, C. F. 1942 Laboratory diagnosis of protozoan diseases. Lea and Febiger, Philadelphia. 349 pp.

Cross, J. B. 1941 A study of *Oxymonas minor* Zeliff from the termite *Kalotermes minor* Hagen. Univ. Calif. Pub. Zool., **43**, 379–404.

Cuboni, G., and Garbini, A. 1890 Sopra una malattia del gelso in rapporto colla flaccidezza del baco de seta. R. Acad. Lincei, **6**, Ser. 4, 26–27.

Cuénot, L. 1896 Études physiologiques sur les orthoptères. Arch. Biol., **14**, 293–341.

Culbertson, J. T. 1942 Medical parasitology. Columbia Univ. Press, New York. 285 pp.

Cunha, A. M. da. 1934 Sur la culture des rickettsia du typhus exanthématique de São Paulo dans la membrane chorio-allantoide de l'embryon de Paulet. Compt. Rend. Soc. Biol., **117**, 392–394.

Currie, D. G. 1910 Mosquitoes in relation to the transmission of leprosy. Flies in relation to the transmission of leprosy. Pub. Health Repts. Bull., 39. 42 pp.

Currie, J. R. 1936 Laboratory practice. William Wood and Co., Baltimore. 378 pp. (See pp. 308–321.)

Damon, S. R. 1926 A note on the spirochaetes of termites. J. Bact., **11**, 31–36.

Dangeard, P.-A. 1902 Sur les caryophysème des eugleniens. Compt. Rend. Acad. Sci., Paris, **134**, 1365–1366.

Daubney, R. 1930 Natural transmission of heartwater of sheep by *Amblyomma variegatum*. Parasitology, **22**, 260–267.

Davaine, C. 1868 Expériences rélatives à la durée de l'incubation des maladies charbonneuses. Bull. Acad. Méd., **33**, 816.

Davaine, C. 1870a Etudes sur la contagion du charbon chez les animaux domestiques. Bull. Acad. Méd., **35**, 215–235.

Davaine, C. 1870b Etudes sur la genèse et la propagation du charbon. Bull. Acad. Méd., **35**, 471–498.

Davis, G. E. 1935 Tularemia. Susceptibility of the white-tailed prairie dog, *Cynomys leucurus* Merriam. Pub. Health Repts., **50**, 731–732.

Davis, G. E. 1939a The Rocky Mountain spotted fever rickettsia in the tick genus *Ornithodoros*. Proc. 6th Pacific Sci. Congr., **5**, 577–579.

Davis, G. E. 1939b *Rickettsia diaporica*: recovery of three strains from *Dermacentor andersoni* collected in southeastern Wyoming: their identity with Montana strain 1. Pub. Health Repts., **54**, 2219–2227.

Davis, G. E. 1940a *Bacterium tularense*: its persistence in the tissues of the argasid ticks *Ornithodoros turicata* and *O. parkeri*. Pub. Health Repts., **55**, 676–680.

Davis, G. E. 1940b *Rickettsia diaporica*: its persistence in the tissues of *Ornithodoros turicata*. Pub. Health Repts., **55**, 1862–1864.

Davis, G. E. 1941 *Ornithodoros turicata*: the male; feeding and copulation habits, fertility, span of life, and the transmission of relapsing fever spirochetes. Pub. Health Repts., **56**, 1799-1802.

Davis, G. E. 1942 Species unity or plurality of the relapsing fever spirochetes. Pub. Am. Assoc. Advancement Sci., 18, 41-47.

Davis, G. E. 1943a Relapsing fever: the tick *Ornithodoros turicata* as a spirochetal reservoir. Pub. Health Repts., **58**, 839-842.

Davis, G. E. 1943b The tick *Ornithodoros rudis* as a host to the rickettsiae of the spotted fevers of Colombia, Brazil, and the United States. Pub. Health Repts., **58**, 1016-1020.

Davis, G. E. 1943c Experimental transmission of the rickettsiae of the spotted fevers of Brazil, Colombia, and the United States by the argasid tick *Ornithodoros nicollei*. Pub. Health Repts., **58**, 1742-1744.

Davis, G. E. 1943d Further attempts to transmit *Pasteurella tularensis* by the bedbug (*Cimex lectularius*). J. Parasitol., **29**, 395-396.

Davis, G. E., and Cox, H. R. 1938 A filter-passing infectious agent isolated from ticks. I. Isolation from *Dermacentor andersoni*, reactions in animals and filtration experiments. Pub. Health Repts., **53**, 2259-2267.

Davis, G. E., and Kohls, G. M. 1937 *Ixodes ricinus californicus* (Banks) as a possible vector of *Bacterium tularense*. Pub. Health Repts., **52**, 281-282.

Davis, G. E., and Parker, R. R. 1934 Comparative experiments on spotted fever and boutonneuse fever (I). Pub. Health Repts., **49**, 423-428.

Davis, J. J., and Luginbill, P. 1921 The green June beetle or fig-eater. North Carolina Agr. Exptl. Sta. Bull., 242. 35 pp.

Davis, N. C. 1932 The effect of various temperatures in modifying the extrinsic incubation period of the yellow fever virus in *Aëdes aegypti*. Am. J. Hyg., **16**, 163-176.

Davis, N. C. 1933a Attempts to transmit yellow fever virus with *Triatoma megista* (Burmeister). J. Parasitol., **19**, 209-214.

Davis, N. C. 1933b The survival of yellow fever virus in ticks. Am. J. Trop. Med., **13**, 547-554.

Davis, N. C. 1934 Attempts to determine the amount of yellow fever virus injected by the bite of a single infected stegomyia mosquito. Am. J. Trop. Med., **14**, 343-354.

Davis, N. C., Frobisher, M., Jr., and Lloyd, W. 1933 The titration of yellow fever virus in stegomyia mosquitoes. J. Exptl. Med., **58**, 211-226.

Davis, N. C., and Shannon, R. C. 1930 The location of yellow fever virus in infected mosquitoes and the possibility of hereditary transmission. Am. J. Hyg., **11**, 335-344.

Dean, R. W. 1933 Morphology of the digestive tract of the apple maggot fly, *Rhagoletis pomonella* Walsh. N.Y. (Geneva) Agr. Expt. Sta. Tech. Bull., 215.

Dean, R. W. 1935 Anatomy and postpupal development of the female reproductive system in the apple maggot fly, *Rhagoletis pomonella* Walsh. N.Y. (Geneva) Agr. Expt. Sta. Tech. Bull., 229.

De Bach, P. H., and McOmie, W. A. 1939 New diseases of termites caused by bacteria. Am. Entomol. Soc. Am., **32**, 137–146.

De Bary, A. 1879 Die Erscheinung der Symbiose. Karl J. Trübner, Strassburg. 30 pp.

Delpy, L., and Kaweh, M. 1937 Transmission de *Bacillus anthracis* à l'homme par *Argas persicus* Oken 1818. Rev. Path. Comparée, **37**, 1229–1234.

Dennis, E. W. 1931 The life history of *Babesia bigemina* in the North American fever tick. Science, **73**, 620–621.

Dennis, E. W. 1932 The life-cycle of *Babesia bigemina* (Smith and Kilbourne) of Texas cattle-fever in the tick *Margaropus annulatus* (Say). Univ. Calif. Pub. Zool., **36**, 263–298.

Derrick, E. H. 1937 "Q" fever. A new fever entity. Clinical features and laboratory investigation. Med. J. Australia, Aug. 21, 281–299.

Derrick, E. H. 1939 *Rickettsia burneti:* the cause of "Q" fever. Med. J. Australia, Jan. 7, 14.

Derrick, E. H. 1944 The epidemiology of Q fever. J. Hyg., **43**, 357–361.

Derrick, E. H., Johnson, D. W., Smith, D. J. W., and Brown, H. E. 1938 The susceptibility of the dog to Q fever. Australian J. Exptl. Biol. Med. Sci., **16**, 245–248.

Derrick, E. H., Smith, D. J. W., and Brown, H. E. 1942 Studies in the epidemiology of Q fever. IX. The role of the cow in the transmission of human infection. Australian J. Exptl. Biol. Med. Sci., **20**, 105–110.

Dewèvre, L. 1892a Note sur le rôle des pediculi dans la propagation de l'impétigo. Compt. Rend. Soc. Biol., **44**, 232–234. (Quoted from Nuttall, 1899.)

Dewèvre, L. 1892b Note sur la transmissibilité de la tuberculose par la punaise des lits. Rev. Méd., **12**, 291–294. (Quoted from Nuttall, 1899.)

Dias, E. 1937 Sobre a presença de symbiontes em hemipteros hematophagos. Mem. Inst. Oswaldo Cruz, **32**, 165–168.

Dias, E., and Martins, A. V. 1937 Aspectos do typho exanthematico em Minas Geraes. Brasil Medico, **51**, 431–441.

Dickman, A. 1933 Studies on the waxmoth, *Galleria mellonella,* with particular reference to the digestion of wax by the larvae. J. Cell. and Comp. Physiol., **3**, 223–246.

Difco manual of dehydrated culture media and reagents. 1943 7th ed. Difco Laboratories, Inc., Detroit. 239 pp.

Dobell, C. 1910 On some parasitic protozoa from Ceylon. Spolia Zeylanica, **7**, 65–87.

Dobell, C. 1911 On *Cristispira veneris* n. sp., and the affinities and classification of spirochaetes. Quart. J. Microscop. Sci., **56**, 507–540.

Dobell, C. 1912 Researches on the spirochaetes and related organisms. Arch. Protistenk, **26**, 117–240.

Doflein, F. 1911 Probleme der Protistenkunde. II. Die Natur der Spirochäten. G. Fischer, Jena. 36 pp.

Donatien, A., and Gayot, G. 1942a Conjonctivité rickettsiene du porc. Bull. Soc. Path. Exotique séance de novembre, 1942.

Donatien, A., and Gayot, G. 1942b Rickettsiose générale du porc. Bull. Soc. Path. Exotique séance de novembre, 1942.

Donatien, A., and Lestoquard, F. 1935 Existence en Algérie d'une rickettsia du chien. Bull. Soc. Path. Exotique, 28, 418-419.

Donatien, A., and Lestoquard, F. 1936a Recherches sur *Rickettsia canis*. Comparaison avec *Rickettsia conori*. Bull. Soc. Path. Exotique, 29, 1052-1056.

Donatien, A., and Lestoquard, F. 1936b *Rickettsia bovis*, nouvelle espèce pathogène pour le boeuf. Bull. Soc. Path. Exotique, 29, 1057-1061.

Donatien, A., and Lestoquard, F. 1937 Etat actuel des connaissances sur les rickettsioses animales. Arch. Inst. Pasteur Algérie, 15, 142-187.

Donatien, A., and Lestoquard, F. 1938 Les rickettsioses des animaux domestiques. Acta Conv. Tertii Trop. Malariae Morbis, Pars I, 564.

Donisthorpe, H. 1945 Ants as carriers of diseases. Entomol. Monthly Mag., 81, 185.

Donovan, C. 1903 The aetiology of one of the heterogenous fevers in India. Brit. Med. J., 2, 1401.

Donovan, C. 1909 Kala-azar. Ann. Rept. Statist. Gov. Gen. Hosp., Madras, 1908, p. 28.

Dougherty, E. C. 1942 Unpublished manuscript.

Douglas, J. R. 1943 The internal anatomy of *Dermacentor andersoni* Stiles. Univ. Calif. Pub. Entomol., 7, 207-272.

Douglas, J. R., and Wheeler, C. M. 1943 Sylvatic plague studies. II. The fate of *Pasteurella pestis* in the flea. J. Infectious Diseases, 72, 18-30.

Dove, W. E., and Shelmire, B. 1931 Tropical rat mites, *Liponyssus bacoti* Hirst, vectors of endemic typhus. J. Am. Med. Assoc., 97, 1506-1510.

Dowden, P. B. 1934 Recently introduced parasites of three important forest insects. Ann. Entomol. Soc. Am., 27, 599-603.

Dowson, W. J. 1939 On the systematic position and generic names of the gram negative bacterial plant pathogens. Zentr. Bakt. Parasitenk. Infekt., II, 100, 177-193.

Drbohlav, J. J. 1925 I-III. Studies on the relation of insect herpetomonad and crithidial flagellates to leishmanias. Am. J. Hyg., 5, 580, 599, 611.

Drobotjko, V., Martchouk, P., Eisenman, B., and Sirotskaya, S. 1938 An experiment on combating caterpillars by microbiological methods. Mikrobiologichnii Zhurnal, 5, 11-26. (Russian with English summary.)

Dropkin, V. H. 1941 Host specificity relations of termite protozoa. Ecology, 22, 200-202.

Dschunkowsky, E. 1913 Deut. Med. Wochschr., 39, 419.

Duboscq, O., and Grassé, P. 1927 Flagellés et schizophytes de *Calotermes* (*Glyptotermes*) *iridipennis* Frogg. Arch. Zool. Exptl. Gen., 66, 451-496.

Duboscq, O., and Grassé, P. 1934 Notes sur les Protistes parasites des termites de France. 9. L'enkystement des flagellés de *Calotermes flavicollis*. Arch. Zool. Exptl. Gen., **76,** 66–73.

Dudgeon, L. S. 1919 The dysenteries: bacillary and amoebic. Bacillary. Brit. Med. J., **1,** no. 3041, 448–451.

Dufrenoy, J. 1919 Les formes de dégénérescence des chenilles de *Cnethocampa pityocampa* parasitées. Compt. Rend. Soc. Biol., **82,** 288–289.

Duggar, B. M. 1896 On a bacterial disease of the squashbug (*Anasa tristis* DeG.). Bull. Illinois State Lab. Natural History, **4,** 340–379.

Duncan, J. T. 1926 On a bactericidal principle present in the alimentary canal of insects and arachnids. Parasitology, **18,** 238–252.

Du Porte, E. M. 1915 Two bacterial diseases of injurious insect larvae. 7th Ann. Rept. Quebec Soc. Prot. Plants Insects Fung. Diseases (1914–1915), 81–85.

Du Porte, E. M., and Vanderleck, J. 1917 Studies on *Coccobacillus acridiorum* D'Herelle, and on certain intestinal organisms of locusts. Ann. Entomol. Soc. Am., **10,** 47–62.

Durand, P. 1932 Rôle du chien comme réservoir de virus dans la fièvre boutonneuse. Arch. Inst. Pasteur Tunis, **21,** 239–250.

Durand, P., and Conseil, E. 1930 Transmission de la fièvre exanthématique par la tique du chien. Compt. Rend. Acad. Sci., Paris, **190,** 1244–1246.

Durand, P., and Conseil, E. 1931 Transmission expérimentale de la fièvre boutonneuse par *Rhipicephalus sanguineus*. Arch. Inst. Pasteur Tunis, **20,** 54–55.

Durand, P., and Sparrow, H. 1939 Innocuité pour l'homme des rickettsias du type *R. rocha limae*. Arch. Inst. Pasteur Tunis, **28,** 74–81. (Also Bull. Soc. Path. Exotique, **32,** 258–261.)

Dutky, S. R. 1940 Two new spore-forming bacteria causing milky diseases of Japanese beetle larvae. J. Agr. Research, **61,** 57–68.

Dutky, S. R. 1941 Susceptibility of certain scarabaeid larvae to infection by Type A milky disease. J. Econ. Entomol., **34,** 215–216.

Dutton, J. E., and Todd, J. L. 1907 A note on the morphology of *Spirochaeta duttoni*. Lancet, pt. II, 1523–1525.

Dyer, R. E. 1934a Effect of flea passage on epidemic typhus virus. Pub. Health Repts., **49,** 224–225.

Dyer, R. E. 1934b Endemic typhus fever: susceptibility of woodchucks, house mice, meadow mice, and white-footed mice. Pub. Health Repts., **49,** 723–724.

Dyer, R. E. 1938 A filter-passing infectious agent isolated from ticks. IV. Human infection. Pub. Health Repts., **53,** 2277–2282.

Dyer, R. E. 1939 Similarity of Australian "Q" fever and a disease caused by an infectious agent isolated from ticks in Montana. Pub. Health Repts., **54,** 1229–1237.

Dyer, R. E. 1944 The rickettsial diseases. J. Am. Med. Assoc., **124,** 1165–1172.

Dyer, R. E., Rumreich, A., and Badger, L. F. 1931a Typhus fever. A virus of the typhus type derived from fleas collected from wild rats. Pub. Health Repts., **46,** 334–338.

Dyer, R. E., Rumreich, A., and Badger, L. F. 1931b The typhus-Rocky Mountain spotted fever group in the United States. J. Am. Med. Assoc., **97,** 589–595.

Dyer, R. E., Workman, W. G., Ceder, E. T., Badger, L. F., and Rumreich, A. 1932 Typhus fever. The multiplication of the virus of endemic typhus in the rat flea *Xenopsylla cheopis*. Pub. Health Repts., **47,** 987–994.

Dyson, J. E. B., and Lloyd, L. 1933 Remarks on the flies breeding in the bacteria beds at the Knostrop Sewage Works, Leeds. Surveyor, **84,** 335–337.

Eckstein, K. 1894 Untersuchungen über die in Raupen vorkommenden Bakterien. Z. Forst-u. Jagdwesen, **26,** 3–20, 228–241, 285–298, 413–424.

Ekblom, T. 1931 I. Cytological and biochemical researches into the intracellular symbiosis in the intestinal cells of *Rhagium inquisitor* L. Skand. Arch. Physiol., **61,** 35–48. II. *Ibid.*, 1932, **64,** 279–298.

Elford, W. J., and Van den Ende, M. 1944 Studies on the viability and filterability of typhus rickettsiae. Brit. J. Exptl. Path., **25,** 213–220.

Ellinger, T., and Chorine, V. 1930a Note on the bacteria isolated from *Ephestia kühniella* Zell. Intern. Corn Borer Invest., Sci. Repts., **3,** 37–38.

Ellinger, T., and Chorine, V. 1930b Sur les microbes d'*Ephestia kühniella* Zell. Compt. Rend. Soc. Biol., **103,** 401–402.

Elliott, C. 1930 Manual of bacterial plant pathogens. Williams and Wilkins Co., Baltimore. 349 pp.

Elliott, C., and Poos, F. W. 1940 Seasonal development, insect vectors, and host range of bacterial wilt of sweet corn. J. Agr. Research, **60,** 645–686.

Elson, J. A. 1937 A comparative study of Hemiptera. Ann. Entomol. Soc. Am., **30,** 579–597.

Eltringham, H. 1930 Histological and illustrative methods for entomologists. Oxford Univ. Press, London. 139 pp.

Emik, L. O. 1941 Ingestion of food by *Trichonympha*. Trans. Am. Microscop. Soc., **60,** 1–6.

Engelhardt, V. M. 1914 New work on foulbrood. Russian Beekeeping Gaz., nos. 1–6, Jan.–June, 1914, pp. 12–16, 46–49, 84–86, 126–130, 162–165, 195–200.

Enigk, K. 1942 Eine Rickettsieninfektion beim Bison. Berlin u. Münch. Tierärztl. Wochschr. Jan. 23, 25–27. (Abstract in Vet. Bull., **12,** 449.)

Erikson, D. 1935 The pathogenic aerobic organisms of the actinomyces group. Spec. Rept. Ser. Med. Research Council, London, no. 203. 61 pp.

Escherich, K. 1900 Ueber das regelmässige Vorkommen von Sprosspilzen in dem Darmepithel eines Käfers. Biol. Zentr., **20,** 350–358.

Escherich, K. 1913 Neues über Polyederkrankheiten. Naturw. Z. Forst- u. Landw., **11,** 86–97.

Escherich, K., and Miyajima, M. 1911 Studien über die Wipfelkrankheit der Nonne. Naturw. Z. Forst- u. Landw., **9,** 381–402.

Eskey, C. R. 1938 Fleas as vectors of plague. Am. J. Pub. Health, **28,** 1305–1310.

Eskey, C. R., and Haas, V. H. 1940 Plague in the western part of the United States. U.S. Pub. Health Service, Pub. Health Bull., 254. 83 pp.

Evlakhova, A. A. 1939 A new yeast-like fungus *Blastodendrion pseudococci* nov. sp. pathogenic for mealy bugs. Bull. Plant Protection, no. 1 (20), 79-84. (In Russian.)

Ewing, H. E., Jr. 1942 The relation of flies (*Musca domestica* Linnaeus) to the transmission of bovine mastitis. Am. J. Vet. Research, **3**, 295-299.

Eyer, H., and Ruska, H. 1944 Ueber den Feinbau der Fleckfieber-Rickettsiae. Z. Hyg. Infektionskrankh., **125**, 483-492.

Eyre, J. W. H. 1930 Bacteriological technique. Baillière, Tindall and Cox, London. 619 pp.

Faasch, W. J. 1935 Darmkanal und Blutverdauung bei Aphanipteren. Z. Morphol. Ökol. Tiere, **29**, 559-584.

Faddeeva, T. 1932 The role of ticks in the transmission and preservation of plague virus. Rev. Microbiol. Epidemiol. Parasitol., **12**, 279.

Faichnie, N. 1909a Fly-borne enteric fever; the source of infection. J. Roy. Army Med Corps, **13**, 580-584.

Faichnie, N. 1909b *Bacillus typhosus* in flies. J. Roy. Army Med. Corps, **13**, 672.

Fantham, H. B. 1912 Some insect flagellates and the problem of the transmission of *Leishmania*. Brit. Med. J., **2**, 1196-1197.

Fantham, H. B., and Porter, A. 1913 The pathogenicity of *Nosema apis* to insects other than hive bees. Ann. Trop. Med., **7**, 569-579.

Farner, D. S., and Katsampes, C. P. 1944 Tsutsugamushi disease. U.S. Naval Medical Bull., **43**, 800-836.

Faust, E. C. 1944 Tropical diseases. Sommer Memorial Lecture. Metropolitan Press Printing Co., Seattle. 49 pp.

Fawcett, H. S. 1929 *Nematospora* on pomegranates, citrus, and cotton in California. Phytopathology, **19**, 479-482.

Fawcett, H. S. 1944 Fungus and bacterial diseases of insects as factors in biological control. Botan. Rev., **10**, 327-348.

Fejgin, B. 1927 Sur le rapport des variations du *Proteus* X-19 avec les cultures isolées par Kuczynski de cas de typhus exanthématique. Compt. Rend. Soc. Biol., **96**, 339-341.

Ferris, G. F. 1931 The louse of elephants. Parasitology, **23**, 112-127.

Ficker, M. 1903 Typhus und Fliegen. Arch. Hyg., **46**, 274-283.

Fielding, J. W. 1943 Typhus: modified Breinl method for staining rickettsiae and other inclusions. Med. J. Australia, May 15, 435-436.

Fildes, P. 1931 Anaerobic cultivation. *In* A system of bacteriology, vol. IX, chap. 6, pp. 92-99. Medical Research Council, London.

Finlay, C. J. 1881 *et seq.* Trabajos selectos. República de Cuba. Secretaria de Sanidad y Beneficencia, Havana, 1912. xxxiv + 657 pp.

Fitzpatrick, F. K.   1939   Vaccination against spotted fever with agar-tissue cultures. Proc. Soc. Exptl. Biol. Med., **42**, 219–220.

Fitzpatrick, H. M.   1930   The lower fungi—Phycomycetes. McGraw-Hill Book Co., Inc., New York.   331 pp.

Flexner, S., and Clark, P. F.   1911   Contamination of the fly with poliomyelitis virus. J. Am. Med. Assoc., **56**, 1717–1718.

Flögel, J. H. L.   1905   Monographie der Johannisbeerblattlaus (*Aphis ribis* L.). Z. Wiss. Insektenbiol., **1**, 97–106.

Florence, L.   1924   An intracellular symbiont of the hog louse. Am. J. Trop. Med., **4**, 397–410.

Foley, H., and Parrot, L.   1937   Rickettsia du trachome. Arch. Inst. Pasteur Algérie, **15**, 339–351.

Forbes, S. A.   1882   Bacterium a parasite of the chinch bug. Am. Naturalist, **16**, 824–825.

Forbes, S. A.   1886   Studies on the contagious diseases of insects. Bull. Illinois State Lab. Natural History, Art. 4, 257–321.

Forbes, S. A.   1892   Bacteria normal to digestive organs of Hemiptera. Bull. Illinois State Lab. Natural History, **4**, art. 1.

Forbes, S. A.   1895   Experiments with the muscardine disease of the chinch-bug and with the trap and barrier method for the destruction of that insect. Illinois Agr. Expt. Sta. Bull., **38**, 25–86.

Forbes, S. A.   1896   On contagious diseases in the chinch-bug. Illinois Agr. Expt. Sta., 19th Rept. Entomol., 16–176.

Fox, H.   1944   Verruga péruana. *In* Clinical tropical medicine, ed. Z. T. Bercovitz, chap. 30, pp. 371–381. Paul B. Hoeber, Inc., New York.   957 pp.

Fraenkel, G., and Blewett, M.   1943   Intracellular symbionts of insects as a source of vitamins. Nature, **152**, 506.

Fraenkel, H.   1921   Die Symbionten der Blattiden im Fettgewebe und Ei insbesondere von *Periplaneta orientalis*. Z. Wiss. Zool., **119**, 53–66.

França, C.   1919   L'insecte transmetteur de *Leptomonas davidi*. (Note préliminaire.) Bull. Soc. Path. Exotique, **12**, 513–514.

França, C.   1920   La flagellose des euphorbes. Ann. Inst. Pasteur, **34**, 432–464.

Francis, E.   1927   Microscopic changes of tularaemia in the tick *Dermacentor andersoni* and the bedbug, *Cimex lectularius*. Pub. Health Repts., **42**, 2763–2772.

Francis, E.   1938   Longevity of the tick *Ornithodoros turicata* and of *Spirochaeta recurrentis* within this tick. Pub. Health Repts., **53**, 2220–2241.

Francis, E., and Lake, G. C.   1921   Experimental transmission of tularaemia in rabbits by the rabbit louse, *Haemodipsus ventricosus* (Denny). Pub. Health Repts., **36**, 1747–1753.

Francis, E., and Lake, G. C.   1922a   Transmission of tularaemia by the bedbug, *Cimex lectularius*. Pub. Health Repts., **37**, 83–95.

Francis, E., and Lake, G. C.   1922b   Transmission of tularaemia by the mouse louse *Polyplax serratus* (Burm.). Pub. Health Repts., **37**, 96–101.

Francis, E., and Mayne, B. 1921 Experimental transmission of tularaemia by flies of the species *Chrysops discalis*. Pub. Health Repts., **36**, 1738–1746.

Francis, T., and Rendtorff, R. C. 1943 Survival of the Lansing strain of poliomyelitis virus in the common house fly, *Musca domestica* L. J. Infectious Diseases, **73**, 198–205.

Fransen, J. J. 1939 Iepenziekte, Iepenspintkevers en Beider Bestrijding. Dissertation, Landbouwhoogeschool, Wageningen. (Not seen.)

Frazer, H. L. 1944 Observations on the method of transmission of internal boll disease of cotton by the cotton stainer-bug. Ann. Appl. Biol., **31**, 271–290.

Fred, E. B., Baldwin, I. L., and McCoy, E. 1932 Root nodule bacteria and leguminous plants. Univ. Wisconsin Studies in Science, 5, Madison. 343 pp.

Fred, E. B., and Waksman, S. A. 1928 Laboratory manual of general microbiology. McGraw-Hill Book Co., New York. 145 pp.

Fresenius, G. 1856 Notiz. Insekten-Pilze betreffend. Botan. Z., **14**, 882.

Fricks, L. D. 1916 Rocky Mountain spotted fever. A report of laboratory investigations of the virus. Pub. Health Repts., **31**, 516–521.

Frost, F. M., Herms, W. B., and Hoskins, W. M. 1936 The nutritional requirements of the larva of the mosquito, *Theobaldia incidens* (Thom.). J. Exptl. Zool., **73**, 461–479.

Frye, W. W., and Meleney, H. E. 1936 The viability of *Endameba histolytica* cysts after passage through the cockroach. J. Parasitol., **22**, 221–222.

Galli-Valerio, B. 1907 Les insects comme propagateurs des maladies: le rôle des arthropodes dans la dissémination des maladies. Ber. XIV Intern. Kongr. Hyg. Demogr., Berlin, 189–194.

Gambetta, L. 1926 Ricerche sulla simbiosi ereditaria di alcuni coleotteri silofagi. Ricerche Morfol. Biol. Anim. Napoli, **1**, 105–119.

Gäumann, E., and Dodge, C. W. 1928 Comparative morphology of fungi. McGraw-Hill Book Co., Inc., New York. 701 pp.

Gay, D. M., and Sellards, A. W. 1927 The fate of *Leptospira icteroides* and *Leptospira icterohaemorrhagiae* in the mosquito, *Aëdes aegypti*. Ann. Trop. Med., **21**, 321–342.

Gear, J., and De Meillon, B. 1941 The hereditary transmission of the rickettsiae of tick-bite fever through the common dog-tick, *Haemaphysalis leachi*. S. African Med. Jr., **15**, 389–392.

Geiman, Q. M. 1932 The intestinal protozoa of the larvae of the crane fly, *Tipula abdominalis*. J. Parasitol., **19**, 173.

Gier, H. T. 1936 The morphology and behavior of the intracellular bacteroids of roaches. Biol. Bull., **71**, 433–452.

Gier, H. T. 1937 Growth of the intracellular symbionts of the cockroach, *Periplaneta americana*. Anat. Record, **70**, 69.

Gieszczykiewicz, M. 1939 Zur Frage der Bakterien-Systematik. Bull. Intern. Acad. Polon. Sci., Classe Sci. Math. Nat., B I, 9–30.

Gildemeister, E., Haagen, E., and Waldmann, O. 1939 Handbuch der Viruskrankheiten. G. Fischer, Jena. 2 vols., 652 and 768 pp.

Gildemeister, E., and Peter, H. 1943 Fleckfieberstudien. III. Ueber das Vorkommen und den Nachweisder *Rickettsia prowazeki* im bebruteten und infizierten Huhnerei. Zentr. Bakt. Parasitenk. Infekt., Orig., **149**, 425–428.

Gillette, G. P. 1896 The grasshopper disease in Colorado. U.S. Bur. Entomol., Bull., **6**, 89–93.

Girard, G. 1935 Présence d'un bactériophage antipesteux chez la *Xenopsylla cheopis* au cours d'une petite épidémie de peste à Tananarive. Compt. Rend. Soc. Biol., **120**, 333–334.

Giroud, P., and Panthier, R. 1942 L'évolution des rickettsies des fièvres exanthématiques et fonction de leur végétabilité dans les tissus qu'elles parasitent. Bull. Soc. Path. Exotique, **35**, 6–8.

Gjullin, C. M., Hegarty, C. P., and Bollen, W. B. 1941 The necessity of low oxygen concentration for the hatching of *Aëdes* mosquito eggs. J. Cellular Comp. Physiol., **17**, 193–202.

Gjullin, C. M., Yates, W. W., and Stage, H. H. 1939 The effect of certain chemicals on the hatching of mosquito eggs. Science, **89**, 539–540.

Glaser, R. W. 1918a A systematic study of the organisms distributed under the name of *Coccobacillus acridiorum* D'Herelle. Ann. Entomol. Soc. Am., **11**, 19–42.

Glaser, R. W. 1918b On the existence of immunity principles in insects. Psyche, **25**, 39–46.

Glaser, R. W. 1918c A new disease of gypsy moth caterpillars. J. Agr. Research, **13**, 515–522.

Glaser, R. W. 1920 Biological studies on intracellular bacteria. Biol. Bull. Marine Biol. Lab., **39**, 133–145.

Glaser, R. W. 1922 *Herpetomonas muscae domesticae,* its behavior and effect in laboratory animals. J. Parasitol., **8**, 99.

Glaser, R. W. 1924a A bacterial disease of silkworms. J. Bact., **9**, 339–352.

Glaser, R. W. 1924b The relation of microorganisms to the development and longevity of flies. Am. J. Trop. Med., **4**, 85–107.

Glaser, R. W. 1924c A bacterial disease of adult house flies. Am. J. Hyg., **4**, 411–415.

Glaser, R. W. 1925a Acquired immunity in silkworms. J. Immunol., **10**, 651–662.

Glaser, R. W. 1925b Specificity in bacterial diseases with special reference to silkworms and tent caterpillars. J. Econ. Entomol., **18**, 769–771.

Glaser, R. W. 1926 The green muscardine disease in silkworms and its control. Ann. Entomol. Soc. Am., 19, 180–192.

Glaser, R. W. 1928 Virus diseases of insects. *In* Rivers, T. M., Filterable viruses, chap. 8, pp. 301–333. Williams and Wilkins Co., Baltimore. (See p. 308.)

Glaser, R. W. 1930a On the isolation, cultivation and classification of the so-called intracellular "symbiont" or "rickettsia" of *Periplaneta americana*. J. Exptl. Med., **51**, 59–82.

Glaser, R. W. 1930b Cultivation and classification of "bacteroids," "symbionts," or "rickettsiae" of *Blattella germanica*. J. Exptl. Med., **51**, 903–907.

Glaser, R. W. 1930c The intracellular "symbionts" and the "rickettsiae." Arch. Path., **9**, 79–96 and 557–576.

Glaser, R. W. 1932 Studies on *Neoaplectana glaseri,* a nematode parasite of the Japanese beetle (*Popillia japonica*). N.J. Dept. Agr., Bur. Plant Ind. Circ., 211. 34 pp.

Glaser, R. W. 1938a A method for the sterile culture of houseflies. J. Parasitol., **24**, 177–179.

Glaser, R. W. 1938b Test of a theory on the origin of bacteriophage. Am. J. Hyg., **27**, 311–315.

Glaser, R. W., and Chapman, J. W. 1912 Studies on the wilt disease or "flacherie" of the gypsy moth. Science, **36**, 219–224.

Glaser, R. W., and Chapman, J. W. 1913 The wilt disease of gypsy moth caterpillars. J. Econ. Entomol., **6**, 479–488.

Glaser, R. W., and Chapman, J. W. 1916 The nature of the polyhedral bodies found in insects. Biol. Bull., **30**, 367–391.

Glaser, R. W., and Cowdry, E. V. 1928 Experiments on the visibility of the polyhedral viruses. J. Exptl. Med., **47**, 829–834.

Glaser, R. W., and Farrell, C. C. 1935 Field experiments with the Japanese beetle and its nematode parasite. J. N.Y. Entomol. Soc., **43**, 345–371.

Glaser, R. W., and Lacaillade, C. W. 1934 Relation of the virus and the inclusion bodies of silkworm "jaundice." Am. J. Hyg., **20**, 454–464.

Glaser, R. W., and Stanley, W. M. 1943 Biochemical studies on the virus and the inclusion bodies of silkworm jaundice. J. Exptl. Med., **77**, 451–466.

Glasgow, H. 1914 The gastric caeca and the caecal bacteria of the Heteroptera. Biol. Bull., **26**, 101–170.

Godoy, A., and Pinto, C. 1922 Da presenca dos symbiotes nos Ixodidas. Brasil Medico, **2**, 335.

Goetsch, W. 1936 Beiträge zur Biologie des Termitenstaates. Z. Morphol. Ökol. Tiere, **31**, 490–560.

Gohar, M. A. 1942 Bacterial flora in some common insects. Lab. Med. Prog., **3**, 29–31.

Goidanich, G. 1936 Il genera di ascomiceti "Grosmannia" G. Boid. Boll. Staz. Patol. Vegetale, N.S., **16**, 26–60.

Goldsworthy, M. C. 1926 Studies on the spot disease of cauliflower; a use of serum diagnosis. Phytopathology, **16**, 877–884.

Goot, P. van der. 1915 Over eenige Engerlingensoorten, die in Riettuinen voorkomen. Med. Proefstation voor de Java Suikerindustrie, pt. 5, no. 10, 275–316.

Gordan, H. M. 1933 On the presence of a ricksettsia-like organism in the lymphocytes of sheep. Australian J. Exptl. Biol. Med. Sci., **11**, 95–97.

Goss, R. W. 1931 Infection experiments with spindle tuber and unmottled curly dwarf of the potato. Neb. Agr. Expt. Sta. Research Bull., 53.

Graber, V. 1872 Anatomisch-physiologische Studien über *Phthirius inguinalis,* Leach. Z. Wiss. Zool., **22,** 137–167.

Gracian, M. 1942 Ein einfaches Verfahren zur Färbung der Rickettsien. Z. Hyg. Infektionskrankh., **124,** 81–82.

Gradwohl, R. B. H. 1938 Clinical laboratory methods and diagnosis. 2d ed. C. V. Mosby Co., St. Louis. 1607 pp.

Graham, H. 1902 The dengue: a study of its pathology and mode of propagation. Med. Record, **61,** 204–207.

Graham-Smith, G. S. 1909 Preliminary note on examinations of flies for the presence of colon bacilli. Gt. Brit. Local Govt. Bd., Repts. Pub. Health Med. Subjs., N.S., 16, 9–13.

Graham-Smith, G. S. 1910 Observations on the ways in which artificially infected flies (*Musca domestica* and *Calliphora erythrocephala*) carry and distribute pathogenic and other bacteria. Gt. Brit. Local Govt. Bd., Repts. Pub. Health Med. Subjs., 40; 1–40.

Graham-Smith, G. S. 1911 Further observations on the ways in which artificially infected flies (*Musca domestica* and *Calliphora erythrocephala*) carry and distribute pathogenic and other bacteria. Gt. Brit. Local Govt. Bd., Repts. Pub. Health Med. Subjs., 53, 31–48.

Graham-Smith, G. S. 1912 Forty-first Ann. Rept., Local Govt. Bd. (1911–12), Suppl. Rept. Med. Office, 304–329, 330–335.

Graham-Smith, G. S. 1913 Flies in relation to disease. Cambridge Univ. Press, London. 292 pp.

Granovsky, A. A. 1929 Preliminary studies of the intracellular symbionts of *Saissetia oleae* (Bernard). Trans. Wisconsin Acad. Sci., **24,** 445–456.

Grassi, B. 1879 Dei protozoi parassiti e specialmente di quelli che sono nell'uomo. Gaz. Med. Ital. Lombard., **39,** 445–448.

Grassi, B. 1882 Intorno ad alcuni protista endoparassitici ed appartenenti alle classi dei flagellati, Lobosi, sporozoi e ciliati. Memoria di parassitologia comparata. Atti Soc. Ital. Sci. Nat., **24,** 135–224.

Grassi, B. 1883 Les méfaits des mouches. Arch. Ital. Biol., **4,** 205–208.

Grassi, B. 1898 La malaria propagata per mezzo di peculiari insetti. II. Nota preliminare. Atti Accad. Lincei, Classe Sci. Fis. Mat. Nat., **7,** 163–172.

Green, R. G. 1931 The occurrence of *Bact. tularense* in the eastern wood tick, *Dermacentor variabilis.* Am. J. Hyg., **14,** 600–613.

Green, R. G., and Evans, C. A. 1938 Role of fleas in the natural transmission of tularaemia. Minn. Wildlife Dis., Invest., April, 1938, pp. 25–28.

Gregson, J. D. 1938 Cytoplasmic inclusion bodies in the engorging tick. J. Path. Bact., **47,** 143–153.

Gregson, J. D. 1941 The discovery of an ixovotoxin in *Dermacentor andersoni* eggs. Proc. Entomol. Soc. British Columbia, Feb. 22. 2 pp.

Greiff, D., Pinkerton, H., and Moragues, V. 1944 Effect of enzyme inhibitors and activators on the multiplication of typhus rickettsiae. J. Exptl. Med., **80**, 561–574.

Griffitts, S. D. 1942 Ants as probable agents in the spread of *Shigella* infections. Science, **96**, 271–272.

Gropengiesser, C. 1925 Untersuchungen über die Symbiose der Blattiden mit neideren pflanzlichen Organismen. Zentr. Bakt. Parasitenk. Infekt., II, **64**, 495–511.

Grundmann, A. W., Kitselman, C. M., Roderick, L. M., and Smith, R. C. 1943 Studies on the transmission of the western strain virus of equine encephalomyelitis by the American dog tick, *Dermacentor variabilis* Say and by *Triatoma sanguisuga* (Leconte). J. Infectious Diseases, **72**, 163–171.

Guilliermond, A. 1920 The yeasts. Tr. F. W. Tanner. John Wiley and Sons, Inc., New York. 424 pp.

Guilliermond, A. 1940 Sexuality, developmental cycles and phylogeny of the yeasts. Botan. Rev., **6**, 1–24.

Guimarais, A. 1922 Flore microbienne du *Phthirius inguinalis;* remarque sur des éléments de nature rickettsienne. Compt. Rend. Soc. Biol., **87**, 711–713.

Gunderson, M. F. 1935 Insects as carriers of *Clostridium botulinum*. J. Bact., **30**, 333.

Guyénot, E. 1906 Sur le mode de nutrition de quelques larves de mouches. Compt. Rend. Soc. Biol., **61**, 634–635.

Guyénot, E. 1907 L'appareil digestive de quelques larves des mouches. Bull. Sci. France Belg., **41**, 353–370.

Guyénot, E. 1913 Etudes biologiques sur une mouche, *Drosophila ampelophila* Löw. Compt. Rend. Soc. Biol., **74**, 97–99, 178–180, 223–227, 270–272, 332–334, 389–391, 443–445.

Guyénot, E. 1917 Recherches expérimentales sur la vie aseptique et la développement d'un organisme (*Drosophila ampelophila*) en fonction du milieu. Bull. Biol. France Belg., **51**, 1–330.

Guyer, M. F. 1936 Animal micrology. 4th ed. Univ. of Chicago Press. 331 pp.

Gwynne-Vaughan, H. C. I., and Barnes, B. 1937 The structure and development of the fungi. Cambridge Univ. Press, London. 449 pp.

Hagan, W. A. 1943 The infectious diseases of domestic animals. Comstock Publishing Co., Ithaca, N.Y. 665 pp.

Hagen, H. A. 1879 Destruction of obnoxious insects by application of the yeast fungus. Cambridge Univ. Press, Cambridge. 11 pp.

Hamilton, A. 1903 The fly as a carrier of typhoid; an inquiry into the part played by the common house fly in the recent epidemic of typhoid fever in Chicago. J. Am. Med. Assoc., **40**, 576–583.

Hamilton, H. L. 1945 Effect of p-aminobenzoic acid on growth of rickettsiae and elementary bodies, with observations on mode of action. Proc. Soc. Exptl. Biol. Med., **59**, 220–226.

Hammon, W. McD., and Reeves, W. C. 1943 Laboratory transmission of

St. Louis encephalitis virus by three genera of mosquitoes. J. Exptl. Med., **78**, 241–253.

Hammon, W. McD., Reeves, W. C., and Gray, M. 1943 Mosquito vectors and inapparent animal reservoirs of St. Louis and western equine encephalitis viruses. Am. J. Pub. Health, **33**, 201–207.

Hankin, E. H. 1897 Note on the relation of insects and rats to the spread of plague. Zentr. Bakt. Parasitenk. Infekt., I, **22**, 437–438.

Hansen, G. A. 1872 [Leprosy bacillus.] Norsk. Mag. Laegevidensk., Ser. 3, **2**, 1.

Hansen, H. N., and Smith, R. E. 1937 A bacterial gall disease of Douglas fir, *Pseudotsuga taxifolia*. Hilgardia, **10**, 569–577.

Harding, H. A., and Morse, W. J. 1909 The bacterial soft rots of certain vegetables. I. The mutual relationships of the causal organisms. N.Y. (Geneva) Agr. Expt. Sta. Tech. Bull., **11**, 251–287.

Hargett, M. V. 1944 The control of yellow fever. (Seen in manuscript.)

Harrar, J. G., and McKelvey, J. J. 1942 Biological control of the mealybug (*Pseudococcus* spp.). Phytopathology, **32**, 7.

Hartig, T. 1844 Ambrosia des *Bostrychus dispar*. Allgem. Forst- u. Jadg-Z., **13**, 73–75.

Hartman, E. 1931 A flacherie disease of silkworms caused by *Bacillus bombysepticus* n. sp. Lingnan Sci. J., **10**, 279–281.

Hartzell, A. 1937 Movement of intracellular bodies associated with peach yellows. Contrib. Boyce Thompson Inst., **8**, 375–388.

Hass, G. M., and Pinkerton, H. 1936 Spotted fever. II. An experimental study of *fièvre boutonneuse*. J. Exptl. Med., **64**, 601–623.

Hassler, F. R., Sizemore, P., and Robinson, R. A. Manuscript.

Hatcher, E. 1939 The consortes of certain North Carolina blattids. J. Elisha Mitchell Sci. Soc., **55**, 329–334.

Haudroy, P., and Ehringer, G. 1937 Dictionnaire des bactéries pathogènes. Bactériologie humaine. Masson et Cie., Paris. 597 pp.

Hayashi, D., and Sako, W. 1913 Recherches sur la grasserie des vers à soie. Moniteur des Soies, Lyons. 40 pp.

Hayashi, N. 1920 Etiology of tsutsugamushi disease. J. Parasitol., **7**, 53–68.

Heaslip, W. G. 1940 An investigation of the condition known as coastal fever in north Queensland: its separation from scrub-typhus. Med. J. Australia, pt. 2, 555–564.

Heaslip, W. G. 1941 Tsutsugamushi fever in north Queensland, Australia. Med. J. Australia, pt. 1, 380–392.

Hecht, O. 1924 Embryonalentwicklung und Symbiose bei *Camponotus ligniperda*. Z. Wiss. Zool., **122**, 173–204.

Hecht, O. 1928 Ueber die Sprosspilze der Oesophagusausstülpungen und über die Giftwirkung der Splicheldrüsen von Stechmücken. I. Vorläufige Mitteilung. Arch. Schiffs-u. Tropen-Hyg., **32**, 561–575.

Hecht, O.   1943   La simbiosis intracellular de insectos con bacterias y hongos y su significacion fisiologica en el suministro de vitaminas. Bol. Soc. Venezolana Cienc. Nat., **8**, 161–179.

Hegler, C., and Prowazek, S. von.   1913   Untersuchungen über Fleckfieber (vorläufiger Bericht). Berl. Klin. Wochschr., **50**, 2035–2040.

Hegner, R. W.   1915   Studies in germ-cells.   4. Protoplasma—differentiation in the oöcytes of certain Hymenoptera. J. Morphol., **26**, 495–536.

Hegner, R. W.   1938   Big fleas have little fleas or who's who among the protozoa. Williams and Wilkins Co., Baltimore.   285 pp.

Hegner, R. W., Root, F. M., Augustine, D. L., and Huff, C. G.   1938   Parasitology, with special reference to man and domesticated animals. D. Appleton-Century Co., New York.   812 pp.

Heim, F.   1894   Du rôle de quelques coléoptères dans la dissémination de certain cas de charbon. Compt. Rend. Soc. Biol., 58–61.   (Quoted from Nuttall, 1899.)

Heitz, E.   1927   Über intrazelluläre Symbiose bei holzfressenden Käferlarven. I. Z. Morphol. Ökol. Tiere (Abt. A. Z. Wiss. Biol.), **7**, 279–305.

Hendee, E. C.   1933   The association of termites with fungi. Science, **77**, 212–213.

Hendee, E. C.   1935   The role of fungi in the diet of the common damp-wood termite, *Zootermopsis angusticollis*. Hilgardia, **9**, 499–525.

Henderson, J. C.   1941   Studies of some amoebae from a termite of the genus *Cubitermes*. Univ. Calif. Pub. Zool., **43**, 357–378.

Henneguy, L. F.   1904   Les insectes (morphologie; réproduction; embryogénie). Leçons recuillies par A. Lecaillon et C. Poirault. Masson et Cie., Paris.   804 pp.

Henning, M. W.   1932   Heartwater. *In* Animal diseases in South Africa, vol. II, chap. 29, pp. 539–560. Cent. News Agency Ltd., So. Africa.

Henrici, A. T.   1930   Molds, yeasts and actinomycetes. John Wiley and Sons, Inc., New York.   296 pp.

Henrici, A. T.   1939   The biology of bacteria.   2d ed.   D. C. Heath and Co., New York.   494 pp.

Henrici, A. T.   1941   The yeasts, genetics, cytology, variation, classification, and identification. Bact. Revs., **5**, 97–179.

Herelle, F. d'.   1911   Sur une épizootie de nature bactérienne sévissant sur les sauterelles au Mexique. Compt. Rend. Acad. Sci., Paris, **152**, 1413–1415.

Herelle, F. d'.   1915   Sur le procédé biologique de destruction des sauterelles. Compt. Rend. Acad. Sci., Paris, **161**, 503–505.

Hering, M.   1926   Die Oekologie der blattminierenden Insektenlarven. Gebrüder Borntraeger, Berlin. Chap. 10, pp. 100–144.

Herms, W. B.   1939   Medical entomology.   3d ed.   The Macmillan Co., New York.   582 pp.

Hertig, M.   1921   Attempts to cultivate the bacteroides of the Blattidae. Biol. Bull., **41**, 181–187.

Hertig, M. 1923 The normal and pathological histology of the ventriculus of the honey-bee, with special reference to infection with *Nosema apis*. J. Parasitol., **9**, 109–140.

Hertig, M. 1936 The rickettsia, *Wolbachia pipientis* (gen. et sp. n.) and associated inclusions of the mosquito, *Culex pipiens*. Parasitology, **28**, 453–486.

Hertig, M. 1942 Phlebotomus and Carrión's disease. Am. J. Trop. Med., **22**, Suppl., 1–81.

Hertig, M., and Wolbach, S. B. 1924 Studies on rickettsia-like microorganisms in insects. J. Med. Research, **44**, 329–374.

Hesselbrock, W., and Foshay, L. 1945 The morphology of *Bacterium tularense*. J. Bact., **49**, 209–236.

Hewitt, C. G. 1914 The house-fly. Cambridge Univ. Press, London. 382 pp.

Heymons, R. 1892 Die Entwicklung der weiblichen Geschlechtsorgane von *Phyllodromia (Blatta) germanica* L. Z. Wiss. Zool., **53**, 434–536.

Heymons, R. 1895 Die Embryonalentwicklung von Dermapteren und Orthopteren unter besonderer Berücksichtigung der Keimblätterbildung. G. Fischer, Jena. 136 pp.

Hill, S. B., Jr., Yothers, W. W., and Miller, R. L. 1934 Effect of arsenical and copper insecticides on the natural control of white flies and scale insects by fungi on orange trees in Florida. Florida Entomologist, **18**, 1–4.

Hindle, E. 1911 The transmission of *Spirochaeta duttoni*. Parasitology, **4**, 133–149.

Hindle, E. 1912 The inheritance of spirochaetal infection in *Argas persicus*. Proc. Cambridge Phil. Soc., **16**, 457.

Hindle, E. 1921 Notes on *Rickettsia*. Parasitology, **13**, 152–159.

Hindle, E. 1931 A system of bacteriology. Medical Research Council, London. Vol. VIII, chaps. 4, 5, 6, 9.

Hindle, E., and Duncan, J. T. 1925 The viability of bacteria in *Argas persicus*. Parasitology, **17**, 434–446.

Hingston, R. W. G. 1929 Instinct and intelligence. The Macmillan Co., New York. 296 pp.

Hinman, E. H. 1930 A study of the food of mosquito larvae. (*Culicidae*). Am. J. Hyg., **12**, 238–270.

Hinman, E. H. 1932 The presence of bacteria within the eggs of mosquitoes. Science, **76**, 106–107.

Hinman, E. H. 1933 The role of bacteria in the nutrition of mosquito larvae. The growth-stimulating factor. Am. J. Hyg., **18**, 224–236.

Hirst, S. 1915 On the tsutsugamushi (*Microtrombidium akamushi* Brumpt). A carrier of Japanese river fever. J. Econ. Biol., **10**, 79–82.

Hitz-Green, S. 1938 Cultivo de la *Rickettsia prowazeki* in vitro. Anales Esc. Nac. Cienc. Biol., **1**, 7–28.

Hoare, C. A. 1923 An experimental study of the sheep-trypanosome (*T. melo-*

phagium Flu, 1908), and its transmission by the sheep-ked (*Melophagus ovinus* L.). Parasitology, **15**, 365–424.

Hoare, C. A. 1924 A note on the specific name of the herpetomonad of the house fly. Trans. Roy. Soc. Trop. Med. Hyg., **17**, 403–406.

Hoare, C. A. 1931 Studies on *Trypanosoma grayi*. III. Life-cycle in the tsetse-fly and in the crocodile. Parasitology, **23**, 449–484.

Hobson, R. P. 1932 Studies on the nutrition of blow-fly larvae. II. Role of the intestinal flora in digestion. J. Exptl. Biol., **9**, 128–138.

Hobson, R. P. 1933 Growth of blow-fly larvae on blood serum. I. Response of aseptic larvae to vitamin B. Biochem. J., **27**, 1899–1909.

Hobson, R. P. 1935 Growth of blow-fly larvae on blood and serum. II. Growth in association with bacteria. Biochem. J., **29**, 1286–1291.

Hoelling, B. A. 1910 Die Kernverhältnisse von *Fusiformis termitidis*. Arch. Protistenk., **19**, 239.

Hofmann, E. 1888 Ueber die Verbreitung der Tuberculose durch Stubenfliegen. Correspondenzbl. d. arztl. Kreis-u. Bezirksvereine Königr. Sachsen, **44**, 130–133.

Hofmann, O. 1891 Die Schlaffsucht (Flacherie) der Nonne (*Liparis monacha*) nebst einem Anhang. Insektentötende Pilze mit besonderlr. Berucksichtigung der Nonne. P. Weher, Frankfurt a. M. 31 pp.

Höhnel, F. von. 1918 Mykologische Fragmente. Ann. Mycol., **16**, 35–174.

Höhnel, F. von, and Litschauer, V. 1907 Beiträge zur Kenntnis der Corticieen. (II Mitteil.) Sitzber. Akad. Wiss. Wien. Math.—Naturw. Klasse, **116**, 739–852.

Hollande, A. C. 1911 Etude histologique comparée du sang des insectes à hémorrhée et des insectes sans hémorrhée. Arch. Zool. Exptl. Gén., **2**, 271–294.

Hollande, A. C., and Vernier, P. 1920 *Coccobacillus insectorum*, n. sp., variété *malacosomae*, bacille pathogène, du sang de la chenille *Malacosoma castrensis*, L. Compt. Rend. Hebdom. Acad. Sci., **171**, 206–208.

Holloway, J. K., and Young, T. R. 1943 The influence of fungicidal sprays on entomogenous fungi and on the purple scale in Florida. J. Econ. Entomol., **36**, 453–457.

Holmgren, N. 1909 Zur Frage der Inzucht bei Termiten. Biol. Zentr., **29**, 125–128.

Holst, E. C. 1936 *Zygosaccharomyces pini*, a new species of yeast associated with bark beetles in pines. J. Agr. Research, **53**, 513–518.

Honeij, J. A., and Parker, R. R. 1914 Leprosy: flies in relation to the transmission of the disease. J. Med. Research, **30**, 127–130.

Hornibrook, J. W., and Nelson, K. R. 1940 An institutional outbreak of pneumonitis. I. Epidemiological and clinical studies. Pub. Health Repts., **55**, 1936–1944.

Howard, C. W., and Clark, P. F. 1912 Experiments on insect transmission of the virus of poliomyelitis. J. Exptl. Med., **16**, 850–859.

Howard, L. O. 1911 The house fly; disease carrier. 2d ed. Frederick A. Stokes Co., New York. 312 pp.

Howard, L. O., Dyar, H. G., and Knab, F. 1912 The mosquitoes of North and Central America and the West Indies. I. Carnegie Inst., Washington, D.C. 520 pp.

Howard, W. R. 1900 New York bee disease, or black brood. Gleanings in Bee Culture, 28, 121–127.

Howland, L. J. 1930 The nutrition of mosquito larvae, with special reference to their algal food. Bull. Entomol. Research, 21, 431–440.

Huang, C. H., Chang, H. C., and Lieu, V. T. 1937 Salmonella infection. A study of 17 cases of *S. enteritidis* septicemia. Chinese Med. J., 52, 345–366.

Hubbard, H. G. 1897 The ambrosia beetles of the United States. U.S. Dept. Agr., Div. Entomol. Bull., 7, 9–30.

Hucker, G. J. 1932 Studies on the Coccaceae. XVII. Agglutination as a means of differentiating the species of *Streptococcus* and *Leuconastoc*. N.Y. (Geneva) Agr. Expt. Sta. Tech. Bull., 190, 17.

Hudson, N. P. 1940 A macroscopic agglutination test with typhus rickettsiae prepared from infected rodent lungs. J. Infectious Diseases, 67, 227–231.

Huff, C. G. 1929 The effects of selection upon susceptibility to bird malaria in *Culex pipiens* Linn. Ann. Trop. Med., 23, 427–442.

Huff, C. G. 1931 The inheritance of natural immunity to *Plasmodium cathemerium* in two species of *Culex*. J. Prevent. Med., 5, 249–259.

Huff, C. G. 1940 Immunity in invertebrates. Physiol. Rev., 20, 68–88.

Huff, C. G. 1941 Factors influencing infection of *Anopheles* with malarial parasites. Am. Assoc. Advancement Sci., Pub. 15, 108–112.

Humphreys, F. A., and Gibbons, R. J. 1942 Some observations on corynebacterial infections, with special reference to their occurrence in mule deer (*Odocoileus hemionus*) in British Columbia. Can. J. Comp. Med. Vet. Sci., 6, 35-45.

Humphreys, F. A., Helmer, D. E., and Gibbons, R. J. 1944 Studies on a virus disease originating in guinea pig injected with ticks (*Dermacentor andersoni* Stiles). J. Infectious Diseases, 74, 109–120.

Hurst, A. 1942 The cause and prognosis of trench fever. Brit. Med. J., 2, 318–320.

Husz, B. 1927 *Bacillus thuringiensis* Berl., a bacterium pathogenic to corn borer larvae. Intern. Corn Borer Invest., Sci. Repts., 1, 191–193.

Husz, B. 1929 The use of *Bacillus thuringiensis* in the fight against the corn borer. Intern. Corn Borer Invest., Sci. Repts., 2, 99–110.

Husz, B. 1930 Field experiments on the application of *Bacillus thuringiensis* against the corn borer. Intern. Corn Borer Invest., Sci. Repts., 3, 91–98.

Huxley, T. H. 1858 On the agamic reproduction and morphology of aphids. Trans. Linn. Soc. London, 22, 193–236.

Imms, A. D. 1919 On the structure and biology of *Archotermopsis,* together with descriptions of new species of intestinal Protozoa and general observations on the Isoptera. Phil. Trans. Roy. Soc., B, **209,** 75–180.

Imms, A. D. 1929 Some methods of technique applicable to entomology. Bull. Entomol. Research, **20,** 165–171.

Imms, A. D. 1937 Recent advances in entomology. P. Blakiston's Son and Co., Inc., Philadelphia. 431 pp.

Irons, J. V., Bohls, S. W., Thurman, D. C., and McGregor, T. 1944 Probable role of the cat flea, *Ctenocephalides felis,* in transmission of murine typhus. Am. J. Trop. Med., **24,** 359–362.

Ishimori, N. 1924 Sur l'immunisation des chenilles. Compt. Rend. Soc. Biol., **90,** 843–845.

Ishimori, N., and Metalnikov, S. 1924 Immunisation de la chenille de *Galleria mellonella* par des substances non spécifiques. Compt. Rend. Acad. Sci., Paris, **178,** 2136–2138.

Iyengar, M. O. T. 1935 Two new fungi of the genus *Coelomomyces* parasitic in larvae of *Anopheles.* Parasitology, **27,** 440–449.

Jackson, D. D. 1907 Report to committee on pollution of the Merchant's Association of New York. (Cited by Graham-Smith, 1913.)

Jacobi, J. 1942 Neue Beobachtungen ueber Fünftagefieber mit besonderer Berücksichtigung der Differentialdiagnose und Therapie. Münch. Med. Wochschr., **89,** 615–618.

Jacobson, H. P. 1932 Fungous diseases. Charles C. Thomas, Springfield, Ill. 315 pp.

Jacot, A. P. 1934 Acarina as possible vectors of the Dutch elm disease. J. Econ. Entomol., **27,** 858–859.

Jacot, A. P. 1936 Three possible mite vectors of the Dutch elm disease. Ann. Entomol. Soc. Am., **29,** 627–635.

Jaffé, J. 1907 *Spirochaeta culicis* n. sp. Arch. Protistenk., **9,** 100.

Janda, V., and Jírovec, O. 1937 Ueber künstlich hervorgerufenen Parasitismus eines freilebenden Ciliaten *Glaucoma piriformis* und Infektionsversuche mit *Euglena gracilis* und *Spirochaeta biflexa.* Mém. Soc. Zool. Techéc. Prague, **5,** 34–58.

Jaschke, W. 1933 Beiträge zur Kenntnis der symbiotischen Einrichtungen bei Hirudineen und Ixodiden. Z. Parasitenk., **5,** 515–541.

Javelly, E. 1914 Les corps bactéroides de la blatte (*Periplaneta orientalis*) n'ont pas encore été cultivées. Compt. Rend. Soc. Biol., **77,** 413–414.

Jellison, W. L. 1934 Rocky Mountain spotted fever. The susceptibility of mice. Pub. Health Repts., **49,** 363–367.

Jellison, W. L. 1945 The geographical distribution of Rocky Mountain spotted fever and Nuttall's cottontail in the western United States. Pub. Health Repts., **60,** 958–961.

Jellison, W. L., and Philip, C. B. 1933 Technique for routine and experimental feeding of certain ixodid ticks on guinea pigs and rabbits. Pub. Health Repts., **48,** 1081–1082.

Jepson, W. F. 1937 Observations on the morphology and bionomies of *Serica brunnea* L. with notes on allied chafer pests. Part I. The morphology of the larva of *Serica brunnea* L. Bull. Entomol. Research, **28**, 149–165.

Jersin, A. (See Yersin.)

Jírovec, O. 1930 Notizen über parasitische Protisten I. Zentr. Bakt. Parasitenk. Infekt., I Orig., **118**, 77–80.

Johannsen, O. A., and Butt, F. H. 1941 Embryology of insects and myriapods. McGraw-Hill Book Co., New York. 462 pp.

Johnson, D. E. 1930 The relation of the cabbage maggot and other insects to the spread and development of soft rot of Cruciferae. Phytopathology, **20**, 857–872.

Johnson, E. M. 1934 Dissemination of angular leaf spot of tobacco by the southern tobacco worm. Phytopathology, **24**, 1381–1382.

Johnston, T. H., and Hitchcock, L. 1923 A bacteriosis of prickly pear plants (*Opuntia* spp.). Trans. Proc. Roy. Soc. S. Australia, **47**, 162–164.

Joly, P. R. 1898 Importancia de papal de los insectos en la transmission de las enfermedades infeciosas y parasitarias. Tr. L. Gaitan. Bol. Consejo Sup. Solubridad., 329–337.

Joly, P. R. 1898 Importance du rôle des insectes dans la transmission des malades infectieuses et parasitaires. Gaz. Hôpit., 1202–1204.

Jones, H. N. 1910 Further studies on the nature of the wilt disease of the gypsy moth larvae. *In* 7th Ann. Rep. State Forester, Mass., Public Doc., **73**, 101–105.

Jones, L. R. 1901 *Bacillus carotovorus* n. sp. die Ursache einer weichen Faulnis der Mohre. Zentr. Bakt. Parasitenk. Infekt., II, **1**, 12–21, 61–68.

Jordan, E. O., and Burrows, W. 1941 Textbook of bacteriology. 13th ed. W. B. Saunders Co., Philadelphia. 731 pp.

Jungmann, P. 1917 Klinik und Aetiologie des wolhynischen Fiebers. Berlin. Klin. Wochschr., **54**, 147–149.

Jungmann, P. 1918 Untersuchungen über Schaflausrickettsien (*Rickettsia melophagi* Nöller). Deut. Med. Wochschr., **44**, 1346–1348.

Jungmann, P. 1919 Das wolhynische Fieber. Julius Springer, Berlin. 126 pp.

Kamil, S., and Bilal, S. 1938a Recherches expérimentales sur l'étiologie de la tularémie en Turquie. Ann. Parasit. Humaine et Comparée, **16**, 530–542.

Kamil (Tokgöz), S., Bilal (Golem), S. 1938b Laboratoriumsversuche über Tularämie. Turkische Z. Hyg. Exptl. Biol., **1**, 155–157.

Karawaiew, W. 1899 Über Anatomie und Metamorphose des Darmkanals der Larve von *Anobium paniceum*. Biol. Zentr., **19**, 122–130, 161–171, 196–202.

Katić, R. V. 1940 *Rickettsia melophagium* and their spread in sheep and ectoparasites. Vojno-san. Glasnik., **11**, 523–527. (Abstract in Vet. Bull., **12**, 491. 1942.)

Katzin, L. I., and Kirby, H. 1939 The relative weights of termites and their protozoa. J. Parasitol., **25**, 444–445.

Kawamura, R. 1926 Studies on tsutsugamushi disease. Published as a book by The Medical Bulletin, Coll. Med., Univ. Cincinnati. 229 pp.

Kawamura, R. 1934 Article in Japanese, Nisshin Igaku, 23. (Not seen.)

Kawamura, R., and Imagawa, Y. 1931a Die Feststellung des Erregers bei der Tsutsugamushikrankheit. Zentr. Bakt. Parasitenk. Infekt., I, **122**, 253–261.

Kawamura, R., and Imagawa, Y. 1931b Ueber die Proliferation der pathogenen Rickettsia im tierischen Organismus bei der Tsutsugamushi-Krankheit. Trans. Japan. Path. Soc., **21**, 455–461.

Keilin, D. 1914 Une nouvelle schizogrégarine *Caulleryella aphiochoetae,* n. g., n. sp., parasite intestinal d'une larve d'un diptère cyclorhaphe (*Aphiochaeta rufipes* Meig.). Compt. Rend. Soc. Biol., **76**, 768–771.

Keilin, D. 1917 Une nouvelle entamibe, *Entamoeba mesnili* n. sp., parasite intestinale d'une larve d'un diptère. Compt. Rend. Soc. Biol., **80**, 133–136.

Keilin, D. 1920 On a new saccharomycete *Monosporella unicuspidata* gen. n. nom., n. sp., parasitic in the body cavity of a dipterous larva (*Dasyhelea obscura* Winnertz). Parasitology, **12**, 83–91.

Keilin, D. 1921a On the life-history of *Helicosporidium parasiticum,* n. g., n. sp., a new type of protist parasitic in the larva of *Dasyhelea obscura* Winn. (Diptera, Ceratopogonidae) and in some other arthropods. Parasitology, **13**, 97–113.

Keilin, D. 1921b On a new ciliate, *Lambornella stegomyiae* n. g., n. sp., parasitic in the body cavity of the larvae of *Stegomyia scutellaris* Walker (Diptera, Nematocera, Culicidae). Parasitology, **13**, 216–224.

Keilin, D. 1921c On a new type of fungus: *Coleomomyces stegomyiae* n.g., n. sp., parasitic in the body cavity of the larva of *Stegomyia scutellaris* Walker (Siptera, Nematocera, Culicidae). Parasitology, **13**, 226–234.

Keilin, D. 1921d On the life-history of *Dasyhelea obscura* Winnertz (Diptera, Nematocera, Ceratopogonidae) with some remarks on the parasites and hereditary bacterian symbiont of this midge. Ann. Nat. Hist., **8**, 576–590.

Keilin, D. 1927 On *Coelomomyces stegomyiae* and *Zografia notonectae,* fungi parasitic in insects. Parasitology, **19**, 365–367.

Kelser, R. A. 1933 Mosquitoes as vectors of the virus of equine encephalomyelitis. J. Am. Vet. Med. Assoc., **82** (N.S. **35**), 767–771.

Kelser, R. A. 1938 Transmission of the virus of equine encephalomyelitis by *Aëdes taeniorhynchus.* J. Am. Vet. Med. Assoc., **102** (N.S. **45**), 195–203.

Kennedy, C. H. 1932 Methods for the study of the internal anatomy of insects. Mimeo. Dept. Entomol., Ohio State Univ., Columbus. 103 pp.

Kent, W. S. 1880–1882 A manual of Infusoria. D. Bogue, London. 3 vols. 913 pp.

Kidder, G. W. 1937 The intestinal protozoa of the wood-feeding roach *Panesthia.* Parasitology, **29**, 163–204.

Kidder, G. W. 1938 Nuclear reorganization without cell division in *Paracleve-*

landia simplex (family Clevelandellidae) an endocommensal ciliate of the wood feeding roach *Panesthia.* Arch. Protistenk., **91,** 69–78.

King, A. F. A. 1882 Insects and disease, mosquitoes and malaria. Read before Philos. Soc. of Washington, Feb. 10. (See also Popular Science Monthly, Sept., 1883, **23,** 644–658.)

King, K. M., and Atkinson, N. J. 1928 The biological control factors of the immature stages of *Euxoa ochrogaster,* Gn. (Lepidoptera, Phalaenidae) in Saskatchewan. Ann. Entomol. Soc. Am., **21,** 167–188.

Kingsbury, B. F., and Johannsen, O. A. 1927 Histological technique. John Wiley & Sons, Inc., New York. 142 pp.

Kirby, H. 1924 Morphology and mitosis of *Dinenympha fimbriata* sp. nov. Univ. Calif. Pub. Zool., **26,** 199–220.

Kirby, H. 1926 The intestinal flagellates of the termite, *Cryptotermes hermsi* Kirby. Univ. Calif. Pub. Zool., **29,** 103–120.

Kirby, H. 1926 On *Staurojoenina assimilis* sp. nov., an intestinal flagellate from the termite, *Kalotermes minor* Hagen. Univ. Calif. Pub. Zool., **29,** 25–102.

Kirby, H. 1927 Studies on some amoebae from the termite *Microtermes,* with notes on some other protozoa from the Termitidae. Quart. J. Microscop. Sci., **71,** 189–222.

Kirby, H. 1928 A species of *Proboscidiella* from *Kalotermes* (*Cryptotermes*) *dudleyi* Banks, a termite of Central America, with remarks on the oxymonad flagellates. Quart. J. Microscop. Sci., **72,** 355–386.

Kirby, H. 1929 *Snyderella* and *Coronympha,* two new genera of multinucleate flagellates from termites. Univ. Calif. Pub. Zool., **31,** 417–432.

Kirby, H. 1930 Trichomonad flagellates from termites. I. *Tricercomitus* gen. nov., and *Hexamastix alexeieff.* Univ. Calif. Pub. Zool., **33,** 393–444.

Kirby, H. 1931 Trichomonad flagellates from termites. II. *Eutrichomastix,* and the subfamily Trichomonadinae. Univ. Calif. Pub. Zool., **36,** 171–262.

Kirby, H. 1932a Flagellates of the genus *Trichonympha* in termites. Univ. Calif. Pub. Zool., **37,** 349–476.

Kirby, H. 1932b Protozoa in termites of the genus *Amitermes.* Parasitology, **24,** 289–304.

Kirby, H. 1936 Two polymastigote flagellates of the genera *Pseudodevescovina* and *Caduceia.* Quart. J. Microscop. Sci., **79,** 309–335.

Kirby, H. 1937a Host-parasite relations in the distribution of protozoa in termites. Univ. Calif. Pub. Zool., **41,** 189–212.

Kirby, H. 1937b The devescovinid flagellate *Parajoenia grassi* from a Hawaiian termite. Univ. Calif. Pub. Zool., **41,** 213–224.

Kirby, H. 1938a The devescovinid flagellates *Caduceia theobromae* França, *Pseudodevescovina ramosa* new species, and *Macrotrichomonas pulchra* Grassi. Univ. Calif. Pub. Zool., **43,** 1–40.

Kirby, H. 1938b Polymastigote flagellates of the genus *Foaina Janicki,* and two new genera *Crucinympha* and *Bullanympha.* Quart. J. Microscop. Sci., **81,** 1–25.

Kirby, H. 1939 Two new flagellates from termites in the genera *Coronympha* Kirby, and *Metacoronympha* Kirby, new genus. Proc. Calif. Acad. Sci., **22,** 207–220.

Kirby, H. 1941 Organisms living on and in protozoa. *In* Protozoa in biological research, chap. 20. Columbia Univ. Press, New York. 1148 pp.

Kirby, H. 1942a Devescovinid flagellates of termites. II. The genera *Caduceia* and *Macrotrichomonas*. Univ. Calif. Pub. Zool., **45,** 93–166.

Kirby, H. 1942b Devescovinid flagellates of termites. III. The genera *Foaina* and *Parajoenia*. Univ. Calif. Pub. Zool., **45,** 167–246.

Kirby, H. 1944 The structural characteristics and nuclear parasites of some species of *Trichonympha* in termites. Univ. Calif. Pub. Zool., **49,** 185–282.

Kirby, H. 1945 Devescovinid flagellates of termites. IV. The genera *Metadevescovina* and *Pseudodevescovina*. Univ. Calif. Pub. Zool., **45,** 247–318.

Kitajima, T., and Metalnikov, S. 1923 Une maladie mortelle chez les chenilles de *Galleria mellonella*. Compt. Rend. Soc. Biol., **88,** 476–477.

Kitasato, S. 1894 Preliminary note on the bacillus of bubonic plague. Lancet, **2,** 428.

Kitselman, C. M., and Grundmann, A. W. 1940 Equine encephalomyelitis virus isolated from naturally infected *Triatoma sanguisuga* LeConte. Kansas Agr. Expt. Sta. Bull., 50. 15 pp.

Klein, E. 1908 "Flies" as carriers of the *Bacillus typhosus*. Brit. Med. J., **2,** 1150–1151.

Kleine, F. K. 1909 Positive Infektionsversuche mit *Trypanosoma brucei* durch *Glossina palpalis*. Deut. Med. Wochschr., **35,** 469–470.

Klevenhusen, F. 1927 Beiträge zur Kenntnis der Aphidensymbiose. Z. Morphol. Ökol. Tiere (Abt. A. Z. Wiss. Biol.), **9,** 97–165.

Kligler, I. J., and Aschner, M. 1931 Cultivation of rickettsia-like microorganisms from certain blood-sucking pupipara. J. Bact., **22,** 103–118.

Kluyver, A. M., and Niel, C. B. van. 1936 Prospects for a rational system of bacterial classification. Zentr. Bakt. Parasitenk. Infekt., II, **94.** 369 pp.

Knoche, E. 1912 Nonnenstudien. Die Wipfelkrankheit und ihr Erreger. Forstwiss. Centr., **34,** 177–194.

Knowles, R. 1928 An introduction to medical protozoology. Thacker, Spink and Co., Calcutta. 887 pp.

Knowles, R., and Basu, B. C. 1935 A blood-inhabiting spirochaete of the guinea pig. Indian J. Med. Research, **22,** 449–468.

Koch, A. 1930 Über das Vorkommen von Mitochondrien in Mycetocyten. Z. Morphol. Ökol. Titere, **19,** 259–290.

Koch, A. 1931 Die Symbiose von *Oryzaephilus surinamensis* L. Z. Morphol. Ökol. Tiere, **23,** 389–424.

Koch, A. 1933a Ueber das Verhalten symbiontenfreier Sitodrepa-Larven. Biol. Zentr., **53,** 199–203.

Koch, A. 1933b Symbionten und Vitamine. Naturwissenschaften, **21,** 543.

Koch, A. 1936a Symbiosestudien. I. Die Symbiose des Splintkäfers, *Lyctus linearis* Goez. Z. Morphol. Ökol. Tiere, **32**, 92–136.

Koch, A. 1936b Symbiosestudien. II. Experimentelle Untersuchungen an *Oryzaephilus surinamensis* L. (Cucujidae, Coleopt.). Z. Morphol. Ökol. Tiere, **32**, 137–180.

Koch, A. 1936c Bau und Entwicklungsgeschichte der Mycetome von *Lyctus linearis* Goez. Verhandl. Deut. Zool. Ges., **38**, 252–261.

Kodama, M. 1932 On classification of typhus in its epidemiological, clinical and etiological observations. Kitasato Arch. Exptl. Med., **9**, 357–361.

Kodama, M., and Kono, M. 1933 Studies on experimental transmission of virus of "eruptive fever" and "typhus" by several blood sucking insects. Kitasato Arch. Exptl. Med., **10**, 99–112.

Kodama, M., Takahashi, K., and Kono, M. 1931 On experimental observation of the so-called Manchurian typhus and its etiological agent (*Rickettsia manchuriae*). Saitake Gaku Zasshi, No. 426, 427. Aug. and Sept., 1931.

Kohls, G. M. 1937 Tick rearing methods with special reference to the Rocky Mountain wood tick, *Dermacentor andersoni* Stiles. *In* Galtsoff, Paul S., *et al.*, *Culture methods for invertebrate animals*, pp. 246–256. Comstock Publishing Co., Ithaca, N.Y.

Kohls, G. M., Armbrust, C. A., Irons, E. N., and Philip, C. B. 1945 Studies on tsutsugamushi disease (scrub typhus, mite-borne typhus) in New Guinea and adjacent islands: further observations on epidemiology and etiology. Am. J. Hyg., **41**, 374–396.

Kohls, G. M., and Steinhaus, E. A. 1943 Tularemia: spontaneous occurrence in shrews. Pub. Health Repts., **58**, 842.

Kolmer, J. A., and Boerner, F. 1941 Approved laboratory technic. 3d ed. D. Appleton-Century Co., New York. 921 pp.

Komárek, J., and Breindl, V. 1924 Die Wipfelkrankheit der Nonne und der Erreger derselben. Z. Angew. Entomol., **10**, 99–162.

Königsberger, J. C., and Zimmermann, A. 1901 De dierlijke vijanden der Koffieculture op Java. Med. Plantentium Java, **44**, 7–113.

Korke, V. T. 1916 On a *Nosema* (*Nosema pulicis*) parasitic in the dog flea (*Ctenocephalus felis*). Indian J. Med. Research, **3**, 725–730.

Korshunova, O. S. 1943 Etiology of tick spotted typhus in Krasnoyarsk province. Zhur. Mikrobiol., 1–2, 59–64. (In Russian.)

Kouwenaar, W., and Wolff, J. W. 1942 Rickettsia infections in Sumatra. Proc. 6th Pacific Sci. Congress (1939), **5**, 633–637.

Krassilstschik, I. M. 1893 La graphitose et la septicémie chez les insectes. Mém. Soc. Zool. France, **6**, 245–285.

Krassilstschik, I. M. 1896 Sur une nouvelle propriété du corpuscule (microsporidium) de la pébrine. Compt. Rend. Acad. Sci., Paris, **123**, 427–429.

Krassilstschik, I. M. 1916 Report on the work of the Bio-Entomological Station for Bessarabia in 1914–1915. Kishinew. 96 pp.

Krassilstschik, J. 1886 De insectorum morbis qui fungis parasitis efficiunter. Mém. Soc. Nat. Nouv. Russie, Odessa. 97 pp.

Krassilstschik, M. J. 1889 Sur les bactéries biophytes. Note sur la symbiose de pucerons avec des bactéries. Am. Inst. Pasteur, **3**, 465–472.

Krassilstschik, M. J. 1890 Ueber eine neue Kategorie von Bacterien (Biophyten), die im Innern eines Organismus leben und ihm Nutzen bringen. Biol. Zentr., **10**, 421.

Krassilstschik, M. J. 1896 Sur les parasites des vers à soie sains et malades. Contribution à l'étude de la flacherie, de la grasserie et de la pébrine. (Communication préliminaire.) Mém. Soc. Zool. France, **9**, 513–522.

Ksenjoposky, A. V. 1916 Review of the pests of Volhynia and report of the work of Volhynia Entomological Bureau for 1915 (translation). Published by the Zemstvo of Volhynia, Jitomir. 24 pp.

Kuczynski, M. H. 1927 Die Erreger des Fleck- und Felsenfiebers. Julius Springer, Berlin. 256 pp.

Kudo, R. R. 1924 A biologic and taxonomic study of the Microsporidia. Illinois Biol. Mono., **9**. 268 pp.

Kudo, R. R. 1939 Protozoology. Charles C. Thomas, Baltimore. 689 pp.

Kudo, R. R. 1942 On the microsporidian, *Duboscqia legeri* Perez 1908, parasitic in *Reticulitermes flavipes*. J. Morphol., **71**, 307–333.

Kudo, R. R. 1943 *Nosema termitis* n. sp., parasitic in *Reticulitermes flavipes*. J. Morphol., **73**, 265–279.

Kudo, R. R., and De Coursey, J. D. 1940 Experimental infection of *Hyphantria cunea* with *Nosema bombycis*. J. Parasitol., **26**, 123–125.

Kuffernath, H. 1921 Microbe pathogène pour les sauterelles et d'autres insectes, *Micrococcus* (*Staphylococcus*) *acridicida,* Kuff. nov. spec. Ann. Gembloux, Brussels, **27**, 253–257.

Kuhns, D. M., and Anderson, T. G. 1944 A fly-borne bacillary dysentery epidemic in a large military organization. Am. J. Pub. Health, **34**, 750–755.

Kumm, H. W., and Frobisher, M., Jr. 1932 Attempts to transmit yellow fever with certain Brazilian mosquitoes (Culicidae) and with bedbugs (*Cimex hemipterus*). Am. J. Trop. Med., **12**, 349–361.

Kumm, H. W., and Turner, T. B. 1936 The transmission of yaws from man to rabbits by an insect vector, *Hippelates pallipes* Loew. Am. J. Trop. Med., **16**, 245–271.

Kumm, H. W., Turner, T. B., and Peat, A. A. 1935 The duration of motility of the spirochaetes of yaws in a small West Indian fly—*Hippelates pallipes* Loew. Am. J. Trop. Med., **15**, 209–223.

Kunkel, L. O. 1933 Insect transmission of peach yellows. Contrib. Boyce Thompson Inst., **5**, 19–28.

Kunkel, L. O. 1937 Effect of heat on ability of *Cicadula sexnotata* (Fall.) to transmit aster yellows. Am. J. Botan., **24**, 316–327.

Kuskop, M. 1924 Bakteriensymbiosen bei Wanzen. Arch. Protistenk., **47**, 350–385.

Küster, H. A. 1902 Über den Durchgang von Bakterien durch Insektendarm. Inaugural Dissertation, Univ. Heidelberg, Rössler. 44 pp.

Lafar, F. 1904–1914 Handbuch der technischen Mykologie. Fischer, Jena. 5 vols., 2903 pp.

Lal, R. B., Ghosal, S. C., and Mukherji, B. 1939 Investigations on the variation of vibrios in the house fly. Indian J. Med. Research, **26**, 597–609.

Lamborn, W. A. 1921 A protozoon pathogenic to mosquito larvae. Parasitology, **13**, 213.

Landois, L. 1864 Untersuchungen über die auf dem menschen schmarotzenden Pediculiden. 1. Anatomie des *Phthirius inguinalis* Leach. Z. Wiss. Zool., **14**, 27–41.

Laveran, A. 1891 De l'étiologie du paludisme. Int. Congr. Hyg. Trans., **2**, 10–20.

Laveran, A. and Franchini, G. 1919 Infection des souris blanches à l'aide des cultures de *Herpetomonas ctenocephali*. Bull. Soc. Path. Exotique, **12**, 379–383.

Laveran, A., and Franchini, G. 1920 Infections expérimentales de chiens et de cobayes à l'aide de cultures d'*Herpetomonas* d'insectes. Bull. Soc. Path. Exotique, **13**, 569–576.

Lavier, G. 1921 Les parasites des invertébrés hématophages. Trav. Lab. Parasitol. Faculté Méd. Paris. (Quoted from Wenyon, Protozoology, **2**, 1249, 1510.)

Lavier, G. 1942–1943 L'évolution de la morphologie dans le genre *Trypanosoma*. Ann. Parasitol. Humaine et Comparée, **19**, 168–200.

Leach, J. G. 1925 The seed-corn maggot and potato blackleg. Science, **61**, 120.

Leach, J. G. 1926 The relation of the seed-corn maggot (*Phorbia fusciceps* Zett.) to the spread and development of potato blackleg in Minnesota. Phytopathology, **16**, 149–176.

Leach, J. G. 1927 The relation of insects and weather to the development of heart rot of celery. Phytopathology, **17**, 663–667.

Leach, J. G. 1930 The identity of the potato blackleg pathogen. Phytopathology, **20**, 743–751.

Leach, J. G. 1931 Further studies on the seed-corn maggot and bacteria with special reference to potato blackleg. Phytopathology, **21**, 387–406.

Leach, J. G. 1933 Methods of survival of bacteria in puparia of the seed corn maggot (*Hylemyia cilicrura* Rond.). Z. Angew. Entomol., **20**, 150–161.

Leach, J. G. 1940 Insect transmission of plant diseases. McGraw-Hill Book Co., New York. 615 pp.

Leach, J. G., and Dosdall, L. 1938 Observations on the dissemination of fungi by ants. Phytopathology, **28**, 444–446.

Leach, J. G., Hodson, A. C., Chilton, St. J. P., and Christensen, C. M. 1940 Observations on two ambrosia beetles and their associated fungi. Phytopathology, **30**, 227–236.

Lebert, H. 1858 Berlin. Entomol. Z. (From Northrup, 1914a.)

Ledingham, J. C. G. 1911 On the survival of specific microorganisms in pupae and imagines of *Musca domestica,* raised from experimentally infected larvae. Experiments with *B. typhosus.* J. Hyg., **11**, 333–340.

Ledingham, J. C. G. 1940 Sex hormones and the Foa-Kurloff cell. J. Path. Bact., **50**, 201–219.

Lee, A. B. 1937 The microtomist's vade mecum. 10th ed. P. Blakiston's Son and Co., Philadelphia. 784 pp.

Lee, H. A. 1924 Dry rot of citrus fruits caused by a *Nematospora* species. Philippine J. Sci., **24**, 719–733.

Léger, L. 1902a Sur un flagellé parasite de l'*Anopheles maculipennis.* Compt. Rend. Soc. Biol., **54**, 354–356.

Léger, L. 1902b Bactéries parasites de l'intestin des larves de chironome. Compt. Rend. Acad. Sci., Paris, **134**, 1317–1319.

Léger, L. 1908 Mycétozoaires endoparasites des insectes. I. *Sporomyxa scauri* nov. gen., nov. spec. Arch. Protistenk., **12**, 109–130.

Léger, L., and Duboscq, O. 1909 Protistes parasites d'une larve de *Ptychoptera* et leur action sur l'hôte. Bull. Acad. Roy. Belg. Sci., 8, 831, 885.

Lehmann, K. B., Neumann, R. O., and Breed, R. S. 1931 Determinative bacteriology, vol. II. G. E. Stechert and Co., New York. 868 pp.

Leidy, J. 1856 A synopsis of Entozoa and some of their ecto-congeners observed by the author. Proc. Acad. Natural Sci., Phila., **8**, 42–58.

Leidy, J. 1879 On *Amoeba blattae.* Proc. Acad. Natural Sci., Phila., **31**, 204–205.

Leidy, J. 1881 Parasites of the termites. J. Acad. Natural Sci., Phila., **8**, 426–448.

Leishman, W. B. 1903 On the possibility of the occurrence of trypanosomiasis in India. Brit. Med. J., **1**, 1252.

Leishman, W. B. 1918 A note on the "granule clumps" found in *Ornithodorus moubata* and their relation to the spirochaetes of African relapsing fever (tick fever). Ann. Inst. Pasteur, **32**, 49.

Leishman, W. B. 1920 The Horace Dobell Lecture on an experimental investigation of *Spirochaeta duttoni,* the parasite of tick fever. Lancet, pt. II, 1237.

Leon, A. P. 1942 Precipitacion de sueros anti-tifo por la orina de enfermos de tifo exantematico. Una nueva reaccion serologica para el diagnostico del tifo. Rev. Inst. Salubridad Enfermedades Trop. (Mex.), **3**, 201–208.

Lepesme, P. 1937a Action de *Bacillus prodigiosus* et *Bacillus pyocyaneus* sur le criquet pèlerin (*Schistocerca gregaria,* Forsk.). Compt. Rend. Soc. Biol., **125**, 492–494.

Lepesme, P. 1937b Sur la présence du *Bacillus prodigiosus* chez le criquet pèlerin (*Schistocerca gregaria* Forsk.) Bull. Soc. Hist. Afr. N., **28**, 406–411.

Lepesme, P. 1938 Recherches sur une aspergillose des acridiens. Bull. Soc. Hist. Nat. Agr., **29**, 372–384.

Lépine, P., Mathis, M., and Sauter, V. 1941 Infestation expérimentale de *Triatoma infestans* par le virus de l'encéphalomyélite équine américaine, type Venezuela. Bull. Soc. Path. Exotique, **34**, 115.

Lespès, C. 1856 Note sur un nématoïde (*Isakis migrans*) parasite des termites. Ann. Sci. Nat., **5**, 335-336.

Lestoquard, F., and Donatien, A. 1936 Sur une nouvelle *Rickettsia* du mouton. Bull. Soc. Path. Exotique, **29**, 105-108.

Levaditi, C., and Lépine, P. 1938 Les ultravirus des maladies humaines. Librairie Maloine, Paris. 2 vols., 1182 pp.

Lévi, M. L. 1940 Contribution à l'étude des conjunctivités rickettsiennes des petits ruminants. Ann. Inst. Pasteur, **64**, 542-557.

Levine, M., and Schoenlein, H. W. 1930 A compilation of culture media for the cultivation of microorganisms. Williams and Wilkins Co., Baltimore. 969 pp.

Levkovich, E. N., and Skrinnik, A. N. 1940-1941 An experimental and immunological analysis of natural immunization of the population in the foci of tick-borne encephalitis. Arch. Biol. Nauk., **59**, 118-121.

Lewis, E. A., and Fotheringham, W. 1941 The transmission of *Theileria parva* by ticks. Parasitology, **33**, 251-277.

Lewis, W. W. 1933 New species of *Proboscidiella* and *Devescovina* from *Kalotermes occidentis* Walker, a termite of lower California. Univ. Calif. Pub. Zool., **39**, 77-96.

Lewthwaite, R. 1930 Clinical and epidemiological observations on tropical typhus in the Federated Malay States. Bull. Inst. Med. Research Federated Malay States, **1**, 1-42.

Lewthwaite, R., and Savoor, S. R. 1938 An instance of the mutation of a murine strain of typhus from the X-19 type to the XK type. Acta Conv. Tertii Trop. Malariae Morbis, **1**, 520-525.

Lewthwaite, R., and Savoor, S. R. 1940 Rickettsia diseases of Malaya. Identity of tsutsugamushi and rural typhus. Lancet, pt. I., pp. 255, 305.

Leydig, F. 1850 Einige Bemerkungen über die Entwicklung der Blattläuse. Z. Wiss. Zool., **2**, 62-66.

Leydig, F. 1854 Zur Anatomie von *Coccus hesperidum*. Z. Wiss. Zool., **5**, 1-12.

Leydig, F. 1857 Lehrbuch der Histologie des Menschen und der Tiere. Von Meidinger Sohn and Co., Frankfurt a. M. 551 pp.

Lien-teh, W., Chun, J. W., Pollitzer, R., and Wu, C. Y. 1936 Plague. A manual for medical and public health workers. The Mercury Press, Shanghai. 547 pp.

Light, S. F., and Sanford, M. F. 1927 Are protozoan faunae of termites specific? Proc. Soc. Exptl. Biol. Med., **25**, 95-96.

Light, S. F., and Sanford, M. F. 1928 Experimental transformation of termites. Univ. Calif. Pub. Zool., **31**, 269-274.

Lilienstern, M. 1932 Beiträge zur Bakteriensymbiose der Ameisen. Z. Morphol. Ökol. Tiere (Abt. A. Z. Wiss. Biol.), **26**, 110-134.

Lillie, R. D. 1930 Psittacosis: rickettsia-like inclusions in man and in experimental animals. Pub. Health Repts., **45**, 773-778.

Lilly, J. H. 1931 A preliminary study of the presence of bacteria in the blood of the house fly *Musca domestica*. Univ. of Wisconsin, B. S. Thesis. 30 pp.

Lindeijer, E. J. 1932 De bacterie-ziekte van den wilg veroorzakt door *Pseudomonas saliciperda* n. sp. Univ. of Amsterdam, Baarn, Thesis, pp. 1–82.

Lindner, P. 1895 Ueber eine in *Aspidiotus nerii* parasitisch lebende Apiculatushefe. Zentr. Bakt. Parasitenk. Infekt., II, **1**, 782–787.

Link, G. K. K., and Taliaferro, W. H. 1928 Further agglutination tests with bacterial plant pathogens. Botan. Gaz., **85**, 198–207.

List, G. M., and Kreutzer, W. A. 1942 Transmission of the causal agent of the ring-rot disease of potatoes by insects. J. Econ. Entomol., **35**, 455–456.

Livingston, S. K., and Prince, L. H. 1932 The treatment of chronic osteomyelitis. With special reference to the use of maggot active principle. J. Am. Med. Assoc., **98**, 1143.

Lochhead, A. G. 1928a Studies on the etiology of European foulbrood of bees. Intern. Congr. Entomol., **2**, 1005–1009.

Lochhead, A. G. 1928b Cultural studies of Bacillus larvae (White). Sci. Agr. **9**, 80–89.

Lochhead, A. G. 1928c The etiology of European foulbrood of bees. Science, **67**, 159–160.

Lochhead, A. G. 1937 The nitrate reduction test and its significance in the detection of *Bacillus larvae*. Can. J. Research, **15**, 79–86.

Lochhead, A. G. 1942 Growth factor requirements of *Bacillus larvae*, White. J. Bact., **44**, 185–189.

Loeb, J., and Northrop, J. H. 1917a On the influence of food and temperature upon the duration of life. J. Biol. Chem., **32**, 103–121.

Loeb, J., and Northrop, J. H. 1917b What determines the duration of life in Metazoa? Proc. Nat. Acad. Sci. U.S., **3**, 382–386.

Löffler, W., and Mooser, H. 1942 Zum Vebertragungsmodus des Fleckfiebers. Beobachtungen anlässlich einer Laboratoriums-Gruppeninfektion. Schweiz. Med. Woch., **72**, 755–761.

Lohde, G. 1872 Insectenepidemien, welche durch Pilze hervorgerufen werden. Berlin. Entomol. Z., **16**, 17–44.

Löhnis, F., and Smith, N. R. 1916 Life cycles of the bacteria. J. Agr. Research, **6**, 675–702.

Long, E. C. 1911a Note on the transmission of leprosy. J. Trop. Med. Hyg., **14**, 254.

Long, E. C. 1911b Note on the transmission of leprosy. Brit. Med. J., **2**, 470.

Longfellow, R. C. 1913 The common house roach as a carrier of disease. Am. J. Pub. Health, **3**, 58–61.

Lounsbury, C. P. 1900 Heartwater in sheep and goats. Agr. J. Cape of Good Hope, **16**, 682–687.

Lounsbury, C. P. 1902 Heartwater in sheep and goats. Special tick investigations. Agr. J. Cape of Good Hope, **21**, 315.

Lounsbury, C. P. 1913 Locust bacterial disease. Agr. J. Union S. Africa, **5**, 607–611.

Lucas, C. L. T. 1927 Two new species of amoeba found in cockroaches: with notes on the cysts of *Nyctotherus ovalis* Leidy. Parasitology, **19**, 223–235.

Lwoff, A. 1924 Infection expérimentale à *Glaucoma piriformis* (infusoire) chez *Galleria mellonella* (lépidoptère). Compt. Rend. Acad. Sci., Paris, **178**, 1106–1108.

Lysaght, A. M. 1936 A note on an unidentified fungus in the body cavity of two thysanopterous insects. Parasitology, **28**, 293–294.

Maassen, A. 1907 Über die sogenannte Faulbrut der Honigbienen. Mitt. Kaiserl. Biol. Anstalt Land- u. Forst-w., **4**, 51–53.

Maassen, A. 1908 Zur Etiologie der sogenannten Faulbrut der Honigbienen. Arb. Kaiserl. Biol. Anstalt Land- u. Forst-w., **6**, 53–70.

Maassen, A. 1913 Weitere Mitteilungen über die seuchenhaften Brutkrankheiten der Bienen insbesondere übr die Faulbrut. Mitt. Kaiserl. Biol. Anstalt Land- u. Forst-w., 8th Ann. Rept., pt. 14, 48–58.

Macalister, C. J. 1912 A new cell proliferant. Its clinical application in the treatment of ulcers. Brit. Med. J., **1**, 10–12.

MacArthur, W. P. 1922 A holotrichus ciliate pathogenic to *Theobaldia annulata* Schrank. J. Roy. Army Med. Corps, **38**, 83–92.

Macchiavello, A. 1938 Sistematica del grupo Rickettsiae. Rev. Chilena Hig. Med. Prevent., **1**, 297–330.

Macchiavello, A. 1939 Investigations of Chilean exanthematicus typhus and experimental exanthematous typhus. Proc. 6th Pacific Sci. Congr., 1942, **5**, 649–675.

McClung, C. E. (editor). 1937 Handbook of microscopical technique. Paul Hoeber, Inc., New York. 698 pp.

McClung, L. S., and McCoy, E. 1941 The anaerobic bacteria and their activities in nature and disease. A subject bibliography. Univ. Calif. Press, Berkeley. 244 pp.

McCoy, E., and McClung, L. S. 1939 The anaerobic bacteria and their activities in nature and disease. A subject bibliography. Univ. Calif. Press, Berkeley. 2 vols., 295, 602 pp.

McCoy, E. E., and Glaser, R. W. 1936 Nematode culture for Japanese beetle control. N.J. Dept. Agr. Circ., 265. 9 pp.

McCoy, G. W. 1911 A plague-like disease of rodents. U.S. Pub. Health Service, Pub. Health Bull., 43, 53–71.

McCoy, G. W., and Chapin, C. W. 1912 *Bacterium tularense,* the cause of a plaguelike disease of rodents. Hyg. Lab., U.S. Pub. Health Service Bull., 53 (January). (See also J. Infectious Diseases, 1912, **10**, 61.)

McCoy, G. W., and Clegg, M. T. 1912 A note on acid-fast bacilli in head lice (*Pediculus capitis*). Pub. Health Repts., **27**, 1464–1465.

McCray, A. H. 1917 Spore-forming bacteria of the apiary. U.S. Dept. Agr., J. Agr. Research, **8**, No. 11, p. 399.

McCulloch, I. 1917 *Crithidia euryophthalmi,* sp. nov., from the hemipteran bug, *Euryophthalmus convivus* Stål. Univ. Calif. Pub. Zool., **18,** 75–88.

MacGregor, M. E. 1917 A summary of our knowledge of insect vectors of disease. Bull. Entomol. Research, **8,** 155–163.

MacGregor, M. E. 1929 The significance of the pH in the development of mosquito larvae, chiefly *Aëdes argenteus* Poir. Parasitology, **21,** 132–157.

Mackehenie, D., and Coronado, D. 1933 Plantas reservoérios de virus: contribución al conocimiento de la fitopatogenesis peruana. Rev. Med. Peruana, **5,** 803.

McKenzie, M. A., and Becker, W. B. 1937 The Dutch elm disease: a new threat to elm. Mass. Agr. Expt. Sta. Bull., 343.

Mackie, D. B. 1913 The Philippine locust (*Pachytylus migratoroides,* R. & F.); natural influences affecting its propagation and distribution. Philippine Agr. Rev., **6,** 538.

Mackinnon, D. L. 1910 New protist parasites from the intestine of Trichoptera. Parasitology, **3,** 245–254.

Mackinnon, D. L. 1912 Protists parasitic in the larva of the crane-fly, *Tipula* sp. Parasitology, **5,** 175–189.

Mackinnon, D. L. 1913 Studies on parasitic protozoa. II. Quart. J. Microscop. Sci., **59,** 459–470.

MacLeod, J. 1932 The bionomics of *Ixodes ricinus* L., the "sheep tick" of Scotland. Parasitology, **24,** 382–400; also 1934, **26,** 282–305.

Macrae, R. 1895 Flies and cholera diffusion. Indian Med. Gaz., **29,** 407–412.

Maddox, R. L. 1885 Experiments on feeding some insects with the curved or "comma" bacillus, and also with another bacillus (*subtilis?*). J. Roy. Microscop. Soc., Ser. 2, pt. II, **5,** 602–607, 941–952.

Magalhães, O., and Rocha, A. 1942 Tifo exantemático do Brasil; paper do cão (*C. familiaris*) na constituição dos fócos da moléstia. Brazil Medico, **56,** 370–377.

Mahdihassan, S. 1924 Lac-secretion and symbiotic fungi. In Some studies in biochemistry by some students of Dr. G. J. Fowler, pp. 136–141. The Phoenix Printing House, Bangalore.

Mahdihassan, S. 1928 Symbionts specific of wax and pseudo lac insects. Arch. Protistenk., **63,** 18–22.

Mahdihassan, S. 1929 The microorganisms of red and yellow lac insects. Arch. Protistenk., **68,** 613–624.

Mahdihassan, S. 1932 The symbiote of *Lakshadia communis.* Arch. Protistenk., **78,** 514–521.

Mahdihassan, S. 1935 Further studies on the symbiotes of scale insects. Arch. Protistenk., **85,** 61–73.

Mahdihassan, S. 1939 Pigmentbildende Bakterien aus einer entsprechend gefärbten Cicade. Zool. Anz. suppl. (Deut. Zool. Gesell.), **12,** 420–430.

Mahdihassan, S. 1940 Diagenesis *versus* mutation. Current Sci., **9,** 495–496.

Mahdihassan, S. 1941 Insect tumours of bacterial origin. Deccan Med. J., pp. 1–19. (Seen in reprint form only.)

Mains, E. B. 1934 The genera *Cordyceps* and *Ophiocordyceps* in Michigan. Proc. Am. Philosophical Soc., **74,** 263–271.

Mains, E. B. 1939a *Cordyceps* from the mountains of North Carolina and Tennessee. Elisha Mitchell Sci. Soc., **55,** 117–129.

Mains, E. B. 1939b *Cordyceps* species from Michigan. Papers of the Mich. Acad. Sci., Arts, & Lett., **25** (1940), 79–84.

Malbrant, R. 1939a Existence de *Rickettsia conjunctivae* du mouton au Congo français. Bull. Soc. Path. Exotique, **32,** 907–908.

Malbrant, R. 1939b Rickettsiose canine au Congo français. Bull. Soc. Path. Exotique, **32,** 908–913.

Mallman, W. L. 1930 The interagglutinability of members of the *Brucella* and *Pasteurella* genera. J. Am. Vet. Med. Assoc., **77,** 636.

Mallory, F. B. 1938 Pathological technique. W. B. Saunders Co., Philadelphia. 274 pp.

Manson, P. 1896 On the life-history of the malaria germ outside the human body. Lancet, I, 1896, pp. 751–754, 831–833. (See also Lancet, II, 1896, pp. 1715–1716.)

Manson, P. 1898 The mosquito and the malaria parasite. Brit. Med. J., Sept. 24, 849–853.

Manson, P. 1907 [On *Borrelia carteri* (*Spiroschaudinnia carteri*).] (See Breed et al., Bergey's Manual, 5th ed., p. 953.)

Manson-Bahr, P. 1919 Bacillary dysentery. Trans. Roy. Soc. Trop. Med. Hyg., **13,** 64–72.

Mansour, K. 1930 Preliminary studies on the bacterial cell-mass (accessory cell-mass) of *Calandra oryzae* (Linn.): the rice weevil. Quart. J. Microscop. Sci., **73,** 421–436.

Mansour, K. 1934a On the intracellular micro-organisms of some bostrychid beetles. Quart. J. Microscop. Sci., **77,** 243–254.

Mansour, K. 1934b On the so-called symbiotic relationship between coleopterous insects and intracellular micro-organisms. Quart. J. Microscop. Sci., **77,** 255–272.

Mansour, K., and Mansour-Bek, J. J. 1934 On the digestion of wood by insects. J. Exptl. Biol., **11,** 243–256.

Mareschal, P. 1914 Rapport phytopathologique pour l'année 1913. Rev. Phytopath. App., nos. 18–19, pp. 9–13.

Mariani, G. 1940 Caratteristiche del Ceppo Etiopico di *Rickettsia rocha limae*. Ann. Igiene, **50,** 59–66.

Marpmann, G. 1897 Bakteriologische Mitteilungen. III. Ueber den Zusammenhang von pathogenen Bakterien mit Fliegen. Zentr. Bakt. Parasitenk. Infekt., I, **22,** 122–132.

Martchouk, P. 1934 Méthode microbiologique de laboratoire pour combattre les nuisibles à l'économie rurale: chenilles de *Phlyctaenodes sticticalis, Malacosoma neustria* L., *Hyponomeuta malinellus* Zell. et *Pieris brassicae*. Mikrobiologichnii Zhurnal, **1**, 20–21.

Martinaglia, G.   1932   The fate of anthrax bacilli in ticks from an anthrax carcass. J. Am. Vet. Med. Assoc., **80**, 805–806.

Marzocchi, V.   1908   Sul parassita del giallume del *Bombyx mori*, etc. Arch. Parasitol., **12**, 456–466.

Masera, E.   1934a   Il *Bacterium prodigiosum* L. et N. nella patologia del baco da seta. R. Staz. Bacologica Sper. Padova, Ann., **47**, 90–98.

Masera, E.   1934b   Il *Bacillus prodigiosus* nella patologia del baco da seta. R. Staz. Bacologica Sper. Padova, Ann., **47**, 99–102.

Masera, E.   1936a   Il *Bacillus prodigiosus* Flügge nella patologia del baco da seta e degli insetti. R. Staz. Bacologia Sper. Padova, Ann., **48**, 409–416.

Masera, E.   1936b   Compartemente del *Bombyx mori* L. alla infezione sperimentale del *Bacterium prodigiosum* L. et N.   R. Staz. Bacologica Sper. Padova, Ann., **48**, 417–422.

Masera, E.   1936c   Fenomeni de antagonismo et antibiosi fra *Bacillus prodigiosus* Flügge e *Beauveria bassiana* Vuill. R. Staz. Bacologica Sper. Padova, Ann., **48**, 423, 458.

Masera, E.   1936d   Flora microbica nelle uova di *"Bombyx mori."* R. Staz. Bacologica Sper. Padova, Ann., **48**, 459–476.

Mason, J. H., and Alexander, R. A.   1939   Studies of the rickettsias of the typhus-Rocky-Mountain-spotted-fever group in South Africa. IV. Discussion and classification. Onderstepoort J. Vet. Sci. Animal Ind., **13**, 67–76.

Mattes, O.   1927   Parasitäre Krankheiten der Mehlmottenlarven und Versuche über ihre Verwendbarkeit als biologisches Bekämpfungsmittel. Sitzber. Ges. Beforder. ges. Naturw. Marburg, **62**, 381–417.

Maxcy, K. F.   1926   An epidemiological study of endemic typhus (Brill's disease) in the southeastern United States. With special reference to its mode of transmission. Pub. Health Repts., **41**, 2967–2995.

May, E.   1941   The behavior of the intestinal protozoa of termites at the time of the last ecdysis. Trans. Am. Microscop. Soc., **60**, 281–292.

Megaw, J. W. D.   1935   Typhus fevers in the tropics. Trans. Roy. Soc. Trop. Med. Hyg., **29**, 105–110.

Megaw, J. W. D.   1943   (See abstract of paper by C. B. Philip, Trop. Diseases Bull., **40**, 828–830.)

Meillon, B. de, and Muspratt, J.   1943   Germination of the sporangia of *Coelomomyces* Keilin.   Nature, **152**, no. 3861, 507.

Melampy, R. M., and MacLeod, G. F.   1938   Bacteria isolated from the gut of the larval *Agriotes mancus* (Say). J. Econ. Entomol., **31**, 320.

Mera, B.   1943   Present status of human bartonellosis.   Bol. Oficina Sanit. Panamericana, **22**, 304–309.

Mercier, L.   1906   Les corps bactéroides de la blatte (*Periplaneta orientalis*): *Bacillus cuenoti* n. sp.   Compt. Rend. Soc. Biol., **58**, 682–684.

Mercier, L.   1907   Recherches sur les bactéroides des blattides.   Arch. Protistenk., 9, 346–358.

Mereshkovsky, S.   1925   Ann. State Inst. Exptl. Agron., Leningr., 3, 7.

Merrill, M. H., Lacaillade, C. W., Jr., and Ten Broeck, C.   1934   Mosquito transmission of equine encephalomyelitis.   Science, 80, 251–252.

Merrill, M. H., and Ten Broeck, C.   1934   Multiplication of equine encephalomyelitis virus in mosquitoes.   Proc. Soc. Exptl. Biol. Med., 32, 421–423.

Merrill, M. H., and Ten Broeck, C.   1935   The transmission of equine encephalomyelitis virus by *Aëdes aegypti*.   J. Exptl. Med., 62, 687–695.

Messerlin, A., and Couzi, G.   1942   Epidémie infantile grave à *Salmonella suipestifer*.   Bull. Inst. Hyg. Maroc, 2, 15–33.   (Published 1943.)

Metalnikov, S.   1914   De la tuberculose chez les insectes.   Compt. Rend. Soc. Biol., 76, 95–96.

Metalnikov, S.   1920a   Immunité de la chenille contre divers microbes.   Compt. Rend. Soc. Biol., 83, 119–121.

Metalnikov, S.   1920b   Sur la digestion des bacilles tuberculeux dans le corps des chenilles des mites des abeilles (*Galleria mellonella*).   Compt. Rend. Soc. Biol., 83, 214–215.

Metalnikov, S.   1922   Une épizootie chez les chenilles de *Galleria mellonella*.   Compt. Rend. Hebdom. Acad. Sci., Paris, 175, 68–70.

Metalnikov, S.   1924   Sur l'hérédité de l'immunité acquise.   Compt. Rend. Acad. Sci., Paris, 179, 514–516.

Metalnikov, S.   1927   L'infection microbienne et l'immunité chez la mite des abeilles *Galleria mellonella*.   Monog. Inst. Pasteur.   Masson et Cie., Paris. 140 pp.

Metalnikov, S.   1930   Utilisation des microbes dans la lutte contre *Lymantria* et autres insectes nuisibles.   Compt. Rend. Soc. Biol., 105, 535–537.

Metalnikov, S., and Chorine, V.   1928a   Maladies bactériennes chenilles de la pyrale du maïs.   (*Pyrausta nubilalis* Hb.).   Compt. Rend. Acad. Sci., Paris, 186, 546–549.

Metalnikov, S., and Chorine, V.   1928b   The infectious diseases of *Pyrausta nubilalis* Hb.   Intern. Corn Borer Invest., Sci. Repts., 1, 41–69.

Metalnikov, S., and Chorine, V.   1929a   L'utilisation des microbes dans la lutte contre la pyrale du maïs *Pyrausta nubilalis* Hb.   Ann. Inst. Pasteur, 43, 1391–1395.

Metalnikov, S., and Chorine, V.   1929b   On the natural and acquired immunity of *Pyrausta nubilalis*.   Intern. Corn Borer Invest., Sci. Repts., 2, 22–38.

Metalnikov, S., and Chorine, V.   1929c   Experiments on the use of bacteria to destroy the corn borer.   Intern. Corn Borer Invest., Sci. Repts., 2, 54–59.

Metalnikov, S., and Chorine, V.   1929d   On the infection of the gypsy moth and certain other insects with *Bacterium thuringiensis*.   Intern. Corn Borer Invest., Sci. Repts., 2, 60–61.

Metalnikov, S., and Chorine, V.   1929e   Maladies microbiennes chez les chenilles de *Pyrausta nubilalis* Hb. Ann. Inst. Pasteur, **43**, 136-151.

Metalnikov, S., Ellinger, T., and Chorine, V. 1928 A new yeast species, isolated from diseased larvae of *Pyrausta nubilalis* Hb. Intern. Corn Borer Invest., Sci. Repts., **1**, 70-71.

Metalnikov, S., Ermolaev, J., and Skobaltzyn, V. 1930 New bacteria pathogenic to the larvae of *Pyrausta nubilalis* Hb. Intern. Corn Borer Invest., Sci. Repts., **3**, 28-36.

Metalnikov, S., and Gaschen, H. 1921 Sur la rapidité d'immunisation chez la chenille de *Galleria mellonella*. Compt. Rend. Soc. Biol., **85**, 224-226.

Metalnikov, S., Kostritsky, L., and Toumanoff, H. 1924 *Bacterium tumefaciens* chez les chenilles de *Galleria mellonella*. Compt. Rend. Acad. Sci., Paris, **179**, 225-227.

Metalnikov, S., and Meng, L. Y. 1935 Utilisation des microbes contre les courtilières. Compt. Rend. Acad. Sci., Paris, **201**, 367-368.

Metalnikov, S., and Metalnikov, S. S. 1932 Maladies des vers du caton (*Gelechia gossypiella* et *Prodenia litura*). Compt. Rend. Acad. Agr., **18**, 203-207.

Metalnikov, S., and Metalnikov, S. S. 1933 Utilisation des bactéries dans la lutte contre les insectes nuisibles aux catonniers. Compt. Rend. Soc. Biol., **113**, 169-172.

Metalnikov, S., and Toumanoff, K. 1928 Experimental researches on the infection of *Pyrausta nubilalis* by entomophytic fungi. Intern. Corn Borer Invest., Sci. Repts., **1**, 72-73.

Metchnikoff, E. 1866 Embryologische Studien an Insekten. Vorl. Mitt. Z. Wiss. Zool., **16**, 23-32.

Metchnikoff, E. 1879 Diseases of the larva of the grain weevil. Insects harmful to agriculture [series]. Issue III, The grain weevil. Published by the Commission attached to the Odessa Zemstvo office for the investigation of the problem of insects harmful to agriculture. Odessa. 32 pp. [In Russian.]

Metchnikoff, E. 1905 Immunity in infective diseases. Tr. F. G. Binnie. Cambridge Univ. Press, Cambridge. pp. 40-41.

Meves, F. 1918 Die Plastosomentheorie der Vererbung. Arch. Mikrskop. Anat., **92**, 41-136.

Meyer, K. F. 1925 The "bacterial symbiosis" in the concretion deposits of certain operculate land mollusks of the families Cyclostomatidae and Annulariidae. J. Infectious Diseases, **36**, 1-107.

Meyer, K. F. 1938 The rôle of the infected and the infective flea in the spread of sylvatic plague. Vierteljahrsschr. Naturforsch. Ges. Zürich, **83**, 160-169.

Meyer, K. F. 1942 The ecology of plague. Medicine, **21**, 143-174.

Miller, E. D. 1937 A study of the bacterial and alleged mitochondrial content of the cells of the clover nodule. Biol. Bull., **73**, 112-125.

Milovidov, P. F. 1928 A propos des bactéroides des blattes (*Blattella germanica*). Compt. Rend. Soc. Biol., **99**, 127-128.

Milzer, A. 1942 Studies on the transmission of lymphocytic choriomeningitis virus by arthropods. J. Infectious Diseases, **70**, 152-172.

Minchin, E. A., and Thomson, J. D. 1915 The rat-trypanosome, *Trypanosoma lewisi,* in its relation to the rat-flea, *Ceratophyllus fasciatus.* Quart. J. Microscop. Sci., **60,** 463–692.

Mingazinni, P. 1889 Richerche sul tubo digerente dei lamellicorni fitofagi. Boll. Soc. Nat. Napoli, **3,** 24–30.

Mingazinni, P. 1889 Richerche sul canale digerente delle larve dei lamellicorni fitofagi. Mitt. Zool. Stat., **9,** 1–112.

Mitchell, E. 1907 Mosquito life. G. P. Putnam and Sons, New York. 145 pp.

Mitscherlich, E. 1941 Die aetiologische Bedeutung von *Rickettsia conjunctivae* (Coles, 1931) für die spezifische Kerato-Konjunktivitis der Schafe in Deutsch-Südwestafrika. Z. Infektionskrankh. Parasit. Krankh. Hyg. Haustiere, **57,** 271–287.

Mitscherlich, E. 1943 Die Uebertragung der Kerato-Conjunctivitis infectiosa des Rindes durch Fliegen und die Tenazität von *Rickettsia conjunctivae* in der Aussenwelt. Deut. Tropenmed. Z., **47,** 57–64.

Mitzmain, M. B. 1914 Summary of experiments in the transmission of anthrax by biting flies. Hyg. Lab. Bull., **94,** 41–48.

Mohamed, Z. 1939 The discovery of *Rickettsia* in a fish. Ministry Agr., Egypt, Tech. Sci. Service Bull., 214. 6 pp.

Möller, A. 1893 Die Pilzgärten einiger südamerikanischer Ameisen. Schlimper's Botan. Mitt. Trop., **6,** 1–127.

Möller, A. 1901 Phycomyceten und Ascomyceten. Untersuchungen aus Brasilien. Schlimper's Botan. Mitt. Trop., **9,** 1–319.

Moniez, R. 1887 Sur un champignon parasite du *Lecanium hesperidum* (*Lecaniascus polymorphus nobis*). Bull. Soc. Zool., **12,** 150–152.

Monteiro, J. L. 1931 Estudos sobre o typho exanthematico de São Paulo. Mem. Inst. Butantan, **6,** 5–135.

Monteiro, J. L., and Fonseca, F. 1932 Typho exanthematico de São Paulo. XII. Sobre um "virus" isolado de ratos da zona urbana da cidade e suas relacoes com o do typho exanthematico de São Paulo. Brasil Medico, **46,** 1029–1033.

Mooser, H. 1928a An American variety of typhus. Trans. Roy. Soc. Trop. Med. Hyg., **22,** 175–176.

Mooser, H. 1928b Reaction of guinea-pigs to Mexican typhus (tabardillo). J. Am. Med. Assoc., **91,** 19–20.

Mooser, H. 1929 Tabardillo, an American variety of typhus. J. Infectious Diseases, **44,** 186–193.

Mooser, H., Castañeda, M. R., and Zinsser, H. 1931a Rats as carriers of Mexican typhus fever. J. Am. Med. Assoc., **97,** 231–232.

Mooser, H., Castañeda, M. R., and Zinsser, H. 1931b The transmission of the virus of Mexican typhus from rat to rat by *Polyplax spinulosus.* J. Exptl. Med., **54,** 567–575.

Mooser, H., and Dummer, C. 1930a Experimental transmission of endemic typhus of the southeastern Atlantic states by the body louse. J. Infectious Diseases, **46,** 170–172.

Mooser, H., and Dummer, C. 1930b On the relation of the organisms in the tunica vaginalis of animals inoculated with Mexican typhus to *Rickettsia prowazeki* and to the causative agent of that disease. J. Exptl. Med., **51**, 189–207.

Mooser, H., Varela, G., and Pilz, H. 1934 Experiments on conversion of typhus strains. J. Exptl. Med., **59**, 137–157.

Moragues, V., Pinkerton, H., and Greiff, D. 1944 Therapeutic effectiveness of penicillin in experimental murine typhus infection in dba mice. J. Exptl. Med., **79**, 431–438.

Morgan, H. de R., and Ledingham, J. C. G. 1909 The bacteriology of summer diarrhoea. Proc. Roy. Soc. Med., **2**, 133–158.

Morgan, J. C., and Harvey, D. 1909 The bacteriology of summer diarrhoea. Proc. Roy. Soc. Med., **11** (Epidem. Sect.), 133–158.

Morris, H. 1918 Blood-sucking insects as transmitters of anthrax or charbon. Louisiana Agr. Expt. Sta. Bull., 163. 15 pp.

Moshkovsky, Sh. D. [Mochovski, Ch.] 1937 Sur l'existence, chez le cobaye, d'une rickettsiose chronique déterminée par *Ehrlichia* (*Rickettsia*) *kurlovi* subg. nov. sp. nov. Compt. Rend. Soc. Biol., **126**, 379–382.

Moshkovsky, Sh. D. 1945 Cytotropic agents of infections and the place of Rickettsia among Chlamydozoa. Advances in Modern Biol. (Uspekhi Souremennoi Biologii), **19**, 1–44.

Mosing, H. 1936 Une nouvelle infection à *Rickettsia*: *Rickettsia weigli* nov. sp. Arch. Inst. Pasteur Tunis, **25**, 373–387.

Mudrow, E. 1932 Über die intrazellulären Symbionten der Zecken. Z. Parasitenk., **5**, 138–183.

Mühlens, P. 1913 XVI. Rückfallfieber-Spirochäten. XVII. Andere, zum Teils als pathogenen geltende Spirochäten. Kolle & Wassermann's Handb. Path. Mikroorg. 2d ed. G. Fischer, Jena. VII, 864, 921.

Müller, J. 1915 Zur Naturgeschichte der Kleiderläuse. Oesterr. Sanitätswesen, **27**, 1–63.

Munk, F., and Rocha-Lima, H. da 1917 Klinik und Aetiologie des sogennanten "wolhynischen Fiebers" (Werner-Hissche Krankheit). II. Ergebnis der aetiologischen Untersuchungen und deren Beziehungen zur Fleckfieberforschung. Münch. Med. Wochschr., **64**, 1422–1426.

Munro, J. A. 1936 Disease checks Bertha army worms. J. Econ. Entomol., **29**, 218.

Muttkowski, R. A. 1923 Studies on the blood of insects. I. The composition of the blood. Bull. Brooklyn Entomol. Soc., **18**, 127–136.

Muttkowski, R. A. 1924 Studies on the blood of insects. II. The structural elements of the blood. Bull. Brooklyn Entomol. Soc., **18**, 127–136.

Nagayo, M., Tamiya, T., Mitamura, T., and Sato, K. 1930 On the virus of tsutsugamushi disease and its demonstration by a new method. Japan. J. Exptl. Med., **8**, 309–318.

Nauck, E. G., and Weyer, F. 1941a Erfahrungen bei der Zucht von Kleiderläusen und der kunstlichen Infektion von Läusen mit Fleckfieber. Zentr. Bakt. Parasitenk. Infekt., I Orig., **147**, 353.

Nauck, E. G., and Weyer, F. 1941b Versuche zur Züchtung von Rickettsien in explantiertem Läusegewebe. Zentr. Bakt. Parasitenk. Infekt., I, **147**, 365–376.

Nauck, E. G., and Zumpt, F. 1941 Versuche zur Uebertragung des epidemischen Fleckfiebers durch die Wanzen *Cimex lectularius* L. und *Triatoma rubrofasciata* De Geer. Zentr. Bakt. Parasitenk. Infekt., I, **147**, 376–381.

Needham, J. G., Galtsoff, P. S., Lutz, F. E., and Welch, P. S. (editors). 1937 Culture methods for invertebrate animals. Comstock Publishing Co., Ithaca, N.Y. 590 pp.

Needham, N. Y. 1937 A bacterial disease of *Aphis rumicus* Linn., apparently caused by *Bacillus lathyri* Manns and Taubenhaus. Ann. Applied Biol., **24**, 144–147.

Neger, F. W. 1908 Ambrosiapilze. I. Ber. Deut. Botan. Ges., **26**, 735–754.

Neger, F. W. 1909 Ambrosiapilze. II. Die Ambrosia der Holzbohrkafer. Ber. Deut. Botan. Ges., **27**, 372–389.

Neger, F. W. 1910 Ambrosiapilze. III. Weitere Beobachtungen an Ambrosiagallen. Ber. Deut. Botan. Ges., **28**, 455–480.

Neger, F. W. 1911 Zur Ueberträgung des Ambrosiapilzes von *Xyleborus dispar*. Naturwiss. Z. Forst- u. Landw., **9**, 223–225.

Neill, M. H. 1917 Experimental typhus in guinea pigs. Pub. Health Repts., **32**, 1105–1108.

Neitz, W. O. 1943 The aetiology of East Coast fever. J. S. African Vet. Med. Assoc., **14**, 39–46.

Neitz, W. O., Alexander, R. A., and Du Toit, P. J. 1934 *Eperythrozoon ovis* (sp. nov.), infection in sheep. Onderstepoort J. Vet. Sci. Animal Ind., **3**, 263–271.

Neitz, W. O., Alexander, R. A., and Mason, J. H. 1941 The transmission of tick-bite fever by the dog tick *Rhipicephalus sanguineus*, Labr. Onderstepoort J. Vet. Sci. Animal Ind., **16**, 9–17.

Neitz, W. O., and Thomas, A. D. 1938 Rickettsiosis in the dog. J. S. African Vet. Med. Assoc., **9**, 166–174.

Neukomm, A. 1927a Action des rayons ultra-violets sur les bactéroides des blattes (*Blattela germanica*). Compt. Rend. Soc. Biol., **96**, 1155–1156.

Neukomm, A. 1927b Sur la structure des bactéroides des blattes (*Blattela germanica*). Compt. Rend. Soc. Biol., **96**, 306–308.

Neukomm, A. 1932 La réaction de la fixation du complément apliquée à l'étude des bactéroides des blattes ("*Blatella germanica*"). Compt. Rend. Soc. Biol., **111**, 928–929.

Nicholas, G. E. 1873 The fly in its sanitary aspect. Lancet, pt. II, 724.

Nicholls, L. 1912 The transmission of pathogenic micro-organisms by flies in Saint Lucia. Bull. Entomol. Research, **3**, 251–267.

Nicoll, W. 1911 On the varieties of *Bacillus coli* associated with the housefly (*Musca domestica*). J. Hyg., **11**, 381–389.

Nicolle, C., and Anderson, C. 1928 Un nouveau spirochète recurrent pathogène pour le cobaye, *Spirochaeta sogdianum,* transmis par *Ornithodorus papillipes.* Arch. Inst. Pasteur Tunis, **17,** 295–309.

Nicolle, C., Comte, C., and Conseil, E. 1909 Transmission expérimentale du typhus exanthématique par le pou du corps. Compt. Rend. Acad. Sci., Paris, **149,** 486–489.

Nieschulz, O. 1929 Über die mechanische Übertragung von einigen Bakterienkrankheiten durch blutsaugende Insekten. Arch. Schiffs- u. Tropen-Hyg., **33,** 282–287.

Nieschulz, O. 1935 Uebertragungsversuche mit Milzbrand und Bettwanzen (*Cimex lectularius*). Zentr. Bakt. Parasitenk. Infekt., I, **135,** 228–229.

Nieschulz, O., and Bos, A. 1940 Versuche mit Mücken und Geflügelspirochäten. Zentr. Bakt. Parasitenk. Infekt., I Orig., **145,** 258–261.

Nieschulz, O., and Kranefeld, F. C. 1929 Experimentelle Untersuchungen über die Uebertragung der Büffelseuche durch Insekten. Zentr. Bakt. Parasitenk. Infekt., I Orig., **113,** 403–417.

Noc, F. 1920 Les spirochétoses humaines à Dakar (Senegal). Bull. Soc. Path. Exotique, **13,** 672.

Noguchi, H. 1926a Cultivation of rickettsia-like microorganisms from the Rocky Mountain spotted fever tick, *Dermacentor andersoni.* J. Exptl. Med., **43,** 515–532.

Noguchi, H. 1926b A filter-passing virus obtained from *Dermacentor andersoni.* J. Exptl. Med., **44,** 1–10.

Noguchi, H., and Tilden, E. B. 1926 Comparative studies of herpetomonads and leishmanias. I. Cultivation of herpetomonads from insects and plants. J. Exptl. Med., **44,** 307–325.

Nöller, W. 1912 Ueber Blutprotozoen einheimischer Nagetiere und ihre Übertragung. Berlin. Klin. Wochschr., 49, **1,** 524–525.

Nöller, W. 1916 Beitrag zur Flecktyphusübertragung durch Läuse. Berlin. Klin. Wochschr., **53,** 778–780.

Nöller, W. 1917a Blut- und Insektenflagellaten Züchtung auf Platten. Arch. Schiffs- u. Tropen-Hyg., **21,** 53.

Nöller, W. 1917b Neue Zuchtungsergebnisse bei Blut- und Insektemparasiten. Berlin Klin. Wochschr., **54,** 346–348.

Nöller, W., and Kuchling, M. 1923 Zur Züchtung des Schaftrypanosomas und der Schaflausrickettsia aus dem Schalblute. Zentr. Bakt. Parasitenk. Infekt., I Ref., **75,** 237–238.

Nolte, H.-W. 1937 Beiträge zur Kenntnis der Symbiontischen Einrichtungen der Gattung *Apion* Hrbst. Z. Morphol. Ökol. Tiere, **33,** 165–200.

Nolte, H.-W. 1938 Die Legeapparate der Dorcatominen (Anobiidae) unter besonderer Berücksichtigung der symbiontischen Einrichtungen. Zool. Anz. Suppl. (Verhandl. Deut. Zool. Ges.), **11,** 147–154.

Northrop, J. H. 1917 The role of yeast in the nutrition of an insect (*Drosophila*). J. Biol. Chem., **30,** 181–187.

Northrup, Z. 1914a A bacterial disease of June beetle larvae, *Lachnosterna* sp. Mich. Agr. Coll. Exptl. Sta. Tech. Bull., 18. 36 pp.

Northrup, Z. 1914b A bacterial disease of the June beetles, *Lachnosterna*. Zentr. Bakt., Parasitenk. Infekt., II, 41, 321-339.

Novy, F. G. 1907 The role of Protozoa in pathology. Proc. Path. Soc., Phila., 1, 1.

Novy, F. G., and Knapp, R. E. 1906 Studies on *Spirillum obermeieri* and related organisms. J. Infectious Diseases, 3, 291.

Nowakowski, L. 1884 Entomophthoreae. Przycrynek doznajómości pasorzytnych grzybków sprariajacyck pomór owadów. Pamietnik Akad. Umiejejnosci Krakau, 8, 153-183.

Nuttall, G. H. F. 1898 Zur Aufklärung der Rolle, welche stechende Insekten bei der Verbreitung von Infektionskrankheiten spielen. Zentr. Bakt. Parasitenk. Infekt., I, 23, 625-635.

Nuttall, G. H. F. 1899 On the role of insects, arachnids and myriapods, as carriers in the spread of bacterial and parasitic diseases of man and animals. A critical and historical study. Johns Hopkins Hosp. Repts., 8, 1-155.

Nuttall, G. H. F. 1908 Spirochaetosis in man and animals. J. Roy. Inst. Pub. Health, 16, 419-464.

Nyka, W. 1944 A method for staining rickettsiae of typhus in histological sections. J. Path. Bact., 56, 264.

Obermeier, O. 1873 Vorkommen feinster, eine eigen Bewegung zeigender Fäden im Blue von Recurrenskranken. Centr. Med. Wiss., 11, 145.

Odriozola, E. 1898 La maladie de Carrión ou la verruga péruvienne. G. Carre et C. Naud, Paris. 217 pp.

O'Farrell, W. R. 1913 Hereditary infection with special reference to its occurrence in *Hyalomma aegyptium* infected with *Crithidia hyalommae*. Ann. Trop. Med. Paras. 7, 545-562.

Ogata, M. 1897 Ueber der Pestepidemie in Formosa. Zentr. Bakt. Parasitenk. Infekt., 21, 769-777.

Ogata, N. 1930 Aetiology der Tsutsugamushikrankheit: *Rickettsia tsutsugamushi*. Trans. Congr. Far East. Assoc. Trop. Med., 8th Congr., 1930, 2, 167-171. (Published June, 1932.)

Ogata, N. 1931 Aetiologie der Tsutsugamushikrankheit: *Rickettsia tsutsugamushi*. Zentr. Bakt. Parasitenk. Infekt., Orig., I, 122, 249-253.

Olarte, J. 1942 *Klebsiella ozaenae* aislada del cobayo. Rev. Inst. Salubridad Enfermedades Trop. (Mex.), 3, 135-138.

Olmer, D. 1930 La fièvre exanthématique. Bull. Mens., Office Intern. Hyg. Publ., 22, 1494-1521.

Olsufiev, N. G. 1939a The role of *Stomoxys calcitrans* L. in the transmission and preservation of tularemia infection. Arkh. Biol. Nauk., 58, 25-31. (In Russian with English summary.)

Olsufiev, N. G. 1939b The role of mosquitoes in the transmission and retention of tularemia. Problems Regional Parasitol., **3**, 213–246. (In Russian with English summary.)

Olsufiev, N. G. 1939c The specific composition and seasonal dynamics of the blood-sucking diptera numbers in the delta of the Volga and their possible role in the epidemiology of tularemia. Zool. zhur., **18**, 786, 798. (In Russian with English summary.)

Olsufiev, N. G. 1940a Nouvelles données expérimentales sur la transmission de l'infection tularemique par les taons (*Tabanus*). Med. parasitol. i parazit. bol., **9**, 260–270. (In Russian with French summary.)

Olsufiev, N. G. 1940b Results of studies of transmitters of tularemia in U.S.S.R. Proc. All-Union Congr. Microbiol., Epidemiol. Infectionists, 1939 (Moscow, 1940), pp. 247–253.

Olsufiev, N. G., and Golov, D. A. 1936 Horse flies as transmitters and conservators of tularemia. Animaux pathogenes, **2**, 187–226. (In Russian with English summary.)

Osborn, H. 1896 Insects affecting domestic animals. U.S. Dept. Agr., Bur. Entomol. Bull., 5.

Ostrolenk, M., and Hunter, A. C. 1939 Bacteria of the colon-aerogenes group on nut meats. Food Research, **4**, 453–460.

Ostrolenk, M., and Welch, H. 1942 The common house-fly (*Musca domestica*) as a source of pollution in food establishments. Food Research, **7**, 192–200.

Otto, R., and Bickhardt, R. 1941 Ueber das Gift der Fleckfieberrickettsien. Z. Hyg. Infektionskrankh., **123**, 447–462.

Otto, R., and Dietrich, S. 1917 Beiträge zur "Rickettsien"-Frage. Deut. Med. Wochschr., **43**, 577–580.

Packchanian, A. A. 1944 Malaria thick films contaminated with excretions of flies containing flagellates (*Herpetomonas*). Am. J. Trop. Med., **24**, 141–143.

Paillot, A. 1913 Coccobacilles parasites d'insectes. Compt. Rend. Acad. Sci., Paris, **157**, 608–611.

Paillot, A. 1916 Les coccobacilles du hanneton. Action pathogène sur quelques chenilles de macrolépidoptères. Compt. Rend. Soc. Biol., **79**, 1102–1103.

Paillot, A. 1917 Microbes nouveau parasites des chenilles de *Lymantria dispar*. Compt. Rend. Hebdom. Acad. Sci., **164**, 525–527.

Paillot, A. 1918 Coccobacillus nouveaux parasites du hanneton. Compt. Rend. Hebdom. Acad. Sci., **167**, 1046–1048.

Paillot, A. 1919a La pseudograsserie, maladie nouvelle des chenilles de *Lymantria dispar*. Compt. Rend. Acad. Sci., Paris, **168**, 258–260.

Paillot, A. 1919b Coccobacilles parasites des chenilles de *Pieris brassicae*. Compt. Rend. Acad. Sci., Paris, **168**, 476–478.

Paillot, A. 1919c La Karyokynetose, nouvelle réaction d'immunité naturelle observée chez les chenilles de macrolépidoptères. Compt. Rend. Hebdom. Acad. Sci., **169**, 396–398.

Paillot, A. 1919d Contribution à l'étude des parasites microbiens des insectes. Etude de *Bacillus hoplosternus* (Paillot). Ann. Inst. Pasteur, **33**, 403–419.

Paillot, A. 1920 L'immunité acquise chez les insectes. Compt. Rend. Soc. Biol., **83**, 278–280.

Paillot, A. 1922 Les maladies bactériennes des insectes. Utilisation en agriculture des bactéries entomophytes. Ann. Epiphyt. phytogénét., **8**, 95–291.

Paillot, A. 1923 Les caractères de l'immunité chez les insectes. Célébration du 75me Anniversaire Societé de Biologie, Paris, 120–123.

Paillot, A. 1924a Sur deux bactéries parasites des larves de *Neurotoma nemoralis*. Compt. Rend. Hebdom. Acad. Sci., **178**, 246–249.

Paillot, A. 1924b Sur une nouvelle maladie des chenilles de *Pieris brassicae* et sur les maladies du noyau chez les insectes. Compt. Rend. Acad. Sci., Paris, **179**, 1353.

Paillot, A. 1926 Sur l'étiologie et l'épidémiologie de la gattine du ver à soie ou "maladie des têtes claires." Compt. Rend. Acad. Sci., Paris, **183**, 251.

Paillot, A. 1928a On the natural equilibrium of *Pyrausta nubilalis* Hb. Intern. Corn Borer Invest., Sci. Repts., **1**, 77–106.

Paillot, A. 1928b Les maladies du ver à soie grasserie et dysenteries. Editions du Service Photographique, de l'Université, Lyon. 328 pp.

Paillot, A. 1929 Sur la spécificité parasitaire des bactéries infectant normalement les pucerons. Compt. Rend. Soc. Biol., **103**, 89–91.

Paillot, A. 1930a Les réactions cellulaires et humorales d'immunité antimicrobienne dans le phenomene de la symbiose chez *Macrosiphum jaceae*. Compt. Rend. Acad. Sci., Paris, **190**, 330–332.

Paillot, A. 1930b Traité des maladies du ver à soie. G. Doin et Cie., Paris. 279 pp.

Paillot, A. 1931a Parasitisme et symbiose chez les aphides. Compt. Rend. Acad. Sci., Paris, **193**, 300–301.

Paillot, A. 1931b Les variations morphologiques du bacille symbiotique de *Macrosiphum tanaceti*. Compt. Rend. Acad. Sci., Paris, **193**, 1222–1224.

Paillot, A. 1932 Les variations du parasitisme bactérien normal chez la *Chaitophorus lyropictus* Kessl. Compt. Rend. Acad. Sci., Paris, **194**, 135–137.

Paillot, A. 1933 L'infection chez les insectes. G. Patissier, Trévoux. 535 pp.

Paillot, A. 1936 Contribution à l'étude des maladies à ultravirus des insectes. Ann. Epiphyt. Phytogénét., **2**, 341–379.

Paillot, A. 1940 Existence d'une septicémie à spirochètes chez les chenilles de *Pieris brassicae*. Compt. Rend. Acad. Sci., Paris, **210**, 615–616.

Paillot, A. 1942 Un nouveau bacille sporulé pathogène pour le bombyx du murier: *Bacillus bombycoides* nov. spec. Compt. Rend. Acad. Agr. France, **28**, 158–161.

Paillot, A. 1943 Sur une nouvelle mycose des chenilles de *Pieris brassicae* L. Compt. Rend. Acad. Sci., Paris, **217**, 383–384.

Pammel, L. H. 1895 Bacteriosis of rutabaga (*Bacillus campestris*, n. sp.). Iowa Agr. Expt. Sta. Bull., **27**, 130–134.

Pang, K. H. 1941 Isolation of typhus rickettsia from rat mites during epidemic in an orphanage. Proc. Soc. Exptl. Biol. Med., **48**, 266–267.

Parker, H. B., Beyer, G. E., and Pothier, O. L. 1903 A study of the etiology of yellow fever. U.S. Pub. Health Marine Hosp. Service, Yellow Fever Inst. Bull., 13. 48 pp.

Parker, R. R. 1933 Recent studies of tick-borne diseases made at the United States Public Health Service Laboratory at Hamilton, Montana. Proc. 5th Pacific Sci. Congr., **6**, 3367–3374.

Parker, R. R. 1938 Rocky Mountain spotted fever. J. Am. Med. Assoc., **110**, 1185–1188, 1273–1278.

Parker, R. R. 1942 *Ornithodoros* ticks as a medium for the transportation of disease agents. Pub. Health Repts., **57**, 1963–1966.

Parker, R. R., Brooks, C. S., and Marsh, H. 1929 The occurrence of *Bacterium tularense* in the wood tick, *Dermacentor occidentalis*, in California. Pub. Health Repts., **44**, 1299–1300.

Parker, R. R., and Davis, G. E. 1938 A filter-passing infectious agent isolated from ticks. II. Transmission by *Dermacentor andersoni*. Pub. Health Repts., **53**, 2267–2270.

Parker, R. R., and Kohls, G. M. 1943 American Q fever: the occurrence of *Rickettsia diaporica* in *Amblyomma americanum* in eastern Texas. Pub. Health Repts., **58**, 1510–1511.

Parker, R. R., Kohls, G. M., Cox, G. W., and Davis, G. E. 1939 Observations on an infectious agent from *Amblyomma maculatum*. Pub. Health Repts., **54**, 1482–1484.

Parker, R. R., Kohls, G. M., and Steinhaus, E. A. 1943a *Amblyomma americanum*, a vector of Rocky Mountain spotted fever. Pub. Health Repts., **58**, 491.

Parker, R. R., Kohls, G. M., and Steinhaus, E. A. 1943b Rocky Mountain spotted fever: spontaneous infection in the tick *Amblyomma americanum*. Pub. Health Repts., **58**, 721–729.

Parker, R. R., and Spencer, R. R. 1924 Tularaemia. XI. Tularaemia infection in ticks of the species *Dermacentor andersoni* Stiles in the Bitterroot Valley, Mont. Pub. Health Repts., **39**, 1057–1073.

Parker, R. R., and Spencer, R. R. 1926 Hereditary transmission of tularaemia infection by the wood tick, *Dermacentor andersoni* Stiles. Pub. Health Repts., **41**, 1403–1407.

Parker, R. R., Spencer, R. R., and Francis, E. 1924 Tularaemia infection in ticks of the species *Dermacentor andersoni* Stiles in the Bitterroot Valley, Mont. Pub. Health Repts., **39**, 1057–1073.

Parker, R. R., and Steinhaus, E. A. 1943a Rocky Mountain spotted fever: duration of potency of tick-tissue vaccine. Pub. Health Repts., **58**, 230–232.

Parker, R. R., and Steinhaus, E. A. 1943b American and Australian Q fevers: persistence of infectious agents in guinea pig tissues after defervescence. Pub. Health Repts., **58**, 523–527.

Parker, R. R., and Steinhaus, E. A. 1943c *Salmonella enteritidis:* experimental transmission by the Rocky Mountain wood tick *Dermacentor andersoni* Stiles. Pub. Health Repts., **58**, 1010–1012.

Patiño-Camargo, L. 1941 Nuevas observaciones sobre un tercer foco di fiebre petequial (maculosa) en el hemisferio americano. Bol. Oficiana Sanit. Panamericana, **20**, 1112–1124.

Patton, W. S. 1908 The life-cycle of a species of *Crithidia* parasitic in the intestinal tract of *Gerris fossarum* Fabr. Arch. Protistenk., **12**, 131–146.

Patton, W. S. 1909 The life cycle of a species of *Crithidia* parasitic in the intestinal tracts of *Tabanus hilarius* and *Tabanus* sp. Arch. Protistenk., **15**, 333–362.

Patton, W. S. 1912 *Spirochaeta ctenocephali*, sp. nov., parasitic in the alimentary tract of the Indian dog flea, *Ctenocephalus felis.* Ann. Trop. Med., **6**, 357.

Patton, W. S. 1921 Studies on the flagellates of the genera *Herpetomonas, Crithidia,* and *Rhynchoidomonas.* No. 7. Some miscellaneous notes on insect flagellates. Indian J. Med. Research, **9**, 230–239.

Patton, W. S. 1931 Insects, ticks, mites, and venomous animals of medical and veterinary importance. Part II. Public health. H. R. Grubb, Ltd., Croydon. 740 pp.

Patton, W. S., and Cragg, F. W. 1913 A textbook of medical entomology. Christian Literature Society for India, London, Madras, and Calcutta. 764 pp.

Patton, W. S., and Evans, A. M. 1929 Insects, ticks, mites and venomous animals. Part I. Medical. H. R. Grubb, Ltd., Croydon. 786 pp.

Paul, J. R., Trask, J. D., Bishop, M. B., Melneck, J. L., and Casey, A. E. 1941 The detection of poliomyelitis virus in flies. Science, **94**, 395–396.

Pavlovsky, E. N., and Soloviev, V. D. 1940 Experimental study of the circulation of the encephalitis virus within the organism of the tick *Ixodes persulcatus.* Arch. Sci. Biol., **59**, 116–117.

Payne, N. M. 1933 A parasitic hymenopteron as a vector of an insect disease. Entomol. News, **44**, 22.

Peck, C. 1879 *Massospora* gen. nov. Thirty-first Rept., N.Y. State Mus. Nat. Hist., p. 44. *Massospora cicadina,* n. g. et sp. Ann. Rept. N.Y. State Museum, **31**, 44.

Peklo, J. 1912 Ueber symbiontische Bakterien der Aphiden. Ber. Deut. Botan. Ges., **30**, 416–419.

Peklo, J. 1916 Sur le puceron lanigère. Zemedelskeho Archivu., **7**. 18 pp.

Peppler, H. J. 1944 Usefulness of microorganisms in studying dispersion of flies. Bull. U.S. Army Med. Dept., **75**, 121–122.

Perroncito, E. 1899 Sopra una speciale forma di micosi delle Zanzare. Boll. Acad. Med. Torino. (Not seen; taken from Keilin, 1921.)

Petch, T. 1923a The genus *Trichosterigma* Petch. Trans. Brit. Mycol. Soc., **9**, 93–94.

Petch, T. 1923b Studies in entomogenous fungi. III. *Torrubiella.* Trans. Brit. Mycol. Soc., **9**, 108–128.

Petch, T. 1924a Studies in entomogenous fungi. IV. Some Ceylon *Cordyceps*. Trans. Brit. Mycol. Soc., **10**, 28–45.

Petch, T. 1924b Studies in entomogenous fungi. V. *Myriangium*. Trans. Brit. Mycol. Soc., **10**, 45–80.

Petch, T. 1925a Studies in entomogenous fungi. VI. *Cephalosporium* and associated fungi. Trans. Brit. Mycol. Soc., **10**, 152–182.

Petch, T. 1925b Studies in entomogenous fungi. VII. *Spicaria*. Trans. Brit. Mycol. Soc., **10**, 183–189.

Petch, T. 1925c Entomogenous fungi and their use in controlling insect pests. Ceylon Dept. Agr. Bull., **71**, 1–40.

Petch, T. 1926 Studies in entomogenous fungi. VIII. Notes on *Beauveria*. Trans. Brit. Mycol. Soc., **10**, 244–271.

Peterson, A. 1934, 1937 A manual of entomological equipment and methods. Part I. Edwards Bros., Inc., Ann Arbor, Mich. 180 pp. Part II. J. S. Swift Co., Inc., St. Louis. 334 pp.

Petri, L. 1905 Ulteriori richerche sopra i batteri che si trovani nell' intestino della larva della *Mosca olearia*. Atti. Accad. Lincei, **14**, 399–404.

Petri, L. 1909 Richerche sopra i batteri intestinali della *Mosca olearia*. Mem. Staz. Patol. Vegetale, Roma, **4**, 1–130.

Petri, L. 1910 Utersuchungen über die Darm-bakterien der Olivenfliege. Zentr. Bakt. Parasitenk. Infekt., II, **26**, 357–367.

Peus, F. 1938 Die Flöhe hygienische Zoologie Monographien. Paul Schöps, Leipzig. Vol. V, 106 pp.

Pfeiffer, H. 1931 Beiträge zu der Bakteriensymbiose der Bettwanze (*Cimex lectularius*) und der Schwalbenwanze (*Oeciacus hirundinis*). Zentr. Bakt. Parasitenk. Infekt., Orig., I, **123**, 151–171.

Pfeiffer, H., and Stammer, H.-J. 1930 Pathogenes Leuchten bei Insekten. Z. Morphol. Ökol. Tiere, **20**, 136–171.

Philip, C. B. 1929 Possibility of hereditary transmission of yellow fever virus by *Aëdes aegypti* (Linn.). J. Exptl. Med., **50**, 703–708.

Philip, C. B. 1930 Possibility of mechanical transmission by insects in experimental yellow fever. Ann. Trop. Med. and Parasitol., **24**, 493–501.

Philip, C. B. 1937 The transmission of disease by flies. Pub. Health Repts., Suppl. no. 29. 22 pp.

Philip, C. B. 1943 Nomenclature of the pathogenic rickettsiae. Am. J. Hyg., **37**, 301–309.

Philip, C. B., Davis, G. E., and Parker, R. R. 1932 Experimental transmission of tularaemia by mosquitoes. Pub. Health Repts., **47**, 2077–2088.

Philip, C. B., and Dias, E. 1938 Rocky Mountain spotted fever. Failure of triatomid bugs to transmit the virus experimentally. Mem. Inst. Oswaldo Cruz, **33**, 469–476.

Philip, C. B., and Jellison, W. L. 1934 The American dog tick, *Dermacentor variabilis,* as a host of *Bacterium tularense*. Pub. Health Repts., **49**, 386–392.

Philip, C. B., and Parker, R. R. 1933 Rocky Mountain spotted fever: investigation of sexual transmission in the wood tick *Dermacentor andersoni*. Pub. Health Repts., 48, 266–272.

Philip, C. B., and Parker, R. R. 1938 The persistence of the viruses of endemic (murine) typhus, Rocky Mountain spotted fever, and boutonneuse fever in tissues of experimental animals. Pub. Health Repts., 53, 1246–1251.

Phillips, E. F. 1907 Wax moths and American foul brood. U.S. Dept. Agr., Bur. Entomol. Bull., 75, 19–22.

Picard, F. 1913 Le *Cleonus mendicus* et le *Lixus scabricollis* charançons nuisibles à la betterave dans le midi de la France. Bull. Soc. Etude Vulgarisation Zool. Agr., 12, 129–137.

Picard, F. 1914 Les insectes nuisibles à la betterave dans la midi de la France. Vie Agr. Rur., 3, 390–391.

Picard, F., and Blanc, G. R. 1913a Sur une septicémie bacillaire des chenilles d'*Arctia caja* L. Compt. Rend. Acad. Sci., Paris, 156, 1334–1336.

Picard, F., and Blanc, G. R. 1913b Les infections à coccobacilles chez les insectes. Compt. Rend. Hebdom. Acad. Sci., 157, 79–81.

Picard, F., and Blanc, G. R. 1914 Les chelonies à chenilles barrués. Progrès Agr. Vit., 31, 261–266.

Pierantoni, U. 1909 L'origine di alcuni organi d'*Icerya purchasi* e la simbiosi ereditaria. Nota preliminare. Boll. Soc. Nat. Napoli, 23, 147–150.

Pierantoni, U. 1910a Origine e struttura del corpo ovale del *Dactylopius citri* e del corpo verde dell'*Aphis brassicae*. Boll. Soc. Nat. Napoli, 24, 1–4.

Pierantoni, U. 1910b Ulteriori osservazioni sulla simbiosi ereditaria degli omotteri. Zool. Anz., 36, 96–111.

Pierantoni, U. 1911 Sul corpo ovale del *Dactylopius citri*. Boll. Soc. Nat. Napoli, 24, 303–304.

Pierantoni, U. 1912 Studi sullo sviluppo d'*Icerya purchasi* Mask. Parte 1. Origine ed evoluzione degli elementi sessuali femminili. Arch. Zool. Ital., 5, 321–400.

Pierantoni, U. 1913 Struttura ed evoluzione dell'organo simbiotico di *Pseudococcus citri* Risso, e ciclo biologico del *Coccidomyces dactylopii* Buchner. Arch. Protistenk., 31, 300–316.

Pierantoni, U. 1929 L'organo simbiotico di *Silvanus surinamensis* (L.). Rend. Accad. Lincei, 9, 451–455.

Pierantoni, U. 1930 Origine e sviluppo degli organi symbiotici *Oryzaephilus* (*Silvanus*) *surinamensis* L. Atti Accad. Sci. Fis. Mat. Napoli, 18, 1–16.

Pierantoni, U. 1936 La simbiosi fisiologica nei termitidi xilofagi e nei loro flagellati intestinali. Arch. Zool. Ital., 22, 135–173.

Pierce, W. D. 1921 Sanitary entomology. Gorham Press, Boston. 518 pp.

Pinkerton, H. 1934 The study of typhus and Rocky Mountain spotted fever by the tissue culture method. Arch. Exptl. Zellforsch. Gewebezücht., 15, 425–430.

Pinkerton, H. 1936 Criteria for the accurate classification of the rickettsial diseases (rickettsioses). Parasitology, 28, 172–189.

Pinkerton, H. 1942 The pathogenic rickettsiae with particular reference to their nature, biologic properties, and classification. Bact. Rev., **6**, 37–78.

Pinkerton, H., and Hass, G. M. 1932 Spotted fever. I. Intranuclear rickettsiae in spotted fever studied in tissue culture. J. Exptl. Med., **56**, 151–156.

Pinkerton, H., and Haas, G. M. 1937 Spotted fever. III. The identification of *Dermacentroxenus rickettsi* and its differentiation from non-pathogenic rickettsia in ticks. J. Exptl. Med., **66**, 729–739.

Pinto, C. 1930 Arthropodes parasitos e transmissores de doencas. Pamenta de Mella e Co., Rio de Janeiro. Vol. I, 395 pp. (See p. 367.)

Plotz, H. 1943 Complement fixation in rickettsial diseases. Science, **97**, 20–21.

Plotz, H., Reagan, R. L., and Wertman, K. 1944 Differentiation between fièvre boutonneuse and Rocky Mountain spotted fever by means of complement fixation. Proc. Soc. Exptl. Biol. Med., **55**, 173–176.

Plotz, H., Smadel, J. E., Anderson, T. F., and Chambers, L. A. 1943 Morphological structure of rickettsiae. J. Exptl. Med., **77**, 355–358.

Poos, F. W., and Elliott, C. 1936 Certain insect vectors of *Aplanobacter stewarti*. J. Agr. Research, **52**, 585–608.

Porter, A. 1909 *Amoeba chironomi*, nov. sp., parasitic in the alimentary tract of the larva of *Chironomus*. Parasitology, **2**, 32–41.

Porter, A. 1910 The structure and life-history of *Crithidia melophagia* (Flu), an endoparasite of the sheep ked, *Melophagus ovinus*. Quart. J. Microscop. Sci., **55**, 189–224.

Portier, P. 1911 Passage de l'asepsie à l'envahissement symbiotique humoral et tissulaire par les microorganismes dans la série des larves des insectes. Compt. Rend. Soc. Biol., **70**, 914–917.

Portier, P. 1918 Les symbiotes. Masson et Cie., Paris. 315 pp.

Portier, P. 1919 Développement complet des larves de *Tenebrio molitor*, obtenu au moyen d'une nourriture sterilisée à haute température (130°). Compt. Rend. Soc. Biol., **82**, 59–60.

Pospelov, V. P. 1913 *Bothynoderes punctiventris*, Germ. and methods of fighting it. 2d ed. Central Board of Land Administration and Agriculture, Dept. of Agr., St. Petersburg. 116 pp.

Pospelov, V. P. 1926 The influence of temperature on the maturation and general health of *Locusta migratoria*, L. Bull. Entomol. Research, **16**, 363–367.

Pospelov, V. P. 1927 Flacherie (septicaemia) of the larvae of *Agrotis segetum*, Schiff. Rept. Bur. Appl. Entomol., **3**, 1–23. (In Russian with English summary.)

Pospelov, V. P. 1929 Intracellular symbiosis and its relation to insect diseases. Ann. State Inst. Exptl. Agronomy (Inst. Opytnoi Agronomii; Leningrad. Izvestiia.), **7**, 551–568.

Pospelov, V. P. 1936 Summary of the scientific research work of the institute of plant protection for the year 1935. Lenin Acad. Agr. Sci., pp. 318–321. (In Russian.)

Pospelov, V. P., and Noreiko, E. S. 1929 Wilt disease (Polyederkrankheit) of the caterpillars and yeasts *Debaryomyces tyrocola* Kon. as its virus. Repts. Applied Entomol., **4**, 167–183.

Prenant, A., Bouin, P., and Maillard, L. 1904 Traité d'histologie. Tome I. Cytologie générale et spéciale. Schleicher, Paris. 977 pp.

Pringault, E. 1921 Présence de spirochètes chez *Phlebotomus perniciosus* Newstead. Compt. Rend. Soc. Biol., **84**, 209.

Profft, J. 1937 Beiträge zur Symbiose der Aphiden und Psylliden. Z. Morphol. Ökol. Tiere, **32**, 289–326.

Proust, A. 1894 Pustule maligne transmise par des peaux de chèvres venant de chine, etc. Bull. Acad. Méd., **30**, 57–66.

Prowazek, S. von. 1907 Chlamydozoa. II. Gelbsucht der Seidenraupen. Arch. Protistenk., **10**, 358–364.

Prowazek, S. von. 1910 Parasitische Protozoen aus Japan, gesammelt von Herrn Dr. Mine in Fukuoka. Arch. Schiffs- u. Tropen-Hyg., **14**, 297.

Prowazek, S. von. 1912 Untersuchungen über die Gelbsucht der Seidenraupen. Zentr. Bakt. Parasitenk. Infekt., I Orig., **67**, 268–284.

Pshenichnov, A. V. 1943 A universal method for studying infections transmitted to man by blood-sucking insects and a new vaccine against spotted typhus. Zhur. Mikrobiol., Epidemiol. i Immunobiol., Nos. 1–2, 43–48.

Putnam, J. D. 1880 Biological and other notes on Coccidae. Proc. Davenport Acad., **2**, 293–347.

Raimbert, A. 1869 Recherches expérimentales sur la transmission du charbon par les mouches. Compt. Rend. Acad. Sci., Paris, **69**, 805–812. (See also Paris Acad. Méd. Bull., 1870, **35**, 50, 215, 471; and Union Méd. Paris, 1870, **9**, 209, 350, 507, 709.)

Raistrick, H. 1932 Biochemistry of the lower fungi. *In* Ergebnisse der Enzymforschung, I, 345–363. Akad. Verlagsgesellschaft m.b.h., Leipzig. 369 pp.

Rand, F. V. 1915 Dissemination of bacterial wilt of cucurbits. J. Agr. Research, **5**, 257–260.

Rand, F. V. 1923 Bacterial wilt or Stewart's disease of corn. Canner, **56**, 164–165.

Rand, F. V., and Cash, L. C. 1920 Some insect relations of *Bacillus tracheiphilus*. Phytopathology, **10**, 133–140.

Rand, F. V., and Cash, L. C. 1933 Bacterial wilt of corn. U.S. Dept. Agr. Tech. Bull., 362. 31 pp.

Rand, F. V., and Enlows, E. M. A. 1916 Transmission and control of bacterial wilt of cucurbits. J. Agr. Research, **6**, 417–434.

Rand, F. V., and Pierce, W. D. 1920 A coordination of our knowledge of insect transmission in plant and animal diseases. Phytopathology, **10**, 189–231.

Ratsburg, J. T. C. 1839 Die Forstinsekten. Berlin. (Quoted by Leach, 1940.)

Rawlins, T. E. 1933 Phytopathological and botanical research methods. John Wiley and Sons, New York. 156 pp.

Ray, H. 1933 On the gregarine, *Lankesteria culicis* (Ross), in the mosquito, *Aëdes (Stegomyia) albopictus* Skuse. Parasitology, **25**, 392-396.

Readio, P. A. 1935 The entomological phases of the Dutch elm disease. J. Econ. Entomol., **28**, 341-353.

Rebagliati, R. 1940 Verruga péruana (enfermedad de Carrión). T. Aguirre, Lima. 204 pp.

Reed, W. 1900 The etiology of yellow fever. Phila. Med. J., **6**, 790-796.

Reed, W., Vaughan, V. C., and Shakespeare, E. O. 1904 Report on the origin and spread of typhoid fever in U.S. military camps during the Spanish war of 1898. Govt. Print. Office, Washington. 2 vols.

Regendanz, P. 1936 Über den Entwicklungsgang von *Babesia bigemina* in der Zecke *Boöphilus microplus*. Zentr. Bakt. Parasitenk. Infekt., I Orig., **137**, 423-428.

Reichenow, E. 1922 Intracelluläre Symbionten bei blutsaugenden Milben und Egeln. Arch. Protistenk., **45**, 95-116.

Reichenow, E. 1935 Uebertragungsweise und Entwicklung der Piroplasmen. Zentr. Bakt. Parasitenk. Infekt., I Orig., **135**, 108-119.

Reichenow, E. 1940 Der Entwicklungsgang des Küstenfiebererregers im Rinde und in der übertragenden Zecke. Arch. Protistenk., **94**, 1-56.

Reiff, W. 1909 Einige Flacherie—Experimente mit der "Gypsy moth" (*Liparis dispar*). Soc. entomol., **24**, 178-181.

Reiff, W. 1911 The "wilt disease," or "flacherie," of the gypsy moth. Contrib. Entomol. Lab. Bussey Inst., Harvard Univ., No. 36. 60 pp.

Rendtorff, R. C., and Francis, T. 1943 Survival of the Lansing strain of poliomyelitis virus in the common house fly, *Musca domestica* L. J. Infectious Diseases, **73**, 198-205.

Richards, O. W. 1941 The staining of acid-fast tubercle bacteria. Science, **93**, 190.

Ricketts, H. T. 1906 The transmission of Rocky Mountain spotted fever by the bite of the wood tick (*Dermacentor occidentalis*). J. Am. Med. Assoc., **47**, 358.

Ricketts, H. T. 1907a The role of the wood tick (*Dermacentor occidentalis*) in Rocky Mountain spotted fever, and the susceptibility of local animals to this disease; a preliminary report. J. Am. Med. Assoc., **49**, 24-27.

Ricketts, H. T. 1907b Observations on the virus and means of transmission of Rocky Mountain spotted fever. J. Infectious Diseases, **4**, 141-153.

Ricketts, H. T. 1909 A micro-organism which apparently has a specific relationship to Rocky Mountain spotted fever. A preliminary report. J. Am. Med. Assoc., **52**, 379-380.

Ricketts, H. T. 1911 Contributions to medical science. Univ. of Chicago Press, Chicago. pp. 373-408.

Ricketts, H. T., and Wilder, R. M. 1910 Further investigations regarding the etiology of tabardillo, Mexican typhus fever. J. Am. Med. Assoc., **55**, 309-311.

Rickter, G. 1928 Untersuchungen an Homopterensymbionten. Z. Morphol. Ökol. Tiere, **10**, 174-206.

Ries, E. 1931 Die Symbiose der Läuse und Federlinge. Z. Morphol. Ökol. Tiere, **20**, 233–367.

Ries, E. 1932a Experimentelle Symbiosestudien. Z. Morphol. Ökol. Tiere (Abt. A. Z. Wiss. Biol.), **25**, 184–234.

Ries, E. 1932b Die prozesse der Eibildung und des Eiwachstums bei Pediculiden und Mallophagen. Z. Zellforsch. Mikroskop. Anat., **16**, 314–388.

Ries, E. 1935 Über den Sinn der erblichen Insektensymbiose. Naturwissenschaften, **23**, 744–749.

Riley, W. A., and Johannsen, O. A. 1932 Medical entomology. 1st ed. McGraw-Hill Book Co., New York. 476 pp. (See p. 121.)

Riley, W. A., and Johannsen, O. A. 1938 Medical entomology. 2nd. ed. McGraw-Hill Book Co., New York. 483 pp. (See p. 260.)

Ripper, W. 1930 Zur Frage des Celluloseabbaus bei der Holzverdauung xylophager Insektenlarven. Z. vergleich. Physiol., **13**, 314–333.

Rivas, G. M. 1942 Transmission of leprosy by fleas. Rev. Facultad Med., **10**, 635 pp. (See in J. Am. Med. Assoc., 1943, **30**, 378.)

Roberg, D. N. 1915 I. The rôle played by the insects of the dipterous family Phoridae in relation to the spread of bacterial infections. II. Experiments on *Aphiochaeta ferruginea* Brunetti with the cholera vibrio. Philippine J. Sci., **10**, Ser. B, 309–336.

Roberts, J. L. 1935 A new species of the genus *Bacillus* exhibiting mobile colonies on the surface of nutrient agar. J. Bact., **29**, 229–238.

Robinson, L. E., and Davidson, J. 1913 The anatomy of *Argas persicus* (Oken 1818). Parasitology, **6**, pt. I, 20–48; pt. II, 217–256; pt. III, 382–424.

Robinson, W. 1935 Stimulation of healing in nonhealing wounds by allantoin occurring in maggot secretions and of wide biological distribution. J. Bone Joint Surg., **17**, 267–271.

Robinson, W. 1937 The healing properties of allantoin and urea discovered through the use of maggots in human wounds. Smithsonian Inst. Pub., Repts., **3471**, 451–461.

Robinson, W. 1940 Ammonia as a cell proliferant and its spontaneous production from urea by the enzyme urease. Am. J. Surg., **49**, 319–325.

Robinson, W., and Norwood, V. H. 1933 The role of surgical maggots in the disinfection of osteomyelitis and other infected wounds. J. Bone Joint Surg., **15**, 409–412.

Robinson, W., and Norwood, V. H. 1934 Destruction of phygenic bacteria in the alimentary tract of surgical maggots implanted in infected wounds. J. Lab. Clin. Med., **19**, 581–586.

Roca-Garcia, M. 1944 The isolation of three neurotropic viruses from forest mosquitoes in eastern Colombia. J. Infectious Diseases, **75**, 160–169.

Rocha-Lima, H. da. 1916 Zur Aetiologie des Fleckfiebers. Berlin Klin. Wochschr., **53**, 567–569.

Rocha-Lima, H. da. 1930 Rickettsien. *In* Kolle, W., and Wasserman, A. V.,

Handbuch der pathogenen Mikroorganismen, vol. VIII, 1350 pp. G. Fischer, Jena.

Rooseboom, M. 1937 Contribution à l'étude de la cytologie du sang de certains insectes, avec quelques considérations générales. Joh. Enschedé, Haarlem, Holland. 135 pp.

Rorer, J. B. 1915 Report on the inoculation of locusts with *Coccobacillus acridiorum*. Bull. Dept. Agric., Trinidad Tobago, **14**, pt. 6, 197–198.

Rosenau, M. J. 1927 Preventive medicine and hygiene. 5th ed. D. Appleton and Co., New York. 1458 pp. (See page 420.)

Rosenau, M. J., and Brues, C. T. 1912 Tran. 15th Intern. Cong. Hyg. Dermography, 1913, **1**, pt. 2. 616 pp.

Rosenholz, H. P. 1927 Die Rolle der Wanzen in Epidemiologie des Rückfallfiebers. Zentr. Bakt. Parasitenk. Infekt., I Orig., **102**, 179–213.

Rosenkranz, W. 1939 Die Symbiose der Pentatomiden (Hemiptera Heteroptera). Z. Morphol. Ökol. Tiere, **36**, 279–309.

Rosenow, E. C., South, L. H., and McCormack, A. T. 1937 Bacteriologic and serologic studies in an epidemic of poliomyelitis in Kentucky. Kentucky Med. J., **35**, 437–446.

Ross, E. S., and Roberts, H. R. 1943 Mosquito atlas. Part II. Eighteen old world anophelines important to malaria. Am. Entomol. Soc., Acad. Natural Sci., Sept. 22. 44 pp.

Ross, R. 1895 The crescent-sphere-flagella metamorphosis of the malarial parasite in the mosquito. Trans. S. Indian Branch Brit. Med. Assoc., **6**, 334–338.

Ross, R. 1897 On some peculiar pigmented cells found in two mosquitoes fed on malarial blood. Brit. Med. J., Dec. 18, pp. 1786–1788. (See also Brit. Med. J., Feb. 26, 1898, pp. 550–551.)

Ross, R. 1898 Report on the cultivation of *Proteosoma* Labbe, in grey mosquitoes. Office of Supt. Govt. Printing, India. 21 pp.

Ross, R. 1903 [I.] Note on the bodies recently described by Leishman and Donovan. [II.] Further notes on Leishman's bodies. Brit. Med. J., **2**, 1261, 1401.

Rossi, P., and Detroyat, C. 1938 Existence en France de *Rickettsia conjunctivae* du boeuf. Bull. Soc. Path. Exotique, **31**, 788–790.

Roubaud, E. 1911 *Cercoplasma* (n. g.) *caulleryi* (n. sp.); nouveau flagellé à formes trypanosomiennes de l'intestin d'*Auchmeromyia luteola* Fabr. (Muscide). Compt. Rend. Soc. Biol., **71**, 570–573.

Roubaud, E. 1919 Les particularités de la nutrition et la vie symbiotique chez les mouches tsétsés. Ann. Inst. Pasteur, **33**, 489–536.

Roubaud, E., and Colas-Belcour, J. 1927 Action des diastases dans le determinisme d'éclosion de l'oeuf chez le moustique de la fièvre jaune. Compt. Rend. Acad. Sci., Paris, **184**, 244–249.

Roubaud, E., and Descazeaux, J. 1923 Sur un agent bactérien pathogène pour les mouches communes: *Bacterium delendaemuscae* n. sp. Compt. Rend. Hebdom. Acad. Sci., **177**, 716–717.

Rozeboom, L. E. 1934 The effect of bacteria on the hatching of mosquito eggs. Am. J. Hyg., **20**, 496–501.

Rozeboom, L. E. 1935 The relation of bacteria and bacterial filtrates to the development of mosquito larvae. Am. J. Hyg., **21**, 167–179.

Ruhland, H. H., and Huddleson, I. F. 1941 The role of one species of cockroach and several species of flies in the dissemination of *Brucella*. Am. J. Vet. Research, **2**, 371–372.

Rumbold, C. T. 1941 A blue stain fungus, *Ceratostomella montium* n. sp., and some yeasts associated with two species of *Dendroctonus*. J. Agr. Research, **62**, 589–601.

Sabin, A. B., and Ward, R. 1941 Flies as carriers of poliomyelitis virus in urban epidemics. Science, **94**, 590–591.

Sabin, A. B., and Ward, R. 1942 Insects and epidemiology of poliomyelitis. Science, **95**, 300–301.

St. John, J. H., Simmons, J. S., and Reynolds, F. H. K. 1930 The survival of various microorganisms within the gastro-intestinal tract of *Aëdes aegypti*. Am. J. Trop. Med., **10**, 237–241.

Sanders, M. 1939 Cultivation of the viruses. Arch. Path., **28**, 541–586.

Sartirana, S., and Paccanaro, A. 1906 Der *Streptococcus bombycis* in Bezug auf die Aetiologie der Auszehrung und Schlaffsucht der Seidenraupen. Zentr. Bakt. Parasitenk. Infekt., I Orig., **40**, 331–336.

Sassuchin, D., and Tichomirova, M. 1936 De la conservation des *Pasteurella pestis* dans les larves et les nymphes des tiques *Dermacentor silvarum* Olen. Rev. Microbiol. Epidemiol. Parasitol., **15**, 362.

Saunders, D. A. 1940a *Musca domestica*, a vector of bovine mastitis (preliminary report). J. Am. Vet. Med. Assoc., **97**, 120–123.

Saunders, D. A. 1940b Hippelates flies as vectors of bovine mastitis (preliminary report). J. Am. Vet. Med. Assoc., **97**, 306–308.

Sawamura, S. 1903 Investigations on flacherie. Tokyo Imp. Univ., Coll. Agr. Bull., **5**, 403–448.

Sawamura, S. 1905 On the large bacillus observed in flacherie. Tokyo Imp. Univ., Coll. Agr. Bull., **6**, 375–386.

Sawamura, S. 1906 Note on bacteria pathogenic to silk-worm. Tokyo Imp. Univ. Coll. Agr. Bull., **7**, 105.

Sawyer, W. A., and Frobisher, M., Jr. 1929 The filterability of yellow fever virus as existing in the mosquito. J. Exptl. Med., **50**, 713–718.

Sawyer, W. H. 1929 Observations on some entomogenous members of the Entomophthoraceae in artificial culture. Am. J. Botan., **16**, 87–121.

Schaudinn, F. 1902 Beiträge zur Kenntnis der Bakterien und verwandter Organismen. I. *Bacillus butschlii* n. sp. Arch. Protistenk., **1**, 306–343.

Schaudinn, F. 1904 Generations- und Wirtswechsel bei Trypanosoma und Spirochaete. Arb. Kaiserl. Gesundh., **20**, 387–439.

Scheinert, W. 1933 Symbiose und Embryonalentwicklung bei Rüsselkäfern. Z. Morphol. Ökol. Tiere (Abt. A. Z. Wiss. Biol.), **27**, 76–128.

Schellack, C. 1907 Morphologische Beiträge zur Kenntnis der europäischen, amerikanischen und afrikanischen Rekurrens-spirochaeten. Arb. Kaiserl. Gesundh., **27**, 364–387.

Schmidberger, J. 1836 Naturgeschichte des Appelborkenkäfers, *Apate dispar*. Beitr. zur Obstbaumzucht und zur Naturgeschichte der Obstbäumen schädlichen Insekten., **4**. (Quoted by Leach, 1940.)

Schminke, A. 1917 Histopathologischer Befund in Roseolen der Haut bei wolhynischem Fieber. Münch. Med. Wochschr., **64**, 961.

Schneider, G. 1940 Beiträge zur kenntnis der symbiontischen Einrichtungen der Heteropteren. Z. Morphol. Ökol. Tiere, **36**, 595–644.

Schneider, K. C. 1902 Lehrbuch der vergleichenden Histologie der Tiere. G. Fischer, Jena. 988 pp.

Schneider-Orelli, O. 1911 Die Ueberträgung und Keimung des Ambrosipilzes von *Xyleborus* (*Anisandrus*) *dispar* F. Naturw. Z. Forst- u. Landw., **9**, 186–192.

Schneider-Orelli, O. 1913 Untersuchungen über den pilzzüchtenden Obstbaumborkenkäfer *Xyleborus* (*Anisandrus*) *dispar* und seinen Nährpilz. Zentr. Bakt. Parasitenk. Infekt., II, **38**, 25–110.

Schölzel, G. 1937 Die Embryologie der Anopluren und Mallophagen. Z. Parasitenk., **9**, 730–770.

Schomann, H. 1937 Die Symbiose der Bockkäfer. Z. Morphol. Ökol. Tiere, **32**, 542–612.

Schrader, F. 1923 The origin of the mycetocytes in *Pseudococcus*. Biol. Bull. Marine Biol. Lab., **45**, 279–302.

Schuberg, A., and Böing, W. 1914 Ueber die Uebertragung von Krankheiten durch einheimische stechende Insekten. Arb. Kaiserl. Gesundh., **47**, 491–512.

Schuberg, A., and Kuhn, P. 1912 Ueber die Uebertragung von Krankheiten durch einheimische stechende Insekten. Arb. Kaiserl. Gesundh. **40**, 209–234.

Schuhardt, V. T. 1940 A "ticktorium" for the propagation of a colony of infected *Ornithodoros turicata*. J. Parasitol., **26**, 201–206.

Schultz, E. S., and Folsom, D. 1923 A "spindling tuber" disease of Irish potatoes. Science, **57**, 149.

Schulz, K. 1939 A rickettsiosis new to South Africa. Onderstepoort J. Vet. Sci., Animal Ind., **13**, 287–289. (Also J. S. African Vet. Med. Assoc., **10**, 176–178.)

Schütte, L. 1921 Das Tönnchen der Musciden. Zool. Anz., **53**, 49–51.

Schwartz, W. 1924 Untersuchungen über die Pilzsymbiose der Schildläuse. Biol. Zentr., **44**, 487–527.

Schwarz, I. 1929 Untersuchungen an mikrosporidien minierender Schmelterlingsraupen, den "Symbionten" Portiers. Z. Morphol. Ökol. Tiere, **13**, 665–705.

Schwerdtfeger, F. 1936 Zur Kenntnis der roten Kiefernbuschhornblattwespe, *Diprion sertifer* Geoffr. (*Lophyrus rufus* Panz.). Z. Pflanzenkrankh. **46**, 513–534. (See also Rev. Applied Entomol. A, **25**, 54.)

Scott, J. R. 1917 Studies upon the common house-fly (*Musca domestica* Linn.)

II. Isolation of *B. cuniculicida,* a hitherto unreported isolation. J. Med. Research, **37,** 121–124.

Seargent, A. 1938 Fièvre récurrente Hispano-Nord-Africaine. Arch. Inst. Pasteur Algérie, **16,** 403–415.

Sellards, A. W. 1923 The cultivation of a *Rickettsia*-like micro-organism from tsutsugamushi disease. Am. J. Trop. Med., **3,** 529–547.

Semans, F. M. 1936 Protozoan parasites of the Orthoptera, with special reference to those of Ohio. I. Introduction and methods. Ohio J. Sci., **36,** 315–320.

Semans, F. M. 1939 Protozoan parasites of the Orthoptera with special reference to those of Ohio. II. Description of the protozoan parasites recognized in this study. Ohio J. Sci., **39,** 157–181.

Semans, F. M. 1941 Protozoan parasites of the Orthoptera, with special reference to those of Ohio. III. Protozoan parasites in relation to the host and host ecology. Ohio J. Sci., **41,** 457–464.

Semans, F. M. 1943 Protozoan parasites of the Orthoptera, with special reference to those of Ohio. IV. Classified list of the protozoan parasites of the Orthoptera of the world. Ohio J. Sci., **43,** 221–234, 271–276.

Senior-White, R. 1928 Algae and the food of anopheline larvae. Indian J. Med. Research, **15,** 969–988.

Serbinow, I. L. 1912a Chernaia cherva. (Blackbrood in bees.) Věst. Obsč. Pceloved., no. 11, 426–429.

Serbinow, I. L. 1912b A new epizootic of bees in North European Russia. Věst. Obsč. Pceloved. (Messager de la société russe d'apiculture), no. 3, 1912.

Serbinow, I. L. 1913 On the etiology of foulbrood in bees. Selisk. Choz. Lesovodstvo, **242,** 367–382.

Serbinow, I. L. 1915 Contribution to the etiology of infectious diarrhea of bees caused by *Bacterium coli apium,* n. sp., and *Proteus alveicola,* n. sp. Zhur. Microbiol., Petrograd, **2,** 19–44.

Sergent, E., Donatien, A., Parrot, L., and Lestoquard, F. 1936 Cycle évolutif de *Theileria dispar* du boeuf chez la tique *Hyalomma mauritanicum.* Arch. Inst. Pasteur Algérie, **14,** 259–294.

Sergent, E., and Foley, H. 1908 Fièvre récurrente du Sud-Oranais et *Pediculus vestimenti.* Note préliminaire. Bull. Soc. Path. Exotique, **1,** 174–176.

Sergent, E., and Foley, H. 1910 Recherches sur la fièvre récurrente et sa mode de transmission, dans une épidémie algérienne. Ann. Inst. Pasteur, **24,** 337–373.

Sergent, E., Foley, H., and Vialette, C. 1914 Sur des formes microbiennes abondantes dans le corps de poux infectés par le typhus exanthématique, et toujours absentes dans poux témoins, non typhiques. Mém. Soc. Biol., **77,** 101–103.

Sergent, E., and Sergent, E. 1906 Sur un flagellé nouveau de l'intestin des *Culex* et des *Stegomyia, Herpetomonas algeriense.* Sur un autre flagellé et sur des *Spirochaete* de l'intestin des larves de moustiques. Compt. Rend. Soc. Biol., **60,** 291.

Settle, E. B., Pinkerton, H., and Corbett, A. J. 1945 A pathologic study of tsutsugamushi disease (scrub typhus) with notes on clinicopathologic correlation. J. Lab. Clin. Med., **30**, 639–661.

Seymour, A. B. 1929 Host index of the fungi of North America. Harvard Univ. Press, Cambridge, Mass. 732 pp.

Shannon, R. C. 1929 Entomological investigations in connection with Carrión's disease. Am. J. Hyg., **10**, 78–111.

Shaughnessy, H. J., and Milzer, A. 1939 Experimental infection of *Dermacentor andersoni* Stiles with the virus of lymphocytic choriomeningitis. Am. J. Pub. Health, **29**, 1103–1108.

Sheldon, J. L. 1903 Cultures of *Empusa*. J. Applied Microscop. Lab. Methods, **6**, 2212–2221.

Shepherd, D. 1924 Life history and biology of *Echocerus cornutus* (Fab.). J. Econ. Entomol., **17**, 572–577.

Shinji, G. O. 1919 Embryology of coccids, with special reference to the ovary, origin and differentiation of the germ cells, germ layers, rudiments of the midgut, and the intracellular symbiotic organism. J. Morphol., **33**, 73–167.

Shiperovich, V. 1925 A sawfly injurious to the pine and its control. (In Russian.) Protect. Plants Ukraine, 1925, 41–46. (See also Rev. Applied Entomol., A, **14**, 209.)

Shkorbatov, V. I. 1943 Experiments with urine in quick diagnosis of typhus exanthematicus. J. Microbiol., Epidemiol., Immunol., nos. 1–2, 23–25.

Shooter, R. A., and Waterworth, P. M. 1944 A note on the transmissibility of hemolytic streptococcal infection by flies. Brit. Med. J., no. 4337, 247–248.

Shope, R. E. 1927 Bacteriophage isolated from the common house fly (*Musca domestica*). J. Exptl. Med., **45**, 1037–1044.

Shule, P. A. 1928 Dengue fever: transmission by *Aëdes aegypti*. Am. J. Trop. Med., **8**, 203–213.

Shulguina, O. G., and Kalinicker, P. A. 1927 Experimental infection of *Locusta migratoria* with bacterial disease. Rept. Bur. Applied Entomol., **3**, 99–104. Leningrad. (English summary.)

Sieyro, L. 1942 Die Hausfliege (*Musca domestica*) als Ueberträger von *Entamoeba histolytica* und anderen Darmprotozoen. Deut. Trop. Z., **46**, 361–372.

Sikora, H. 1916 Beiträge zur Anatomie, Physiologie und Biologie der Kleiderlaus (*Pediculus vestimenti* Nitzsch). 1. Anatomie des Verdauungstraktes. Arch. Schiffs- u. Tropen-Hyg., **20**, Beiheft 1, 1–76.

Sikora, H. 1918 Beiträge zur Kenntnis der Rickettsien. Arch. Schiffs- u. Tropen-Hyg., **22**, 442–446.

Sikora, H. 1919 Vorläufige Mitteilungen über Mycetome bei Pediculiden. Biol. Zentr., **39**, 287–288.

Sikora, H. 1920 Beobachtungen an Rickettsien, besonders zur Unterscheidung der *R. prowazeki* von *R. pedikuli*. Arch. Schiffs- u. Tropen-Hyg., **24**, 347–353.

Sikora, H. 1922 Neue Rickettsien bei Vogelläusen. Arch. Schiffs- u. Tropen-Hyg., **26**, 271–272.

Sikora, H. 1942 Zur Morphologie der Rickettsien. Z. Hyg. Infektionskrankh., **124**, 250-270.

Siler, J. F., Hall, M. W., and Hitchens, A. P. 1926 Dengue: its history, epidemiology, mechanism of transmission, etiology, clinical manifestations, immunity and prevention. Philippine J. Sci., **29**, 1-302.

Silva, P. da. 1916 Exper. sur la trans. de la leishmaniose infantile par les puces (*Pulex irritans*). Arq. Inst. Camara Pestana, **4**, 26-27. (Not seen.)

Simmonds, M. 1892 Fliegen und Choleraübertragung. Deut. Med. Wochschr., no. 41, 931.

Simmons, J. S. 1941 The transmission of malaria by the *Anopheles* mosquitoes of North America. Pub. Am. Assoc. Advancement Sci., **15**, 113-130.

Simmons, J. S., St. John, J. H., and Reynolds, F. H. K. 1931 Experimental studies of dengue. Philippine J. Sci., **44**, 1-252.

Simmons, S. W. 1935a Adequacy of nutritional retardation in culture of sterile maggots for surgical use. Arch. Surg., **30**, 1014, 1025.

Simmons, S. W. 1935b The bactericidal properties of excretions of the maggot of *Lucilia sericata*. Bull. Entomol. Research, **26**, 559-563.

Simmons, S. W. 1935c A bactericidal principle in excretions of surgical maggots which destroys important etiological agents of pyogenic infections. J. Bact., **30**, 253-267.

Simmons, S. W. 1939 Digestive enzymes of the larva of the cattle grub *Hypoderma lineatum* (De Villiers). Ann. Entomol. Soc. Am., **32**, 621-627.

Simond, P. L. 1898 La propagation de la peste. Ann. Inst. Pasteur, **12**, 625.

Singer, E. 1941 Experiments on the survival of rickettsiae in cell-free media. Australian J. Exptl. Biol. Med. Sci., **19**, 123-124.

Sinton, J. A., and Shute, P. G. 1939 Spirochaetal infections of mosquitoes. J. Trop. Med. Hyg., **42**, 125-126.

Smart, J. 1943 A handbook for the identification of insects of medical importance. British Museum, London. 269 pp.

Smith, D. J. W. 1940a Studies in the epidemiology of Q fever. 3. The transmission of Q fever by the tick *Haemaphysalis humerosa*. Australian J. Exptl. Biol. Med. Sci., **18**, 103-118.

Smith, D. J. W. 1940b Studies in the epidemiology of Q fever. 4. The failure to transmit Q fever with the cat-flea *Ctenocephalides felis*. Australian J. Exptl. Biol. Med. Sci., **18**, 119-123.

Smith, D. J. W. 1941 Studies in the epidemiology of Q fever. 8. The transmission of Q fever by the tick *Rhipicephalus sanguineus*. Australian J. Exptl. Biol. Med. Sci., **19**, 133-136.

Smith, D. J. W. 1942a Studies in the epidemiology of Q fever. 10. The transmission of Q fever by the tick *Ixodes holocyclus* (with notes on tick paralysis in bandicoots). Australian J. Exptl. Biol. Med. Sci., **20**, 213-217.

Smith, D. J. W. 1942b Studies in the epidemiology of Q fever. 11. Experimental infection of the ticks *Haemaphysalis bispinosa* and *Ornithodorus* sp. with *Rickettsia burneti*. Australian J. Exptl. Biol. Med. Sci., **20**, 295-296.

Smith, D. J. W., and Derrick, E. H. 1940 Studies in the epidemiology of Q fever. 1. The isolation of six strains of *Rickettsia burneti* from the tick *Haemaphysalis humerosa*. Australian J. Exptl. Biol. Med. Sci., **18**, 1–8.

Smith, E. F. 1893 Two new and destructive diseases of cucurbits. Botan. Gaz., **18**, 339.

Smith, E. F. 1896 A bacterial disease of the tomato, eggplant, and Irish potato (*Bacillus solanacearum* n. sp.). U.S. Dept. Agr., Div. Physiol. Path. Bull., 12.

Smith, E. F. 1897 *Pseudomonas campestris* (Pammel), the cause of a brown rot in cruciferous plants. Zentr. Bakt. Parasitenk. Infekt., II, **3**, 284–291, 408–415, 478–486.

Smith, G. 1938 An introduction to industrial mycology. Edward Arnold and Co., London. 302 pp.

Smith, K. M. 1937 A textbook of plant virus diseases. P. Blakiston's Son and Co., Philadelphia. 615 pp.

Smith, M. G., Blattner, R. J., and Heys, F. M. 1944 The isolation of the St. Louis encephalitis virus from chicken mites (*Dermanyssus gallinae*) in nature. Science, **100**, 362–363.

Smith, N. R., and Clark, F. E. 1938 Motile colonies of *Bacillus alvei* and other bacteria. J. Bact., **35**, 59–60.

Smith, R. C. 1933 Fungous and bacterial diseases in the control of grasshoppers and chinch bugs. 28th Biennial Rept. Kansas State Board Agr., 44–58.

Smith, T., and Kilbourne, F. E. 1893 Investigations into the nature, causation, and prevention of Texas or southern cattle fever. U.S. Dept. Agr., Bur. Animal Ind. Bull., 1. 177 pp.

Smyth, E. G. 1917 The white grubs injuring cane in Porto Rico. I. Life histories of May-beetle. J. Dept. Agr., Porto Rico, **1**, no. 3, 141–169.

Smyth, E. G. 1920 The white grubs injuring sugar cane in Porto Rico. II. The rhinoceros beetles. J. Dept. Agr., Porto Rico, **4**, no. 2, 3–29.

Snodgrass, R. E. 1935 Principles of insect morphology. McGraw-Hill Book Co., Inc., New York. 667 pp.

Snow, F. W. 1890 Experiments for the destruction of chinch-bugs. 21st Rept. Entomol. Soc. Ontario, 93–97.

Snyder, J. C., and Wheeler, C. M. 1945 The experimental infection of the human body louse, *Pediculus humanus corporis*, with murine and epidemic louse-borne typhus strains. J. Exptl. Med., **82**, 1–20.

Sokoloff, V. P., and Klotz, L. J. 1942 Mortality of the red scale on citrus through infection with a spore-forming bacterium. Phytopathology, **32**, 187–198.

Soper, F. L. 1936 Jungle yellow fever, a new epidemiological entity in South America. Rev. Hyg. Saude Pub., **10**, 107–144.

Souza-Araujo, H. C. 1943 Verificação, em condições naturais, da infecção de mais três Hematófagos (Anophelineos, Flebótomos e Simulídeos) em leprosos. Mem. Inst. Oswaldo Cruz, **39**, 167–176.

Spaet, G. 1944 Investigations into the aetiology of exanthematic typhus. J. Trop. Med. Hyg., **47**, 28-30.

Sparrow, F. K. 1944 Personal Communication.

Sparrow, H. 1939 Infection spontanée des poux d'élevage par une rickettsia du type *Rickettsia rocha limae*. Arch. Inst. Pasteur Tunis, **28**, 64-73.

Sparrow, H. 1940 Abondance des rickettsias du typhus murin cultivée dans les poux. Arch. Inst. Pasteur Tunis, **29**, 250-261.

Sparrow, H., and Mareschal, P. 1942 Innocuité pour l'homme de la piqûre du pou typhique et données expérimentales sur les conditions de l'infection typhique. Compt. Rend. Acad. Sci., Paris, **215**, 389-391.

Speare, A. T. 1912 *Entomophthora* disease of the sugar cane mealy bug. Rept. Work Exp. Sta. Hawaiian Sugar Planters' Assoc. Bull. 12. (See Fungi parasitic upon insects injurious to sugar cane. Hawaiian Sugar Planters' Assoc. Exptl. Sta., Path. Physiol. Bull., 12. 62 pp.)

Speare, A. T. 1920 Further studies of *Sorosporella uvella*, a fungous parasite of noctuid larvae. J. Agr. Research, **18**, 399-439.

Speer, A. J. 1927 Compendium of the parasites of mosquitoes (Culicidae). U.S. Pub. Health Service, Hygienic Lab. Bull., 146. 36 pp.

Spegazzini, C. 1899 Fungi argentini novi v. critici. Ann. museo nac. Buenos Aires, **6**, 81-365.

Spencer, R. R., and Parker, R. R. 1923 Rocky Mountain spotted fever: infectivity of fasting and recently fed ticks. Pub. Health Repts., **38**, 333-339.

Spencer, R. R., and Parker, R. R. 1924a Rocky Mountain spotted fever: viability of the virus in animal tissues. Pub. Health Repts., **39**, 55-57.

Spencer, R. R., and Parker, R. R. 1924b Rocky Mountain spotted fever: experimental studies on tick virus. Pub. Health Repts., **39**, 3027-3040.

Spencer, R. R., and Parker, R. R. 1924c Rocky Mountain spotted fever: nonfilterability of tick and blood virus. Pub. Health Repts., **39**, 3251-3255.

Spencer, R. R., and Parker, R. R. 1926 Rocky Mountain spotted fever: certain characteristics of blood virus. Pub. Health Repts., **41**, 1817-1822.

Spencer, R. R., and Parker, R. R. 1930 Studies on Rocky Mountain spotted fever. Hyg. Lab. Bull., 154. 116 pp.

Spielman, M., and Haushalter, M. 1887 Dissémination du bacilli de la tuberculose par la mouch. Compt. Rend. Acad. Sci., Paris, **105**, 352-353.

Stammer, H.-J. 1929a Die Bakteriensymbiose der Trypetiden (Diptera). Z. Morphol. Ökol. Tiere (Abt. A. Z. Wiss. Biol.) **15**, 481-523.

Stammer, H.-J. 1929b Die Symbiose der Lagriiden (Coleoptera). Z. Morphol. Ökol. Tiere, **15**, 1-34.

Stammer, H.-J. 1935 Studien an Symbiosen zwischen Käfern und Mikroorganismen. I. Die Symbiose der Donaciinen (Coleopter. Chrysomel.) Z. Morphol. Ökol. Tiere, **29**, 585-608.

Stammer, H.-J. 1936 Studien an Symbiosen zwischen Käfern und Mikroorganismen. II. Die Symbiose des *Bromius obscurus* L. und der Cassida-Arten

(Coleopt. Chrysomel.) Z. Morphol. Ökol. Tiere (Abt. A. Z. Wiss. Biol.), **31**, 682–697.

Statelov, N. 1932 Ein pathogenischer Brazillus auf den Larven der *Barbitistes amplipennis*. Mitt. Bulg. Entomol. Ges. (Bulg. Entomol. Druzhestvo, Sofia), **7**, 56–61.

Stein, F. 1878 (1867–1883) Der Organismus der Infusionstiere. Abt. II and III. W. Engelmann, Leipzig. (Quoted from Wenyon, 1926.)

Steinhaus, E. A. 1939 Unpublished data.

Steinhaus, E. A. 1940a The microbiology of insects—with special reference to the biologic relationships between bacteria and insects. Bact. Revs., **4**, 17–57.

Steinhaus, E. A. 1940b Unpublished data.

Steinhaus, E. A. 1940c A discussion of the microbial flora of insects. J. Bact., **40**, 161–162.

Steinhaus, E. A. 1941a A study of the bacteria associated with thirty species of insects. J. Bact., **42**, 757–790.

Steinhaus, E. A. 1941b Unpublished data.

Steinhaus, E. A. 1942a Rickettsia-like organism from normal *Dermacentor andersoni* Stiles. Pub. Health Repts., **57**, 1375–1377.

Steinhaus, E. A. 1942b The microbial flora of the Rocky Mountain wood tick, *Dermacentor andersoni* Stiles. J. Bact., **44**, 397–404.

Steinhaus, E. A. 1942c Note on a toxic principle in the eggs of the tick *Dermacentor andersoni* Stiles. Pub. Health Repts., **57**, 1310–1312.

Steinhaus, E. A. 1942d Catalogue of bacteria associated extracellularly with insects and ticks. Burgess Publishing Co., Minneapolis, Minn. 206 pp.

Steinhaus, E. A. 1942e Unpublished data.

Steinhaus, E. A. 1943 A new bacterium, *Corynebacterium lipoptenae,* associated with the louse fly *Lipoptena depressa* Say. J. Parasitol., **29**, 80.

Steinhaus, E. A. 1944 Unpublished data.

Steinhaus, E. A., and Birkeland, J. B. 1939 Studies on the life and death of bacteria. I. The senescent phase in aging cultures and the possible mechanisms involved. J. Bact., **38**, 249–261.

Steinhaus, E. A., and Parker, R. R. 1943 Experimental Rocky Mountain spotted fever: results of treatment with certain drugs. Pub. Health Repts., **58**, 351–352.

Steinhaus, E. A., and Parker, R. R. 1944 The isolation of a filter-passing agent from the rabbit tick *Haemaphysalis leporis-palustris* Packard. Pub. Health Repts., **59**, 1528–1529.

Stelling-Dekker, N. M. 1931 Die Hefesammlung des "Centraalbureau voor Schimmelcultures." I. Teil. Die Sporogenen Hefen. Amsterdam.

Stempell, W. 1909 Über *Nosema bombycis* Nageli. Arch. Protistenk., **16**, 281–358.

Stévenel, L. 1913 Quelques observations et examens microbiologiques faits à Pointe à Pitre. Bull. Soc. Path. Exotique, **6**, 356.

Stevens, N. M.   1905   A study of the germ cells of *Aphis rosae* and *Aphis oenotherae*.   J. Exptl. Zool., **2**, 313–334.

Stewart, M. A.   1934   The rôle of *Lucilia sericata* Meig. larvae in osteomyelitis wounds.   Ann. Trop. Med. Paras., **28**, 445–454.

Stiles, C. W.   1944   Isolation of the *Bacillus anthracis* from spinose ear ticks *Ornithodorus megnini*.   Am. J. Ves. Res., **5**, 318–319.

Storey, H. H.   1933   Investigations of the mechanism of the transmission of plant viruses by insect vectors. I.   Proc. Roy. Soc. (London) B, **113**, 463–485.

Storey, H. H.   1939   Transmission of plant viruses by insects.   Botan. Rev., **5**, 240–272.

Stratman-Thomas, W. K.   1940   The influence of temperature on *Plasmodium vivax*.   Am. J. Trop. Med., **20**, 703–715.

Strickland, C.   1912   *Agrippina bona* nov. gen. et nov. sp. representing a new family of gregarines.   Parasitology, **5**, 97–108.

Strindberg, H.   1913   Embryologische Studien an Insekten.   Z. Wiss. Zool., **106**, 1–227.

Strohmeyer, H.   1911   Die biologische Bedeutung sekundarer Geschlectscharaktere am Kopfe weiblicher Platypodiden.   Entomol. Blätter, **7**, 103–108.

Strong, R. P.   1924   Investigations upon flagellate infections.   Am. J. Trop Med., **4**, 345–385.

Strong, R. P.   1940   Progress in the study of infections due to *Bartonella* and *Rickettsia*, with special reference to the work performed at Harvard University.   Am. J. Trop. Med., **20**, 13–46.

Strong, R. P.   1942   Stitt's diagnosis, prevention and treatment of tropical diseases. 6th ed.   The Blakiston Co., Philadelphia.   2 vols., 1747 pp.

Strong, R. P., Tyzzer, E. E., Brues, C. T., Sellards, A. W., and Gastiaburú, J. C.   1915   The Harvard School of Tropical Medicine. Report of the first expedition to South America, 1913.   Cambridge, Mass.   220 pp.

Stuhlmann, F.   1907   Beiträge zur Kenntnis der Tsetsefliege.   Arb. Kaiserl. Gesundh., **26**, 301–383.

Sturtevant, A. P.   1923   Les maladies du couvain d'abeilles telles qu'on les connaît aux États-Unis.   Congr. Intern. Apiculture Marseille (1922), **6**, 168–177.

Sturtevant, A. P.   1924   The development of American foulbrood in relation to the metabolism of its causative organism.   J. Agr. Research, **28**, 129–168.

Sturtevant, A. P.   1925   The relation of *Bacillus alvei* to the confusing symptoms in European foulbrood.   J. Econ. Entomol., **18**, 400–405.

Stutzer, M. J., and Wsorow, W. J.   1927   Über Infectionen der Raupen der Wintersaateule (*Euxoa segetum* Schiff).   Zentr. Bakt. Parasitenk. Infekt., **71**, 113–129.

Subramaniam, M. K., and Naidu, M. B.   1944   On a new plerocercoid from a sand fly.   Current Sci., **13**, 260–261.

Šulc, K.   1906   *Kermincola kermesina* n. g. n. sp. und *physokermina* n. sp., neue Mikroendosymbiontiker der Cocciden.   Sitzber. Böhm. Ges. Wiss., Art. XIX. 6 pp.

Šulc, K. 1910a "Pseudovitellus" und ähnliche Gewebe der Homopteren sind Wohnstätten symbiotischer Saccharomyceten. Sitzber. Böhm. Ges. Wiss. Math.- Naturw. Classe, **3**, 1–39, art. III.

Šulc, K. 1910b Symbiotische Saccharomyceten der echten Cicaden (Cicadidae). Sitzber. Böhm. Ges. Wiss. Math.- Naturw. Classe, **3**, 1–6, art. XIV.

Sulkin, S. E. 1945 Recovery of equine encephalomyelitis virus (western type) from chicken mites. Science, **101**, 381–383.

Sweetman, H. L. 1936 The biological control of insects. Comstock Publishing Co., Ithaca, N.Y. 461 pp.

Swezy, O. 1923 The pseudopodial method of feeding by trichonymphid flagellates parasitic in wood eating termites. Univ. Calif. Pub. Zool., **20**, 391–400.

Swezy, O., and Severin, H. P. 1930 A rickettsia-like microorganism in *Eutettix tenellus* (Baker), the carrier of curly top of sugar beets. Phytopathology, **20**, 169–178.

Sydow, H., and Sydow, P. 1919 Mykologische Mitteilungen. Ann. Mycol., **17**, 33–47.

Sylla, A. 1942 Ueber die Wolhynische Krankheit. Med. Klin., **38**, 726–729.

Syverton, J. T., and Berry, G. P. 1941 Hereditary transmission of the western type of equine encephalomyelitis virus in the wood tick, *Dermacentor andersoni* Stiles. J. Exptl. Med., **73**, 507–530.

Syverton, J. T., and Thomas, L. 1944 A method for staining *Rickettsia orientalis* in yolk sac and other smear preparations. Nat. Research Council, Office of Sci. Research Development, Trop. Disease Rept. no. 35. Mimeographed. 3 pp.

Tannreuther, G. W. 1907 History of the germ cells and early embryology of certain aphids. Zool. Jahrb., Abt. Anat. Ontog., **24**, 609–642.

Tanquary, M. C. 1913 Biological and embryological studies on formicidae. Bull. Illinois State Lab. Natural History, **9**, 417–479.

Tarr, H. L. A. 1935 Studies on European foul brood of bees. I. A. description of strains of *Bacillus alvei* obtained from different sources, and of another species occurring in larvae affected with this disease. Ann. Applied Biol., **22**, 709–718.

Tate, P. 1940 On *Mycetosporidium jacksonae* n. sp. parasitic in species of *Sitona* weevils. Parasitology, **32**, 462–469.

Tatlock, H. 1944 A rickettsia-like organism recovered from guinea pigs. Proc. Soc. Exptl. Biol. Med., **57**, 95–99.

Tatum, E. L. 1939 Development of eye-colors in *Drosophila*: bacterial synthesis of V + hormone. Proc. Nat. Acad. Sci. U.S., **25**, 486–490.

Tauber, O. E. 1940 Mitotic response of roach hemocytes to certain pathogenes in the hemolymph. Ann. Entomol. Soc. Am., **83**, 113–119.

Tauber, O. E., and Griffiths, J. T. 1942 Isolation of *Staphylococcus albus* from hemolymph of the roach *Blatta orientalis*. Proc. Soc. Exptl. Biol. Med., **51**, 45–47.

Taylor, A. B., and King, R. L. 1937 Further studies on the parasitic amebae found in grasshoppers. Trans. Am. Microscop. Soc., **56**, 172–176.

Taylor, A. W. 1930 The domestic mosquitoes of Gadau, Northern Nigeria, and their relation to malaria and filariasis. Ann. Trop. Med. Paras., 24, 425–435.

Tebbutt, H. 1913 On the influence of the metamorphosis of *Musca domestica* upon bacteria administered in the larval stage. J. Hyg., 12, 516–526.

Tejera, E. 1926 Les blattes envisagées comme agentes de dissémination des germes pathogènes. Compt. Rend. Soc. Biol., 95, 1382–1384. (See also Rev. Soc. Argentina Biol., 2, 243–256.)

Teodoro, G. 1918 Alcune osservazioni sui saccaromicete del *Lecanium persicae* Fab. Redia Firenze, 13, 1–5.

Thaxter, R. 1888 The Entomophthoreae of the United States. Mem. Bost. Soc. Natural Hist., 4, 133–201.

Thaxter, R. 1896 Contribution towards a monograph of the Laboulbeniaceae. Am. Acad. Arts Sci., (N.S.) 12, 187–429.

Thaxter, R. 1908 Contribution towards a monograph of the Laboulbeniaceae. Part II. Am. Arts Sci., (N.S.) 13, 219–469.

Thaxter, R. 1920 Second note on certain peculiar fungus-parasites of living insects. Botan. Gaz., 69, 1–27.

Thomas, H. D. 1943 Preliminary studies on the physiology of *Aëdes aegypti* (Diptera: Culicidae). I. The hatching of eggs under sterile conditions. J. Parasitol., 29, 324–328.

Thomas, L. J., and Cahn, A. R. 1932 A new disease of moose. I. Preliminary report. J. Parasitol., 18, 219–231.

Thompson, W. R. 1930 The biological control of insect and plant pests. His Majesty's Stationery Office, London, E.M.B. 29. 124 pp.

Thorpe, W. A. 1930 The biology of the petroleum fly (*Psilopa petrolii*, Coq.). Trans. Entomol. Soc. London, 78, 331–344.

Tizzoni, G., and Cattani, J. 1886 Untersuchungen über Cholera. Centr. Med. Wiss., 769–771.

Toda, T. 1923 Cholera and the ship cockroach. J. Hyg., 21, 359–361.

Todd, J. C., and Sanford, A. H. 1943 Clinical diagnosis by laboratory methods. 10th ed. Saunders Co., Philadelphia. 911 pp.

Toomey, J. A., Takacs, W. S., and Tischer, L. A. 1941 Poliomyelitis virus from flies. Proc. Soc. Exptl. Biol. Med., 48, 637–639.

Töpfer, H. 1916a Zur Aetiologie des "Febris Wolhynica." Berlin Klin. Wochschr., 53, 323.

Töpfer, H. 1916b Zur Ursache und Uebertragung des Wolhynischen Fiebers. Münch. Med. Wochschr., 63, 1495–1496.

Töpfer, H., and Schüssler, H. 1916 Zur Aetiologie des Fleckfiebers. Deut. Med. Wochschr., 42, 1157–1158.

Topley, W. W. C., and Wilson, G. S. 1936 The principles of bacteriology and immunity. William Wood and Co., Baltimore. 1645 pp.

Topping, N. H., Heilig, R., and Naidu, V. R. 1943 A note on the rickettsioses in India. Pub. Health Repts., 58, 1208–1210.

Torrey, J. C. 1912 Numbers and types of bacteria carried by city flies. J. Infectious Diseases, **10**, 166–177.

Tóth, L. 1937 Entwicklungszyklus und Symbiose von *Pemphigus spirothecae* Pass. (Aphidina). Z. Morphol. Ökol. Tiere, **33**, 412–437.

Toumanoff, K. 1927 La flore microbienne d'un couvain malade. Recueil de Médecine Vétérinaire (Bull. Soc. Cent. de Méd. Vétér.), **103**, 367–374.

Toumanoff, K. 1928 On the infection of *Pyrausta nubilalis* Hb. by *Aspergillus flavus* and *Spicaria farinosa*. Intern. Corn Borer Invest., Sci. Repts., **1**, 74–76.

Toumanoff, K. 1933 Action des champignons entomophytes sur la pyrale du maïs (*Pyrausta nubilalis*). Ann. Parasitol. Humaine et Comparée, **11**, 129–143.

Townsend, C. H. T. 1913 The transmission of verruga by phlebotomus. J. Am. Med. Assoc., **61**, 1717–1718.

Townsend, C. H. T. 1915 Two years' investigation in Peru of verruga and its insect transmission. Am. J. Trop. Diseases, **3**, 16–32.

Trager, W. 1934 The cultivation of a cellulose-digesting flagellate, *Trichomonas termopsidis*, and of certain other termite protozoa. Biol. Bull., **66**, 182–190.

Trager, W. 1935a The culture of the mosquito larvae free from living microorganisms. Am. J. Hyg., **22**, 18–25.

Trager, W. 1935b On the nutritional requirements of mosquito larvae (*Aëdes aegypti*). Am. J. Hyg., **22**, 475–493.

Trager, W. 1935c Cultivation of the virus of grasserie in silkworm tissue cultures. J. Exptl. Med., **61**, 501–513.

Trager, W. 1937 The hatching of spores of *Nosema bombycis* Nägeli and the partial development of the organism in tissue cultures. J. Parasitol., **23**, 226–227.

Trager, W. 1939 Intracellular microorganism-like bodies in the tick *Dermacentor variabilis* Say. J. Parasitol., **25**, 233–239.

Trager, W., Miller, D. K., and Rhoads, C. P. 1938 The absence from the urine of pernicious anemia patients of a mosquito growth factor present in normal urine. J. Exptl. Med. **67**, 469–480.

Trager, W., and Subbarow, Y. 1938 The chemical nature of growth factors required by mosquito larvae. I. Riboflavin and thiamin. Biol. Bull., **75**, 75–84.

Trask, J. D., Paul, J. R., and Melnick, J. L. 1943 The detection of poliomyelitis virus in flies collected during epidemics of poliomyelitis. I. Methods, results, and types of flies involved. J. Exptl. Med., **77**, 531–544.

Travassos, J., and Biocca, E. 1942 Açáo da prata eletrolisada sôbre certas toxinas, venenos, protozóarios, rickettsias, virus filtráveis e bacteriófagos. Mem. Inst. Butantan, **16**, 309–314.

Travis, B. V. 1932 A discussion of synonymy in the nomenclature of certain insect flagellates, with the description of a new flagellate from the larvae of *Lygy rodes relectus* Say. Iowa State Coll. J. Sci., **6**, 317–323.

Travis, B. V., and Becker, E. R. 1931 A preliminary report on intestinal protozoa of white grubs (*Phyllophaga* spp.—Coleoptera). Iowa State Coll. J. Sci., **5**, 223–235.

Treillard, M., and Lwoff, A. 1924 Sur un infusoire parasite de la cavité générale des larves de chironomes. Sa sexualité. Compt. Rend. Acad. Sci., Paris, **178**, 1761–1764.

Trotter, A. 1934 Il fungo-Ambrosia delle gallerie die un Xyloborino di Ceylon. Ann. Ist. Super. Agrar. Portici, Ser. III, **6**, 256–275.

Tschirch, A. 1922–1924 Handbuch der Pharmakognosie. C. H. Tauchnitz, Leipzig. 252 pp.

Tubeuf, C. von. 1892a Die Krankheiten der Nonne (*Liparis monacha*). Forstl. Naturw. Z., **1**, 34–47, 62–79.

Tubeuf, C. von. 1892b Weitere Beobachtungen über die Krankheiten der Nonne. Forstl. Naturw. Z., **1**, 277–279.

Tubeuf, C. von. 1911 Zur Geschichte der Nonnenkrankheit. Naturw. Z. Forst- u. Landw., **9**, 357–377.

Uichanco, L. B. 1924 Studies on the embryogeny and postnatal development of the Aphididae with special reference to the history of the "symbiotic organ" o. "mycetom." Philippine J. Sci., **24**, 143–247.

Uribe, C. 1926 *Crithidia ortheae* n. sp. from reduviids of the genus Orthea. J. Parasitol., **12**, 199–202.

Usinger, R. L. 1944 The Triatominae of North and Central America and the West Indies and their public health significance. U.S. Pub. Health Service, Pub. Health Bull., no. 288. 83 pp.

Uvarov, B. P. 1928 Insect nutrition and metabolism. Trans. Entomol. Soc. London, **76**, 255–343.

Valleau, W. D., and Johnson, E. M. 1936 *Physalis subglabrata:* a natural host of *Bacterium angulatum.* Phytopathology, **26**, 388–390.

Vaney, C., and Contè, A. 1903 Sur un diptére (*Degeeria funebris* Mg.) parasite de l'altise de la vigne (*Haltica ampelophaga* Guer.) Compt. Rend. Acad. Sci., Paris, **136**, 1275–1276.

van Rooyen, C. E., and Rhodes, A. J. 1940 Virus diseases of man. Oxford Univ. Press, London. 932 pp.

Vaternahm, T. 1924 Zur Ernährung und Verdauung unserer einheimsichen Geotrupesarten. Z. Wiss. Insektenbiol., **19**, 20–27.

Veeder, M. A. 1898 Flies as spreaders of disease in camps. N.Y. Med. Record, Sept. 17, 429–430.

Vejdovsky, F. 1906 Bemerkungen zum Aufsatze des Herrn Dr. K. Šulc über *Kermicola kermesina,* etc. Sitzber. Böhm. Ges. Wiss., Math.- Naturw. Classe (České spolecnosti Nauk), art. XX, 6–12.

Verjbitski, D. T. 1908 The part played by insects in the epidemiology of plague. J. Hyg., **8**, 162–208.

Vervoort, H. 1938 (See discussion at end of paper by Donatien and Lestoquard, 1938.)

Vincent, M. 1922 On the life-history of a new gregarine: *Pyxinia anobii* n. sp., intestinal parasite of *Anobium paniceum* L. (Coleoptera). Parasitology, **14**, 299–306.

REFERENCES

Vincent, M. 1927 On *Legerella hydropori* n. sp., a coccidian parasite of the malpighian tubes of *Hydroporus palustris* L. (Coleoptera). Parasitology, **19**, 394–400.

Volferz, A. A., Kolpakov, S. A., and Flegontoff, A. A. 1934 The role of ectoparasites in the tularaemic epizootic of the ground squirrels. Rev. Microbiol. Epidemiol. Parasitol., **13**, 103–116.

Wachtl, F. H., and Kornauth, K. 1893 Additions to the knowledge of the morphology, biology, and pathology of the Nonne moth (*Psilura monacha*), and experiments with methods for destroying the caterpillars. Mitt. Ver. Förd. Landw. Versuchsw. Oesterr., **16**. 38 pp.

Wadsworth, A. B. 1939 Standard methods of the division of laboratories and research of the New York State Department of Health. 2d ed. Williams and Wilkins Co., Baltimore. 681 pp.

Wahl, B. 1909–1912 Über die Polyderkrankheit der Nonne (*Lymantria monacha* L.). Centr. Ges. Forstwesen, **35**, 164, 212; **36**, 193, 377; **37**, 247; **38**, 355.

Waite, M. B. 1891 Results from recent investigations in pear blight. Botan. Gaz., **16**, 259.

Walczuch, A. 1932 Studien an Coccidensymbionten. Z. Morphol. Ökol. Tiere, **25**, 623–729.

Wallace, F. G. 1943 Flagellate parasites of mosquitoes with special reference to *Crithidia fasciculata* Léger, 1902. J. Parasitol., **29**, 196–205.

Wallace, G. B. 1932 Coffee bean disease. Relation of *Nematospora gossypii* to the disease. Trop. Agr. (Trinidad), **9**, 127.

Wallace, G. I., Cahn, A. R., and Thomas, L. J. 1933 *Klebsiella paralytica* a new pathogenic bacterium from "moose disease." J. Infectious Diseases, **53**, 386–414.

Wallace, J. M., and Murphy, A. M. 1938 Studies on the epidemiology of curly top in southern Idaho, with special reference to sugar beets and weed hosts of the vector *Eutettix tenellus*. U.S. Dept. Agr. Tech. Bull., 624.

Wallengren, H., and Johansson, R. 1929 On the infection of *Pyrausta nubilalis* Hb. by *Metarrhizium anisopliae* (Metsch.). Intern. Corn Borer Invest., Sci. Repts., **2**, 131–145.

Waller, E. F. 1940 Tularemia in Iowa cottontail rabbits (*Sylvilagus floridanus mearnsi*) and in a dog. Vet. Student, **2**, 54, 55, 73.

Wallin, I. E. 1922 On the nature of mitochondria. I. Observations on mitochondria staining methods applied to bacteria. II. Reactions of bacteria to chemical treatment. III. The demonstration of metochondria by bacteriological methods. IV. A comparative study of the morphogenesis of root nodule bacteria and chloroplasts. Am. J. Anat., **30**, 203–229, 451–466.

Wallin, I. E. 1927 Symbioticism and the origin of species. Williams and Wilkins Co., Baltimore. 171 pp.

Wallin, I. E. 1943 Personal communication.

Wardle, R. A. 1929 The problems of applied entomology. Manchester University Press, England. 588 pp.

Watson, J. R. 1912 Utilization of fungous parasites of Coccidae and Aleurodidae in Florida. J. Econ. Entomol., **5**, 200–204.

Webb, J. L. 1940 The occurrence of rickettsia-like bodies in the reduviid bug, *Triatoma rubrofasciata,* and their transmission to laboratory animals. Parasitology, **32**, 355–360.

Weber, G. F. 1933 Pecans infected with *Nematospora coryli* Pegl. Phytopathology, **23**, 1000–1001.

Weber, N. A. 1937 The biology of the fungus-growing ants. II. Nesting habits of the Bachac (*Atta cephalotes* L.). Trop. Agr. (Trinidad), **14**, 223–226.

Weber, N. A. 1938 The biology of the fungus-growing ants. III. The sporophore of the fungus grown by *Atta cephalotes* and a review of other reported sporophores. Rev. Entomol., **8**, 265–272.

Weigl, R. 1920 Studies on *Rickettsia prowazeki.* Przglad. Epidemj., **1**, 15.

Weigl, R. 1921 Further studies on *Rickettsia rocha-limae* (n. s.). Przglad. Epidemj., **1**, 375.

Weigl, R. 1923 Die Bezietungen der X Stämme zur *Rickettsia prowazeki.* Z. Hyg. Infektionskrankh., **99**, 303.

Weigl, R. 1924 Further studies on *Rickettsia rocha-limae* (nov. spec.). J. Trop. Med. Hyg., **27**, 14–15.

Weigl, R. 1939 Trachomstudium. I. *Rickettsia rocha-limae* (Weigl) ist nicht der Trachomerreger. Zentr. Bakt., Parasitenk. Infekt., I, **143**, 291–298.

Weinland, E. 1907 Weitere Beobachtungen an *Calliphora.* IV. Über chemische Momente bei der Metamorphose (und Entwicklung). Z. Biol., **49**, 486–493.

Weinland, E. 1908 Über die Bildung von Fett aus eiweissartiger Substanz im Brei der Calliphoralarven. Z. Biol., **51**, 197–278.

Weinman, D. 1944 Infectious anemias due to *Bartonella* and related red cell parasites. Trans. Am. Phil. Soc., **33**, 243–350.

Weiss, L. J. 1943 Electron-micrographs of rickettsiae of typhus fever. J. Immunol., **47**, 353–357.

Welander, E. 1896 Wien. Klin. Wochschr., no. 52. (Quoted by Pierce, 1921, p. 108.)

Welch, D. S., Herrick, G. W., and Curtis, R. W. 1934 The Dutch elm disease. Cornell Univ. Extension Bull., 290.

Wenrich, D. H. 1937 Protozoological methods. *In* Microscopical Technique, ed. C. E. McClung, chap. 8. Paul Hoeber, Inc., New York. 698 pp. (See pp. 522–551.)

Wenyon, C. M. 1911 Oriental sore in Bagdad, together with observations on a gregarine in *Stegomyia fasciata,* the haemogregarine of dogs and the flagellates of house flies. Parasitology, **4**, 273–344.

Wenyon, C. M. 1926 Protozoology. William Wood and Co., New York. 2 vols. 1563 pp.

Werner, E. 1926a Die Ernährung der Larve von *Potosia cuprea,* Frb. (*Cetonia floricola* Hbst.) Ein Beitrag zum Problem der Celluloseverdauung bei Insectenlarven. Z. Morphol. Ökol. Tiere (Abt. A. Z. Wiss. Biol.) **6**, 150–206.

Werner, E. 1926b *Bacillus cellulosam fermentans.* Zentr. Bakt. Parasitenk. Infekt., II, **67**, 297.

Werner, H. 1939 Neuere Ergebnisse der Fünftagefieberforschung. Deut. Med. Wochschr., **65**, 174–176.

Wharton, D. R. A. 1928 Etiology of European foul-brood of bees. Science, **66**, 451–452.

Wheeler, C. M. 1942 The distribution of the spirochete of California relapsing fever within the body of the vector, *Ornithodoros hermsi.* A symposium on relapsing fever in the Americas. Pub. Am. Assoc. Advancement Sci., 18, 89–99.

Wheeler, C. M., and Douglas, J. R. 1945 Sylvatic plague studies. V. The determination of vector efficiency. J. Infectious Diseases, **77**, 1–12.

Wheeler, W. M. 1889 The embryology of *Blatta germanica* and *Doryphora decemlineata.* J. Morphol., **3**, 291–386.

Wheeler, W. M. 1907 The fungus-growing ants of North America. Bull. Am. Mus. Natural Hist., **23**, 669–807.

Wheeler, W. M. 1914 Ants and bees as carriers of pathogenic organisms. Am. J. Trop. Med. Prev. Med., **2**, 160–168.

Wheeler, W. M. 1923 Social life among the insects. Harcourt, Brace and Co., New York. 375 pp.

Wheeler, W. M. 1937 Mosaics and other anomalies among ants. Harvard Univ. Press, Cambridge. 95 pp.

White, G. F. 1904 The further investigation of the diseases affecting the apiaries in the State of New York. New York Dept. Agr., 11th Ann. Rept. Com. Agr. for 1903, Jan. 15, pp. 103–114.

White, G. F. 1905 The bacterial flora of the apiary with special reference to bee diseases. Cornell University Thesis.

White, G. F. 1906 The bacteria of the apiary with special reference to bee diseases. U.S. Dept. Agr., Bur. Entomol. Tech. Bull., 14. 50 pp.

White, G. F. 1912 The cause of European foulbrood. U.S. Dept. Agr., Bur. Entomol. Circ., 157. 15 pp.

White, G. F. 1913 Sacbrood, a disease of bees. U.S. Dept. Agr., Bur. Entomol. Circ., 169. 5 pp.

White, G. F. 1917 Sacbrood. U.S. Dept. Agr. Bull., 431. 54 pp.

White, G. F. 1920a American foulbrood. U.S. Dept. Agr., Bur. Entomol. Bull., 809. 46 pp.

White, G. F. 1920b European foulbrood. U.S. Dept. Agr. Bull., 810. 39 pp.

White, G. F. 1920c Some observations on European foulbrood. Am. Bee J., **60**, 225–227, 266–268.

White, G. F. 1923a Hornworm septicemia. J. Agr. Research, **26**, 477–486.

White, G. F. 1923b Cutworm septicemia. J. Agr. Research, **26**, 487–496.

White, G. F. 1927 A protozoan and a bacterial disease of *Ephestia kühniella* Zell. Proc. Entomol. Soc. Wash., **29**, 147-148.

White, G. F. 1928 Potato beetle septicemia, with the proposal of a new species of bacterium. Proc. Entomol. Soc. Wash., **30**, 71-72.

White, G. F. 1935 Potato beetle septicemia. J. Agr. Research, **51**, 223-234.

White, G. F., and Noble, L. W. 1936 Notes on pink bollworm septicemia. J. Econ. Entomol., **29**, 122-124.

White, P. B. 1933 The O-receptor complex of *B. proteus* X-19. Brit. J. Exptl. Path., **14**, 145.

White, R. T. 1940 Survival of Type A milky disease of Japanese beetle larvae under adverse field conditions. J. Econ. Entomol., **33**, 303-309.

White, R. T. 1941 Development of milky disease of Japanese beetle larvae under field conditions. J. Econ. Entomol., **34**, 213-215.

White, R. T., and Dutky, S. R. 1942 Cooperative distribution of organisms causing milky disease of Japanese beetle grubs. J. Econ. Entomol., **35**, 679-682.

Whitman, L. 1937 The multiplication of the virus of yellow fever in *Aëdes aegypti*. J. Exptl. Med., **66**, 133-143.

Whitman, L., and Antunes, P. C. A. 1938 Studies on *Aëdes aegypti* infected in the larval stage with the virus of yellow fever. Proc. Soc. Exptl. Biol. Med., **37**, 664-666.

Widmann, E. 1915 Zur Frage der Übertragung von Bakterien durch Läuse. Munch. Med. Wochschr., **62**, 1336-1338.

Wigglesworth, V. B. 1927 Digestion of the cockroach. I. The hydrogen ion concentration in the alimentary canal. Biochem. J., **21**, 791-796.

Wigglesworth, V. B. 1929 Digestion in the tsetse-fly: a study of structure and function. Parasitology, **21**, 288-321.

Wigglesworth, V. B. 1936 Symbiotic bacteria in a blood-sucking insect, *Rhodnius prolixus* Stål. (Hemiptera, Triatomidae). Parasitology, **28**, 284-289.

Wigglesworth, V. B. 1939 The principles of insect physiology. E. P. Dutton and Co., Inc., New York. 434 pp.

Will, L. 1889 Entwicklungsgeschichte der viviparen Aphiden Zool. Jahrb. Abt. Anat. Ontog., **3**, 201-286.

Williams, R. W. 1944 A check list of the mite vectors and animal reservoirs of tsutsugamushi disease. Am. J. Trop. Med., **24**, 355-357.

Wingard, S. A. 1925 Studies on the pathogenicity, morphology and cytology of *Nematospora phaseoli*. Bull. Torrey Botan. Club, **52**, 249-290.

Witlaczil, E. 1882 Zur Anatomie der Aphiden. Arb. Zool. Inst. Wien, **4**, 397-441.

Witlaczil, E. 1884 Entwicklungsgeschichte der Aphiden. Z. Wiss. Zool., **40**, 559-696.

Wolbach, S. B. 1919 Studies on Rocky Mountain spotted fever. J. Med. Research, **41**, 1-197.

Wolbach, S. B., Pinkerton, H., and Schlesinger, M. J. 1923 The cultivation of the organisms of Rocky Mountain spotted fever and typhus in tissue cultures. Proc. Soc. Exptl. Biol. Med., **20,** 270–273.

Wolbach, S. B., and Schlesinger, M. J. 1923 The cultivation of the microorganisms of Rocky Mountain spotted fever (*Dermacentroxenus rickettsi*) and of typhus (*Rickettsia prowazeki*) in tissue plasma cultures. J. Med. Research, **44,** 231–256.

Wolbach, S. B., and Todd (misspelled Tood), J. L. 1920 Note sur l'étiologie et l'anatomie pathologique du typhus exanthématique au Mexique. Ann. Inst. Pasteur, **34,** 153–158.

Wolbach, S. B., Todd, J. L., and Palfrey, F. W. 1922 The etiology and pathology of typhus. Harvard Univ. Press, Cambridge. 222 pp.

Wolf, J. 1924a Contribution à la morphologie des bactéroides dans les blattes (*Periplaneta orientalis* L.). Compt. Rend. Soc. Biol., **91,** 1180–1182.

Wolf, J. 1924b Contribution à la localisation des bactéroides dans les corps adipeux des blattes (*Periplaneta orientalis*). Compt. Rend. Soc. Biol., **91,** 1182–1183.

Wolff, M. 1910 Ueber eine neue Krankheit der Raupe von *Bupalus piniarius* L. Mitt. Kaiser-Wilhelm Inst. Landw. Bromberg, **3,** 69–92.

Wollman, E. 1911 Sur l'élevage des mouches steriles. Contribution à la connaissance du rôle des microbes dans les voies digestives. Ann. Inst. Pasteur, **25,** 79–88.

Wollman, E. 1921 La méthode des élevages aseptiques en physiologie. Arch. Intern. Physiol., **18,** 194–199.

Wollman, E. 1926 Observations sur une lignée aseptique de blattes (*Blattella germanica*) datant de cinq ans. Compt. Rend. Soc. Biol., **95,** 164–165.

Wolzogen Kühr, C. A. H. von. 1931–1932 Über eine Gärungsmikrobe in Fäkalien von Mückenlarven. Zentr. Bakt. Parasitenk. Infekt., II, **85,** 223–250.

Womersley, H., and Heaslip, W. G. 1943 The Trombiculinae (Acarina) or Itchmites of the Austro-Malayan and Oriental regions. Trans. Roy. Soc. S. Australia, **67,** 68–142.

Wood, S. F. 1942 The persistence of *Trypanosoma cruzi* in dead cone-nosed bugs (Hemiptera, Reduviidae). Am. J. Trop. Med., **22,** 613–621.

Woodcock, H. M. 1923 "Rickettsia" bodies as a result of cell digestion or lysis. J. Roy. Army Med. Corps, **40,** 81–91, 241–269.

Woodland, J. C., McDowell, M. M., and Richards, J. T. 1943 Bullis fever (Lone Star fever—tick fever). J. Am. Med. Assoc., **122,** 1156–1160.

Wright, E. 1935 *Trichosporium symbioticum*, n. sp.: a wood-staining fungus associated with *Scolytus ventralis*. J. Agr. Research, **50,** 525–538.

Wright, E. 1938 Further investigations of brown-staining fungi associated with engraver beetles (*Scolytus*) in white fur. J. Agr. Research, **57,** 759–774.

Yakimoff, W. L. 1930 Zur Frage über Parasiten bei Protozoa. Arch. Protistenk., **72,** 135–138.

Yamasaki, M. 1931 Studies on the intestinal protozoa of termites. II. Oxygenation experiments under the influence of temperature. Mem. Coll. Sci. Kyoto Imp. Univ., Ser. B, **7**, 179–188.

Yao, H. Y., Yuan, I. C., and Huie, D. 1929 The relation of flies, beverages and well water to gastro-intestinal diseases in Peiping. Nat. Med. J. China, **15**, 410–418.

Yeager, J. F. 1945 The blood picture of the southern armyworm (*Prodenia eridania*). J. Agr. Res., **71**, 1–40.

Yeager, J. F., and Knight, H. H. 1933 Microscopic observations on blood coagulation in several different species of insects. Ann. Entomol. Soc. Am., **26**, 591–602.

Yeager, J. F., McGovran, E. R., Munson, S. C., and Mayer, E. L. 1942 Effect of blocking hemocytes with Chinese ink and staining nephrocytes with trypan blue upon the resistance of the cockroach *Periplaneta americana* (L.) to sodium arsenite and nicotine. Ann. Entomol. Soc. Am., **35**, 23–40.

Yersin, A. 1894 La peste bubonique à Hongkong. Ann. Inst. Pasteur, **8**, 652–667.

Yolles, S. F., and Knigin, T. D. 1943 Note on a new transparent cage for collecting and feeding insects. Am. J. Trop. Med., **23**, 465–469.

Yoshida, S. 1935 On the tissue culture of tsutsugamushi virus (*Rickettsia tsutsugamushi*). Kitasato Arch. Exptl. Med., **12**, 324–337.

Young, M. D. 1937 Cockroaches as carriers of *Giárdia* cysts. J. Parasitol., **23**, 102–103.

Zacharias, A. 1928 Untersuchungen über die intrazellulare Symbiose bei den Pupiparen. Z. Morphol. Ökol. Tiere, **10**, 676–737.

Zasukhin, D. N. 1936a The ticks (Ixodidae) and their role in epizootology and epidemiology of tularaemia in the southeast of the U.S.S.R. Rev. Microbiol. Epidemiol. Parasitol., **15**, 461–470. (In Russian.)

Zasukhin, D. N. 1936b Transovarial transmission of the causative agents of protozoan, spirochetal, bacterial, and virus diseases in ticks. Rev. Microbiol. Epidemiol. Parasitol., **15**, 457–460.

Zeliff, C. C. 1933 A new protozoan from the larva of the beetle *Osmoderma scabra*. Proc. U.S. Nat. Museum, **82**, 1–3.

Zernoff, V. 1928a Sur la spécificité de l'immunité passive chez *Galleria mellonella*. Compt. Rend. Soc. Biol., **98**, 1500–1502.

Zernoff, V. 1928b Sur la nature de l'immunité passive chez les chenilles de *Galleria mellonella*. Compt. Rend. Soc. Biol., **99**, 315–317.

Zernoff, V. 1929 Essai de sérothérapie chez *Galleria mellonella*. Compt. Rend. Acad. Sci., Paris, **188**, 1321–1323.

Zernoff, V. 1930 L'immunité passive et la sérothérapie chez les insectes (chenilles de *Galleria mellonella*). Ann. Inst. Pasteur, **44**, 604–618.

Zernoff, V. 1931 Microbes virulents pour les chenilles (*Galleria mellonella* et *Pyrausta nubilalis*). Compt. Rend. Soc. Biol., **106**, 543–546.

Zernoff, V. 1934 Influence des différentes concentrations des vaccins dans l'immunisation de *Galleria mellonella*. Compt. Rend. Soc. Biol., **116**, 304.

Zernoff, V., and Ajolo, J. 1939 Chimiotherapie chez les insectes: action du para-aminophenylsulfamide (1162 F.) dans l'infection expérimentale chez de *Galleria mellonella*. Compt. Rend. Soc. Biol., **131**, 232–234.

Zilber, L. A., and Dosser, E. M. 1933 Culture of virus of exanthematous typhus. Sovet. Vrach. Gaz., April 15, 209–210. (See also Zentr. Bakt. Parasitenk. Infekt., 1934, **131**, 222–232.)

Zinsser, H. 1934 Rats, lice and history. Blue Ribbon Books, Inc., New York. 301 pp.

Zinsser, H. 1937 The rickettsia diseases. Varieties, epidemiology and geographical distribution. Am. J. Hyg., **25**, 430–463.

Zinsser, H. 1940 Epidemiology and immunity in the rickettsial diseases. *In* Gordon, J. E., *et al.,* Virus and rickettsial diseases. Harvard Univ. Press, Cambridge. 907 pp. (See pp. 872–907.)

Zinsser, H., and Castañeda, M. R. 1931 Studies on typhus fever. VIII. Ticks as a possible vector of the disease from animals to man. J. Exptl. Med., **54**, 11–21.

Zinsser, H., and Castañeda, M. R. 1932 A method of obtaining large amounts of *Rickettsia provaceki* by x-ray radiation of rats. Proc. Soc. Exptl. Biol. Med., **29**, 840–844.

Zinsser, H., and Macchiavello, A. 1936 Enlarged tissue cultures of European typhus rickettsiae for vaccine production. Proc. Soc. Exptl. Biol. Med., **35**, 84–87.

Zinsser, H., Wei, H., and Fitzpatrick, F. K. 1937 Agar slant tissue cultures of typhus rickettsiae (both types). Proc. Soc. Exptl. Biol. Med., **37**, 604–606.

Zorin, P. V., and Zorina, L. M. 1928 Contributions à la biologie de la *Polia oleracea*. Défense Plantes, no. 5–6, **5**, 475–486.

Zuelzer, M. 1925 Die Spirochäten. *In* Prowazek, S. *Handbuch der pathogenen Protozoen*. Vol. III, lief. XI, pp. 1627–1974. J. A. Barth, Leipzig. 2171 pp.

AUTHOR INDEX

Abe, N., 128
Acqua, C., 418
Adametz, L., 181
Adler, S., 454, 455, 456
Adlerz, G., 247
Agramonte, A., 432
Ahlm, C. E., 291
Ajolo, J., 130, 156
Alexander, R. A., 262, 303, 339, 341, 343
Allen, T. C., 22, 174
Amaral, A. do, 264, 290, 329, 339
Amzel, R., 275
Anderson, C., 453, 455
Anderson, C. R., 277
Anderson, J. F., 441
Anderson, T. F., 275
Anderson, T. G., 132
André, C., 49
Anigstein, L., 275, 291, 292, 313, 314, 330, 339, 345
Antunes, P. C., 435
Aoki, K., 88, 418, 419
Aristotle, 566
Ark, P. A., 117
Arkwright, J. A., 78, 219, 288, 316, 317, 319, 320, 321
Armbrust, C. A., 291
Armer, J. M., 521
Arneth, J., 288
Aschner, M., 201, 205, 211, 212, 213, 221, 245, 246, 276, 277, 584
Ascolese, 134
Ashbel, R., 454, 455, 456
Ashworth, J. H., 579
Atanasoff, D., 412
Atkin, E. E., 22, 23, 28, 270, 316, 317, 319
Atkinson, N. J., 75
Augustine, D. L., 434, 477, 577

Babers, F. H., 58
Bacot, A. W., 14, 19, 22, 23, 28, 78, 163, 169, 270, 288, 316, 317, 319, 320, 575
Bader, M. N., 330, 339, 345
Badger, L. F., 281, 282, 283, 338, 340, 343
Baer, W. S., 35
Bahr, L., 131
Bahr, P. H., 128, 129
Bailey, I. W., 406, 408
Baker, A. C., 259, 261
Balbiani, E. G., 190, 197, 201, 522
Baldwin, I. L., 203
Baltazard, M., 160, 163, 282
Bandelli, 134
Banerji, R. N., 329

Bang, F. B., 441, 442
Barber, M. A., 12, 23, 24, 28, 134, 179, 183
Barnes, B., 590
Barraud, P. J., 575
Bary, A. de (*see* De Bary)
Basile, C., 466
Basu, B. C., 454
Bates, L. B., 453
Baumberger, J. P., 29, 349, 350
Beard, R. L., 83
Beauperthuy, L. D., 3, 432
Becker, E. R., 469, 471, 483
Becker, W. B., 410, 463
Begg, A. M., 275
Beijerinck, M. W., 203
Belding, D. L., 501
Bell, E. J., 290, 291, 300
Bell, J. F., 153, 168
Beltiukova, K. I., 101
Benecke, W., 607
Benedek, T., 71, 366
Bengtson, Ida A., 94, 286, 301, 302, 336, 587
Bennett, C. W., 446
Bensley, R. R., 202
Bequaert, J., 54
Bergey, D. H., 125
Bergold, 416
Berlese, A., 193, 231, 349, 356, 362
Berliner, E., 91, 92
Berry, G. P., 438
Bertarelli, E., 128
Beyer, G. E., 349
Bickhardt, R., 287
Bilal, S., 167, 168, 169
Biocca, E., 337
Bishop, A., 485
Bishop, M. B., 441
Black, L. M., 447
Blake, F. G., 290, 291
Blanc, G. R., 59, 69, 160, 163, 183, 184, 270, 282, 341, 343, 344, 437
Blanchard, C., 47
Blattner, R. J., 438
Blewett, M., 213, 369
Blochmann, F., 190, 215, 217, 247
Boardman, E. T., 574
Bobb, M. L., 400
Bode, H., 204
Boek, 47
Boerner, F., 582
Bogdanow, E. A., 29, 30, 281
Bogoyavlensky, N., 382
Bohls, S. W., 281
Böing, W., 138
Bolle, J., 417

Bollen, W. B., 24
Bollinger, O., 53
Bonde, R., 119
Bongert, 54
Bonnamour, S., 610
Borchert, A., 81
Borzenkov, A., 165
Bos, A., 458
Bouin, P., 216
Bourne, B. A., 78
Boutroux, 348
Bowers, R. E., 32
Bozhenko, V. P., 168
Bradley, G. H., 575
Bradley, W. G., 575
Brain, C. K., 352, 353, 354, 355, 357, 358, 360, 361, 364, 365, 366, 372, 373
Brandes, E. W., 445
Brecher, G., 141, 234, 235
Breed, R. S., 63, 66, 70, 125, 132, 134, 451
Brefeld, O., 378, 380
Breindl, V., 425
Breinl, F., 12, 271
Breitsprecher, E., 366, 368, 369
Brigham, G. D., 281, 285, 286
Brill, N. E., 280
Brinley, F. J., 19, 135
Brooks, C. S., 167
Brown, F. M., 31, 67, 69, 125, 145, 150, 158, 180
Brown, H. E., 301, 308
Brown, J. G., 118
Brown, J. H., 162
Bruce, D., 3, 474
Bruch, A., 340
Brues, C. T., 212, 356, 357, 441
Brumpt, E., 262, 263, 321, 322, 328, 333, 339, 341, 452, 453, 466
Brunchorst, J., 203
Buchanan, D., 174
Buchanan, E. D., 263
Buchanan, R. E., 263, 583
Buchanan, W. D., 405
Buchner, P., 176, 190, 191, 197, 199, 206, 217, 218, 219, 221, 222, 223, 224, 225, 226, 231, 237, 238, 239, 243, 245, 246, 247, 248, 250, 251, 254, 262, 316, 318, 352, 354, 359, 361, 363, 364, 365, 366, 367, 368, 369, 370, 373, 374
Buddington, A. R., 350
Buen, S. de, 453
Bullamore, G. W., 78
Burnet, F. M., 259, 260, 281, 295, 296, 298, 300, 301
Burnett, W. I., 471
Burnside, C. E., 25, 52, 55, 66, 76, 77, 81, 97, 98, 139, 351, 395, 401, 402, 415, 427
Burri, R., 66, 67

Burrill, T. J., 116, 147
Burroughs, A. L., 91, 145, 146, 151, 162
Burrows, W., 113
Busacca, A., 262
Bush, S. F., 494
Butt, F. H., 195, 196, 222
Byam, W., 288

Cahn, A. R., 123, 124
Calkins, G. N., 501
Cameron, G. R., 18, 43, 46, 50, 73, 74, 90, 95, 96, 122, 126, 128, 130, 138, 141, 142, 143, 154, 156, 157, 170, 179, 557, 564, 565, 566, 567, 568, 569
Caminopétros, M. J., 339, 341, 343, 344, 437
Campbell, F. L., 575
Campo, G. C. (see Clavero, G. C.)
Canham, A. S., 311
Cao, G., 12, 19, 46, 49, 53, 55, 60, 71, 80, 83, 85, 87, 90, 93, 95, 96, 121, 122, 124, 126, 132, 133, 138, 151, 152, 154, 155, 156, 159, 160, 164, 169, 170, 171, 172, 178, 179
Carmichael, J., 308
Carpano, M., 310
Carrasquillo, 47
Carter, W., 231, 232, 363
Cartwright, K. St. G., 392
Casey, A. E., 441
Cash, L. C., 120, 177
Caspari, E., 130
Castañeda, M. R., 261, 269, 272, 276, 279, 281, 282, 286, 298, 584
Castellani, A., 349
Cattani, G., 178
Ceder, E. T., 282, 283
Cellí, A., 49, 127, 155
Chahovitch, X., 170
Chalgren, W. S., 153
Chalmers, A. J., 349
Chambers, L. A., 275
Chandler, A. C., 436, 465, 473, 501
Chang, H. C., 129
Chapin, C. W., 165, 166
Chapman, J. W., 146, 186, 413, 421, 423
Chapman, R. N., 12
Charles, V. K., 377, 392, 394
Charrin, 80
Chatton, E., 57, 71, 365
Cheshire, F. R., 25, 52, 63
Cheyne, W. W., 25, 52
Chigasaki, Y., 88, 418, 419
Chilton, St. J. P., 405
Chittenden, F. H., 150
Chodzko, W., 312
Cholodkowsky, N., 190, 215, 221
Chorine, V., 43, 63, 64, 76, 87, 88, 92, 93, 96, 99, 100, 101, 103, 110, 112, 135, 145,

179, 184, 185, 351, 556, 560, 562, 565, 566, 568, 569, 571
Chow, C. Y., 122, 128, 131
Christensen, C. M., 405
Chumakov, M. P., 439
Chun, J. W., 163
Ciferri, R., 184
Clark, P. F., 441
Clavera, G. C., 578, 584
Clegg, M. T., 48
Clements, F. E., 383, 393
Cleveland, L. R., 461, 526, 530, 531, 532, 533, 534, 535, 536, 537
Clinton, G. P., 410
Cochrane, E. W. W., 128
Coggeshall, L. T., 440, 501, 504
Cohn, F., 377
Colas-Belcour, J., 23
Coles, J. D. W. A., 304, 305, 306
Colla, S., 384
Collier, J., 461
Comb, H., 128, 129
Comstock, J. H., 105, 135, 147, 150, 154, 193, 194, 214, 215, 216, 225, 226, 227, 231, 270, 316, 405, 417, 527, 567, 576
Comte, C., 270
Conor, A., 340
Conseil, E., 270, 341
Conté, A., 355, 400
Contos, B., 344
Cook, M. T., 443
Cooley, R. A., 250, 330, 340, 454, 575
Cornalia, E., 413
Coronodo, D., 347
Corson, J. F., 291
Corrodor, 47
Couch, J. N., 381, 389, 390, 391
Couvreur, E., 170
Couzi, 129
Cowan, T. W., 67
Cowdry, E. V., 191, 202, 215, 218, 220, 232, 247, 249, 251, 252, 253, 255, 256, 262, 297, 302, 303, 316, 418, 508, 509, 510, 580
Cox, G. L., 14, 120, 122, 125, 129, 141, 144, 153, 155
Cox, H. R., 276, 295, 296, 299, 300, 302, 343, 344, 442, 584
Cox, J. A., 400
Cragg, F. W., 316
Craig, C. F., 582
Cuboni, C., 60
Cuénot, L., 202, 216
Culbertson, J. T., 501
Cunha, A. M. da, 276, 584
Currie, D. G., 47, 48
Currie, J. R., 579
Curtis, R. W., 410

Damon, S. R., 460
Darling, 128
Daubney, R., 303
Davaine, C., 53
Davis, G. E., 167, 168, 295, 296, 297, 330, 333, 340, 341, 344, 442, 453, 456
Davis, J. J., 148
Davis, N. C., 434, 435
Dean, R. W., 22
De Bach, P. H., 39, 99, 133, 134
De Bary, A., 5, 70, 188, 189
De Coursey, J. D., 514
Delpy, L., 55
Dennis, E. W., 506, 507
Derrick, E. H., 295, 296, 298, 301, 308
Descazeaux, 102
Desportes, C., 333
Detroyat, C., 305
Dewèvre, L., 50, 153
Dias, E., 233, 331
Dickman, A., 30
Dietrich, S., 270
Dobell, C. C., 460, 554, 590
Dodge, C. W., 385
Doflein, F., 460
Donatien, A., 304, 305, 306, 307, 308, 309, 310, 324, 510
Donisthorpe, H., 442
Donovan, C., 472, 473
Donskov, G., 165
Dosdall, L., 408
Dosser, E. M., 276, 584
Dougherty, E. C., 61, 116
Douglas, J. R., 163, 164, 579
Dove, W. E., 282
Dowden, P. B., 575
Dowson, W. J., 173
Drbohlav, J. J., 471
Drobotjko, V., 76
Dropkin, V. H., 530
Dschunkowsky, E., 452
Duboscq, O., 61, 115, 116, 461, 492, 532
Dudgeon, L. S., 131
Dufrenoy, J., 11, 142
Duggar, B. M., 61
Dummer, C., 282, 283, 284, 285
Duncan, J. T., 12, 33, 34, 54, 55, 73, 74, 78, 91, 134, 143, 154, 155, 156, 157, 170, 233, 288
Dunn, R. C., 212
Du Porte, E. M., 148
Durand, P., 320, 341, 343
Dutky, S. R., 67, 68, 82, 83
Dutton, J. E., 456, 457
Dyar, H. G., 187
Dyer, R. E., 270, 271, 280, 281, 282, 283, 285, 286, 295, 301, 336
Dyson, J. E. B., 29

AUTHOR INDEX

Eckstein, K., 52, 55, 56, 58, 59, 60, 62, 63, 68, 73, 74, 75, 87, 88, 91, 93, 109, 147, 151, 424
Ehlers, 47
Ehrlich, P., 325, 592
Eisenman, B., 76
Ekblom, T., 370
Elford, W. J., 277
Ellinger, T., 93, 103, 351
Elliott, C., 177
Eltringham, H., 580
Emerson, A. E., 541
Emik, L. O., 535
Enigk, K., 345
Enlows, E. M. A., 120
Erikson, D., 234
Ermolaev, J., 80, 100, 106, 112
Escherich, K., 366, 369, 413, 425
Eskey, C. R., 162, 163
Evans, A. M., 168, 579
Evlakhova, A. A., 351
Ewing, H. E., Jr., 139
Eyer, H., 275
Eyre, J. W. H., 582

Faasch, W. J., 247
Faddeeva, T., 165
Faichnie, N., 19, 128, 178
Fantham, H. B., 466, 515
Faraday, M., 7
Farner, D. S., 291, 292, 294
Farrell, C. C., 8
Faucheron, L., 355
Faust, E. C., 501
Fawcett, H. S., 350, 399
Fejgin, B., 275
Ferris, G. F., 220
Ficker, M., 127, 129, 131
Fielding, J. W., 261, 584
Fiennes, R. N. T.-W., 308
Fildes, P., 583
Finlay, C. J., 3, 432
Fitzpatrick, F. K., 276, 336, 584
Fitzpatrick, H. M., 377
Flegontoff, A. A., 168
Flexner, S., 441
Flögel, J. H. L., 197
Florence, L., 223, 224, 225
Flügge, 72
Foley, H., 262, 269, 452
Folsom, D., 444
Fonesca, F., 280
Forbes, S. A., 14, 65, 116, 146, 396, 400
Foshay, L., 165
Foster, R. E., 77
Fotheringham, W., 508
Fox, H., 345
Fraenkel, G., 213, 369
Fraenkel, H., 217

França, C., 473
Franchini, G., 467
Francis, E., 166, 168, 454, 455, 456
Francis, T., 442
Fransen, J. J., 410
Frazer, H. L., 412
Fred, E. B., 203, 582
Freeman, M., 295, 296, 298
Fresenius, G., 377
Fricks, L. D., 335
Frobisher, M., Jr., 434, 435
Frost, F. M., 32
Frost, W. H., 441
Frye, W. W., 485
Fulmer, E. I., 583
Fulton, F., 275
Furniss, R. L., 425

Gallardo, F. P., 578, 584
Galli-Valerio, B., 155
Galtsoff, P. S., 575
Gambetta, L., 236
Garbini, A., 60
Gaschen, H., 126, 560, 572
Gäumann, E., 384
Gay, D. M., 434
Gayot, G., 304, 306
Gear, J., 341
Geiman, Q. M., 483
Ghosal, S. C., 26
Gibbons, R. J., 44, 440
Gier, H. T., 60, 203, 204, 206, 207, 208, 216, 217
Gieszczykiewicz, M., 263
Gildemeister, E., 584, 594
Gill, C. A., 178
Gillette, G. P., 113
Girard, G., 450
Giroud, P., 275
Gjullin, C. M., 24
Gladkikh, S., 439
Gladstone, 7
Glaser, R. W., 8, 29, 31, 42, 45, 60, 81, 82, 124, 140, 146, 157, 182, 183, 186, 191, 204, 205, 211, 216, 356, 357, 397, 402, 413, 415, 416, 418, 419, 420, 421, 423, 429, 441, 442, 449, 450, 472, 560, 562, 565, 571, 572, 575
Glasgow, H., 14, 15, 16
Glynn, E. E., 14, 120, 122, 125, 129, 141, 144, 153, 155
Godoy, A., 251
Goetsch, W., 532
Gohar, M. A., 631
Goidanich, G., 410
Goldsworthy, M. C., 174
Golov, D. A., 167, 169
Goodpasture, 335
Goot, van der, 64

AUTHOR INDEX

Gordon, H. M., 315
Goss, R. W., 444
Graber, V., 221
Gracian, M., 261
Gradwohl, R. B. H., 582
Graham, H., 436
Graham-Smith, G. S., 12, 19, 43, 49, 54, 55, 125, 128, 129, 131, 132, 133, 178
Granovsky, A. A., 355, 356
Grassé, P., 61, 115, 116, 532
Grassi, B., 471, 501, 538
Gratia, 416
Gray, M., 437, 438
Green, R. G., 167, 168
Greiff, D., 632
Gregson, J. D., 253
Griffiths, J. T., 18
Griffitts, D., 132
Gropengiesser, C., 204, 216, 218
Grundmann, A. W., 439
Guilliermond, A., 348, 365, 588
Guimarais, A., 223
Gunderson, M. F., 94
Guyénot, E., 30, 349
Guyer, M. F., 580
Gwynne-Vaughan, H. C. I., 590

Haagen, E., 594
Haas, V. H., 162
Hadley, C. H., 82
Hagan, W. A., 54, 138, 157, 481, 506, 508, 510
Hagen, H. A., 351
Hall, M. W., 436
Hall, S. R., 461
Ham, A. W., 508, 509
Hamilton, A., 127, 129, 131, 178
Hamilton, H. L., 336
Hammon, W. McD., 437, 438, 439
Hankin, E. H., 164
Hansen, G. A., 47
Hansen, H. N., 175
Harding, H. A., 119
Hargett, M. V., 432
Harrar, J. G., 400
Hartig, T., 402, 404
Hartman, E., 57
Hartzell, A., 448
Hass, G. M., 322, 335, 340, 342, 343
Hassler, F. R., 337
Hatcher, E., 59, 90, 98, 149
Haushalter, M., 49
Hayashi, D., 290, 418
Heaslip, W. G., 93, 291
Hecht, O., 247, 349
Hegarty, C. P., 24
Hegler, C., 269
Hegner, R. W., 247, 434, 477, 577
Heilig, R., 329

Heim, F., 55
Heitz, E., 366, 369, 370
Helmer, D. E., 440
Hendee, E. C., 408, 409
Henneguy, L. F., 190, 197, 202, 216
Henning, M. W., 302
Henrici, A. T., 588, 589, 590
Herelle, F. d', 181, 182, 183
Hering, M., 31
Herms, W. B., 32, 154, 155, 178
Herrick, G. W., 270, 316, 410
Hertig, M., 11, 60, 204, 205, 216, 218, 220, 243, 244, 253, 254, 262, 263, 312, 313, 314, 315, 316, 318, 320, 325, 326, 327, 328, 346, 584
Hesselbrock, W., 165
Hewitt, C. G., 14
Heymons, R., 190, 206, 215
Heys, F. M., 438
Hill, S. B., Jr., 400
Hindle, E., 33, 55, 74, 91, 155, 220, 223, 289, 321, 457, 458, 459, 590
Hinman, E. H., 23, 24, 28
Hiratsuka, E., 134
Hitchens, A. P., 436
Hitz-Green, S., 276
Hoare, C. A., 313, 471, 482
Hobson, R. P., 33, 235
Hodson, A. C., 405
Hoelling, B. A., 116
Hofmann, E., 49
Hofmann, O., 56, 74, 424
Höhnel, F. von, 389, 410
Hollande, A. C., 185, 563
Holloway, J. K., 400
Holmgren, N., 536
Holst, E. C., 67, 349
Honeij, J. A., 48
Hooke, 221
Hornibrook, J. W., 296
Hoskins, W. M., 32
Hough, W. S., 400
Howard, C. W., 441
Howard, L. O., 128, 154, 178, 187, 216
Howard, W. R., 73, 91
Huang, C. H., 129
Hubbard, H. G., 402, 405, 406
Hucker, G. J., 139
Huddleson, I. F., 158
Hudson, N. P., 286, 587
Huff, C. G., 43, 130, 434, 477, 504, 505, 557, 558, 568, 570, 577
Huie, D., 9
Humphreys, F. A., 44, 440
Hungate, 534
Hurst, A., 228
Husz, B., 91
Huxley, T. H., 190, 197

AUTHOR INDEX

Imagawa, Y., 290, 291, 292
Imms, A. D., 32, 536, 579
Irons, E. N., 291
Irons, J. V., 281
Ishimori, N., 146, 560
Iyengar, M. O. T., 382, 383

Jackson, D. D., 12
Jacobi, J., 288
Jacobson, H. P., 409
Jacot, A. P., 410
Jaffé, J., 460
Janda, V., 520
Jaschke, W., 251, 253
Javelly, E., 216
Jellison, W. L., 22, 337, 575, 586
Jepson, W. F., 100
Jírovec, O., 460, 520
Johannsen, O. A., 48, 49, 128, 195, 196, 222, 580
Johansson, R., 397
Johnson, D. E., 640
Johnson, D. W., 308
Johnson, E. M., 173
Johnston, T. H., 117
Joly, N., 87, 154, 155
Jones, H. N., 421
Jones, L. R., 118
Jones, O. R., 183
Jordan, E. O., 113
Jungmann, P., 205, 287, 288, 312, 313, 315

Kalinicker, P. A., 84
Kamil, S., 167, 168
Karawaiew, W., 366
Kasarov, 24
Katić, R. V., 313, 315
Katsampes, C. P., 291, 292, 294
Katzin, L. I., 538
Kawamura, R., 290, 291, 292
Kaweh, M., 55
Keilin, D., 187, 243, 351, 381, 382, 485, 497, 518, 519, 520, 521
Kelser, R. A., 437
Kennedy, C. H., 577, 580, 593
Kent, W. S., 471
Kidder, G. W., 537
Kilbourne, F. E., 3, 506, 510
King, A. F. A., 501
King, K. M., 75
King, R. L., 486
Kingsbury, B. F., 580
Kirby, H., 36, 37, 460, 461, 485, 526, 527, 529, 530, 531, 532, 538, 539, 540, 541, 542, 546, 550, 553, 554
Kitajima, T., 103
Kitasato, S., 162
Kitselman, C. M., 439
Klein, E., 128

Kleine, F. K., 474
Klevenhusen, F., 228
Kligler, I. J., 205, 245, 276, 277, 584
Klotz, L. J., 58
Kluyver, A. M., 99
Knab, F., 187
Knapp, R. S., 452, 460
Knight, H. H., 564
Knigin, T. D., 575
Knoche, E., 425
Knowles, R., 454, 590
Koch, A., 200, 201, 209, 213, 237, 241, 242, 243, 369, 457
Koch, R., 53, 178, 191, 246, 252
Kodama, M., 268, 280, 282
Kofoid, 501
Kohls, G. M., 167, 290, 291, 296, 344, 575
Kolmer, J. A., 582
Kolpakov, S. A., 168
Komárek, J., 425
Königsberger, J. C., 355
Kono, M., 280, 282
Korke, V. T., 515
Kornauth, K., 424
Korshunova, O. S., 270
Kostritsky, L., 113
Kouwenaar, W., 290
Kramis, N. J., 45, 49, 90, 92, 107, 108, 109, 128, 133, 166, 178, 228, 233, 272, 294, 299, 305, 307, 309, 314, 324, 334, 342, 450, 453, 469, 476, 502, 503
Kranefeld, F. C., 160
Krassilstschik, I. M., 86, 93, 142, 147, 156, 190, 395, 417
Kreutzer, W. A., 176
Ksenjoposky, A. V., 53
Kuchling, M., 315
Kuczynski, M. H., 275, 331
Kudo, R. R., 484, 489, 506, 511, 512, 514, 517, 518, 521, 552
Kuffernath, H., 153
Kuhn, P., 54
Kuhns, D. M., 132
Kumm, H. W., 435, 459
Kunkel, L. O., 447
Kurloff, 325
Kuskop, M., 16
Küster, H. A., 49, 55, 160

Lacaillade, C. W., 419, 437
Lafar, F., 590
Lake, G. C., 166, 168
Lal, R. B., 26, 178
Lamborn, W. A., 520
Landois, L., 221
Larrey, 34
Laveran, A., 467, 501
Lavier, G., 465

AUTHOR INDEX

Law, 225, 226
Lazear, J., 432
Leach, J. G., 1, 3, 19, 20, 117, 118, 119, 171, 172, 173, 174, 175, 176, 177, 350, 404, 408, 409, 411, 443, 444, 445, 448, 449, 582
Ledingham, J. C. G., 125, 128, 129, 131, 169, 178, 325
Lee, A. B., 580
Léger, L., 20, 381, 461, 492
Lehmann, K. B., 63, 66, 125, 132
Leidy, J., 471, 485, 552
Leishman, W. B., 456, 473
Leon, A. P., 279
Lepesme, P., 134, 170
Lépine, P., 439, 594
Lespès, C., 538
Lestoquard, F., 305, 306, 307, 308, 309, 310, 324, 510
Levaditi, C., 590, 594
Lévi, M. L., 304
Levine, M., 582
Levkovich, E. N., 440
Lewis, E. A., 508
Lewis, F. C., 14, 120, 122, 125, 129, 141, 144, 153, 155
Lewis, W. W., 648
Lewthwaite, R., 283, 291, 292, 294
Leydig, F., 14, 189, 190, 197
Lien-teh, W., 163
Lieu, V. T., 129
Light, S. F., 530
Lilienstern, M., 247, 248, 249
Lillie, R. D., 262
Lindeijer, E. J., 175
Lindner, P., 352, 354
Link, G. K. K., 119
Linnaeus, 47
Lipshutz, J., 291
List, G. M., 176
Litschauer, V., 389
Lloyd, L., 29, 34
Lloyd, W., 434
Lly, J. H., 17
Lochhead, A. G., 52, 67, 80, 81
Lodi, B. de, 396
Loeb, J., 349
Löffler, W., 271
Lohde, G., 381, 389
Löhnis, F., 203
Long, E. C., 47
Longfellow, R. C., 12, 43, 122, 126, 134, 155, 156, 180
Lounsbury, C. P., 183, 303, 304
Lubbock, J., 215
Lucas, C. L. T., 485
Lugger, O., 135
Luginbill, P., 148
Lutz, F. E., 575

Lwoff, A., 520
Lysaght, A. M., 349

Maassen, A., 25, 52, 66, 67, 139
Macalister, C. J., 36
MacArthur, W. P., 520
McCay, C. M., 32
Macchiati, L., 57
Macchiavello, A., 259, 261, 264, 265, 272, 276, 315, 584
Macfie, 48, 49, 125
McClung, C. E., 580, 582
McClung, L. S., 583
McCormack, A. T., 441
McCormick, F. A., 410
McCoy, E. E., 8, 203
McCoy, Elizabeth, 203, 583
McCoy, G. W., 48, 165, 166
McCray, A. H., 77
McCulloch, I., 469, 470
McDowell, M. M., 345
McGovran, E. R., 562
MacGregor, M. E., 24, 158, 575
McGregor, T., 281
McKelvey, J. J., 400
Mackehenie, D., 347
McKenzie, M. A., 410
Mackie, D. B., 183
Mackinnon, D. L., 62, 461, 483
MacLeod, G. F., 11
MacLeod, J., 574
McOmie, W. A., 39, 99, 133, 134
Macrae, R., 178
Maddox, R. L., 178
Magalhães, O., 338
Mahdihassan, S., 205, 352, 375
Maillard, L., 216
Mains, E. B., 388, 579, 582
Malbrant, R., 306
Malden, 78
Mallman, W. L., 165
Mallory, F. B., 335, 582
Manson, P., 452, 501
Manson-Bahr, P., 128
Mansour, K., 199, 200, 208, 209, 236, 237, 238
Mareschal, P., 184, 271
Mariani, G., 319
Marlatt, C. L., 154, 216
Marpmann, G., 99
Marsh, H., 167
Martchouk, P., 76
Martin, C. J., 163
Martinaglia, G., 55
Martins, A. V., 331
Marzocchi, V., 417
Masera, E., 24, 90, 134
Mason, J. H., 339, 341, 343
Mathis, M., 439

Mattes, O., 51, 66, 145
Maxcy, K. F., 281, 290, 291
Mayer, E. L., 562
Mayne, B., 166
Megaw, J. W. D., 280, 287
Meillon, B. de, 341, 382
Melampy, R. M., 11
Meleney, H. E., 485
Melnick, J. L., 441
Meng, L. Y., 64, 104
Mera, B., 345
Mercier, L., 60, 204, 216, 217, 218
Mercurialis, 3
Mereshkovsky, S., 182
Merrill, M. H., 437, 438
Messerlin, A., 129
Metalnikov, S., 43, 48, 49, 64, 76, 80, 87, 88, 90, 92, 95, 96, 100, 103, 104, 106, 110, 111, 112, 113, 122, 126, 128, 130, 131, 134, 135, 138, 142, 143, 146, 154, 156, 157, 170, 179, 184, 351, 393, 397, 555, 556, 560, 561, 562, 565, 566, 568, 569, 571
Metalnikov, S. S., 104, 111, 113
Metchnikoff, E., 86, 197, 201, 396, 555, 559, 562
Meves, F., 202
Meyer, K. F., 162, 164, 188, 210, 211
Miller, D. K., 29
Miller, E. D., 202
Miller, R. L., 400
Milovidov, P. F., 202
Milzer, A., 440
Minchin, E. A., 479
Mitamura, T., 290
Mitchell, E., 28
Mitscherlich, E., 304
Mitzmain, M. B., 54
Miyajima, M., 425
Mochkovski, Ch. P. (*see* Moshkovsky, Sh. D.)
Mohamed, Z., 311
Möller, A., 388, 406
Moniez, R., 352, 354
Monteiro, J. L., 264, 280, 290, 329, 339
Mooser, H., 269, 271, 280, 281, 282, 283, 284, 285
Moragues, V., 632
Morgan, J. C., 125
Morischila, 128
Morris, H., 54
Morse, W. J., 119
Moshkovsky, Sh. D., 189, 263, 290, 302, 304, 306, 308, 310, 324, 325
Mosing, H., 311, 312
Moulton, F. R., 575
Mrack, E. M., 350
Mudrow, E., 191, 250, 251, 252, 253
Mühlens, P., 452
Mukherji, B., 26

Müller, J., 221
Munk, F., 287, 288
Munro, J. A., 123
Munson, S. C., 562
Murphy, A. M., 446
Muspratt, J., 382
Muttkowski, R. A., 563, 564

Nabarro, D., 474
Nagayo, M., 290
Naidu, V. R., 329
Napoleon, 35
Nauck, E. G., 270, 276, 584
Needham, J. G., 575
Needham, N. Y., 119
Neger, F. W., 403, 404
Neill, M. H., 280
Neitz, W. O., 262, 307, 341, 507
Nelson, C. I., 123
Nelson, K. R., 296
Neukomm, A., 203
Neumann, R. O., 63, 66, 125, 132
Nicholas, G. E., 178
Nicholls, L., 155
Nicoll, W., 14, 63, 64, 94, 120, 121, 122, 125, 130, 131, 178
Nicolle, C., 270, 453, 455, 467
Nieschulz, O., 54, 55, 160, 458
Nikanoroff, 164
Noble, L. W., 77
Noc, F., 461
Noguchi, H., 62, 84, 85, 442, 473
Nöller, W., 205, 206, 270, 312, 313, 315, 325, 515, 584
Nolte, H.-W., 368
Noreiko, E. S., 418
Northrop, J. H., 349
Northrup, Z., 86, 148, 149
Norwood, V. H., 12
Nott, J., 3, 432
Nowakowski, L., 378
Nuttall, G. H. F., 12, 14, 34, 47, 50, 53, 54, 164, 178, 452, 501

Obermeier, O., 452
Odriozola, E., 345
O'Farrell, W. R., 469
Ogata, M., 162
Ogata, N., 290, 291
Olarte, J., 123
Olitsky, P. K., 202
Olmer, D., 341
Olsufiev, N. G., 169
Osborn, H., 159
Ostrolenk, M., 122, 129
Otto, R., 270, 287

Packchanian, A. A., 472

AUTHOR INDEX

Paillot, A., 18, 25, 26, 27, 46, 57, 60, 63, 64, 65, 69, 70, 71, 72, 75, 76, 79, 83, 84, 86, 87, 88, 100, 106, 108, 109, 110, 111, 136, 137, 140, 142, 143, 148, 156, 179, 184, 185, 186, 191, 210, 227, 228, 229, 231, 351, 373, 394, 415, 418, 420, 426, 430, 431, 461, 467, 524, 555, 559, 560, 562, 564, 565, 568, 570, 571
Palfrey, F. W., 270, 271, 273, 274, 288
Pammel, L. H., 173
Pang, K. H., 282
Panthier, R., 275
Paré, 34
Parker, H. B., 349
Parker, R. R., 22, 48, 130, 166, 167, 168, 286, 296, 297, 301, 322, 330, 331, 332, 333, 336, 337, 340, 342, 344, 575, 585
Parrott, L., 262, 510
Pasteur, L., 57, 513, 522, 555, 558
Patiño-Camargo, L., 331
Patton, W. S., 101, 159, 316, 460, 468, 469, 472, 579
Paul, J. R., 441
Pavlovsky, E. N., 439
Payne, N. M., 524
Peat, A. A., 459
Peck, C., 380
Peklo, J., 229
Peppler, H. J., 129, 349
Perroncito, E., 134, 187
Petch, T., 376, 387, 388, 392, 394
Peter, H., 584
Peterson, A., 574, 575
Petri, L., 20, 21, 23, 30, 175, 176, 181
Peus, F., 247
Pfeiffer, H., 102, 104, 106, 233, 316, 319
Phaff, H., 350
Philip, C. B., 10, 22, 167, 168, 263, 264, 280, 286, 290, 291, 295, 333, 337, 435, 575, 586
Phillips, E. F., 566
Picard, F., 59, 69, 183, 184
Pierantoni, U., 190, 212, 229, 231, 241, 243, 362, 363, 364
Pierce, W. D., 12, 19, 47, 55, 71, 96, 114, 137, 138, 153, 443
Pilz, H., 269, 281
Pinkerton, H., 256, 259, 262, 263, 268, 275, 276, 277, 280, 289, 322, 332, 333, 335, 340, 341, 342, 343, 584
Pinto, C., 138, 251
Plotz, H., 259, 275, 281, 284, 340
Pollini, 134
Pollitzer, R., 163
Ponce, 81
Poos, F. W., 177
Porchet, 46
Porter, A., 460, 486, 515
Porthier, O. L., 349
Portier, P., 31, 32, 201, 202

Pospelov, V. P., 51, 57, 80, 134, 151, 171, 182, 418
Prenant, A., 216
Prince, L. H., 34
Pringault, E., 460
Profft, J., 230
Proust, A., 54
Prowazek, S. von, 256, 269, 413, 417, 418, 460
Pshenichnov, A. V., 585
Putnam, J. D., 189, 357

Raimbert, A., 3, 53
Raistrick, H., 590
Rand, F. V., 120, 177, 443
Ratsburg, J. T. C., 402
Rawlins, T. E., 590
Ray, H., 489
Re, G., 134
Readio, P. A., 410
Reagan, R. L., 340
Rebagliati, R., 345
Reed, W., 3, 127, 432
Rees, C. W., 510
Reeves, W. C., 437, 438, 439
Regendanz, P., 507
Reichenow, E., 254, 255, 507, 508, 510
Reiff, W., 421
Rendtorff, R. C., 442
Reynolds, F. H. K., 48, 128, 135, 156, 436
Rhoads, C. P., 29
Rhodes, A. J., 594
Rice, L., 436
Richards, J. T., 345
Richards, O. W., 583
Ricketts, H. T., 256, 257, 269, 274, 322, 330, 332, 336
Rickter, G., 374
Ries, E., 200, 211, 213, 219, 220, 221, 222, 223, 225, 226, 227
Riker, A. J., 22, 174
Riley, W. A., 47, 48, 49, 128
Rivas, G. M., 48
Roberg, D. N., 179
Roberts, H. R., 504
Roberts, J. L., 85
Robinson, R. A., 337
Robinson, W., 12, 35, 36
Roca-Garcia, M., 436
Rocha, A., 338
Rocha-Lima, H. da, 256, 263, 268, 269, 270, 272, 287, 288
Roderick, L. M., 439
Rolander, 47
Rolfs, 401
Rolleston, 576
Romanevich, B. V., 101
Rondelli, 251
Rooseboom, M., 564

Root, F. M., 434, 477, 577
Rorer, J. B., 183
Rosenau, M. J., 47, 441
Rosenholz, H. P., 455
Rosenow, E. C., 441
Ross, E. S., 504
Ross, R., 3, 473, 501
Rossi, P., 305
Roubaud, E., 23, 31, 32, 102, 245, 246, 472
Rozeboom, L. E., 23, 24, 28
Rozenkranz, W., 16, 17
Rozier, 134
Ruhland, H. H., 158
Rumbold, C. T., 349
Rumreich, A., 281 282, 283
Ruska, H., 275

Sabin, A. B., 441
Sadusk, J. F., 290, 291
St. John, J. H., 48, 128, 135, 156, 436, 453
Sako, W., 418
Sanders, 47
Sanders, E. P., 461
Sanders, M., 594
Sanford, A. H., 582, 583
Sanford, M. F., 530
Sartirana, S., 165
Sato, K., 290
Saunders, D. A., 138
Sauter, V., 439
Savoor, S. R., 283, 291, 292, 294
Sawamura, S., 61, 62, 63, 70, 74, 85, 94, 122, 155, 170
Sawyer, W. A., 435
Sawyer, W. H., 378
Schaudinn, F., 58, 325, 461
Scheinert, W., 238
Schellack, C., 452
Schlesinger, M. J., 275, 584
Schmidberger, J., 402
Schminke, A., 287
Schneider, G., 17, 235, 236
Schneider, K. C., 261
Schneider-Orelli, O., 403, 404
Schoenlein, H. W., 582
Schölzel, G., 222
Schomann, H., 370, 371, 372
Schuberg, A., 54, 138
Schuhardt, V. T., 575
Schultz, E. S., 444
Schulz, K., 310
Schüssler, H., 270
Schütte, L., 31
Schwammerdam, 221
Schwartz, W., 352
Schwerdtfeger, F., 86
Scott, J. R., 122, 126, 129, 141, 143, 144, 154, 155, 156, 161

Seargent, A., 454
Seitlenok, N. A., 439
Sellards, A. W., 290, 293, 434
Semans, F. M., 463, 468, 491, 492, 493, 494, 521, 591
Serbinow, I. L., 53, 57, 58, 67, 72, 101, 124, 152
Sergent, Ed., 269, 452, 461, 510
Sergent, Et., 461
Severin, H. H. P., 226, 227
Seymour, A. B., 376
Shakespeare, E. O., 127
Shannon, R. C., 346, 434, 435
Shaughnessy, H. J., 440
Shear, C. L., 383, 393
Sheldon, J. L., 378
Shelmire, B., 282
Shepherd, D., 91
Shinji, G. O., 374
Shiperovich, V., 86
Shkorbatov, V. I., 279
Shooter, R. A., 143
Shope, R. E., 442, 449
Shule, P. A., 436
Shulguina, O. G., 84
Shute, P. G., 461
Sieyro, L., 485
Sikora, H., 218, 219, 221, 223, 245, 257, 271, 312, 313, 320, 321, 325
Siler, J. F., 436
Silva, G. R., 279, 281
Silva, P. da, 138
Simmonds, M., 178
Simmons, J. S., 48, 128, 135, 156, 435, 504
Simmons, S. W., 31, 35, 575
Simond, P. L., 3, 162, 163
Singer, E., 276, 300, 584
Sinton, J. A., 461
Sirotskaya, S., 76
Sizemore, P., 337
Skobaltzyn, V., 80, 100, 106, 112
Skorodumoff, 164
Skrinnick, A. N., 440
Smadel, J. E., 275
Smith, D. J. W., 296, 297, 298, 301, 308
Smith, E. F., 120, 173, 175, 176
Smith, G., 590
Smith, K. M., 444, 594
Smith, M. G., 438
Smith, N. R., 203
Smith, R. C., 113, 147, 439
Smith, R. E., 175
Smith, R. F., 425
Smith, T., 3, 4, 506, 510
Smyth, E. G., 148
Snodgrass, R. E., 192, 196
Snow, F. W., 400
Snyder, J. C., 278, 585

AUTHOR INDEX

Sokoloff, V. P., 58
Soloviev, V. D., 439
Soper, F. L., 432
South, L. H., 441
Souza-Araujo, H. C., 48
Spaet, G., 584
Sparrow, F. K., 381
Sparrow, H., 271, 283, 319, 320
Speare, A. T., 378, 395
Specht, G., 71, 366
Speer, A. J., 516
Spegazzini, C., 406
Spencer, R. R., 22, 166, 167, 322, 331, 332, 333, 336, 337, 585
Spielman, M., 49
Stage, H. H., 24
Stammer, H.-J., 11, 17, 22, 29, 102, 104, 106, 242, 243
Stanley, W. M., 416, 419, 420
Statelov, N., 56
Stein, F., 471
Steinhaus, E. A., 11, 14, 19, 22, 23, 44, 51, 59, 60, 65, 90, 97, 98, 104, 105, 106, 108, 109, 112, 113, 114, 115, 120, 121, 122, 125, 126, 127, 130, 135, 141, 144, 145, 146, 148, 149, 169, 172, 180, 191, 205, 206, 233, 245, 254, 301, 315, 319, 322, 324, 335, 349, 393, 441, 450, 585
Stelling-Dekker, N. M., 588
Stempell, W., 512, 513
Stévenel, L., 461
Stevens, N. M., 197
Stewart, M. A., 35
Stieben, 24
Storey, H. H., 443, 447, 594
Stratman-Thomas, W. K., 504
Strickland, C., 495
Strindberg, H., 247
Strohmeyer, H., 404
Strong, R. P., 262, 287, 345, 473, 582
Stuhlmann, F., 246, 252
Sturtevant, A. P., 52, 55, 66, 67, 77, 427
Stutzer, M. J., 98, 110, 111, 113, 120, 130, 152, 172, 186
Subramaniam, M. K., 8
Šulc, K., 190, 196, 197, 352, 358, 359, 360, 361, 362, 364, 365, 374
Sulkin, S. E., 438
Summers, F. M., 501
Sweetman, H. L., 8, 112, 148, 351, 380, 396, 398, 401, 413, 467, 492, 522, 523, 525
Swezy, O., 226, 227, 535
Sydow, H., 410
Sydow, P., 410
Sylla, A., 288
Syverton, J. T., 292, 438

Takacs, W. S., 441

Takahashi, K., 280
Taliaferro, W. H., 119
Tamiya, T., 290
Tannreuther, G. W., 190, 197
Tanquary, M. C., 247
Tarr, H. L. A., 81
Tate, P., 485, 486
Tateiwa, 135
Tatlock, H., 345
Tatum, E. L., 33
Tauber, O. E., 18, 154
Taylor, A. B., 486
Taylor, A. W., 461
Tebbutt, H., 50, 128
Tejera, E., 48, 49, 460, 468
Ten Broeck, C., 4, 437, 438
Teodoro, G., 355
Thaxter, R., 378, 379, 384, 385, 386, 393
Theiler, 441
Thomas, A. D., 307
Thomas, H. D., 24
Thomas, H. E., 117
Thomas, L., 292
Thomas, L. J., 123, 124
Thompson, W. R., 575
Thomson, J. D., 479
Thorpe, W. A., 11
Thurman, D. C., 281
Tichomirova, M., 164, 165
Tilden, E. B., 473
Tischer, L. A., 441
Tizzoni, G., 178
Toda, T., 179
Todd, J. C., 582, 583
Todd, J. L., 270, 271, 273, 274, 280, 288, 329, 456, 457
Toit, P. J. du, 262
Toomey, J. A., 441
Töpfer, H., 270, 288, 289
Topley, W. W. C., 583
Topping, N. H., 329
Torrey, J. C., 10, 18, 120, 122, 130, 141, 142, 143, 146, 153, 154, 180
Toumanoff, H., 113
Toumanoff, K., 51, 123, 147, 393, 397
Townsend, C. H. T., 346
Trager, W., 24, 29, 253, 421, 422, 513, 532, 575
Trask, J. D., 441
Travassos, J., 337
Travis, B. V., 483
Treillard, M., 520
Troitsky, 457
Trotter, A., 405
Tschirch, A., 375
Tsuchimochi, 128
Tubeuf, C. von, 56, 74, 108, 109, 424
Turner, T. B., 459

AUTHOR INDEX

Uichanco, L. B., 190, 197, 198, 206
Uribe, C., 470
Usinger, R. L., 477

Valleau, W. D., 173
van den Ende, M., 275, 277
Vaney, C., 400
van Niel, C. B., 99
van Rooyen, C. E., 594
Varela, G., 269, 281
Vargas-Curiel, J., 286
Vasco, 134
Vaughan, V. C., 127
Vedder, 47
Veeder, M. A., 128
Veillon, 149
Vejdovsky, F., 358
Verjbitski, D. T., 163
Vernier, P., 185
Vervoort, H., 290
Vialette, C., 269
Vincent, M., 495, 499
Volferz, A. A., 168

Wachtl, F. H., 424
Wadsworth, A. B., 582
Wahl, B., 109, 413, 424
Waite, M. B., 3, 116
Waksman, S., 582
Waldmann, O., 594
Walker, 164
Wallace, F. G., 468
Wallace, G. B., 350
Wallace, G. I., 124
Wallace, H. E., 446
Wallace, J. M., 446
Wallengren, H., 397
Waller, E. F., 168
Wallin, I. E., 202
Ward, R., 441
Wardle, R. A., 137, 140
Waterworth, P. M., 143
Watson, J. R., 400
Wayer, F., 584
Webb, J. L., 235, 345
Weber, G. F., 350
Weber, N. A., 406
Wei, H., 276, 336, 584, 585
Weigl, R., 270, 275, 276, 288, 311, 319, 320, 585
Weinland, E., 29, 30
Weinman, D., 345
Weiss, L. J., 275
Welander, E., 157
Welch, D. S., 410
Welch, H., 129
Welch, P. S., 575
Wenrich, D. H., 591
Wenyon, C. M., 452, 464, 466, 467, 468, 471, 472, 473, 475, 477, 478, 489, 490, 491, 497, 500, 520, 590, 593
Werner, E., 31, 32, 96, 97
Werner, H., 288
Wertman, K., 340
Weyer, F., 276
Wharton, D. R. A., 80
Wheeler, C. M., 163, 164, 278, 456, 585
Wheeler, W. M., 128, 190, 215, 403, 406, 407, 408
Wherry, 47
White, G. F., 10, 25, 50, 52, 56, 61, 63, 66, 67, 68, 73, 74, 75, 76, 77, 80, 88, 89, 97, 99, 102, 109, 121, 122, 129, 144, 171, 426, 427, 428, 429, 430
White, R. T., 68
Whitman, L., 434, 435
Widmann, E., 138, 153
Wigglesworth, V. B., 32, 33, 141, 212, 213, 233, 234, 235, 246
Wilcocks, C., 291
Wilder, R. M., 269, 274
Will, L., 197
Williams, R. W., 291
Wilson, G. S., 583
Wingard, S. A., 350
Witlaczil, E., 197, 201
Wolbach, S. B., 205, 218, 220, 243, 244, 253, 254, 262, 263, 270, 271, 273, 274, 275, 280, 288, 290, 312, 313, 314, 315, 316, 318, 320, 322, 325, 328, 329, 330, 331, 333, 334, 335, 336, 337
Wolf, J., 217
Wolff, J. W., 290
Wolff, M., 417
Wollman, E., 29, 30, 32, 101, 159, 216
Wolzogen Kühr, C. A. H. von, 29
Womersley, H., 291
Wood, S. F., 477
Woodcock, H. M., 315
Woodland, J. C., 345
Woodward, T. E., 270
Workman, W. G., 283
Wright, E., 411
Wsorow, W. J., 98, 110, 111, 113, 120, 130, 152, 172, 186
Wu, C. Y., 163

Yakimoff, W. L., 37
Yamasaki, M., 534
Yao, H. Y., 9
Yates, W. W., 24
Yeager, J. F., 562, 564
Yersin, A., 162
Yolles, S. F., 575
Yoshida, S., 293
Yothers, W. W., 400
Young, G., 339

Young, M. D., 484
Young, T. R., 400
Yuan, I. C., 9

Zacharias, A., 245
Zasukhin, D. N., 167, 457
Zeliff, C. C., 522
Zernoff, V., 130, 134, 156, 561, 562
Ziklauri, 24

Zilber, L. A., 276, 584
Zimmermann, A., 110, 355
Zinsser, H., 269, 276, 280, 281, 282, 294, 336, 584, 587
Zorin, P. V., 53
Zorina, L. M., 53
Zuber, 149
Zuelzer, M., 460
Zumpt, F., 54, 270

SUBJECT INDEX

Those pages which contain the most important or the most complete discussion of a subject are indicated by **boldface type**. In the case of most of the insects, the scientific name is followed by the name or the abbreviation of the name of the authority for that species.

Abraxas grossulariata (Linn.), 515
Acallomyces, 386
Acanthochermes quercus Koll., 230
Acanthoscelides obtectus (Say), 91
Acanthosominae, 16, 17
Acanthospora polymorpha, 494
Acanthotermes ochraceus, 550
Acarina, 215, **249-255**, 462
Accessory food substance, 29
Accessory glands, 192, 193, 194
Accessory lobe, 316
Acephalina, 488-491
Aceratagallia sanguinolenta (Prov.), 447
Acherontia atropos (Linn.), 394
Acheta campestris (Linn.):
  associated protozoa–*Gregarina macrocephala*, 493; *G. oblonga*, 493; *G. ovata*, 493
Achraca grissella, 48
Achroia grisella (Fabr.), 556
Achromobacter, 41, **97**; *A. delicatulum*, 97, 98; *A. eurydice*, 25, 52, 81, **97-98**; *A. hyalinum*, 98; *A. larvae*, 98; *A. superficiale*, 98
Acid-fast bacteria, 583, 597
Acleris variana, 414
Acnidosporidia, 487, 511
Acompsomyces, 385
Acquired immunity (*see* Immunity in insects)
Acridium aegyptium Linn., 183, 184
Acromobacter hyalinum, 98
Acromyrmex (*Atta*) *disciger* Mayr:
  associated fungi, 408–*Pholiota gongylophora*, 406
Acromyrmex (*Atta*) *lundi* (Guér.):
  associated fungi, 408–*Xylaria micrura*, 406
Acrostalagmus, 393; *A. aphidium*, 400
Acrotrophic type of ovariole, 193
Acrydium arenosum angustum Hanc., 492
Actinocephalus acutispora, 494; *A. amphoriformis*, 494; *A. elongatus*, 494; *A. fimbriatus*, 494; *A. giganteus*, 494
Actinomyces necrophorus, 114; *A. rhodnii*, 234, 235
Actinomycetaceae, 262
Actinomycetales, 40, 42
Actinomycetes, 262
Actinopoda, 484
Active immunity (*see* Immunity in insects, acquired, active)
Adelea, 498

Adeleidea, 498
Adelges laricis, 229
Adelgidae, 229-230, 498
Adelina cryptocerci, 498; *A. mesnili*, 498; *A. tribolii*, 498
Adelphocoris rapidus (Say), 116
Adoretus compressus Web., 64
Adosomus roridus Pallas, 239
Aëdes abnormalis, 442
Aëdes aegypti (Linn.):
  accessory food substances, requirement of, 29
  associated bacteria–*Eberthella typhosa*, 127, 128; *Escherichia coli*, 24; *Malleomyces pseudomallei*, 160; *Mycobacterium leprae*, 46, 48; *Pasteurella bollingeri*, 160, 161; *P. pestis*, 162; *P. tularensis*, 166; *Pseudomonas aeruginosa*, 24; *Sarcina lutea*, 24; *Serratia marcescens*, 24, 133, 135; *Staphylococcus aureus*, 155, 156
  associated molds, 24
  associated protozoa–*Crithidia fasciculata*, 486; *Lankestria culicis*, 488; *Myxococcidium stegomyiae*, 349; *Nosema stegomyiae*, 515; *Plistophora stegomyiae*, 516; *Trypanosoma congolense*, 480
  associated spirochete, 461–*Borrelia anserina*, 458
  associated viruses–dengue, 436; encephalitis, 437; fowlpox, 442; lymphocytic choriomeningitis, 440; yellow fever, 3, 432-435
  associated yeasts, 24, 349
  food for, microorganisms, 28, 29
  hatching of eggs, effect of bacteria on, **22-24**
Aëdes africanus Theobald, 433
Aëdes albopictus (Skuse):
  associated fungus–*Coelomomyces stegomyiae*, 381
  associated protozoan–*Lankestria culicis*, 489
  associated viruses–dengue, 436; equine encephalomyelitis, 437; lymphocytic choriomeningitis, 440; yellow fever, 433
Aëdes aldrichi Dyar & Knab (see *Aëdes lateralis*)
Aëdes argenteus, 433 (see *Aëdes aegypti*)
Aëdes caballus, 442
Aëdes calopus, 433 (see *Aëdes aegypti*)
Aëdes canadensis (Theobald):
  associated bacterium–*Pasteurella tularensis*, 166, 168

SUBJECT INDEX

Aëdes cantans Meig., 515
Aëdes caspius Pallas, 166
Aëdes cinereus Meig.:
  associated bacteria–*Pasteurella tularensis*, 166; *Staphylococcus albus*, 153, 154; *S. aureus*, 156
  bactericidal substance in, 34
Aëdes dorsalis (Meig.):
  associated bacterium–*Pasteurella tularensis*, 166, 168
  associated virus–equine encephalomyelitis, 437, 439
Aëdes fluviatilis (Lutz), 433
Aëdes geniculatus Oliv:
  associated viruses–equine encephalomyelitis, 437; yellow fever, 433
Aëdes lateralis Meig.:
  associated viruses—equine encephalomyelitis, 437; St. Louis encephalitis, 437, 439
  eggs, hatching of, 24
Aëdes leucocelaenus (D. & S.), 433
Aëdes lineatopennis, 442
Aëdes luteocephalus Newstead, 433
Aëdes metallicus Edwards, 433
Aëdes nearcticus Dyar, 166, 168
Aëdes nemorosus (= *Aëdes punctor* Kirby):
  associated protozoa–*Nosema* sp., 515; *Thelohania* sp., 516
Aëdes nigromaculis (Ludlow), 439
Aëdes scapularis (Rondani), 433
Aëdes scutellaris (Walk.), 8, 520
Aëdes scutellaris hebrideus Edwards, 436
Aëdes simpsoni Theobald, 433
Aëdes sollicitans (Walk.):
  associated virus–equine encephalomyelitis, 437
  eggs, hatching of, 23
Aëdes spp.:
  associated protozoa–*Plasmodium* spp., 505
  associated viruses, 442
Aëdes stimulans (Walk.):
  associated bacterium–*Pasteurella tularensis*, 166, 168
  associated virus–fowlpox, 442
Aëdes stokesi Evans, 433
Aëdes sylvestris (see *Aëdes vexans*)
Aëdes taeniorhynchus (Wied.):
  associated viruses–St. Louis encephalitis, 437, 439; yellow fever, 433
Aëdes taylori Edwards, 433
Aëdes triseriatus (Say), 433
Aëdes vexans (Meig.):
  associated bacteria–*Bacillus anthracis*, 53, 54; *Pasteurella tularensis*, 166, 168
  associated viruses–equine encephalomyelitis, 437; fowlpox, 442; St. Louis encephalitis, 439
  eggs, hatching of, 24

Aëdes vittatus (Bigot), 433
Aegeria exitiosa Say, 412
Aegerita, 393
Aegerita webberi, 400
Aepyprymnus rufescens, 300
Aerobacter, 41, 120
Aerobacter aerogenes, 28, 120; *A. bombycis*, 125 (see also *Proteus bombycis*); *A. cloacae*, 120-121
Aerobic bacteria, **581-583**, 596
Aeschna constricta Say, 495; *A. grandis* Linn., 515
African sleeping sickness, 474, 476
Agallia sticticollis, 446
Agaricales, 392-393
Agglutination, 323
Agglutinins, 562, 571, 572
Aggregata, 523
Agriolimax agrestis Linn., 173
Agrion puella, 494
Agriotes mancus (Say), 11
Agrippina bona, 495
Agrotis ashworthi Dbld., 18; *A. pronubana* Linn., 147, 148; *A. segetum* Shiff. (see *Euxoa segetum*); *Agrotis* sp., 75, 104; *A. ypsilon* (Rott.), 395
Agyriales, 383
Akimerus schäfferi, 371
Alabama argillacea (Hbn.), 414
Albococcus pyogenes, 153 (see also *Staphylococcus albus*); *Albococcus* sp., 153
Alcaligenes, 42, **180**; *A. ammoniagenes*, 180; *A. bronchisepticus*, 180; *A. faecalis*, 180; *A. stevensae*, 67, 180
Alces americana var. *americana*, 124
Aleurodes, 364
Aleurodomyces, 364
Aleurodomyces signoretti, 364
Aleyrodes citri, 412; *A. variabilis* Quaint, 394
Aleyrodids: mycetome in, 199; symbiosis in, 190
Algae, 28
Alimentary tract of insects, 10, 11, 12, 576, 578
Allantocystis dasyhelei, 490
Allantoin, 35, 36
Allodermanyssus sanguineus (Hirst), 291
Allorrhina nitida (see *Cotinis nitida*); *Allorrhina* spp., 148
Alophus pericarpus, 240; *A. triguttatus* Fabr., 240
Alpine rock crawler (see *Grylloblatta campodeiformis*)
Amara augustata Say, 494
Amblyomma agamum (see *A. rotundatum*)
Amblyomma americanum (Linn.):
  associated bacteria-*Bacillus cereus*, 13, 58, **59**; *Pasteurella tularensis*, 166, 167
  associated rickettsiae–*Dermacentroxenous*

Amblyomma americanum (Linn.) (*Cont.*)
  rickettsi, 330; *Rickettsia burneti*, 296; *Rickettsia* sp., 345
  bacterial flora of, 13, 14
  intracellular symbiotes in, 251
Amblyomma brasiliensis, 331
Amblyomma cajennense (Fabr.):
  associated bacterium—*Mycobacterium leprae*, 47, 48
  associated rickettsia—*Dermacentroxenous rickettsi*, 330, 331, 333
  associated virus—yellow fever, 436
  bacterial flora of, 13
  intracellular symbiotes in, 251
Amblyomma hebraeum Koch:
  associated rickettsiae—*Dermacentroxenous conori*, 340, 341; *Rickettsia ruminantium*, 297, 303
Amblyomma maculatum (Linn.):
  associated rickettsia—rickettsia, unnamed, 344
  bacterial flora of, 13
  intracellular symbiotes in, 251
Amblyomma ovale Koch, 331
Amblyomma rotundatum Koch, 251
Amblyomma spp.:
  associated bacterium—*Actinomyces necrophorus*, 114
  associated protozoan—*Leptomonas* sp., 468
  associated rickettsia—*Rickettsia prowazeki* var. *typhi*, 282; *R. tsutsugamushi*, 292
Amblyomma striatum Koch, 330
Amblyomma tuberculatum Marx, 251
Amblyomma variegatum (Fabr.):
  associated rickettsia—*Rickettsia ruminantium*, 303
  associated virus—Nairobi disease of sheep, 442
Ambrosia, 402
Ambrosia beetles, 402-406; breeding habits of, 403-404; fungi of, 404-406; galleries of, 403-404
Ambrosia fungi (*see* Fungi, ambrosia)
Ambrosiamyces zeylanicus, 405
Amebocytes, 200, **564**
Ameletus ludeus, 516
American cockroach (see *Periplaneta americana*)
American foul brood, 66, 67
Amitermes beaumonti Banks:
  associated protozoa—*Chilomastix minuta*, 538; *Endamoeba beaumonti*, 553; *Retortamonas termitis*, 538; *Trichomonas lighti*, 540
Amitermes coachellae Light:
  associated protozoa—*Endamoeba beaumonti*, 553; *Trichomonas lighti*, 540
Amitermes emersoni Light:
  associated protozoa—*Nyctotherus silvestrianus*, 550; *Trichomonas lighti*, 540
Amitermes medius Banks, 553

Amitermes minimus Light:
  associated protozoa—Coccidia, 553; *Endamoeba beaumonti*, 553; *Nyctotherus silvestrianus*, 554; *Trichomonas lighti*, 540
Amitermes silvestrianus Light:
  associated protozoa—*Nyctotherus silvestrianus*, 554; *Trichomonas lighti*, 540
Amitermes wheeleri Desneux:
  associated protozoa—*Endamoeba beaumonti*, 553; *Trichomonas lighti*, 540
Amniotic cavity, 196
Amoeba chironomi, 486
Amoebida, 484
Amoebidae, 484
Amoebina, 553
Amorphomyces, 385
Amphibia, 520
Amphigonous, 191, 198
Ampullae, 226
Anabolia, 483
Anacanthotermes, 528
Anacanthotermes macrocephalus Desneux, 548
Anacanthotermes murgabicus Vasiljev:
  associated protozoa—*Devescovina elongata*, 540; *Monocercomonas termitis*, 539; *Trichonympha turkestanica*, 548; *T. vermiformis*, 539
Anacanthotermes ochraceus Burm.:
  associated protozoa—*Deltotrichonympha numidis*, 550; *Devescovina elongata*, 540; *Foaina pectinata*, 544; *Rostronympha magna*, 551
Anacanthotermes vagans Hagen, 548
Anacanthotermes viarum (Koenig):
  associated protozoa—*Trichonympha fletcheri*, 548; *T. turkestancia*, 548
Anaerobic bacteria, 581, **583**, 597
Ananas virus *1*, 448
Anaphylaxis in insects, 572
Anaplasma, 262; *A. centrale*, 510; *A. marginale*, 510, 511
Anaplasmata, 510-511
Anasa tristis (DeG.):
  associated bacteria—*Bacillus entomotoxicon*, 61; *Erwinia tracheiphila*, 119
  ceca, 15; bacteria of, 16
Anaxipha exigua Say, 493
Ancyrophora gracilis, 494; *A. uncinata*, 494
Anguillulinidae, 7
Anisolobus dacnecola, 493
Anisoplia austriaca Hbst.:
  associated bacterium—*Bacillus salutarius*, 86
  associated fungus—*Metarrhizium anisopliae*, 396
Anobiidae, 236, 369, 370
Anobium paniceum Linn. (see *Stegobium paniceum*); *A. striatum* Oliv., 367, 369
Anomala frischii Fabr., 394

SUBJECT INDEX

Anomala orientalis Wtrh.:
  associated bacterium–*Bacillus popilliae*, 82, 83
  associated protozoa, 483
Anopheles aconitus Dönitz:
  associated fungi–*Coelomomyces anophelesica*, 382; *C. indiana*, 382
Anopheles albimanus Wied., 504
Anopheles albitarsis Lynch:
  associated bacterium–*Mycobacterium leprae*, 46, 48
  associated protozoan, 504
Anopheles annularis Wulp.:
  associated fungi–*Coelomomyces anophelesica*, 382; *C. indiana*, 382
  associated protozoan–*Thelohania legeri*, 515
Anopheles argyritarsis Rob.-Desv., 504
Anopheles barbirostris Wulp.:
  associated fungi–*Coelomomyces anophelesica*, 382; *C. indiana*, 382
  associated protozoan–*Thelohania legeri*, 515
Anopheles bellator Dyar & Knab, 504
Anopheles bifurcatus Linn., 34 (see also *Anopheles claviger*)
Anopheles boliviensis, 436
Anopheles claviger Meig.:
  associated bacteria–*Staphylococcus albus*, 153, 154; *S. aureus*, 155, 156
  associated protozoa–*Caulleryella anophelis*, 497; *Thelohania legeri*, 515
  bactericidal substance in, 34
Anopheles crucians Wied.:
  associated protozoa–malarial, 504; *Thelohania legeri*, 515; *T. pyriformis*, 516
Anopheles culcifacies Giles, 504
Anopheles darlingi Root, 504
Anopheles elutus Edwards, 504
Anopheles fuliginosus Giles, 160, 161 (see also *Anopheles annularis*)
Anopheles funestus Giles, 504
Anopheles gambiae Giles, 504, 516
Anopheles hectoris Giaquinto, 504
Anopheles hyrcanus (Pallas):
  associated bacterium–*Pasteurella tularensis*, 166
  associated protozoa–malarial, 504
Anopheles hyrcanus var. *nigerrimus* Giles:
  associated fungi–*Coelomomyces anophelesica*, 382; *C. indiana*, 382
  associated protozoa–*Thelohania indica*, 516; *T. legeri*, 515
Anopheles jamesii Theobald:
  associated fungi–*Coelomomyces anophelesica*, 382; *C. indiana*, 382
Anopoheles larvae, source of food, 28
Anopheles ludlowii Theobald, 504
Anopheles maculipennis Meig.:
  associated bacteria–*Bacillus thuringiensis*, 91, 92; *Leptotrix buccalis*, 187

associated protozoa–*Crithidia fasciculata*, 468; malarial, 504; *Thelohania legeri*, 515
associated spirochetes, 461
associated viruses–fowlpox, 442; western equine encephalomyelitis, 439
Anopheles maculipennis var. *atroparvus* Van Thiel, 461
Anopheles maculipennis var. *freeborni* Aitken, 439
Anopheles minimus Theobald, 504
Anopheles pseudopunctipennis Theobald, 504
Anopheles punctimacula Dyar & Knab, 504
Anopheles punctipennis (Say):
  associated protozoa–malarial, 504; *Thelohania legeri*, 515
Anopheles quadrimaculatus Say:
  associated protozoa–malarial, 504; *Nosema anophelis*, 515; *Thelohania legeri*, 515; *T. obesa*, 516; *T. pyriformis*, 516
Anopheles ramsayi Covell:
  associated fungi–*Coelomomyces anophelesica*, 382; *C. indiana*, 382
  associated protozoa–*Thelohania anomola*, 516; *T. indica*, 516; *T. legeri*, 515
Anopheles stephensi Liston, 504
Anopheles spp., 512
Anopheles subpictus Grassi:
  associated fungi–*Coelomomyces anophelesica*, 382; *C. indiana*, 382
  associated protozoan–*Thelohania legeri*, 516
Anopheles superpictus Grassi, 504
Anopheles tarsimaculatus Goeldi:
  associated bacterium–*Mycobacterium leprae*, 46, 48
  associated protozoa–malarial, 504
Anopheles vagus Dönitz:
  associated fungi–*Coelomomyces anophelesica*, 382; *C. indiana*, 382
  associated protozoan–*Thelohania legeri*, 516
Anopheles varuna Iyen.:
  associated fungi–*Coelomomyces anophelesica*, 382; *C. indiana*, 382
  associated protozoa–*Thelohania legeri*, 516; *T. obscura*, 516
Anoplura, 199, 206, 215, **220-226**
Anoxia australis Gyll.:
  associated bacteria–*Bacillus lymantriae*, 69; *Coccobacillus cajae*, 183, 184
Ant (*see also* specific names):
  associated bacteria–*Coccobacillus acridiorum*, 183, 558; *Eberthella typhosa*, 127, 128; *Pasteurella pestis*, 162, 164; *Shigella paradysenteriae*, 131-132
  associated fungi, 376, 385, **406-408**
  associated virus–smallpox, 443
  intracellular symbiotes in, 190, **247-249**
Antelope, 304
Antestia lineaticollis Stål:

Antestia lineaticollis Stål (*Cont.*)
  associated yeasts–*Nematospora coryli*, 350; *N. gossypii*, 350
Antherea mylitta Dru. [*A. paphia* (Linn.)], 414; *A. pernyi* Guér., 414; *A. yamamaï* Guér., 414
Anthonomus pomorum Linn., 394
Anthrax, 53; bacillus, 3, 12, **53-55** (see also *Bacillus anthracis*)
Anthrenus museorum Linn.:
  associated bacterium–*Bacillus anthracis*, 53, 55
  associated virus, 413
Anthrenus sp., 116
Anthrenus verbasci (Linn.), 495
Antibodies in insects, 562; agglutinins, 562, **571**; antitoxins, 562, **571**; bactericidins, 571; bacteriolysins, 562, **571**; complement, absence of, 572; opsonins, absence of, 572; precipitins, absence of, 571
Anticarsia gemmatilis (Hbn.), 394
Antitoxins, 562, 571
Ants (*see* Ant)
Aonidiella aurantii (Mask.), 58
Apanteles glomeratus (Linn.), 79, 137, 179, 186, 524
Apantesis virgo (see *Callarctia virgo*)
Apes, 277
Aphalara, 230
Aphalara calthae (Linn.):
  associated yeasts–*Aphidomyces aphalarae calthae*, 364; *Cicadomyces aphalarae calthae*, 361
Aphididae, 414; intracellular symbiotes in, 227-229; yeastlike symbiotes in, 372-373
Aphidomyces, 364; *A. aphalarae calthae*, 364; *A. aphidis*, 364; *A. drepanosiphi*, 364; *A. psyllae forsteri*, 365; *A. sulcii*, 365
Aphids, 227-229 (*see also* specific names *and* Aphididae):
  associated bacterium–*Xanthomonas coronofaciens*, 173
  associated fungi, 412–*Empusa aphidis*, 380
  associated yeasts, 372-373
  associated viruses, 444, 445
  bacteria in blood of, 18
  intracellular symbiotes of, 227-229
  mycetome of, 190, 197-199
  symbiosis in, 190
Aphiochaeta ferruginea, 178, 179; *A. rufipes* Meig., 497
Aphis amenticola, Kalt., 364
Aphis atriplicis Linn., 228 (see also *Hyalopterus atriplicis*)
Aphis avenae Fabr., 116
Aphis brassicae Linn. (see *Brevicoryne brassicae*)
Aphis forbesi Weed, 228

Aphis gossypii Glov.:
  associated bacterium–*Erwinia tracheiphila*, 119
  associated fungi, 400
  associated virus, 445
Aphis helianthi Monell, 444
Aphis maidis Fitch, 445
Aphis pomi DeG.:
  associated bacterium–*Erwinia amylovora*, 116
  intracellular symbiotes in, 228
Aphis rhamni Kalt., 445
Aphis rumicis Linn.:
  associated bacterium–*Erwinia lathyri*, 119
  associated viruses, 444, 445
  intracellular symbiotes in, 228
Aphis spiraecola Patch, 401
Aphrophora (*Cercopis*) *alni*, 360
Aphrophora (*Cercopis*) *salicis* (DeG.):
  associated yeasts–*Cicadomyces aphrophorae salicis*, 361; *C. rubricinctus*, 361
Apidae, 414
Apiomerus nalipa, 470; *A. pilipes*, 477
Apion, 199, 206; *A. pisi* Fabr., 240
Apioninae, 238, 240
Apis florea Fabr., 514
Apis mellifera Linn.:
  associated bacteria–*Achromobacter eurydice*, 25; *Aerobacter cloacae*, 120, 121; *BACILLUS* A White, 50; *B. agilis*, 51; *B. agilis larvae*, 51; *B. alvei*, 25, 52-53; *B. alveolaris*, 53; *B. apisepticus*, 55; *Bacillus* B White, 56; *B. butlerovii*, 57-58; *Bacillus* E White, 61; *B. gaytoni*, 63; *B. lanceolatus*, 66; *B. larvae*, 66-67; *B. mesentericus*, 72; *B. milii*, 73; *B. mycoides*, 74; *B. orpheus*, 76, 77; *B. paraalvei*, 77; *B. pestiformis apis*, 78; *B. pluton*, 25, **80-82**; *B. subgastricus*, 89; *B. thoracis*, 91; *BACTERIUM acidiformans*, 99; *B. coli apium*, 101; *B. cyaneus*, 102; *Bacterium* D White, 102; *B. mycoides*, 109; *B. noctuarum*, 110; *Erwinia tracheiphila*, 119; *Escherichia coli*, 121, 122; *E. paradoxa*, 123; *Micrococcus* C, 144; *M. luteus liquefaciens*, 147; *Proteus alveicola*, 124; *Pseudomonas fluorescens*, 170, 171; *Salmonella choleraesuis*, 129; *S. schottmülleri* var. *alvei*, 131; *Sarcina aurantiaca*, 152; *S. lutea*, 152; *Streptococcus apis*, 25, **139**
  associated fungi–*Aspergillus flavus*, 401; *A. fumigatus*, 401; *A. glaucus*, 401; *A. nidulans*, 401; *A. niger*, 401; *A. ochraceus*, 401; *Aspergillus* spp., 395; *Mucor hiemalis*, 402; *Penicillium* spp., 395
  associated protozoa–*Entamoeba apis*, 485; *Malpighamoeba mellificae*, 486; *Nosema apis*, 514; *Nosema bombi*, 514
  associated viruses–paralysis, 415; sacbrood, 414, 415, **426-430**

immunization of, 560
Aplanobacter stewarti, 177 (see also *Xanthomonas stewarti*)
Apodermus agrarius, 295
Aponida, 238
Aporia crataegi Linn.:
  associated bacteria—*Bacillus thuringiensis*, 91, 92; *Bacterium cazaubon*, 100; *B. pyrenei* Nos. *1–3*, 112
Apple aphid (see *Aphis pomi*)
Apple leafhopper (see *Empoasca maligna*)
Apple maggot (see *Rhagoletis pomonella*)
Aproophyla nigra, 18
Aptinothrips rufus (Gmelin), 349
Arachnida, 215, **249-255**, 385, 577
Arachnolysin in spiders and their eggs, 572
Araujia angustifolia, 473
Archotermopsis, 528
Archotermopsis wroughtoni Desneux:
  associated protozoa—*Joenopsis cephalotricha*, 548; *J. polytricha*, 548; *Microjoenia axostylis*, 551; *Pseudotrichonympha pristina*, 550; *Trichomonas termitis*, 539
Arctia caja Linn.:
  associated bacteria—*Bacillus hoplosternus*, 65; *Coccobacillus cajae*, 59, **183**
  associated protozoan—*Nosema bombycis*, 513
Arctiidae, 414
Arctocorixa interrupta (Say), 472
Arcyptera flavicosta, 181
Argas miniatus Koch, 253, 457 (see also *Argas persicus*)
Argas persicus Oken.:
  anatomy of, 577
  associated bacteria—*Bacillus anthracis*, 33, **34**, 53, **55**; *B. mesentericus*, **34**, 72, **73**; mycoides, 33, **74**; *B. subtilis*, 33, **34**, 89, **91**; *B. vulgatus*, 34; *Brucella abortus*, 34; *Clostridium tetani*, 96; *Eberthella typhosa*, 34; *Neisseria catarrhalis*, 34, **157**; *Pasteurella pestis*, 33, 162, **165**; *Serratia marcescens*, 34; *Staphylococcus albus*, 153, 154; *S. aureus*, 33, 155, 156; *Streptococcus faecalis*, 34; *S. hemolyticus*, 34; *S. pyogenes*, 143
  associated spirochetes—*Borrelia anserina*, 457, 458; *Borrelia cobayae*, 454
  associated virus—yellow fever, 436
  bacterial flora of, 13, 14
  bactericidal substance in, 33, 34
  intestinal tract, sterile, 12
  intracellular symbiotes in, 206, **253**, 254
Argas reflexus (Fabr.):
  associated bacterium—*Staphylococcus aureus*, 155
  associated spirochete—*Borrelia anserina*, 457
Argasidae, 250, **253-254**, 577
Armigeres obturbans Walk., 160, 161

Armyworms (see specific names)
Arphia sulphurea (Fabr.):
  associated protozoa—*Actinocephalus elongatus*, 494; *Gregarina rigida columna*, 492; *G. rigida rigida*, 491
Arthropod-borne virus encephalitides, 437
Arthropods (see Insects *and* Ticks)
Arthrorhynchus, 385
Arytaena, 230
Aschersonia, 395; *A. aleyrodis*, 400; *A. goldiana*, 400
Ascia rapae (Linn.) (see *Pieris rapae*)
Ascobacterium, 42, 181; *A. luteum*, 21, 175, 181
Ascomycetes, 348, 376, **382-388**, 393
Asemini, 370, 372
Asiatic garden beetle (see *Autoserica castanea*)
Asida opaca Say, 494; *Asida* sp., 494
Asopinae, 16
Aspergillus, 393, 395; *A. flavus*, 134, 170, 394, 398, 401; *A. fumigatus*, 401, *A. glaucus*, 401; *A. nidulans*, 401; *A. niger*, 24, 394, 411; *A. ochraceus*, 401; *A. parasiticus*, 401
Asphondylia coronillae Vall., 395
Aspidapion, 239; *A. aeneum* Fabr., 240
Aspidiotus, 391; *A. ancylus* (Putn.), 391; *A. cyanophylli* Sign., 374; *A. forbesi* Johns., 391; *A. hederae* (Vallot), 374; *A. juglansregiae* Comst., 391; *A. latiniae* Sign., 374; *A. nerii* Bouché, 374; *A. osborni* Newell & Ckll., 389, 391, 392; *A. pini* Comst., 374
Assassin bugs, 233-235
Aster yellows, 446
Asterolecaniae, 374
Asterolecanium aurem (Bdv.), 374; *A. variolosum* (Ratz.), 232, 374
Asterophora elegans, 494; *A. philica*, 494
Atebrin, 288
Atomus, 255
Atricholaelaps glasgowi, 440
Atropidae, 218
Atta cephalotes (Linn.), 406
Atta sexdens Linn.:
  associated bacterium—*Coccobacillus ocridiorum*, 181
  associated fungi, 408
Atta texana Buckley, 408
Atta vollenweideri Forel, 408
Attagenus pellio Linn.:
  associated bacterium—*Bacillus anthracis*, 53, 55
  associated protozoan—*Pyxinia frenzeli*, 495
Attagenus piceus (Oliv.), 116
Attini, 406
Aulacophora foveicollis (Lucas), 495
Aureogenus vastans, 447
Aurococcus, 153
Australian typhus, 291
Autoecomyces, 385

SUBJECT INDEX

Autographa brassicae (Riley):
associated bacterium—*Xanthomonas campestris*, 173
associated virus, 414
Autographa gamma californica (Speyer), 414
Autoserica castanea (Arrow), 82, 83
Axolotl, 483
Azygous mycetomes (*see* Mycetome, azygous)

Babesia berbera, 507; *B. bigemina*, 3, **506-507**;
B. bovis, 507; *B. caballi*, 507; *B. canis*, 507;
B. equi, 507; *B. felis*, 507; *B. gibsoni*, 507;
B. motasi, 507; *B. ovis*, 507; *B. sergenti*, 507
Babesiidae, 500, **506-511**
Bacillaceae, 41, 50
Bacillus, 20, 41, **50-94**, 103, 106, 112; *Bacillus* A Ledingham, 50; *Bacillus* A White, 10, **50**; *B. acidi lactici*, 122 (see also *Escherichia coli*); *B. acridiorum*, 75 (see also *Coccobacillus acridiorum*); *B. aegypticus*, 101, 159 (see also *Bacterium conjunctivitides*); *B. aerifaciens*, 51; *B. agilis* Hauduroy *et al.*, 51; *B. agilis* Mattes, **51**, 66; *B. agilis larvae*, 51; *B. agrotidis typhoides*, 51; *B. alacer*, 52; *B. albolactis*, 59 (see *B. cereus*); *B. alvei*, 25, **52-53**, 58, 73, 80, 81, 139; *B. alveolaris*, 53; *B. anthracis*, 12, 33, 34, **53-55**, 97, 561; *B. apisepticus*, 55; *B. aureus*, 55-56; *Bacillus* B Hofmann, **56**, 74, 108, 109, 424; *Bacillus* B White, 56; *B. barbitistes*, 56; *B. bombycis*, 56-57, 415; *B. bombycis non liquefaciens*, 57; *B. bombycoides*, 57; *B. bombysepticus*, 57; *B. brandenburgiensis*, 67 (see *B. larvae*); *B. bruneus*, 111; *B. burrii*, 67 (see *B. larvae*); *B. butlerovii*, 57-58; *B. bütschlii*, 58; *Bacillus* C, 58; *B. cacticidus*, 117 (see also *Erwinia cacticida*); *B. cajae*, 59, 184 (see also *Coccobacillus cajae*); *B. campestris*, 173 (see also *Xanthomonas campestris*); *B. canadensis*, 100, 560; *B. canus*, 58; *B. carbonchio*, 53 (see also *B. anthracis*); *B. carotovorus*, 118 (see also *Erwinia carotovora*); *B. cellulosam fermentans*, 97 (see also *Clostridium werneri*); *B. cereus*, 13, **58-59**, 61; *B. cholerae suis*, 129 (see also *Salmonella choleraesuis*); *B. circulans*, 59; *B. cleoni*, 59; *B. cloacae*, 121 (see also *Aerobacter cloacae*); *B. coeruleus*, 59; *B. coli*, 33, **121**, 122 (see also *Escherichia coli*); *B. coli communior*, 122 (see also *Escherichia coli*); *B. coli communis*, 122 (see also *Escherichia coli*); *B. coli mutabilis*, 122 (see also *Escherichia coli*); *B. coli-simile*, 60; *B. cubonianus*, 60; *B. cuenoti*, **60**, **204**, 216; *B. cuniculicida*, 161 (see also *Pasteurella cuniculicida*); *B. decolor*, 60; *B. della barbone bufalino* (see *B. anthracis*); *B. diphtheriae* (see *Corynebacterium diphtheriae*); *B. dobelli*, 61; *B. dysenteriae* (Flexner), 132; *Bacillus* E White, 61; *B. ellenbachensis*, 61; *B. ellenbachi*, 61; *B. enteritidis* (see *Salmonella enteritidis*); *B. entomotoxicon*, 61-62; *B. ephestia*, 76; *B. equidistans*, 62; *B. faecalis alkaligenes* (see *Alcaligenes faecalis*); *B. ferrugenus*, 62; *B. flavus*, 62; *B. flexilis*, 61, **62-63**; *B. fluorescens liquefaciens*, 84, 170, **171** (see also *Pseudomonas fluorescens*); *B. fluorescens non-liquefaciens*, 171 (see also *Pseudomonas non-liquefaciens*); *B. fluorescens septicus*, 72 (see also *Pseudomonas septica*); *B. foetidus*, 63; *B. fuchsinus*, 63; *B. gasoformans nonliquefaciens*, 63; *B. gastricus*, 89; *B. gaytoni*, 63; *B. gibsoni*, 63-64; *B. gigas*, 64; *B. gortynae*, 64; *B. graphitosis* (see *B. tracheïtis sive graphitosis*); *B. gruenthali*, 64; *B. grünthal*, 64; *B. gryllotalpae*, 64; *B. hoplosternus*, **65**, 100; *B. hyalinus*, 98; *B. immobilis*, 65; *B. insectorum*, 147 (see also *Micrococcus insectorum*); *B. intrapallens*, 65; *B. lactis acidi*, 142; *B. lactis aerogenes*, 120 (see also *Aerobacter aerogenes*); *B. lanceolatus*, 51, 66; *B. larvae*, 66-67, antibiotic from, 67; *B. lasiocampa*, 67; *B. lathyri*, 119 (see also *Erwinia lathyri*); *B. lentimorbus*, 67-68, 82, 83; *B. leptinotarse*, 68; *B. lineatus*, 68; *B. liparis*, 27, **69**; *B. lutzae*, **69**, 150, 158; *B. lymantriae*, **69**, 136, alpha, 69, 70, beta, 69, 70; *B. lymantricola adiposus*, 69, 70; *B. megaterium*, 70; *B. megaterium* [*megatherium*] *bombycis*, 70-71; *B. megaterium* De Bary, 70, **71**, 87, 100, 366; *B. megaterium* Ravenel, 71; *B. melolonthae*, 71; *B. melolonthae liquefaciens*, 71, alpha, 71, beta, 72, gamma, 72; *B. melolonthae non liquefaciens*, **72**, 560, 570, 571; alpha, 72; beta, 72; gamma, **72**, 565; delta, 72; epsilon, 72; *B. mesentericus*, 24, 34, **72-73**, 152; *B. mesentericus ruber*, 24; *B. milii*, **73**, 91; *B. minimus*, 73; *B. monachae*, **73-74**, 108, 109; Bacillus of Morgan No. *1* (see *Proteus morgani*); *B. mycoides*, 28, 33, **74**, 109, 558, 567; *B. necrophorus*, 114; *B. neurotomae*, **74-75**, 110; *B. noctuarum*, 68, **75**, 89, 110; *B. novus*, 24; *B. oblongus*, 75; *B. ochraceus*, 110; *B. ocniriae*, 76; *B. ontarioni*, **76**, 110; *B. orpheus*, **76-77**, 81; *B. oxytocus perniciosus*, 120 (see also *Aerobacter aerogenes*); *B. para-alvei*, 77-78; *B. paratyphi* B, 130 (see also *Salmonella schottmülleri*); *B. paratyphi alvei*, 131 (see also *Salmonella schottmülleri* var. *alvei*); *B. para-typhosus*, 130 (see also *Salmonella schottmülleri*); *B. paratyphus* Type A, 130 (see also *Salmonella paratyphi*); *B. pectinophorae*, 77-78; *B. pediculi*, 78; *B. pestiformis apis*, 78; *B. pestis*, 78-79 (see also *Pasteurella pestis*); *B. pieris agilis*, **79**, 137, 186; *B.*

SUBJECT INDEX

pieris fluorescens, 79; *alpha*, 79; *beta*, 79; *B. pieris non liquefaciens alpha*, 79; *beta*, 79; *B. piocianemus*, 80; *B. pirenei*, 80; *B. pluton*, 25, 52, 58, 66, **80-81**, 97, 139; *B. poncei*, **81-82**, 565, 571, 572; *B. popilliae*, 67, 68, **82-83**; *B. prodigiosus*, 101, **133, 134**, 273 (see also *Serratia marcescens*); *B. proteidis*, 79, 83; *B. proteisimile*, 83-84; *B. proteus*, 126 (see also *Proteus vulgaris*); *B. proteus fluorescens*, 171 (see also *Pseudomonas jaegeri*); *B. proteus vulgaris*, 126 (see also *Proteus vulgaris*); *B. pseudo-difterico* (see *Corynebacterium pseudodiphthericum*); *B. punctatus*, 84; *B. pyocyaneus*, 169, 171 (see also *Pseudomonas aeruginosa*); *B. pyrameis* I, 84; II, 84; *B. radiciformis*, 84-85; *B. rickettsiformis*, 85; *B. rotans*, 85; *B. rubefaciens*, 85; *B. salutarius*, 86; *B. schafferi*, 122 (see also *Escherichia coli*); *B. septicaemiae lophyri*, 86; *B. septicus agrigenus*, 99 (see *Bacterium agrigenum*); *B. septicus insectorum*, 86, 148; *B. similcarbonchio*, 86-87; *B. similis*, 87; *B. similtubercolari*, 87; *B. simplex*, 24; *B. solanacearum*, 176 (see also *Xanthomonas solanacearum*); *B. sotto*, 57, **87-88**, 100; *B. spermatozoides*, 88; *B. sphingidis*, 68, 75, **88-89**, 110; *B. subgastricus*, 89; *B. subtilis*, 24, 28, 33, 65, **89-91**, 350, 556, 557, 560; *B. tenax*, 91; *B. thoracis*, 91; *B. thuringiensis*, **91-93**, 100, 560, 566; *B. tifosimile*, 93; *B. tingens*, 93; *B. tracheiphilus* (see *Erwinia tracheiphila*); *B. tracheïtis sive* [or] *graphitosis*, 93; *B. tropicus*, 93-94; *B. typhosimile* (see *B. tifosimile*); *B. verrucosus*, 63 (see also *B. foetidus*); *B. violaceus*, 180 (see also *Chromobacterium violaceum*); *B. viridans*, 94; *B. visiculosis*, 94; *B. vulgatus*, 34, 73 (see also *B. mesentericus*); *Bacillus* X (see *B. larvae*); *Bacillus* Y (see *B. pluton*)

Bacteria: acid-fast, examination for, 580, **583**, 597; aerobic, examination for, 580, **581-583**, 596; anaerobic, examination for, 580, **583**, 597; associated with termite flagellates, 36-37; bactericidal substance of insects against, 34; in blood of insects, 17, 18; cecal (see Ceca, bacteria in); cellulose-fermenting, 31; coliform, 182; culture methods for, 583; effect of, on hatching of mosquito eggs, 22, 23; examination for, 580, **581-583**, 598-599; external, on insects, 9; extracellular, **9-37, 43-187**; as food, 28-29; *genera incertae sedis vel dubia*, 181-187; identification of, 583, **598-599**; and immunity of insects, 556-558; and insect digestion, 30-32; and insect eggs, 22-23; in insect eggs, 24; in insects, alimentary tract, 10-14; in insects, blood, 17, 18; in insects, ceca, 14-17; in insects, factors influencing flora, 18-19; in insects, fate of during metamorphosis, 19-20; in insects, percentage of bacteria, 9; in insects, regional sterility, 11; nomenclature of, 38-40; role in insect nutrition, 27-33; as source of vitamins and growth accessory substances, 28, **32-33**; synthesis of hormone by, 33; in ticks, 13, 14 (*see also* specific names); variation in, 24-27

Bacteriaceae, 41, 97
Bacterial crypts, 14
Bactericidal action of insect blood plasma, 571
Bactericidal substance in insects and ticks, 33-34
Bactericidins, 571
Bacteridium luteum, 181
Bacteriocytes, 190, 196
Bacteriolysins, 562, 571
Bacteriophage, 13, **449-450**
Bacteriotomes, 190
Bacterium, 41, 99, 105; *B. acidi lactici*, 122 (see also *Escherichia coli*); *B. acidiformans*, 99; *B. aerogenes*, 120 (see also *Aerobacter aerogenes*); *B. agrigenum*, 99; *B. angulatum*, 173; *B. bombycis* (see *Bacillus bombycis*); *B. bombycivorum*, 125 (see also *Aerobacter bombycis*); *B. canadensis*, 99, 101; *B. cazaubon*, 99; *B. cellulosum*, 99; *B. cholerae gallinarum*, 160, 161 (see *Pasteurella cuniculicida* and *P. avicida*); *B. cleonusum*, 101; *B. coli*, 122 (see *Escherichia coli*); *B. coli anaerogenes*, 126 (see *Eberthella belfastiensis*); *B. coli apium*, 101, 124; *B. conjunctivitidis*, 101-102; *B. coronofaciens*, 173 (see also *Xanthomonas coronofaciens*); *B. cyaneus*, 10, **102**; *Bacterium* D, 102; *B. delendae-muscae*, 102; *B. elbvibrionen*, 102; *B. ephestiae* No. *1* and No. 2, 92, 93, **103**; *B. eurydice*, 25, **97**, 139; *B. fermentationis*, 24; *B. galleriae*, 103; No. 2 and No. *3*, 101, **103**, 560, 561; *B. gelechiae*, No. *1*, No. 2, and No. *5*, 104; *B. gryllotalpae*, 104; *B. hebetisiccus*, 104; *B. hemophosphoreum*, 104; *B. imperiale*, 105; *B. incertum*, 105, 107; *B. insectiphilium*, **105-106**, 107; *B. intrinsectum*, 106; *B. italicum* No. 2, 106; *B. kiliense*, 132; *B. knipowitchii*, 106; *B. lactis acidi*, 142; *B. luteum*, 181; *B. lymantriae* (see *Bacillus lymantriae*); *B. lymantricola adiposus*, 25, 26, 27, **106**; *B. melolonthae liquefaciens*, **108**, 184; *B. melolonthae liquefaciens alpha*, 184; *gamma*, 25, **108**, 570; *B. minutiferula*, 108; *B. monachae*, 56, 74, **108-109**, 424; *B. mutabile*, 27, **109**; *B. mycoides*, 74, **109**; *B. neapolitanum*, 99, 122; *B. neurotomae*, 27, **109-110**; *B. noctuarum*, 110, 184; *B. ochraceum*, 110; *B. ontarioni*, 76, 101, **110**; *B. paracoli*, 111; *B. paratyphi*, 130 (see also *Salmonella paratyphi*); *B. pieris*

Bacterium (Cont.)
  liquefaciens alpha, 25, 26, **111**; *B. pityocampae*, **111**, 142; *B. prodeniae*, 111-112; *B. prodigiosum*, 135 (see also *Serratia marcescens*); *B. prodigiosus* (see *Serratia marcescens*); *B. pseudotsugae*, 175 (see also *Xanthomonas pseudotsugae*); *B. pyraustae* Nos. *1-7*, 112; *B. pyrenei*, 80; Nos. *1-3*, 112; *B. qualis*, 112; *B. rubrum*, 113; *B. salicis*, 119 (see also *Erwinia salicis*); *B. savastanoi*, 175 (see also *Xanthomonas savastanoi*); *B. sotto* (see *Bacillus sotto*); *Bacterium* sp., **99**, 134; *B. sphingidis*, 184; *B. suipestifer* (see also *Salmonella choleraesuis*); *B. tegumenticola*, 113; *B. termo*, 113; *B. thuringiensis*, 92 (see also *Bacillus thuringiensis*); *B. tularense*, 165, 166, 167 (see also *Pasteurella tularensis*); *B. tumefaciens*, 113; *B. turcosum*, 24; *B. vesiculosum*, 94; *B. viscosum non liquefaciens*, 113
Bacteriumlike symbiotes (see Intracellular symbiotes)
Bacteroides, 114
Bacteroids, 41, 60, **203**, 204, 208, 262
Baëtis pygmaea Hagen, 516
Baëtis rhodani, 516
Baëtis sp.:
  associated protozoa—*Nosema baëtis*, 514; *Ophryoglena collini*, 521
Bagous binodulus Hbst., 240
Bagworm (see *Thyridopteryx ephemeraeformis*)
Balanius glandium Mrsh., 240; *B. nucorum* Linn., 240
Balantidium blattarum, 521; *B. coli*, 521; *B. knowlesi*, 521; *B. orchestium*, 521; *B. ovatum*, 521; *B. praenucleatum*, 521
Bandicoots, 295, 298, 300
Bangasternus orientalis Cap., 240
Barathrax configurata (Walk.), 123
Barbitistes amplipennis (see *Isophya amplipennis*)
Barbylanympha coahoma, 536; *B. estoboga*, 536; *B. laurabuda*, 536; *B. ufalula*, 484, 536
Baris chlorizans Germ., 240; *B. laticollis* Mrsh., 240; *B. maculipennis* (Mansour), 240; *B. morio* Boh., 240
Bark beetles, 349 (see also specific names)
Barroella zeteki, 547
Barrouxia ornata, 498
Bartonella, 262, **345-347**; *B. bacilliformis*, 299, **346**; *B. muris*, 345
Basidiobolus, 377; *B. ranarum*, 377
Bats, 325, 337
Beauveria, 393; *B. bassiana*, 393, 394, 398; *B. densa*, 394; *B. globulifera*, 394, 396, 400; *B. stephanoderis*, 394
Basidiomycetes, 376, **388-393**
Basidiospores, 388
Basidium, 388

Bedbug, 12, 47, **168**, 169, **233**, 282, **316-319**, 440, 454, 468 (see also *Cimex lectularius*)
Bees, 3, 10, 116, 117, 412, 515 (*see also* specific names); black brood of, 53, 73; foul brood of (*see* Foul brood); Isle of Wight disease, 78; New York disease, 73, 91; paralysis of, 63
Beet weevil, 101
Beetles, 162, 163, 377 (*see also* specific names); ambrosia beetles, 402-406; bostrychid, 206, 207; darkling, 243; flat, 241-242; leaf, 242-243; wood-boring, 376
Beloides firmus, 495
Belostoma, 482
Bemisia gossypiperda Misra & Lamba, 449; *B. gossypiperda* var. *mosaicivectura*, 449; *B. nigeriensis* Corb., 449
Berkelella, 387
Berlese's organ, 317
Beschmierorgan, 17
Beta virus *1*, 446; Beta virus *3*, 448; Beta virus *5*, 448
Bibio albipennis Say, 116; *B. marci* Linn., 495
Biologic relationships, general, 4, 5
Birds, rickettsiae from, 310, 311
Bison, rickettsia from, 345
Biting midges (see Ceratopogonidae)
Blaberus, 217, 468
Blaberus atropos:
  associated protozoan—*Entamoeba nana*, 485
  associated spirochete—*Spirochaeta blattae*, 460
Blaberus cranifer, 212
Black brood of bees, 53, 73
Black carpet beetle (see *Attagenus piceus*)
Blackleg of cattle, 95
Blaps mucronata Latr.:
  associated bacteria—*Bacillus anthracis*, 53, 55; *B. proteisimile*, 83; *B. radiciformis*, 84, 85; *Clostridium chauvoei*, 95; *C. sporogenes*, 95; *C. tetani*, 96; *Corynebacterium pseudodiphthericum*, 46; *Diplococcus pneumoniae*, 137, 138; *Klebsiella pneumoniae*, 124; *Malleomyces mallei*, 159; *Mycobacterium tuberculosis*, 48, 49; *Pseudomonas fluorescens*, 170; *P. non-liquefaciens*, 171, 172; *Sarcina alba*, 151; *Vibrio colerigenes*, 177
Blaptica dubia Serv., 494
Blasteme stage, 195
Blastocladiales, 377, 381
Blastodendrion pseudococci, 351
Blastophaga psenes Linn., 350, 411
Blatta, 207
Blatta aethiopica Sauss., 216, 217
Blatta lapponica Linn.:
  associated protozoan—*Gamocystis tenax*, 493
  intracellular symbiote in, 216
Blatta orientalis Linn.:

anatomy of, 576
associated bacteria–*Bacillus anthracis*, 53, 55;
B. bütchlii, 58; *B. cuenoti*, **60**, 204; *B. megatherium*, 71; *B. proteisimile*, 83; *B. radiciformis*, 84, 85; *B. similcarbonchio*, 86, 87; *B. subtilis*, 89, 90; *B. tifosimile*, 93; *Chromobacterium violaceum*, 180; *Clostridium chauvoei*, 95; *C. sporogenes*, 95; *C. tetani*, 96; *Coccobacillus cajae*, 183, 184; *Corynebacterium diphtheriae*, 43; *C. pseudodiptherrcum*, 46; *Diplococcus pneumoniae*, 137, 138; *Eberthella typhosa*, 127, 128; *Escherichia coli*, 121, 122; *Klebsiella pneumoniae*, 124; *Malleomyces mallei*, 159; *Mycobacterium tuberculosis*, 48, 49; *Pasteurella avicida*, 160; *Proteus vulgaris*, 126; *Pseudomonas aeruginosa*, 170; *P. fluorescens*, 170; *P. nonliquefaciens*, 171, 172; *Sarcina alba*, 151; *S. aurantiaca*, 152; *S. lutea*, 152; *Serratia marcescens*, 132, 134; *Staphylococcus albus*, 153; *S. aureus*, 18, **155**; *S. citreus*, 156; *Vibrio colerigenes*, 177, 178; *V. metschnikovii*, 179
associated protozoa–*Balantidium praenucleatum*, 521; *Diplocystis schneideri*, 490; *Endamoeba blattae*, 485; *Endolimax blattae*, 485; *Entamoeba thomsoni*, 485; *Gregarina blattarum*, 491, 493; *Hartmanella blattae*, 486; *Nycototherus ovalis*, 521; *Plistophora kudoi*, 516; *P. periplanetae*, 517; *Protomagalhaensia serpentula*, 493; *Retortamonas blattae*, 482; *R. orthoperorum*, 482; *Tetratrichomastix blattidarum*, 483
immunization of, 560
intracellular symbiotes in, 190, 203, 207, **215-217**
mycetocytes of, 200
Blattella, 207
Blattella germanica (Linn.):
associated bacteria–*Aerobacter cloacae*, 120, 121; *Corynebacterium blattellae*, 42, **204, 205**, 216; *Escherichia coli*, 121, 122; *Staphylococcus albus*, 153, 154; *S. aureus*, 155; *Streptococcus faecalis*, 141
associated fungus–*Cordyceps blattae*, 387
associated protozoa–*Entamoeba thomsoni*, 485; *Gregarina blattarum*, 481; *Nyctotherus ovalis*, 521; *Plistophora periplanetae*, 517; *Tetratrichomastix blattidarum*, 483
associated virus–lymphocytic choriomeningitis, 441
associated yeast, 349
complement fixation tests for symbiotes of, 203
intracellular symbiotes in, 190, **215-217**
mycetocytes of, 200
vitamins, importance of in diet, 349
Blattidae, **215-218**, 536

Blattids, symbiosis in, 190 (*see also* specific names)
Blissus leucopterus (Say):
associated bacteria–*Bacillus entomotoxicon*, 61, 62; *Micrococcus insectorum*, 146-147
associated fungus–*Beauveria globulifera*, 394, 400
ceca, 15; bacteria of, 16
Blood, insect, 563, 567; bacteria in, 17, 18; clotting properties of, 564; constituents of, 563; hemocytes of, 563; hemolymph of, 563; virus in, 426
Blood cells (*see* Cells, insect, blood)
"Blood virus," 331
Bluetongue of sheep, 442
Bodo muscarum, 471 (see also *Herpetomonas muscarum*)
Bombus agrorum Fabr., 514; *B. hortorum* Linn., 514, 515; *B. lapidarium* (Linn.), 514, 515; *B. latreillelus*, 514, 515; *B. sylvarum* (Linn.), 514; *B. terrestris* (Linn.), 514, 515; *B. venustus* (Radokowsky), 515
Bombycidae, 414, 514
Bombyx mori (Linn.):
associated bacteria–*Bacillus bombycis*, **56, 57**, 415; *B. bombycoides*, 57; *B. bombysepticus*, 57; *B. cubonianus*, 60; *B. ellenbachi*, 61; *B. ferrugenus*, 62; *B. fuchsinus*, 63; *B. intrapallens*, 65; *B. lymantricola adiposus*, 70; *B. megaterium*, 70; *B. megaterium [megaterium] bombycis*, 70, 71; *B. melolonthae*, 71; *B. mycoides*, 74; *B. noctuarum*, 75; *B. orpheus*, 76; *B. rubefaciens*, 85; *B. sotto*, 87; *B. sphingidis*, 88, 89; *B. viridans*, 94; *Bacterium lymantricola adiposus*, 106; *Coccobacillus acridiorum*, **181**, 558; *C. cajae*, 183, 184; *Diplobacillus melolonthae*, 185; *Diplococcus bombycis*, 136; *D. melolonthae*, 137; *Escherichia coli*, 121, 122; *Micrococcus lardarius*, 147; *M. ovatus*, 49; *Proteus bombycis*, 124-125; *Pseudomonas aeruginosa*, 169, 170; *Sarcina lutea*, 152; *Serratia marcescens*, 122, **134**, 135; *Staphylococcus aureus*, 155; *S. insectorum*, 156; *Streptococcus bombycis*, **139**, 142, 415; *S. disparis*, 140, 141; *S. pastorianus*, 142
associated fungi–*Beauveria bassiana*, 394; *B. densa*, 394
associated protozoa–*Chlamydozoa bombycis*, 417; *Nosema bombycis*, 512, 513-514
associated spirochete–*Spirochaeta pieridis*, 461
associated viruses–414, 415, 416, **417-421**
bacteria as aid to digestion in, 31
flacherie in, **57**, 60, 61, 70, 88, 90, 135, 140, **415**
"flacherie-like" disease of, 560
gattine in, 140, 415

SUBJECT INDEX

Bombyx mori (Linn.) (*Cont.*)
 grasserie in, 70, 147 (see also *Bombyx mori*, jaundice in)
 jaundice in, 415, 416, 417-421, 422
 pébrine in, 513, 522, 555, 558
Bont tick (see *Amblyomma hebraeum*)
Book lice (*see* Atropidae)
Boöphilus, 252
Boöphilus annulatus (Say), 3:
 associated protozoa–*Anaplasma centrale*, 511; *A. marginale*, 510; *Babesia bigemina*, 506, 507
 associated rickettsia–*Rickettsia burneti*, 297, 298
 intracellular symbiotes in, 251, 252
Boöphilus annulatus australis Fuller, 251
Boöphilus annulatus microplis Canestrini:
 associated rickettsia–*Rickettsia burneti*, 297
 intracellular symbiotes in, 251
Boöphilus australis (see *Margaropus australis*)
Boöphilus decoloratus (Koch):
 associated bacterium–*Bacillus anthracis*, 53, 55
 associated rickettsia–*Dermacentroxenous conori*, 341
 associated spirochete–*Borrelia theileri* (see *Margaropus decoloratus*, 459)
 intracellular symbiotes in, 251
Boöphilus microplus (Canestrini):
 associated bacterium–*Mycobacterium leprae*, 47, 48
 associated protozoan–*Babesia bigemina*, 507
 associated virus–yellow fever, 436
Borborus, 471
Boris granulipennis, 238
Borrelia, 451; *B. anserina*, 454, 457-458, 459; *B. berberum*, 452; *B. carteri*, 452; *B. cobayae*, 454; *B. duttoni*, 452, 456, 457; *B. gallinarum*, 457 (see also *B. anserina*); *B. glossinae*, 460; *B. hermsi*, 453; *B. hispanicum*, 453, 454; *B. kochi*, 452; *B. neotropicalis*, 453; *B. novyi*, 452; *B. parkeri*, 453; *B. persica*, 452, 454, 455; *B. recurrentis*, 452; *B. rossi*, 452; *B. sogdianum*, 453; *B. theileri*, 458-459, *B. turicatae*, 453; *B. venezuelense*, 452
Borrellina bombycis, 418
Bostrichidae, 236
Bostrychoplites zickeli (Marseul): intracellular symbiotes in, 236; mycetome of, 199; symbiote transmission by, 208, 209
Bothriopsis histrio, 495
Bothynoderes angullicollis Chevt., 239
Bothynoderes meridionalis Chevt., 239
Bothynoderes punctiventris Germ.:
 associated bacterium–*Bacillus bombycis*, 56, 57
 intracellular symbiotes in, 239

Botrytis, 393; *B. anthophila*, 412; *B. bassiana*, 394 (see also *Beauveria bassiana*)
Boutonneuse fever (see *Fièvre boutonneuse*)
Brachycerinae, 240
Brachycerus apterus Linn., 238, 240; *B. undatus* Fabr., 238, 240
Brachyrhinus cardiniger (H.J.), 239; *B. corruptor* (H.J.), 239; *B. gemmatus* (Scop.), 238, 239; *B. inflatus* (Gyll.), 238, 239; *B. niger* (Fabr.), 239; *B. sulcatus* (Fabr.), 239; *B. sulphurifer* (Oliv.), 239; *B. ventricola* (Wse.), 239
Brachystola magna (Gir.), 491
Brassolis astyra Godart, 514
Brazilian spotted fever, 260, **329**
Breakbone fever, 436
Bremus (see *Bombus*)
Brevicoryne brassicae (Linn.), 228
Brill's disease, 280
Bromius, 242
Bromius obscurus Linn.: cecal bacteria in, **17**; intracellular symbiotes of, 242
Brotolomia meticulosa Linn., 394
Brown fungus (see *Aegerita webberi*)
Brown-tail moth (see *Nygmia phaeorrhoea*)
Brownian movement, 273
Brucella, 42, **158**, 165; *B. abortus*, 158-159; *B. tularensis*, 165
Brucelleae, 42, **158**
Brucellosis, 158, 159
Bruchus obtectus (see *Acanthoscelides obtectus*)
Bubonic plague (see *Pasteurella pestis* and Plague)
Bulbocephalus elongatus, 494
Bullanympha silvestrii, 545
Bullfinch, 310
Bullis fever, 345
Bupalus piniarius Linn., 414, 417
Burrowing beetle (see *Melolontha vulgaris*)
Bursa copulatrix, 193, 208, 209
Bürstenbesatzes, 314

Cabbage looper (see *Autographa brassicae*)
Cabbage worm (see *Autographa brassicae* and *Pieris rapae*)
Cactoblastis bucyrus, 117; *C. cactorum* (Berg.), 117
Cactobrosis fernaldiales Hulst., 118
Caduceia bugnioni, 36, 460, 542; *C. kalshoveni*, 542; *C. monile*, 542; *C. nova*, 542; *C. pruvoti*, 542; *C. theobromae*, 542
Caeculia, 514
Caenis, 490
Caenomyces, 385
Calandra granaria Linn. (see *Sitophilus granarius*); *C. oryzae* Linn. (see *S. oryzae*); *Calandra* spp., 199
Calandrinae, 240

SUBJECT INDEX

Calcaritermes brevicollis (Banks):
  associated protozoa–*Calonympha* sp., 546; *Devescovina lepida*, 541; *Foaina reflexa*, 544; *Hexamastix claviger*, 539; *Macrotrichomonas procera*, 542; *Microrhopalodina kofoidi*, 547; *Oxymonas ovata*, 547
Calcaritermes emarginicollis Snyder:
  associated protozoa–*Devescovina lepida*, 541; *Foaina reflexa*, 544; *Macrotrichomonas procera*, 542
Calcaritermes nearcticus Snyder:
  associated protozoa–*Devescovina lepida*, 541; *Foaina reflexa*, 544; *Macrotrichomonas procera*, 543
Calcaritermes parvinotus Light:
  associated protozoa–*Devescovina lepida*, 541; *Foaina reflexa*, 544; *Macrotrichomonas procera*, 543
Calcaritermes spp., 528
Callarctia virgo (Linn.), 414
Callimorpha jacobeae (Linn.), 515
Calliphora, 158
Calliphora erythrocephala (Meig.):
  associated bacteria–*Bacillus anthracis*, 53, 54; *Salmonella enteritidis*, 129
  associated protozoa–*Herpetomonas muscarum*, 471; *Nosema apis*, 515
  associated virus–poliomyelitis, 441
Calliphora vomitoria (Linn.):
  associated bacteria–*Bacillus anthracis*, 53; *B. colisimile*, 60; *B. radiciformis*, 84, 85; *B. similcarbonchio*, 86, 87; *B. tifosimile*, 93; *Escherichia coli*, 121, 122; *Proteus vulgaris*, 126; *Pseudomonas aeruginosa*, 169; *P. fluorescens*, 170, 171; *P. jaegeri*, 171; *P. nonliquefaciens*, 171, 172; *Sarcina aurantiaca*, 152; *Serratia kielensis*, 132; *S. marcescens*, 133; *Staphylococcus albus*, 153, 154; *S. citreus*, 156; *Vibrio comma*, 178
  associated virus, 414
Calliphoridae, 441
Calliptamus italicus (Linn.):
  associated bacterium–*Bacillus thuringiensis*, 91, 92
  associated protozoan–*Gregarina acridiorum*, 493
Callistephus virus 1, 446; *Callistephus virus 1a*, 446
Calocoris bipunctatus Fabr., 445
Calonympha, 546; *C. grassi*, 546
Calonymphidae, 531
Caloptenus, 181, 182
Calotermes iridipennis, 116
Calves, 301
Calyx, 192
Camponotus, 199, 206, 209, **247, 249**; *C. atramentarius* Forel var. *liocnemis* Emery, 248; *C. brutus*, 407; *C. ligniperdus* Latr. [= *C.*

herculeanus (Linn.)], 248; *C. maculatus* (Fabr.), 248; *C. rectangularis* Emery, 248; *C. rubroniger* Forel, 248; *C. senex* Smith, 248
Camptomyces, 386
Campylomma verbasci (Meyer), 116
Candida, 349; *C. albicans*, 375
Canker of apple trees, 412
Cantharomyces, 386
Cantharosphaeria chilensis, 384
Capnodium citri, 412
Capybara, 337
Carabid beetles, 14 (*see* specific names)
Carabidae, 385
Carabus, 494; *C. auratus* Heuer, 494; *C. violaceus* Linn., 494
Carausius morosus Brunn., 560, 561
Carbohydrates used in identifying bacteria, 599
Carolinaia cyperi Ainslie, 445
Carpocapsa pomonella (Linn.), 116
Carpophilus hemipterus (Linn.), 350, 411
Carrión's disease, 345, 346
Carrot rust fly (see *Psila rosae*)
Caryococcus cretus, 37; *C. dilatator*, 37; *C. hypertrophicus*, 37; *C. invadens*, 37; *C. nucleophagus*, 37
Caryoletira, 37
Casinaria infesta (Cress.), 249
Cassava mosaic, 449
Cassida, 242; *C. flaveola* Thunb., 242; *C. hemisphaerica* Hbst., 242; *C. nebulosa* Linn., 242; *C. nobilis*, Linn., 242; *C. rubiginosa* Müll., *C. vibex* Linn., 242; *C. viridis* Linn., 242
Cat, 286
Catalogue, method and plan of, 40
Catalpa sphinx (see *Ceratomia catalpae*)
Catapion seniculus Kirby, 240
Catarrhal fever of sheep, 442
Catopsilia eubule Scudd., 514
Cattle, 302, 308
Caulleryella annulatae, 497; *C. anophelis*, 497; *C. aphiochaetae*, 497; *C. maligna*, 497; *C. pipientis*, 497; *C. crenata*, 495
Ceca, 14, **15**, 21, 22, 237, 238, 242, 243:
  bacteria in, **14-17**, 147; generation-to-generation transmission of, 15; species specificity of, 14
  function of, 16
Cecal bacteria (*see* Ceca)
Cecropia moth (see *Samia cecropia*)
Cediopsylla simplex (Baker), 165, 168
Celerio, 414; *C. lineata* (Fabr.), 61
Cellia allopha, 497
Cells, insect:
  blood–amebocytes, 200, **564**; leucocytes, 563, 564; macronucleocytes, 564; macrophages, 562; micronucleocytes, 564; microphages, 562; nephrocytes, 562; oenocytoids, 563,

Cells, insect (*Cont.*)
564; phagocytic, 562, 563, **564-569**; proleucocytes, 563; spherules, **564**, 568
other types of cells–cleavage, 194; fat, 247; fixed, 562; germ, 194; giant, 246, **569**; nurse, 192; pericardial, 244, **564**, 567; primordial germ, 192; wandering, 562; yolk, 194
Cellular immunity (*see* Immunity in insects, cellular)
Cellulose digestion by bacteria, 30, 31
Cephalina, 491-495
Cephalosporium, 393
Ceraiomyces, 385
Cerambycidae, 236, 370
Cerambycini, 371
Cerambyx scopolii Füssl., 370
Ceramica picta (Harr.), 65
Ceratitis capitata (Wied), 184
Ceratomia catalpae (Bdv.):
associated bacteria–*Bacillus immoblis*, 65; *B. noctuarum*, 75; *B. sphingidis*, 88, 89; *B. subtilis*, 89, 90; *Flavobacterium maris*, 115
Ceratomyces, 385
Ceratomycetaceae, 385
Ceratonyssus musculi, 254
Ceratophyllus acutus, 166 (see also *Diamanus montanus*)
Ceratophyllus anisus (see *Monopsyllus anisus*)
Ceratophyllus columbae Walk. & Ger., 498
Ceratophyllus fasciatus Bosc. (see *Nosopsyllus fasciatus*)
Ceratophyllus gallinae (Schrank):
associated bacterium–*Pasteurella pestis*, 162
associated protozoan–*Legerella parva*, 498
Ceratophyllus hirundinis Curtis:
associated protozoa–*Trypanosoma duttoni*, 479; *T. lewisi*, 478
Ceratophyllus laverani Roth., 479
Ceratophyllus lucifer Roth., 478
Ceratophyllus rectangulatus Wahlgren, 165
Ceratophyllus sciurorum Schrank, 466
Ceratophyllus sp., 466
Ceratophyllus tesquorum Wagner, 161, 164
Ceratopogon solstitialis Winn.:
associated protozoan–*Schizocystis gregarinoides*, 495, 497; *Taeniocystis mira*, 495
Ceratopogon sp.:
associated protozoa–*Spiroglugea octospora*, 518; *Toxogulgea vibrio*, 518
Ceratopogonidae, 167, 243
Ceratostomella ips, 410; *C. penicillata*, 410; *C. piceaperda*, 410; *C. pilifera*, 410; *C. pluriannulata*, 405; *C. pseudotsugae*, 410; *C. ulmi*, 405, 410
Cercomonas muscae domesticae, 471 (see also *Herpetomonas muscarum*)
Cercoplasma, 472
Cerecoccus, 391

Ceroplastes, 394; *C. egbarum fulleri* Ckll., 356; *C. pallidus* Brain, 356; *C. rusci* Linn., 356
Cetonia, 32, 483
Cetonia aurata (Linn.):
associated bacterium–*Coccobacillus cajae*, 183, 184
associated fungus–*Beauveria densa*, 394
Cetonia floricola Hbst., 100
Ceuthophilus brevipes Scudd., 492; *C. divergens*, 492; *C. gracilipes* (Hald.), 492; 493; *C. meridionalis* Scudd., 493; *C. nigricans*, 493; *C. pallidipes*, 493; *C. uhleri*, 494; *C. uniformis* Scudd., 493
Ceutorhynchus alliariae Bris., 240; *C. constrictus* Mrsh., 240; *C. pleurostigma* Mrsh., 240; *C. punctiger* Gyll., 239, 240; *C. sulcicollis*, 240; *C. symphyti* Bedel., 240
Chaetocnema confinis Crotch, 177; *C. denticulata* (Ill.), 177; *C. pulicaria* Melsh., 177
Chaetomyces, 385
Chagas' disease, 474
Chagasella alydi, 498; *C. hartmanii*, 498
Chaitophorus aceris, 228; *C. lyropictus*, 228
Chariesterus cuspidatus Dist., 473
Cheimatobia brumata Linn., 394
Chermes, 391; *C. abietis* Linn., 365; *C. cooleyi* Gill, 175; *C. strobilobius* Kalt., 365
Chermidae, 230
Chermomyces chermetis abietis, 365; *C. chermetis strobilobii*, 365
Chicken eggs, cultivation of microorganisms in, 293, 299, 300, 314, 315, 319, 323, 328, 334, 336, 343, 584
Chilo simplex (Butl.), 514
Chilomastix minuta, 538
Chimabache fagella, 413
Chinch bug, 396, 400 (see *Blissus leucopterus*)
Chionaspis, 391; *C. salicis* Linn., 374
Chipmunk, 286, 324, 337 (see also *Tamias striatus striatus*)
Chironomids (see *Chironomus*)
Chironomus, 19:
associated bacterium–*Serratia marcescens*, 133, 135
associated protozoan–*Nosema chironomi*, 515
Chironomus dorsalis Meig., 486
Chironomus plumosus Burrill:
associated bacteria–*Bacillus* sp., 20; *Pseudomonas fermentans*, 29; *Streptothrix* sp., 20
associated protozoa–*Amoeba chironomi*, 486; *Crithidia campanulata*, 470; *Glaucoma pyriformis*, 520; *Mrazekia brevicauda*, 518
associated spirochetes, 210, 461
Chitonomyces, 386
Chlamydozoa bombycis, 417; *C. prowazeki*, 417
Chlamydozoon conjunctivae, 304

Chlorops (*Musca*) *leprae* (Linn.), 46, 47; *C. vomitoria*, 46, 47
Choix fever, 329
Cholera, 178-179 (see also *Vibrio comma*)
Chorion, 193, 195
Chorthippus biguttulus Linn., 91, 93; *C. curtipennis*, 491 (see also *C. longicornis*); *C. curtipennis curtipennis*, 491; *C. dorsatus*, 91, 92; *C. longicornis* (Latr.), 492; *C. pulvinatus*, 91, 92
Chortophaga viridifasciata (DeG.), 492, 494; *C. viridifasciata* var. *austratior*, 412
Chromobacterium, 42, 180; *C. violaceum*, 180
Chromoderus fasciatus Müll., 239
Chromophile leucocytes (see Leucocytes)
Chromosomes, reduction of, 194
Chrysomela sanguinolenta Linn., 183, 184
Chrysomelidae:
  associated bacteria–*Bacillus subtilis*, 89, 91; *Bacterium intrinsectum*, 106
  intracellular symbiotes of, 236, 242-243
  mycetome of, 199
Chrysomelids, 17
Chrysomphalus, 391; *C. dictyospermi* (Morg.), 374
Chrysomyia megacephala (Fabr.):
  associated bacteria–*Eberthella typhosa*, 127, 128; *Escherichia coli*, 121, 122; *Shigella dysenteriae*, 131
Chrysopa oculata Say, 218
Chrysopidae, 218
Chrysops caecutiens Linn., 53; *C. discalis* Willist., 166; *C. dispar* Fabr., 160, 161; *C. noctifer* O.S., 166, 167; *C. relicta* Meig., 166
Chrysozona turkestanica, 166
Chytridiales, 377, 381-382
Cicada: symbiosis in, 191; mycetome of, 199
Cicada orni Linn.:
  associated yeasts–*Cicadocola cicadarum*, 359; *Cicadomyces cicadarum*, 361, 362
  mycetome of, 196
Cicadella viridis, 352
Cicadellidae, 226-227
Cicadellids, 190
Cicadocola, 359; *C. cicadarum*, 359, 360
Cicadomyces, 353, **359**; *C. aphalarae calthae*, 361; *C. aphrophorae alni*, 360; *C. aphrophorae salicis*, 360-361; *C. cicadarum*, 361, 362; *C. dubius*, 361, 362; *C. liberiae*, 361; *C. minimus*, 361; *C. minor*, 360; *C. ptyeli lineati*, 359, 360; *C. rubricinctus*, 361; *C. sulcii*, 365 (see also *Aphidomyces sulcii*)
Cicadula sexnotata Fall., 446
Cicadulina mbila (China), 447; *C. storeyi* China, 447; *C. zeae* China, 447
Cicindelidae, 385
Cidnorrhinus quadrimaculatus Linn., 240
Ciliata, 463, **520-522**, 537, 554

Ciliophora, 463, **519-522**
Cimex, 199, 206, 316
Cimex hemipterus (Fabr.):
  associated protozoan–*Leishmania donovani*, 473
  associated virus–yellow fever, 435
Cimex hirundinis Jenyns, 316
Cimex lectularius Linn.:
  associated bacteria–*Bacillus anthracis*, 53, 54; *B. mesentericus*, 72, 73; *B. mycoides*, 74; *B. subtilis*, 89, 91; *Bacterium tegumenticola*, 113; *Corynebacterium paurometabolum*, **44-45**, 233; *Micrococcus conglomeratus*, 145; *Mycobacterium leprae*, 46, 47; *Pasteurella pestis*, 164; *P. tularensis*, 48, 50, **165, 166**; *Salmonella paratyphi*, 130; *Sarcina flava*, 152; *Staphylococcus albus*, 153, 154; *S. aureus*, 155, 156
  associated protozoa–*Leptomonas ctenocephali*, 468; *L. pulicis*, 468
  associated rickettsiae–*Dermacentroxenus rickettsi*, 331; *Rickettsia lectularia*, 316-319; *R. prowazeki* var. *prowazeki*, 270; *R. prowazeki* var. *typhi*, 282
  associated spirochetes, 454
  associated viruses–lymphocytic choriomeningitis, 440; poliomyelitis, 441; yellow fever, 435
  bactericidal substance in, 34
  intracellular symbiotes in, 233, 235
  mycetome in, 199
  NR bodies in, 328
  sterile gut of, 12
Cimex pipistrelli Jenyns:
  associated protozoan–*Trypanosoma vespertilionis*, 479
  NR bodies in, 328
Cimex rotundatus, 316, 331 (see also *Cimex hemipterus*)
Cimicidae, 233
Cionus hortulanus Geoffr., 240; *C. scrophulariae* (Linn.), 240; *C. thaspi* Gyll., 240
Circulatory system of insects, 562-563; dorsal vessel of, 563; hemocele of, 562
Cirphis unipuncta (Haw.):
  associated bacterium–*Streptococcus disparis*, 140, 141
  associated viruses, 414, 425
Cis bidentatus Oliv., 492
Cissococcomyces, 362; *C. natalensis*, 362
Cissococcus fulleri Ckll., 362
Citellophilus tesquorum (Wagner), 161
Citellus, 270; *C. beecheyi*, 166; *C. citellus*, 343
Cladosporium, 393
Classification of rickettsiae, 262–268; of termites, 528
Claviceps purpurea, 411
Cleavage, 195

Cleavage cells, 194, 195
Clematomyces, 385
Cleoninae, 239
Cleonus clathratus Oliv., 239; *C. exanthineaticus,* 239; *C. maculatus,* 239; *C. mendicus* Gyll. (see *Temnorrhinus mendicus*); *C. ophthalmicus* Rossi, 239; *C. piger* Scop., 239; *C. punctiventris* Germ., 395, 396; *C. strabus* Gyll., 239; *C. verrucosus* Gebl., 239
Cletus varius Kl., 470
Clidiomyces, 386
Climatic bubo, 262
Clostridium, 41, 94; *C. botulinum* Type C, 94-95; *C. chauvoei,* 95; *C. luciliae* (see *C. botulinum* Type C); *C. novyi,* 95; *C. oedematiens,* **95,** 557; *C. perfringens,* **95,** 558, 567; *C. septicum,* 557; *C. sporogenes,* **95, 96,** 557; *C. tetani,* **96,** 557; *C. welchii,* 35, **95;** *C. werneri,* 96-97
Clotting of insect blood, 564
Clovia bipunctata, 374
Cnethcampa pityocampa:
  associated bacteria—*Bacterium pityocampae,* 111; *Streptococcus pityocampa alpha, beta,* 111, **142-143**
Cnidiosporidia, 487, 511-519
Coastal fever, 94, 291
Coati, 337
Coccidia, 487, **497-500,** 553
Coccidiascus legeri, 365
Coccidioides immitis, 409
Coccidioidomycosis, 409
Coccidomyces, 362; *C. dactylopii,* 362-363; *C. pierantonii,* 364 (see also *Icerymyces pierantonii*); *C. rosae,* 354 (see also *Lecaniocola rosae*)
Coccids, 199 (see also specific names)
Coccinella novemnotata Hbst.:
  associated bacteria—*Escherichia coli,* 121, 122; *Flavobacterium chlorum,* 114; *F. devorans,* 115; *Micrococcus ochraceus,* 149
Coccobacillus, 42, 137, **181;** *C. acridiorum,* 71, 81, 84, **181-183,** 561; *C. cajae,* 183-184; *C. ellingeri,* 110, **184,** 560, 566; *C. gibsoni,* 64, **185;** *C. insectorum* var. *malacosomae,* 185; *C. lymantriae,* 136, 137
Cocconema, 517 (see *Coccospora*)
Cocconematidae, 523
Coccospora micrococcus, 517; *C. octospora,* 517; *C. polyspora,* 517
Coccosporidae, 513, 517
Coccus hesperidum, 354 (see also *Lecanium hesperidum*)
Cochliomyia macellaria (Fabr.), 471
Cockchafers, 387 (see specific names)
Cockroaches, 12, 14, 32, 48, 162, 206, 207, **215-218,** 460, 484, 486, 511, 522, 576 (see also specific names)

Codling moth (see *Carpocapsa pomonella*)
Coelomyces, 381, 382; *C. anophelesica,* 382, 383; *C. indiana,* 382; *C. notonecte,* 382; *C. stegomyiae,* 381
Coelomomycetacea, 381
Coelosporidium blattellae, 511; *C. periplanetae,* 511
Coenagrionidae, 149
Coleoptera, 17, 138, 378, 385, 387, 393, 413, 443, 563; symbiosis in, 191, 206, 214, 215, **236-243**
Coleorhynchus heros, 495
Colepismatophila watsonae, 494
Colias eurytheme Bdv., 414, 425; *C. philodice* Godart, 414
Colibacillus paradoxus, 123
Collection of insects and ticks, 514
Colleterial glands, 193
Colombian spotted fever, 329
Colorado potato beetle (see *Leptinotarsa decemlineata*)
Colorado tick fever, 442
Columba livia, 311, 506
Columbicola columbae (Linn.), 219
Colymbetes, 495
Cometoides capitatus, 494
Commensalism, 5, 188
Commensals, 210
Complement, absence of in insects, 572
Complement fixation test for symbiotes, 203
Compsolixus ascanii Linn., 239
Compsomyces, 385
Conchylis ambiguella: associated fungus, 394; associated virus, 414
Conifers, blue stain of, 410-411
Coniocleonus glaucus Fahr. var. *turbatus* Fahr., 239
Conjunctivitis, 102, 304, 305
Conocephalus fasciatus var. *fasciatus* DeG.
  associated bacteria—*Aerobacter aerogenes,* 120; *Bacillus subtilis,* 89, 90; *Eberthella insecticola,* 127; *Flavobacterium acidificum,* 114
Conomelus limbatus Fabr., 358
Conorrhynchus mendicus, 59
Convergent lady beetle (see *Hippodamia convergens*)
Coptosoma scutellatum Geoffr., 17
Coptotermes, 528, 547, 554
Coptotermes ceylonicus Escherich, 550
Coptotermes corvignathus Kalshoven, 550
Coptotermes formosanus Shiraki:
  associated protozoa—*Holomastigotoides hartmanni,* 551; *Pseudotrichonympha grassii,* 550; *Spirotrichonympha leidyi,* 551
Coptotermes hartmanni Holmgren:
  associated protozoa—*Holomastigotoides hertwigi,* 551; *Pseudotrichonympha hertwigi,* 550

Coptotermes heimi Wasm., 554
Coptotermes lacteus Froggatt:
  associated protozoa–*Holomastigotoides hemigymnum*, 557; *Pseudotrichonympha hertwigi*, 550; *Spirotrichonympha flagellata* var. *coptotermitis lactei*, 551
Coptotermes sjöstedti Holmgren:
  associated protozoa–*Holomastigotoides mirabile*, 551; *Pseudotrichonympha hertwigi*, 550
Coptotermes travians Haviland, 550
Copulation, 193
Cordyceps, 387; *C. barnesii*, 387; *C. blattae*, 387; *C. clavulata*, 400; *C. coccinea*, 387; *C. cristata*, 388; *C. dipterigena*, 387; *C. flacata*, 387; *C. gonylepticida*, 388; *C. militaris*, 387, 389; *C. rhynchoticola*, 387; *C. robertsi*, 398; *C. thyrsoides*, 388
Coreidae, 14, 16
Coreomyces, 385
Corethra, 516; *C. plumicornis* Fabr., 516
Corethromyces, 385
Corn borer (see *Pyrausta nubilalis*)
Corn streak virus, 447
Cornfield ant (see *Lasius niger americanus*)
Coronilla emerus, 395
Coronympha clevelandi, 545; *C. octonaria*, 545
Corrodentia, 214, 218
Corthylus, 405; *C. columbianus* Hopk., 404; *C. punctatissimus* Zimm., 404
Corycella armata, 494
Corynebacterium, 41, 42; *C. blattellae*, 42, **216**; *C. diphtheriae*, **43**, 556, 558; *C. lipoptenae*, **43**, 206, 245; *C. ovis*, 44; *C. paurometabolum*, **44-45**, 233; *C. periplaneta* var. *americana*, 45, **216**; *C. pseudodiphthericum*, **46**, 319, 557; *C. sepedonicum*, 176
Coryssomerus capucinus Beck., 240
Cossus, 394
Cotinis nitida (Linn.):
  associated bacteria–*Bacillus popilliae*, 82, 83; *Micrococcus nigrofaciens*, 48
Cotton: internal boll disease, 412; wilt, 412
Cotton mouse, 285
Cotton rat, 285
Coxiella, 263; *C. burneti*, 295 (see also *Rickettsia burneti*)
Crapium, 410
Cricetus frumentarius, 499
Crickets, 183, 379 (see also specific names)
Criocephalus, 371
Cristospira, 451
Crithidia, 464, 465, **468-470**; *C. anophelis*, 469; *C. campanulata*, 470; *C. christophersi*, 470; *C. cleopatrae*, 470; *C. cleti*, 470; *C. ctenopthalmi*, 470; *C. euryophthalmus*, 469; *C. fasciculata*, 468; *C. gerridis*, 469; *C. haemaphysalidis*, 470; *C. haematopini*, 470; *C. hystrichopsyllae*, 470; *C. liturae*, 470; *C.*

minuta, 470; *C. nalipi*, 470; *C. nycteribiae*, 470; *C. ortheae*, 470; *C. oxycareni*, 470; *C. pulicis*, 470; *C. simuliae*, 470; *C. subulata*, 470; *C. tabani*, 470; *C. tenuis*, 470; *C. tullochi*, 470; *C. vacuolata*, 470; *C. xenopsyllae* (see *C. cleopatrae*)
Cross-immunity tests for rickettsiae, 587
Cryptocercus, 30, 207, 530, 531, 535, 536, 537
Cryptocercus punctulatus Scudd., 463:
  associated bacterium–*Bacillus subtilis*, 89, 90
  associated protozoa, 484, 498, 531, **535-537**; comparison of with termites, 537
  associated spirochetes, 461
Cryptorhynchus lapathi Linn., 175 (see also *Sternochetus lapathi*)
Cryptotermes, 528
Cryptotermes breviarticulatus Snyder:
  associated protozoa–*Devescovina lemniscata*, 541; *Foaina reflexa*, 544
Cryptotermes brevis Walk.:
  associated protozoa–*Devescovina striata*, 541; *Foaina humilis*, 543; *Hexamastix disclaviger*, 539
Crytotermes cavifrons Banks, 544
Cryptotermes cynocephalus Light:
  associated protozoa–*Devescovina parasoma*, 541; *D. transita*, 542; *Foaina nana*, 543
Cryptotermes darwini Light:
  associated protozoa–*Devescovina striata*, 541; *Foaina nana*, 543
Cryptotermes dolei Light, 550
Cryptotermes domesticus (Haviland):
  associated protozoa–*Devescovina lemniscata*, 541; *Foaina nana*, 543
Cryptotermes dudleyi Banks:
  associated protozoa–*Devescovina cometoides*, 540; *D. lemniscata*, 541; *Foaina nana*, 543; *F. solita*, 543; *Hexamastix claviger*, 539; *Microrhopalodina kofoidi*, 547; *M. multinucleata*, 547; *Pentatrichomonoides scroa*, 540; *Tricercomitus divergens*, 539
Cryptotermes fatulus Light:
  associated protozoa–*Devescovina striata*, 541; *Foaina nana*, 543
Cryptotermes grassii Silvestri:
  associated protozoa–*Devescovina striata*, 541; *Calonympha grassii*, 546; *Foaina humilis*, 543
Cryptotermes havilandi (Sjöstedt):
  associated protozoa–*Devescovina glabra*, 540; *Foaina reflexa*, 544; *Stephanonympha silvestrii* var. *cryptotermitis havilandi*, 546
Cryptotermes hermsi Kirby:
  associated protozoa–*Devescovina lemniscata*, 541; *Foaina nana*, 543; *Oxymonas parvula*, 546; *Stephanonympha nelumbium*, 546
Crytotermes longicollis (Banks):
  associated protozoa–*Devescovina lepida*, 541; *Foaina reflexa*, 544; *F. solita*, 543; *Penta-*

Cryptotermes longicollis (Banks) (*Cont.*)
  trichomonoides scroa, 540; *Snyderella tobogae*, 546; *Tricercomitus divergens*, 539
Cryptotermes merwei Fuller:
  associated protozoa–*Devescovina coghilli*, 541; *Foaina nucleoflexa*, 544
Cryptotermes piceatus Snyder:
  associated protozoa–*Devescovina striata*, 541; *Foaina humilis*, 543; *Hexamastix conclaviger*, 539
Cryptotermes queenslandis Hill:
  associated protozoa–*Devescovina lemniscata*, 541; *Foaina funifera*, 544; *F. solita*, 543
Cryptotermes spp.:
  associated protozoa–*Devescovina cometoides*, 540; *D. glabra*, 540; *D. lemniscata*, 541; *D. parasoma*, 541; *D. similis*, 542; *D. striata*, 541; *Foaina acontophora*, 544; *F. hamata*, 544; *F. nana*, 543; *F. nucleoflexa, F. parvula*, 543; *F. ramulosa*, 544; *F. solita*, 543
Ctenocephalides canis (Curtis):
  associated bacterium–*Pasteurella pestis*, 162
  associated protozoa–*Leptomonas ctenocephali*, 466; *Nosema ctenocephali*, 515; *N. pulicis*, 515; *Trypanosoma lewisi*, 478; *T. rabinowitschi*, 479
  associated rickettsia–*Rickettsia prowazeki* var. *typhi*, 281
  intracellular symbiotes in, 247
Ctenocephalides felis (Bouché):
  associated bacterium–*Pasteurella pestis*, 162
  associated protozoan–*Crithidia* sp., 470
  associated rickettsiae–*Rickettsia burneti*, 297; *R. ctenocephali*, 267, 320; *R. prowazeki* var. *typhi*, 281
  associated spirochete–*Spirochaeta ctenocephali*, 460
Ctenocephalus = *Ctenocephalides* (which see)
Ctenophthalmus agyrtes (Heller):
  associated bacterium–*Pasteurella pestis*, 161
  associated protozoa–*Crithidia ctenophthalmi*, 470; *Herpetomonas ctenophthalmi*, 466, 472; *Trypanosoma lewisi*, 478
Ctenophthalmus assimilis (Taschenb.):
  associated bacterium–*Pasteurella tularensis*, 165
  associated protozoan–*Trypanosoma rabinowitschi*, 479
Ctenophthalmus orientalis Wagner, 165, 168
Ctenophthalmus pollex, 165, 168
Ctenopsylla musculi, 466 (see *Leptopsylla segnis*)
Cubitermes spp.:
  associated protozoa–*Endamoeba granosa*, 554; *E. lutea*, 554; *E. pellucida*, 554; *Endolimax suggrandis*, 554
Cucujidae, 236, 241-242, 494
Cucumber mosaic, 445

Cucumis virus, 445
Culex apicalis Adams, 166, 168
Culex coronator Dyar & Knab, 438, 439
Culex fatigans, 321, 322, 436, 468 (see also *C. quinquefasciatus*)
Culex leprincei Dyar & Knab:
  associated protozoa–*Thelohania minuta*, 516; *T. rotunda*, 516
Culex nebulosus Theobald, 461
Culex pipiens Linn.:
  associated bacteria–*Bacillus thuringiensis*, 91, 92; *Pasteurella pestis*, 162
  associated protozoa–*Caulleryella pipientis*, 497; *Crithidia fasciculata*, 468; *Nosema culicis*, 515; *Plasmodium cathemerium*, 558; *P. relictum*, 558; *Stempellia magna*, 516; *Thelohania* sp., 516
  associated rickettsia–*Wolbachia pipientis*, 325, 327, 328
  associated spirochetes–*Borrelia anserina*, 458; unnamed spirochete, 461
  associated viruses–fowlpox, 442; lymphocytic choriomeningitis, 440; St. Louis encephalitis, 438, 439
Culex quinquefasciatus Say:
  associated nematode–*Wuchereria bancrofti*, 8
  associated protozoan–*Leptomonas culicis*, 468
  associated viruses–dengue fever, 436; St. Louis encephalitis, 439; western equine encephalomyelitis, 439
Culex restuans Theobald:
  associated protozoa–*Stempellia magna*, 516; *Thelohania opacita*, 516
Culex spp.:
  associated fungus–*Empusa culicis*, 380
  associated protozoa, 504, 505, 506; *Thelohania opacita*, 516
  associated spirochete–*Spirochaeta culicis*, 460
Culex stigmatosoma Dyar, 439
Culex tarsalis Coq.:
  associated bacterium–*Pasteurella tularensis*, 166, 168
  associated viruses–encephalitis, 438; St. Louis encephalitis, 438, 439; western equine encephalitis, 438, 439
Culex territans Walk., 328 (see also *C. restuans*)
Culex testaceus Wulp., 516
Culex thalassius Theobald, 433
Culicoides, 442, 474; *C. peregrinus*, 521; *C. sanguisugus* Malloch, 243
Culiseta (see *Theobaldia*)
Culiseta inornata, 438, 439
Cultivation of intracellular symbiotes, 204-206
Culture methods, 44, 204-206, 582, 583, 584
Curculionae, 240
Curculionidae:
  associated bacterium–*Bacillus subtilis*, 89, 91
  intracellular symbiotes in, 236, 237-240

mycteome in, 199, 237, 238
Curly top of beets, 445, 446
Cutworm septicemia, 75
Cutworms, 88 (*see also* specific names)
Cybister, 183, 184
Cycadomyces ptyeli lineati, 359
Cyclic endosymbioses, 206
Cyclocephala borealis, 82, 83
Cyclopodia sykesi, 470
Cynomya cadaverina Desvoidy, 116
Cyphocleonus tigrinus Panz., 239
Cyphomyrmex comalensis Wheeler, 406; *C. rimosus* Spinola, 408
Cyphon pallidulus Bohem., 494
Cyrtacantharcris ruficornis Fabr., 492
Cystocephalus algerianus, 494
Cystotrypanosoma, 472
Cytotrope, definition, 189
Cytotropic agent, 189

Dacne rufifrons Fabr., 493
Dactylopius, 362
Dactylosoma, 506
Dactylosphaera vitifoli, 230
Dacus oleae (Gmelin):
 associated bacteria–*Ascobacterium luteum*, 21, 181; *Xanthomonas savastanoi*, 20, 175
 bacteria as aid in digestion, 30
 ceca of, 21-22
 generation-to-generation transmission of bacteria in, 22
 head of, 23
 ovipositor of, 21
Danais (Danaus) erippus Dbld., 514; *D. gilippus* Cramer, 514
Danaus archippus = *Danaus plexippus*, 145, 146
Danysz bacillus, 130, 561 (see also *Salmonella enteritidis*)
Darkling beetles (*see* Tenebrionidae)
Dasyhelea brevitibialis Goetgr., 243; *D. flavifrons* Guér., 243; *D. longipalpis* Kieffer, 243; *D. obscura* Baum., 243, 351, 490, 518; *D. versicolor* Winn., 243
Datana angusii Grote & Robinson, 65; *D. ministra* Drury, 65
Debaryomyces tyrocola, 418
Decticus (Tettigonia) verrucivorus Linn., 394
Deer fly (see *Chrysops discalis*)
Defaunation of termites, 533-534
Deilephila lineata (Fabr.) (see *Celerio lineata*)
Deltatrichonympha numidica, 550; *D. operculata*, 550
Deltocephalus dorsalis Motsch, 447
Dematium, 357
Demodex, 46; *D. folliculorum*, 114
Dendroctonus monticolae Hopk., 349; *D. piceaperda* Hopk., 410; *D. ponderosae* Hopk., 349; *D. pseudotsugae* Hopk., 410

Dendrorhunchus systeni, 494
Dengue fever, 436-437
Depot mycetocytes (*see* Mycetocyte)
Dermacentor, 252
Dermacentor albipictus Pack.:
 associated bacteria, 13–*Corynebacterium ovis*, 44; *Klebsiella paralytica*, 123-124; *Pasteurella tularensis*, 166
 bacteriophage from, 450
 intracellular symbiotes in, 251
Dermacentor andersoni Stiles:
 associated bacteria, 13, 14–*Bacillus equidistans*, 62; *B. pseudoxerosis*, 84; *B. rickettsiformis*, 85; *Pasteurella tularensis*, 22, **166**, **167**; *Salmonella enteritidis*, 22, **129**, 130, 556; *Serratia marcescens*, 133, 135
 associated protozoan–*Anaplasma marginale*, 510, 511
 associated rickettsiae–*Dermacentroxenus rickettsi*, 330, 331, 332, 333; *Rickettsia burneti*, 259, **296**, 297; *R. dermacentrophila*, 322, 323, 324; *R. prowazeki* var. *typhi*, 282; *R. tsutsugamushi*, 292
 associated spirochetes–relapsing fever, 454
 associated viruses–Colorado tick fever, 442; equine encephalomyelitis, 438; lymphocytic choriomeningitis, 440
 associated yeasts, 349
 bacteriophage from, 450
 collection of 574
 intracellular symbiotes in, 251, 253
Dermacentor auratus, 292
Dermacentor modestus, 330 (see *D. andersoni*)
Dermacentor nitens Neum. (see *Otocentor nitens*)
Dermacentor nuttalli, 270
Dermacentor occidentalis Neum.:
 associated bacterium–*Pasteurella tularensis*, 166, 167
 associated protozoan–*Anaplasma marginale*, 510
 associated rickettsiae–*Dermacentroxenus rickettsi*, 330; *Rickettsia burneti*, 296
Dermacentor parumapertus Neum.:
 associated bacterium–*Pasteurella tularensis*, 166, 167
 associated rickettsia–*Dermacentroxenus rickettsi*, 330
Dermacentor parumapertus margenatus (see *D. parumapertus*)
Dermacentor reticulatus (Fabr.):
 associated protozoa–*Babesia caballi*, 507; *B. canis*, 507; *B. equi*, 507
Dermacentor silvarum Olen.:
 associated bacteria–*Pasteurella pestis*, 162, 165, 166, 167; *P. tularensis*, 166, 167
 associated rickettsia, 270
Dermacentor variabilis (Say):

Dermacentor variabilis (Say) (*Cont.*)
associated bacterium—*Pasteurella tularensis*, 22, 166, 167, 168
associated rickettsia—*Dermacentroxenus rickettsi*, 330, 333
associated viruses—St. Louis encephalitis, 438; western equine encephalomyelitis, 439
intracellular symbiotes in, 251, 253
Dermacentor venustus Banks, 330 (see *Dermacentor andersoni*)
Dermacentroxenus, 257, 262, 263, 264, 265, 280, 331
Dermacentroxenus conori, 267, 308, 329; cultivation of, 342-343; disease caused by, 339-340; immunologic aspects, 343-344; generic relations to *D. rickettsi*, 340; morphology of, 342; pathogenicity of, 343; staining, 342; synonyms, 339; ticks associated with, 340-342
Dermacentroxenus orientalis, 290
Dermacentroxenus rickettsi, 249, 263, 265, 267, 292, 297, 322, **328**, 340, 342; blood virus, 331; cultivation of, 335-336; disease caused, 260, 329; functional phases, 332; generation-to-generation transmission, 333; growth of, 335-336; immunity of ticks to, 556; immunologic aspects, 339; longevity of, 336-337; morphology, 334-335; pathogenicity, 210, 337-338; properties of, 336-337; serum virus, 331; staining, 334-335; synonyms, 328; tick virus, 331; ticks and insects associated with, and biologic relationships involved, 330-334; toxin of, 339
Dermacentroxenus rickettsi var. *conori*, 339 (see *D. conori*)
Dermacentroxenus rickettsi var. *pijperi*, 339 (see *D. conori*)
Dermacentroxenus typhi, 339 (see *Rickettsia prowazeki* var. *typhi*)
Dermanyssidae, 254
Dermanyssus avium Gervais, 255, 457; *D. gallinae*, 438; *Dermanyssus* sp., 255, 499
Dermatophilus penetrans (see *Tunga penetrans*)
Dermestes lardarius Linn.:
associated protozoa—*Beloides firmus*, 495; *Pyxinia bulbifera*, 495; *P. rubecula*, 495
associated virus, 413
Dermestes peruvianus Cast., 495
Dermestes vulpinus Fabr.:
associated bacterium—*Bacillus anthracis*, 53, 54
associated protozoa—*Pyxinia cystalligera*, 495; *P. rubecula*, 495
Dermestidae, 413
Derocalymma, 217; *Derocalymma stigmosa*, 216
Desmometopa nigra Zett., 29
Desvoidea obturbans, 436

Deuteromycetes, 376, 393-396
Deutomerite, 491, 497
Devescovina, 529; *D. arta*, 541; *D. coghilli*, 541; *D. cometoides*, 540; *D. elongata*, 540; *D. exiiis*, 541; *D. fissa*, 540; *D. glabra*, 540; *D. hawaiensis*, 541; *D. insolita*, 541; *D. lemniscata*, 529, 541; *D. lepida*, 541; *D. parasoma*, 541; *D. robusta*, 541; *D. similis*, 542; *D. striata*, 541; *D. tendicula*, 540; *D. transita*, 541-542; *D. uniflexura*, 541; *D. vestita*, 541; *D. vittata*, 541
Devescovininae, 531
Diabrotica duodecimpunctata (Fabr.):
associated bacteria—*Erwinia tracheiphila*, 119; *Xanthomonas stewarti*, 177
Diabrotica longicornis (Say), 177
Diabrotica soror LeC., 116
Diabrotica vittata (Fabr.):
associated bacterium—*Erwinia tracheiphila*, 119, 120
associated protozoan—*Gregarina diabrotica*, 493
associated virus, 445
Dialeges pauper, 371
Dialeurodes citri (Riley & Howard) (see *Aleurodes citri*)
Diamanus montanus (Baker):
associated bacteria—*Pasteurella pestis*, 161, 163; *P. tularensis*, 166
Diapheromera femorata (Say):
associated bacteria—*Bacterium hebetisiccus*, 104; *Flavobacterium rheni*, 115
associated yeasts, 349
Diaspidinae, 373
Dichomyces, 386
Dichromorpha viridia, 494
Dicnidea, 513, 518
Didelphis virginiana, 285
Dieuches humilis, 473
Dilixellus barbanae Fabr., 239
Dimorormyces, 386
Dimorphomyces, 386
Dinenympha, 548; *D. exilis*, 548; *D. fimbriata*, 548; *D. gracilis*, 548; *D. leidyi*, 548; *D. nobilis*, 548; *D. parva*, 548; *D. porteri*, 548; *D. rugosa*, 548
Dinopsyllus lypusus Jordan & Roth., 161, 164
Dioecomyces, 385
Dione juno Cramer, 514; *D. vanillae* (Linn.), 514
Dioptidae, 414
Diplobacillus, 42, **185**; *D. melolonthae*, 185; *D. pieris*, 79, 186
Diplococcus, 41; *D. bombycis*, 136; *D. gonorrhoeae*, 157 (see also *Neisseria gonorrhoeae*); *D. intracellularis*, 158 (see also *Neisseria intracellularis*); *D. liparis*, 136; *D. lymantriae*, 136-137; *D. melolonthae*, 137; *D. pemphigi*

SUBJECT INDEX

contagiosi, 137; *D. pemphigocontagiosus*, 137; *D. pieris*, 79, **137**; *D. pluton*, 81 (see also *Bacillus pluton*); *D. pneumoniae*, **137-138**, 557, 567
Diplocystis major, 490; *D. minor*, 490; *D. schneideri*, 490
Diplodia griffoni, 412
Diplomyces, 385
Diplonympha parvulus, 546
Diprion rufus, 144; *D. sertifer* Geoffr., 86
Diprionidae, 414
Diptera, 215, **243-246**, 346, 350, 378, 385, 387, 414
Discorhynchus truncatus, 495
Diseases of insects: bacterial, see ch. III, 43-187; fungous, 396-402; protozoan, 522-525; virus, 413-432
Disinfectants, 579
Disonycha glabrata (Fabr.), 177; *D. triangularis* (Say), 444
Dissection of insects and ticks, 576-579
Dissosteira carolina (Linn.):
  associated fungus–*Fusarium vasinfectum*, 412
  associated protozoa–*Actinocephalus fimbriatus*, 494; *Gregarina locustae*, 491; *G. rigida columna*, 492; *G. rigida rigida*, 491
Distichomyces, 385
Dixippus morosus, 520
Dobellina mesnili, 485 (see also *Entamoeba mesnili*)
Dociostaurus maroccanus Thunb., 181
Docophorus icterodes Nitz., 219; *D. leontodon*, 219
Dog, 300, 306, 307, 308, 325, 337, 338, 343, 345
Dog louse (see *Linognathus piliferus*)
Dolichopsyllidae, 247
Donacia, 199, 242
Donkey, 277
Dorcus, 32
Dorsal vessel, 563
Dorypteryx pallida, 218
Dorytomus longimanus Forst., 240; *D. melanophthalmus* Payk., 240
Dothideales, 383, 387
Downy mildew of lima beans, 412
Drepanosiphum, 346; *D. platanoides* Schrank, 210, 228, 229
Dreyfusia nordmannianae Eckstein, 229; *D. picea* (Ratz.), 229
Drone fly (see *Eristalis tenax*)
Drosophila, 33, 349, 350, 471
Drosophila ampelophila Loew:
  associated bacterium–*Erwinia amylovora*, 116, 117
  associated fungus–*Aspergillus niger*, 411
  associated protozoan–*Herpetomonas ampelophilae*, 472

associated yeasts, 349, 350
Drosophila confusa Staeger:
  associated protozoa–*Herpetomonas drosophilae*, 472; *H. roubaudi*, 472; *Octosporea monospora*, 518; *O. muscae-domesticae*, 518
  associated spirochete–*Spirochaeta drosophilae*, 460
Drosophila funebris (Fabr.):
  associated bacterium–*Erwinia amylovora*, 116
  associated yeast, 365
Drosophila melanogaster (see *D. ampelophila*)
Drosophila plurilineata:
  associated protozoa–*Octosporea monospora*, 518; *O. muscae–domesticae*, 518
Drosophila rubrostriata, 472
Drosophilidae, 117
Drug store weevil (see *Stegobium paniceum*)
Duboscqia legeri, 552
Ductus ejaculatorius, 193
Dumdum fever (*see* Kala-azar)
Dutch elm disease, 405, 409-410
Dwarf disease of rice, 447
Dysdercus, 350, 412; *D. cingulatus* (Fabr.), 350; *D. fasciatus* Signoret, 350; *D. intermedius*, 350; *D. nigrofasciatus* Stål, 350; *D. ruficollis* (Linn.), 498; *D. superstitiosus*, 358; *D. suturellus* Scha., 15
Dysentery, bacillary, 131, 132
Dytiscus, 183, 184

Eacles imperialis (Dru.):
  associated bacteria–*Alcaligenes ammoniagenes*, 180; *Bacterium imperiale*, 105; *Micrococcus ochraceus*, 149
East Coast fever of cattle, 507
Eastern spotted fever, 329
Eastern tent caterpillar (see *Malacosoma americana*)
Eberthella, 41, 126; *E. belfastiensis*, 126; *E. insecticola*, 127; *E. pyogenes*, *E. typhosa*, 12, 19, 127-128, 131, 449, 557, 567
Echidnophaga gallinaceae (Westwood), 281
Echocerus cornutus (Fabr.), 91
Ectinomyces, 385
Ectobia lapponica Linn., 482
Ectobia livida Fabr., 216
Egg, insect, 222, 235, 237, 245, 250, 253, 323, 327 (*see also* Chicken egg); bacteria in, 24; cells of, 191, 248; development of, 193-196; fertilization of, 191, 193; hatching of, 22, 23; holoblastic, 194; intracellular symbiotes in, 190, 206, 235; micropyle of, 193; nucleus of, 193, 194, 195; toxic agent in, 572; tube, 191, 192, 193; yolk of, 194, 198
Ehrlichia, 263; *E. bovis*, 308; *E. canis*, 306; *E. (Rickettsia) kurlovi*, 266, 325; *E. ovina*, 310
Eimeridea, 498
Ejaculatory duct, 194

SUBJECT INDEX

Elachiptera costata Loew, 118
Electrocollargol, 288
Electron microscope, 275, 419, 420
Eleodes, 494
Elephant louse (see *Haematomyzus elephantis*)
Elleschus bipunctatus Linn., 240
Ellopia fiscellaria lugubrosa, 414, 425
Enbadomonas agilis (see *Retortamonas agilis*);
  E. alexeieff, 483
Empoasca maligna (Walsh), 116
Empusa, 376, **377-380**; *E. aphidis*, 380; *E. aulicae*, 183; *E. conglomeratua*, 380; *E. culicis*, 380; *E. delpiniana*, 380; *E. fresenii*, 401; *E. grylli*, 379; *E. muscae*, 377, **379**, 380, 381; *E. radicans*, 380; *E. sciarae*, 380
Empusaceae, 377
Enarthromyces, 386
Encephalitides, 437-440 (*see also* Encephalitis)
Encephalitis, 437-440; equine, 437; Japanese B, 437; mosquitoes associated with, 437; Russian spring-summer, 437; St. Louis, 437; ticks associated with, 438; Venezuelan type, 437; verno-aestival (*see* Russian spring-summer)
Encephalomyelitis, 437
Encoptolophus sordidus (Burm.):
  associated bacterium–*Bacillus poncei*, 81, 82
  associated protozoa–*Gregarina nigra*, 491; *G. rigida columna*, 492; *G. rigida rigida*, 491
Encoptolophus texensis, 412
Endamoeba, 485; *E. beaumonti*, 553; *E. blattae*, 485; *E. disparata*, 553; *E. granosa*, 554; *E. javanica*, 537; *E. lutea*, 554; *E. majestas*, 553; *E. pellucida*, 554; *E. philippinensis*, 537; *E. sabulosa*, 553; *E. simulans*, 553; *E. thomsoni* (see *Entamoeba thomsoni*)
Endamoebidae, 484, 485, 553
Endolimax, 485; *E. goheeni*, 554; *E. suggrandis*, 554; *E. termitis*, 554
Endosclerotium pseudococci, 400
Endosepsis of figs, 411
Endosymbiote, 202
Endromis versicolora Linn., 18
Endrosis fenestrella Zell., 494
Ennomos autumnaria Möschler, 18
Enochrus hamiltoni (Horn), 94
Ensign coccids (*see* Orthezeiidae)
Entamoeba, 485; *E. apis*, 485; *E. histolytica*, 485; *E. mesnili*, 485; *E. minchini*, 485; *E. nana*, 484; *E. thomsoni*, 485
Enterobacillus larvae, 98
Enterobacteriaceae, 41, 116
Enterococcus, 42, 186; *E. citreus*, 186
Enterocystis ensis, 490
Entomophthora, 377, **378-380**; *E. sphaerosperma*, 380
Entomophthorales, 377-381
Eomenacanthus stramineus (Nitz.), 200

Epacromia strepens, 183, 184
Epeorus torrentium, 516
Eperythrozoön, 262
Ephemera, 514
Ephemera vulgata Linn.:
  associated protozoa–*Nosema schneideri*, 514; *Stempellia mutabilis*, 516; *Telomyxa glugeiformis*, 518
Ephemerella ignita, 515
Ephestia kühniella Zell.:
  associated bacteria–*Bacillus agilis*, 51; *B. thuringiensis*, 91, 92, 93; *Bacterium canadensis*, 99, 100; *B. cazaubon*, 100; *B. ephestiae*, 103; *B. galleriae*, 103; *B. ontarioni*, 110; *B. pyrenei*, 112; *Micrococcus curtissi*, 145; *M. ephestiae*, 145
  associated protozoa–*Mattesia dispora*, 497; *Theohania ephestiae*, 516, 524
  mycetome, transplanted, 200
Ephialtes angulosa, 514
Epicalotermes aethiopicus (see *Kalotermes aethiopicus*)
Epicauta cinerea marginata (Fabr.), 121
Epicauta pennsylvanica (DeG.):
  associated bacteria–*Aerobacter cloacae*, 120, 121; *Alcaligenes ammoniagenes*, 180; *Xanthomonas sepedonica*, 176
Epidemic series, steps in, 260
Epididymis, 193
Epilachna borealis (Fabr.), 119
Epilampra grisea DeG., 216
Epimerite, 489
Epitrix cucumeris (Harris):
  associated bacterium–*Edwinia tracheiphila*, 119
  associated virus, 444
Equine infectious anemia, 442
Eratyrus cuspidatus Stål, 477
Ergot, 411
Eriococcidae, 231-232
Eriogaster lanestris Linn.:
  associated bacteria–*Diplococcus bombycis*, 136; *D. melolonthae*, 137
Eriophyes ribis Nal, 449
Eriosoma lanigerum (Hausm.), 227
Eristalis tenax (Linn.), 178
Ernobius abietis Fabr., 367, 369; *E. mollis* Linn., 367, 369
Eruptive fever (see *Fièvre boutonneuse*)
Erwineae, 41, **116**
Erwinia, 41, 116; *E. amylovora*, 116; *E. aroideae*, 118, 119; *E. cacticida*, 117-118; *E. carnegieana*, 118; *E. carotovora*, 20, **118-119**; *E. lathyri*, 119; *E. salicis*, 119; *E. tracheiphila*, 119-120
Erytrapion, 238; *E. miniatum* Germ., 240
Eschericheae, 41, 120
Escherichia, 41, 121

Escherichia coli, 59, 556, 557, 558; associated insects, 121-122; bacteriophage against, 449; effect on hatching of insect eggs, 24; as food for mosquitoes, 28; as immunizing agent, 560, 561; var. *acidilactici*, 122; var. *communior*, 122; var. *neapolitana*, 122
Escherichia coli mutabile, 122
Escherichia ellingeri, 184 (see also *Coccobacillus ellingeri*)
Escherichia freundii, 122
Escherichia noctuarii, 75 (see also *Bacillus noctuarum*)
Escherichia paradoxa, 123
Escherichia sphingidis, 89 (see also *Bacillus sphingidis*)
Espejoia mucicola, 521
Espundia (see *Leishmania braziliensis*)
Eubacteriales, 41, 50
Eucantharomyces, 386
Euchlaena mexicana, 177
Euchloris vernaria Hbst., 18
Eucomonympha imla, 484, 536
Eucorethromyces, 385
Eucryptotermes, 528
Euglena deses, 37
Eugregarinaria, **488-495**, 496, 523
Euhaplomyces, 386
Eulixus iridis Oliv., 239; *E. myagri* Oliv., 239; *E. subtilis* Strm., 239
Eumonoecomyces, 386
Eupedicinus longiceps, 440
Euphorbia hypericifolia, 472; *E. pilulifera*, 472; *E. thymifolia*, 472
Euproctis chrysorrhoea Linn., 25 (see also *Nygmia phaeorrhoea*); *E. flava* Bremer, 394
Eupterix auratus, 445
Euricania ocellus Walk., 374
European corn borer (see *Pyrausta nubilalis*)
European foul brood (see Foul brood, European)
Eurycotis, 207, 216; *E. floridana* Walk., 484
Eurydema ornata Linn., 183, 184
Eurygaster maura (Linn.), 17
Eurymus eurytheme (see *Colias eurytheme*); *E. philodice* (see *Colias philodice*)
Euryophthalmus convivus (Stål):
associated bacterium–*Xanthomonas maculicola*, 174
associated protozoan–*Crithidia euryophthalmus*, 469
Eusattus, 494
Euscelis bicolor, 177; *E. striatulus* (Fall.), 448
Euschistus servus Say, 16
Eutettix tenellus (Bak.): associated viruses, 446; intracellular symbiotes in, 226, 227
Eutriatoma flavida Neiva, 233; *E. uhleri*, 331
Eutrichapion loti Kirby, 240; *E. virens* Hbst., 240

Eutrichomastix axostylis, 539; *E. passali*, 483; *E. phyllophagae*, 483; *E. termitis*, 539; *E. trichopterorum*, 483
Euura atra, 119
Euxoa, 570
Euxoa ochrogaster (Guen.):
associated bacteria–*Bacillus noctuarum*, 75; *Bacterium ochraceum*, 110
Euxoa segetum Schiff.:
associated bacteria–*Achromobacter larvae*, 98; *Aerobacter aerogenes*, 120; *Bacillus agrotidis typhoides*, 51; *B. melolonthae liquefaciens*, 71, 72; *B. melolonthae non liquefaciens*, **72**, 560, 571; *Bacterium lymantricola adiposus*, 26, 106; *B. melolonthae liquefaciens*, 108; *B. ochraceum*, 110; *B. paracoli*, 111; *B. viscosum non liquefaciens*, 113; *Enterococcus citreus*, 186; *Micrococcus neurotomae*, 147, 148; *M. saccatus*, 151; *Pseudomonas fluorescens*, 170, 171; *P. non-liquefaciens*, 171; *P. septica*, 172; *Salmonella paratyphi*, 130; *Sarcina flava*, 152
associated fungus–*Sorosporella uvella*, 395
associated viruses, 414–polyhedral diseases, 415, 426; pseudojaundice, 415, **430-432**
Euxoa tessellata Harr., 395
Euzodiomyces, 385
Exosymbiote, 202
Expaion fuscirostre Fabr., 240
Extracellular bacteria (see Bacteria, extracellular)

False blossom of cranberry, 448
Fannia canicularis Linn., 471; *F. scalaris* (Fabr.), 516, 518
Fat body, 199, 207, 208, 219, 229, 230, 237, 243
Fat cells (see Cells, insect, fat)
Feltia, 75; *F. jaculifera*, 395; *F. subgothica* (Haw.), 395
Fermentation chambers, 31, 32
Fettraupen, 417
Fettsucht, 417 (see Jaundice)
Fibrin in insect blood, 563
Fièvre boutonneuse, 329, **339**, 340, 344
Fièvre escharonodulaire, 340
Figs, diseases of, 411
Fiji disease of sugar cane, 448
Filariasis, 8
Filippia chilianthi Brain, 357
Fire ant (see *Solenopsis geminata*)
Fire blight, 3, 116, 117
Firefly (see *Photinus pyralis*)
Fish, 311; rickettsia in, 311
Fission, multiplication by, 227
"Five-day fever," 287 (see *Rickettsia wolhynica*)
Fixed cells (see Cells, insect, fixed)

Flacherie, **57,** 60, 61, 70, 88, 90, 135, 140, **415,** 424
Flagellates, 463 (*see also* Mastigophora); associated bacteria, 36-37; cellulose digestion in termites, 30; occurrence in termites, 538-552
Flat beetles (*see* Cucujidae)
Flavobacterium, 41, 69, 114; *F. acidificum*, 114; *F. chlorum*, 114; *F. devorans*, 115; *F. fermentans* (see *Pseudomonas fermentans*); *F. maris*, 115; *F. ochraceum*, 110; *F. rheni*, 115
Flea beetle, 400 (see *Haltica*)
Flea larva, 19
Flea typhus, 264
Fleas, 3, 161, 167, 260, 270, 281, 441 (*see also* specific names); blocking mechanism in, 163-164; vector efficiency of, 163
Flebotomus (see *Phlebotomus*)
Flexilis, 61
Flies, 3, 53, 161, 163, 169, 379, 385, 387, 388 (*see also* specific names); blowfly, 102, 379; fleshfly, 102; Ichneumon, 249; Hippelates, 138; horsefly, 169, 243, 244; housefly, 9, 10, 49, 349, 379; stablefly, 169; whitefly, 400
Flood fever, 290 (see *Rickettsia tsutsugamushi*)
Fluorescent microscopy, 584
Flying squirrel (see *Glaucomys volans saturatus*)
Foa-Kurloff bodies, 325
Foaina, 529; *F. acontophora*, 544; *F. appendicula*, 544; *F. decipiens*, 544; *delicata*, 544; *F. dogieli*, 544; *F. duo*, 544; *F. exempta*, 543; *F. falcifera*, 544; *F. funifera*, 544; *F. gracilis*, 543; *F. grassii*, 543; *F. hamata*, 544; *F. hilli*, 544; *F. humilis*, 543; *F. minuscula*, 544; *F. nana*, 543; *F. nucleoflexa*, 544; *F. ovata*, 543; *F. parvula*, 543; *F. pectinata*, 544; *F. ramulosa*, 544; *F. reflexa*, 544; *F. signata*, 544; *F. solita*, 543; *F. taeniola*, 543
Follicle cells, 230
Follicular epithelium, 192, 197, 198
Forest tent caterpillar (see *Malacosoma disstria*)
Forficula auricularia Linn., 493
Formica, 248, 249; *F. fusca* Latr., 199, 209, 248, 249; *F. fusca glebaria* Nyl. Em., 248; *F. fusca subsericea* (Say), 116; *F. pallidefulva schaufussi* var. *incerta*, 116; *F. rufa* Linn., 248; *F. sanguinea* Latr., 248
Formicidae, 247-249
Fossilis quintana, 287
Foul brood: American, 66-67; European, 25, 66, 67, 76, **80-81,** 97, 98; para-, **77,** 139
Fowlpox, 442
Fox, gray, 286
Frankel's diplo bacillus, 138
Frankliniella lycopersici Andrewarther, 448; *F. moultoni* Hood, 448; *F. occidentalis* (Perg.), 448

Friedländer's bacillus, 124
Frogs, 325
Fungi, **375-412,** 570; ambrosia, 402-406; and ants, 406-408; diseases of insects caused by, 396-402; examination of insects and ticks for, 589; and termites, 408-409
Fungi Imperfecti, 367, **393-396**
Fungous disease of insects, 396-402; optimum conditions for, 397-398; practical use of, 398; symptoms of, 396-397
Fusarium, 393; *F. moniliforme*, 24; *F. moniliforme* var. *fici*, 411; *F. vasinfectum*, 412
Fusiformis, 114, 115; *F. hilli*, 61, 115; *F. termitidis*, 61, 116
Fusobacterium, 41, 115

Gaffkya, 41, 144; *G. tetragena*, 144
Galleria mellonella (Linn.):
 associated bacteria–*Bacillus anthracis*, 561; *B. megatherium*, 557; *B. mesentericus*, 72, 73; *B. mycoides*, **74,** 558, 567; *B. subtilis*, 89, **90,** 557; *Bacterium canadensis*, 99, 100; *B. galleriae*, **103,** 561; *B. ontarioni*, 110; *B. tumefaciens*, 113; *Clostridium novyi*, 95; *C. oedematiens*, **95,** 557; *C. perfringens*, **95,** 557, 567; *C. septicum*, 557; *C. sporogenes*, 95, **96,** 557; *C. tetani*, **96,** 557; *C. welchii*, 95; *Coccobacillus acridiorum*, 561; *C. ellingeri*, 184; *Corynebacterium diphtheriae*, **43,** 557, 558; *C. pseudodiphthericum*, **46,** 557; *Diplococcus pneumoniae*, 137, **138,** 557, 567; *Eberthella typhosa*, 127, **128,** 557, 567; *Escherichia coli*, 121, **122,** 557, 558, 560, 561; *E. paradoxa*, 123; *Hemophilus influenzae*, 557, 567; *Micrococcus curtissi*, 145; *M. galleriae*, 146, 561; *Mycobacterium avium*, 557; *M. lacticola*, **46,** 557, 567; *M. tuberculosis*, 48, **49, 50,** 557, 567; *M. tuberculosis* var. *bovis*, 557; *M. tuberculosis* var. *hominis*, 557; *Neisseria gonorrhoeae*, **157,** 557; *Pasteurella avicida*, 557; *P. aviseptica*, 557; *P. pestis*, 557; *Proteus vulgaris*, **126,** 557, 558, 567; *Pseudomonas aeruginosa*, 169, **170,** 557, 558; *Salmonella enteritidis*, 129, **130,** 561; *S. schottmülleri*, **130,** 557; *Serratia marcescens*, 132, **134,** 557; *Shigella dysenteriae*, **131,** 557, 567; *Staphylococcus albus*, 153, **154,** 557, 567; *S. aureus*, 155, **156,** 557, 567; *Streptococcus faecalis*, **141,** 558; *S. galleriae*, 142; *S. pyogenes*, **143,** 558; *Vibrio comma*, 178, **179,** 557, 560, 561, 567, 572; *V. leonardi*, 179
 associated fungi–*Aspergillus flavus*, 393; *A. niger*, 394; *Beauveria bassiana*, 393; *Spicaria farinosa*, 394
 associated protozoa–*Glaucoma pyriformis*, 520; *Trypanosoma brucei*, 557
 beeswax, utilization of, 30
 blood cells of, **564**

characteristics of, 566
immunity in, 556, 557, 559, 560, 561, 562, 564, 566, 567, 569, 571, 572
microbe-free, 29
Gamasidae, 166, 168 (*see also* specific names)
Gametocyst, 496, 497
Gametocyte, 488, 490, 496, 497, 498, 501, 502, 503
Gamocystis polymorpha, 493; *G. steini*, 493; *G. tenax*, 493
Gastroenteritis, 129
Gattine, 140, 415
Gecko (see *Tarentola mauritanica*)
Gelbsucht, 417 (*see* Jaundice)
Gelechia gossypiella, 104 (see also *Platyedra gossypiella*)
Geneiorhynchus aeschanae, 495
Generation-to-generation transmission: of bacteria in insects, 15, 16, 20-22, 44, 130, 167; of intracellular symbiotes in insects, **206-209**, 218, 219, 222, 223, 225, 226, 230, 231, 232, 236, 237, 241, 242, 245, 247, 250, 254, 265; of protozoa in insects and ticks, 506, 507, 511, 524; of rickettsiae in insects and ticks, 264, 292, 297, 306, 314, 317, 327, 333, 341, 345; of spirochetes in ticks, 457; of viruses in insects, 438, 439, 440, 447; of yeasts in insects, 356, 363, 368, 372, 374
Genital ducts, 191
Geometridae, 414
Gerbil (*Gerbillus indicus*), 164, 277, 343, 499
Germ band, 194
Germ cell, 194, 195, 196
German cockroach (see *Blattella germanica*)
Germarium, 192
Germinal disc, 194
Gerris fossarum, 469; *G. marginatus* Say, 469; *G. paludum* Fabr., 469; *G. remiges* Say, 469; *G. rufoscutellatus*, 469
Giant cell area, 569
Giant cells (*see* Cells, insect, giant)
Giardia lamblia, 484
Gilpinia hercyniae (Htg.), 414
Gland: milk, 206, 245; rectal, 244; salivary, 247, 250
Glanders, 159
Glaucoma pyriformis, 520
Glaucomys volans saturatus, 286
Gliricola, 219
Glischrochilus emarginatus, 411; *G. fasciatus*, 116; *G. grandicollis*, 410; *G. integer*, 411; *G. oregoni*, 411; *G. pini*, 410
Gloeosporium perennans, 412
Glomerella grossypii, 24
Glossina, 206, 212, 246
Glossina austeni, 480
Glossina brevipalis Newstead:
associated protozoa–*Trypanosoma caprae*, 481;

T. congolense, 480; *T. rhodesiense*, 476; *T. simiae*, 481
Glossina fusca (Walk.):
associated protozoa–*Crithidia* sp., 470; *Trypanosoma gambiense*, 474
Glossina longipalpis (Wied.):
associated protozoa–*Trypanosoma congolense*, 480; *T. vivax*, 481
Glossina morsitans Westwood:
associated protozoa–*Crithidia* sp., 470; *Trypanosoma caprae*, 481; *T. congolense*, 480; *T. gambiense*, 474; *T. rhodesiense*, 476; *T. simiae*, 481; *T. vivax*, 481
Glossina pallidipes Austen, 474
Glossina palpalis (Rob.-Desv.):
associated protozoa–*Crithidia tullochi*, 470; *Hepatozoon* sp., 499; *Trypanosoma congolense*, 480; *T. gambiense*, 414; *T. grayi*, 482; *T. rhodesiense*, 476; *T. uniforme*, 481; *T. vivax*, 481
associated spirochete–*Borrelia glossinae*, 460
intracellular symbiotes in, 246
Glossina spp.:
associated protozoa–*Herpetomonas grayi*, 472; *Trypanosoma evansi*, 481
mycetome of, 199
Glossina submorsitans Newstead:
associated protozoan–*Trypanosoma gambiense*, 474
intracellular symbiotes in, 246
Glossina swynnertoni Austen, 476
Glossina tachinoides Westwood:
associated protozoa–*Trypanosoma congolense*, 480; *T. gambiense*, 474; *T. vivax*, 481
Glycogenase, 31
Glyptotermes, 528
Glyptotermes angustus Snyder:
associated protozoa–*Devescovina arta*, 451; *Foaina solita*, 543; *Metadevescovina modesta*, 545; *Oxymonas barbouri*, 547; *Tricercomitus divergens*, 539; *Trichomonas barbouri*, 540
Glyptotermes barbouri, 539, 540
Glyptotermes borneensis Haviland, 550
Glyptotermes brevicaudatus Haviland:
associated protozoa–*Macrotrichomonas ramosa*, 542; *Trichonympha chattoni*, 549
Glyptotermes brevicornis Froggatt:
associated protozoa–*Foaina delicata*, 544; *F. hilli*, 544; *Macrotrichomonas pulchra*, 542; *Trichonympha chattoni*, 549
Glyptotermes caudomunitis Kemner:
associated protozoa–*Devescovina glabra*, 540; *Foaina hamata*, 544; *Macrotrichomonas unguis*, 542
Glyptotermes ceylonicus Holmgren:
associated protozoa–*Macrotrichomonas pulchra*, 542; *Trichonympha chattoni*, 549
Glyptotermes contracticornis (Snyder):
associated protozoa–*Foaina falcifera*, 544;

Glyptotermes contracticornis (Snyder) (*Cont.*)
Foaina parvula, 543; Macrotrichomonas pulchra, 542; Oxymonas spp., 547; Trichomonas cartagoensis, 540; Trichonympha chattoni, 549; Trichonympha sp., 550; Tricercomitus divergens, 539

Glyptotermes dilatatus (Bugnion & Popoff):
associated protozoa–Devescovina cometoides, 540; D. transita, 541; Foaina nana, 543; F. solita, 543

Glyptotermes dubius Hill:
associated protozoa–Foaina delicata, 544; F. hilli, 544; Macrotrichomonas pulchra, 542

Glyptotermes iridipennis Froggatt:
associated bacteria–Bacillus dobelli, 61; Fusiformis hilli, 116; F. termitidis, 116
associated protozoa–Foaina delicata, 544; F. hilli, 544; Macrotrichomonas pulchra, 542; Trichonympha chattoni, 549

Glyptotermes minutus Kemner:
associated protozoa–Devescovina arta, 541; Macrotrichomonas unguis, 542

Glyptotermes montanus Kemner:
associated protozoa–Foaina flacifera, 544; F. parvula, 543; Macrotrichomonas pulchra, 542; Trichonympha chattoni, 549

Glyptotermes neotuberculatus Hill:
associated protozoa–Foaina delicata, 544; F. parvula, 543; Macrotrichomonas pulchra, 542; Trichonympha chattoni, 549

Glyptotermes niger Kemner:
associated protozoa–Devescovina vestita, 541; Foaina duo, 544

Glyptotermes nigrolabrum Hill, 550

Glyptotermes parvulus Sjöstedt:
associated protozoa–Diplonympha parvulus, 546; Foaina hamata, 544; F. solita, 543; Macrotrichomonas pulchra, 542; Metadevescovina stereociliata, 545; Trichonympha chattoni, 549

Glyptotermes perangustus Hill:
associated protozoa–Foaina delicata, 544; F. hilli, 544; Macrotrichomonas pulchra, 542

Glyptotermes satsumensis Matsumura, 550

Glyptotermes spp.:
associated protozoa–Devescovina cometoides, 540; D. glabra, 540; D. lemniscata, 541; D. transita, 541; Foaina hamata, 544; F. solita, 543; Macrotrichomonas pulchra, 542; M. unguis, 542; Metadevescovina mediocris, 545; M. stereociliata, 545; Trichonympha chattoni, 549

Glyptotermes taveuniensis Hill:
associated protozoa–Foaina parvula, 543; Macrotrichomonas pulchra, 36, 542; Trichonympha chattoni, 549

Glyptotermes trilineatus Mjöberg, 550
Glyptotermes tuberculatus (Froggatt):
associated protozoa–Devescovina lemniscata, 541; Foaina funifera, 544; F. solita, 543

Gnathotrichus asperulus (LeC.), 404; G. materiarius (Fitch), 404, G. retusus (LeC.), 404

Gnorimoschema operculella (Zell.), 132

Goat, 278, 302, 304, 343

Goat louse (see Linognathus stenopsis)

Gonads, insect, 194, 217, 237

Gonepteryx rhamni (Linn.), 18

Goniocotes compar Nitz., 219; G. hologaster Nitz., 219

Goniodes, 219; Goniodes damicornis Nitz., 219

Gonopore, 193

Gopher, pocket, 337

Gordiaceae, 7

Goriidae, 7

Gortyna ochracea, 64

Gossypium virus 1, 449

Gracilaria syringella (Fabr.), 31

Grahamella, 262

Granular leucocytes (see Leucocytes)

Graphiphora agathina, 18; G. triangulum, 18

Graphitose, 93

Graphium ulmi (see Ceratostomella ulmi)

Grasserie, 70, 147, 416, **417** (see also Jaundice of silkworms)

Grasshoppers, 14, 75, 88, 89, 182, 379, 494, 558 (see also specific names)

Green body, 190

Green June beetle (see Cotinis nitida)

Green stink bug (see Nezara hilaris)

Gregarina acridiorum, 493; G. acrydiinarum, 492; G. blattarum, 491, 493; G. ceuthophili, 492; G. chagasi, 493; G. conica, 493; G. consobrina, 493; G. cuneata, 491, 492, 493; G. davini, 493; G. diabrotica, 493; G. galliveri, 492; G. hadenoeci, 492; G. hyalocephala, 493; G. indianensis, 492; G. kingi, 492; G. legeri, 491; G. locustae, 491; G. longiducta, 493; G. macrocephala, 493; G. mirotermitis, 553; G. mystacidarum, 492; G. neglecta, 493; G. nigra, 492; G. oblonga, 493; G. ohioensis, 492; G. ovata, 493; G. oviceps, 491; G. paranensis, 493; G. parcoblattae, 492; G. platydema, 493; G. prima, 492; G. proteocephala, 492; G. rigida columna, 492; G. rigida rigida, 491; G. segmatata, 492; G. sphaerulosa, 493; G. stygia, 493; G. termitis, 552; G. thomasi, 492; G. udeopsylla, 493

Gregarines, 487-497

Gregarinida, **487-497**, 523

Grosmannia, 410

Growth forms, 26, 227

Gryllidae, 493

Grylloblatta campodeiformis campodeiformis Walk.:
associated bacteria–Bacillus subtilis, 89, 91;

Micrococcus epidermidis, 145; *M. freudenreichii*, 146; *M. subflavus*, 151
Gryllomorpha dalmatina, 493
Gryllotalpa gryllotalpa (Linn.):
  associated bacteria–*Bacillus gryllotalpae*, 64, 104; *Bacterium gryllotalpae*, 104
  associated protozoa–*Nyctotherus ovalis*, 521; *Retortamonas gryllotalpae*, 482; *R. orthopterorum*, 482
Gryllotalpa spp.:
  associated protozoa–*Gregarina sphaerulosa*, 493; *Hirmocystis gryllotalpae*, 493; *Lophomonas blattarum*, 484; *L. striata*, 484
Gryllus abbreviatus Serville:
  associated protozoa–*Gregarina oviceps*, 491; *Leidyanna erratica*, 493
Gryllus americanus Blatchl., 491
Gryllus assimilis (Fabr.):
  associated bacterium–*Bacillus poncei*, 81, 82
  associated protozoa–*Gregarina galliveri*, 492; *G. kingi*, 492; *Leidyanna erratica*, 493; *L. gryllorum*, 493
Gryllus domesticus Linn.:
  associated protozoa–*Diplocystis major*, 490; *D. minor*, 490; *Gregarina macrocephala*, 493; *Leidyanna gryllorum*, 493
Gryllus pennsylvanicus Burm. (see *G. assimilis*)
Guinea pigs, 277, 278, 282, 283, 285, 286, 289, 294, 297, 298, 300, 301, 306, 308, 315, 319, 320, 323, 324, 325, 331, 334, 336, 338, 343, 344, 345, 437, 585, 586, 587
Gurleya francottei, 515; *G. légeri*, 515
Gymnetron antirrhini Payk., 240; *G. collinum* Gyll., 240; *G. villosulum* Gyll., 240
Gymnoascales, 383
Gypsy moth (see *Porthetria dispar*)
Gyrinidae, 385
Gyrinus natator Linn., 494
Gyrococcus, 42, 186; *G. flaccidifex*, 145, 146, 186, 421
Gyropus, 219

Hadenoecus puteanus Scudd., 492
Hadrotettix trifasciatus, 486
Haemagogus capricorni (Lutz), 433
Haemaphysalis bispinosa Neum., 297, 298
Haemaphysalis cinnabarina Koch, 166, 167
Haemaphysalis concinna Koch:
  associated rickettsia, 270
  associated virus–Russian spring-summer encephalitis, 442
Haemaphysalis flava, Neum., 470
Haemaphysalis humerosa Warburton & Nuttall:
  associated bacterium–*Bacillus tropicus*, 93-94
  associated rickettsia–*Rickettsia burneti*, 296, 297, 298
Haemaphysalis leachi (Audouin):
  associated protozoan–*Babesia canis*, 507
  associated rickettsia–*Dermacentroxenus conori*, 340, 341
  intracellular symbiotes in, 251
Haemaphysalis leporis-palustris Pack.:
  associated bacterium, 13–*Pasteurella tularensis*, 166, 167
  associated rickettsia–*Dermacentroxenus rickettsi*, 330, 333
  associated virus–rabbit papilloma, 442
  intracellular symbiotes in, 251
Haematobia irritans Linn., 53, 54; *H. serrata* Rob.-Desv., 439; *Haematobia* sp., 442
Haematomyzidae, 220
Haematomyzus elephantis Piaget, 220
Haematopinidae, 223-226
Haematopinus, 226; *H. asini* (Linn.), 225; *H. columbianus* Osborn, 162; *H. eurysternus* Nitz., 225; *H. macrocephalus* Burm. (see *H. asini*); *H. spinulosus* Burm., 470, 479; *H. stephensi* Christ. & Newst., 499; *H. suis* (Linn.), 223, 224, 225
Haematopota duttoni Newstead, 470; *H. itlaico* (Meig.), 470; *H. pluvialis* (Linn.), 53; *H. vandenbrandeni*, 470
Haemocoel (see Hemocele)
Haemocytes (see Hemocytes)
Haemodipsus ventricosus (Denny), 165, 166
Haemogregarina triatomae, 498
Haemolymph (see Hemolymph)
Haemoproteidae, 505-506
Haemoproteus, 506; *H. columbae*, 506; *H. lophortyx*, 506
Haemosporidia, 487, **500-511**
Hairworms, 7
Halesidotis, 514
Halo blight, 174
Haltica, 400
Hamster, 294, 499
Handling of insects and ticks, 574-575
Hanseniaspora, 351
Hansenula, 351
Hapithus agitator agitator:
  associated protozoa–*Gregarina galliveri*, 492; *Leidyanna erratica*, 493
Haplomyces, 386
Haplosporidia, 486, 511
Hare, snowshoe, 168
Harpalus pennsylvanicus erythropus, 493
Harpyia bifida, 414
Hartmannella blattae, 486
Harvest mites (see Trombidiidae)
Hatching of mosquito eggs, effect of bacteria on, 22-24
Heart rot of celery, 118
Heartwater disease, **302-304**, 310
Heartwater fever, 302
Helicosporidia, 518-519

Helicosporidium parasiticum, 518, 519
Heliothis armigera (Hbn.):
 associated fungus–*Spicaria heliothis*, 394
 associated protozoan–*Nosema heliotidis*, 514
 associated virus, 414
Heliothis obsoleta (Fabr.) (see *H. armigera*)
Heliothrips femoralis (see *Hercinothrips femoralis*)
Helops striatus Oliv., 494
Hemerocampa leucostigma (S. & A.), 414, 425
Hemiderma perspicillatum, 337
Hemileuca maia (Dru.), 414; *H. oliviae* Ckll., 414, 425
Hemiptera, 14, 16, 215, **233-236**, 443
Hemocele, 520, 562, 563
Hemocytes, 563, 567, 568 (*see also* Cells)
Hemolymph, 563 (*see also* Blood, insect)
Hemolysis in insects, 567
Hemophileae, 42, 159
Hemophilus, 42, **159**; *H. duplex*, 101, **159**; *H. influenzae*, 557, 567
Hepatozoon, 499; *H. balfouri*, 499; *H. canis*, 499, 500; *H. criceti*, 499; *H. gerbilli*, 499; *H. muris*, 498; 499
Hercinothrips femoralis (Reut.), 174
Hericia hericia, 518
Herpetomonas, 464, 465, 466, 469, **470-472**; *H. algeriense*, 472; *H. ampelophilae*, *H. aretocorixae*, 472; *H. calliphorae*, 471 (see also *H. muscarum*): *H. caulleryi*, 472; *H. ctenocephali*, 466; *H. ctenophthalmi*, 466, 472; *H. ctenopsyllae*, 466; *H. culicidarum*, 468 (see also *Crithidia fasciculata*); *H. culicis*, 468, 472 (see also *Crithidia fasciculata*); *H. debreuli*, 466; *H. drosophilae*, 472; *H. grayi*, 472; *H. jaculum*, 472; *H. luciliae*, 471 (see also *H. muscarum*); *H. mesnili*, 472; *H. mirabilis*, 472; *H. muscae domesticae*, 470 (see also *H. muscarum*); *H. muscarum*, 466, 471, 472; *H. myzomyiae*, 472; *H. pattoni*, 466; *H. pessoai*, 472; *H. pseudoleishmania*, 466; *H. pyraustae*, 47, 472; *H. roubaudi*, 472; *H. rubrostriatae*, 472; *H. sarcophagae*, 471 (see also *H. muscarum*); *H. siphunculinae*, 472; *H. vespae*, 472
Herpomyces, 385
Hesperotettix viridis pratensis, 491
Heterocampa guttivitta, 414
Heterogomia, 217; *H. aegiptica*, 216
Heteromastix batrachorum var. *tipulae*, 484; *H. claviger*, 593; *H. conclaviger*, 539; *H. disclaviger*, 539; *H. laticeps*, 539; *H. termitis*, 539; *H. termopsidis*, 539
Heterotricha, 554
Hexamastix, 539
Hexamita cryptocerci, 536, 537; *H. periplanetae*, 484
Hexapoda, 214, 215, 462, 513

Hippelates flavipes Loew, 459; *H. pallipes* Loew, 459
Hippobosca capensis Olfers, 205, 245; *H. equina* Linn., 205, 245
Hippoboscidae, 244
Hippodamia convergens Guér., 116
Hirmocystis gryllotalpae, 493; *H. hartpali*, 493; *H. termitis*, 552 (see also *Gregarina termitis*)
Hirsutella arachnophila, 392; *H. citriformis*, 392; *H. entompohila*, 392; *H. floccosa*, 392; *H. saussurei*, 392; *H. subulata*, 392
Histiogaster fungivorax Jacot, 410
Histiotus velatus, 337
Histological methods, 579
History: general, 3, 4; on intracellular symbiotes, 189-191; on viruses, 413, 432
Hodotermes, 528
Hodotermes mossambicus Hagen:
 associated protozoa–*Holomastigotes crassum*, 551; *H. elongatum*, 551; *Joenia intermedia*, 548; *Spirotrichonympha africana*, 551; *Trichomonas macrostoma*, 539
Hodotermitidae, 527, 528
Hodotermopsis, 528; *H. japonicus* Holmgren, 549
Hoffman's bacillus (see *Corynebacterium pseudodiphthericum*)
Hog louse (see *Haematopinus suis*)
Holomastigotes crassum, 531; *H. elongatum*, 551
Holomastigotoides hartmanni, 551; *H. hemigymnum*, 551; *H. hertwigi*, 551; *H. mirabile*, 551
Holotricha, 520-521
Homalo, 217; *H. demascruralis*, 216
Homalomyia scalaris (see *Fannia scalaris*); *Homalomyia* sp., 471
Homona coffearia, 394
Homoptera, 215, **226-232**, 373, 387, 393, 395, 443
Honeybee, 11, 351, 426-430 (see also *Apis mellifera* and Bees)
Hoplonympha natator, 548
Hopper, garden flea, 173
Hormetica laevigata Burm., 522
Horse, 277, 286, 343, 437
Horse louse (see *Haematopinus asini*)
Horsefly (see Flies, horsefly)
Host-parasite specificity, 529-531
Housefly (see Flies, housefly, and *Musca domestica*)
Humoral immunity (see Immunity in insects, humoral)
Hyalomma aegyptium (Linn.):
 associated protozoa–*Crithidia hyalommae*, 469
 associated rickettsiae–*Dermacentroxenus conori*, 341; *Rickettsia tsutsugamushi*, 292
Hyalomma impressum, 508

Hyalomma mauritanicum, 510
Hyalomma spp., 308
Hyalomma volgense (*H. uralense*), 162, 165
Hyalopterus atriplicis (Linn.), 444
Hyalospora affinis, 493
Hydaticus, 495
Hydraeomyces, 386
Hydria, 514
Hydromys chrysogaster, 300
Hydromyza livens, 31
Hydrophilidae, 385
Hydrophilomyces, 385
Hydrophilus, 183, 184
Hydrophilus piceus Linn.:
  associated protozoa–*Phialoides ornata,* 495; *Rhynchophrya palpans,* 522
Hydroporus palustris Linn., 498, 499
Hydrous caraboides Linn., 494; *Hydrous* sp., 494
Hyla arborea, 184
Hylemya antiqua Meig., 116
Hylemya brassicae (Bouché):
  associated bacterium–*Erwinia carotovora,* 20, 118
  associated fungus–*Phoma lingam,* 412
Hylemya cilicrura Rondani, 19:
  associated bacteria–*Erwinia carotovora,* 20, **118**; *Pseudomonas fluorescens,* 20, 169, **170**; *P. non-liquefaciens,* 20, 171, **172**
Hylemya lipsia Walk., 116
Hylemya trichodactyla Rondani, 118
Hylobius, 238; *H. abietis* Linn., 237, 238, 240
Hylurgopinus rufipes (Eich.), 410
Hymenoptera, 385, 387, 414; symbiosis in, 191, 215, **247-249**
Hypermastigida, 484, 536, 538, **548-552**
Hypermastigotes, 531
Hypersensitivity, 572
Hyphantria cunea (Dru.):
  associated bacterium–*Streptococcus faecalis,* 141
  associated protozoan–*Nosema bombycis,* 514
  associated viruses, 414
Hyphomycetes, 395
Hypocrella, 387
Hypoderma lineata (DeVill.), 31
Hyponomeuta evonymella Linn.:
  associated bacteria–*Bacillus monachae,* 73, 74; *B. oblongus,* 75; *B. spermatozoides,* 88
Hyponomeuta sp., 147
Hysteroneura setariae (Thos.), 445
Hystrichopsylla talpae (Curtis):
  associated protozoan–*Crithidia hystrichopsyllae,* 470
  intracellular symbiotes in, 247
Hystrichopsyllidae, 247

Icerya purchasi Mask., 364

Icerymyces pierantonii, 364
Ichneumonidae, 249
Identification of bacteria, 583, **598-599**; of protozoa, 590; of rickettsiae, 586-587; yeasts, 588
Idiomyces, 385
Idionympha perissa, 484, 536
Illinoia solanifolii (see *Macrosiphum solanifolii*)
Immunity in insects, 126, **555-572**; acquired, 559-562; acquired, active, 559; acquired, passive, 278, 561; antibodies (agglutinins, 562, 571; antitoxins, 562, 571; bactericidins, 571; bacteriolysins, 562, 571; complement, 572; lysins, 562; opsonins, 572; precipitins, 571); celluar, 210, **562-570**; humoral, 210, **570-572**; natural, 556-559; nonspecific, 561; phagocytosis, 210; race, 556; species, 556; theories of, 555; in ticks, 555
Imperial moth (see *Eacles imperialis*)
Impetigo contagiosa, 153
Indian mealworm (see *Plodia interpunctella*)
Infectious ophthalmia (*see* Conjunctivitis)
Insect microbiology, applications of, 6, 7
Insects (*see* specific names): collecting methods for, 574; dissection methods, 576-579; handling methods for, 574-575; histologic methods, 579; and immunity, 555-572; microbiological examination of, 580-601; rearing of, 575; reproductive system (female, 191-193; male, 193); shipping and storing methods, 574-575
Intracellular bacteria (*see* Intracellular symbiotes)
Intracellular symbiotes: arthropod hosts of, 213-255; bacteroids, 203; cultivation of, 190, 204-206; generation-to-generation transmission of, 206-209; history, 189-191; and insect host, 191; in insects, examples of, 213-249 (*see also* specific insect names); life cycles in insects, 217-218; mitochondria, comparison with, 202; nature of, 201-204; origin of, 209-210; penicillin and, 212; role of, 210-213
Invertase, 31
Ipidae, 403
Isaria, 387, 393, 400
Ischnodemus, 236; *I. sabuleti* Fall., 235, 236
Ischnorrhynchus, 236; *I. ericae* Horv., 236; *I. resedae,* 236
Isle of Wight disease, 78
Isoodon torosus, 94, 300
Isophya amplipennis Brunn., 56
Isoptera, 249, 385, 462, 526-529, 536
Isotrich caulleryi, 520
Ixodes, 250, 251, 437
Ixodes autumnalis (see *I. hexagonus*)
Ixodes dentatus Marx:

SUBJECT INDEX

Ixodes dentatus Marx (*Cont.*)
associated bacteria, 13, 152, 153
associated rickettsia–*Dermacentroxenus rickettsi*, 330
Ixodes hexagonous Leach:
associated bacterium–*Pasteurella pestis*, 162, 165
intracellular symbiotes in, 250, 251, 252
Ixodes holocyclus Neum., 287, 298
Ixodes pacificus, 166
Ixodes persulcatus, 439, 440
Ixodes ricinus (Linn.):
associated bacteria–*Bacillus anthracis*, 53, 55; *Pasteurella tularensis*, 166; *Staphylococcus aureus*, 155
associated protozoa–*Anaplasma marginale*, 510; *Babesia bovis*, 507; *Trypanosoma melophagium*, 480
associated rickettsia–*Dermacentroxenus conori*, 341
associated virus, 442
intracellular symbiotes in, 250, 251
Ixodes ricinus californicus Banks, 166
Ixodes texanus, 13
Ixodidae, 250, 253, 254
Ixodisymbionte, 251
Ixodisymbiotes, 263
Ixodoidea, 215, 249

Jackal, 307
Janthinosoma sayi (see *Psorophora sayi, P. ferox*)
Japanese B virus, 437
Japanese beetle (see *Popillia japonica*)
Japanese cicada, 365
Jaundice of silkworms, 415, 416, **417**; cultivation of, 421, 422; history of, 417-418; pathology of, 420-421; virus of, 418-420
Joenia annectens, 548; *J. intermedia*, 548
Joenina pulchella, 548
Joenopsis cephalotricha, 548; *J. polytricha*, 548
"Jolly bodies," 510
Jumping plant lice (*see* Chermidae)

Kala azar, **473**
Kalotermes aethiopicus (Silvestri), 548
Kalotermes barbouri Snyder, 547
Kalotermes castaneiceps Sjöstedt:
associated protozoa–*Devescovina lemniscata*, 544; *Foaina acontophora*, 544; *F. parvula*, 543; *Trichonympha corbula*, 549
Kalotermes clevelandi Snyder:
associated protozoa–*Coronympha clevelandi*, 545; *Oxymonas clevelandi*, 547; *Tricercomitus divergens*, 539; *Trichonympha subquasilla*, 549
Kalotermes condonensis Hill, 545
Kalotermes condonensis var. *chryseus* Hill, 545
Kalotermes durbanensis Haviland, 545

Kalotermes emersoni Light:
associated protozoa–*Coronympha octonaria*, 545; *Metacoronympha senta*, 545; *Tricercomitus divergens*, 539; *Trichonympha lighti*, 549
Kalotermes flavicollis (Fabr.), 538:
associated protozoa–*Foaina dogiela*, 544; *F. grassii*, 543; *Hexamastix termitis*, 539; *Joenia annectens*, 548; *Microrhopalodina enflata*, 547; *Prolophomonas kalotermitis*, 548
transmission of protozoa in, 532
Kalotermes hubbardi Banks, 530:
associated protozoa–*Metadevescovina debilis*, 545; *Staurojoenina* sp., 548
Kalotermes immigrans Snyder:
associated protozoa–*Coronympha clevelandi*, 545; *Trichonympha subquasilla*, 549
Kalotermes jeannelanus Sjöstedt:
associated protozoa–*Caduceia theobromae*, 542; *Devescovina tendicula*, 540; *Foaina reflexa*, 544; *F. solita*, 543
Kalotermes jouteli Banks:
associated protozoa–*Macrotrichomonas restis*, 542; *Metadevescovina turbula*, 545
Kalotermes lighti Snyder:
associated protozoa–*Coronympha octonaria*, 545; *Metacoronympha senta*, 545
Kalotermes longus (Holmgren):
associated protozoa–*Foaina acontophora*, 544; *F. parvula*, 543; *Trichonympha corbula*, 549
Kalotermes madagascariensis Wasm., 545
Kalotermes magninotus, 547
Kalotermes marginipennis (Latr.):
associated protozoa–*Hexamastix claviger*, 539; *Metadevescovina modica*, 545; *M. magna*, 545
Kalotermes milleri Emerson, 549
Kalotermes minor Hagen:
associated fungi, 408
associated protozoa–*Metadevescovina minor*, 545; *Oxymonas minor*, 547; *Staurojoenina assimilis*, 548
Kalotermes nigriceps, 547
Kalotermes nocens Light, 547
Kalotermes obscurus (Walk.):
associated protozoa–*Trichonympha* sp., 550; *T. zeylanica*, 549
Kalotermes occidentis Walk.:
associated protozoa–*Metadevescovina polyspira*, 545; *Microrhopalodina occidentis*, 547; *Trichonympha ampla*, 549
Kalotermes pacificus Banks:
associated protozoa–*Coronympha octonaria*, 545; *Metacoronympha senta*, 545
Kalotermes perezi Holmgren:
associated protozoa–*Devescovina uniflexura*, 541; *Foaina reflexa*, 544; *Hexamastix claviger*, 539; *Trichonympha quasilla*, 549
Kalotermes perparvus, 546

Kalotermes platycephalus:
  associated protozoa—*Coronympha octonaria,* 545; *Metacoronympha senta,* 545
Kalotermes repandus Hill:
  associated protozoa—*Devescovina cometoides,* 540; *D. transita,* 542; *Foaina nana,* 543; *F. solita,* 543
Kalotermes rufinotum Hill, 550
Kalotermes schwartzi Banks, 550
Kalotermes snyderi Light:
  associated protozoa—*Oxymonas panamae,* 547; *Tricercomitus divergens,* 539
Kalotermes spp., 528, 529, 530:
  associated protozoa—*Caduceia pruvoti,* 542; *C. theobromae,* 542; *Devescovina cometoides,* 540; *D. glabra,* 540; *D. insolita,* 541; *D. lemniscata,* 541; *D. vittata,* 541; *Foaina nana,* 543; *F. Parvula,* 543; *F. reflexa,* 544; *F. signata,* 544; *F. solita,* 543; *Metadevescovina fulleri,* 545; *Oxymonas brevis,* 547; *O. hubbardi,* 547; *O. janicki,* 547; *O. jouteli,* 547; *O. snyderi,* 547; *Pseudodevescovina ramosa,* 545; *Trichonympha corbula,* 37, **549**; *T. divexa,* 549
Kalotermes tabogae Snyder:
  associated protozoa—*Coronympha octonaria,* 545; *Metacoronympha senta,* 545, 546; *Oxymonas clevelandi,* 547; *Tricercomitus divergens,* 538; *Trichonympha tabogae,* 549
  defaunation of, 533
Kalotermes taylori Light, 550
Kalotermitidae, 527, 528, 531
Kalotermitinae, 541
Karyokinetosis, 568
Karyolysus lacertarum, 499
Kedani fever, 290
Kenya typhus, 329, 339
Kermes quercus Linn., 358
Kermicola, 359; *K. kermesina,* 358, 359; *K. tenuis,* 359
Kirbyella zeteki, 547
Klebsiella, 41, **123**; *K. capsulata,* 123; *K. ozaenae,* 123; *K. paralytica,* 123-124; *K. pneumoniae,* 124
Kofoidia loriculata, 548
Kofoidina ovata, 553
Kräuselkrankheit of sugar beet, 448
Kurloff bodies, 325
Kusanoöpis, 388

Laboulbenia, 385; *L. cristata,* 386; *L. pheropsophi,* 386; *L. variabilis,* 386
Laboulbeniaceae, 367, 385-386
Laboulbeniales, 383, 384-387
Laccifer albizziae, 375; *L. lacca* (Kerr), 375
Laccomyces symbioticus, 375
Lacerta muralis, 499
Lachnosterna smithi, 78, 79

Lactase, 31
Lactobacillus lactis, 142
Lactobacteriaceae, 41, 136
Lady beetle (see *Coccinella novemnotata*)
Laelaps echidninus, 499
Lagenidiales, 377
Lakshadia, 375; *L. communis* Mahd., 375
Lambornella stegomyiae, 520, 521
Lamellicorn larva, 31
Lamia, 378
Lankestria culicis, 488-490
Laphygma frugiperda (A. & S.), 414
Larinus cynarae Fabr., 239; *L. flavescens* Germ., 239; *L. jaceae* Fabr., 240; *L. latus* Hbst., 239; *L. planus* Fabr., 240; *L. scolymus,* 239; *L. sturnus* Schaller, 240; *L. turbinatus* Gyll., 240
Lasiocampidae, 414
Lasioderma redtenbacherei Bach, 367; *L. serricorne* (Fabr.), 213, 369
Lasiodiplodia triflorae, 412
Lasius niger, 407; *L. n. americanus* Emery, 116
Latent (X) virus of potato, 449
Lateral ducts, 193
Lateral oviduct, 192
Leaf beetles (see Chrysomelidae)
Leaf curl of cotton, 449
Leaf curl of sugar beet, 448
Leaf-cutting ants, 406
Leafhoppers, 392 (see Cicadellidae)
Lecaniascus polymorphus, 345 (see *Lecaniocola parasitica*)
Lecaniidae, 71
Lecaniocola, 353, 354; *L. ceroplastidis pallidi,* 356; *L. conomeli limbati,* 358; *L. contei,* 355; *L. egbarum fulleri,* 356; *L. filippiae,* 357; *L. inglisiae,* 357; *L. lecanii viridi,* 355; *L. macropsidis lanionis,* 358; *L. parasitica,* 354; *L. proteae,* 354-355; *L. protopulvinariae,* 357; *L. pulvinariae,* 357; *L. putnami,* 357; *L. rosae,* 354; *L. saccardiana,* 356; *L. saissetiae,* 355-356
Lecaniodiaspis pruninosa Hunter, 374
Lecanium corni Bouché:
  associated bacterium—*Bacillus megatherium,* 71, 366
  associated yeasts—*Lecaniocola rosae,* 354; *Torula lecanii corni,* 366
Lecanium hemisphericum Targ., 355
Lecanium hesperidum Burm.:
  associated yeasts—*Lecaniocola parasitica,* 354; *L. contei,* 355
  intracellular symbiotes in, 189
Lecanium longulum, 355
Lecanium oleae Bernard, 355
Lecanium persicae (Fabr.), 355
Lecanium proteae Brain, 354
Lecanium viride Green, 355

Legerella grassii, 498; *L. hydropori*, 498, 499; *L. parva*, 498
Legeria agilis, 495
Leidyanna erratica, 493; *L. gryllorum*, 493; *L. tinei*, 494; *L. xylocopae*, 493
Leidyopsis, 534
Leiognathus laverani, 479
Leishmania, 464, 465, 473-474; *L. braziliensis*, 473, 474; *L. donovani*, 466, 473-474; *L. donovani* var. *infantum* (*L. infantum*), 473; *L. infantum*, 473; *L. tarentolae*, 474; *L. tropica*, 473, 474
Lentinus, 389; *L. atticolus*, 406
Leogorrus litura, 470
Lepidoptera, 112, 214, 378, 387, 388, 393, 395, 413, 415, 563, 566
Lepidosaphes, 374, 391; *L. gloverii* (Pack.), 374; *L. ulmi* (Linn.), 374
Lepisma saccharina Linn., 215
Lepismatidae, 215
Lepismatophila thermobiae, 494
Leptinotarsa decemlineata (Say):
  associated bacteria–*Achromobacter delicatulum*, 97; *Aerobacter aerogenes*, 120; *Bacillus leptinotarse*, 68; *Flavobacterium acidificum*, 114; *Pseudomonas ovalis*, 172; *P. septica*, 172; *Xanthomonas sepedonica*, 176; *X. solanacearum*, 176
  associated virus, 444
Leptocoris trivittatus (Say), 61
Leptoglossus balteatus, 350; *L. zonatus* (Dallas), 350
Leptomonas, 464, 465, 466-468, 473; *L. agilis*, 468; *L. blaberae*, 468; *L. bütschlii*, 466; *L. ctenocephali*, 466, 467; *L. culicis*, 468; *L. familiaris*, 468; *L. fasciculata*, 468; *L. foveati*, 468; *L. jaculum*, 468; *L. michiganensis*, 468; *L. mirabilis*, 468; *L. naucoridis*, 468; *L. pangoniae*, 468; *L. pulicis*, 467; *L. pyraustae*, 467, 468; *L. vespae*, 467, 468
Leptopsylla musculi (see *L. segnis*)
Leptopsylla segnis (Schönherr):
  associated bacterium–*Pasteurella pestis*, 161, 162
  associated protozoa–*Herpetomonas ctenopsyllae*, 466; *Trypanosoma lewisi*, 478
  associated rickettsia–*Rickettsia prowazeki* var. *typhi*, 281
Leptosphaeria coniothyrium, 412
Leptospira, 451
Leptospironympha eupora, 484, 536; *L. rudis*, 536; *L. wachula*, 536
Leptothrix, 187; *L. buccalis*, 187
Leptothrichia buccalis, 187
Leptotrix, 42, 187; *L. buccalis*, 187
Leptura cerambyciformis Schrank, 370; *L. rubra* Linn., 370, 371, 372
Lepturini, 371

Lesser grain borer (see *Rhizopertha dominica*)
Leucochromus imperialis Zoubk., 239; *L. lehmanni* Menet., 239
Leucocytes, 563, 564, 565, 567, 568, 569, 570
Leucocytozoon anatis, 506
Leucosomus pedestris Poda, 239
Leucotermes, 528; *L. aureus*, 550; *L. flaviceps*, 551; *L. indicola*, 554; *L. tenuis*, 551
Lice, 48, 137, 270, 271, 281, 288, 289, 321, 578 (see also specific names)
Ligyrodes relectus, 483
Limber neck of chickens, 94
Limnaeomyces, 386
Limnophilus flavicornis, 483
Limnophilus rhombicus:
  associated protozoa–*Entamoeba* sp., 485; *Eutrichomastix trichopterorum*, 483; *Thelohania janus*, 516
Lincus spathyliger, 473
Linognathoides citelli, 162
Linognathus, 226; *L. piliferus* (Burm.), 226; *L. stenopsis* (Burm.), 267, 321; *L. tenuirostris*, 226; *L. vituli*, 226
Linostoma, 410
Liparidae, 414
Liparis auriflua, 55, 56; *L. germanus* Linn., 240; *L. salicis* (see *Stilpnotia salicis*)
Lipase, 30, 31
Lipeurus, 219; *L. baculus* Nitz. (see *Columbicola columbae*); *L. frater* Giebel, 219; *L. lacteus* Giebel., 219
Liponyssus arcuatus:
  associated protozoa–*Hepatozoon criceti*, 499; *Trypanosoma vespertilionis*, 479
Liponyssus bacoti (Hirst):
  associated rickettsiae–*Rickettsia burneti*, 296; *R. prowazeki*, 260; *R. prowazeki* var. *typhi*, 282
Liponyssus nagayoi Yamada, 282
Liponyssus saurarum Oudemans:
  associated protozoan–*Karyolysus lacertarum*, 499
  intracellular symbiotes in, 254
Lipoptena caprina Austen, 205, 245; *L. cervi* (Linn.), 245; *L. depressa* (Say), 44, 266, 245
Lipothropha macrospora, 497; *L. microspora*, 497
Listerella, 105
Listeria, 105; *L. monocytogenes*, 105
Lixochelus elongatus Goeze, 239
Lixus bidentatus, 239; *L. bisoleatus*, 239; *L. cavipennis*, 239; *L. coarctatus*, 239; *L. fimbriolatus*, 239; *L. frater*, 239; *L. massaicus*, 239; *L. nebulofasciatus*, 239; *L. nitidicollis*, 239; *L. paraplecticus*, 239; *L. plicaticollis*, 239; *L. praegrandus*, 239; *L. rhomboidalis*, 239; *L. sturmi*, 239; *L. truncatulus*, 239

Lobitermes longicollis Banks, 539
Locusta migratoria Linn.:
  associated bacteria, 84–*Bacillus fluorescens liquefaciens*, 84; *B. punctatus*, 84
  associated protozoan–*Gregarina oblonga*, 493
Locusta migratoroides R. & F., 181, 183
Locusta viridissima Linn.:
  associated bacterium–*Staphylococcus acridicida*, 153
  associated fungus–*Beauveria densa*, 394
Locustidae, 182
Loemopsylla cleopatrae, 470
Lophocampa flavostica, 514
Lophocephalus insignis, 494
Lophomonas blattarum, 484; *L. striata*, 484
Lophotryx spp., 506
Louping ill of sheep, 442
Louse, 129, 137, 210, 260, 261, 270, 271, 288, 289, 321, 578 (see also specific names)
Louse typhus, 264
Loxa variegata Dist.:
  associated bacteria–*Aerobacter aerogenes*, 120; *Eberthella insecticola*, 127
Loxostege sticticalis:
  associated bacteria–*Bacillus ocniriae*, 76; *Bacterium cazaubon*, 100; *Serratia marcescens*, 133
Lucilia argyricephala, 471, 472
Lucilia caesar Linn.:
  associated bacteria–*Bacillus anthracis*, 53, 54; *B. colisimile*, 60; *B. radiciformis*, 84, 85; *B. similcarbonchio*, 86, 87; *B. tifosimile*, 93; *Clostridium botulinum* Type C, 94; *Escherichia coli*, 121, 122; *Proteus vulgaris*, 126; *Pseudomonas aeruginosa*, 169; *P. fluorescens*, 170, 171; *P. jaegeri*, 171; *P. non-liquefaciens*, 171, 172; *Sarcina aurantiaca*, 152; *Serratia kielensis*, 132; *S. marcescens*, 133; *Staphylococcus albus*, 153, 154; *S. citreus*, 156
  associated protozoa, 471
  associated virus–poliomyelitis, 441
  osteomyelitis, treatment of with, 35
Lucilia craggi, 471
Lucilia sericata Meig.:
  associated bacteria, 11–*Bacillus lutzae*, 69; *Erwinia amylovora*, 116, 117; *Escherichia coli*, 33; *Micrococcus rushmori*, 150; *Neisseria luciliarum*, 158
  associated protozoan–*Herpetomonas muscarum*, 471
  bacteria as source of growth factors in, 33
  osteomyelitis, treatment of with, 35
Lucilia sp.:
  associated bacteria–*Brucella abortus*, 158, 159; *Mycobacterium leprae*, 46, 48
Lucoppia curviseta, 255
Lupinus arboreus, 469
Lycopersicum virus *3*, 448
Lyctidae, 236, 237

Lyctus, 199, 206, 209; *L. linearis* Goeze, 236, 237
Lygaeidae, 14, 16
Lygaeus familiaris, 468; *L. kalmii* Stål, 473
Lygus pratensis (Linn.):
  associated bacteria–*Bacillus subtilis*, 89, 90; *Bacterium qualis*, 112; *Erwinia amylovora*, 116; *E. carotovora*, 118, 119
  associated virus, 444
Lymantria dispar (Linn.) (see *Porthetria dispar*)
Lymantria monacha Linn., 60:
  associated bacteria–*Bacillus alacer*, 52; *Bacillus* B Hofmann, 56; *B. canus*, 58; *B. coeruleus*, 59; *B. lineatus*, 68; *B. minimus*, 73; *B. monachae*, 73, 74; *B. tenax*, 91; *B. similis*, 87; *B. tingens*, 93; *Bacterium monachae*, 108; *Micrococcus major*, 147; *M. vulgaris*, 151
  associated viruses, 414; *Wipfelkrankheit*, 415, **424-425**
Lymphocytes, 565, 567, 568, 569
Lymphocytic choriomeningitis, 440-441
Lymphogranuloma venereum, 262
Lynchia brunea Oliv., 506; *L. capensis*, 506; *L. fusca* (Macq.), 506; *L. hirsuta* Ferris, 506; *L. lividicolor*, 506; *L. maura* Bigot, 506
Lyperosia exigua (De Meij.), 160, 161
Lyreman cicada (see *Tibicen linnei*)
Lysins, 562

Macacus inuus, 307
Machadoella triatomae, 497
Machilis cylindrica, 493
Macilenze, 140
Macrodactylus subspinosus (Fabr.), 82, 83
Macronucleocytes, 564
Macrophages, 210, 562
Macrophoma coronillae, 395
Macropsis lanio Linn., 358; *M. trimaculata* (Fitch), 447
Macrosiphum gei Koch, 444, 445 [see *Macrosiphum solanifolii* (Ashm.)]
Macrosiphum jaceae (Linn.), 228, 570
Macrosiphum onobrachis (B. deF.) = *M. pisi* Kalt.
Macrosiphum pisi Kalt., 444
Macrosiphum rosae (Linn.), 228
Macrosiphum solanifolii (Ashm.):
  associated bacterium–*Xanthomonas stewarti*, 177
  associated viruses–Cucumis virus, 445; Solanus virus *12*, 444; Solanus virus *14*, 445
Macrosiphum tanaceti Linn.: intracellular symbiotes of, 198; mycetome of, **198**, 228
Macrospironympha xylopletha, 484, 536
Macrosteles divisus Uehler, 445
Macrosymbiote, 5
Macrotrichomonas lighti, **542**

Macrotrichomonas procera, 542
Macrotrichomonas pulchra: associated bacteriumlike organisms, 36; associated termites, 542
Macrotrichomonas ramosa, 542
Macrotrichomonas restis, 542
Macrotrichomonas unguis, 542
Macrotrichomonas virgosa: associated bacteriumlike organisms, 36; associated termites, 542
Maculatum disease, 315, 325, 344
Magdalis armigera Geoffr.: associated fungi, 410; intracellular symbiotes of, 240
Magenscheibe, 221
Maggot fly (see *Calliphora vomitoria*)
Maggots, surgical, 34-36
Magicicada septendecim (Linn.):
associated fungus–*Massospora cicadina*, 380-381
intracellular symbiotes of, 232
Malacosoma americana (Fabr.):
associated bacteria–*Alcaligenes stevensae,* 180; *Bacillus lasiocampa,* 67; *Micrococcus nigrofaciens,* 148
associated viruses, 414, 425
Malacosoma castrensis, Linn., 185
Malacosoma disstria Hbn., 414, 425
Malameba locustae, 486
Malaraeus telchinum, 162
Malaria, 3, **500-505;** biologic relationships, 504; types of, 500
Malleomyces, 42; 159; *M. mallei,* 159; *M. pseudomallei,* 160
Mallophaga, 215, 219, 321
Malpighamoeba locustae, 486; *M. mellificae,* 486
Malpighian tubes, 201, 218, 220, 232, 235, 237, 239, 242, 243, 244, 245, 247, 249, 250, 251, 252, 253, 254, 317, 327, 331, 334, 456, 457, 458, 466, 482, 486, 488, 490, 511, 514, 523
Malpighiella refringens, 485; *M. locustae,* 486
Maltase, 31
Mamestra brassicae Linn., 394
Mamestra oleraceae:
associated bacterium–*Bacterium hemophosphoreum,* 104
associated protozoan–*Microklossia* sp., 525
Mamestra picta Linn., 65
Manihot virus *1,* 449
Mansonia africana (Theobald) (see *Taeniorhynchus africanus*): *M. (Taeniorhynchus) fuscopennata* (Theobald), 442; *M. (T.) microannulata* (Theobald), 442; *M. (T.) uniformis* Theobald, 474; *M. (T.) versicolor* Edwards, 442
Margaronia pyloalis, 514
Margarodes, 374; *M. polonicus* (Linn.), 374
Margarodinae, 374

Margaropus australis, 459 (see also *Boöphilus australis*); *M. decoloratus,* 459 (see also *Boöphilus decoloratus*); *M. microplis,* 251 (see also *Boöphilus annulatus microplis*); *M. winthemi* Korsch, 251
Marmot, 164, 337
Marmota monax monax, 285
Marseilles fever, 340 (see also *Fièvre boutonneuse*)
Masse polaire, 197
Massospora, 377, 380; *M. cicadina,* 380
Mastigophora, 462, **463-484,** 523, 537; in termites, 538-552
Mastitis, 138
Mastotermes, 528, 531
Mastotermes darwiniensis Froggatt:
associated protozoa–*Deltatrichonympha operculata,* 550; *Metadevescovina extranea,* 545; *Mixotrichia paradoxa,* 550
Mastotermitidae, 527, 528
Mattesia dispora, 497
Maturation, 194, 195
May beetle (see *Phyllophaga vaudinei*)
Mealworm (see *Tenebrio molitor*)
Mealy bugs, 231, 232 (*see also* specific names)
Mecaspis alternans Hbst., 239
Mechanites lysimnia, 514
Megabothris lucifer (see *Ceratophyllus lucifer*); *M. rectangulatus* (see *Ceratophyllus rectangulatus*)
Megninietta ulmi, 410
Melanoplus angustipennis (Dodge):
associated protozoan–*Gregarina rigida rigida,* 491
Melanoplus atlantis Riley, 181, 183 (see also *Melanoplus mexicanus*)
Melanoplus bivittatus (Say):
associated bacteria–*Bacterium termo,* 113; *Coccobacillus acridiorum,* 181, 183
associated protozoa–*Gregarina indianensis,* 492; *G. rigida columna,* 492; *G. rigida rigida,* 491
Melanoplus differentialis (Thos.):
associated bacterium–*Xanthomonas sepedonica,* 176
associated fungus–*Fusarium vasinfectum,* 412
associated protozoa–*Gregarina indianensis,* 492; *G. nigra,* 491; *G. rigida columna,* 492; *G. rigida rigida,* 491; *Malameba locustae,* 486
Melanoplus femur-rubrum (DeG.):
associated bacteria–*Bacillus poncei,* 81, 82, 565, 571, 572; *Bacterium termo,* 113; *Coccobacillus acridiorum,* 181, 183
associated fungus–*Fusarium vasinfectum,* 412
associated protozoan–*Malameba locustae,* 486
Melanoplus femur-rubrum femur-rubrum (DeG.):

associated protozoa–*Gregarina nigra*, 492; *G. rigida rigida*, 491
Melanoplus keeleri luridus Thos.:
associated protozoa–*Gregarina rigida columna*, 492; *G. rigida rigida*, 491
Melanoplus mexicanus (Sauss.) (= *M. atlantis* Riley):
associated bacterium–*Coccobacillus acridiorum*, 181, 183
associated fungus–*Fusarium vasinfectum*, 412
associated protozoa–*Malameba locustae*, 486
Melanoplus mexicanus mexicanus (Sauss.):
associated protozoa–*Gregarina indianensis*, 492; *G. rigida columna*, 492; *G. rigida rigida*, 491
Melanoplus obovatipennis Blatchl.:
associated protozoa–*Gregarina rigida columna*, 492; *G. rigida rigida*, 491
Melanoplus scudderi scudderi Uehler, 492
Melanoplus spp., 444
Melanotus oregonensis (LeC.), 116
Melitara prodenialis, 117
Melolontha melolontha (Linn.):
associated bacteria–*Bacillus hoplosternus*, 65; *B. melolonthae*, 71; *B. melolonthae liquefaciens alpha, beta, gamma*, 71, 72; *B. melolonthae nonliquefaciens alpha, beta, gamma, delta, epsilon*, 72; *B. septicus insectorum*, 86; *B. tracheitis sive graphitosus*, 93; *Bacterium melolontha liquefaciens*, 108; *gamma*, 108; *Diplococcus melolonthae*, 137, 185; *Micrococcus nigrofaciens*, 148
associated protozoan–*Polymastix melolonthae*, 483
Melolontha sp., 495
Melolontha vulgaris Fabr.:
associated bacteria–*Bacillus similcarbonchio*, 86, 87; *B. similtubercolari*, 87; *B. tifosimili*, 93; *Bacterium melolonthae liquefaciens*, 108; *Clostridium sporogenes*, 95, 96; *C. tetani*, 96; *Coccobacillus cajae*, 183, 184; *Pseudomonas aeruginosa*, 169; *P. fluorescens*, 170, 171; *Serratia marcescens*, 133; *Staphylococcus albus*, 153, 154; *S. aureus*, 155; *S. citreus*, 156
Melomys littoralis, 300
Melophagus, 246
Melophagus ovinus (Linn.):
associated bacterium–*Bacillus anthracis*, 53, 54
associated protozoa–*Nosema apis*, 515; *Trypanosoma melophagium*, 480
associated rickettsia–*Rickettsia melophagi*, 44, 244, 312-315
associated spirochete–*Spirochaeta melophagi*, 460
extracellular symbiotes of, 205
intracellular symbiotes in, 244, 245
Menopon biseriatum Nitz., 200; *M. gallinae* Linn., 220; *M. pallidum* Nitz., 220; *Menopon* sp., 219
Menoponidae, 219
Menospora polyacantha, 494
Mephitis elongata, 286
Mermithidae, 7
Meroblastic cleavage, 194
Merozoites, 487, 496, 497, 498, 501, 505, 519
Mesoderm, 194
Metacoronympha senta, 545, 546
Metadevescovina carina, 545; *M. cristata*, 545; *M. cuspidata*, 545; *M. debilis*, 545; *M. extranea*, 545; *M. fulleri*, 545; *M. magna*, 545; *M. mediocris*, 545; *M. modesta*, 545; *M. modica*, 545; *M. nitida*, 545; *M. nudula*, 545; *M. patula*, 545; *M. polyspira*, 545; *M. stereociliata*, 545; *M. turbula*, 545
Metamorphosis, survival of bacteria through, 12, 14, **19-20**, 50, 54, 121, 125, 127, 132, 133, 135, 152, 156, 169, 171, 179
Metarrhizium anisopliae, 396, 397, 400, 402
Metatachardia conchiferata, 375
Methods and procedures for microbiological examinations of insects and ticks, 573-601
Miarus, 210; *M. campanulae* Linn., 238, 240
Mice, 285, 289, 315, 319, 320 (*see also* Mouse)
Micrattacus nanus, 514
Microbiologic examination of insects and ticks, 580-601; preparation of specimens, 580-581; record keeping and forms, 594-601
Microbracon hebetor Say, 524
Micrococcaceae, 37, 41, 144
Micrococcus, 41, **144**, 450; *M. acridicida*, 153; *M. amylovorus*, 116; *M. batrochorum*, 37; *Micrococcus* C, 10, 144; *M. candidus*, 24; *M. catarrhalis*, 157; *M. chersonesia*, 144; *M. cinnebareus*, 24; *M. conglomeratus*, 45; *M. curtissi*, 145; *M. ephestriae*, 145; *M. epidermidis*, 145; *M. flaccidifex danai*, 145-146; *M. flavus*, 146; *M. freudenreichii*, 146; *M. galleriae* No. 3, **146**, 561; *M. insectorum*, 146-147; *M. lardarius*, **147**, 417; *M. luteus*, 181; *M. luteus liquefaciens*, 147; *M. major*, 147; *M. melitensis*, 273; *M. neurotomae*, 147-148; *M. nigrofaciens*, 148; *M. nitrificans*, 86, **148**; *M. nonfermentans*, 149; *M. ochraceus*, 149; *M. ovatus*, 149; *M. parvulus*, 149; *M. perflavus*, 24; *M. pieridis*, 150; *M. pyogenes aureus*, 155; *M. rushmori*, 69, **150**; *M. saccatus*, 151; *M. subflavus*, 151; *M. tetragenus*, 144; *M. vulgaris*, 151
Microfilaria bancrofti, 321
Microgamete, 502
Microhodotermes, 528
Microjoenia hexamitoides, 551; *M. pyriformis*, 551; *M. ratcliffei*, 551
Microklossia sp., 525
Microlynchia pusilla (Speiser), 506

Micronucleocytes, 564, 570; with spherules, 564
Microphages, 562
Micropyle, 21, 193, 195, 208, 209
Microrhopalodina enflata, 547; *M. kofoidi*, 547; *M. multinucleata*, 547; *M. occidentis*, 547
Microsporidia, 511-518, 523
Microsporidian, 425
Microsporidium polyedricum, 417
Microsymbiote, 5, 189, 201, 214
Microthyriales, 383
Microtus montebelli, 294; *M. pennsylvanicus pennsylvanicus*, 285
Microvelia, 469; *M. americana* (Uehler), 469
Midges, biting, 243
Milk glands, 206, 245, 246
Milkweed bug (see *Oncopeltus fasciatus*)
Milky disease, 67, 68, 82, 83
Mimorista flavidissimalis, 117
Minas Geraes typhus, 329
Mirotermes hispaniolae Banks: associated protozoa—*Endamoeba disparata*, 553; *E. majestas*, 553; *Endolimax termitis*, 554; *Trichomonas labelli*, 540
Mirotermes panamaensis: associated protozoa—*Endamoeba sabulosa*, 553; *E. simulans*, 553; *Gregarina mirotermitis*, 553
Mite typhus, 264
Mites, 162, 215, 254, 282, 292, 385, 410, 499; harvest, 255 (*see also* Acarina *and* specific names)
Mitochondria, 202, 204, 367
Mitotic division, 194
Mixotricha paradoxa, 550
Miyagawanella, 263; *M. lymphogranulomatosis*, 262, 268
Moellerius heyeri, 407, 408
Mogannia hebes, 374
Molds, examination of insects and ticks for, 580, 589
Molytes germanus Linn., 238
Monarthrum, 405; *M. dentiger* (LeC.), 404; *M. fasciatum* (Say), 404; *M. mali* (Fitch), 404, 405; *M. scutellare* (LeC.), 404
Moneilema spp., 117
Monieziella arborea Jacot, 410
Monilia, 405; *M. candida*, 376, 402, 405; *M. niger*, 375
Moniliales, 393
Monkey, 277, 278, 283, 285, 294, 300, 307, 310, 315, 320, 324, 334, 338, 343, 345, 346
Monkey louse (see *Pedicinus eurygaster*)
Monoblepharidales, 377
Monocercomonas axostylis, 539; *M. mackinnoni*, 484; *M. melolonthae*, 484; *M. termitis*, 539
Monocercomonoides globus, 536, 546; *M. ligyrodis*, 483; *M. panesthiae*, 537

Monochamus notatus (Dru.), 412; *M. scutellatus* (Say), 412
Monocnidea, 512
Monoecomyces, 386
Monomita muscarum, 471
Monomorium vastator, 162, 164
Monopsyllus anisus (Roth.), 161
Monospora, 351
Monosporella, 351; *M. unicuspidata*, 351
Moose disease, 123-124
Morax-Axenfeld bacillus, 101, 159
Morreira odorata, 473
Moschomyces, 385
Mosquitoes, 3, 32, 162, 164, 169, 472, 556
Mouse, 300; cotton, 285; deer, 285, 337; field, 346; golden, 285; house, 285; meadow, 285, 337; oldfield, 285; white, 286, 324, 338, 342; white-footed, 285
Mouse louse (see *Polyplax serratus*)
Mrazekia bacilliformis, 518; *M. brevicauda*, 518; *M. tetraspora*, 518
Mrazekiidae, 513, 517, 523
Mucidus sp., 382
Mucor hiemalis, 402
Mucorales, 377
Mud-dauber wasp (see *Sceliphron caementarium*)
Murgantia histrionica (Hahn), 16
Mus concolor, 294; *M. diardii*, 294; *M. musculus musculus*, 285; *M. rattus rufescens*, 294
Musca, 350
Musca domestica Linn.: associated bacteria—*Aerobacter aerogenes*, 120; *A. cloacae*, 120-121; *Alcaligenes faecalis*, 180; *BACILLUS A*, 50; *B. anthracis*, 12, 53; *B. colisimile*, 60; *B. gasoformans non-liquefaciens*, 63; *B. grünthal*, 64; *B. lutzae*, 69; *B. piocianemus*, 80; *B. radiciformis*, 84, 85; *B. similcarbonchio*, 86, 87; *B. tifosimile*, 93; *B. vesiculosis*, 94; *BACTERIUM agrigenum*, 99; *B. conjunctivitidis*, 101; *B. delendaemuscae*, 102; *Brucella abortus*, 158, 159; *Corynebacterium diphtheriae*, 43; *Eberthella belfastiensis*, 126; *E. typhosa*, 12, 127, **128**; *Erwinia amylovora*, 116, 117; *Escherichia coli*, 121, 122; *Gaffkya tetragena*, 144; *Hemophilus duplex*, 159; *Micrococcus flavus*, 146; *Mycobacterium leprae*, 46, 47, 48; *M. tuberculosis*, 48, 49; *Neisseria gonorrhoeae*, 157; *N. intracellularis*, 158; *Pasteurella cuniculicida*, 161; *P. pestis*, 162; *P. tularensis*, 166; *Proteus morganii*, 125; *P. vulgaris*, 126; *Pseudomonas aeruginosa*, 14, 19, **169**; *P. fluorescens*, 170, 171; *P. jaegeri*, 171; *P. nonliquefaciens*, 171, 172; *Salmonella choleraesuis*, 129; *S. enteritidis*, 12, **129**; *S. paratyphi*, 130; *S. schottmülleri*, 130; *Sarcina aurantiaca*, 152; *Serratia kielensis*, 132; *S. marces-*

SUBJECT INDEX

cens, 12, **133**; *Shigella dysenteriae*, 12; *S. paradysenteriae*, 131, 132; *Shigella* sp., 131; *Staphylococcus albus*, 153, 154; *S. aureus*, 155, 156; *S. citreus*, 156; *S. muscae*, 157; *Staphylococcus* spp., 152, 153; *Streptococcus agalactiae*, 138; *S. equinus*, 141; *S. faecalis*, 141; *S. lactis*, 142; *S. pyogenes*, 143; *S. salivarius*, 143, 144; *Vibrio comma*, 27, 178
associated fungus–*Empusa muscae*, 379, 380
associated protozoa–*Endamoeba histolytica*, 485; *Herpetomonas muscarum*, 466; *Octosporea muscae-domesticae*, 518
associated rickettsia–*Rickettsia conjunctivae*, 304
associated spirochete–*Treponema pertenue*, 459
associated virus–poliomyelitis, 441, 442
bacterial flora of, 12, 18
bactericidal substance in, 34
bacteriophage in, 449
survival of bacteria during metamorphosis, 12, 19
Musca humilis, 471
Musca inferior, 160, 161
Musca nebulo Wied., 471
Musca leprae Linn., 46, 47
Muscardine diseases, 376, 387, 394, 397
Muscidae, 246
Muscina assimilis (Fall.), 116
Muscina stabulans Fall.:
associated bacteria–*Brucella abortus*, 158, 159; *Erwinia amylovora*, 116
associated virus–poliomyelitis, 441
Mutualism, 5, 210, 211, 212; definition, 188; facultative, 210
Mutualist, 260
Mycetobia pallipes Meig., 518
Mycetoblasts, 198, 199
Mycetocyte, 190, **196**, 197, 200, 210, 216, 217, 219, 220, 223, 225, 226, 229, 230, 231, 238, 241, 243, 244, 246, 248, 249, 362, 364, 367, 368, 371, 373, 570; depot, 225, 226; size, 198
Mycetom, 190, 222, 223
Mycetome, 190, **196-201**, 209, 211, 212, 220, 221, 225, 230, 236, 238, 243, 244, 262, 264, 265, 354, 357, 359, 360, 361, 364, 365, 371, 373, 374, 570:
azyqous, 225, 226
functions of, 201
in–Aleurodids, 199; Anoplura, 199; *Apion*, 199; *Blatta orientalis*, 200; *Blattella germanica*, 200; *Bostrychoplites zickeli*, 199; *Calandra*, 199; *Camponotus*, 199; Chermidae, 230; *Cicada orni*, 196; cicadas, 199; *Cimex lectularius*, 199, 233, 316, **317**, 318; coccids, 199; *Donacia*, 199; *Eomenacanthus stramineus*, 200; *Ephestia kuhniella*, 200; *Formica*

fusca, 199; *Haematopinus suis*, 199; *Hylobius abietis*, 237; *Icerya purchasi*, 364; lice, 221, 222; *Lyctus*, 199; *Lyctus lineans*, 237, 238; *Macrosiphum tanaceti*, 198; *Margarodes polonicus*, 374; *Menopon biseriatum*, 200; mites, 254; *Oryzaephilus surinamensis*, 199, 200; *Pseudococcus adonidum*, 200, 231; *P. brevipes*, 231; *P. citri*, 197, 200, 231; *Psylla buxi*, 197, 200; *Rhizopertha*, 199; *R. dominica*, 199, 200; *Rhynchophorus ferrugineus*, 238; *Sinoxylon ceratoniae*, 199; *Sitophilus granarius*, 200, 238; *S. oryzae*, 238; *Stegobium paniceum*, 199, 200, **367**
tracheoles of, 198
transplantation of, 200
Mycetosporidium jacksonae, 486; *M. talpa*, 486
Mycobacteriaceae, 40, 42
Mycobacterium, 41, 46
Mycobacterium aquae, 46
Mycobacterium avium, 557
Mycobacterium lacticola, 46, 557, 567
Mycobacterium leprae, 46-48
Mycobacterium piscium, 556
Mycobacterium smegmatis (see *M. lacticola*)
Mycobacterium tuberculosis: associated insects, 48-50; immunity of insects to, 556; phagocytosis of, 568, 569; strains of, 48, 50
Mycobacterium tuberculosis var. *bovis*, 557
Mycobacterium tuberculosis var. *hominis*, 557
Mycoderma clayi, 351
Mydaea, 387
Myodocha serripes Oliv., 15
Myriangiaceae, 387, 388
Myriangium acaciae, 388; *M. brasiliensis*, 388; *M. curtissi*, 388; *M. dolichosporum*, 388; *M. duriaei*, 388; *M. montagnei*, 388; *M. philippinense*, 388
Myrmicinae, 406
Myxococcidium, 349; *M. stegomyiae*, 349
Myzus circumflexus (Buckt.), 445
Myzus persicae (Sulz.):
associated viruses–Cucumis virus, 445; Solanum virus *12*, 444; Solanum virus *14*, 445
Myzus pseudosolani Theobald, 445

Nagana, 476
Nairobi disease of sheep, 442
Naucoris maculatus Fabr., 468
Nauphoeta cinerea (Oliv.), 216
Necrophorus dauricus Motsch., 162
Nectria, 387; *N. galligena*, 412
Necydaliini, 371
Necydalis major Linn., 370
Neisseria, 41, 148, 157; *N. catarrhalis*, 34, 157; *N. gonorrhoeae*, 157, 557; *N. intracellularis*, 158, 258; *N. luciliarum*, 69, 158; *N. perflava*, 158
Neisseriaceae, 41, 157

Nemathelminthes, 7
Nematoda, 7
Nematodes, parasitic, **7,** 570
Nematomorpha, 7
Nematospora, 350; *N. coryli,* 350; *N. gossypii,* 350, 410; *N. phaseoli,* 350
Nemobius fasciatus var. *fasciatus* DeG., 135:
associated bacteria–*Aerobacter aerogenes,* 120; *Eberthella pyogenes,* 127; *Escherichia coli,* 121, 122; *Serratia plymouthensis,* 135
associated protozoa–*Leidyanna erratica,* 493; *L. gryllorum,* 493
Nemobius fasciatus var. *socius* DeG., 493
Nemobius sylvestris Fabr.:
associated protozoa–*Gregarina macrocephala,* 493; *Leidyanna gryllorum,* 493
Neoaplectana glaseri, 8
Neocleonus vittiger Fabr., 239
Neoconocephalus fratellus, 493
Neocurtilla hexadactyla, 522
Neohaematopinus laeviusculus (Grube), 165, 167, 168
Neotermes, 528
Neotermes aburiensis Sjöstedt:
associated protozoa–*Caduceia theobromae,* 542; *Foaina decipiens,* 544; *Parajoenia decipiens,* 544; *Pseudodevescovina brevirostris,* 545; *P. punctata,* 545
Neotermes amplus (Sjöstedt), 549
Neotermes castaneiceps Sjöstedt (see *Kalotermes castaneiceps*)
Neotermes castaneus Burm.:
associated protozoa–*Devescovina arta,* 541; *D. lepida,* 541; *Foaina reflexa,* 544; *F. solita,* 543; *Stylorhynchus* sp., 553
Neotermes connexus Snyder:
associated protozoa–*Devescovina exilis,* 541; *D. hawaiensis,* 541; *Foaina gracilis,* 543; *F. nana,* 543; *F. solita,* 543; *Oxymonas granulosa,* 546; *Parajoenia grassii,* 544; *Stephanonympha silvestrii,* 546
Neotermes dalbergiae Kalshoven:
associated protozoa–*Caduceia kalshoveni,* 542; *Devescovina parasoma,* 541; *Foaina nana,* 543; *Oxymonas grandis,* 547
Neotermes desneuxi (Sjöstedt):
associated protozoa–*Foaina ovata,* 543; *Trichonympha peplophora,* 549
Neotermes erythraeus Silvestri:
associated protozoa–*Bullanympha silvestrii,* 545; *Caduceia nova,* 542; *Devescovina robusta,* 541; *Foaina parvula,* 543; *F. reflexa,* 544; *Stephanonympha silvestrii* var. *neotermitis erythraei,* 546
Neotermes europae (Wasm.):
associated protozoa–*Foaina reflexa,* 544; *F. signata,* 544
Neotermes gestri Silvestri, 542

Neotermes gracilidens (Sjöstedt):
associated protozoa–*Foaina ovata,* 543; *Trichonympha peplophora,* 549
Neotermes greeni Holmgren:
associated protozoa–*Caduceia bugnioni,* 36, 460, 542; *Foaina nana,* 513; *F. solita,* 543
Neotermes holmgreni Banks:
associated protozoa–*Devescovina lepida,* 541; *Foaina reflexa,* 544; *F. solita,* 543; *Oxymonas,* spp., 547; *Tricercomitus divergens,* 538; *Tritrichomonas divergens,* 538
Neotermes howa (Wasm.):
associated protozoa–*Foaina ovata,* 543; *Trichonympha peplophora,* 549
Neotermes howa var. *mauritiana* Sjöstedt, 543
Neotermes insularis (White):
associated protozoa–*Devescovina lemniscata,* 541; *Foaina exempta,* 543; *Pseudodevescovina uniflagellata,* 545
Neotermes larseni Light:
associated protozoa–*Devescovina lemniscata,* 541; *Foaina nana,* 543
Neotermes longus (Holmgren) (see *Kalotermes longus*)
Neotermes meruensis Sjöstedt:
associated protozoa–*Devescovina glabra,* 540; *Foaina reflexa,* 544; *F. signata,* 544; *Trichonympha teres,* 549
Neotermes militaris (Desneux):
associated protozoa–*Nyctotherus termitis,* 554; *Trichonympha zeylanica,* 549
Neotermes minutus (Kemner), 544
Neotermes sonneratiae Kemner:
associated protozoa–*Devescovina parasoma,* 541; *Foaina nana,* 543; *F. solita,* 543
Neotermes spp.:
associated protozoa–*Devescovina glabra,* 540; *D. insolita,* 541; *D. lemniscata,* 541; *Foaina parvula,* 543; *F. reflexa,* 544
Neotermes tectonae Dammerman:
associated protozoa–*Caduceia monile,* 542; *Devescovina parasoma,* 541; *Foaina nana,* 548; *F. solita,* 543; *Oxymonas grandis,* 547
Neotermes zuluensis Holmgren:
associated protozoa–*Devescovina glabra,* 540; *Foaina reflexa,* 544; *F. signata,* 544
Nepa, 183, 184
Nepa cinerea Linn.:
associated protozoa–*Barrouxia ornata,* 498; *Coleorhynchus heros,* 495; *Leptomonas jaculum,* 468; *Syncystis mirabilis,* 497
Nepa jaculum, 472
Nephelodes emmedonia (Cramer), 414
Nephotettix apicalis var. *cincticepts,* 447
Nephrocytes, 562
Nepticula malella, 31
Neuroctena anilis Fall., 471
Neuroptera, 214, 218, 385

SUBJECT INDEX

Neurotoma nemoralis Linn.:
  associated bacteria–*Bacillus neurotomae*, 74; *Bacterium neurotomae*, 27, 109; *Micrococcus neurotomae*, 147-148
Neutralization test, 587
New York bee disease, 73, 91
Nezara hilaris Say (= *Acrosternum hilaris*), 350; *N. viridula* (Linn.), 350
Nicotiana virus *1*, 445
"Nine mile fever," 295, 301 (see also *Rickettsia burneti*)
Nirmus, 219; *N. fuscus* Nitz., 219; *N. merulensis* Denny, 219; *N. subtilis* Giebel, 219
Nitzschia sp., 219
Noctua clandestina Harris, 414
Noctuidae, 414, 425
Nomenclature: of bacteria, 38-40; of rickettsia, 262-264
Nonpolyhedral diseases, 426-432
Nonscutate ticks (see Argasidae)
Nosema adiei, 515; *N. anophelis*, 512; *N. apis*, 514; *N. astyrae*, 514; *N. auriflammae*, 514; *N. baëtis*, 514; *N. bombi*, 514; *N. bombycis*, 511, 512, 513-514, 522, 523, 524, 558; *N. caeculiae*, 514; *N. chironomi*, 515; *N. ctenocephali*, 515; *N. culicis*, 515; *N. ephemerae*, 514; *N. ephialtis*, 514; *N. erippi*, 514; *N. eubules*, 514; *N. halesidotidis*, 514; *N. heliotidis*, 514; *N. hydriae*, 514; *N. junonis*, 514; *N. longifilum*, 515; *N. lophocampae*, 514; *N. lysimniae*, 514; *N. micrattaci*, 514; *N. pulicis*, 515; *N. sabaunae*, 514; *N. schneideri*, 514; *N. stegomyiae*, 515; *N. termitis*, 552; *N. vanillae*, 514
Nosematidae, 513-517, 523
Nosopsyllus fasciatus (Bosc d'Antic):
  associated bacteria–*Pasteurella pestis*, 161, 163; *P. tularensis*, 165
  associated protozoa–*Agrippina bona*, 495; *Legerella grassii*, 498; *Malpighiella refringens*, 485; *Trypanosoma lewisi*, 478; *T. rabinowitschi*, 470
  associated rickettsia–*Rickettsia prowazeki* var. *typhi*, 281
  intracellular symbiotes in, 247
Notaris acridulus Linn., 240
Notocyrtus foveatus, 468
Notodontidae, 414
Notolophus antiqua (Linn.), 414
Notonecta sp.:
  associated bacterium–*Coccobacillus cajae*, 183, 184
  associated fungus–*Coelomomyces notonectae*, 382
NR bodies, 328
Nucleophaga, 37
Nucleus: mitotic division of, 194; parasitization of, 37; zygote, 194

Nun moth (see *Lymantria monacha*)
Nurse cells, 192, 206
Nycteribia biarticulata Herman, 245, 246; *N. blainvillei* Leach, 246; *N. blasii* Kol., 245, 246; *N. kollari*, 245
Nycteribiidae, 244, 245
Nycteribosca kollari Frfld. (see *Nycteribia kollari*)
Nyctobates pennsylvanica DeG., 494
Nyctotherus buissoni, 521; *N. fletcheri*, 554; *N. hormeticae*, 522; *N. militaris*, 554; *N. neocurtillae*, 522; *N. osmodermae*, 522; *N. ovalis*, 521; *N. panesthiae*, 522; *N. pintoi*, 522; *N. silvestrianus*, 554; *Nyctotherus* spp., 521, 554; *N. termitis*, 554; *N. uichancoi*, 522, 537
Nygmia phaeorrhoea (Donov.):
  associated bacteria–*Bacillus hoplosternus*, 65; *B. lymantricola adiposus*, 70; *B. melolonthae liquefaciens*, 71, 72; *B. melolonthae non liquefaciens*, 72, 565; *Bacterium melolonthae liquefaciens gamma*, 108; *B. pieris liquefaciens, alpha*, 26, 111; *Diplobacillus melolonthae*, 185; *Diplococcus bombycis*, 136
  associated fungus–*Beauveria densa*, 394
  phagocytosis in, 565
Nymphalidae, 414
Nysius, 236; *N. euphorbiae* Horvath, 473; *N. lineolatus* Costa, 236; *N. punctipennis* H.S., 236; *N. senecionis* Schill., 236; *N. thymi* Wolff, 236

Oak caterpillar (see *Phryganidia californica*)
Octosporea monospora, 518; *O. muscae-domesticae*, 518
Oculium, 393
Odioporus glabricollus, 238
Oecanthus angustipennis Fitch, 412; *O. niveus* (DeG.), 412
Oecophoridae, 413
Oedalens nigrofasciatus, 181, 183
Oedipoda, 493; *O. coerulescens* Linn., 493
Oenocytes, 229, 429, 564, 565, 568
Oenocytoids, 563, 564
Oliarus horisanus Mats., 374
Oligomerus brunneus Stim., 367
Olive fly (see *Dacus oleae*)
Olyca junctoliniella, 117
Omophlus brevicollis:
  associated protozoa–*Ophryocystis hessei*, 496; *Thelohania cepedei*, 516
Omphalapion laevigatum Payk., 238, 240
Oncopeltus fasciatus Say:
  associated bacteria–*Eberthella insecticola*, 11, **127**; *Proteus insecticolens*, 11, **125**; *P. recticolens*, 11, **126**; *Serratia marcescens*, 133, 135; *Streptococcus faecalis*, 11, **141**

Oncopeltus fasciatus Say (Cont.)
  associated protozoa–*Phytomonas elmassiani*, 473
  rearing of, 575
Onion maggot (see *Hylemya antiqua*)
Oöcyst, 487, 489, 490, 496, 497, 500, 504
Oöcyte, 192, 206, 207, 208, 250, 498
Oögonia, 192
Oökinete, 487, 502, 507
Oösome, 194, 195
Oöspora, 393; *O. saccardiana*, 356 (see also *Lecaniocola saccardiana*)
Oötheca, 193
Opalinata, 520
Opartum sabulosum, 183, 184
Ophionectria, 387
Ophiostoma, 410; *O. ulmi*, 410
Ophryocystidae, 496
Ophryocystis hessei, 496; *O. mesnili*, 496
Ophryoglena collini, 521
Opossum, 285, 337 (see also *Didelphis virginiana*)
Opsonins, absence of in insects, 572
Opuntia, 117
Orchestia agilis, 521
Oregon wireworm (see *Melanotus oregonensis*)
Organ of synthesis, 201
Orgyia pudibunda (Linn.):
  associated bacteria–*Bacillus mycoides*, 74; *B. tingens*, 93
Oriental beetle (see *Anomala orientalis*)
Oriental cockroach (see *Blatta orientalis*)
Oriental sore (see *Leishmania tropica*)
Ormensis, 374
Ornithodoros, 453
Ornithodoros amblus, 477
Ornithodoros coprophilus, 13
Ornithodoros erraticus Lucas:
  associated rickettsia–*Dermacentroxenus conori*, 341
  associated spirochete–relapsing fever, 454
Ornithodoros furcosus Neum.:
  associated bacterium–*Klebsiella ozaenae*, 123
  associated protozoan–*Trypanosoma cruzi*, 477
Ornithodoros gurneyi Warburton, 297
Ornithodoros hermsi Wheeler, Herms, and Meyer:
  associated rickettsiae–*Dermacentroxenus rickettsi*, 330; *Rickettsia burneti*, 297
  associated spirochete–relapsing fever, 454, 456, 457
Ornithodoros lahorensis Neum.:
  associated bacterium–*Pasteurella tularensis*, 166, 167
  associated protozoan–*Trypanosoma evansi*, 481
  associated spirochete–relapsing fever, 454
Ornithodoros marocanus, 454

Ornithodoros moubata (Murray):
  associated bacteria–*Bacillus mesentericus*, 72, 73; *B. mycoides*, 74; *B. subtilis*, 89, 91; *Pasteurella tularensis*, 166; *Staphylococcus albus*, 153, 154; *S. aureus*, 155; *Streptococcus pyogenes*, 143
  associated rickettsiae–*Dermacentroxenus conori*, 341; *Rickettsia burneti*, 297; *R. tsutsugamushi*, 292
  associated spirochetes–*Borrelia anserina*, 457; *B. duttoni*, 456; *B. sogdianum*, 454; *B. recurrentis*, 454
  bactericidal substance in, 34
  intracellular symbiotes of, 253
  sterile gut of, 12
Ornithodoros nereensis, 454
Ornithodoros nicollei Mooser, 330
Ornithodoros normandi Larrousse, 454
Ornithodoros papillipes, 454, 457
Ornithodoros parkeri Cooley:
  associated bacterium–*Pasteurella tularensis*, 166, 168
  associated protozoan–*Trypanosoma cruzi*, 477
  associated rickettsiae–*Dermacentroxenus rickettsi*, 330, 331; *Rickettsia burneti*, 297
  associated spirochete–relapsing fever, 454
Ornithodoros rudis Karsch:
  associated rickettsia–*Dermacentroxenus rickettsi*, 330
  associated spirochete–relapsing fever, 454
Ornithodoros savignyi (Audouin), 454
Ornithodoros talaje (Guerin-Méneville), 454
Ornithodoros tartakovskyi, 454
Ornithodoros tholozani (Laboulbene & Megnin):
  associated protozoan–*Trypanosoma evansi*, 481
  associated spirochetes–*Borrelia persica*, 454; *B. sogdianum*, 453
Ornithodoros turicata (Duges):
  associated bacteria–*Pasteurella tularensis*, 166, 168
  associated rickettsiae–*Dermacentroxenus rickettsi*, 330, 331, 333; *Rickettsia burneti*, 297
  associated spirochete–relapsing fever, 454, 456, 457
Ornithodoros venezuelensis, 454
Ornithomyia avicularis Linn., 245
Orobitis cyaneus Linn., 240
Oropsylla silantiewi Wagner, 161, 163
Oroya fever, 345 (see also Carrión's disease)
Orsodacne atra (Ahrens), 116
Orthaea bilobata (Say), 470; *O. vincta* (Say), 470
Orthezia insignis Dougl., 231
Ortheziidae, 231-232
Orthocladius, 518
Orthognathotermes wheeleri Snyder, 539

SUBJECT INDEX

Ortholixus sanguineus Rossi, 239
Orthoptera, 214, 215, 378, 380, 385, 387, 443, 463, 563, 591
Orthotylus flavosparsus (Sahlbg.), 116
Oryctes nasicornis, 32, 483
Oryza virus *I*, 447
Oryzaephilus, 206, 209
Oryzaephilus surinamensis (Linn.), 213; mycetome of, 199, 200, 201, 241, 242, 443, 463, 563
Oryzaephilus unidentatus, 242
Oryzomys palustris palustris, 285
Osmoderma, 32; *O. scabra* (Beauvois), 522
Otiorhynchinae, 239
Otiorhynchus fuscipes Oliv.:
 associated protozoa–*Mycetosporidium talpa*, 486; *Nosema longifilum*, 515
Otiorhynchus gemmatus Scop. (see *Brachyrhinus gemmatus*)
Otiorhynchus inflatus Gyll. (see *Brachyrhinus inflatus*)
Otobius megnini (Duges), 13, 53
Otocentor nitens (Neum.):
 associated rickettsiae–*Dermacentroxenus rickettsi*, 331; *Rickettsia prowazeki* var. *typhi*, 282
Ovarian ampules, 221, 222, 225
Ovaries, **191, 192**, 194, 207, 208, 217, 218, 222, 233, 235, 249, 250, 251, 253, 254, 317, 325, 434, 456
Ovarioles, **191-193**, 208, 219, 374; types, 192, 193
Oviduct, 192, 193, 220, 221, 223, 225, 226
Oviposition, 193
Ovum, 194, 197
Oxycareni lavaterae, 470
Oxymirus, 372; *O. cursor* Linn., 370
Oxymonadidae, 531
Oxymonas, 547, 553; *O. barbouri*, 547; *O. brevis*, 547; *O. clevelandi*, 547; *O. dimorpha*, 546; *O. gracilis*, 547; *O. grandis*, 547; *O. granulosa*, 546; *O. hubbardi*, 547; *O. janicki*, 547; *O. jouteli*, 547; *O. kirbyi*, 547; *O. minor*, 547; *O. ovata*, 547; *O. panamae*, 547; *O. parvula*, 546; *O. pediculosa*, 547; *O. projector*, 546; *O. synderi*, 547
Oxystoma craccae Linn., 240
Oxytoma, 238
Oxyuridae, 7

P factor, 279
Pachycerus madidus Oliv., 239
Paederus, 386
Palaeopsylla gracilis, 479
Palomena prasina, 17
Pamphagus, 493
Panesthia angustipennis, 522
Panesthia javanica, 537:

 associated protozoa–*Endamoeba javanica*, 537; *E. philippinensis*, 537; *Hexamita cryptocerci*, 537; *Monocercomonoides panesthiae*, 537; *Nyctotherus uichancoi*, 522, 537; *Paraclevelandia brevis*, 537; *P. constricta*, 537; *P. contorta*, 537; *P. elongata*, 537; *P. hastula*, 537; *P. panesthiae*, 537; *P. parapanesthiae*, 537; *P. simplex*, 537
Panesthia spadica:
 associated protozoa–*Endamoeba javanica*, 537; *Hexamita cryptocerci*, 537; *Monocercomonoides panesthiae*, 537; *Nyctotherus uichancoi*, 537; *Paraclevelandia brevis*, 537; *P. constricta*, 537; *P. contorta*, 537; *P. nipponensis*, 537; *P. panesthiae*, 537; *P. simplex*, 537
Panesthiinae, 536
Pangonia neavei, 468
Panstrongylus megistus (Burm.)
Papainase, 23
Papilio pompejus Jablonsky & Hbst., 514
Pappataci fever, 442
Para-aminobenzoic acid, 277, 336
Para-aminophenylsulfonamide, 130, 156
Paraclevelandia brevis, 537; *P. constricta*, 537; *P. contorta*, 537; *P. elongata*, 537; *P. hastula*, 537; *P. nipponensis*, 537; *P. panesthiae*, 537; *P. parapanesthiae*, 537; *P. simplex*, 537
Parajoenia decipiens, 544; *P. grassii*, 544
Paraneotermes, 528
Paraneotermes simplicicornis Banks:
 associated protozoa–*Foaina taeniola*, 543; *Hoplonympha natator*, 548; *Kofoidia loriculata*, 548; *Macrotrichomonas lighti*, 542; *Oxymonas dimorpha*, 546; *Spirotrichonympha bispira*, 551
Parasite, 203, 210, 260
Parasitism, 5, 188, 203, 211
Parasol ants, 406
Paratetranychus yothersi (Burm.), 492
Parcoblatta, 207
Parcoblatta lata, 468
Parcoblatta pennsylvanica (DeG.):
 associated protozoa–*Gregarina blattarum*, 491; *G. parcoblattae*, 492; *G. thomasi*, 492; *Leptomonas* sp., 468; *Nyctotherus ovalis*, 521
 intracellular symbiotes in, 217
Parcoblatta uhleriana Sauss., 492
Parcoblatta virginica Brunner:
 associated protozoa–*Gregarina ohioensis*, 492; *Leptomonas* sp., 468
 intracellular symbiotes in, 217
Pardalophora apiculata (Harr.):
 associated protozoa–*Actinocephalus* sp., 494; *Gregarina rigida columna*, 492
Paria canella (Fabr.) var. *gilvipes* Horn:
 associated bacteria–*Escherichia coli*, 121, 122; *E. freundii*, 122

Parlatorea olea (Colvée), 374
Parrhinotermes, 529
Parthenogenesis, 191, 198, 201
Parvobacteriaceae, 41, 158
Passive immunity (*see* Immunity in insects, passive)
Pasteurella, 42, 160, 161, 165
Pasteurella avicida, 160, 161
Pasteurella aviseptica, 160 (see also *Pasteurella avicida*)
Pasteurella bollingeri, 160
Pasteurella bubaliseptica, 160
Pasteurella cuniculicida, 161
Pasteurella pestis, 33, 79, **161-165**, 556, 557; bacteriophage against, 450; blocking action of in fleas, 163-164
Pasteurella tularensis, 165-169, 262
Pasteurelleae, 42, 159
Pastorianus, 349
Pathogenicity of various bacteria for *Galleria mellonella*, 557-558
Peach yellows, 447
Pébrine, **513**, 522, 555, 558
Pectinophora gossypiella (Saunders):
  associated bacteria—*Bacillus pectinophorae*, 77, 78; *Bacterium ephestiae* Nos. *1-2*, 103
Pedicel, 192
Pedicinus albidus Rudow, 270; *P. eurygaster* (Furm.), 456; *P. longiceps* Piaget, 270, 454; *P. rhesi* Fahr., 223
Pediculidae, 221-223
Pediculus, 201, 219, 223, 226, 321
Pediculus capitis (see *Pediculus humanus capitis*)
Pediculus corporis (see *Pediculus humanus corporis*)
Pediculus humanus Linn., 271:
  associated bacteria—*Bacillus pediculi*, 78; *Pasteurella pestis*, 162
  associated rickettsiae—*Rickettsia prowazeki* var. *typhi*, 283; *R. rocha-limae*, 319; *R. tsutsugamushi*, 292; *R. weigli*, 312
  associated virus—poliomyelitis, 441
  role of symbiotes in life of, 211, 212
Pediculus humanus capitis DeG.:
  associated bacteria—*Eberthella typhosa*, 127, 128; *Mycobacterium leprae*, 46; *Staphylococcus* sp., 152, 153
  associated rickettsiae—*Rickettsia prowazeki* var. *prowazeki*, 270
  intracellular symbiotes in, 221-223
Pediculus humanus corporis DeG.:
  associated bacteria—*Eberthella typhosa*, 127, 128; *Staphylococcus* sp., 152, 153
  associated rickettsiae—*Rickettsiae prowazeki* var. *prowazeki*, 270; *R. prowazeki* var. *typhi*, 282, 283; *R. wolhynica*, 288
  associated spirochete—relapsing fever, 454

  intracellular symbiotes in, 223
Pegomyia calyptrata Zetterstedt, 116
Penicillin, 212, 277
Penicillium, 24, 387, 393, 395
Peniophora gigantia, 412
Penis, 193
Pentatomidae, 14, 16, 17, 120
Pentatrichomonoides scroa, 540
Pepsin, 23
Perapion curtirostre Germ., 240
Perapion, 239; *P. violaceum* Kirby, 240
Perezia aeschnae, 515; *P. legeri*, 515, 524; *P. mesnili*, 515, 523, 524, 525; *P. pieris*, 515, 523, 525; *P. pyraustae*, 515, 524, 525
Peribalus limbolarius Stål, 15, 16
Pericardial cell (see Cells, insect, pericardial)
Periplasm, 195
Periplaneta, 207
Periplaneta americana (Linn.):
  associated bacteria—*Achromobacter hyalinum*, 98; *Bacillus albolactis*, 59; *B. cereus*, 58, 59; *Brucella abortus*, 158, 159; *Corynebacterium periplaneta* var. *americana*, 45, **204**, 216; *Eberthella typhosa*, 127, 128; *Micrococcus nigrofaciens*, 148; *M. parvulus*, 149; *Mycobacterium leprae*, 46; *M. tuberculosis*, 48, 49; *Vibrio comma*, 178
  associated protozoa—*Balantidium blattarum*, 521; *B. ovatum*, 521; *Diplocystis schneideri*, 490; *Endamoeba blattae*, 485; *E. thomsoni*, 485; *Endolimax blattae*, 485; *Entomoeba histolytica*, 485; *Giardia lamblia*, 484; *Gregarina blattarum*, 491; *G. legeri*, 491; *Isotrich caulleryi*, 521; *Nyctotherus ovalis*, 521; *Plistophora periplanetae*, 517; *Tetratrichomastix blattidarum*, 483
Periplaneta australasiae (Fabr.), 485:
  associated protozoa—*Nyctotherus ovalis*, 521
  intracellular symbiotes in, 216
Periplaneta brunea Burm., 484
Perisporiales, 383
Perittopus, 468
Perkinsiella saccharicida Kirk., 448; *P. vastatrix* (Bred.), 448
Peromyscus gossypinus gossypinus, 285; *P. leukopus noveboracensis*, 285; *P. nuttalli aureolus*, 285; *P. polionotus polionotus*, 285
Peronosporales, 377
Pestalozzia, 405
Peyritschiella, 386
Peyritschiellaceae, 385, **386-387**
Peziotrichum, 393
Pezizales, 383
Pezomachus botrana, 181
Pfeifferella, 165
Phacidiales, 383
Phagocytes, 562, 563, **564-569**, 570
Phagocytic blood cells (see Cells, insect, blood)

Phagocytic response of larvae: to bacteria, 565, 566; to foreign cells, 567; to foreign particles, 567
Phagocytosis, **564-569**, 570; conditions affecting, 568; disposal of engulfed substances, 568; manner of 567, 568; specialized, 570
Phallus, 193
Phenacoccus aceris (Sign.), 363; *P. piceae*, 363
Pheropsophus, 385
Phialoides ornata, 494
Philopteridae, 219
Philosamia cynthia (Dru.), 414
Phithrogena semicolorata, 516
Phlebotomus (= *Flebotomus*), 346
Phlebotomus argentipes Annandale & Burnetti, 474
Phletbotomus intermedius, 46, 48, 474
Phlebotomus longicuspis, 474
Phletbotomus major, 474
Phletbotomus major var. *chinensis*, 474
Phlebotomus minutus, 474
Phlebotomus noguchii Shannon, 346
Phlebotomus papatasii Scopoli:
 associated protozoa–*Hepatozoon* sp., 499; *Leishmania braziliensis*, 474
 associated virus–pappataci fever, 442
Phlebotomus perniciosus:
 associated protozoan–*Leishmania donovani*, 474
 associated spirochete–*Spirochaeta phlebotomi*, 460
Phlebotomus sergenti Parrot, 474
Phlebotomus verrucarum Townsend, 346
Phlebotomus vexator, 328
Phlegethontius sexta (see *Protoparce sexta*)
Pholiota, 389; *P. gongylophora*, 406
Phoma lingam, 412; *P. (Isaria) stenobothri*, 400
Phomales, 395
Phorbia fusciceps Zettersted, 118 (see also *Hylemya cilicrura*)
Phormia regina (Meig.), 35, 471
Photinus pyralis (Linn.), 172
Photosynthesis, 412
Photuris pennsylvanicus (DeG.), 125
Phryganidia californica Pack., 414, 425
Phthia picta (Dru.), 350
Phthirius pubis (Linn.):
 associated bacterium–*Mycobacterium leprae*, 46, 48
 intracellular symbiotes in, 223
Phthorimaea operculella (see *Gnorimoschema operculella*)
Phycomycetes, 376, 377-382, 393
Phyllophaga:
 associated bacteria–*Micrococcus nigrofaciens*, 148; *Xanthomonas stewarti*, 177
 associated protozoa–*Polymastix melolonthae*, 483; *P. phyllophagae*, 483; *Retortamonas phyllophagae*, 482; *Vahlkampfia* spp., 486
Phyllophaga anzia (LeC.), 82, 83
Phyllophaga bipartita Horn, 82, 83
Phyllophaga ephilida (Say), 82, 83
Phyllophaga fusca (Frölich):
 associated bacteria–*Bacillus entomotoxicon*, 61, 62; *B. popilliae*, 82, 83
Phyllophaga rugosa (Melsh.), 82, 83
Phyllophaga vandinei (Smyth), 148
Phyllopsis, 230
Phylloxeridae, 230
Phymatosphaeria brasiliensis, 388
Phymatotrichum, 393
Physokermes abietis Mod., 359
Physokermincola physokermina, 359
Physostomum, 219
Phytomastigina, 463
Phytomonas (bacteria), 173 (see also *Xanthomonas*); *P. savastanoi*, 20
Phytomonas (protozoa), 464, 465, **472-473**; *P. bordasi*, 473; *P. davidi*, 472; *P. elmassiani*, 472; *P. leptovasorum*, 473; *P. lygaeorum*, 473
Phytonomus arator Linn., 240
Phytophthora phaseoli, 412
Pidonia lurida, 371
Pieridae, 414
Pieris brassicae (Linn.):
 associated bacteria–*Bacillus monachae*, 73, 74; *B. pieris agilis*, 79; *B. pieris fluorescens*, 79; *B. pieris liquefaciens alpha* and *beta*, 79; *B. pieris non liquefaciens alpha* and *beta*, 79; *B. proteidis*, 83; *B. similis*, 87; *B. thuringiensis*, 91, 92; *Bacterium pieris liquefaciens alpha*, 25, 26, 111; *B. pyrenei*, Nos. 1–3, 112; *Diplobacillus pieris*, 186; *Diplococcus pieris*, 137; *Micrococcus vulgaris*, 151; *Vibrio pieris*, 179
 associated fungus–*Empusa conglomerata*, 380, 381
 associated protozoa–*Nosema apis*, 515; *Perezia legeri*, 515; *P. mesnili*, 515, 525; *P. pieris*, 515; *Thelohania mesnili*, 524
 associated spirochete–*Spirochaeta pieridis*, 461
 associated virus, 414
 immunization of, 560
Pieris rapae (Linn.) = *Ascia rapae*:
 associated bacteria–*Aerobacter aerogenes*, 120; *A. cloacae*, 120-121; *Bacillus aerifaciens*, 51; *B. intrapallens*, 65; *Bacterium hemophosphoreum*, *Flavobacterium acidificum*, 114; *Micrococcus pieridis*, 150; *Staphylococcus acridicida*, 153
 associated viruses, 414
Piesma cinerea (Say), 448; *P. quadrata* Fieb., 448
Pig, 343

SUBJECT INDEX

Pigeon, 343
Pilagra fusiformis, 374
Pileocephalus, 491; *P. blaberae*, 494; *P. chinensis*, 494; *P. heerii*, 494; *P. striatus*, 494; *P. tachycines*, 494
Pimelia bifurcata:
  associated bacteria—*Bacillus anthracis*, 53, 55; *B. proteisimile*, 83; *B. radiciformis*, 84, 85; *Clostridium chauvoei*, 95; *C. sporogenes*, 95; *C. tetani*, 96; *Corynebacterium pseudodiphthericum*, 46; *Diplococcus pneumoniae*, 137, 138; *Klebsiella pneumoniae*, 124; *Malleomyces mallei*, 159; *Mycobacterium tuberculosis*, 48, 49; *Pseudomonas fluorescens*, 170; *P. non-liquefaciens*, 171, 172; *Sarcina alba*, 151; *Vibrio colerigenes*, 177, 178
Pimelia sardea Sol.:
  associated bacteria—*Bacillus anthracis*, 53, 55; *B. proteisimile*, 83; *B. radiciformis*, 84, 85; *Clostridium chauvoei*, 95; *C. sporogenes*, 95; *C. tetani*, 96; *Corynebacterium pseudodiphthericum*, 46; *Diplococcus pneumoniae*, 137, 138; *Klebsiella pneumoniae*, 124; *Malleomyces mallei*, 159; *Mycobacterium tuberculosis*, 48, 49; *Pseudomonas fluorescens*, 170; *P. non-liquefaciens*, 171, 172; *Sarcina alba*, 151; *Vibrio colerigenes*, 177, 178
Pimelia sp., 494
Pineus orientalis, 229; *P. pineoides*, 229; *P. pini*, 229; *P. strobus* Hgt., 229
Pink bollworm (see *Pectinophora gossypiella*)
Pionea forficalis Linn., 394
Piroplasma, 506 (see *Babesia*)
Pissodes notatus Fabr., 238, 240; *P. pini* Linn., 238
Plagiognathus politus Uehler, 116
Plague bacillus, 161-165 (see *Pasteurella pestis*); disease, 3
Plants: bacterial diseases of, 116-120, 173-177; virus diseases of, 443-449
Plasmal, 205
Plasmochin, 288
Plasmodiidae, 500-505
Plasmodium agamae, 505; *P. brodeni*, 505; *P. bubalis*, 505; *P. cathemerium*, 505, 558; *P. circumflexum*, 505; *P. elongatum*, 505; *P. falciparum*, 500, 502, 503; *P. inui*, 505; *P. kochi*, 505; *P. mabuiae*, 505; *P. malariae*, 500, 502, 503; *P. melanipherum*, 505; *P. minasence*, 505; *P. murinum*, 505; *P. nucleophilum*, 505; *P. ovale*, 500, 502; *P. polare*, 505; *P. praecox*, 505; *P. pteropi*, 505; *P. reichenowi*, 505; *P. relictum*, 558; *P. tropiduri*, 505; *P. vassali*, 505; *P. vaughni*, 505; *P. vivax*, 500, **501-502**, 503, 504
Plasmodroma, 463-519
Platydema excavatum (Say), 493

Platyedra gossypiella (Saunders), 104, 113
Platypleura kaempferi, 374
Platypus, 405; *P. compositus* Say, 404; *P. flavicornis* (Fabr.), 414; *P. quadridentatus* (Oliv.), 404
Platysamia cecropia (Linn.) (see *Samia cecropia*)
Platyzosteria, 217; *P. armata*, 217
Plerocercoid, 8
Plinia ampla, 374
Plistophora kudoi, 516; *P. periplaneta*, 517; *P. simulii*, 516; *P. stegomyiae*, 516; *P. vayssierei*, 516
Plodia interpunctella (Hbn.):
  associated bacterium—*Bacillus cereus*, 58, 59
  associated protozoan—*Mattesia dispora*, 497
Plusia brassicae Riley (see *Autographa brassicae*)
Pneumococco sp., 138
Podonectria coccicola, 400, 401
Poecilus koyi, 183, 184
Polar bodies, 194, 195
Polia oleracea Linn., 52, 53
Poliomyelitis, 441-442
Polistes sp., 116
Polyascomyces, 386
Polychrosis, 400
Polyete lardaris, 380
Polyhedra, 417, 418, 419, 421, 423, 424, 425, 426; chemical structure of, 416; morphology of, **415-416**; staining of, 416
Polyhedral bodies (see Polyhedra)
Polyhedral diseases, 415-426
"Polyhedrikrankheit" (see Viruses, diseases, insects)
Polymastigida, 463, **483-484**, 536
Polymastix melolonthae, 483; *P. phyllophagae*, 483
Polyphylla fullo Linn., 394
Polyplax, 226; *P. serrata* Burm., 165, 166; *P. spinulosus* (Burm.), 226, 270, 281
Pomace fly (see *Drosophila funebris*)
Popillia japonica Newm.:
  associated bacteria—*Bacillus lentimorbus*, 67-68, 82-83; *B. popilliae*, 67-68, 82-83
  associated nematode—*Neoaplectana glaseri*, 8
  associated protozoa—*Actinocephalus*, 494; *Endolimax*, 485; *Endomoeba*, 485; *Monocercomonas mackinnoni*, 484; *M. melolonthae*, 484; *Retortamonas phyllophagae*, 482; *Tetratrichomastix mackinnoni*, 483
Poroniopsis bruchi, 408
Porotermes, 528
Porotermes adamsoni (Froggatt):
  associated protozoa—*Joenina pulchella*, 548; *Pseudotrypanosoma giganteum*, 540; *P. minimum*, 540; *Spirotrichonympha mirabilis*,

SUBJECT INDEX

551; *S. pundibunda*, 551; *Trichonympha magna*, 549
Porotermes grandis Holmgren:
 associated protozoa–*Pseudotrypanosoma giganteum*, 540; *P. minimum*, 540; *Spirotrichonympha grandis*, 551
Porthesia auriflua:
 associated bacteria–*Bacillus lineatus*, 68; *B. monachae*, 73, 74; *B. similis*, 87
Porthetria dispar (Linn.):
 associated bacteria–*BACILLUS bombycis non liquefaciens*, 57; *B. gortynae*, 64; *B. hoplosternus*, 65; *B. liparis*, 69; *B. lymantriae alpha* and *beta*, 69, 70; *B. lymantricola adiposus*, 70; *B. melolonthae*, 71; *B. melolonthae liquefaciens alpha, beta, gamma*, 71, **72**, 570; *B. melolonthae non liquefaciens alpha, beta, gamma, delta, epsilon*, 72; *B. thuringiensis*, 91, 92; *BACTERIUM cazaubon*, 100; *B. lymantricola adiposus*, 26, **106**; *B. melolonthae liquefaciens gamma*, 25, **108**; *B. pieris liquefaciens alpha*, 25, 26, **111**; *B. pyrenei*, Nos. *1, 2, 3*, 112; *Diplobacillus melolonthae*, 185; *Diplococcus bombycis*, 136; *D. liparis*, 136; *D. lymantriae*, 136; *D. melolonthae*, 137; *Gyrococcus flaccidifex*, 186; *Serratia marcescens*, 132, 134; *Streptococcus disparis*, 140
 associated viruses, 414, 425; wilt of, 415, 416, 421, 422
Potato: leaf roll of, 445; ring rot of, 176; yellow dwarf of, 447
Potato flea beetle (see *Epitrix cucumeris*)
Potosia, 32; *P. cuprae* Fabr., 31, 96-97
Powder-post beetles (*see* Bostrichidae)
Prairie dogs, 168
Prenolepis imparis (Say), 116
Preparation of insect specimens, 574-579, 580-581
Pretibial fever, 345
Primary epithelium, 194
Primitive streak, 194
Primordial germ cells, 192
Prismatospora evansi, 494
Procedures and methods, 573-601
Procryptotermes, 528:
 associated protozoa–*Devescovina fissa*, 540; *Foaina minuscula*, 544; *F. nana*, 543; *F. solita*, 543; *Macrotrichomonas virgosa*, 36, **542**; *Metadevescovina cristata*, 545; *M. nitida*, 545; *Trichonympha corbula*, 37, 549
Procryptotermes paradoxus, 545
Prodenia, 75
Prodenia eridania (Cramer), 58, 59
Prodenia litura (Fabr.):
 associated bacteria–*Bacterium gellechiae* Nos. *1-3*, 104; *B. prodeniae*, 111-112
 associated viruses, 414

Proleucocytes, 563
Prolophomonas kalotermitis, 548; *P. tocopola*, 484, 563
Prorhinotermes, 528
Protapion, 238; *P. aestivum* Germ., 240; *P. assimile* Kirby, 240; *P. flavipes* Payk., 240; *P. nigritarse* Kirby, 240; *P. varipes* Germ., 240
Proteae, 41, 124
Protection test, for rickettsiae, 587
Proteus, 41, **124**, 279, 450; *P. alveicola*, 101, 124; *P. ammoniae*, 125; *P. bombycis*, 124-125; *P. insecticolens*, 11, 125; *P. morgani*, 125; *P. nocturarum*, 75; *Proteus* OX-K, 279, 295, 344; *Proteus* OX-2, 279, 344; *Proteus* OX-19, 279, 286, 295, 344; *Proteus* X-K, 283; *Proteus* X-19, 283, 286; *P. photuris*, 125; *P. recticolens*, 11, 126; *P. sphingidis*, 89 (see also *Bacillus sphingidis*); *P. vulgaris*, 113, **126**, 279, 556, 558, 566, 567
Protomagalhaensia serpentula, 493
Protomerite, 491-497
Protomonadida, 463, **464-483**, 538
Protoparce, 61, 62
Protoparce quinquemaculata (Haw.):
 associated bacteria–*Bacillus nocturarum*, 75; *B. sphingidis*, 88, 89
Protoparce sexta (Johan.):
 associated bacteria–*Bacillus nocturarum*, 75; *B. sphingidis*, 88, 89; *Xanthomonas angulata*, 173
Protoplattoidae, 536
Protopulvinaria pyriformis Ckll., 357
Protozoa, 462-554; in arthropod hosts, 462-463; diseases caused by, 522-525; examination of insects and ticks for, 590-594; in termites, 526-554; termites and, physiologic aspects, 533-535
Protozoan disease of insects, 522-525
Prunus virus *1*, 448
Psammolestes coreodes Bergroth, 233
Pseudocleonus cinereus Schrank, 239
Pseudococcus adonidum (Linn.): intracellular symbiotes in, 231; mycetome of, 200; yeast of, 363
Pseudococcus brevipes (Ckll.), 231, 232, 363
Pseudococcus citri (Risso):
 associated bacterium–*Serratia marcescens*, 132, 134
 associated yeasts, 351, 362
 intracellular symbiotes in, 231
 mycetome of, 197, 200
Pseudococcus comstocki (Kuwani), 400
Pseudococcus diminutus Leon., 362
Pseudococcus medanieli, 362
Pseudodevescovina brevirostris, 545; *P. punctata*, 545; *P. ramosa*, 545; *P. uniflagellata*, 545

SUBJECT INDEX

Pseudograsserie (pseudojaundice) No. *1*, 415, 430-431
Pseudograsserie (pseudojaundice) No. 2, 415, 431-432
Pseudolynchia canariensis (Macq.), 245
Pseudomonadaceae, 42, 169
Pseudomonadeae, 42, 169
Pseudomonas, 42, **169**; *P. aeruginosa*, 14, 19, 24, 28, **169-170**, 556, 558, 560; *P. fermentans*, 29 (see also *Flavobacterium fermentans*); *P. fluorescens*, 20, 28, **170-171**; *P. jaegeri*, 171; *P. non-liquefaciens*, 171-172; *P. ovalis*, 172; *P. saliciperda*, 175; *P. septica*, 172
Pseudonuclei, 227
Pseudoparlatoria parlatorioides (Comst.), 374
Pseudotarsonemoides innumerabilis Vitzth., 410
Pseudotrichonympha grassii, 550; *P. hertwigi*, 550; *P. introflexibilis*, 550; *P. magnipapillosa*, 550; *P. parvipapillosa*, 550; *P. pristina*, 550; *P. sphaerophora*, 550
Pseudotrypanosoma giganteum, 540; *P. minimum*, 540
Pseudotsuga taxifola, 175
Pseudovitellus, 190, 197
Psila rosae (Fabr.), 118, 119
Psilopa petrolei Coq., 11
Psittacosis, 262
Psocidae, 218
Psocus, 218
Psorophora ciliata (Fabr.), 439; *P. columbiae* Dyar & Knab, 442; *P. ferox* (Humb.) 433; *P. sayi* Dyar & Knab (= *P. ferox*), 53, 54
Psychoda, 346
Psychodidae, 346
Psylla, 230; *P. buxi* (Linn.), 197, 200; *P. forsteri* Fl., 365
Psyllidae, 230 (*see also* Chermidae)
Psyllidomyces tenuis, 359
Psyllids, symbiosis in, 190
Psylliodes affinis Payk., 445
Ptelobius kroatzi, 410; *P. vittatus*, 410
Pterocallis juglandicola Buckt., 228
Ptinus, 53, 55
Ptychoptera contaminata Linn.:
associated protozoa–*Crithidia campanulata*, 470; *Gurleya francottei*, 515; *Phileocephalus striatus*, 494
associated spirochetes, 461
Ptyelus lineatus (Stål), 361:
associated yeasts–*Cicadomyces minor*, 360; *C. ptyeli lineati*, 359
Pubic louse (see *Phthirius pubis*)
Pulex brasiliensis, 478
Pulex canis, 47, 48
Pulex cleopatrae, 499
Pulex irritans Linn.:

associated bacteria–*Diplococcus pneumoniae*, 137, 138; *Mycobacterium leprae*, 47, 48; *Pasteurella pestis*, 162; *Salmonella choleraesuis*, 129
associated protozoa–*Crithidia pulicis*, 470; *Leptomonas pulicis*, 467; *Trypanosoma lewisi*, 478
associated rickettsia–*Rickettsia prowazeki* var. *typhi*, 281
intracellular symbiotes in, 247
Pulex spp., 466
Pulicidae, 247
Pulvinaria floccifera, 355
Pulvinaria innumerabilis Rath.:
associated yeast–*Lecaniocola putnami*, 356, 357
intracellular symbiotes in, 190
Pulvinaria mesembryanthemi Sign., 357
Pupipara, 206, 244-246
Pyrameis cardui (Linn.) (see *Vanessa cardui*)
Pyrausta nubilalis (Hbn.):
associated bacteria–*Bacillus gibsoni*, 63, 185; *B. ontarioni*, 76; *B. pirenei*, 80; *B. sotto*, 87, 88; *B. thuringiensis*, 91, **92**, 566; *Bacterium canadensis*, 99; *B. cazaubon*, 100; *B. christiei*, 101; *B. ephestiae*, Nos. *1*, 2, 103; *B. galleriae* Nos. 2, *3*, 103; *Bacterium gelechiae* Nos. *1*, 2, 5, 104; *B. italicum* No. 2, 106; *B. ontarioni*, 110; *B. pyraustae* Nos. *1-7*, 112; *B. pyrenei* Nos. *1-3*, 112; *Coccobacillus ellingeri*, **184**, 566; *C. gibsoni*, 64, 185; *Micrococcus curtissi*, 145; *Mycobacterium tuberculosis*, 569; *Proteus vulgaris*, 126; 566; *Serratia marcescens*, 132, 134; *Vibrio leonardi*, **179**, 566
associated fungi–*Aspergillus flavus*, 393; *A. niger*, 393; *Beauveria bassiana*, 393; *Metarrhizium anisopliae*, 400; *Spicaria farinosa*, 394
associated protozoa–*Herpetomonas pyraustae*, 472; *Leptomonas pyraustae*, 467, 468; *Perezia pyraustae*, 515; *Thelohania mesnili*, 516
immunization of, 556, 560, 565, 566
Pyrrhula europae, 310
Pyrsonympha affinis, 547; *P. elongata*, 547; *P. grandis*, 547; *P. granulata*, 548; *P. major*, 548; *P. minor*, 547; *P. modesta*, 547; *P. vacuolata*, 547; *P. vertens*, 547
Pyrosonimphinae, 531
Pyxinia anobii, 495; *P. bulbifera*, 495; *P. crystalligera*, 495; *P. frenzeli*, 495; *P. möbuszi*, 495; *P. rubecula*, 495

Q fever, American and Australian, 258, 263, 295-302, 308, 315, 324 (see also *Rickettsia burneti*)
Queensland fever, 295

SUBJECT INDEX

Quercus palustris, 391
Quinine, 288

Rabbit, 277, 285, 294, 300, 315, 319, 322, 324, 325, 334, 337, 338, 343, 345; cottontail, 165, 168, 286, 337; jack, 337; swamp, 286; snowshoe, 337
Rabbit flea (see *Spilopsyllus cuniculi*)
Rabbit louse (see *Haemodipsus ventricosus*)
Rabbit papilloma, 442
Raccoon, 286
Race immunity, 556
Ranatra, 183, 184
Range caterpillar (see *Hemileuca oliviae*)
Rapid plant bug (see *Adelphocoris rapidus*)
Rat, 161, 162, 260, 277, 278, 282, 285, 286, 289, 295, 315, 345; albino, 300; bushy-tailed wood, 324; cotton, 277, 285; field, 294; house, 294; rice, 285; sewer, 183; white, 286, 338; wood, 285, 337
Rat flea (see *Xenopsylla cheopis*)
Rat louse (see *Polyplax spinulosus*)
Rattus assimilis, 300; *R. conatus*, 300; *R. concolor browni*, 295; *R. culorum culorum*, 300; *R. culorum youngi*, 300; *R. decumanus*, 478; *R. losea*, 295; *R. lutreolus*, 300; *R. norvegicus*, 499; *R. rattus*, 478; *R. rattus alexandrinus*, 286
Rearing of insects and ticks, 575
Receptaculum seminis, 193
Record keeping, 594-595; forms for, 596-601
Red bacillus of Kiel, 132
Red-backed cutworm (see *Euxoa ochrogaster*)
Red-bordered stinkbug (see *Euryophthalmus convivus*)
Red fungus (see *Aschersonia aleyrodis*)
Red-headed fungus (see *Sphaerostible auranticola*)
Red-legged grasshopper (see *Melanoplus femur-rubrum*)
Red-water fever of cattle, 506
Reduviidae, 233-235
Relapsing fever, 261, **452-457**; means of differentiation, 453
Reticulitermes, 528, 529, 530, 531, 534, 547
Reticulitermes flaviceps Oshima:
 associated protozoa–*Dinenympha exilis*, 548; *D. leidyi*, 548; *D. parva*, 548; *D. porteri*, 548; *Duboscqia legeri*, 552; *Holomastigotes elongatum*, 551; *Pyrsonympha grandis*, 547; *Teratonympha mirabilis* var. *formosana*, 552
Reticulitermes flavipes Kollar, 535, 538:
 associated protozoa–*Dinenympha gracilis*, 548; *Duboscqia legeri*, 552; *Gregarina termitis*, 552; *Holomastigotes elongatum*, 551; *Nosema termitis*, 552; *Pyrsonympha vertens*, 547; *Spirotrichonympha kofoidi*, 551;

Trichonympha agilis, 549; *Teratonympha mirabilis* var. *formosana*, 552
Reticulitermes fukienensis Light, 549
Reticulitermes hageni Banks:
 associated protozoa–*Microjoenia pyriformis*, 551; *Spirotrichonympha gracilis*, 551; *S. kofoidi*, 551; *S. pulchella*, 551
Reticulitermes hesperus Banks:
 associated fungi, 408
 associated protozoa–*Dinenympha fimbriata*, 548; *Microjoenia ratcliffei*, 551; *Pyrsonympha granulata*, 548; *P. major*, 548; *P. minor*, 547; *Spironympha ovalis*, 551; *Torquenympha octoplus*, 552; *Trichonympha agilis*, 549
Reticulitermes lucifugus Rossi, 538:
 associated protozoa–*Dinenympha gracilis*, 548; *Duboscqia legeri*, 552; *Hexamastix termitis*, 539; *Holomastigotes elongatum*, 551; *Microjoenia hexamitoides*, 551; *Pyrsonympha elongata*, 547; *P. vacuolata*, 547; *Spirotrichonympha flagellata*, 551; *S. segmentata*, 551; *Trichomonas trypanoides*, 539; *T. agilis*, 549; *T. minor*, 549; *T. serbica*, 549
Reticulitermes speratus Kolbe, 534:
 associated protozoa–*Dinenympha exilis*, 548; *D. leidyi*, 548; *D. nobilis*, 548; *D. parva*, 548; *D. porteri*, 548; *D. rugosa*, 548; *Holomastigotes elongatum*, 551; *Pyrsonympha grandis*, 547; *P. modesta*, 547; *Teratonympha mirabilis*, 552; *Trichonympha agilis*, 549
Reticulitermes tibialis Banks, 549
Reticulo-endothelial system, 562
Retortamonas agilis, 483; *R. belostomae*, 482; *R. blattae*, 482; *R. caudacus*, 483; *R. gryllotalpae*, 482; *R. orthopterorum*, 482; *R. phyllophagae*, 482; *R. termitis*, 538; *R. wenrichi*, 482
Reversion of black currants, 449
Rhabditidae, 7
Rhabdophaga saliciperda (Duf.), 119
Rhachomyces, 385
Rhadinomyces, 385
Rhagium bifasciatum Fabr., 370, 371, 372; *R. inquisitor* (Linn.), 370, 371, 372
Rhagoletis pomonella (Walsh), 22, 174
Rhamnusium bicolor Schrank, 370
Rhinoceros beetle (see *Strategus utanus*)
Rhinocoris albopilosus, 470; *R. iracundus*, 468
Rhinocyllus comicus Frölich, 240; *R. olivieri* Mgl., 240
Rhinoncus bruchoides, 240
Rhinotermes, 529; *R. nasutus* (Perty), 550
Rhinotermitidae, 527, 528
Rhinotrichum depauperatum, 396
Rhipicephalus, 191, 252
Rhipicephalus appendiculatus Neum.;

Rhipicephalus appendiculatus Neum. (*Cont.*)
associated protozoan–*Theileria parva*, 507, 508, 509
associated rickettsia–*Dermacentroxenus conori*, 341
associated viruses–louping ill of sheep, 442; Nairobi disease of sheep, 442; Rift Valley fever, 442
intracellular symbiotes in, 251
Rhipicephalus bursa Canestrini & Fanzago:
associated protozoan–*Babesia motasi*, 507
associated rickettsia–*Rickettsia ovina*, 310
Rhipicephalus capensis Koch:
associated protozoan–*Theileria parva*, 507
intracellular symbiotes in, 251
Rhipicephalus evertsi Neum.:
associated protozoa–*Babesia caballi*, 507; *B. equi*, 507; *Theileria parva*, 507
associated spirochete–*Borrelia theileria*, 459
intracellular symbiotes in, 251
Rhipicephalus nitens, 507
Rhipicephalus sanguineus (Latr.):
associated bacterium–*Pasteurella tularensis*, 166, 167
associated protozoa–*Babesia canis*, 507; *Crithidia christophersi*, 470; *Hepatozoon canis*, 499, 500
associated rickettsiae–*Dermacentroxenus conori*, 340, 341; *D. rickettsi*, 330; *Rickettsia burneti*, 297, 298; *R. canis*, 306; *R. tsutsugamushi*, 292; unnamed rickettsia, 345
associated spirochete–relapsing fever, 454
associated virus–yellow fever, 436
intracellular symbiotes in, 251, 252, 253
Rhipicephalus schulzei Olen., 162, 165
Rhipicephalus simus Koch:
associated protozoa–*Babesia felis*, 507; *Theileria parva*, 507
associated rickettsia–*Dermacentroxenus conori*, 341
Rhizobiaceae, 42, 180
Rhizobium, 203
Rhizomastigida, 463, 483
Rhizomastix gracilis, 483
Rhizomyces, 385
Rhizopertha dominica (Fabr.): intracellular symbiotes in, 236; mycetome in, 199, 200; symbiote transmission, 208, 209
Rhizopertha spp., 199
Rhizopoda, 484
Rhizopus nigricans, 24
Rhizotrogus, 495
Rhizotrogus solstitialis Linn.:
associated bacterium–*Diplococcus melolonthae*, 137
associated fungus–*Beauveria densa*, 394
Rhodnius pictipes Stål, 476, 477
Rhodnius prolixus Stål:

associated bacteria–*Actinomyces rhodnii*, 234, 235; *Bacillus mesentericus*, 72, 73; *B. mycoides*, 74; *B. subtilis*, 89, 91; *Klebsiella ozaenae*, 123; *Staphylococcus albus*, 153, 154, *S. aureus*, 155, 156
associated protozoan–*Tryponasoma cruzi*, 476
associated rickettsia–*Dermacentroxenus rickettsi*, 331
bacteria as source of vitamins for, 33
bactericidal substance of, 34
intracellular symbiotes in, 233
Rhopalosiphum prunifoliae (Fitch), 444
Rhopalosiphum ribis Koch, 228
Rhynchoidomonas, 472
Rhynchonympha tarda, 484, 536
Rhynchophoromyces, 385
Rhynchophorus, 238
Rhynchophorus ferrugineus Oliv., 238, 240
Rhynchophrya palpans, 522
Rhyncolus lignarius Mrsh., 239, 240
Rhyparobia maderae, 134
Ribes virus *1*, 449
Rickettsia, 202, **256**, 257, 263, 264, 265, 280, 303 (see also Rickettsiae); culture methods, 584-585; examination of insects and ticks for, 580, 584-587; experimental animal tests, 585; identification of, 586; staining methods, 584
Rickettsia akamushi, 290
Rickettsia akari, 291
Rickettsia avium, 266, **310-311**
Rickettsia blanci, 339 (see also *Dermacentroxenus conori*)
Rickettsia bovis, 266, 305, **308**; disease caused, 308-309; immunologic aspects, 310; morphology, 308-309; pathogenicity, 309-310; staining, 309; tick associated with, 308
Rickettsia brasiliensis, 265, **329** (see also *Dermacentroxenus rickettsi*)
Rickettsia burneti, 258, 267, **295-302**; biologic relationships, 297-298; cultivation of, 300; diseases caused, 295-296; filterability of, 261, 300; generation-to-generation transmission, 297; growth of, 300; immunologic aspects, 301-302; insects and ticks associated with, 296-297; morphology, 298-299; pathogenicity, 300-301; resistance of, 300; staining, 298-299
Rickettsia cairo, 320
Rickettsia canis, 266, **306,** 324; comparison with *Rickettsia burneti*, 308; disease caused, 306; morphology, 306-307; pathogenicity, 307-308; staining, 307; tick associated with, 306
Rickettsia conjunctivae, 265, 266, **304**; disease caused, 304; morphology, 305; pathogenicity, 306; staining, 305-306; suspected vectors, 304; varieties of, 304-305

SUBJECT INDEX

Rickettsia conjunctivae bovis, 305
Rickettsia conjunctivae-bovis, 266, **305**
Rickettsia conjunctivae galli, 305
Rickettsia conjunctivae-galli, 266, **305**
Rickettsia conori, 265, **339** (see also *Dermacentroxenus conori*)
Rickettsia cowdria ruminantium, 302 (see also *Rickettsia ruminantium*)
Rickettsia ctenocephali, 267, **320**
Rickettsia culicis, 267, **321-322**
Rickettsia dermacentrophila, 249, 268, **322**, 335; cultivation, 323; and *Dermacentor andersoni,* 322; immunity aspects, 324; morphology, 323; pathogenicity, 323-324; staining, 323
Rickettsia dermacentroxenus, 328 (see also *Dermacentroxenus rickettsi*)
Rickettsia diaporica, **295**, 296, 300 (see also *Rickettsia burneti*)
Rickettsia exanthematojebri, 280 (see also *Rickettsia prowazeki*)
Rickettsia fletcheri, 280 (see also *Rickettsia prowazeki* var. *typhi*)
Rickettsia hirundinis, 316
Rickettsia kurlovi, 266, **324**
Rickettsia lectularia, 265, 267, **316**; and biologic relationship with *Cimex lectularius,* 233, **316-317**; cultivation of, 319; cycle of, 317-318; morphology, 317-319; pathogenicity, 319; staining, 317, 318
Rickettsia lestoquardi, 267, **306**
Rickettsia linognathi, 220, 267, **321**
Rickettsia manchuriae, 280 (see also *Rickettsia prowazeki* var. *typhi*)
Rickettsia megawi, 290 (see also *Rickettsia tsutsugamushi*)
Rickettsia megawi var. *breinli,* 290 (see also *Rickettsia tsutsugamushi*)
Rickettsia megawi var. *fletcheri,* 290 (see also *Rickettsia tsutsugamushi*)
Rickettsia megawi var. *pijperi,* 339 (see also *Dermacentroxenus conori*)
Rickettsia melophagi, 265, 267, **312**, 320, 321, 480; cultivation, 205, 261, 315; growth, 315; generation-to-generation transmission, 314; and *Melophagus ovinus,* 44, 244, 245, 312-313; morphology, 314-315; pathogenicity, 315; staining, 314-315
Rickettsia mooseri, 280 (see also *Rickettsia prowazeki* var. *typhi*)
Rickettsia muricola, 280 (see also *Rickettsia prowazeki* var. *typhi*)
Rickettsia murina, 280 (see also *Rickettsia prowazeki* var. *typhi*)
Rickettsia nipponica, 290 (see also *Rickettsia tsutsugamushi*)
Rickettsia orientalis, 290 (see also *Rickettsia tsutsugamushi*)
Rickettsia orientalis var. *schüffneri,* 290 (see also *Rickettsia tsutsugamushi*)
Rickettsia ovina, 266, **310**
Rickettsia pediculi, 221, 287, **288**, 289, 311, 320 (see also *Rickettsia wolhynica*)
Rickettsia pisces, 266, 311
Rickettsia prowazeki, 210, 220, 256, 261, 263, 265, **268**, 269, 270, 292, 297, 312, 319, 335, 556
Rickettsia prowazeki var. *mooseri,* 268, **280** (see also *Rickettsia prowazeki typhi*)
Rickettsia prowazeki var. *prowazeki,* 266, 282, 283, 285; biologic relationships, 270-271; cultivation of, 275-276; cycle of development, 274; disease caused, 269-270; immunologic aspects (agglutination, 279; complement fixation, 279; "dry-blood" test for, 279; relationship to spotted fever, 279; toxin of, 279; vaccine for, 278; Weil-Felix reaction, 279); insects associated with, 270-279; modes of transmission, 271; morphology and staining, 271-275; pathogenicity, 277-278; physiology of, 276-277; varieties, 268-269
Rickettsia prowazeki var. *typhi,* 266, 268, **280**; biologic relationships, 282-283; disease caused, 280-281; guinea pig scrotal involvement in, 285; immunologic aspects, 286-287; insects and ticks associated with, 281-282; morphology and staining, 283-286; pathogenicity, 284-286; synonyms for, 280
Rickettsia pseudotyphi, 290 (see also *Rickettsia tsutsugamushi*)
Rickettsia psittaci, 262, 268, 310
Rickettsia quintana, 287 (see also *Rickettsia wolhynica*)
Rickettsia rickettsi, 328 (see also *Dermacentroxenus rickettsi*)
Rickettsia rocha-limae, 220, 261, 265, 267, 312, **319**; comparison with *Rickettsia prowazeki,* 319-320; insect associated with, 319; pathogencity, 320
Rickettsia ruminantium, 265, 266, **302**, 310; biologic relationships, 303; cultivation, 303; disease caused, 302; immunologic aspects, 304; morphology, 303; pathogenicity, 304; ticks concerned, 297, 303
Rickettsia suis, 266, **304**
Rickettsia sumatrana, 290
Rickettsia trachomae, 262
Rickettsia trachomitis, 262, 268
Rickettsia trichodectae, 220, 267, **321**
Rickettsia tsutsugamushi, 263, 266, **290-295**; disease caused, 290-291; cultivation, 293-294; immunity aspects, 295; insects and arachnids associated with, 291-292; morphology, 292-293; pathogenicity, 294-295; staining, 292-293; synonyms, 289

SUBJECT INDEX

Rickettsia tsutsugamushi-orientalis, 290 (see also *Rickettsia tsutsugamushi*)
Rickettsia typhi, 329 (see also *Dermacentroxenus rickettsi*, 329)
Rickettsia weigli, 267, **311-312**; characteristics of, 311-312; pathogenicity, insect associated with, 312
Rickettsia wolhynica, 265, 267, **287**, 311; biologic relationships, 288-289; disease caused, 287-288; immunologic aspects, 289-290; insects associated with, 288-289; morphology, 289; pathogenicity, 289; physiologic properties, 289; staining, 289
Rickettsiaceae, 220, 256, 257, 259, 262, 263, 264, 265; key to, 266-268
Rickettsiae, 205, **256-347** (*see also* specific names); characteristics of, 261; classification of, 262-268; cross-immunity tests, 587; cultivation of, 584-585; definitions of, 256-259; evolution of pathogenicity, 259-261; examination of insects and ticks for, 580, **584-587**, 597; experimental animal tests, 585; identification of, 586; nature of, 259; neutralization test, 587; nomenclature of, 262-268; protection test, 587; stages of, 257; taxonomic arrangement of, Macchiavello's, 264-265; Steinhaus', 266-268
Rickettsiaemia Weigli disease, 312
Rickettsial diseases, groups of, 263
Rickettsiales, 263
Rickettsialike bacteria, 262
Rickettsialike organisms, 264 (*see also* Intracellular symbiotes)
Rickettsias, 256 (*see also* Rickettsiae)
Rickettsioides, 263
Rickia, 385
Rift Valley fever, 442
Rihana ocrea, 374
Roaches (*see* Cockroaches *and* specific names)
Rocky Mountain spotted fever, 244, 269, 323, 324, **328-339**, 344
Ropalosiphum persicae, 400
Rostronympha magna, 551
Roundworms (*see* Nemathelminthes)
Rozites gongylophora, 406
Rugitermes, 528
Rugitermes kirbyi Snyder:
associated protozoa—*Foaina nana*, 543; *F. solita*, 543; *Oxymonas kirbyi*, 547; *Tricercomitus divergens*, 539; *Trichonympha saepicula*, 549
Rugitermes panamae Snyder:
associated protozoa—*Foaina nana*, 543; *F. solita*, 543; *Oxymonas kirbyi*, 547; *Tricercomitus divergens*, 539; *Trichonympha saepicula*, 549
Russian spring-summer encephalitis, 437, 439

Sabethini, 433
Sabimamorpha, 374
Sacbrood, 413, 415, **426-430**; geographical distribution, 429; histopathology, 429; stages in post-mortem appearance, 427, **428-429**; symptoms, 427; transmission, 430; the virus, 429
Saccharomyces, 351, 353, 358, 366; *S. anobii*, 366-370; *S. apiculatus* var. *parasiticus*, 354; *S. aphalarae calthae*, 364; *S. aphidis*, 364; *S. cerevisae*, 24, 349, 350; *S. cicadarum*, 359; *S. ellipsoideus*, 349, 351
Saccharomycetaceae, 365, 384
Saccharomycetales, 376
Saccharum virus *I*, 445; Saccharum virus 2, 448
Sacchiphantes abietis, 229; *S. viridis*, 229
Saccinobaculus ambloaxostylus, 536; *S. doroaxostylus*, 536; *S. minor*, 536
Saissetia hemisphaerica Targ., 355; *S. oleae* (Bern.), 355, 356
Salivary glands, 235, 271, 283, 292, 331, 334, 434, 457, 505, 507, 508, 510
Salix spp., 175
Salmonella, 41, 129; *S. choleraesuis*, 129; *S. enteritidis*, 12, 19, 22, **129-130**, 556, 560; *S. enteritidis* var. *Danysz*, 561; *S. paratyphi*, **130**, 449; *S. schottmülleri*, 130; *S. schottmülleri* var. *alvei*, **131**, 557; *S. suipestifer*, 129
Salmonelleae, 41, 126
Salvarsan, 288
Samia cecropia (Linn.):
associated bacteria—*Bacillus circulans* "Group," 59; *B. subtilis*, 89, 90; *Escherichia coli*, 121, 122; *E. freundii*, 122
Sandfly, 8, 346 (*see also* specific names)
Sandpipers, 95
São Paulo typhus, 329
Saperda punctata Linn., 410; *S. tridentata* Oliv., 410
Saphanini, 371, 373
Saprolegniales, 377
Saprospira, 451
Sarcina, 41; *S. alba*, 151; *S. aurantiaca*, 152; *S. flava*, 152; *S. lutea*, 24, 28, 152
Sarcodina, 463, 484-486, 537, 553
Sarcophaga barbata, 46, 48
Sarcophaga bullata Park, 471
Sarcophaga carnaria (Linn.):
associated bacteria—*Bacillus anthracis*, 53; *B. colisimile*, 60; *B. radiciformis*, 84, 85; *B. similcarbonchio*, 86, 87; *Escherichia coli*, 121, 122; *Proteus vulgaris*, 126; *Pseudomonas aeruginosa*, 169; *P. fluorescens*, 170, 171; *P. jaegeri*, 171; *P. non-liquefaciens*, 171, 172; *Sarcina aurantiaca*, 152; *Serratia kielensis*, 132; *S. marcescens*, 132; *Staphylo-*

SUBJECT INDEX

coccus albus, 153; *S. citreus*, 156
Sarcophaga haemorhoidalis (Fall.), 441
Sarcophaga melanura Meig., 468
Sarcophaga pallinervis Thomson, 46, 48
Sarcophagula sp., 155
Sarcoptes scabiei DeG., 46, 47
Sarcosporidia, 511
Saturnia pavonia Linn. var. *major*, 414
Saturniidae, 414, 425
Savoy of beets, 448
Scale, in sacbrood, 428
Scale insects (*see also* specific names):
  associated fungi–*Aegerita webberi*, 400; *Aschersonia goldiana*, 400; *A. aleyrodis*, 400; *Myriangium* spp., 388; *Podonectria coccicola*, 400; *Septobactium* spp., 389; *Spaerostilbe auranticola*, 400; *Torrubiella* spp., 388
Scaptomyza graminum (Fall.), 118, 119
Scaurus tristis, 381
Scea auriflamma, 514
Sceliphron caementarium (Dru.), 108
Schedoacercomonas muscae domesticae, 471
Schedorhinotermes, 529
Schedorhinotermes intermedius Brauer:
  associated protozoa–*Pseudotrichonympha parvipapillosa*, 550; *Spirotrichonympha elongata*, 551; *S. flagellata* var. *schedorhinotermitis intermedii*, 551
Schedorhinotermes putorius Sjöstedt:
  associated protozoa–*Pseudotrichonympha introflexibilis*, 550; *P. magnipapillosa*, 550
Schistocerca americana (Dru.), 412
Schistocerca americana americana Scudd., 491
Schistocerca gregaria Forsk.:
  associated bacteria–*Pseudomonas aeruginosa*, 169, 170; *Serratia marcescens*, 132, 134
Schistocerca obscura Fabr., 412
Schistocerca pallens Thunb., 181, 182
Schistocerca paranensis Burm.:
  associated bacterium–*Coccobacillus acridiorum*, 181, 182, 183
  associated protozoan–*Gregarina paranensis*, 493
Schistocerca peregrina Oliv.:
  associated bacterium–*Coccobacillus acridiorum*, 181, 183
  associated fungus–*Beauveria densa*, 394
Schizocystidae, 496-497
Schizocystis gregarinoides, 497
Schizocystis legeri, 497
Schizogony, 487, 496, 497, 500, 501, 505, 512, 513, 518
Schizogregarinaria, 496-497, 523
Schizogregarines, 496
Schizomycetes, 40, 42, 263
Schizosaccharomyces chermetis abietis, 365; *S. chermetis strobilobii*, 365; *S. psyllae forsteri*, 365; *S. sulcii*, 365

Schlaffsucht, 424
Schneiderai mucronata, 495
Sciara, 350, 380; *S. thomae* Linn., 411
Scieroptera splendidula, 374
Sciurus carolinensis, 286, *S. niger niger*, 286
Scolytidae, 403
Scolytus affinis (Escherich), 410; *S. multistriatus* (Mrsh.), 410; *S. praeceps* LeC., 411; *S. pygamaens* Fabr., 410; *S. rugulosus* Ratz., 116; *S. scolytus* Fabr., 410; *S. subscaber* LeC., 411; *S. sulcifrons* Rey, 410; *S. ventralis* LeC., 411
Scrub typhus (see *Rickettsia tsutsugamushi* and tsutsugamushi disease)
Scutate ticks (*see* Ixodidae)
Scutelleridae, 14, 16, 17
Seed-corn maggot (see *Hylemya cilicrura*)
Seminal vesicles, 194
Semliki forest virus, 442
Septicemia, 69, 71, 72, 86, 90, 123, 171, 566; cutworm, 75; hornworm, 88, 89; pink bollworm, 77-78; rabbits, 161
Septobasidium, 376, 389-392; *S. alni*, 391; *S. alni* var. *squamosum*, 391; *S. apiculatum*, 390, 392; *S. burtii*, 389, 390, 392; *S. canescens*, 391; *S. castaneum*, 391; *S. curtisii*, 391; *S. grandisporum*, 391; *S. pseudopedicellatum*, 391; *S. sabalis*, 391; *S. sinuosum*, 391
Sericostoma, 483, 495
Serrateae, 41, 132
Serratia, 41, 132
Serratia fuchsina, 63
Serratia kielensis, 132
Serratia marcescens: and *Dermacentor andersoni*, 135, 555; insects associated with, 12, 24, 99, 128, 132-135; survival during metamorphosis, 135
Serratia plymouthensis, 135
Serratia plymuthicum = S. plymouthensis
Serritermes, 529
"Serum virus," 331
Sex organs, morphology of, 191-193
Sheep, 277, 302, 304, 310, 315, 343
Sheep ked, 312, 313, 337 (see also *Melophagus ovinus*)
Shigella, 41, 131; *S. dysenteriae*, 12, 19, 128, 131, 557, 567; *S. flexner* V, 132; *S. paradysenteriae*, 131-132
Shinbone fever, 287
Shipping of insects, 574-575
Short-nosed cattle louse (see *Haematopinus eurysternus*)
Shot-hole borer (see *Scolytus rugulosus*)
Sibinia pellucens Scop., 238, 240
Sigmodon hispidus hispidus, 285
Silkworm, 394, 396, 397, 400, 511, 555 (see also *Bombyx mori*)
Silky ant (see *Formica fusca subsericea*)

SUBJECT INDEX

Silpha laevigata Fabr., 494; *S. thoracica,* 494
Silverfish (see *Lepisma saccharina*)
Simulium bracteatum Coq.:
 associated protozoa–*Thelohania bracteata,* 516; *T. fibrata,* 516; *T. multispora,* 516
Simulium columbaczense (Schönb.), 470
Simulium decorum katmai Walk., 167
Simulium hirtipes Fries:
 associated protozoa–*Thelohania bracteata,* 516; *T. fibrata,* 516
Simulium maculatum Meig.:
 associated protozoa–*Plistophora simulii,* 516; *Thelohania bracteata,* 516; *T. fibrata,* 516; *T. multispora,* 516
Simulium noelleri, 460
Simulium ochraceum Walk.:
 associated protozoa–*Plistophora simulii,* 516; *Thelohania bracteata,* 516; *T. fibrata,* 516
Simulium ornatum, 516
Simulium pertinax, 46, 48
Simulium reptans Linn.:
 associated protozoan–*Thelohania varians,* 516
 associated spirochete–*Spirochaeta noelleri,* 460
Simulium sp., 511
Simulium venustum Say:
 associated protozoa–*Leucocytozoon anatis,* 516; *Plistophora simulii,* 516; *Thelohania bracteata,* 516; *T. fibrata,* 516
Simulium vittatum, 516
Sinea diadema (Fabr.), 89, 90
Sinoxylon ceratoniae: intracellular symbiotes in, 236, 237; mycetome in, 199; symbiote transmission, 208, 209
Siphonaptera, 215, 247
Siphunculina funicola De Meij., 472
Sirex cyaneus, 393; *S. gigas* Linn., 393
Sirocalus pulvinatus Gyll., 240
Sitodrepa panicea (see *Stegobium paniceum*)
Sitona sp., 486
Sitophilus granarius (Linn.), 200, 209, 238, 240; *S. oryzae* (Linn.), 238, 240
Skunk (see *Mephitis elongata*)
Sleeping sickness (see *Trypanosoma gambiense*)
Smallpox, 443
Smeringomyces, 383
Smerinthus (Sphinx) atlanticus Aust., 414; *S. ocellatus* Linn., 18
Smicronyx cuscutae Bris., 239; *S. jungermanniae* Reich., 240
Smut of figs, 411
Snout beetles (see *Curculionidae*)
Snyderella tabogae, 546
Soft rot of crucifers (see *Erwinia carotovora*)
Solanum virus *12,* 444; Solanum virus *14,* 445; Solanum virus *16,* 447
Solenopsis gemminata (Fabr.):
 associated bacteria–*Coccobacillus acridiorum,* 181; *Shigella paradysenteriae,* 131, 132

Sorosporella uvella, 395
Souring of figs, 411
South African horse sickness, 442
South African tick-bite fever, 329, 339, 340
Sparaganum proliferum, 8
Species immunity, 556
Sperm, 193, 195
Spermatheca, 192, 193
Spermatocytes, 327
Spermatogonia, 327
Spermatozoa, 191, 193, 194, 327
Spermophiles, 343
Sphaeriales, 383, 387-388
Sphaerocystis simplex, 494
Sphaeropsis coronillae, 395
Sphaerostilbe, 387
Spharagemon bolli Scudd.:
 associated protozoa–*Gregarina rigida columna,* 492; *G. rigida rigida,* 491
Spharagemon collare (Scudd.) *collare,* 492
Spharagemon cristatum, 412
Sphingidae, 414, 425
Sphingonotus sp., 493
Sphinx ligustri (Linn.), 394
Spicaria, 393; *S. aleyrodis,* 394; *S. anomala,* 411; *S. arneae,* 394; *S. cossus,* 394; *S. farinosa,* 394, 398, 400; *S. heliothis,* 394; *S. javanica,* 394; *S. prasina,* 394; *S. rileyi,* 394; *S. verticillioides,* 394
Spiders, 388, 392
Spilopsyllus cuniculi, 165, 168, 479
Spiral filament, 519
Spirilleae, 42, 177
Spirillum, 451; *S. minus,* 451
Spirochaeta, 451; *S. anserina,* 20; *S. blattae,* 460; *S. cobayae,* 454; *S. ctenocephali,* 460; *S. culicis,* 460, 461; *S. drosophilae,* 460; *S. glossinae,* 460; *Spirochaeta grassii,* 460; *S. hispanica,* 455; *S. melophagi,* 466; *S. minei,* 46; *S. morsus-muris,* 451; *S. noelleri,* 460; *S. obermeieri,* 456; *S. phlebotomi,* 460; *S. pieridis,* 461; *S. recurrentis,* 452; *S. sogdianum,* 453; *S. termitis,* 460. (See also *Borrelia.*)
Spirochaetaceae, 451
Spirochetes, 451-461; associated insects and ticks, 454, 455, 456; biologic relationships, 455; examination of for, 580, 590, 597; generation-to-generation transmission of in, 457; relapsing fever group, 452-457; types of behavior in lice, 455
Spiroglugea octospora, 518
Spironympha elegans, 551; *S. ovalis,* 551; *S. porteri,* 551
Spirotricha, 520, 521-522
Spirotrichonympha africana, 551; *S. bispira,* 551; *S. clevelandi,* 551; *S. elongata,* 551; *S. flagellata,* 551; *S. flagellata* var. *coptotermitis*

lactei, S. flagellata var. *schedorhinotermitis intermedii,* 551; *S. gracilis,* 551; *S. grandis,* 551; *S. kofoidi,* 551; *S. leidyi,* 551; *S. mirabilis,* 551; *S. polygyra,* 551; *S. pulchella,* 551; *S. pundibunda,* 551; *S. segementata,* 551
Spirotrichosoma capitata, 552; *S. magna,* 552; *S. obtusa,* 552
Spodoptera, 394
Spondylini, 370, 372
Sporadin, 484, 491
Sporocyst, 498, 499, 500
Sporogony, 487, 496, 512
Sporokinetes, 499, 507
Sporomyxa scauri, 381
Sporotrichum, 393; *S. globuliferum* (see *Beauveria globulifera*)
Sporozoa, 463, **487-519**, 523, 537, 552-553
Sporozoites, 487, 488, 489, 496, 497, 498, 499, 500, 501, 502, 503, 504, 505, 507, 508, 519
Spotted fever, 84, 263, 285, 291, 295, 300, 301, 303, 315, 343, 570
Spotted wilt of tomatoes, 448
Spring disease, 172
Squash beetle (see *Epilachna borealis*)
Squash bug (see *Anasa tristis*)
Squirrel, 286; Columbian, 324; flying, 285, 324; fox, 286; gray, 286; ground, 164, 166, 337; pine, 324; side-striped, 324; tree, 337
Stablefly (see *Stomoxys calcitrans*)
Stains: aniline, 256, 261, 262, 264, 303, 582; carbol fuchsin, 273; Castañeda's, 261, 272, 298, 307, 309, 310, 325, 342, 584; Fielding's, 584; gentian violet, 273; Giemsa's, 223, 227, 248, 253, 261, 271, 272, 283, 289, 292, 293, 298, 303, 305, 307, 310, 311, 314, 317, 321, 323, 325, 326, 334, 335, 342, 373, 584; Goodpasture's, 335; Gram's, 273; hematoxylin-eosin, 298; iron hematoxylin, 227, 373; Löffler's methylene blue, 303; Macchiavello's, 44, 261, 272, 298, 315, 323, 335, 342, 584; Mann's, 298; Pappenheim's, 325; for rickettsiae, 584; thionin, 273; Wright's, 223
Staphylinidae, 385
Staphylococcus, 41, **152**, 153, 562; *S. acridicida,* 153; *S. albus,* 18, 19, 34, **153-154**, 155, 557, 567; *S. aureus,* 19, 33, 34, 35, 153, **155-156**, 557, 567; *S. citreus,* 156; *S. insectorum,* 156; *S. muscae,* **157**, 449; *S. parvulus,* 149; *S. pyogenes,* 153; *S. pyogenes albus,* 154; *S. pyogenes aureus,* 155; *S. pyogenes citreus,* 156; *S. septicaemia,* 153
Stauroderus biguttulus (see *Chorthippus biguttulus*)
Staurojoenina assimilis, 548; *S. mirabilis,* 548
Stauronotus maroccanus Stål, 181, 182
Stegobium paniceum (Linn.):
  associated protozoan—*Pyxina anobii,* 495

  associated yeast—*Saccharomyces anobii,* 366-370; role of, 369
  generation-to-generation transmission in, 368
  mycetome in, 199, 200
Stegomyia fasciata, 433 (see also *Aëdes aegypti*)
Steinina ovalis, 493; *S. rotunda,* 494
Stempellia magna, 512, 516, 517; *S. mutabilis,* 516
Stenobothrus, 394
Stenocorus cardui Hbst., 240; *S. meridianus,* 371; *S. quercus,* 371
Stenocephalus agilis Scop., 473
Stenopelmatus fuscus Hald., 493; *S. intermedius,* 493; *S. pictus* Scudd., 493
Stenophylax, 483
Stephanocleonus superciliosus Gebl., 239
Stephanoderes hampei (Ferrari), 394
Stephanonympha nelumbium, 546; *S. silvestrii,* 546; *S. silvestrii* var. *cryptotermitis havilandi,* 546; *S. silvestrii* var. *neotermitis erythraei,* 546
Stereum sanguineolentum, 372
Sternochetus lapathi (Linn.), 238, 240
Stethotorax ater var. *brasiliensis,* 522
Stibula apicalis (Melsh.), 177
Stichomyces, 386
Stictospora provincialis, 495
Stigmatomyces, 385
Stilpnotia salicis (Linn.):
  associated bacteria—*Bacillus aureus,* 55, 56; *B. lineatus,* 68; *B. minimus,* 73; *B. monachae,* 73, 74; *Bacterium cazaubon,* 100; *B. pyrenei,* 112; *Micrococcus vulgaris,* 151
St. Louis virus, 437
Stolotermes, 528; *S. ruficeps,* 552; *S. victoriensis,* 552
Stomach disk, 221-222, 223
Stomoxys, 442
Stomoxys calcitrans (Linn.):
  associated bacteria—*Bacillus anthracis,* 34, 53, **54**; *B. mesentericus,* 34, 72, **73**; *B. mycoides,* 74; *B. subtilis,* 34, 89, **91**; *B. vulgatus,* 34; *Bacterium delendae-muscae,* 102; *Brucella abortus,* 158, 159; *Malleomyces mallei,* 159; *Pasteurella tularensis,* 166; *Pseudomonas aeruginosa,* 169, 170; *Serratia marcescens,* 133, 134; *Staphylococcus albus,* 34, 153, **154**; *S. aureus,* 34, 155, **156**; *Streptococcus,* 138
  associated protozoan—*Trypanosoma congolense,* 480
  associated rickettsia—*Rickettsia conjunctivae,* 304
  associated viruses, 439; fowlpox, 442; poliomyelitis, 441
  bactericidal substance of, 34, 54
Storing of insects, 575
Strangalia maculata Poda, 370
Strategus utanus, 148

Streblidae, 244
Streblomastix strix, 534, 547
Streptococceae, 41, 135
Streptococcus, 41, **138**, 143; *S. agalactiae*, 138-139; *S. apis*, 66, 80, 81, 97, **139**; *S. bombycis*, 139-140; *S. disparis*, 140-141; *S. equinus*, 141; *S. faecalis*, 11, 33, **141**, 557; *S. galleriae*, 142; *S. haemolyticus*, 34, 143; *S. lactis*, 142; *S. liquefaciens*, 139; *S. pastorianus*, 139, 140, **142**; *S. pityocampae*, 142-143; *S. pyogenes*, 143, 557; *S. salivarius*, 143-144
Streptothrix, 20
Strigoderma arboricola (Fabr.), 82, 83
Strigomonas fasiculata, 469
Strophingia, 230; *S. ericae*, 230
Stylocephalus giganteus, 494
Stylorhynchus, 553
Stylotermes, 528
Subulitermes kirbyi Snyder, 539
Suctoria, 463, 522
Sugar-beet leafhopper (see *Eutettix tenellus*)
Sugar-cane mosaic, 445
Sulphonamides, 288, 308
Surgical maggots, 34-36
Swallow flea (see *Ceratophyllus hirundinis*)
Swine, 304, 306
Sylvilagus, 165; *S. aquaticus aquaticus*, 286; *S. floridanus mallurus*, 286; *S. nuttalli*, 337
Symbiont, 188, 189, 203
Symbiosis, 5, 203, 211, 212, 228; definition of, **188**
Symbiotes, 5, **198**; definition of, 188, **189**, 261 (*see also* Intracellular symbiotes)
Symbiotic organs, 190
Sympetrum rubicundulum (Say), 494
Symplectromyces, 385
Symptomatic anthrax, 95
Syncystis mirabilis, 497
Synopsyllus fonquernii, 162
Systena blanda Melsh., 444; *S. taeniata* (Say), 444
Systenus adpropinquans, Loew, 497
Systenus scholtzii:
  associated protozoa—*Dendrorhynchus systeni*, 494; *Schizocystis legeri*, 497
Systenus spp.:
  associated protozoa—*Lipotropha macrospora*, 497; *L. microspora*, 497

Tabanidae, 243-244, 511
Tabanus:
  associated bacteria—*Bacillus anthracis*, 53; *Staphylococcus albus*, 153; *S. aureus*, 155, 156
  associated protozoan—*Trypanosoma evansi*, 481
Tabanus abdominalis Fabr., 243

Tabanus agrestis Wied., 166
Tabanus atratus Fabr., 53, 54
Tabanus autumnalis Linn., 166
Tabanus bovinus Linn., 53, 54
Tabanus bromius Linn., 166
Tabanus costalis Wied. (= *T. vicarius* Walk.), 244
Tabanus erberi Brauer, 166
Tabanus flavoguttatus, 166
Tabanus glaucopis Meig., 470
Tabanus hilaris Walk., 470
Tabanus karybenthinus, 166
Tabanus lineola Fabr., 243
Tabanus peculiaris, 166
Tabanus pumilus Macq., 244
Tabanus punctifer O.S., 439
Tabanus rubidus Wied.:
  associated bacteria—*Bacillus anthracis*, 53, 54; *Pasteurella bollingeri*, 160, 161
Tabanus rupestris, 166, 167
Tabanus septentrionalis Loew, 166, 167
Tabanus solstitialis Meig., 166
Tabanus striatus Fabr.:
  associated bacteria—*Bacillus anthracis*, 53, 54; *Pasteurella bollingeri*, 160
Tabanus tergestinus Egg., 470
Tabanus trispilus Wied., 244
Tabanus turkestanus, 166
Tabardillo, 281
Tachardina lobata Chamb., 375; *T. sylvestrii* Mahd., 375
Tachycines asynamorus Adel., 494
Taeniocystis mira, 495
Taeniorhynchus africanus Theobald, 433
Talorchestia longicornis, 521
Tambina, 374
Tamias striatus striatus, 286
Tanypus, 517, 518; *T. setiger*, 517; *T. varius*, 516
Tanytarsus, 517
Tapeworm, 8
Tarbagan, 164, 165
Tarentola mauritanica, 474
Tarnished plant bug (see *Lygus pratensis*)
Tartessus, 374
Telomyxa glugeiformis, 518
Telomyxidae, 523
Telosporidia, 487
Temnorrhinus mendicus (Gyll.):
  associated bacteria—*Bacillus cleoni*, 59; *Coccobacillus cajae*, 183, 184
Tenebrio molitor Linn.:
  associated bacteria—*Bacillus subtilis*, 89, 90; *Bacterium elbvibrionen*, 102; *B. hemophosphoreum*, 104; *B. knipowitchii*, 106; *Serratia marcescens*, 132, 134
  associated fungus—*Beauveria densa*, 394
  associated protozoa—*Gamocystis polymorpha*,

SUBJECT INDEX

493; *G. steini*, 493; *Gregarina cuneata*, 491, 492, 493; *Ophryocystis mesnili*, 496; *Steinina ovalis*, 493
mycetome, transplanted, 200
Tenebrionidae, 236, 243
Tentyria sp.:
associated bacteria—*Bacillus anthracis*, 53, 55; *B. proteisimile*, 83; *B. radiciformis*, 84, 85; *Clostridium chauvoei*, 95; *C. sporogenes*, 95; *C. tetani*, 96; *Corynebacterium pseudodiphthericum*, 46; *Diplococcus pneumoniae*, 136, 137; *Klebsiella pneumoniae*, 124; *Malleomyces mallei*, 159; *Mycobacterium tuberculosis*, 48, 49; *Sarcina alba*, 151; *Vibrio colerigenes*, 177, 178
Tephritis conura Loew, 11; *T. heiseri* Frfld., 22
Teratomyces, 385
Teratonympha mirabilis, 552; *T. mirabilis* var. *formosana*, 552
Termes (see *Reticulitermes*); *T. lucifugus* (Rossi), 460
Terminal filament, 191, 192
Termitaria, 408
Termites, 376, **526-554;** associated bacterium, 85; castes of, 527; classification of, 528; cross infection of, 530; defaunation, 533-534; diet of, 534-535; evolutionary development, 531-532; flagellates, bacteria of, 36-37; fungi "gardens" of, 408-409; host-parasite specificity, 529-531; protozoa of, 526-554; spirochetes of, 461; transmission of, 530, 532
Termitidae, 527, 529, 531, 553, 554
Termitogeton, 528
Testes, insect, **193,** 194, 207, 209. 217, 247, 275, 317, 325
Testicular follicles, 193
Tetraneura ulmi DeG., 227
Tetraodon fahaka, 311
Tetratrichomastix blattidarum, 483; *T. mackinnoni*, 483
Tetropium castaneum Payk., 370, 371
Tettigoniidae, 144
Tettigoniella ferruginea, 374; *T. viridis*, 374
Tettigonospora stenopelmati, 493
Texas cattle fever, 3, 506 (see also *Babesia bigemina*)
Thamnotettix nigrifrons (Forbes), 177
Theileria, 308, 506; *T. annulata*, 510; *T. dispar*, 308, 510; *T. kochi*, 507; *T. mutans*, 510; *T. parva*, 507, 508, 509; *T. tsutsugamushi*, 290
Thelohania, 516; *T. anomola*, 516; *T. baetica*, 516; *T. bracteata*, 516; *T. brasiliensis*, 516; *T. cepedei*, 516; *T. corethrae*, 516; *T. ephestiae*, 516, 524; *T. fibrata*, 516; *T. indica*, 516; *T. janus*, 516; *T. legeri*, 512, 515; *T. mesnili*, 516, 523, 524; *T. minuta*, 512, 516;

T. multispora, 516; *T. mutabilis*, 516; *T. obesa*, 512; *T. obscura*, 516; *T. opacita*, 512, 516; *T. ovata*, 516; *T. pinguis*, 516; *T. pyriformis*, 516; *T. rotunda*, 512, 516; *T. varians*, 516
Theobaldia annulata Schrank:
associated protozoa—*Caulleryella annulatae*, 497; *Glaucoma pyriformis*, 520
associated spirochete—*Borrelia anserina*, 458
associated virus—fowlpox, 442
Theobaldia incidens (Thomson):
associated bacterium—*Pasteurella tularensis*, 166, 168
associated virus—St. Louis encephalitis, 438, 439
Theobaldia inornata (Williston), 439
Theobaldia spp.:
associated protozoa—*Plasmodium* spp., 505; *Haemoproteus* spp., 506
Thermesia gemmatilis, 394
Thermobia domestica Pack.:
associated protozoa—*Colepismatophila watsonae*, 494; *Lepismatophila thermobiae*, 494
Thetomys gracilicaudatus, 300
Thompsoniella porresta, 374
Thos mesomelas, 307
Thrips, 174, 448
Thrips tabaci Lind, 448
Thyreocoris unicolor Pal. Beauv., 15
Thyridopteryx ephemeraeformis (Haw.):
associated bacteria—*Bacterium insectiphilium*, 105-106; diphtheroid, 45; *Micrococcus freundenreichii*, 146; *Streptococcus faecalis*, 141
Thysanura, 214, 215, 443
Tibicen linnei (Smith and Gross):
associated bacteria—*Bacterium incertum*, 105; *B. mutabile*, 27, 109; *Micrococcus nitrificans*, 148; *M. nonfermentans*, 149; *Streptococcus faecalis*, 141
Tibicina septendecim (see *Magicicada septendecim*)
Tick-bite fever, 340
Tick typhus, 264, 270
Tick virus, 264, 270
Ticks (see also specific names and Ixodoidea), 215, 256, 282; argasid, 333; bacteria in, 9, **13,** 14; bactericidal substance in, 33-44; collecting methods for, 574; dissection of, 577; generation-to-generation transmission in, 22; handling methods, 574; histologic methods, 579; immunity to *Dermacentroxenus rickettsi*, 556; intracellular symbiotes of, 249-254; ixodid, 333; microbiologic examination of, 580-601; nonscutate, 253; protozoa of, 473; rearing of, 575; scutate, 250; shipping and storing methods, 575
Tillomorphini, 371

Timber beetles (see Ambrosia beetles)
Tineidae, 414
Tineola biselliella (Hum.):
  associated protozoan–*Adelina mesnili*, 498
  associated viruses, 414
Tipula, 62, 483; *T. abdominalis* Say, 482, 484; *T. oleracea* Linn., 515; *T. paludosa* Meig., 445
Tipulae, 380
Tmetis muricatus, 181
Tobacco hornworm (see *Protoparce sexta*)
Tobacco mosaic, 445
Tobia fever, 329
Tomato hornworm (see *Protoparce quinquemaculata*)
Tomonatus aztecus, 412
Torquenympha octoplus, 552
Torrubiella, 388; *T. aranicida*, 388; *T. barda*, 388; *T. lecana*, 388; *T. luteorostrata*, 388; *T. ochracea*, 388; *T. rostrata*, 388; *T. ruba*, 388; *T. sericicola*, 388; *T. tenuis*, 388; *T. tomentosa*, 388
Tortricidae, 414
Torula lecanii cornii, 71, 366
Torula sp., 218, 375
Torulopsis, 366
Toxin, 562; diphtheria, 43
Toxoglugea vibrio, 518
Toxoid, 562
Toxoptera graminum Rong., 445
Toxostege sticticalis, 525
Toxotus schaumi, 371; *T. vestitus*, 371, *T. vittiger*, 371
Trachia piniperda, 73, 74
Trachoma, 262
Trachymyrmex septentrionalis, 408
Trachysphaera fructigena, 408
Tramea lacerata (see *Trapezostigma lacerata*)
Transovarial transmission (see Generation-to-generation transmission)
Trapezostigma lacerata (Hagen), 494
Trench fever, 78, 287-290 (see also *Rickettsia wolhynica*)
Treponema, 451; *T. parvum*, 460; *T. pertenue*, 459; *T. stylopygae*, 460
Treponemaceae, 451
Triatoma brasiliensis Neiva, 233
Triatoma chagasi, 476
Triatoma dimidiata (Latr.):
  associated protozoa–*Machadoella triatomae*, 497; *Trypanosoma cruzi*, 476
Triatoma geniculata = *Panstrongylus geniculatus* (Latr.), 476
Triatoma gerstaekeri Stål, 477
Triatoma hegneri Mazzotti, 477
Triatoma heidemanni Neiva = *T. leticularius* (Stål), 477
Triatoma infestans Klug:
  associated bacterium–*Streptococcus faecalis*, 141
  associated protozoan–*Trypanosoma cruzi*, 476
  associated rickettsia–*Dermacentroxenus rickettsi*, 331
  associated viruses, 439
  intracellular symbiotes in, 233, 235
Triatoma longipes Barber, 477
Triatoma megista (Burm.):
  associated protozoan–*Trypanosoma cruzi*, 476, 477
  intracellular symbiotes in, 233
Triatoma protracta Uehler:
  associated protozoa, 477
  associated rickettsia–*Dermacentroxenus rickettsi*, 331
  intracellular symbiotes in, 233
Triatoma rubida (Uehler), 477
Triatoma rubrofasciata (DeG.): associated protozoa, 479; associated rickettsiae, 345; intracellular symbiotes in, 233, 235
Triatoma rubrovaria (Blanchard), 498
Triatoma sanguisuga (LeC.), 439, 477
Triatoma sordida Stål:
  associated protozoan–*Trypanosoma cruzi*, 476
  intracellular symbiotes in, 233
Triatoma spp., 474
Triatoma vitticeps (Stål), 477
Tribolium castaneum (Hbst.) (= *T. ferrugineum* Fabr.), 498; *T. confusum* Duv., 12, 243
Tricercomitus divergens, 529, 538
Trichocera, 485; *T. hiemalis* DeG., 485
Trichochermes, 230
Trichodectes caprae Gurlt: associated rickettsia, 321; intracellular symbiotes, 220
Trichodectes climax Nitz. (see *Trichodectes caprae*)
Trichodectes pilosus Giebel:
  associated rickettsia–*Rickettsia trichodectae*, 267, 321
  intracellular symbiotes in, 220
Trichodectidae, 219-220
Trichoderma, 393
Trichoduboscqia epori, 516
Trichomastix trichopterorum (see *Eutrichomastix trichopterorum*)
Trichomesiini, 371
Trichomitus termitidis, 539 (see *Trichomonas termopsidis*)
Trichomonas, 534; *T. barbouri*, 540; *T. brevicollis*, 540; *T. cartagoensis*, 540; *T. labelli*, 540; *T. lighti*, 540; *T. linearis*, 539; *T. macrostoma*, 539; *T. termitis*, 539; *T. trypanoides*, 539; *T. vermiformis*, 539
Trichonympha, 531, 534, 535, 537, 549-550; *T. acuta*, 536; *T. agilis*, 529, 549; *T. algoa*, 536; *T. ampla*, 549; *T. campanula*, 533, 535,

549; *T. chattoni*, 549; *T. chula*, 536; *T. collaris*, 549, 550; *T. corbula*, 37, 549; *T. divexa*, 549; *T. fletcheri*, 548; *T. grandis*, 484, 536; *T. lata*, 536; *T. lighti*, 549; *T. magna*, 549; *T. minor*, 549; *T. okolona*, 536; *T. parva*, 536; *T. peplophora*, 37, 549; *T. quasilla*, 549; *T. saepicula*, 549; *T. serbica*, 550; *T. sphaerica*, 533, 549, 550; *T. subquasilla*, 549; *T. tabogae*, 549; *T. teres*, 549; *T. termopsidis*, 533, 534; *T. turkestanica*, 548; *T. zeylanica*, 549
Trichoptera, 494
Trichosporium symbioticum, 411
Trichosporon, 353
Tridactylus spp., 493
Trilobus gracilis, 466
Trimentropis citrina, 412
Trinoton, 219
Trioza, 230
Tripsacum dactyloides, 177
Tritrichomonas batrachorum, 37; *T. holmgreni*, 540
Trombicula akamushi (Brumpt), 291, 292, 294; *T. buloloensis*, 291; *T. deliensis* Walch, 291; *T. fletcheri* Wormersley & Heaslip, 291; *T. hirtsi*, 291; *T. minor*, 291; *T. schüffneri*, 291; *T. walchi*, 291
Trombidiidae, 255
Trombidium akamushi, 291
Trophozoites, 487, 489, 498, 518
Tropical typhus (see *Rickettsia tsutsugamushi*)
Tropidacris dux (Dru.), 181, 183
Tryidiomyces formicarum, 406
Trypanosoma, 464, 465, 474-482; *T. blanchardi*, 479; *T. brucei*, 474, 476, 481; *T. caprae*, 481; *T. congolense*, 480; *T. conorrhini*, 477; *T. cruzi*, 474, **476-478**; *T. duttoni*, 479; *T. evansi*, 481; *T. gambiense*, 474-475; *T. grayi*, 481; *T. kochi*, 482; *T. lewisi*, 478-479; *T. melophagium*, 480, 481; *T. nabiasi*, 479; *T. rabinowitschi*, 470; *T. rhodesiense*, 474, **475-476**; *T. simiae*, 481; *T. talpae*, 479; *T. theileri*, 479; *T. uniforme*, 481; *T. vespertilionis*, 479; *T. vivax*, 481
Trypanosomidae, 464-482, 523
Trypetidae, 22, 29
Trypodendron betulae Sw., 405
Trypodendron bivittatum (Kirby), 404; *T. politus*, 404; *T. restusum* (Lex.), 404, 405; *T. scabricollis* (LeC.), 404
Trypopitys carpini Redtb., 367
Trypsin, 23, 31
Tryxalis, 493
Tsetse flies, 3, 474, 475 (see also specific names)
Tsutsugamushi disease, 263, 279, **290-295** (see also *Rickettsia tsutsugamushi*)
Tuberales, 383
Tuberculariella ips, 410

Tularemia, 166 (see also *Pasteurella tularensis*)
Tunga penetrans (Linn.), 96
Tunica propria, 192, 197
Tunica vaginalis, 283, 284
Tupinambis teguixin, 498
Tussock moth (see *Hemerocampa leucostigma*)
Tychius flavicollis Steph., 240
Tychius meliloti Steph., 240
Typhlocyba ulmi, 445
Typhoid fever, 127, 269
Typhus, 94, 256, 260, 263, 279, 288, 291, 295, 300, 301, 303, 339; Australian, 291; classical, 268, 270, 280; endemic, **280-287**, 325; epidemic, **269-279**, 325; European, 269, 270, 278, 280; human, 269; Mexican, 280; Mexican, endemic, 281; Mexican, epidemic, 281; Minas Geraes, 329; murine, 268, 269, **280-287**; rural, 291; São Paulo, 329; scrub, 291, 294; sporadic, 291, 329; tick, 291; tropical, 291
Tyrosinase, 23

Udeopsylla robusta Haldeman, 493
Urinympha talea, 484, 536
Urographis fasciata DeG.:
  associated bacteria—*Achromobacter superficiale*, 98; *Aerobacter aerogenes*, 120; *Alcaligenes ammoniagenes*, 180
Ustilaginales, 393
Ustilago violaceae, 393

Vacca, 417
Vaccinium virus *1*, 448
Vagina, 192, 193
Vahlkampfia mellificae, 486; *Vahlkampfia* spp., 486
Vanessa cardui (Linn.), 84
Vanessa polychloros (Linn.):
  associated bacteria—*Bacillus aureus*, 55; *B. flavus*, 62; *Bacterium pieris liquefaciens alpha*, 25, 26, 111
Vanessa urticae (Linn.):
  associated bacteria—*Bacillus aureus*, 55; *B. decolor*, 60; *B. foetidus*, 63; *B. hoplosternus*, 65; *B. lineatus*, 63, 68; *B. lymantricola adiposus*, 70; *B. melolonthae*, 71; *B. monachae*, 73, 74; *B. similis*, 63, 87; *B. thuringiensis*, 91, 92; *Bacterium cazaubon*, 100; *B. melolonthae liquefaciens*, 108; *B. pieris liquefaciens*, 25, 26, **111**; *B. pyrenei*, 112; *Coccobacillus insectorum* var. *malacosomae*, 185; *Diplobacillus pieris*, 185; *Diplococcus bombycis*, 136; *D. melolonthae*, 137; *D. pieris*, 137; *Micrococcus vulgaris*, 151
  associated viruses, 414—polyhedral disease, 415, **426**
Variation of bacteria in insects, 24-27

SUBJECT INDEX

Vasa deferentia, 193, 194
Veillonella parvula, 149
Verno-aestival encephalitis, 439
Verruga peruana, 345
Verticillium, 387, 393
Vesca penis, 78
Vesicula seminalis, 193
Vespa crabro Linn.:
  associated protozoa—*Herpetomonas vespae,* 472; *Leptomonas vespae,* 468
  associated yeast—*Saccharomyces apiculatus,* 348
Vespa germanica Fabr., 515
Vesperus, 371
Vespula, 116
Vibrio, 42, **177**; *V. Aglaiae,* 87; *V. albensis,* 102; *V. colerigenes,* 177-178; *V. comma,* 28, **178-179**, 557, 560, 561, 567; *V. leonardi,* **179**, 560, 566; *V. metschnikovii,* 179; *V. pieris,* 179
Viruses, 259, 262:
  diseases—animals (including man), 432-443; insects, 413-432; plants, 443-449
  examination of insects and ticks for, 594, 597, **600-601**
  insect relationships, 413-450
Vitamins, 213, 235, 237
Vitellarium, 192
Vitelline membrane, 194
Vitellus, 194, 197
Vole, 294
Volucella obesa, 46, 48
Volvaria, 389, 409
Vulva, 193

Walnut caterpillar (see *Danta angusi*)
Wasps, 3, 117, 190
Watermark disease of willows, 119
Wax moth (see *Galleria mellonella*)
Weasel, 337
Weigl's typhus vaccine, 288
Weil-Felix reaction, 278, **279**, 286, 295, 301, 312, 338, 339, 344
Western celery mosaic, 445
Western equine virus, 437
Wheat mosaic, 449
White ant, 408
White-headed fungus (see *Podonectria coccicola*)
Whitmore's bacillus, 160
Wilt: of insects, 140, 145, 186, 415, 421-424 (see also Virus diseases of insects); of plants, 119, 120, 177
Wipfelkrankheit, 56, 108, 109, 415, **424-425**
Wohlfahrtia nuba, 35
Wolbachia, 263, 264, 265
Wolbachia pipientis, 265, 267; and *Culex pipiens,* 325, **327-328**; cultivation, 328;

morphology, 326; NR bodies, 328; **patho**genicity, 328; staining, 326
Wolhynia, 287
Wolhynian fever, 287
Wood-digesting roach (see *Cryptocercus punctulatus*)
Woodchuck, 285
Woodtick, 259 (see also *Dermancentor andersoni* and *D. variabilis*)
Woolly aphis, 412
Wuchereria bancrofti, 8
Wyeomyia melanocephala, 436

X factor of Castañeda, 279
Xanthomonas, 42, 173; *X. angulata,* 173; *X. campestris,* 173; *X. coronofaciens,* 173-174; *X. maculicola,* 174; *X. medicaginis* var. *phaseolicola,* 174; *X. melophthora,* 174; *X. pseudotsugae,* 175; *X. saliciperda,* 175; *X. savastanoi,* 175-176; *X. sepedonica,* 176; *X. solanacearum,* 176; *X. stewarti,* 177; *X. tumefaciens,* 113
Xenopsylla astia Roth.:
  associated bacterium—*Pasteurella pestis,* 161
  associated rickettsia—*Rickettsia prowazeki* var. *typhi,* 281
Xenopsylla brasiliensis (Baker), 161, 163
Xenopsylla cheopis (Roth.):
  associated bacteria—*Malleomyces pseudomallei,* 160; *Pasteurella pestis,* 161, 162, 163
  associated protozoan—*Trypanosoma lewisi,* 478
  associated rickettsiae—*Rickettsia prowazeki* var. *prowazeki,* 270, 271; *R. prowazeki* var. *typhi,* 281, 283; *R. tsutsugamushi,* 292
Xenopsylla cleopatrae Roth.:
  associated protozoa—*Crithidia cleopatrae,* 470; unnamed species, 470
Xestobium rufovillosum (DeG.), 367, 369
Xylaria, 409; *X. micrura,* 406
Xyleborus celsus Eichh., 403; *X. dispar* Fabr., 402, 404; *X. fuscatus* Eich., 404; *X. germanus* Bldfd., 405; *X. obesus,* 404; *X. pubescens* Zimm., 404; *X. tachygraphus,* 404; *X. xylographus* (Say), 404
Xylocopa aestuans (Linn.), 493
Xylosandrus germanus, 405

Yaws, 459
Yeasts, 196, **348-375**; disease-producing carried by insects, 350-351; effect on hatching of mosquitoes, 23; examination of insects and ticks for, 580, **588-589**, 597; extracellular, 348-351; as food for insects, 33, 349, 350; insects associated with, 348-349; intracellular, 351-375
Yellow fever, 3, 261, **432-436**; biologic relationship with mosquitoes, 433-435; **and** ticks, 436; types of, 432

SUBJECT INDEX

Yellow fungus (see *Aschersonia goldiana*)
Yellow spot of pineapple, 448
Yolk, 194, 195, 198; secondary, 197
Yolk cells, 194, 195, 196
Yponomeuta (see *Hyponomeuta*)

Zea mays, 177, 447
Zea virus 2, 447
Zodiomyces, 385
Zografia notonectae, 382
Zonocerus elegans:
  associated bacterium–*Coccobacillus acridiorum*, 181, 183
  associated protozoan–*Gregarina nigra*, 492
Zoomastigina, 463
Zoötermopsis, 528, 530, 534, 535, 538
Zoötermopsis angusticollis Hagen, 530:
  associated bacteria–*Bacterium neapolitanum*, 99; *Serratia marcescens*, 99, 132, 134
  associated fungi, 408, 409
  associated protozoa–*Gregarina termitis*, 552; *Hexamastix termopsidis*, 539; *Kofoidina ovata*, 553; *Streblomastix strix*, 547; *Trichomonas termopsidis*, 539; *Trichonympha campanula*, 549, 550; *T. collaris*, 549, 550; *T. sphaerica*, 549, 550
Zoötermopsis laticeps Banks:
  associated protozoa–*Hexamastix laticeps*, 539; *Trichomonas termopsidis*, 539; *Trichonympha campanula*, 549
Zoötermopsis nevadensis Hagen, 529:
  associated protozoa–*Gregarina termitis*, 552; *Hexamastix termopsidis*, 539; *Kofoidina ovata*, 553; *Streblomastix strix*, 533; *Trichomonas termopsidis*, 533, 539; *Trichonympha campanula*, 533, 549; *T. collaris*, 549; *T. sphaerica*, 533, 549
defaunation of, 533, 534
Zygosaccharomyces pini, 349
Zygote, 194